**Analogue-based
Drug Discovery**

*Edited by
János Fischer
and C. Robin Ganellin*

Further Titles of Interest

Buschmann, H., Guitart, X., Torrens, A., Holenz, J., Quintana, J.

Antidepressants, Antipsychotics, Antianxiolytics

From Chemistry and Pharmacology to Clinical Application

2006, ISBN 3-527-31058-4

Langer, T., Hoffmann, R. D. (eds.)

Pharmacophores and Pharmacophore Searches

2006, ISBN 3-527-31250-1

Kubinyi, H., Müller, G. (eds.)

Chemogenomics in Drug Discovery

A Medicinal Chemistry Perspective

2004, ISBN 3-527-30987-X

Nielsen, P. E. (ed.)

Pseudo-Peptides in Drug Discovery

2004, ISBN 3-527-30633-1

Wong, C.-H. (ed.)

Carbohydrate-based Drug Discovery

2003, ISBN 3-527-30632-3

Buschmann, H., Christoph, T., Friderichs, E., Maul, C., Sundermann, B. (eds.)

Analgesics

From Chemistry and Pharmacology to Clinical Application

2002, ISBN 3-527-30403-7

Analogue-based Drug Discovery

Edited by
János Fischer and C. Robin Ganellin

WILEY-VCH Verlag GmbH & Co. KGaA

Editors

Dr. János Fischer
Gedeon Richter Ltd.
P.O. Box 27
H-1475 Budapest 10
Hungary

Prof. Dr. C. Robin Ganellin
University College London
Department of Chemistry
20 Gordon Street
London WC1H OAJ
UK

Supported by
The International Union of Pure
and Applied Chemistry (IUPAC)
Chemistry and Human Health Division
PO Box 13757
Research Triangle Park, NC 27709-3757
USA

■ All books published by Wiley-VCH are
carefully produced. Nevertheless, authors,
editors, and publisher do not warrant the
information contained in these books,
including this book, to be free of errors.
Readers are advised to keep in mind that
statements, data, illustrations, procedural
details or other items may inadvertently
be inaccurate.

Library of Congress Card No.: applied for
British Library Cataloguing-in-Publication Data
A catalogue record for this book is available
from the British Library.

**Bibliographic information published by
Die Deutsche Bibliothek**
Die Deutsche Bibliothek lists this publication
in the Deutsche Nationalbibliografie; detailed
bibliographic data is available in the Internet at
<http://dnb.ddb.de>.

© 2006 WILEY-VCH Verlag GmbH & Co. KGaA,
Weinheim

Printed in the Federal Republic of Germany.
Printed on acid-free paper.

Typesetting   Kühn & Weyh, Satz und Medien,
Freiburg
Printing   Strauss GmbH, Mörlenbach
Bookbinding   Schäffer GmbH, Grünstadt
Cover-Design   Grafik Design Schulz, Fußgönheim

ISBN-13:   978-3-527-31257-3
ISBN-10:   3-527-31257-9

Contents

Preface *XVII*

Introduction *XIX*

List of Contributors *XXV*

Abbreviations *XXIX*

Part I   General Aspects of Analogue-Based Drug Discovery   *1*

1   Analogues as a Means of Discovering New Drugs   *3*
            Camille G. Wermuth

1.1   Designing of Analogues   *3*
1.1.1   Analogues Produced by Homologous Variations   *3*
1.1.1.1   Homology Through Monoalkylation   *3*
1.1.1.2   Polymethylenic Bis-Ammonium Compounds: Hexa- and
            Decamethonium   *3*
1.1.1.3   Homology in Cyclic Compounds   *4*
1.1.2   Analogues Produced by Vinylogy   *4*
1.1.2.1   Zaprinast Benzologues   *5*
1.1.3   Analogues Produced by Isosteric Variations   *5*
1.1.3.1   The Dominant Parameter is Structural   *5*
1.1.3.2   The Dominant Parameter is Electronic   *6*
1.1.3.3   The Dominant Parameter is Lipophilicity   *7*
1.1.4   Positional Isomers Produced as Analogues   *7*
1.1.5   Optical Isomers Produced as Analogues   *8*
1.1.5.1   Racemic Switches   *8*
1.1.5.2   Specific Profile for Each Enantiomer   *8*
1.1.6   Analogues Produced by Ring Transformations   *9*
1.1.7   Twin Drugs   *9*
1.2   The Pros and Cons of Analogue Design   *10*
1.2.1   The Success is Almost Warranted   *10*

Analogue-based Drug Discovery. IUPAC, János Fischer, and C. Robin Ganellin (Eds.)
Copyright © 2006 WILEY-VCH Verlag GmbH & Co. KGaA, Weinheim
ISBN: 3-527-31257-9

1.2.2   The Information is Available   *11*
1.2.3   Financial Considerations   *11*
1.2.4   Emergence of New Properties   *12*
1.3   Analogue Design as a Means of Discovering New Drugs   *12*
1.3.1   New Uses for Old Drugs   *12*
1.3.2   The PASS Program   *14*
1.3.3   New Leads from Old Drugs: The SOSA Approach   *14*
1.3.3.1 Definition   *14*
1.3.3.2 Rationale   *15*
1.3.3.3 Availability   *15*
1.3.3.4 Examples   *15*
1.3.3.4 Discussion   *18*
1.4   Conclusion   *20*

2   **Drug Likeness and Analogue-Based Drug Discovery**   *25*
         John R. Proudfoot

3   **Privileged Structures and Analogue-Based Drug Discovery**   *53*
         Hugo Kubinyi

3.1   Introduction   *53*
3.2   Drugs from Side Effects   *54*
3.3   Agonists and Antagonists   *55*
3.4   Privileged Structures   *57*
3.5   Drug Action on Target Classes   *58*
3.5.1   GPCR Ligands   *59*
3.5.2   Nuclear Receptor Ligands   *61*
3.5.3   Integrin Ligands   *61*
3.5.4   Kinase Inhibitors   *63*
3.5.5   Phosphodiesterase Inhibitors   *64*
3.5.6   Neurotransmitter Uptake Inhibitors   *65*
3.6   Summary and Conclusions   *65*

Part II   **Selected Examples of Analogue-Based Drug Discoveries**   *69*

1   **Development of Anti-Ulcer H$_2$-Receptor Histamine Antagonists**   *71*
         C. Robin Ganellin

1.1   Introduction   *71*
1.2   The Prototype Drug, Burimamide, Defined Histamine H$_2$-Receptors   *71*
1.3   The Pioneer Drug, Cimetidine: A Breakthrough for Treating Peptic Ulcer
         Disease   *72*
1.4   Ranitidine: The First Successful Analogue of H$_2$ Antagonists   *73*
1.5   The Discovery of Tiotidine and Famotidine   *76*

1.6   Other Compounds   *77*
1.7   The Use of H$_2$-Receptor Histamine Antagonists as Medicines   *78*

**2   Esomeprazole in the Framework of Proton-Pump Inhibitor
          Development**   *81*
          Per Lindberg and Enar Carlsson

2.1   Towards Omeprazole: The First Proton-Pump Inhibitor   *81*
2.2   The Treatment of Acid-Related Disorders Before Losec®   *81*
2.3   Pioneer Research at Hässle during the 1960s and 1970s   *83*
2.3.1   Toxicological Challenges   *86*
2.3.2   Discovery of H$^+$, K$^+$-ATPase: The Gastric Proton Pump   *87*
2.3.3   Analogue Optimization   *87*
2.4   The Development of Omeprazole   *89*
2.4.1   Further Toxicological Challenges and the Halt of the Clinical
        Program   *89*
2.4.2   Resumption of Clinical Studies   *90*
2.4.3   Omeprazole Reaches the Market and Supersedes H$_2$-Receptor
        Antagonists   *90*
2.5   The Unique Action of Omeprazole   *91*
2.5.1   Inhibition of the Final Step   *91*
2.5.2   Omeprazole Binds Strongly to the H$^+$, K$^+$-ATPase   *91*
2.5.3   Inhibition of Acid Secretion and H$^+$, K$^+$-ATPase Activity   *92*
2.5.4   Omeprazole Concentrates and Transforms in Acid   *93*
2.5.5   Disulfide Enzyme–Inhibitor Complex on the Luminal Side   *93*
2.5.6   Short Half-Life in Plasma and Long Half-Life for Enzyme–Inhibitor
        Complex   *93*
2.5.7   Mechanism at the Molecular Level   *94*
2.5.8   The "Targeted Prodrug" Omeprazole means a Highly Specific
        Action   *96*
2.6   pH-Stability Profile   *97*
2.7   Omeprazole Analogues Synthesized by Other Companies   *98*
2.8   Omeprazole: A Need for Improvement?   *103*
2.8.1   The Omeprazole Follow-Up Program   *103*
2.8.2   No Good Alternative to the Omeprazole Structural Template   *103*
2.8.3   Chemical Approach   *104*
2.8.4   Synthesis and Screening   *105*
2.8.5   Isomers Seemed Unattractive   *106*
2.8.6   Isomer Pharmacokinetics and Pharmacodynamics in Animals   *106*
2.8.7   The Key Experiment in Man   *107*
2.8.8   Production of Esomeprazole (Mg Salt)   *109*
2.8.9   Omeprazole Isomers: Differences in Clearance and Metabolic
        Pattern   *109*
2.9   Summary   *111*

3 **The Development of a New Proton-Pump Inhibitor: The Case History of Pantoprazole** *115*
Jörg Senn-Bilfinger and Ernst Sturm

3.1 Introduction *115*
3.2 History of Gastrointestinal Research at Byk Gulden *117*
3.2.1 The Antacids and Cytoprotectives Projects and the Set-Up of In-Vivo Ulcer Models *117*
3.2.2 Decision to Concentrate on Anti-Secretory Treatments and the Study of Compounds with an Unknown Mechanism of Action *118*
3.3 Identification of the First PPI Project Candidates *121*
3.3.1 Optimizing the Benzimidazole Moiety *121*
3.3.2 Impact of the First PPI Project Compounds *123*
3.4 Elucidation of the Mechanism of Action of PPIs *125*
3.4.1 A Surprising Interrelationship Between Stability and Activity *125*
3.4.2 Isolation and Identification of the Active Principle of the PPIs *125*
3.5 Identification of Pantoprazole as a Candidate for Development *128*
3.5.1 Optimizing the Pyridine Moiety and the First Synthesis of Pantoprazole *128*
3.5.2 Selection Criteria *129*
3.5.3 The Selection of Pantoprazole and Internal Competition with SK&F95601 *130*
3.5.4 Toxicological Problems: Project Development at Risk *131*
3.5.5 Benefits of Pantoprazole for the Patient *132*
3.5.6 Summary *132*
3.6 Outlook on Further Developments *133*

4 **Optimizing the Clinical Pharmacologic Properties of the HMG-CoA Reductase Inhibitors** *137*
Sándor Kerpel-Fronius and János Fischer

4.1 Introduction *137*
4.2 Medicinal chemistry of the Statins *138*
4.3 Clinical and Pharmacologic Properties of the Statin Analogues *142*
4.3.1 Fibrate Coadministration *148*
4.4 Clinical Efficacy of the Statins *149*

5 **Optimizing Antihypertensive Therapy by Angiotensin Receptor Blockers** *157*
Csaba Farsang and János Fischer

5.1 Medicinal Chemistry *157*
5.2 Clinical Results with Angiotensin II Antagonists *160*
5.2.1 Mechanisms of Action *160*
5.2.1.1 Other Effects of ARBs *161*
5.2.2 Target Organ Protection *162*

5.2.2.1   Left Ventricular Hypertrophy   *162*
5.2.2.2   Diabetic and Nondiabetic Nephropathy   *162*
5.2.2.3   Diabetes Prevention   *162*
5.2.2.4   Coronary Heart Disease (CHD)   *162*
5.2.2.5   Congestive Heart Failure   *162*
5.2.2.6   Stroke Prevention and Other CNS Effects   *163*
5.3   Differences Among Angiotensin AT_1 Receptor Blockers   *163*
5.3.1   Structural Differences   *163*
5.3.2 AT_1 Receptor Antagonism   *164*
5.3.3   Pharmacokinetics/Dosing Considerations   *164*
5.3.4   Drug Interactions/Adverse Effects   *165*
5.3.5   Efficacy in Hypertension   *165*
5.4   Summary   *166*

6   **Optimizing Antihypertensive Therapy by Angiotensin-Converting Enzyme Inhibitors**   *169*
         Sándor Alföldi and János Fischer

6.1   Medicinal Chemistry of ACE-inhibitors   *169*
6.2   Clinical Results with ACE-Inhibitors   *173*
6.2.1   Hemodynamic Effects   *173*
6.2.2   Effects of ACE-Inhibitors   *174*
6.2.2.1   Hypotension   *174*
6.2.2.2   Dry Cough   *174*
6.2.2.3   Hyperkalemia   *174*
6.2.2.4   Acute Renal Failure   *175*
6.2.2.5   Proteinuria   *175*
6.2.2.6   Angioedema   *175*
6.2.2.7   Teratogenic Effects   *175*
6.2.2.8   Other Side Effects   *175*
6.2.3   Contraindications   *176*
6.2.4   Drug Interactions   *176*
6.3   Differences Among ACE-Inhibitors   *177*
6.4   Summary and Outlook   *179*

7   **Case Study of Lacidipine in the Research of New Calcium Antagonists**   *181*
         Giovanni Gaviraghi

7.1   Introduction   *181*
7.2   Dihydropyridine Calcium Channel-Blocking Agents   *182*
7.2.1   Nifedipine   *182*
7.2.2   Felodipine   *183*
7.2.3   Isradipine   *183*
7.2.4   Nimodipine   *184*
7.2.5   Nisoldipine   *184*

7.2.6   Amlodipine   *185*
7.2.7   Lacidipine   *185*
7.2.8   Lercanidipine   *185*
7.2.9   Manidipine   *186*
7.3   Lacidipine: A Long-Lasting Calcium Channel-Blocking Drug: Case Study   *187*
7.3.1   The Lacidipine Project   *188*
7.3.2   Synthesis   *190*
7.3.3   The Pharmacological Profile of Lacidipine   *190*
7.4   Conclusion   *191*

8   Selective Beta-Adrenergic Receptor-Blocking Agents   *193*
       Paul W. Erhardt and Lajos Matos

8.1   Introduction   *193*
8.2   Beta-1 Selective Blockers   *201*
8.2.1   Atenolol   *201*
8.2.1.1   Discovery   *201*
8.2.1.2   Synthesis   *203*
8.2.1.3   Clinical Pharmacology   *203*
8.2.2   Betaxolol   *206*
8.2.2.1   Discovery   *206*
8.2.2.2   Synthesis   *209*
8.2.2.3   Clinical Pharmacology   *210*
8.2.3   Celiprolol   *211*
8.2.3.1   Discovery   *211*
8.2.3.2   Synthesis   *214*
8.2.3.3   Clinical Pharmacology   *215*
8.2.4   Nebivolol   *217*
8.2.4.1   Discovery   *217*
8.2.4.2   Synthesis   *218*
8.2.4.3   Clinical Pharmacology   *220*
8.3   Accumulated Structure–Activity Relationships   *222*
8.4   Summary   *226*

9   Case Study: "Esmolol Stat"   *233*
       Paul W. Erhardt

9.1   Introduction   *233*
9.2   Pharmacological Target   *234*
9.3   Chemical Target   *234*
9.3.1   Internal Esters   *234*
9.3.2   External Esters   *236*
9.3.3   Structure–Activity Relationships   *237*
9.4   Chemical Synthesis   *240*
9.5   Pharmacology and Clinical Profile   *241*

9.6   Summary and Some Lessons for Today   *243*
9.6.1   Compound Libraries   *243*
9.6.2   Biological Testing   *244*
9.6.3   SAR   *244*

10   **Development of Organic Nitrates for Coronary Heart Disease**   *247*
      László Dézsi

10.1   Introduction   *247*
10.2   Empirical Observations Leading to the Therapeutic Use of Classic Nitrovasodilators   *247*
10.3   Isoamyl Nitrite: The Pioneer Drug   *248*
10.4   Nitroglycerin (Glyceryl Trinitrate): The Most Successful Analogue   *248*
10.5   Isosorbide Dinitrate: A Viable Analogue with Prolonged Action   *249*
10.6   Isosorbide Mononitrate: The Metabolite of Isosorbide Dinitrate   *250*
10.7   Nicorandil: The Potassium Channel Opener Analogue with a Broad Cardiovascular Spectrum   *251*
10.8   Cardiovascular Efficacy of Organic Nitrates   *252*
10.9   Mechanism of Action of Organic Nitrates   *253*
10.10   Tolerance to Organic Nitrates   *255*
10.11   Concluding Remarks   *256*

11   **Development of Opioid Receptor Ligands**   *259*
      Christopher R. McCurdy

11.1   Introduction   *259*
11.2   Pharmacology Related to the Classic Opioid Receptors.   *261*
11.3   Alkaloids from the Latex of *Papaver somniferum* Initiate Research   *261*
11.4   Morphine: The Prototype Opioid Ligand   *262*
11.4.1   Initial Studies of Morphine Analogues   *263*
11.5   Structure–Activity Relationships of Morphine Derivatives   *265*
11.6   Synthetic Analogues of Thebaine Further Define Morphinan SAR   *266*
11.7   Compounds of the Morphinan Skeleton Produce New Agents   *269*
11.8   Further Reduction of the Morphinan Skeleton Produced the Benzomorphans   *270*
11.9   Another Simplified Version of Morphine Creates a New Class of Opioid Ligand   *271*
11.10   A Breakthrough in the Structural Design of Opioid Ligands   *271*
11.11   Discovery of the 4-Anilidopiperidines   *273*
11.12   Phenylpropylamines: The Most Stripped-Down Opioids Still Related to Morphine   *273*
11.13   The Use of Opioid Analgesics in Clinical Practice: Hope of the Future   *274*

12   Stigmines   *277*
Zeev Tashma

12.1   Historical Background   *277*
12.2   Pharmacological Activities of Physostigmine   *278*
12.3   Chemistry and Biochemistry of Physostigmine   *279*
12.4   Interaction of Acetylcholinesterase with Carbamates   *280*
12.5   SAR of the Eseroline Moiety, and the Development of Miotine   *282*
12.6   The Development of Quaternary Carbamates for Myasthenia
         Gravis   *283*
12.7   Carbamates as Pre-Exposure Treatment against Organophosphate
         Intoxication   *284*
12.8   Carbamates as Insecticides   *286*
12.8.1   Structural Features   *287*
12.9   Carbamates in the Treatment of Alzheimer's Disease   *287*
12.9.2   Close Derivatives of Physostigmine   *288*
12.9.3   Rivastigmine   *289*

13   Structural Analogues of Clozapine   *297*
Béla Kiss and István Bitter

13.1   Introduction   *297*
13.2   Clozapine: The Prototype "Atypical" Antipsychotic; Some Chemical
         Aspects   *299*
13.3   Preclinical Aspects   *300*
13.3.1   Multireceptor Profile: *In-Vitro, In-Vivo* Similarities and Differences   *300*
13.3.2   The Availability of Data   *304*
13.3.3   Dopamine D_2 versus Serotonin 5-HT$_{2A}$ Affinity   *304*
13.3.4   Affinity to other Receptors   *306*
13.3.5   Inverse Agonism   *306*
13.3.6   Receptor Affinity of Metabolites and Clinical Action   *307*
13.4   Clinical Aspects   *307*
13.4.1   Terminology   *307*
13.4.2   Indications   *308*
13.4.3   Dosage   *308*
13.4.4   Clinical Efficacy in Schizophrenia   *308*
13.4.5   Clinical Efficacy in Bipolar Disorder (Especially in Mania)   *309*
13.4.6   Adverse Events   *310*
13.5   Summary and Conclusions   *310*

14   Quinolone Antibiotics: The Development of Moxifloxacin   *315*
Uwe Petersen

14.1   Introduction   *315*
14.2   Aims   *320*
14.3   The Chemical Evolution of Moxifloxacin   *321*

14.4   Synthetic Routes   *338*
14.4.1   *S,S*–2,8-Diazabicyclo[4.3.0]nonane   *338*
14.4.2   Preparation of BAY X 8843 36 and its Analogues   *339*
14.4.3   Preparation of Moxifloxacin Hydrochloride 47   *339*
14.5   The Physicochemical Properties of Moxifloxacin   *342*
14.6   The Microbiological and Clinical Properties of Moxifloxacin   *344*
14.6.1   *Mycobacterium tuberculosis*   *347*
14.6.2   Skin Infections   *347*
14.6.3   Ophthalmology   *348*
14.6.4   Dental Medicine   *348*
14.7   Pharmacokinetics/Pharmacodynamics of Moxifloxacin   *348*
14.8   Development of Resistance to Moxifloxacin   *350*
14.9   Safety and Tolerability of Moxifloxacin   *352*
14.10   Metabolism, Excretion and Biodegradability of Moxifloxacin   *353*
14.11   Future Prospects for Quinolones   *355*
14.12   Summary and Conclusions   *356*

15   The Development of Bisphosphonates as Drugs   *371*
          Eli Breuer

15.1   Historical Background   *371*
15.2   Discovery of the Biological Activity of Pyrophosphate and of
          Bisphosphonates   *372*
15.3   Bone-Related Activity of Bisphosphonates   *372*
15.3.1   Overview   *372*
15.3.2   Osteolytic Bone Diseases   *373*
15.3.2.1 Osteoporosis   *373*
15.3.2.2 Osteolytic Tumors   *373*
15.3.2.3 Paget's Disease   *375*
15.3.3   Structure–Activity Relationships   *375*
15.3.3.1 The Molecular Skeletons of Bisphosphonates   *375*
15.3.3.2 Phosphonic Acid Groups   *375*
15.3.3.3 The Geminal Hydroxy Group   *375*
15.3.3.4 Nitrogen-Containing Side Chain   *375*
15.3.3.5 Structure–Activity Relationships of BPs: A Summary   *376*
15.3.4   Inhibition of Bone Resorption: The Mechanisms of Action   *377*
15.3.5   Clinical Pharmacology of Bisphosphonates   *378*
15.3.6   Bisphosphonates in Clinical Use   *379*
15.4   Miscellaneous Biological Aspects of Bisphosphonates   *379*
15.4.1   Bisphosphonates as Vehicles for Delivering Drugs to Bone   *379*
15.4.2   Bisphosphonates as Potential Drugs for other Indications   *379*
15.4.2.1 Antiviral Drugs   *381*
15.4.2.2 Bisphosphonate Inositol-Monophosphatase Inhibitor: A Potential Drug
          for Bipolar Disorders   *381*

15.4.2.3 Hypocholesterolemic Bisphosphonates (Squalene Synthase Inhibitors) *381*

15.4.2.4 Antiparasitic Drugs *381*

15.4.2.5 Anti-Inflammatory and Anti-Arthritic Bisphosphonates *382*

15.4.2.6 Cardiovascular Applications of Bisphosphonates *382*

15.5   Conclusions *382*

16   Cisplatin and its Analogues for Cancer Chemotherapy *385*
        Sándor Kerpel-Fronius

16.1   Introduction *385*

16.2   Cisplatin *385*

16.2.1   Discovery *385*

16.2.2   Structure *386*

16.2.3   Mechanism of Action *386*

16.2.4   Pharmacokinetics *387*

16.2.5   Clinical Efficacy *387*

16.2.6   Adverse Effects *388*

16.3   Carboplatin *389*

16.3.1   Development *389*

16.3.2   Administration and Pharmacokinetics *389*

16.3.4   Adverse Effects *390*

16.3.5   Clinical Efficacy *390*

16.4   Oxaliplatin *390*

16.4.1   Development *390*

16.4.2   Cellular Resistance to Various Pt Analogues *391*

16.4.3   Metabolism and Pharmacokinetics *392*

16.4.4   Adverse Effects *392*

16.4.5   Clinical Efficacy *392*

16.5   Summary *393*

17   The History of Drospirenone *395*
        Rudolf Wiechert

17.1   General Development *395*

17.2   Syntheses *397*

18   Histamine H$_1$ Blockers: From Relative Failures to Blockbusters Within Series of Analogues *401*
        Henk Timmerman

18.1   Introduction *401*

18.2   The First Antihistamines *402*

18.3   Diphenhydramine as a Skeleton for Antihistamines *403*

18.3.1   The Diaryl Group *404*

18.3.2   The Linker *406*

18.3.3   The Basic Group   *406*
18.3.4   The Analogue Principle   *407*
18.3.5   The Analogue Principle in Perspective   *409*
18.4   The New Antihistamines   *411*
18.5   Antihistamines for Which the Analogue Principle Does not Seem to Work   *415*
18.6   Inverse Agonism   *415*
18.7   A Further Generation of Antihistamines?   *416*
18.8   Conclusions   *417*

19   Corticosteroids: From Natural Products to Useful Analogues   *419*
           Zoltán Tuba, Sándor Mahó, and Csaba Sánta

19.1   Introduction   *419*
19.2   Corticosteroid Analogues   *420*
19.2.1   Cortisone   *422*
19.2.2   Hydrocortisone   *423*
19.2.3   Prednisone and Prednisolone   *424*
19.2.4   Fludrocortisone   *424*
19.2.5   Triamcinolone and Triamcinolone Acetonide   *425*
19.2.6   Dexamethasone   *426*
19.2.7   Betamethasone   *427*
19.2.8   Beclomethasone Dipropionate   *428*
19.2.9   Methylprednisolone   *429*
19.2.10   Fluocinolone Acetonide, Flunisolide, Fluocortin-21-Butylate and Flumetasone   *429*
19.2.11   Budesonide   *431*
19.2.12   Halobetasol Propionate   *432*
19.2.13   Mometasone Furoate   *433*
19.2.14   Fluticasone Propionate   *434*
19.2.15   Loteprednol Etabonate   *435*
19.2.16   Ciclesonide   *436*
19.3   Summary   *437*

Part III Table of Selected Analogue Classes   *441*
           Erika M. Alapi and János Fischer

Index   *553*

Preface

The International Union of Pure and Applied Chemistry (IUPAC) is the global civil organization of chemists. The Union is organized into Divisions, with Division VII being devoted to Chemistry and Human Health. The latter incorporates the Subcommittee for Medicinal Chemistry and Drug Development which has projects in various stages of completion. One of these projects, which is devoted to "Analogue-Based Drug Discovery", was initiated in 2003.

The goal of the project is to study the role of analogue drugs for medicinal chemistry, and in this respect two interesting points have come to light:

1. Statistically, every second drug is an analogue.
2. The market value of analogue drugs amounts to approximately two-thirds of that for all small-molecule drugs.

Clearly, in order to have reached this level of importance, analogue drugs must have special value.

Today, it is not too difficult to identify analogues among the most frequently prescribed drugs, on the basis of their similarities in structure and biological properties. In the present book, analogue drugs have, for the first time, been collected systematically on the basis of two sources:

- by using actual data from the Anatomical-Therapeutic Chemical (ATC) System of the World Health Organization (WHO); and
- by using the most recently available data of IMS (the former Intercontinental Marketing Services) Health.

In this way, among the Top 500 most frequently used drugs, 67 analogue classes and 306 analogue drugs have been identified.

This book focuses on both structural and pharmacological analogues – that is, those analogues which have similar chemical and biological properties – although some examples are also included where the analogue is derived purely on a similar chemical or a similar biological basis (but not both).

Within the book, it is shown how analogues play an important role in medicinal chemistry and, more importantly, how they optimize drug therapies. Hence, it was for this reason that we sought to select diverse fields of drug research and medicinal chemistry.

Analogue-based Drug Discovery. IUPAC, János Fischer, and C. Robin Ganellin (Eds.)
Copyright © 2006 WILEY-VCH Verlag GmbH & Co. KGaA, Weinheim
ISBN: 3-527-31257-9

The aim of the book was not to provide a comprehensive review, but rather to describe selected analogue classes in a more detailed manner. In support of this aim, we should point out that nine of the authors have played key roles as co-inventors in the discovery of some of the very important drugs detailed in the book.

This book should serve as a useful reference for experts in medicinal chemistry and also for students of this field. Moreover, it will also be of interest to a wide range of scientists, including organic chemists, biochemists, pharmacologists and clinicians, who are interested in drug research.

We extend our sincere thanks to many people involved in the book's preparation. For data collection, we thank our co-workers at Richter Ltd. (Budapest, Hungary): Ildikó Balló, Andrea Donát, Péter Erdélyi, Dr. Tamás Fodor, Sándor Lévai, György Szabó, Dr. Attila Szemző, Katalin Szőke, and Krisztina Vukics.

We are also very much obliged to all project members of the IUPAC Medicinal Chemistry Subcommittee: Prof. Eli Breuer, Prof. Giovanni Gaviraghi, Prof. Per Lindberg, Dr. John Proudfoot, Prof. Jörg Senn-Bilfinger, Prof. Henk Timmerman, and Prof. Camille G. Wermuth, each of whom has contributed chapters. We are also grateful to Prof. Paul W. Erhardt who, as President of the Division of Chemistry and Human Health of the IUPAC, helped in the development of the book, not only in several committee discussions but also as an author. Our thanks are also extended to all co-author experts.

Special thanks are due to Dr. Tom Perun of the IUPAC Medicinal Chemistry Subcommittee, Dr. Hanns Wurziger (Merck KGaA, Darmstadt), Dr. Derek Buckle (DRB Associates), and Dr. Stefan Jaroch (Schering AG, Berlin) for their help in the final preparation and review of the manuscript.

We also received outstanding help from Prof. Sándor Kerpel-Fronius who, as a clinical pharmacologist, helped to create a bridge between medicinal chemistry and clinical pharmacology.

We hope that this book will be a useful reading for all experts participating in drug discovery, both in the industry and in academia.

And last – but not least – we welcome comments from readers, and assure them that these will be taken on board if we are fortunate enough to run to a second edition!

János Fischer and C. Robin Ganellin
Budapest and London
May 2005

Introduction

János Fischer and C. Robin Ganellin

The discovery and development of new drugs to provide medicines for treating diseases is the main role of the pharmaceutical industry. The impact of this process on the well-being of society is considerable, but it is a difficult and costly procedure to conduct. Biological organisms – and especially human beings – are extraordinarily complex, and our understanding of how they function at the molecular level remains rudimentary, although considerable advances in knowledge have been made in recent decades. Whilst an advanced industrial society was able to plan and deliver a man to the moon following a 10-year program, almost 40 years on we are still only able to treat about 60% of cancer patients effectively, and do not understand how to correct most mental diseases.

In order to treat a disease, an attempt must be made to put right something that has gone wrong. However, because of our limited understanding of normal state functions at the molecular level, we must work empirically and in doing so resort to much trial and error. In general, the same situation applies to new drug discovery, with the sources of new drugs falling into three main categories:

- Existing drugs
- Screening against a physiological target
- Structure-based drug design

Existing Drugs

The most fruitful basis to discover a new drug is to start with an old drug, and this has been the most common and reliable route to new products. Thus, existing drugs may need to be improved, for example to develop a better dosage form, to improve drug absorption or duration, to increase potency and reduce the daily dose, or to avoid certain side effects. On occasion, the existing drug is a natural product, but more often than not it is a synthetic compound, and many such examples are provided later in this book.

Analogue-based Drug Discovery. IUPAC, János Fischer, and C. Robin Ganellin (Eds.)
Copyright © 2006 WILEY-VCH Verlag GmbH & Co. KGaA, Weinheim
ISBN: 3-527-31257-9

Natural Products

Historically, natural products have formed the oldest basis for new medicines, and natural selection during evolution and competition between the species has produced powerful, biologically active natural products. These can serve as chemical leads, to be refined by the chemist by creating analogues that will provide a more specifically acting drug, or perhaps avoid a delivery problem or an unwanted adverse side effect.

For example, molds and bacteria produce substances that prevent other organisms from growing in their vicinity. The famous *Penicillium* mold led, via the pharmaceutical industry, to penicillin. However, penicillin was not stable in the acidic environment of the stomach, and so compounds were synthesized by chemists to produce a range of useful semisynthetic penicillin analogues. An example of the use of analogues to develop new antibacterial antibiotics is provided in Chapter II-14, on the development of moxifloxacin.

Another fruitful means of identifying pharmacologically active natural products has been that of folk law remedies, many of which are plant products. Typical examples include alkaloids, such as atropine (from plants of the Solanaceae family, known to the ancient Greeks) and reserpine (from *Rauwolfia serpentina*, the snakeroot), which is popular in India as a herbal remedy for use as a tranquilizer or antihypertensive. Other chapters in the book relate to stigmines (based on physostigmine, an anticholinesterase alkaloid from the Calabar bean in West Africa) that are used to treat Alzheimer's disease (Chapter II-12), and opioid receptor ligands (based on morphine, the most important alkaloid of the opium poppy) for pain relief and as antitussives (Chapter II-11).

Synthetic Drugs

A very important type of synthetic drug is one that opens up a new therapeutic treatment, and this is referred to as a *pioneer drug* because it is used to pioneer a new type of therapy, or to make a marked improvement over what was previously possible. Such a drug is often referred to as "First in class", and might arise through the observation of a side effect of a drug that is already in use but can then be exploited by making an analogue. The side effect may be the result of an astute observation made during pharmacological studies in animals, or from clinical investigations in patients. An example is the discovery of sulfonamide diuretics during the 1950s following the observation that the antibacterial drug sulfanilamide made the urine alkaline by inhibiting the enzyme, carbonic anhydrase. This rendered the then-used toxic organomercurials obsolete, and so constituted a considerable improvement in therapy.

Another example is the discovery made during the use of the antihistamine promethazine to treat surgical shock. In order to improve potency, a chlorine atom was incorporated into the drug molecule. Subsequently, when the patients seemed to be unconcerned about undergoing surgery, chlorpromazine – the first phenothiazine antipsychotic drug – was born, thereby opening up a new era in the treatment of mental disease.

A more recent example is that of sildenafil which, as a result of observations made during Phase I studies in male volunteers, is now used to treat erectile dysfunction. Sildefanil had originally been designed as an analogue of zaprinast, and a more selective phosphodiesterase inhibitor (PDE5) for use as a cardiovascular agent (see Chapter I-1).

Although such drugs have not been "designed" for their newly observed action, they usually trigger the start of an analogue program to improve upon their activity.

Physiological Targets

Pioneer drugs may also arise more deliberately as a result of ligand design, for example, to block an enzyme, or to block a biogenic amine receptor. Examples are the first inhibitor of the angiotensin-converting enzyme (ACE), captopril, the first H_1-receptor antihistamine, phenbenzamine, the first β-blocker, pronethalol, and the first histamine H_2-receptor antagonist, cimetidine. These have all given rise to analogue programs described later in the book (Chapters II-6, II-18, II-8 and II-1 respectively).

More recently, as a result of the considerable explosion in scientific knowledge of cellular biochemistry and cell biology, a large number of physiological targets have been considered for drug action. A major problem however in this regard is to relate the isolated target to the physiology of the whole animal. Furthermore, it is never clear at the outset of a research program that the particular target will determine the outcome of a disease, as Nature rarely issues monopolies to the putative transmitters. Another problem is that physiological targets may not have a known specific messenger substance, and so there may not be a chemically based lead. Hence, attempts must be made to identify a lead through the high-throughput screen (HTS) of a chemical library.

Structure-Based Drug Design

On occasion – although still rather rarely – sufficient molecular information is known about the physiological target (e.g., it may be the crystal structure of an enzyme showing the precise geometry of the active site) that an attempt can be made to design directly a drug molecule to fit. Although this situation is rather rare, it is clearly intellectually satisfying if success is achieved. In reality, this situation should perhaps be termed "structure-assisted drug design", and examples to date have occurred in the anti-HIV field of drug research and recently in discovery of a promising renin inhibitor (aliskerin).

The foregoing discussion should have provided the reader with the impression that new drug discovery is a very difficult and risky business, and indeed, there are very few industries prepared to invest in research programs where the likelihood of success is so low, especially if the intended outcome is a pioneer drug. Typically, the time scale is likely to be 10–20 years, and there is no guarantee that the product will have an adequate impact on the disease process, or be sufficiently safe and free of unwanted side effects.

Analogue-Based Drug Design

The development of a pioneer drug is extraordinarily uncertain because its therapeutic use has not yet been validated clinically. On the other hand, preparing an analogue of an established drug has the considerable advantage that the predicted therapeutic use of the analogue has already been proved. This removes a major uncertainty from the overall risk of success. Nonetheless, there are still many hurdles to overcome, notably with regard to pharmacokinetic behavior and safety. It is important, however, to identify at the outset of the research program the expected clinical advantage of the analogue over the established drug which is being used as a lead. The aim must be to provide an improvement in the use as a medicine.

Very often, a pioneer drug which demonstrates its success will stimulate many companies to seek an improved analogue. Since the identification of success will be a clinical publication or annual report of sales, companies tend to start their research programs at about the same time. Many different new potential analogue drugs may therefore appear within a year or two of each other, and all must undergo extensive preclinical laboratory studies for safety, as well as clinical trials to prove their advantage over the lead medicine. Eventually, the potential new drug must be marketed in order to gain access to a sufficiently wide patient population and to reveal its advantages and confirm its safety and lack of adverse side effects.

Thus, companies have exchanged the uncertainty of having to determine the clinical use of a pioneer product for the uncertainty which arises through competition with other companies to demonstrate a clinical advantage for their analogue product. The latter is usually a lesser uncertainty and, indeed, there is always the likelihood that a competitive product may run into difficulties – for example, unexpected side effects, a poorer clinical performance than initially predicted, or inadequate marketing. Thus, there are powerful reasons for continuing to develop new analogue compounds, and this leads to an apparent proliferation of new products which, for some time, may lead to confusion about which is the most suitable drug treatment.

From an outsider's view of the pharmaceutical industry, this proliferation of products may be seen as an example of rampant commercialization. The analogue products are often regarded as "me-too" compounds – a term used pejoratively, notably by those who indulge in attacking the pharmaceutical industry, to suggest that because a pioneer medicine has proved to be commercially valuable, other companies want to share in the commercial success. Of course, companies must be commercially successful to stay in existence, as with any other business enterprise, but this cannot be the only basis for a new drug. As has been explained above, the new drug must demonstrate an advantage if it is to succeed. In effect, the application of new analogue products is the historical means by which a particular drug therapy becomes optimized, and this is clearly demonstrated by the many examples provided in the present book.

Defining Analogues

The term "analogue" is used in its chemical sense and is defined in the IUPAC medicinal chemistry glossary as "... a drug whose structure is related to that of another drug but whose chemical and biological properties may be quite different."

It is useful for present purposes to have a broader definition of analogues, where not only the chemical relationships, but also the pharmacological properties are considered.

In the present context, we consider an analogue drug to be one that has a chemical and/or pharmacological relationship to another drug. We have used a classification of analogue drugs according to three categories in which we have defined chemistry in terms of chemical structure.

Structural and Pharmacological Analogues

Structural and pharmacological analogues are drugs which have similar chemical structures, and a similar main pharmacological activity. One special class of these is considered to be *direct analogues* if they have identical pharmacophores – that is, they can be described by a general structure which includes most of the chemical skeleton. Many analogues, however, share only a small part of the skeleton, and in this case they are not direct analogues.

It is sometimes difficult to discern any true chemical similarity, but the development of an analogue may have started with the lead structure and passed through a series of iterative structure–activity relationships (SARs). Where there is such a historical development, the analogues are still considered to have a structural and pharmacological relationship.

The histamine H_2 antagonist anti-ulcer drugs (see Chapter II-1) illustrate the difference between the definitions. Thus, burimamide provided proof of principle, while metiamide, a *direct analogue*, went into clinic trials and validated the target by demonstrating the therapeutic use. However, metiamide had an unacceptable side effect (granulocytopenia), and so another *direct analogue*, cimetidine, became the *pioneer drug*. With ranitidine came a change in the ring structure (imidazole replaced by dimethylaminomethyl-furan); this does not have the same SAR pattern and so must be considered to be a *structural and pharmacological analogue* (i.e., not a *direct analogue*). Nizatidine appears as a *direct analogue* of ranitidine (thiazole replaces furan). Famotidine has several major changes in structure, but still retains the $CH_2SCH_2CH_2$ chain and the heterocycle and polar "end group"; thus, it is also a *structural and pharmacological analogue*. Roxatidine has many differences in its structure and might be considered as a *pharmacological analogue*; however, its SAR development can be directly traced to other analogues with a general structure that fits all of the above compounds, so it must be considered to be a *structural and pharmacological analogue.*

Structural Analogues

Structural analogues are drugs which have a similar chemical structure but have quite different pharmacological properties. These drugs are usually prepared with the intention of being a structural and pharmacological analogue but then, unexpectedly, another pharmacological activity appears. An example is in the case of opiate agonists and antagonists (see Chapter II-11) where morphine, as a prototype opiate agonist, has an N-methyl substituent. The *N*-allyl-normorphine, nalorphine, is used as a morphine antagonist.

In some cases, the simple modification of a drug structure can essentially modify the biological activity profile, whilst preserving some part of the original activity. Drospirenone (see Chapter II-17), an orally active progestin contraceptive with antimineralcorticoid properties, is an example of an analogue-based drug discovery process that started from the diuretic spironolactone.

Pharmacological Analogues

Pharmacological analogues are drugs which have a similar pharmacological activity without having any discernible chemical or structural relationship. This means that different pharmacophores exist to explain the same pharmacological activity of structurally different drugs. Leads for such compounds can arise through screening and/or computer modeling. These analogues are beyond the scope of this book. An example is provided by the three compounds nifedipine, verapamil, and diltiazem, all of which are L-class voltage-gated calcium-channel blockers (see Chapter II-7).

The Content of the Book

This book presents the role of analogues in medicinal chemistry and their contribution to new drug discovery. In Part I, the various chemical approaches to making analogues are reviewed and the general aspects of analogue-based drug discovery (ABDD) are summarized, whilst in Part II are described 19 examples of analogue classes, together with relevant case studies. A table of structural and pharmacological analogues of the most frequently used drugs is provided in Part III.

List of Contributors

Erika M. Alapi
Gedeon Richter, Ltd
P.O. Box 27
1475 Budapest 10
Hungary

Sándor Alföldi
Semmelweis University
First Medical Clinic
Korányi S. u. 2/a
1083 Budapest
Hungary

István Bitter
Semmelweis University
Clinic of Psychiatry and Psychotherapy
Balassa u. 6
1083 Budapest
Hungary

Eli Breuer
The Hebrew University of Jerusalem
School of Pharmacy
Department of Medicinal Chemistry
P.O. Box 12065
91120 Jerusalem
Israel

Enar Carlsson
AstraZeneca R&D Mölndal
431 83 Mölndal
Sweden

László Dézsi
Gedeon Richter Ltd
P.O. Box 27
1475 Budapest 10
Hungary

Paul W. Erhardt
Center for Drug Design and
Development
The University of Toledo
College of Pharmacy
Toledo, OH 43606-3390
USA

Csaba Farsang
Semmelweis University
First Medical Clinic
Korányi S. u. 2/a
1083 Budapest
Hungary

János Fischer
Gedeon Richter, Ltd
P. O. Box 27
1475 Budapest 10
Hungary

C. Robin Ganellin
University College London
Department of Chemistry
20 Gordon Street
London WC1H OAJ
UK

Analogue-based Drug Discovery. IUPAC, János Fischer, and C. Robin Ganellin (Eds.)
Copyright © 2006 WILEY-VCH Verlag GmbH & Co. KGaA, Weinheim
ISBN: 3-527-31257-9

Giovanni Gaviraghi
Sienabiotech SpA
Via Fiorentina, 1
53100 Siena
Italy

Sándor Kerpel-Fronius
Semmelweis University
Department of Pharmacology
Nagyvárad tér 4
1445 Budapest
Hungary

Béla Kiss
Department of Molecular Pharmacology
Pharmacological and Drug Safety
Research
Gedeon Richter Ltd
Gyömrői u. 19-21
1103 Budapest
Hungary

Hugo Kubinyi
retired from BASF AG
Donnersbergstraße 9
67256 Weisenheim am Sand
Germany

Per Lindberg
AstraZeneca R&D Mölndal
431 83 Mölndal
Sweden

Sándor Mahó
Gedeon Richter Ltd
P.O. Box 27
1475 Budapest 10
Hungary

Lajos Matos
Szent János Hospital
Department of Cardiology
Andrássy út 49
1061 Budapest
Hungary

Christopher R. McCurdy
Department of Medicinal Chemistry
The University of Mississippi
P.O. Box 1848
University, MS 38677
USA

Uwe Petersen
retired from Bayer AG
Auf dem Forst 4
51375 Leverkusen
Germany

John R. Proudfoot
Medicinal Chemistry Department
Boehringer Ingelheim
Pharmaceuticals Inc.
900 Ridgebury Rd
Ridgefield, CT 06877-0368
USA

Csaba Sánta
Gedeon Richter Ltd
P.O. Box 27
1475 Budapest 10
Hungary

Jörg Senn-Bilfinger
Altana Pharma AG
Byk-Gulden-Str. 2
78467 Konstanz
Germany

Ernst Sturm
Altana Pharma AG
Byk-Gulden-Str. 2
78467 Konstanz
Germany

Zeev Tashma
The Hebrew University of Jerusalem
Department of Medicinal Chemistry
P.O. Box 12065
91120 Jerusalem
Israel

Henk Timmerman
Vrije Universiteit
Department of Pharmacochemistry
De Boelelaan 1083
1081 HV Amsterdam
The Netherlands

Zoltán Tuba
Gedeon Richter Ltd
P.O. Box 27
1475 Budapest 10
Hungary

Camille G. Wermuth
Prestwick Chemical Inc.
Boulevard Gonthier d'Andernach
Strasbourg Innovation Park
67400 Illkirch Cedex
France

Rudolf Wiechert
retired from Schering AG
Petzowerstrasse 8a
14109 Berlin
Germany

Abbreviations

ABDD	analogue-based drug discovery
ABS	acute bacterial sinusitis
ACE	angiotensin-converting enzyme
AChE	acetylcholinesterase
ADHD	attention deficit hyperactivity disorder
ADMET	absorption, distribution, metabolism, elimination, toxicity
AECB	acute exacerbation of chronic bronchitis
AHI	antihistaminic index
AHL	amphetamine-induced hyperlocomotion
ANP	atrial natriuretic peptide
APC	apomorphine-induced climbing
ARBs	angiotensin II receptor blockers
AUC	area under the curve
BPs	bisphosphonates
CAD	coronary artery disease
cAMP	cyclic adenosine monophosphate
CAP	community-acquired pneumonia
CAR	conditioned avoidance response
CATL	catalepsy-induced dosing
CCK-B	cholecystokinin-B
cGMP	cyclic guanosine monophosphate
CHD	coronary heart disease
CHF	congestive heart failure
CNS	central nervous system
CoA	coenzyme A
COPD	chronic obstructive pulmonary disease
COX-1	cyclooxygenase 1
COX-2	cyclooxygenase 2
CRF	corticotropin-releasing factor
CWA	chemical warfare agent
DBP	diastolic blood pressure
DOP	delta opioid receptor
ECL	enterochromaffin-like

Analogue-based Drug Discovery. IUPAC, János Fischer, and C. Robin Ganellin (Eds.)
Copyright © 2006 WILEY-VCH Verlag GmbH & Co. KGaA, Weinheim
ISBN: 3-527-31257-9

ED	erectile dysfunction
EDRF	endothelium-derived relaxing factor
EPS	extrapyramidal symptoms
FPP	farnesyl diphosphate
GABA	γ-aminobutyric acid
GERD	gastroesophageal reflux disease
GFR	glomerular filtration rate
GPCRs	G protein-coupled receptors
GRH	gonadotropin-releasing Hormone
HAP	hydroxyapatite
HDL-C	high-density lipoprotein-cholesterol
HERG	human ether-a-go-go related gene
HMG-CoA	hydroxymethyl-glutaryl coenzyme A
5-HT	5-hydroxytryptamine (serotonin)
HTS	high-throughput screening
HTW	head twitch
ICAM	intercellular cell adhesion molecule
IL	interleukin
IMPase	inositol monophosphatase
IND	investigational New Drug
iNOS	inducible nitric oxide synthase
IPP	isopentenyl diphosphate
ISA	intrinsic sympathomimetic activity
KOP	kappa opioid receptor
LDL-C	low-density lipoprotein-cholesterol
LVH	left ventricular hypertrophy
MAO	monoamine oxidase
MDRSP	multidrug-resistant S. pneumoniae
MDRTB	multidrug-resistant tuberculosis
MG	myasthenia gravis
MI	myocardial infarction
MIC	minimum inhibitory concentration
MOP	mu opioid receptor
MPC	mutant prevention concentration
MPP^+	1-methyl-4-phenylpyridinium
MPPP	1-methyl-4-phenyl-4-propionoxypiperidine
MPTP	1-methyl-4-phenyl-1,2,5,6-tetrahydropyridine
MRSA	methicillin-resistant S. aureus
MSA	membrane-stabilizing activity
MSSA	methicillin-sensitive S. aureus
NA/NE	noradrenaline (norepinephrine)
NMDA	N-methyl-D-aspartate
NMJ	neuromuscular junction
NOS	nitric oxide synthase
NSAIDs	nonsteroidal anti-inflammatory drugs

OAD	obstructive airway disease
OATP	organic acid transporter polypeptides
PAA	phosphonoacetic acid
PAF	platelet-activating factor
PASS	Prediction of Activity Spectra for Substances
PCP	phencyclidine
PDE	phosphodiesterase
PET	positron emission tomography
PFA	phosphonoformic acid
Pgp	P-glycoprotein
PK	pharmacokinetic
PKC	protein kinase C
PNS	peripheral nervous system
PPI	proton pump inhibitor
RA	rheumatoid arthritis
RAAS	renin-angiotensin-aldosterone system
RAS	renin-angiotensin system
RGD	arginine-glycine-aspartic acid
SAR	structure-activity relationship
SBP	systolic blood pressure
SHR	spontaneously hypertensive rat
SOSA	Selective Optimization of Side Activities
SSSI	skin and skin structure infections
TNF_a	tumor necrosis factor-alpha
TXA-2	thromboxane-A_2
ULN	upper limit of normal value

Part I
General Aspects of Analogue-Based Drug Discovery

1
Analogues as a Means of Discovering New Drugs

Camille G. Wermuth

1.1
Designing of Analogues

The term analogy, derived from the Latin and Greek *analogia*, is used in natural science since 1791 to describe structural and functional similarity [1]. Extended to drugs, this definition implies that the analogue of an existing drug molecule shares chemical and therapeutic similarities with the original compound. The chemical design of analogues makes use of simple and traditional procedures of medicinal chemistry such as the synthesis of homologues, vinylogues, isosteres, positional isomers, optical isomers, transformation of ring systems, and the synthesis of twin drugs. These approaches are well-documented in textbooks, and have been reviewed extensively [2–7]; thus, they will be considered only briefly hereafter.

1.1.1
Analogues Produced by Homologous Variations

1.1.1.1 Homology Through Monoalkylation

Simple homologous variations applied on a neuramidase-inhibiting lead [8] achieved a 6300-fold increase in potency (Fig. 1.1).

1.1.1.2 Polymethylenic Bis-Ammonium Compounds: Hexa- and Decamethonium
Compounds having the general formula $(CH_3)_3 N^+-(CH_2)_n-N^+(CH_3)_3$ usually show high affinity for the cholinergic receptors. When the values of n are intermediate ($n = 5$ or 6: penta- or hexamethonium), these compounds behave like cholinergic *agonists* (towards the sympathetic ganglia). For higher values ($n = 10$: decamethonium), the compounds become *antagonists* of acetylcholine (at the muscular end-plate). In both cases, increasing acetylcholine levels displace them from their binding sites. When considering neuromuscular blockade, one observes again a curve with an asymmetric profile: sudden changes between $n = 6$ and $n = 8$, and then progressive diminution between $n = 9$ and $n = 12$.

Analogue-based Drug Discovery. IUPAC, János Fischer, and C. Robin Ganellin (Eds.)
Copyright © 2006 WILEY-VCH Verlag GmbH & Co. KGaA, Weinheim
ISBN: 3-527-31257-9

1.1.1.3 Homology in Cyclic Compounds

Homology in cyclic compounds can dramatically change the affinity of a ligand for its target. This is illustrated by a series of cyclic ACE inhibitors related to enalapril (Fig. 1.2).

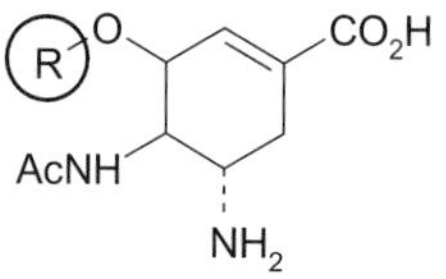

Neuraminidase inhibition

R =	IC_{50} (nM)
H	6.300
CH_3 -	3.700
CH_3 - CH_2 -	2.000
CH3 - CH2 - CH2 -	180
CH_3 - CH_2 - CH_2 - CH_2 -	300
$(CH_3)_2$ -CH_2 - CH_2-	200
CH_3 - CH_2 - $CH(CH_3)$ -	10
$(CH_3$ - $CH_2)_2$ - CH -	1
$(CH_3$ - CH_2 - $CH_2)$ CH -	16
Cyclopentyl	22
Cyclohexyl	60
Phenyl	530

Fig. 1.1 Monoalkylated, cyclohexene-derived, neuraminidase inhibitors.

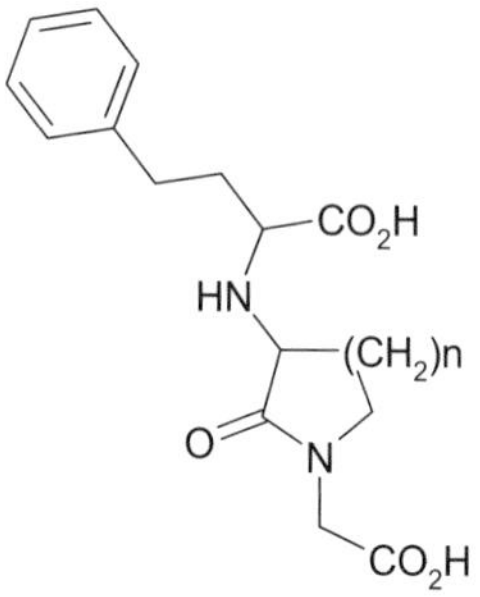

ACE inhibition IC_{50}(nM)

n = 1 :	19,000
n = 2 :	1,700
n = 3 :	19
n = 4 :	4.8

Fig. 1.2 Homology in cyclic compounds.

For these compounds, an almost 4000-fold increase in activity was observed in passing from the five-membered to the eight-membered homologue [9].

1.1.2
Analogues Produced by Vinylogy

Tolcapone (Fig. 1.3) was designed as an inhibitor of the enzyme catechol O-methyltransferase, and is useful in the L-DOPA treatment of Parkinson's disease [10]. In avoiding the methylation of L-DOPA as well as that of dopamine, it prolongs the beneficial activities of these molecules.

Catechol O-methyltransferase inhibition represents therefore a valuable adjuvant to the L-DOPA decarboxylase inhibition. Unfortunately, tolcapone exhibited severe liver damage and had to be removed from the market. The corresponding vinylogue *entacapone* is devoid of this side effect [8–10].

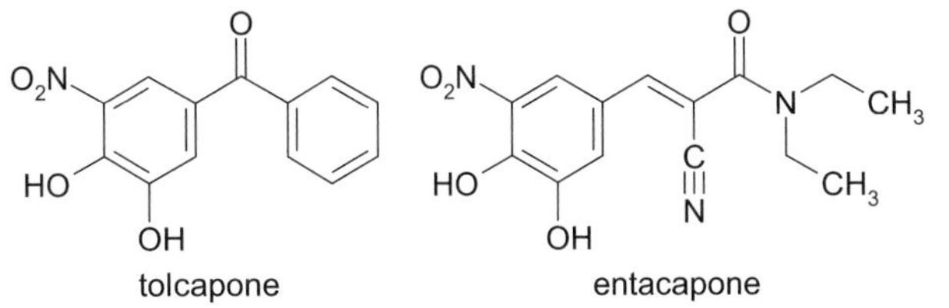

Fig. 1.3 The vinylogy principle applied to the catechol O-methyltransferase inhibitor, tolcapone.

1.1.2.1 Zaprinast Benzologues

A very convincing example of the usefulness of benzologues is provided by the synthesis of compound A, a linear benzologue of the prototypical phosphodiesterase type 5 (PDE$_5$) inhibitor zaprinast, and its optimization to potent and selective PDE$_5$ inhibitors such as compound B [11] (Fig. 1.4).

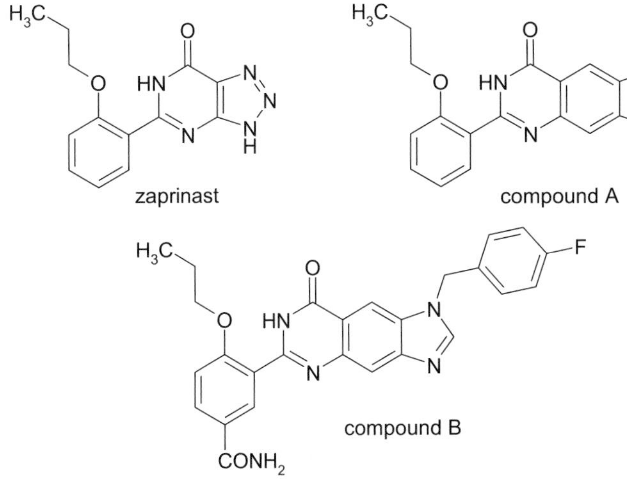

Fig. 1.4 Linear benzologues derived from zaprinast [11].

1.1.3
Analogues Produced by Isosteric Variations

1.1.3.1 The Dominant Parameter is Structural

Structural factors are important when the portion of the molecule involved in the isosteric change serves to maintain other functions in a particular geometry. That is the case for tricyclic psychotropic drugs (Fig. 1.5).

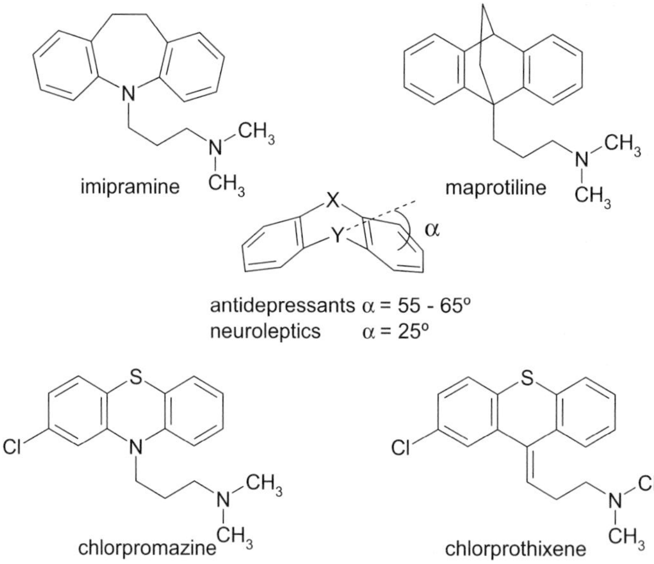

Fig. 1.5 The tricyclic antidepressants (imipramine and maprotiline) are characterized by a dihedral angle of 55° to 65° between the two benzo rings; this angle is only 25° for the tricyclic neuroleptics (chlorpromazine, chlorprothixene) [12].

For the two antidepressants (imipramine and maprotiline), the bioisosterism is geometrical insofar that the dihedral angle a formed by the two benzo rings is comparable: $a = 65°$ for the dibenzazepine and $a = 55°$ for the dibenzocycloheptadiene [12]. This angle is only 25° for the neuroleptic phenothiazines and for the thioxanthenes. In these examples, the part of the molecule modified by isosterism is not involved in the interaction with the receptor. It serves only to position correctly the other elements of the molecule.

1.1.3.2 The Dominant Parameter is Electronic

The electron-attracting nitro group in the antibiotic chloramphenicol (Tab. 1.1) has been replaced by other electron-attracting functions such as methyl-sulfonyl (thiamphenicol) or an acetyl (or cetophenicol). Apparently, the dominant feature is of electronic nature.

Similar electronic effects are found in the benzodiazepines series in which the chlorine atom of diazepam can be exchanged with a bromine (bromazepam) or with a nitro group (nitrazepam).

Tab. 1.1 Isosteric replacements in the amphenicol family.

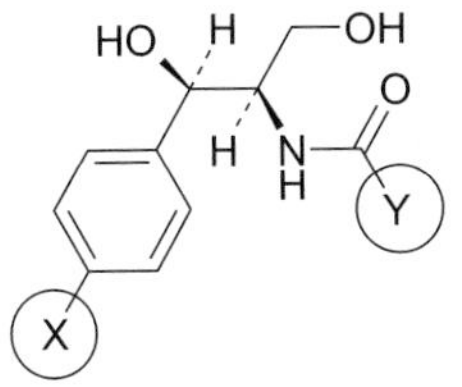

Compound	X	Y
Chloramphenicol	$-NO_2$	$-CH-Cl_2$
Thiamphenicol	CH_3-SO_2-	$-CH-Cl_2$
Cetophenicol	CH_3-CO-	$-CH-Cl_2$

1.1.3.3 The Dominant Parameter is Lipophilicity

Typical examples of lipophilic analogues of prototypes are eptastigmine (*N*-heptyl-physostigmine), derived from physostigmine by replacement of the N-methyl group by a *n*-heptyl group [13,14], and tiagabine, in which a diaryl-butenyl chain is grafted to the nitrogen atom of nipecotic acid in order to yield a compound able to cross the blood–brain barrier [15] (Fig. 1.6).

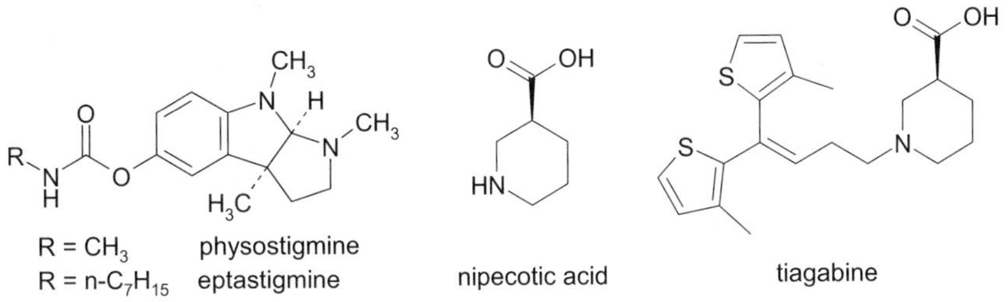

R = CH₃   physostigmine
R = n-C₇H₁₅   eptastigmine

nipecotic acid

tiagabine

Fig. 1.6 Lipophilic analogues of physostigmine and of nipecotic acid.

1.1.4
Positional Isomers Produced as Analogues

In a series of nonpeptide corticotropin-releasing factor 1 (CRF1) antagonists, scientists from Neurocrine [16] observed a dramatic change in affinity simply by shifting one nitrogen atom of the pyrimidine ring (Fig. 1.7).

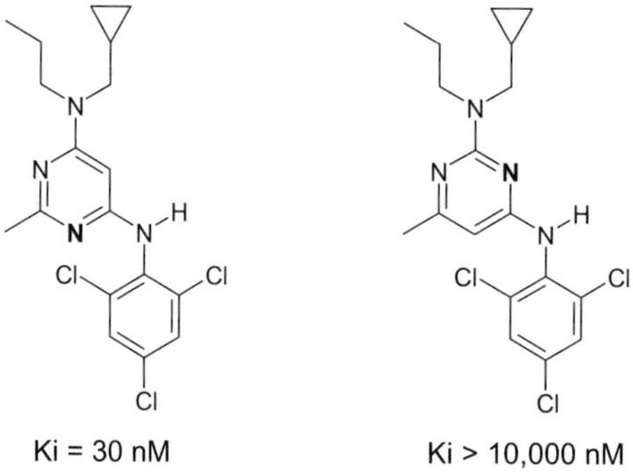

Ki = 30 nM   Ki > 10,000 nM

Fig. 1.7 A particularly striking effect of positional isomerism [K_i values for binding to the human corticotropin-releasing factor 1 (CRF1) receptor] [16].

1.1.5
Optical Isomers Produced as Analogues

Due to their almost identical chemical structure, enantiomers represent a subtle class of analogues. Often in a pair of enantiomers, the desired biological activity is concentrated in only one enantiomer. Then, the passage from a racemic mixture to the pure active eutomer – which is usually termed "racemic switch" – can produce an improved drug. However, in some cases and despite their similar constitution, both enantiomers can have totally different pharmacodynamic or pharmacokinetic profiles.

1.1.5.1 Racemic Switches
A general trend in the pharmaceutical industry is to switch from racemates to single enantiomers. Examples are given by (*R*)-(–)-verapamil, (*S*)-fluoxetin, (*S*)-ketoprofen, (*R*)-albuterol, levofloxacin, esomeprazole (see Chapter II-2), levocetirizine, and many others [17,18]. In addition to the quality improvement of the drug, this switch represents also a way to prolong its life insofar that the isolated eutomer is legally considered as a new drug entity.

1.1.5.2 Specific Profile for Each Enantiomer
The splitting of the anthelmintic drug tetramisole into its two components reveal nematocidal and immunostimulant properties for *S*(–)-levamisole and antidepressant properties for *R*(+)-dexamisole [19,20] (Fig. 1.8).

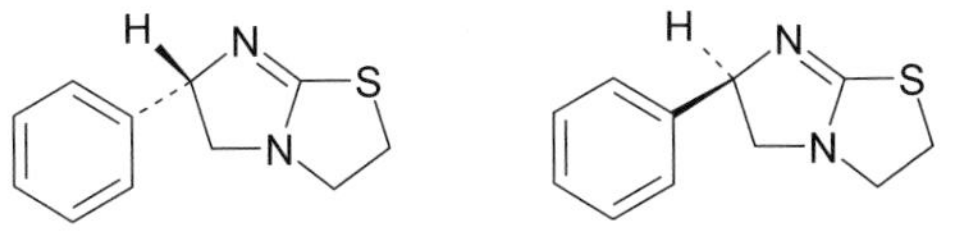

S-(-)-levamisole   R-(+)-dexamisole

Fig. 1.8 The racemate tetramisole can be split into the nematocidal and immunostimulant S(–)-levamisole and into the
antidepressant R(+)-dexamisole [19,20].

1.1.6
Analogues Produced by Ring Transformations

When active molecules contain cyclic systems, these can be opened, expanded,
contracted, and modified in many other ways, or even abolished. Conversely, non-
cyclic molecules can be cyclized, attached to, or included in, ring systems.

Two interesting examples are found in cyclic analogues of β-blockers (Fig. 1.9).
Cyclization to the morpholine yields the antidepressant viloxazine (mode 1),
whereas cyclization to chromanols (mode 2) yields the potassium channel blocker
cromakalim.

viloxazine

cromakalim

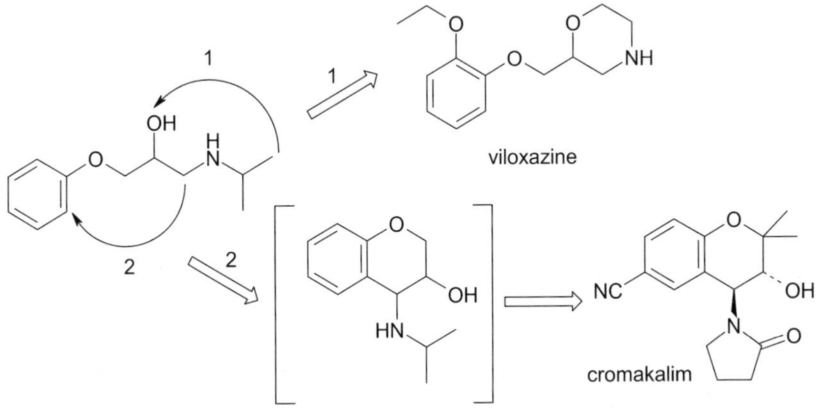

Fig. 1.9 Two modes of cyclization of the side chain in β-blockers yield the noradrenaline (NA) reuptake inhibitor viloxazine
(mode 1) and the potassium channel blocker cromakalim
(mode 2).

1.1.7
Twin Drugs

The search for cationic cholinergic agents has led to numerous twin drugs (Fig. 1.10).
The bis-quaternary ammonium salts hexamethonium and decamethonium are
potent blockers in ganglia and in neuromuscular junctions, respectively. Other

neuromuscular blocking agents such as succinyl and sebacyl dicholines can be regarded as pure acetylcholine twin drugs.

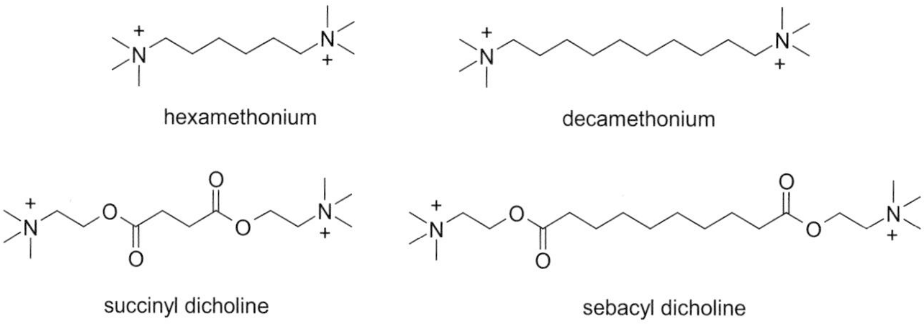

Fig. 1.10 Cholinergic twin drugs.

Other examples of twin drugs are the antioxidant probucol, which lowers the cholesterol level in blood, and cromolyn, a chromone heterocycle, which is useful in the inhalational treatment of bronchial asthma (Fig. 1.11).

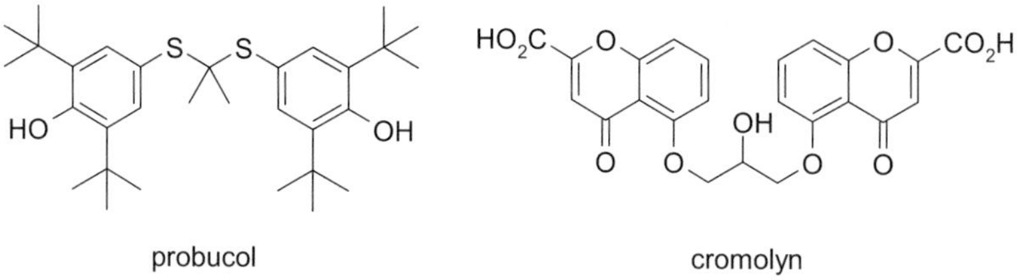

Fig. 1.11 Probucol and cromolyn are also twin drugs.

1.2
The Pros and Cons of Analogue Design

The usual objectives of analogue design are the identification and development of a possibly improved version of a prototype drug. Such compounds are often "direct analogues", and therefore are chemically and pharmacologically similar to the prototype drug. It should also present some advantages over the prototype. However, to achieve these goals is not always an easy task.

1.2.1
The Success is Almost Warranted

A reassuring aspect of making therapeutic copies resides in the quasi-certainty to end with active drugs in the desired therapeutic area. It is indeed extremely rare – and practically improbable – that a given biological activity is unique to a single molecule. Molecular modifications allow the preparation of additional products for which one can expect, if the investigation has been sufficiently prolonged, a comparable activity to that of the copied model, and perhaps even a better one. This factor is comforting for the medicinal chemist, as well as for the financiers that subsidize him or her. It is necessary, however, to bear in mind that the original inventor of a pioneer drug possesses a technological and scientific advantage over analogue producers and, moreover, they are able to design a certain number of analogues of their own drug much earlier than the competitors.

1.2.2
The Information is Available

A second element which favors R&D based on analogues, comes from the information already gained with the original prototype. As soon as the pharmacological models that served to identify the activity profile of a new prototype are known, it suffices to apply them to the analogues. In other words, the pharmacologist will know in advance to what kind of activity he/she will meet and which tests he/she will have to apply to select the desired activity profile. In addition, during clinical investigations, the original studies undertaken with the lead compound, will serve as a reference and can be adopted unchanged to the evaluation of the analogues.

One criticism of this approach is that, in selecting a new active molecule by means of the same pharmacological models as were used for the original compound, one will inevitably end up with a compound presenting an identical activity profile; thus, the innovative character of such a research is very modest.

1.2.3
Financial Considerations

Finally, financial arguments can play in favor of analogue research. Thus, it may be important – and even vital – for a pharmaceutical company or for a national industry, to have its own drugs rather than to subcontract a license. Indeed, in paying license fees, an industry impoverishes its own research. Moreover, the financial profitability of a research based on *direct analogues* can appear to be higher, because no investment in fundamental research is required. The counterpart is that placing the copy on the market will naturally occur later than that of the original drug, and thus it will make it more difficult to achieve a high sales ranking – all the more so because the *direct analogues* will be in competition with other analogues targeting a similar market.

In reality, the situation is more subtle because very often the synthesis of *direct analogues* is justified by a desire to improve the existing drug. Thus, for penicillins the chemical structure that surrounds the beta-lactam ring is still being modified. Current antibiotics that have been derived from this research (e.g., the cephalosporins) are more selective, more active on resistant strains, and can be administered by the oral route. They are as different from the parent molecule as a recent car compared to a 40-year-old model! In other words, innovation can result from the sum of a great number of stepwise improvements, as well as from a major breakthrough.

1.2.4
Emergence of New Properties

It can happen that, during the R&D period of an analogue, a totally new property which was not present in the original molecule appears unexpectedly. Thanks to the emergence of such a new activity, the therapeutic copy becomes in turn a new lead structure. This was the case for imipramine, which was initially synthesized as an analogue of chlorpromazine and presented to the investigators for study of its antipsychotic profile [21]. During its clinical evaluation, this substance demonstrated much more activity against depressive states than against psychoses. Imipramine has truly opened, since 1954, a therapeutic avenue for the pharmacological treatment of depression.

Preparing totally new drugs emerging from already well-known lead structures is certainly one of the most exciting parts of medicinal chemistry, and will form the basis of the following sections.

1.3
Analogue Design as a Means of Discovering New Drugs

1.3.1
New Uses for Old Drugs

In some cases, a new clinical activity observed for an old drug is sufficiently potent and interesting to justify the immediate use of the drug in the new indication, and this is illustrated below.

Amiodarone, for example (Fig. 1.12), was introduced as a coronary dilator for angina, but concern about corneal deposits, discoloration of skin exposed to sunlight and thyroid disorders led to the withdrawal of the drug in 1967. However, in 1974 amiodarone was found to be highly effective in the treatment of a rare type of arrhythmia known as the Wolff–Parkinson–White syndrome. Accordingly, amiodarone was reintroduced specifically for that purpose [22].

Benziodarone was initially used in Europe as a coronary dilator, and proved later to be a useful uricosuric agent. However, it is now withdrawn from the market due to several cases of jaundice associated with its use [22]. The corresponding brominated analogue, benzbromarone was specifically marketed for its uricosuric properties.

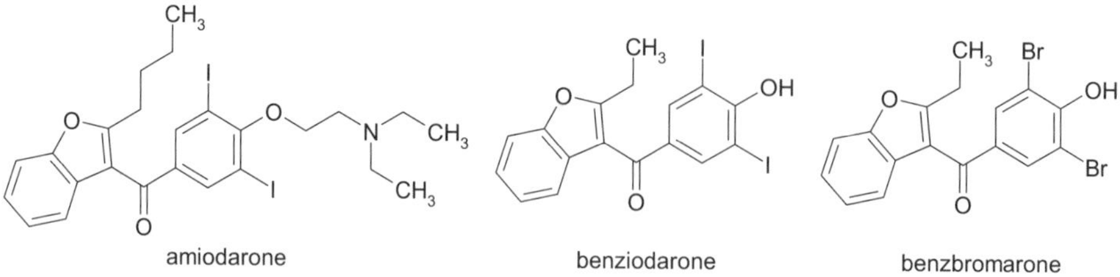

amiodarone   benziodarone   benzbromarone

Fig. 1.12 Structures of the arones.

Thalidomide was initially launched as a sedative/hypnotic drug (Fig. 1.13), but withdrawn because of its extreme teratogenicity. However, under restricted conditions (no administration during pregnancy, or to any woman of childbearing age), it found a new use as an immunomodulator. Thalidomide seems particularly effective in the treatment of erythema nodosum leprosum, a possible complication of the chemotherapy of leprosy [23].

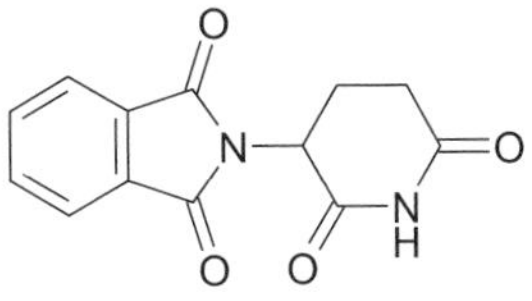

Fig. 1.13 Structure of thalidomide. The marketed compound is the racemate.

In 2001, the antimalarial drug *quinacrine* and the antipsychotic drug *chlorpromazine* (Fig. 1.14) were shown to inhibit prion infection in cells. Prusiner and co-workers [24] identified the drugs independently, and found that they inhibited the conversion of normal prion protein into infectious prions, and also cleared prions from infected cells. Both drugs can cross over from the bloodstream to the brain, where the prion diseases are localized.

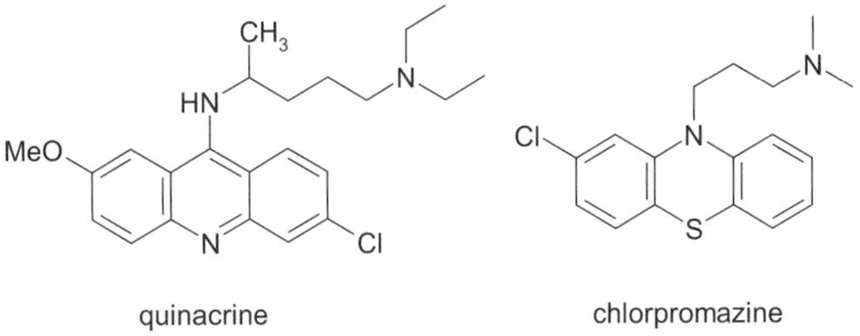

quinacrine   chlorpromazine

Fig. 1.14 Old drugs, new use. The antimalarial drug quinacrine and the antipsychotic drug chlorpromazine are able to inhibit prion infection [24].

Another recent example is provided by the discovery of the use of sildenafil (Viagra®; Fig. 1.15), a phosphodiesterase type 5 (PDE$_5$) inhibitor, as an efficacious, orally active agent for the treatment of erectile dysfunction [25,26]. Initially, this compound was brought to the clinic as an hypotensive and cardiotonic substance, but its usefulness in erectile dysfunction resulted unexpectedly from clinical observations made during a 10-day toleration study in Wales [27].

In many therapeutic families, each generation of compounds induces the birth of the following one. This happened in the past for the sulfamides, penicillins, steroids, conazoles, prostaglandins, and tricyclic psychotropics families, and one can draw real genealogical trees representing the progeny of the discoveries. More recent examples are found in the domain of ACE inhibitors, histaminergic H$_2$ antagonists, angiotensin II receptor antagonists, and HMG-CoA reductase inhibitors.

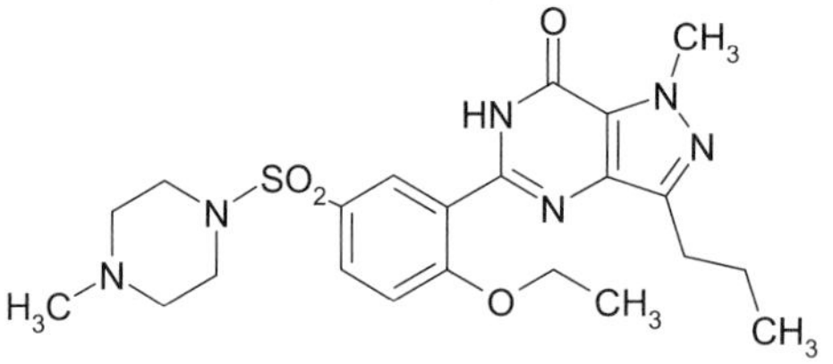

Fig. 1.15 Structure of the phosphodiesterase type 5 (PDE$_5$) inhibitor sildenafil [25,26].

Research programs based on the exploitation of side effects are of great interest in the discovery of new paths, in so far as they depend on information about activities *observed directly in man* and not in animals. On the other hand, they allow the detection of new therapeutic activities *even when no pharmacological models in animals exist.*

1.3.2
The PASS Program

Another very valuable tool, which allows simultaneous evaluation of the "drug-likeness" and prediction of the probable activity profile, is found in the program PASS (Prediction of Activity Spectra for Substances) developed by Poroikov and his team [28]. This approach consists of comparing a newly prepared molecule to a training set of about 35 000 active compounds for which the main and the side pharmacological effects, the mechanism of action, the mutagenicity, the carcinogenicity, the teratogenicity, and the embryotoxicity are (at least partly) known. The program then predicts the potential biological activity of the new molecule. In a published example [28], PASS was applied to a set of 130 pharmaceuticals from the list of the top 200 medicines. The known pharmacological effects were found in the predicted activity spectrum in 93.2% of cases. Additionally, the probability of some additional effects was also predicted to be significant, including angiogenesis inhibition, bone formation stimulation, possible use in cognition disorders treatment, and multiple sclerosis treatment. These predictions, if confirmed experimentally, may provide new leads from drugs that are already on the market. Most of the known side and toxic effects were also predicted by PASS.

1.3.3
New Leads from Old Drugs: The SOSA Approach

1.3.3.1 Definition

The SOSA approach (SOSA = Selective Optimization of Side Activities) represents an original alternative to high-throughput screening (HTS) [29–31]. SOSA consists of two steps:
- Perform screening assays on newly identified pharmacological targets only with a limited set (approximately 1000 compounds)

of well-known drug molecules. For these drugs, bioavailability and toxicity studies have already been performed, and they have proven usefulness in human therapy.

- Once a hit is observed with a given drug molecule, the task is then to prepare analogues of this molecule in order to transform the observed "side activity" into the main effect and to strongly reduce or abolish the initial pharmacological activity.

1.3.3.2   Rationale

The rationale behind the SOSA approach lies in the fact that, in addition to their main activity, almost all drugs used in human therapy show one or several side effects. In other words, if they are able to exert a strong interaction with the main target, they exert also less strong interactions with some other biological targets. Most of these targets are unrelated to the primary therapeutic activity of the compound. The objective is then to proceed to a reversal of the affinities, with the identified side effect becoming the main effect, and vice-versa.

1.3.3.3   Availability

A chemical library available for the SOSA approach is the Prestwick Chemical Library [32]. This contains 1120 biologically active compounds with high chemical and pharmacological diversity, as well as known bioavailability and safety in humans. About 90% of the compounds are well-established drugs, and about 10% are bioactive alkaloids. For scientists interested in drug-likeness, such a library certainly fulfills in the most convincing way the quest for "drug-like" leads!

1.3.3.4   Examples

Antihypertensives
A typical illustration of the SOSA approach is given by the development of *selective ligands for the endothelin ET$_A$ receptors* by scientists from Bristol-Myers-Squibb [28,33]. Starting from an in-house library, the antibacterial compound sulfathiazole (Fig. 11.6) was an initial, but weak, hit (IC$_{50}$ = 69 μM). Testing of related sulfonamides identified the more potent sulfisoxazole (IC$_{50}$ = 0.78 μM). Systematic variations led finally to the potent and selective ligand BMS-182874. *In vivo*, this compound was orally active and produced a long-lasting hypotensive effect.

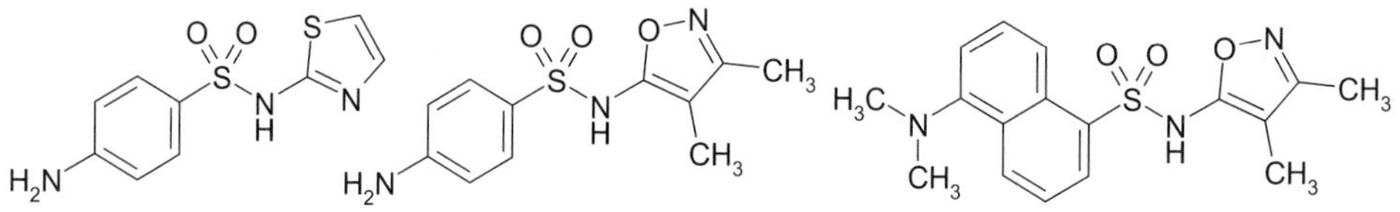

1 sulfathiazole
ET_A IC_{50} = 69 μM

2 sulfisoxazole
ET_A IC_{50} = 0.78 μM

3 BMS-182874
ET_A IC_{50} = 0.15 μM

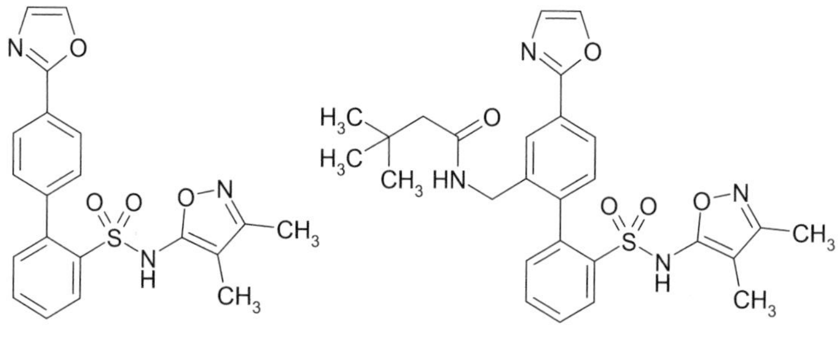

4 BMS-193884
ET_A Ki = 1.4 nM

5 BMS-207940
ET_A Ki = 0.010 nM

Fig. 1.16 A successful SOSA approach allowed the identification of the antibacterial sulfonamide sulfathiazole as a ligand of the endothelin ET_A receptor and its optimization to the selective and potent compounds BMS-182874, BMS-193884, and BMS-207940 [34,35].

Further optimization guided by pharmacokinetic considerations led the Bristol-Myers-Squibb team to replace the naphthalene ring by a diphenyl system [35]. Among the prepared compounds, **4** (BMS-193884, ET_A K_i = 1.4 nM; ET_B K_i = 18 700 nM) showed promising hemodynamic effects in a Phase II clinical trial for congestive heart failure. More recent studies led to the extremely potent antagonist **5** (BMS-207940 ET_A K_i = 10 pM) presenting an 80 000-fold selectivity for ET_A versus ET_B. The bioavailability of **5** is 100% in rats, and it exhibits oral activity already at a 3 μM kg^{-1} dosing [35].

Cholinergic Agonists

In a second example, the starting lead was the antidepressant minaprine (Fig. 1.17). In addition to reinforcing serotonergic and dopaminergic transmission, this amino-pyridazine possesses weak affinity for muscarinic M_1, receptors (K_i = 17 μM). Simple chemical variations allowed the dopaminergic and serotonergic activities to be abolished, and the cholinergic activity to be boosted to nanomolar concentrations [36–38].

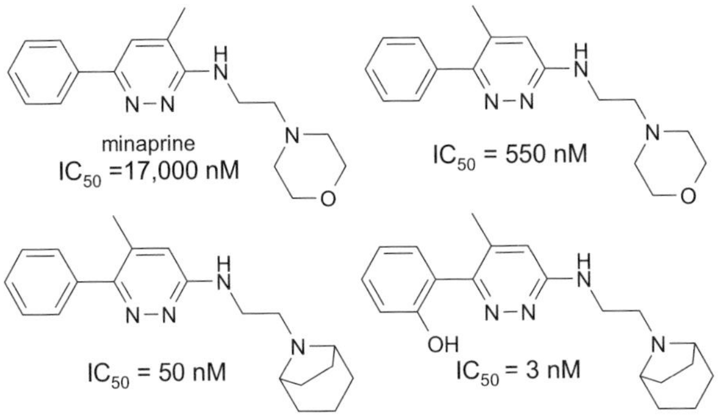

Fig. 1.17 Progressive passage from minaprine to a potent and selective partial muscarinic M_1 receptor agonist [36–38].

Acetylcholinesterase Inhibitors

Starting from the same minaprine lead, it was imagined that this molecule, in being recognized by the acetylcholine receptors, should also be recognized by the acetylcholine enzyme. It transpired that minaprine had only a very weak affinity for acetylcholinesterase (600 μM on electric eel enzyme), but relatively simple modifications (creation of a lipophilic cationic head, increase of the side chain length, and bridging the phenyl and the pyridazinyl rings) led to nanomolar affinities (Fig. 1.18) [39,40].

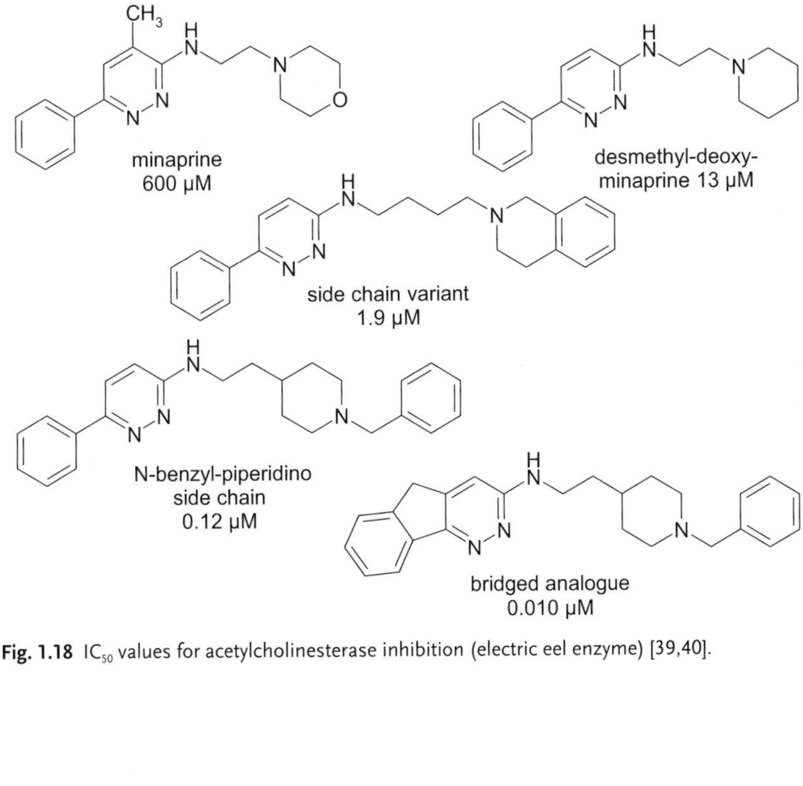

Fig. 1.18 IC_{50} values for acetylcholinesterase inhibition (electric eel enzyme) [39,40].

Corticotropin-Releasing Factor (CRF) Antagonists

Another interesting switch consisted in the progressive passage from desmethyl-minaprine **6** to the bioisosteric thiadiazole **7** (Fig. 1.19), and then to the bioisosteric thiazoles. Tri-substitution on the phenyl ring and replacement of the aliphatic morpholine by a pyridine led to compound **8** which exhibited some affinity for the receptor of the 41 amino-acid neuropeptide CRF. Further optimization led to nanomolar CRF antagonists such as **9** [41,42].

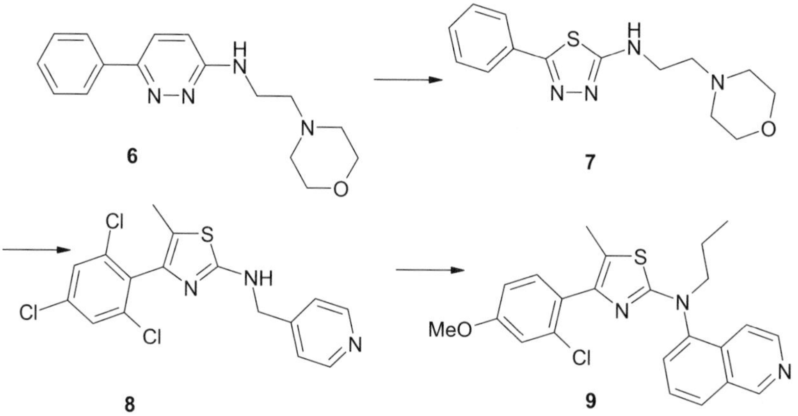

Fig. 1.19 Switch from the antidepressant molecule minaprine to the potent CRF receptor antagonist 9 [41,42].

Subtype-Selective Dopamine D$_3$ Receptor Ligands

In another study, chemical variations of the D$_2$/D$_3$ nonselective antagonist sulpiride (Fig. 1.20) led to the compound Do 897, a selective and potent D$_3$ receptor partial agonist [43].

1.3.3.4 Discussion

The SOSA approach appears to be an efficient strategy for drug discovery, particularly as it is based on the screening of drug molecules and thus automatically yields drug-like hits. Before starting a costly HTS campaign, SOSA can represent a seductive alternative. Once the initial screening has provided a hit, it will be used as the starting point for a drug discovery program. Using traditional medicinal chemistry as well as parallel synthesis, the initial "side activity" is transformed into the main activity and, conversely, the initial main activity is strongly reduced or abolished. This strategy leads with a high probability to *safe, bioavailable, original*, and *patentable* analogues.

Safety and Bioavailability

During years of practicing SOSA approaches, it has been observed at Prestwick that, starting with a drug molecule as a lead substance in performing analogue

sulpiride (1 : 2) nafadotride (1 : 9.6)

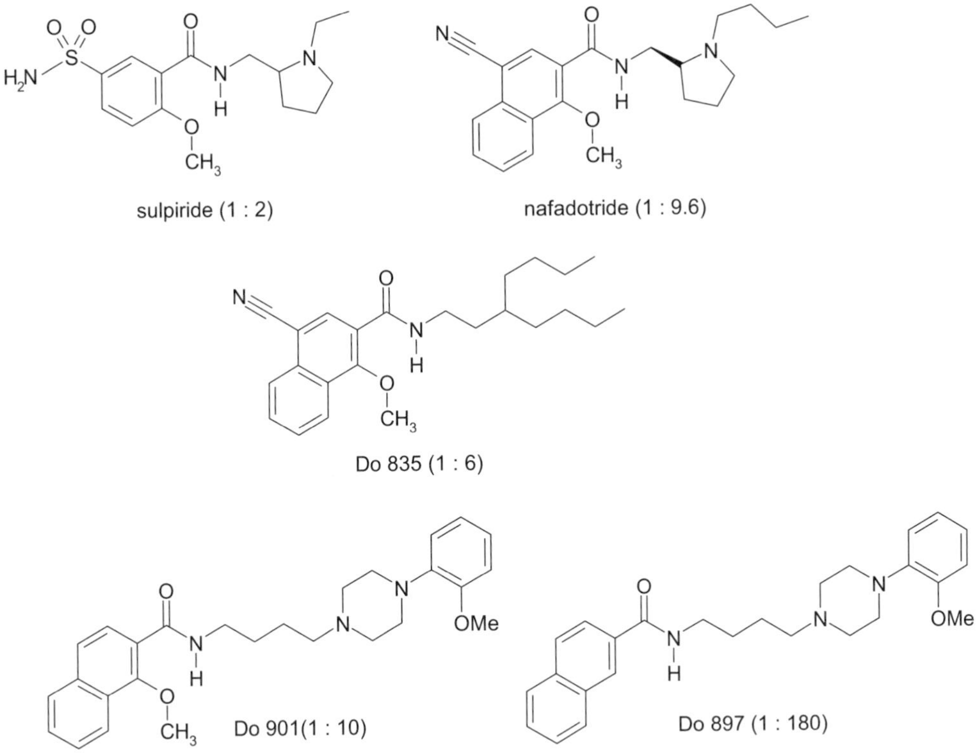

Do 835 (1 : 6)

Do 901 (1 : 10) Do 897 (1 : 180)

Fig. 1.20 The progressive change from the D_2/D_3 receptor nonselective antagonists to the highly D_3-selective compound Do 897 [43]. The numbers in parentheses indicate the D_2/D_3 affinity ratio.

synthesis, has increased notably the probability of obtaining safe new chemical entities. In addition, most of these satisfy Lipinski's [44], Veber's [45], Bergström's [46], and Wenlock's [47] observations in terms of solubility, oral bioavailability, and drug-likeness.

Patentability

When a well-known drug hits with a new target, there is a risk that several hundreds or thousands of analogues of the original drug molecule are already synthesized by the original inventors and their competitors. These molecules are usually protected by patents, or already belong to the public domain. Thus, at a first glance, a high risk of interference appears probable. In fact, in optimizing another therapeutic profile than that of the original inventors, the medicinal chemist will rapidly prepare analogues with chemical structures very different from that of the original hit. As an example, a medicinal chemist interested in a phosphodiesterase (PDE) and using diazepam as lead, will rapidly prepare compounds which are beyond the scope of the original patents, precisely because in becoming PDE inhibitors they present modified (and thus patentable) structures and have lost their affinity for the benzodiazepine receptor.

Originality

On occasion, the screening of a library of several hundred therapeutically diverse drug molecules ultimately produces very surprising results. A nice example of unexpected findings resulting from a systematic screening is found in the tetracyclic compound **11** (BMS-192548) extracted from *Aspergillus niger* WB2346 (Fig. 1.21).

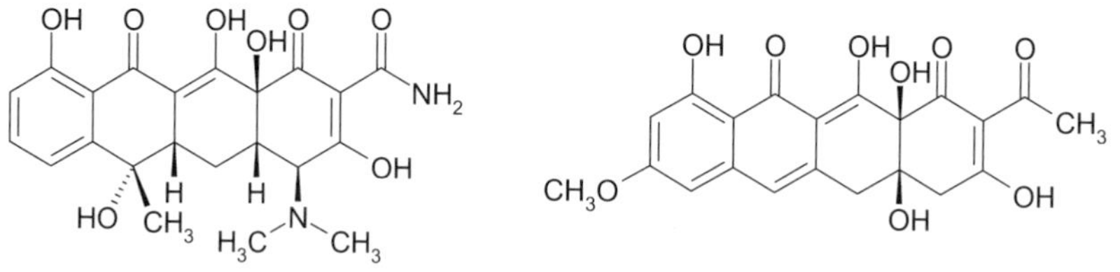

10: tetracycline	**11**: BMS-192548

Fig. 1.21 Unexpected CNS activity of the tetracycline analogue 11 (BMS-192548) [48].

For any medicinal chemist or pharmacologist, the similarity of this compound with the antibiotic tetracycline is striking. However, none of them would *a priori* forecast that BMS-192548 exhibits CNS activities. Actually, the compound turns out to be a ligand for the neuropeptide Y receptor preparations [48].

Orphan Diseases

As mentioned above, a differentiating peculiarity of this type of library is that it is constituted by compounds that have already been safely given to humans. Thus, if a compound were to "hit" with sufficient potency on an orphan target, there is a high chance that it could rapidly be tested in patients for Proof of Principle. This possibility represents another advantage of the SOSA approach.

1.4
Conclusion

Formally, three classes of drug analogues can be considered: (a) analogues presenting chemical and pharmacological similarities (e.g., lovastatin and simvastatin); (b) analogues presenting only structural similarities (e.g., chlorpromazine and imipramine); and (c) chemically different compounds displaying similar pharmacological properties (e.g., chlorpromazine and haloperidol).

The design of analogues of the first category – *direct analogues* – is one of the most rewarding activities of medicinal chemists, and forms part of their daily activity. Indeed, it justifies to a great extent the theme of the present book.

The design of analogues of the second category is sometimes fortuitous, as the result of a pharmacological or a clinical feed-back. For these "structural analogues", the observation of a new activity can also result from a planned and systematic investigation such as provided by the use of the PASS program or the SOSA approach.

Finally, the design of the third category of analogues – "pharmacological analogues" – is not within the scope of this chapter, but is typically relevant from computer-aided drug design (docking, virtual design).

Together, it can be taken for granted that analogue design has been – and continues to be – one of the most fruitful of the methodologies leading to important and useful drug molecules.

References

1 Rey A. *Dictionnaire historique de la langue française*. Dictionnaires Le Robert, Paris, **1992**.

2 Schueller FW. *Chemobiodynamics and Drug Design*. McGraw-Hill, New York, **1960**.

3 Burger A. *Medicinal Chemistry*, 2nd ed. Interscience Publishers, Inc., New York, **1960**.

4 Büchi J. *Grundlagen der Arzneimittelforschung und der Synthetischen Arzneimittel*. Birkhäuser Verlag, Basel, **1963**.

5 Thornber CW. Isosterism and molecular modification in drug design. *Chem. Soc. Rev.*, **1979**, 8, 563–580.

6 Chen X, Wang W. The use of bioisosteric groups in lead optimization. *Annual Reports in Medicinal Chemistry*. Elsevier, Amsterdam, **2003**.

7 Wermuth CG. *The Practice of Medicinal Chemistry*. Academic Press, London, **2003**.

8 Kim CU, Lew W, Williams MA, Wu H, Zhang L, Chen X, Escarpe PA, Mendel DB, Laver WG, Stevens RC. Structure–activity relationship studies of novel carbocyclic influenza neuraminidase inhibitors. *J. Med. Chem.*, **1998**, 41, 2451–2460.

9 Thorsett ED, Harris EE, Aster SD, Peterson ER, Snyder JP, Springer JP, Hirshfield J, Tristram EW, Patchett AA, Ulm EH, Vassil TC. Conformationally restricted inhibitors of angiotensin converting enzyme: synthesis and computations. *J. Med. Chem.*, **1986**, 29, 251–260.

10 Zürcher G, Keller HH, Kettler R, Borgulya J, Bonetti EP, Eigenmann R, Da Prada M. Ro 40–7592, a novel, very potent, and orally active inhibitor of catechol-O-methyltransferase: a pharmacological study in rats. *Adv. Neurol.*, **1990**, 53, 497–503.

11 Rotella DP, Sun Z, Zhu Y, Krupinski J, Pongrac R, Seliger L, Normandin D, Macor JE. N-3-Substituted imidazoquinazolinones: potent and selective PDE5 inhibitors as potential agents for treatment of erectile dysfunction. *J. Med. Chem.*, **2000**, 43, 1257–1263.

12 Wilhelm M. The chemistry of polycyclic psycho-active drugs: serendipity or systematic investigation? *Pharm. J.*, **1975**, 214, 414–416.

13 Brufani M, Marta M, Pomponi M. Anticholinesterase activity of a new carbamate, heptylphysostigmine, in view of its use in patients with Alzheimer-type dementia. *Eur. J. Biochem.*, 986, 157, 115–120.

14 Braida D, Sala M. Eptastigmine: ten years of pharmacology, toxicology, pharmacokinetics, and clinical studies. *CNS Drug Rev.*, **2001**, 7, 369–386.

15 Andersen KE, Braestrup C, Gronwald FC, Jorgensen AS, Nielsen EB, Sonnewald U, Sorensen PO, Suzdak PD, Knudsen LJ. The synthesis of novel GABA uptake inhibitors. 1. Elu-

cidation of the structure–activity studies leading to the choice of (R)-1-[4,4-bis (3-methyl-2-thienyl)-3-butenyl]-3-piperidinecarboxylic acid (tiagabine) as an anticonvulsant drug candidate. *J. Med. Chem.*, **1993**, 36, 1716–1725.

16 Chen C, Dagnino RJ, De Souza EB, Grigoriadis DE, Huang CQ, Kim KI, Liu Z, Moran T, Webb TR, Whitten JP, Xie YF., Jr., Design and synthesis of a series of non-peptide high-affinity human corticotropin-releasing factor 1 receptor antagonists. *J. Med. Chem.*, **1996**, 39, 4358–4360.

17 Stinson ST. Chiral drugs. *Chem. Eng. News*, **1995**, 44–74.

18 Stinson SC. Chiral drug interactions. *Chem. Eng. News*, **1999**, 101–120.

19 Bullock MW, Hand JJ, Waletzky E. Resolution and racemization of dl-tetramisole, dl–6-phenyl-2,3,5,6-tetrahydroimidazo-[2,1-b]thiazole. *J. Med. Chem.*, **1968**, 11, 169–171.

20 Schnieden H. Levamisole: a general pharmacological perspective. *Int. J. Immunopharmacol.*, **1981**, 3, 9–13.

21 Thuilier J. *Les dix ans qui ont changé la folie.* Robert Laffont, Paris, **1981**, pp. 253–257.

22 Sneader W. *Drug Prototypes and their Exploitation.* John Wiley & Sons, Chichester, **1996**, p. 242.

23 Iyer CGS, Languillon J, Ramanujam K. WHO coordinated short-term double-blind trial with thalidomide in the treatment of acute lepra reactions in male lepromatus patients. *Bull. World Health Org.*, **1971**, 45, 719–732.

24 Korth C, May BCH, Cohen FE, Prusiner SB. Acridine and phenothiazine derivatives as pharmacotherapeutics for prion disease. *Proc. Natl. Acad. Sci. USA*, **2001**, 98, 9836–9841.

25 Terret NK, Bell AS, Brown D, Ellis P. Sildenafil (Viagra), a potent and selective inhibitor of type 5 cGMP phosphodiesterase with utility for the treatment of male erectile dysfunction. *Bioorg. Med. Chem. Lett.*, **1996**, 6, 1819–1824.

26 Boolell M, Allen MJ, Ballard SA, Gepi-Attee S, Muirhead GJ, Naylor AM, Osterloh IHC. Sildenafil: an orally active type 5 cyclic GMP-specific phosphodiesterase inhibitor for the treatment of penile erectile dysfunction. *Int. J. Urol. Res.*, **1996**, 8, 47–52.

27 Kling, J. From hypertension to angina to Viagra. Mod. Drug Discovery 1988, 31–38.

28 Poroikov V, Akimov D, Shabelnikova E, Filimonov D. Top 200 Medicines: Can New Actions be Discovered Through Computer-Aided Prediction? *SAR QSAR Environ. Res.*, **2001**, 12, 327–344.

29 Wermuth CG. Search for new lead compounds: the example of the chemical and pharmacological dissection of aminopyridazines. *J. Heterocyclic Chem.*, **1998**, 35, 1091–1100.

30 Wermuth CG, Clarence-Smith K. 'Drug-like' leads: bigger is not always better. *Pharmaceutical News*, **2000**, 7, 53–57.

31 Wermuth CG. The 'SOSA' approach: an alternative to high-throughput screening. *Med. Chem. Res.*, **2001**, 10, 431–439.

32 Prestwick-Chemical-Library. Prestwick Chemical, Inc., Illkirch and Washington DC, www.prestwickchemical.com.

33 Riechers H, Albrecht H-P, Amberg W, Baumann E, Böhm H-J, Klinge D, Kling A, Muller S, Raschack M, Unger L, Walker N, Wernet W. Discovery and optimization of a novel class of orally active non-peptidic endothelin-A receptor antagonists. *J. Med. Chem.*, **1996**, 39, 2123–2128.

34 Stein PD, Hunt JT, Floyd DM, Moreland S, Dickinson KEJ, Mitchell C, Liu EC-K, Webb ML, Murugesan N, Dickey J, McMullen D, Zhang R, Lee VG, Serdino R, Delaney C, Schaeffer TR, Kozlowski M. The discovery of sulfonamide endothelin antagonists and the development of the orally active ETA antagonist 5-(dimethylamino)-N-(3,4-dimethyl-5-isoxazol1)-l-naphthalenesulfonamide. *J. Med. Chem.*, **1994**, 37, 329–331.

35 Murugesan N, Gu Z, Spergel S, Young M, Chen P, Mathur A, Leith L, Hermsmeier M, Liu EC-K, Zhang R, Bird E, Waldron T, Marino A, Koplowitz B, Humphreys WG, Chong S, Morrison RA, Webb ML, Moreland S, Trippodo N, Barrish JC. Biphenylsulfonamide endothelin receptor antagonists. 4. Discovery of N-[[[(4,5-dimethyl-3-isoxazolyl)amino]sulfonyl]-4-(2-oxazo-

lyl)[1,1'-biphenyl]-2-yl]methyl]-N,3,3-tri-methylbutanamide (BMS-207940, a highly potent and orally active ETA selective antagonist. *J. Med. Chem.*, **2003**, 46, 125–137.

36 Wermuth CG, Schlewer G, Bourguignon J-J, Maghioros G, Bouchet M-J, Moire C, Kan J-P, Worms P, Bizière K. 3-Aminopyridazine derivatives with atypical antidepressant, serotonergic and dopaminergic activities. *J. Med. Chem.*, **1989**, 32, 528–537.

37 Wermuth CG. Aminopyridazines – an alternative route to potent muscarinic agonists with no cholinergic syndrome. *Il Farmaco*, **1993**, 48, 253–274.

38 Wermuth CG, Bourguignon J-J, Hoffmann R, Boigegrain R, Brodin R, Kan J-P, Soubrié P. SR 46559A and related aminopyridazines are potent muscarinic agonists with no cholinergic syndrome. Biorg. Med. Chem. Lett., **1992**, 2, 833–836.

39 Contreras J-M, Rival YM, Chayer S, Bourguignon JJ, Wermuth CG. Aminopyridazines as acetylcholinesterase inhibitors. *J. Med. Chem.*, **1999**, 42, 730–741.

40 Contreras J-M, Parrot I, Sippl W, Rival RM, Wermuth CG. Design, synthesis and structure–activity relationships of a series of 3-[2-(1-benzylpiperidin-4-yl)ethylamino]pyridazine derivatives as acetylcholinesterase inhibitors. *J. Med. Chem.*, **2001**, 44, 2707–2718.

41 Gully D, Roger P, Valette G, Wermuth CG, Courtemanche G, Gauthier C. Dérivés alkylamino ramifiés du thiazole, leurs procédés de préparation et les compositions pharmaceutiques qui les contiennent, Elf-Sanofi: French Demande No. 9207736, 24 June, **1992**.

42 Gully D, Roger P, Wermuth CG. 4-Phenyl-aminothiazole derivatives, method for preparing same and pharmaceutical compositions containing said derivatives, Sanofi: World Patent, 09 January, **1997**.

43 Pilla M, Perachon S, Sautel F, Garrido F, Mann A, Wermuth CG, Schwartz J-C, Everitt BJ, Sokoloff P. Selective inhibition of cocaine-seeking behaviour by a partial dopamine D_3 agonist. *Nature*, **1999**, 400, 371–375.

44 Lipinski CA, Lombardo F, Dominy BW, Feeney PJ. Experimental and computational approaches to estimate solubility and permeability in drug discovery and development settings. *Adv. Drug Delivery Rev.*, **2001**, 46, 3–26.

45 Veber DF, Johnson SR, Cheng HY, Smith BR, Ward KW, Kopple KD. Molecular properties that influence the oral bioavailability of drug candidates. *J. Med. Chem.*, **2002**, 45, 2615–2623.

46 Bergström CAS, Strafford M, Lazorova L, Avdeef A, Luthman K, Artursson P. Absorption classification of oral drugs based on molecular surface properties. *J. Med. Chem.*, **2003**, 46, 558–570.

47 Wenlock MC, Austin RP, Barton P, Davis AM, Leeson PD. A comparison of physicochemical property profiles of development and marketed oral drugs. *J. Med. Chem.*, **2003**, 46, 1250–1256.

48 Shu YZ, Cutrone JQ, Klohr SE, Huang S. BMS-192548, a tetracyclic binding inhibitor of neuropeptide Y receptors, from *Aspergillus niger* WB2346. II. Physico-chemical properties and structural characterization. *J. Antibiot.*, **1995**, 48, 1060–1065.

2
Drug Likeness and Analogue-Based Drug Discovery

John R. Proudfoot

Natural products or other drug structures have served as lead structures for the majority of drugs launched during the past 100 years. The process of using other drug structures as leads, or analogue-based drug discovery, leads over time to iterative improvements in the efficacy and safety of therapeutic agents. During the past 15 to 20 years, new technologies such as high-throughput screening (HTS) and combinatorial chemistry have been introduced and have had a major impact on the types of structures now chosen as leads for medicinal chemistry optimization [1,2]. However, a survey of the novel small-molecule therapeutics that were launched or approved for the first time during the period 2000 to 2004 (Q3) [3] (Tab. 2.1) reveals that the majority resulted from the continued exploitation of already well-established structural classes (drug analogues) or the rapid expansion of newer structural classes (early-phase analogues). There are only a few examples of drugs derived from screening leads, and there is no instance disclosed to date where combinatorial chemistry has provided the lead structure for a launched drug.

The continued success of analogue-based approaches hinges on the following: the ongoing improvement in the understanding of drug mechanism of action and *in-vivo* behavior identifies opportunities for improved drug analogues. The discussion below highlights the relationships between some recently launched analogue-based drugs and the structures from which they are derived, particularly in the context of the issues and opportunities that drove the discovery process and the changes in the properties that are typically thought to impact drug-likeness [4]. It is intended to be illustrative rather than comprehensive, and the examples are chosen to convey the range of opportunities available within the context of an analogue-based approach to drug discovery [5].

Among the new structural classes to reach the market recently, the selective cyclooxygenase 2 (COX-2) inhibitors (Fig. 2.1) represent the successful exploitation of a lead structure disclosed in the scientific literature, and constitute an excellent example of the potential in early-phase analogue research. Throughout the 1970s, the mechanism of action of classical nonsteroidal anti-inflammatory drugs (NSAIDs) was linked to inhibition of prostaglandin G/H synthase 1, subsequently known as cyclooxygenase-1 (COX-1) [6]. The discovery of an inducible

Analogue-based Drug Discovery. IUPAC, János Fischer, and C. Robin Ganellin (Eds.)
Copyright © 2006 WILEY-VCH Verlag GmbH & Co. KGaA, Weinheim
ISBN: 3-527-31257-9

Tab. 2.1 Small-molecule therapeutics launched or approved
during the period 2000 to 2004.

Drug	Type	Lead/Lead source	Launched/Approved
Almotriptan	Drug analogue	Sumatriptan	2000
Alosetron	Drug analogue	Ondansetron	2000
Artemotil	Natural product analogue	Artemisinin	2000
Bepotastine		Not disclosed	2000
Bexarotene		Literature	2000
Bulaquine	Drug analogue	Primaquine	2000
Cevimeline	Natural product analogue	Acetylcholine	2000
Dexmedetomidine	Drug enantiomer	Medetomidine	2000
Dofetilide		Not disclosed	2000
Drospirenone	Drug analogue	Spironolactone	2000
Egualen sodium	Drug analogue	Guaiazulene sulfonate	2000
Esomeprazole	Drug enantiomer	Omeprazole	2000
Exemestane		Literature	2000
Frovatriptan		Literature	2000
Lafutidine	Drug analogue	Pibutidine	2000
Levetiracetam	Drug analogue	Piracetam	2000
Levobupivacaine	Drug enantiomer	Bupivacaine	2000
Levosimendan	Drug analogue	Pimobendan	2000
Linezolid	Drug analogue (early phase)	DuP 721/DuP 105	2000
Liranaftate		Tolnaftate/screening	2000
Lopinavir	Drug analogue	Ritonavir/rational drug design	2000
Maxacalcitol	Natural product analogue	1,2-dihydroxy vitamin D_3	2000
Melevodopa	Natural product/Prodrug	Levodopa	2000
Perospirone	Drug analogue	Tiospirone	2000
Ramatroban		Not disclosed	2000
Taltirelin	Natural product analogue	Thyrotropin-releasing hormone	2000

Tab. 2.1 Continued.

Drug	Type	Lead/Lead source	Launched/Approved
Ziprasidone	Drug analogue	Tiospirone	2000
Zofenopril	Drug analogue	Captopril	2000
Zoledronic acid	Drug analogue	Pamidronic acid	2000
Aminolevulinic acid	Natural product		2001
Bimatoprost	Drug analogue	Latanoprost	2001
Bosentan		Screening	2001
Desloratadine	Drug metabolite	Loratadine	2001
Edaravone		Not disclosed	2001
Eletriptan	Drug analogue	Not disclosed	2001
Fudosteine		Not disclosed	2001
Gabexate		Not disclosed	2001
Hexafluorocalcitriol	Natural product analogue	1,2- dihydroxy vitamin D_3	2001
Imatinib		Not disclosed	2001
Levocetirizine	Drug enantiomer	Cetirizine	2001
Tegaserod	Natural product analogue	Serotonin	2001
Tenofovir disoproxil	Drug analogue/Prodrug	Tenofovir	2001
Travoprost	Drug analogue	Not disclosed	2001
Valganciclovir	Drug analogue/Prodrug	Ganciclovir	2001
Adefovir dipivoxil	Drug analogue/Prodrug	Adefovir	2002
Amrubicin	Drug/Natural product analogue	Daunomycin	2002
Aripiprazole		Not disclosed	2002
Balofloxacin	Drug analogue	Not disclosed	2002
Biapenem	Drug analogue	Meropenem	2002
D-threo-Methyl-phenidate	Drug enantiomer	Methylphenidate	2002
Eplerenone	Drug analogue	Spironolactone	2002
Ertapenem	Drug analogue	Meropenem	2002
Escitalopram	Drug enantiomer	Citalopram	2002

Tab. 2.1 Continued.

Drug	Type	Lead/Lead source	Launched/Approved
Etoricoxib	Drug analogue (early phase)	DuP 697	2002
Ezetimibe		Literature/SA 58035	2002
Fulvestrant	Drug analogue	Literature	2002
Gefitinib		Screening	2002
Landiolol	Prodrug	Not disclosed	2002
Neridronic acid	Drug analogue	Pamidronic acid	2002
Nitisinone		None	2002
Olmesartan	Drug analogue/Prodrug	Losartan	2002
Parecoxib	Drug analogue/Prodrug	Valdecoxib	2002
Pazufloxacin	Drug analogue	Ofloxacin	2002
Prulifloxacin	Drug analogue/Prodrug		2002
Sivelestat		Screening	2002
Sodium oxybate	Natural product		2002
Tiotropium	Drug analogue	Oxitropium	2002
Treprostinil	Natural product analogue	Not disclosed	2002
Valdecoxib	Drug analogue	Celecoxib	2002
Voriconazole	Drug analogue	Fluconazole	2002
Aprepitant	Drug analogue	Literature/screening	2003
Atazanavir		Rational drug design	2003
Atomoxetine	Drug analogue	Fluoxetine	2003
Azelnidipine	Drug analogue	Nicardipine	2003
Bortezomib		Literature	2003
Dutasteride	Drug analogue	Finasteride	2003
Emtricitabine	Natural product analogue	Deoxycytidine	2003
Fosamprenavir	Drug analogue/Prodrug	Amprenavir	2003
Gemifloxacin	Drug analogue	Tosufloxacin	2003
Lumiracoxib	Drug analogue	Diclofenac	2003

Tab. 2.1 Continued.

Drug	Type	Lead/Lead source	Launched/Approved
Miglustat		Screening	2003
Mycophenolate	Natural product		2003
Nisvastatin	Drug analogue	Not disclosed	2003
Nitazoxanide	Drug analogue	Tenonitrozole	2003
Norelgestromin	Drug analogue	Norgestrel	2003
Palonosetron	Drug analogue	Tropisetron	2003
Rosuvastatin	Drug analogue	Fluvastatin	2003
Rupatadine	Drug analogue	Literature	2003
Tadalafil		Literature/screening	2003
Vardenafil	Drug analogue	Sildenafil	2003
Cincalcet	Drug analogue	Fendiline	2004
Duloxetine	Drug analogue	Fluoxetine	2004
Eszopiclone	Drug enantiomer	Zopiclone	2004
Fosfluconazole	Drug analogue/Prodrug	Fluconazole	2004
Indisetron	Drug analogue	Granisetron	2004
Mitiglinide		Not disclosed	2004
Pregabalin	Natural product analogue	GABA	2004
Solifenacin		Not disclosed	2004
Ximelagatran	Prodrug	Rational drug design	2004

γ-aminobutyric acid.

form of this enzyme, COX-2, in the late 1980s raised the possibility of safer therapeutics based on selective inhibition of COX-2. COX-1 is expressed constitutively in healthy tissues, whereas COX-2 is induced by inflammatory stimuli. Classical NSAIDS generally inhibit both enzymes to a comparable degree, and the gastrointestinal side effects sometimes seen with NSAIDS are due to inhibition of COX-1 in the stomach where it produces the cytoprotective prostaglandin G_2. Drugs displaying selective inhibition of COX-2 were expected to possess an improved safety profile, particularly with regard to gastrointestinal side effects.

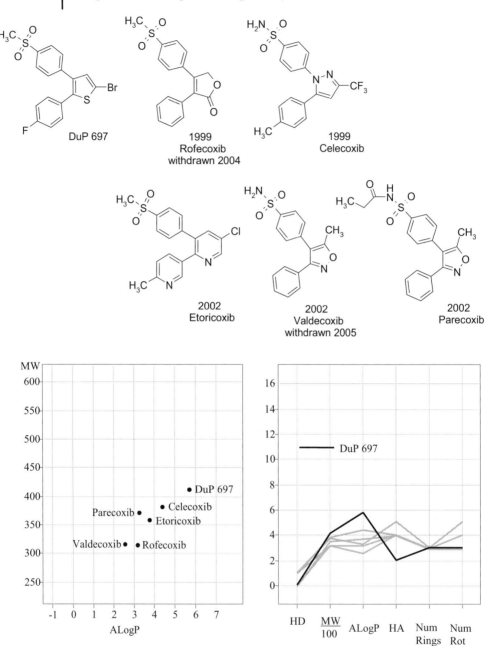

Fig. 2.1 Structures and properties of cyclooxygenase 2 (COX-2) inhibitors.

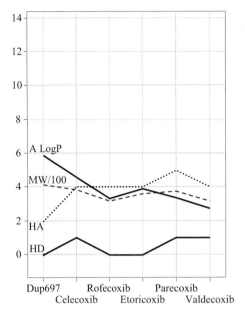

Fig. 2.1 (continued).

The pharmacological profile of the lead structure, DuP 697 [7,8], was consistent with selective inhibition of COX-2 versus COX-1. The successful exploitation of DuP 697 was not accomplished by the organization that disclosed the original discovery – a common theme in analogue-based approaches. Initial optimization, which focused on improving selectivity for the inhibition of COX-2, provided rofecoxib and celecoxib. Etoricoxib [9] arose from an effort to exploit the pyridine ring as a replacement for the central five-membered rings common to the other structures. Research with the goal of identifying an inhibitor with greater potency and selectivity than celecoxib gave valdecoxib [10]. Parecoxib [11] is a prodrug of valdecoxib with improved solubility to enable parenteral administration. This series of drugs represents an analogue-based approach that provided "first in class" representatives, rather than the more common iterative improvement to "best in class" molecules. The drug molecules ultimately derived from DuP 697 all possess a lower molecular weight (mostly a result of replacing the bromo susbtituent) and log P values than the lead structure. These changes are contrary to the trends typically seen when large sets of leads and drugs are examined and increases in molecular weight and log P are generally seen [12].

In contrast to the COX-2 inhibitors above, desloratadine and rupatadine (Fig. 2.2) represent the continued evolution of a drug class which has provided effective therapeutic agents for a half-century. Histamine is a key mediator of allergic inflammation through its activity at H_1 receptors. However, H_1 receptors are located centrally as well as in the periphery, and H_1 receptor antagonists have the potential for sedating side effects if they cross the blood–brain barrier. Since the launch of the first effective antihistamine drugs in the mid-1940s, considerable

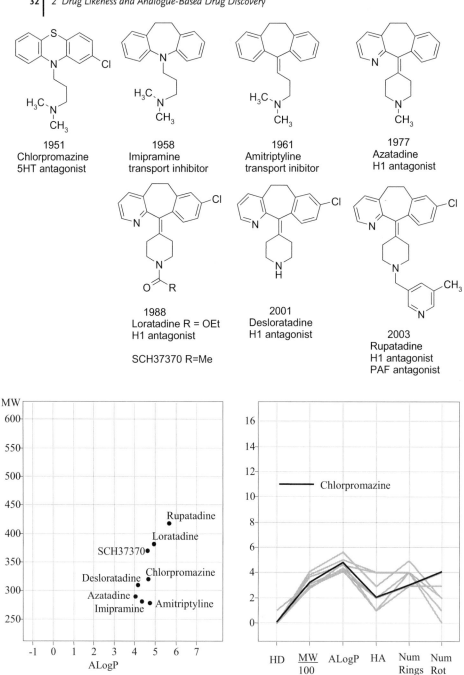

Fig. 2.2 Structures and properties of drugs related to desloratadine.

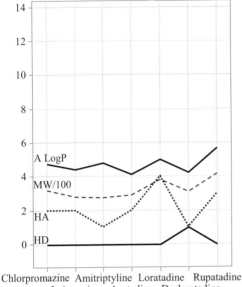

Chlorpromazine Amitriptyline Loratadine Rupatadine
Imipramine Azatadine Desloratadine

Fig. 2.2 (continued).

effort has been devoted to the search for nonsedating antihistamines to avoid the drowsiness side effect commonly seen with these agents – an effort that culminated in the launch of such drugs as loratadine. Desloratadine [13] is the major metabolite of loratadine and displays molecular and pharmacological potency superiority to loratadine. It is also claimed to have a superior side-effect profile over first-generation, nonsedating antihistamines, particularly in regard to drug–drug interactions [14]. Desloratadine illustrates the impact that information on the metabolic fate of a drug, and concerns about the safety profile of a drug class, can have on the discovery of improved therapeutic agents.

Rupatadine represents a novel structural extension of this analogue class, and resulted from a discovery effort focused on the design of dual H_1 and platelet-activating factor (PAF) receptor antagonists [15]. Allergic and inflammatory responses are mediated by histamine and PAF, along with other mediators. These mediators display complimentary effects *in vivo*, so that blockade of both H_1 and PAF receptors may be of greater benefit than the blockade of one alone. The lead structure, SCH37370 – an analogue of loratadine – was reported to have the desired mixed-activity profile. The optimization process focused on improving potency. Again, within this class there is no clear trend to the increasing molecular weight and log P values relative to chlorpromazine, the initial prototype [16]. Relatively minor structural variations have yielded changes in the biological mode of action of the drug molecules.

Atomoxetine is the most recent neurotransmitter reuptake inhibitor to reach the market (Fig. 2.3). It is a selective inhibitor of norepinephrine (noradrenaline) transport, and during the 1980s was – as tomoxetine – evaluated clinically for the

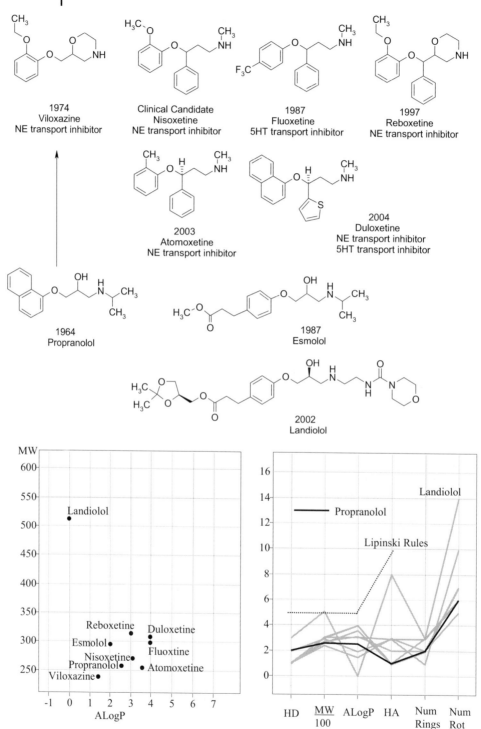

Fig. 2.3 Structures and properties of drugs related to atomoxetine.

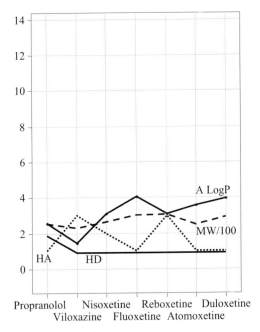

Fig. 2.3 (continued).

treatment of depression [17]. Subsequently, it was patented [18] and approved for the treatment of attention deficit hyperactivity disorder (ADHD). Atomoxetine is an analogue of nisoxetine [19] with the same profile of selective norepinephrine reuptake inhibition, but a longer duration of action [20]. This structural class traces its origin to viloxazine [21], which resulted from attempts to exploit the sedative and anticonvulsant effects observed for propranolol in pharmacological experiments. *Reboxetine* [22], also recently launched, resulted from an exploration of molecules containing structural features of both viloxazine and nisoxetine. Duloxetine, an additional member of this class which displays comparable potencies as a 5-hydroxytryptamine (5HT) and norepinephrine reuptake inhibitor [23], was also recently approved. One additional drug reaching the market in this time period also traces its origin indirectly to propranolol. Landiolol [24] (Fig. 2.3) was designed to be a short-acting, metabolically labile beta blocker modeled after esmolol, but with improved potency. It represents an analogue-based approach where decreased metabolic stability was desired in order to effect a short duration of action. Even though landiolol displays a rather different combination of properties compared to the other members of this class, it is interesting to note that the combination is still consistent with the rule of five formulated by Lipinski [25], even though the drug was designed for intravenous administration. However, aside from landiolol, the property changes seen within this class, relative to propranolol, are consistent with those seen in broader analyses: with a trend towards increasing molecular weight, log P, number of H bond acceptors, number of rings and rotatable bonds [9a].

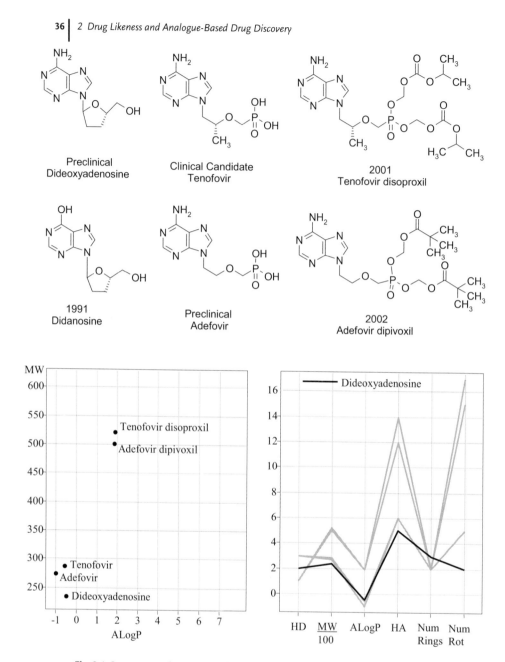

Fig. 2.4 Structures and properties of selected nucleoside and nucleotide analogues.

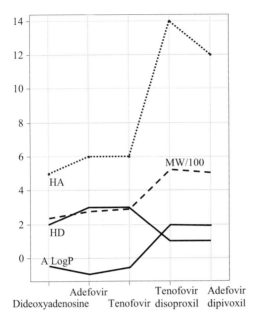

Fig. 2.4 (continued).

Nucleoside analogues have been used as effective therapeutic agents in the anti-infective and anticancer areas for many years. The anti HIV-1 effects of dideoxy-nucleosides such as dideoxyadenosine and didanosine were described in 1985 [26], and subsequently a variety of nucleoside analogues were approved for the treatment of AIDS. A sophisticated prodrug approach led to the nucleotide analogues tenofovir disoproxil [27] (approved for the treatment of HIV-1 infection) and adefovir dipivoxil [28] (approved for the treatment of hepatitis B infection) (Fig. 2.4). These molecules are both designed to deliver the biologically active phosphonates intracellularly. They avoid the slow first phosphorylation step necessary for the activity of nucleosides by incorporation of a metabolically stable phosphonate group, as seen for example in tenofovir. Tenofovir, however, is effective only when administered parenterally – the charges carried by the phosphonate group are detrimental to oral absorption. Masking the phosphonate charges with metabolically labile ester groups allows oral delivery. The successful development of these analogues required an appropriate combination of the structural features desired for efficacy with the physical properties that allowed oral absorption and cell permeability in order to release of the active drug intracellularly. As with the previous case of landiolol, these prodrugs have quite different property profiles from their lead or prototype compounds and are of substantially increased complexity, as evidenced by the large increase in both molecular weight and number of rotatable bonds.

Two new vitamin D analogues, maxacalcitol [29] and hexafluorocalcitriol [30] (Fig. 2.5) were designed to separate the antiproliferative effects of the natural hormone from its role in calcium and phosphate mobilization. There had not been

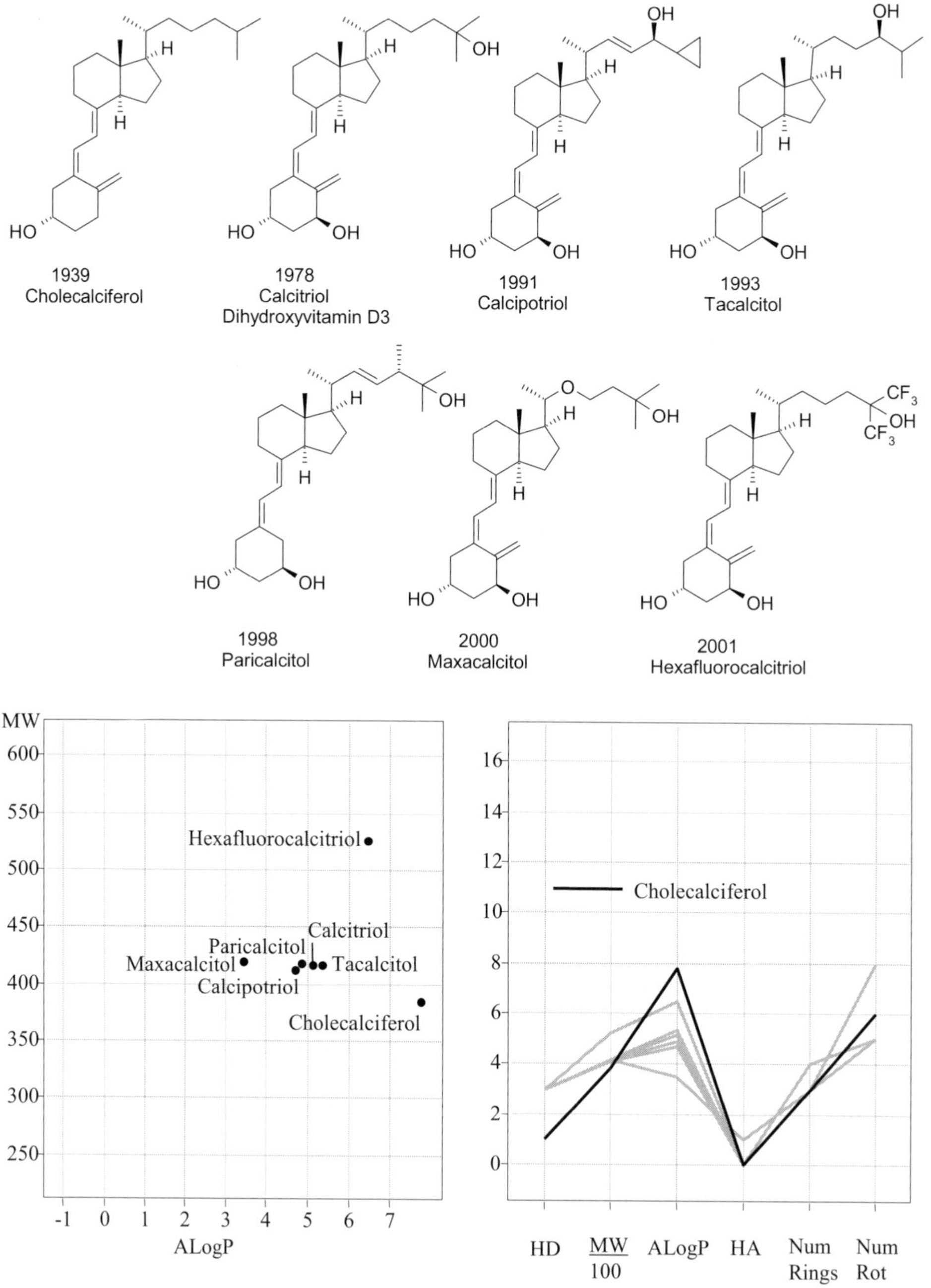

Fig. 2.5 Structures and properties of selected vitamin D analogues.

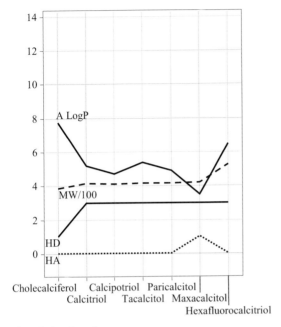

Fig. 2.5 (continued).

much activity within this class for many years, and the recent expansion results in large part from an ongoing improved understanding of the vitamin D nuclear receptor and its mechanism of action. There is also a significant enablement with regard to modern synthetic methodology. The generation of these hormone derivatives is synthetically complex and requires some challenging selective synthetic transformations. There are, no doubt, other examples of modern synthetic methodology impacting recent drug discovery, but this is certainly a noteworthy example.

Angiotensin II is an octapeptide formed in vivo by the action of angiotensin-converting enzyme (ACE) on the decapeptide angiotensin I, and plays an important role in the control of blood pressure. The approval of the first ACE inhibitors (e.g., captopril) in the early 1980s demonstrated that control of angiotensin levels is beneficial in controlling blood pressure, and prompted the search for small molecule antagonists of angiotensin II action at the receptor level. Losartan (Fig. 2.6) was the first angiotensin II receptor antagonist approved for use in the treatment of hypertension, and arose from an exploitation of the structure S-8308, which had been disclosed in the patent literature as the first nonpeptide angiotensin antagonist. Losartan has modest bioavailability (25–35%), and is metabolized to the active 5-carboxylic acid metabolite that is responsible for the long duration of action. Olmesartan is the most recently approved angiotensin II antagonist [31], and resulted from an exploration of the function of substituents at the 4- and 5-positions of the imidazole ring of losartan. This effort identified the hydroxyisopropyl group at the 4-position as a substituent that conferred improved molecular

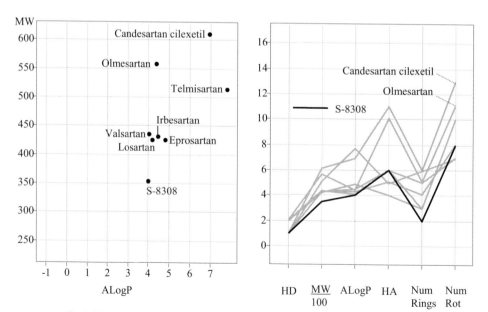

Fig. 2.6 Structures and properties of selected angiotensin II receptor antagonists.

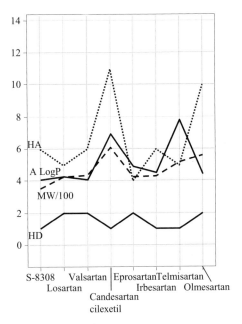

Fig. 2.6 (continued).

potency. This finding, in combination with a prodrug approach to mask the carboxylic acid at the 5-position and allow oral exposure, produced olmesartan. Within this class there is a greater diversity of structure and properties than is commonly seen with analogue-based approaches. The drugs show substantial increases in molecular weight compared to the preclinical lead. The structural variation may be at least partly ascribed to the substantial input from computational chemistry and molecular modeling in the design of telmisartan [32], eprosartan [33], and valsartan [34] in particular. Again, the prodrugs, candesartan and olmesartan, display the largest change in profile with regard to changes in molecular weight and number of rotatable bonds.

Lovastatin, simvastatin, and *pravastatin* (Fig. 2.7) are potent inhibitors of 3-hydroxy-3-methylglutaryl-CoA reductase (HMG-CoA reductase), the rate-limiting enzyme in cholesterol biosynthesis, and are effective cholesterol-lowering agents used in the treatment of hypercholesterolemia. They are active in the beta-hydroxyacid form shown for pravastatin. The disclosure of the biological efficacy of lovastatin [35] opened opportunities for the discovery of similar therapeutic agents based synthetic rather than natural product structures. Rosuvastatin [36,37] is the most recently approved HMG-CoA reductase inhibitor. Pyrimidine replacements for the lower ring system of statins had been disclosed in the literature [38], and the methanesulfonamide-substituted pyrimidine in rosuvastatin was specifically designed to confer physical properties (i.e., log D), preferred for targeting to the liver [39]. This targeting, when combined with greater molecular potency, leads to improved efficacy at lower doses than seen with the earlier statins.

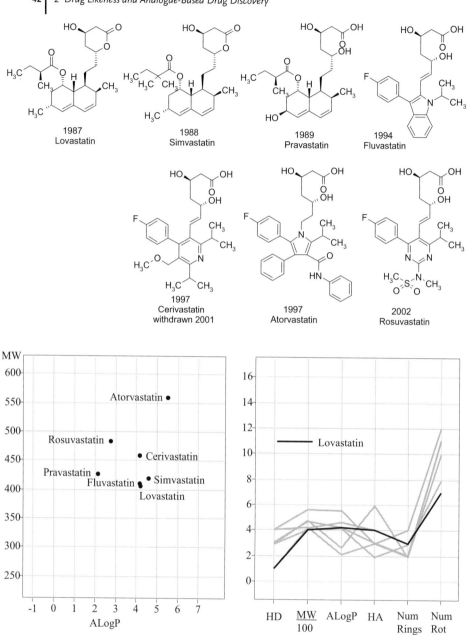

Fig. 2.7 Structures and properties of selected HMG-CoA reductase inhibitors.

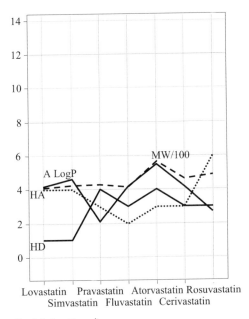

Fig. 2.7 (continued).

Sumatriptan is a selective 5-hydroxytryptamine 1D ($5HT_{1D}$) receptor agonist used for the treatment of migraine [40], and was designed based on the observation that infused 5-HT is effective in alleviating migraine. The low oral bioavailability (14%) of sumatriptan presented an obvious opportunity for improved therapeutic agents within the triptan class (Fig. 2.8). Zolmitriptan [41] (40%), naratriptan (60–70%) and rizatriptan (40%) all display superiority to sumatriptan with regard to oral bioavailability. Eletriptan [42] displays five-fold improved molecular potency over sumatriptan, along with improved (50%) bioavailability. Almotriptan [43] was designed with the intention of having a longer-acting sumatriptan analogue, and achieves 70–80% oral bioavailability.

Buspirone [44] (Fig. 2.9) was discovered through pharmacological screening for sedative agents, and subsequently found to be a dopamine D_2 presynaptic receptor antagonist and a serotonin 1A ($5\text{-}HT_{1A}$) receptor partial agonist. It was originally evaluated in the clinic as an antipsychotic agent, and although it showed disappointing results for this indication, it demonstrated anxiolytic effects and was approved for this indication. Subsequently, tiospirone – a clinical candidate [45] – arose from an effort to impart the antipsychotic activity lacking in buspirone, and tandospirone [46] was designed to engender improved anxiolytic effects by reducing the dopaminergic activity of buspirone. Perospirone [47] (Fig. 2.9) makes use of structural features present in both tiospirone and tandospirone, and its discovery was driven by a desire to generate a drug with $5HT_2$ and D_2 receptor antagonist activity – a profile expected to result in decreased side effects [48]. Ziprasidone [49] also takes advantage of the benzisothiazole ring structure present in tiospirone in combination with a novel right-hand side to achieve a desired $5HT_{2A}$ and

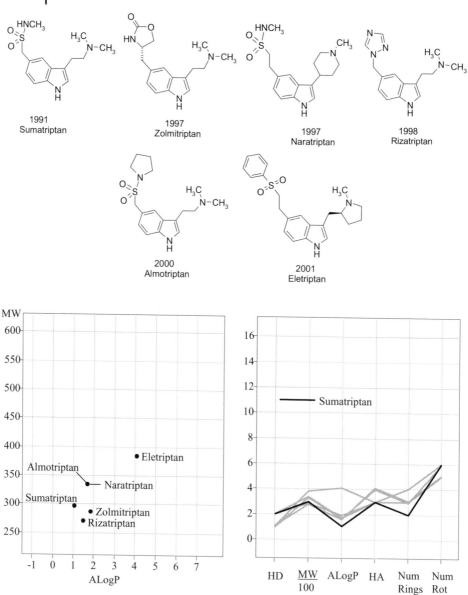

Fig. 2.8 Structures and properties of selected "triptan" drugs.

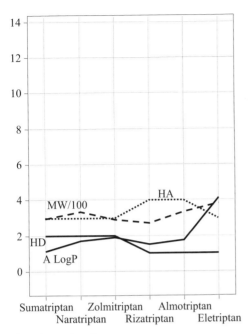

Fig. 2.8 (continued).

D_2 receptor antagonist profile, again targeted at providing an antipsychotic drug with decreased side effects [50].

Three additional "floxacin" or fluoroquinolone antibiotics (Fig. 2.10) were approved, adding to the 20-plus launched members of this class. Pazufloxacin arose from an effort to discover an agent which was less toxic than ciprofloxacin and ofloxacin. The aminocyclopropyl substituent, in combination with the ciprofloxacin quinolone scaffold, was found to confer a favorable toxicity profile, and subsequent combination of this 7-substituent with the ofloxacin scaffold gave pazufloxacin [51]. Prulifloxacin arose from an effort to exploit an underexplored area of the structure–activity relationship (SAR) of the fluoroquinolones and to understand the effects of substitution at the 2-position of the scaffold. These studies led to the discovery that the thiazeto fused ring system confers good antibacterial activity. Subsequent efforts to improve pharmacokinetic properties led to the prodrug prulifloxacin [52]. Balofloxacin resulted from an effort to reduce the photosensitivity-related side effects of fluoroquinolones, and derived from the finding that the methoxy substituent on the scaffold reduced phototoxicity [53]. Interestingly, the methoxy-substituted scaffold is also found in gatifloxacin, and the discovery of this agent was driven by the goal of improving antibacterial efficacy rather than improving the safety profile.

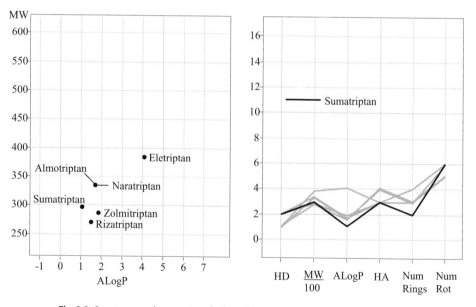

Fig. 2.9 Structures and properties of selected antipsychotic drugs.

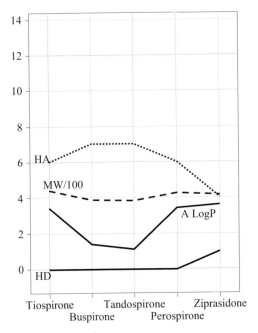

Fig. 2.9 (continued).

In 1983, Alfred Burger wrote "The design of the chemical structure of biologically active molecules is guided by the techniques and working hypotheses of the field of medicinal chemistry." [54]. The techniques and working hypotheses have evolved, been refined and challenged, and increased considerably in sophistication since the early days of medicinal chemistry. The early chemistry-driven exploitation of a limited number synthetically accessible structures has given way to the present-day understanding of how molecular features relate to favorable drug-like properties [25, 55], and this continuous improvement in understanding is essentially what enables successful analogue-based approaches to drug discovery. The discovery of new effective drug prototypes [56] is not a common occurrence. Combining molecular features that balance efficacy and safety and allow clinical success is a challenging task, as evidenced by the perceived low productivity of the pharmaceutical industry in recent years [57]. Despite the contributions from HTS and combinatorial chemistry to the identification of novel lead structures, these techniques have yet to demonstrate a significant impact with regard to the production of successful drugs. The lead structures for most marketed drugs continue to be analogue-based structures, clinical candidates, or structures reported in the scientific or patent literature, or natural products.

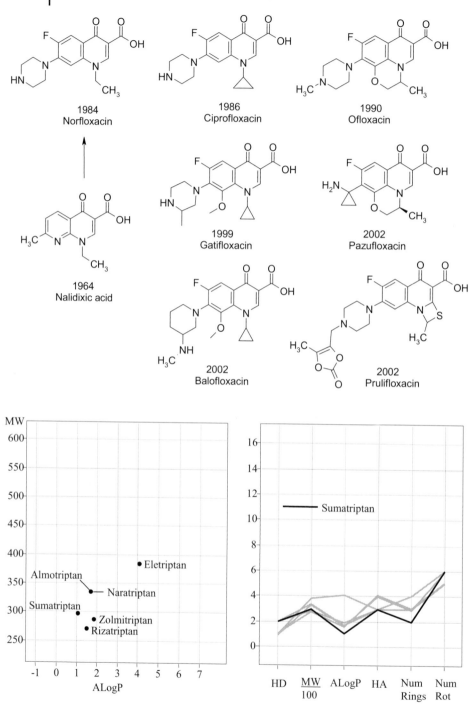

Fig. 2.10 Structures and properties of selected quinolone antibiotics.

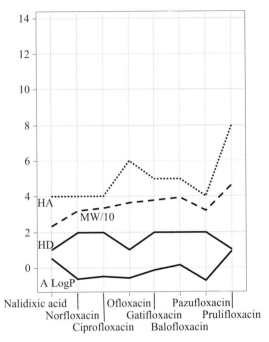

Fig. 2.10 (continued).

References

1 Lipinski CA. *J. Pharmacol. Tox. Methods*, **2000**, 44, 235.

2 Freter K. *Pharm. Res.*, **1988**, 397.

3 (a) Graul A. The Years New Drugs. *Drug News Perspect.*, **2003**, 16, 22–40; (b) Graul A. The Years New Drugs. *Drug News Perspect.*, **2002**, 15, 29–43; (c) Graul A. The Years New Drugs. *Drug News Perspect.*, **2001**, 14, 12–31; (d) Gaudilliere B, Bernardelli P, Berna P. *Annu. Rep. Med. Chem.*, **2001**, 36, 293; (e) Bernardelli P, Gaudilliere B, Vergne F. *Annu. Rep. Med. Chem.*, **2002**, 37, 257–277.

4 A plot of molecular weight versus A log P is depicted for each set of molecules. In addition, comparative profiles are also presented: number of hydrogen bond donors (HD), A Log P, number of hydrogen bond acceptors (HA), number of rings (Num Rings) and number of rotatable bonds (Num Rot). Molecular weight divided by 100 (to allow presentation on the same numerical scale) is also shown. These parameters were calculated within the Scitegic Pipeline software package, with visualization via edited Spotfire charts.

5 Some of these analogue classes have also been reviewed recently: Fischer J, Gere A. Timing of analogue research in drug discovery. *Pharmazie*, **2001**, 56, 9.

6 Flower RJ, Vane JR. Inhibition of prostaglandin synthesis. *Biochem. Pharmacol.*, **1974**, 23, 139–1450.

7 Gans KR, Galbraith W, Roman RJ, Haber SB, Kerr JS, Schmidt WK, Smith C, Hewes WE, Ackerman NR. Anti-inflammatory and safety profile of DuP697, a novel orally effective prostaglandin synthesis inhibitor. *J. Pharmacol. Exp. Ther.*, **1990**, 254, 180–187.

8 For a discussion of the likely origin of DuP697, see: Lednicer D. Tracing the origins of COX-2 inhibitor's structures. *Curr. Med. Chem.*, **2002**, 9, 1457–1461.

9 (a) Chauret N, Yergey JA, Brideau C, Friesen RW, Mancini J, Riendeau D,

Silva J, Styhler A, Trimble LA, Nicoll-Griffith DA. In vitro metabolism considerations, including activity testing of metabolites, in the discovery and selection of the COX-2 inhibitor etoricoxib (MK-0663). *Bioorg. Med. Chem. Lett.*, **2001**, 11, 1059–1062; (b) Friesen RW, Brideau C, Chan CC, Charleson S, Deschenes D, Dube D, Ethier D, Fortin R, Gauthier JY, Girard Y, Gordon R, Greig GM, Riendeau D, Savoie C, Wang Z, Wong E, Visco D, Xu LJ, Young RN. 2-Pyridinyl-3-[4-(methylsulfonyl)phenyl]pyridines: selective and orally active cyclooxygenase-2 inhibitors. *Bioorg. Med. Chem. Lett.*, **1998**, 8, 2777–2782.

10 Talley JJ, Brown DL, Carter JS, Graneto MJ, Koboldt CM, Masferrer JL, Perkins WE, Rogers RS, Shaffer AF, Zhang YY, Zweifel BS, Seibert K. 4-[5-Methyl-3-phenylisoxazol-4-yl]-benzenesulfonamide, Valdecoxib: a p otent and s elective i nhibitor of COX-2. *J. Med. Chem.*, **2000**, 43, 775–777.

11 Talley JJ, Bertenshaw SR, Brown DL, Carter JS, Graneto MJ, Kellogg MS, Koboldt CM, Yuan J, Zhang YY, Seibert K. N-[[(5-Methyl-3-phenylisoxazol-4-yl)-phenyl]sulfonyl]propanamide, sodium salt, parecoxib sodium: a potent and selective inhibitor of COX-2 for parenteral administration. *J. Med. Chem.*, **2000**, 43, 1661–1663.

12 (a) Oprea TI, Davis AM, Teague SJ, Leeson PD. Is there a difference between leads and drugs? A historical perspective. *J. Chem. Inf. Comput. Sci.*, **2001**, 41, 1308–1315; (b) Oprea TI. Current trends in lead discovery: are we looking for the appropriate properties? *J. Comput. Aid. Mol. Des.*, **2002**, 16, 325.

13 (a) Graul A, Leeson P A, Castaner J. Desloratadine: treatment of allergic rhinitis histamine H₁ antagonist. *Drugs of the Future*, **2000**, 25, 339–346; (b) Geha RS, Meltzer EO. Desloratadine: a new, nonsedating, oral antihistamine. *J. Allergy Clin. Immunol.*, **2001**, 107, 752–762.

14 Aberg AKG, McCullough JR, Smith ER. Methods and compositions for treating allergic rhinitis and other disorders using descarbethoxyloratadine. US Patent 595997, Jan 21, **1997**.

15 Carceller E, Merlos M, Giral M, Balsa D, Almansa C, Bartroli J, Garcia-Rafanell J, Forn J. [(3-Pyridylalkyl)piperidylidene]-benzocycloheptapyridine derivatives as dual antagonists of PAF and histamine. *J. Med. Chem.*, **1994**, 37, 2697–2703.

16 Sneader W. *Drug Prototypes and their Exploitation*. John Wiley & Sons, **1996**, p. 677.

17 Zerbe RL, Rowe H, Enas GG, Wong D, Farid N, Lemberger L. Clinical pharmacology of tomoxetine, a potential antidepressant. *J. Pharmacol. Exp. Ther.*, **1985**, 232, 139–143.

18 Heligenstein JH, Tollefson D. Treatment of attention-deficit/hyperactivity disorder. US Patent 5658590, August 19, **1997**.

19 Lemberger L, Rowe H, Carmichael R, Crabtree R. Clinical studies with drugs affecting biogenic amine mechanisms. Catecholamines: Basic clinical frontiers. In: Proceedings, 4th International Catecholamine Symposium, **1979**, part 2, pp. 1923–1924.

20 Wong DT, Threlkeld PG, Best KL, Bymaster FP. A new inhibitor of norepinephrine uptake devoid of affinity for receptors in rat brain. *J. Pharmacol. Exp. Ther.*, **1982**, 222, 61–65.

21 Greenwood DT, Mallion KB, Todd AH, Turner RW. 2-Aryloxymethyl-2,3,5,6-tetrahydro-1,4-oxazines, a new class of antidepressants. *J. Med. Chem.*, **1975**, 18, 573–577.

22 Melloni P, Carniel G, Della Torre A, Bonsignori A, Buonamici M, Pozzi O, Ricciardi S, Rossi AC. Potential antidepressant agents. α-Aryloxy-benzyl derivatives of ethanolamine and morpholine. *Eur. J. Med. Chem.*, **1984**, 19, 235–242.

23 Wong DT, Robertson DW, Bymaster FP, Krushinski JH, Reid A, Leroy R. LY227942, an inhibitor of serotonin and norepinephrine uptake: biochemical pharmacology of a potential antidepressant drug. *Life Sci.*, **1988**, 43, 2049–2057.

24 Iguchi S, Iwamura H, Nishizaki M, Hayashi A, Senokuchi K, Kobayashi K, Sakaki K, Hachiya K, Ichioka Y, Kawamura M. Development of a highly cardioselective ultra short-acting β-blocker, ONO-1101. *Chem. Pharm. Bull.*, **1992**, 40, 1462–146 9.

25 Lipinski CA, Lombardo F, Dominy BW, Feeney PJ. Experimental and computational approaches to estimate solubility and permeability in drug discovery and development settings. *Adv. Drug. Deliv. Rev.*, **1977**, 23, 3–25.

26 Mitsuya H, Broder S. Inhibition of the in vitro infectivity and cytopathic effect of human T-lymphotrophic virus type III/lymphadenopathy-associated virus (HTLV-III/LAV) by 2',3'-dideoxynucleosides. *Proc. Natl. Acad. Sci. USA*, **1986**, 83, 1911–1915.

27 Arimilli MN, Kim CU, Dougherty J, Mulato A, Oliyai R, Shaw JP, Cundy KC, Bischofberger N. Synthesis, in vitro biological evaluation and oral bioavailability of 9-[2-(phosphonomethoxy)propyl] adenine (PMPA) prodrugs. *Antiviral Chem. Chemother.*, **1997**, 8(6), 557–564.

28 Starrett JE, Jr., Tortolani DR, Hitchcock MJM, Martin JC, Mansuri MM. Synthesis and in vitro evaluation of a phosphonate prodrug: bis(pivaloyloxymethyl)-9-(2-phosphonylmethoxyethyl)adenine. *Antiviral Res.*, **1992**, 19, 267–273.

29 Murayama E, Miyamoto K, Kubodera N, Mori T, Matsunga I. Synthetic studies of v itamin D_3 analogues. VIII. Synthesis of 22-oxavitamin D_3 analogues. *Chem. Pharm. Bull.*, **1986**, 34, 4410.

30 Kobayashi Y, Taguchi T, Mitsuhashi S, Eguchi T, Ohshima E, Ikekawa N. Studies on organic fluorine compounds. XXXIX. Studies on steroids. LXXIX. Synthesis of $1a$,25-dihydroxy-26,26,26,27,27,27-hexafluorovitamin D_3. *Chem. Pharm. Bull.*, **1982**, 30, 4297–4303.

31 Yanagisawa H, Amemiya Y, Kanazaki T, Shimoji Y, Fujimoto K, Kitahara Y, Sada T, Mizuno M, Ikeda M, et al. Nonpeptide angiotensin II receptor a ntagonists: synthesis, biological activities, and structure–activity relationships of imidazole-5-carboxylic acids bearing alkyl, alkenyl, and hydroxyalkyl substituents at the 4-position and their related compounds. *J. Med. Chem.*, **1996**, 39, 323–338.

32 Ries UJ, Mihm G, Narr B, Hasselbach KM, Wittneben H, Entzeroth M, van Meel JCA, Wienen W, Hauel NH. 6-Substituted benzimidazoles as new nonpeptide angiotensin II receptor antagonists: synthesis, biological activity, and structure–activity relationships. *J. Med. Chem.*, **1993**, 36, 4040–4051.

33 Samenen JM, Peishoff CE, Keenan RM, Weinstock J. Refinement of a molecular model of angiotensin II (AII) employed in the discovery of potent nonpeptide antagonists. *Bioorg. Med. Chem. Lett.*, **1993**, 3, 909–914.

34 Buhlmayer P, Furet P, Criscione L, de Gasparo M, Whitebread S, Schmidlin T, Lattmann R, Wood J. Valsartan, a potent, orally active angiotensin II antagonist developed from the structurally new amino acid series. *Bioorg. Med. Chem. Lett.*, **1994**, 4, 29–34.

35 Alberts AW, Chen J, Kuron G, Hunt V, Huff J, Hoffman C, Rothrock J, Lopez M, Joshua H, Harris E, Patchett A, Monaghan R, Currie S, Stapley E, Albers-Schonberg G, Hensens O, Hirschfield J, Hoogsteen K, Liech J, Springer J. Mevinolin: a highly potent competitive inhibitor of hydroxymethylglutaryl-coenzyme A reductase and a cholesterol-lowering agent. *Proc. Natl. Acad. Sci. USA*, **1980**, 77, 3957–3961.

36 Davidson MH. Rosuvastatin: a highly efficacious statin for the treatment of dyslipidemia. *Expert Opin. Invest. Drugs*, **2002**, 11, 125–141.

37 Watanabe M, Koike H, Ishiba T, Okada T, Seo S, Hirai K. Synthesis and biological activity of methanesulfonamide pyrimidine- and N-methanesulfonyl pyrrole-substituted 3,5-dihydroxy-6-heptenoates, a novel series of HMG-CoA reductase inhibitors. *Bioorg. Med. Chem.*, **1997**, 5, 437–444.

38 Beck G, Kesseler K, Baader E, Bartmann W, Bergmann A, Granzer E, Jendralla H, Von Kerekjarto B, Krause R, Paulus E, Schubert W, Wess G. Synthesis and biological activity of new HMG-CoA reductase inhibitors. 1. Lactones of pyridine- and pyrimidine-substituted 3,5-dihydroxy-6-heptenoic (-heptanoic) acids. *J. Med. Chem.*, **1990**, 33, 52–60.

39 Roth BD, Bocan TMA, Blankley CJ, Chucholowski AW, Creger PL, Creswell MC, Ferguson E, Newton RS, O' Brien P, Picard JA, Roark WH, Sekerke CS, Sliskovic DR, Wilson MW.

Relationship between tissue selectivity and lipophilicity for inhibitors of HMG-CoA reductase. *J. Med. Chem.*, **1991**, 34, 463–466.

40 Humphrey PPA. The discovery of sumatriptan and a new class of drug for the acute treatment of migraine. *Front. Headache Res.*, **2001**, 10, 3–10.

41 Buckingham J, Glen RC, Hill AP, Hyde RM, Martin GR, Robertson AD, Salmon JA, Woollard PM. Computer-aided design and synthesis of 5-substituted tryptamines and their pharmacology at the 5-HT₁D receptor: discovery of compounds with potential anti-migraine properties. *J. Med. Chem.*, **1995**, 38, 3566–35 80.

42 (a) Rapoport AM, Tepper SJ. All triptans are not the same. *J. Headache Pain*, **2001**, 2, S87–S92; (b) Grujich NN, Gawel MJ. Eletriptan. *Expert Opin. Invest. Drugs*, **2001**, 10, 1869–1874.

43 Palacios JM, Rabasseda X, Castaner J. Almotriptan: antimigraine 5-HT₁D/1B agonist. *Drugs of the Future*, **1999**, 24, 367–374.

44 Wu Y-H, Rayburn JW, Allen LE, Ferguson HC, Kissel JW. Psychosedative agents. 2. 8-(4-substituted 1-piperazinylalkyl)-8-azaspiro[4.5]decane-7,9-diones. *J. Med. Chem.*, **1972**, 15, 477–479.

45 Yevich JP, New JS, Smith DW, Lobeck WG, Catt JD, Minielli JL, Eison MS, Taylor DP, Riblet LA, Temple DL, Jr. Synthesis and biological evaluation of 1-(1,2-benzisothiazol-3-yl)- and (1,2-benzisoxazol-3-yl)piperazine derivatives as potential antipsychotic agents. *J. Med. Chem.*, **1986**, 29, 359–3 69.

46 Ishizumi K, Kojima A, Antoku F. Synthesis and anxiolytic activity of N-substituted cyclic imides (1R*,2S*,3R*,4S*)-N-[4-[4-(2-pyrimidinyl)-1-piperazinyl]butyl]-2,3-bicyclo [2.2.1]heptanedicarboximide (tandospirone) and related compounds. *Chem. Pharm. Bull.*, **1991**, 39, 2288–2300.

47 Ishizume K, Kojima A, Antoku F, Saji I, Yoshigi M. Succinimide derivatives. II. Synthesis and antipsychotic activity of 1-[4-[[4-(1,2-benzisothiazol-3-yl)-1-piperazinyl]butyl]-1,2-cis-cyclohexanedicarboxamide (SM-9018) and related

compounds. *Chem. Pharm. Bull.*, **1995**, 43, 2139.

48 Onrust SV, McClellan K. Perospirone. *CNS Drugs*, **2001**, 15, 329–337.

49 Howard HR, Lowe JA, Seeger TF, Seymour PA, Zorn SH, Maloney PR, Ewing FE, Newman ME, Schmidt AW, Furman JS, Robinson GL, Jackson E, Johnson C, Morrone J. 3-Benzisothiazolylpiperazine derivatives as potential atypical antipsychotic agents. *J. Med. Chem.*, **1996**, 39, 143–148.

50 Hirsch SR, Kissling W, Bauml J, Power A, O'Connor R. A 28-week comparison of ziprasidone and haloperidol in outpatients with stable schizophrenia. *J. Clin. Psychiatry*, **2002**, 63, 516–523.

51 Todo Y, Nitta J, Miyajama M, Fukuoka Y, Yamashiro Y, Nishida N, Saikawa I, Narita H. Pyridonecarboxylic acids as antibacterial agents. VIII. *Chem. Pharm. Bull.*, **1994**, 42, 2063–2070.

52 Segawa J, Kitano M, Kazuno K, Matsuoka M, Shirahase I, Ozaki M, Matsuda M, Tomii Y, Kise M. Studies on pyridonecarboxylic acids. 1. Synthesis and antibacterial evaluation of 7-substituted-6-halo-4-oxo-4H-[1,3]thiazeto [3,2-a]quinoline-3-carboxylic acids. *J. Med. Chem.*, **1992**, 35, 4727–4738.

53 Marutani K, Matsumoto M, Otabe Y, Nagamuta M, Tanaka K, Miyoshi A, Hasegawa T, Nagano H, Matsubara S, et al. Reduced phototoxicity of a fluoroquinolone antibacterial agent with a methoxy group at the 8 position in mice irradiated with long-wavelength UV light. *Antimicrob. Agents Chemother.*, **1993**, 37(10), 2217–2223

54 Burger A. In: *A Guide to the Chemical Basis of Drug Design*. John Wiley & Sons, **1983**, p. v.

55 Wenlock MC, Austin RP, Barton P, Davis AM, Leeson PD. A comparison of physicochemical property profiles of development and marketed oral drugs. *J. Med. Chem.*, **2003**, 46, 1250–1256.

56 Sneader W. *Drug Prototypes and their Exploitation*. John Wiley & Sons, **1996**.

57 Drews J. Strategic trends in the drug industry. *Drug Discovery Today*, **2003**, 8, 411–420.

3
Privileged Structures and Analogue-Based Drug Discovery

Hugo Kubinyi

During the past 100 years, the pharmaceutical industry has achieved great success in the development of drugs to combat many human diseases. Recently, our knowledge of human biology has increased significantly, and currently the sequence of the human genome is known and many signaling pathways are understood. Gene technology provides us with human targets for *in-vitro* screening and for three-dimensional (3D) structure elucidation using X-ray crystallography and nuclear magnetic resonance (NMR) studies. Moreover, it also enables a proof of the therapeutic concept in genetically modified animals. In recent years, new technologies have been developed for drug discovery, amongst which are combinatorial chemistry and automated parallel synthesis, high-throughput screening (HTS) with human targets, structure-based and computer-aided design, and virtual screening. Despite the benefits of all these achievements, there is an increasing gap between the costs of research and development and the number of new drugs introduced into therapy [1].

3.1
Introduction

Chemogenomics [2–5] is a new strategy in drug discovery which, in principle, searches for all molecules that are capable of interacting with a particular cellular function. However, due to the almost infinite number of drug-like organic molecules, this is an impossible task. Therefore, chemogenomics has been defined as the investigation of classes of compounds (congeneric analogues) against families of related biological targets. Whereas such a strategy was first developed in the pharmaceutical industry almost 20 years ago, today it is applied more systematically in the search for target- and subtype-specific ligands. The term "privileged structures" has been defined for scaffolds, like the benzodiazepines, that very often produce biologically active analogues in a target family – in this case in the class of G protein-coupled receptors (GPCRs) [6,7]. By analogy with Emil Fischer's "lock and key" principle [8], Gerhard Müller refers to the chemogenomics approach as a "target family-directed master key concept" [9,10]. In this definition,

Analogue-based Drug Discovery. IUPAC, János Fischer, and C. Robin Ganellin (Eds.)
Copyright © 2006 WILEY-VCH Verlag GmbH & Co. KGaA, Weinheim
ISBN: 3-527-31257-9

chemogenomics deals with the systematic analysis of chemical–biological interactions. Congeneric series of chemical analogues serve as the probes in order to investigate their action on specific target classes, for example GPCRs, kinases, phosphodiesterases, ion channels, serine proteases, and others, in the search for target-specific ligands for every protein encoded by the human genome, as potential lead candidates.

3.2
Drugs from Side Effects

Many drugs have been discovered by chance [11–16], others by the observation and subsequent optimization of side effects [16–18]. Morphine **1** is an opiate-receptor narcotic, which has, in addition to its analgesic activity, also antitussive and constipant properties. In the systematic chemical variation of this lead structure, more specifically acting analogues have been derived, for example the major analgesic fentanyl **2**, the antitussive drug dextromethorphan **3**, the constipating drug loperamide **4**, and even dopamine antagonists (e.g., the neuroleptic haloperidol **5**) (Fig. 3.1).

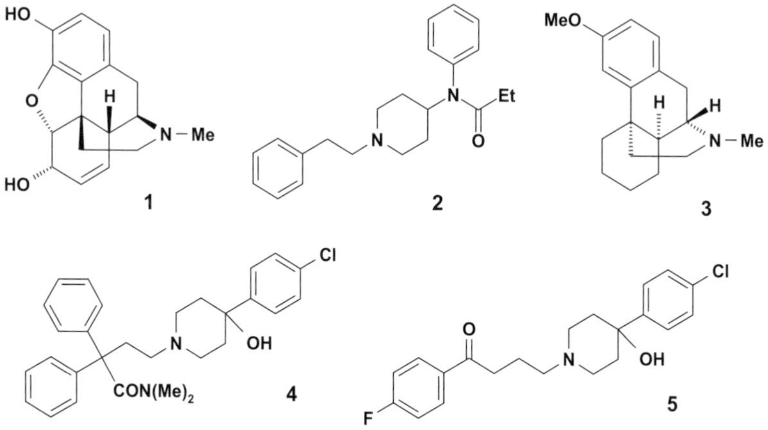

Fig. 3.1 Morphine **1** has been the lead structure for the development of the major analgesic fentanyl **2**, the antitussive drug dextromethorphan **3**, the constipating drug loperamide **4**, and the neuroleptic drug haloperidol **5**.

Another very prominent example of the development of drugs for new indications, from the fortuitous observation of side effects, is that of the sulfonamides. Several sulfonamides of the first generation had, in addition to their antibacterial effect, either diuretic or hypoglycemic activities. Correspondingly, specific diuretics, antiglaucomics, antihypertensives, and antidiabetics could be developed in this group of compounds [11,14,16].

A new strategy in drug discovery, the "selective optimization of side activities" (the SOSA approach), has been formulated by Camille Wermuth [17,18; cf. Chapter I–1]. In this approach, minor side activities of active compounds serve as the starting point for structural variations, ultimately to obtain analogues that now have the former side effect as their main activity. In this manner, nanomolar acetylcholinesterase (AChE) inhibitors as well as nanomolar muscarinic M_1 receptor and 5-HT$_3$ receptor ligands could be derived from the antidepressant minaprine [17,18].

3.3
Agonists and Antagonists

Neurotransmitter agonists and antagonists, as well as other agonists and antagonists of GPCRs, are very often closely related on a chemical basis. Typically, minor chemical changes (often resulting in a more lipophilic molecule) convert an agonist into an antagonist. In this manner, the first beta-blocker dichloroisoprenaline 6 was derived from isoprenaline 7, the opiate antagonist nalorphine 8 from morphine 9, and, for example, the histamine and acetylcholine antagonists 10 and 12 from the neurotransmitters histamine 11 and acetylcholine 13, respectively (Fig. 3.2).

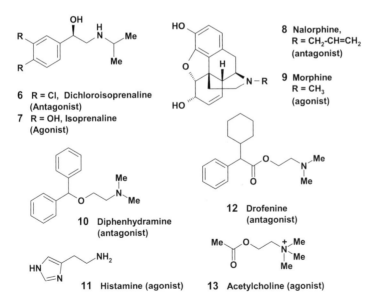

Fig. 3.2 The antagonists dichloroisoprenaline 6, nalorphine 8, diphenhydramine 10, and drofenine 12 are lipophilic analogues of the agonists isoprenaline 7, morphine 9, histamine 11, and acetylcholine 13.

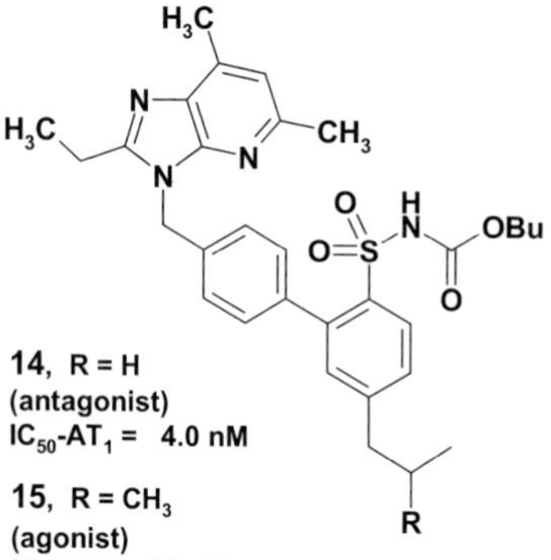

14, R = H
(antagonist)
IC_{50}-AT_1 = 4.0 nM

15, R = CH$_3$
(agonist)
IC_{50}-AT_1 = 25 nM

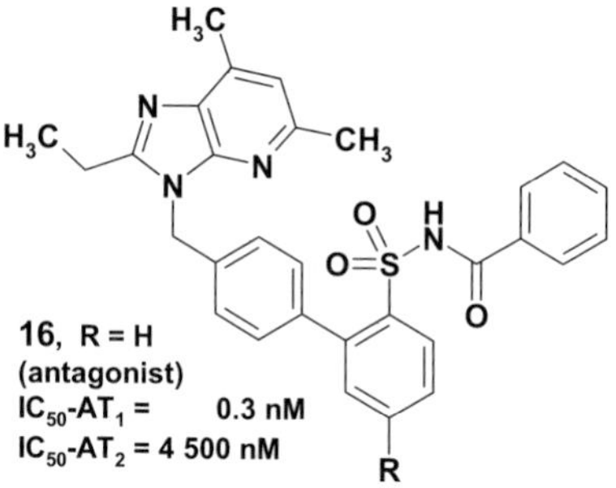

16, R = H
(antagonist)
IC_{50}-AT_1 =   0.3 nM
IC_{50}-AT_2 = 4 500 nM

17, R = CH$_2$CH(CH$_3$)$_2$
(agonist)
IC_{50}-AT_1 = 13 nM
IC_{50}-AT_2 = 10 nM

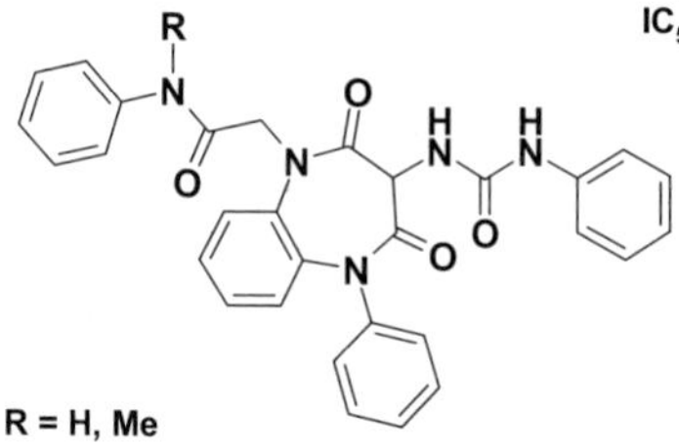

18, R = H, Me
   (CCK$_A$ antagonists)
19, R = Et, Pr, i-Pr
   (CCK$_A$ agonists)

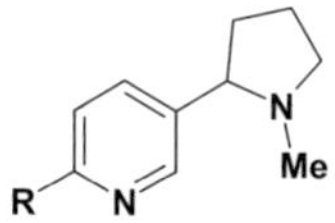

20, R = Et
   (nACh agonist)
21, R = n-Pr
   (nACh antagonist)

Fig. 3.3 In contrast to general experience, the antagonist **14** becomes an agonist **15** by introduction of an additional methyl group. Whereas compound **16** is a subnanomolar and specific AT$_1$ receptor antagonist, the iso-butyl analogue **17** is a nanomolar nonspecific AT$_1$/AT$_2$ agonist. The lower alkyl analogues **18** and **20** are antagonists at their respective receptors, whereas the increase in alkyl chain length to **19** and **21** generates agonists.

In recent times, however, the reverse effect has been observed – that is, the conversion of an antagonist into an agonist – by making it slightly more lipophilic [19–22]. The AT$_1$ receptor antagonist **14** becomes an agonist **15** by the introduction of a single methyl group; the effect is even more pronounced in the case of the AT$_1$-specific antagonist **16**, where the isobutyl analogue **17** is a nonspecific AT$_1$/AT$_2$-agonist. The CCK$_A$-antagonists **18** are also converted into CCK$_A$-agonists **19**, if the N-alkyl residue becomes larger than a methyl group (Fig. 3.3) [19–22]. Some 6-alkylnicotine derivatives discriminate between different activities of nicotine: whereas (–)6-ethylnicotine **20** binds with high affinity to nicotinic acetylcholine receptors (K_i = 5.6 nM) and produces the same actions as (–)nicotine (K_i = 1.2 nM), the 6-n-propyl analogue **21** (K_i = 22 nM) antagonizes the antinociceptive effect of (–)nicotine in the mice tail-flick assay, but not the spontaneous activity or discriminative stimulus effects (Fig. 3.3) [23].

3.4
Privileged Structures

It is common experience in medicinal chemistry that certain structural features produce biologically active compounds more often than others. The histamine antagonist diphenhydramine **10** (see Fig. 3.2) contained a diphenylmethane moiety. However, such a structural feature is present also in several acetylcholine antagonists, in the constipant drug loperamide **4** (see Fig. 3.1), in analgesics (e.g., nefopam), anticonvulsants (e.g., phenytoin), antiparkinson drugs (e.g., budipine), psychostimulants (e.g., modafinil), spasmolytics (e.g., propiverin), and many others.

Another group of such "privileged" structures are the tricyclic antihistaminic, neuroleptic and antidepressant drugs, for example, the anti-allergic H_1 antagonist promethazine **22**, the dopamine antagonist chlorpromazine **23**, and the uptake inhibitor imipramine **24** (Fig. 3.4). Despite their close chemical similarity, these compounds interact with the neurotransmitter binding sites of different GPCRs and transporters.

A closer inspection of the chemical structures of these compounds shows that they contain a modification of the diphenylmethane moiety, a diphenylamine. The large group of benzodiazepines also contains two phenyl rings attached to a carbon atom (1,4-benzodiazepines) or to a nitrogen atom (1,5-benzodiazepines); in addition, they mimic β-turns of proteins, due to their rigid structure. The tranquilizer diazepam **25** is the prototype of a benzodiazepine agonist (Fig. 3.5). On the other hand, flumazenil **26** is a benzodiazepine antagonist that is used as an antidote in benzodiazepine intoxication and after benzodiazepine application in surgery. Compound Ro-15–3505, **27**, is an inverse agonist, which acts as a proconvulsant. However, it was soon discovered that some benzodiazepines act also at other receptors. Tifluadom **28** is a strong kappa opioid receptor agonist [24], whilst devazepide **29** is an orally active cholecystokinin-B (CCK-B) antagonist (Fig. 3.5) [6].

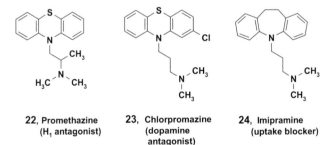

22, Promethazine
(H₁ antagonist)

23, Chlorpromazine
(dopamine
antagonist)

24, Imipramine
(uptake blocker)

Fig. 3.4 Promethazine **22**, chlorpromazine **23** and imipramine **24** are chemically closely related. Nevertheless, they have different mechanisms of action and different therapeutic applications. The antihistamine **22** has been used as an anti-allergic drug, the dopamine antagonist **23** is a neuroleptic, and the uptake inhibitor imipramine **24** is an antidepressant.

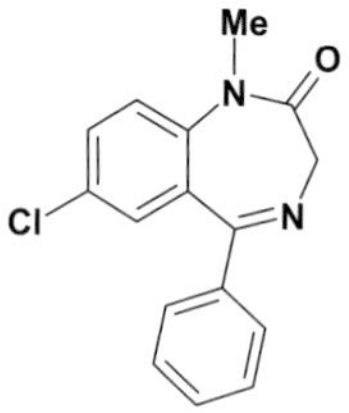

Fig. 3.5 The benzodiazepines are a group of compounds with highly diverse biological activities. At the benzodiazepine receptor, the tranquilizer diazepam **25** is an agonist, flumazenil **26** is an antagonist, and Ro 15–3505, **27**, is an inverse agonist. Tifluadom **28** is an opiate receptor agonist as well as a cholecystokinin (CCK) antagonist; devazepide **29** is a CCK antagonist.

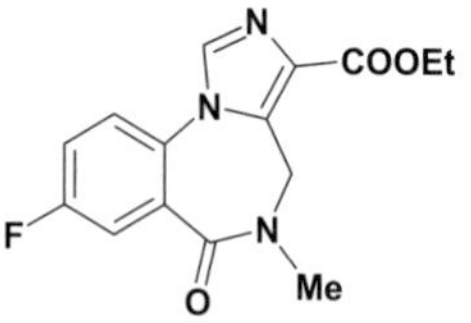

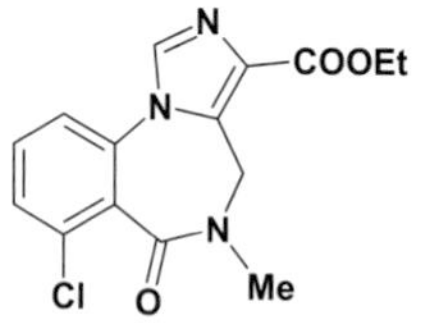

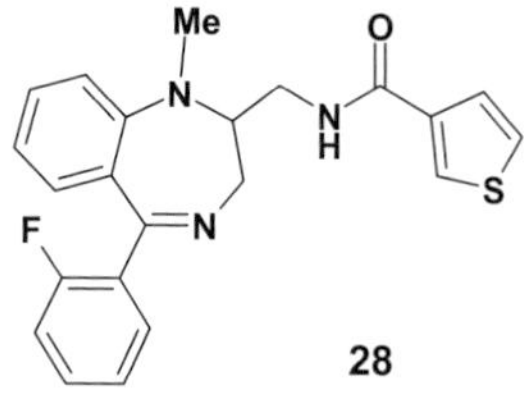

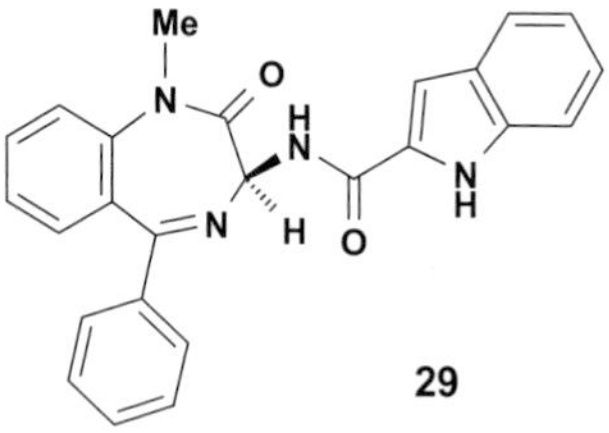

The observation that tifluadom **28** is also a nanomolar cholecystokinin receptor antagonist [25] led Evans et al. [6] to the conclusion that "... *these structures appear to contain common features which facilitate binding to various ... receptor surfaces, perhaps through binding elements different from those employed for binding of the natural ligands ...*". These authors also formulated "... *what is clear is that certain 'privileged structures' are capable of providing useful ligands for more than one receptor and that judicious modification of such structures could be a viable alternative in the search for new receptor agonists and antagonists.*"

This is indeed the case. In addition to gamma-aminobutyric acid (GABA), opiate and cholecystokinin receptor activities, the benzodiazepine scaffold is also found in muscle relaxants, antidepressants, neuroleptics, hypnotics, NK-1 receptor and vasopressin receptor antagonists, integrin antagonists, farnesyl transferase and phosphodiesterase inhibitors, potassium channel modulators, and others.

Some more examples of privileged structures include phenethylamines, steroids, arylpiperazines, arylspiropiperidines, and biphenyltetrazoles (Fig. 3.6) [7,26].

3.5
Drug Action on Target Classes

A very first proof that chemogenomic approaches might significantly increase the chances of success came from the target class of aspartyl proteases. During the 1980s, hundreds of man-years were spent in the search for orally available renin inhibitors, without much success. When HIV protease was recognized to be also

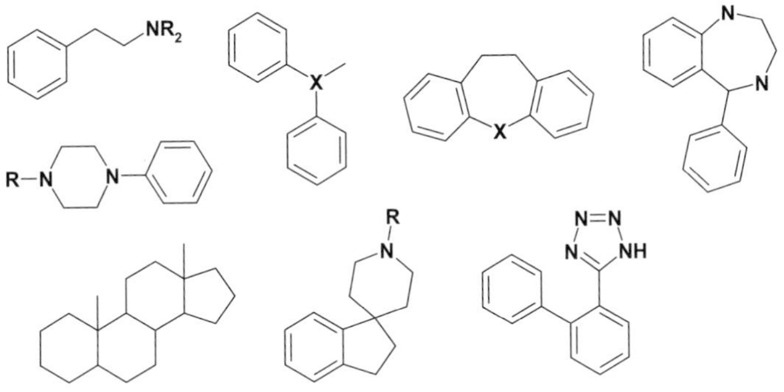

Fig. 3.6 Privileged structures are scaffolds or substituents that often produce biologically active compounds; for example, phenethylamines, diphenylmethyl and diphenylamine compounds (X = C or N, respectively), tricyclic compounds (X = C or N), benzodiazepines, arylpiperidines, steroids, spiropiperidines and tetrazolobiphenyls (from the upper left to the lower right).

an aspartyl protease, the accumulated knowledge could be used to design potent inhibitors for this nonhomologous but inherently related enzyme. The principle of transition state inhibitor interaction with the catalytic aspartates was transferred from renin to HIV protease, and within a relatively short time potent HIV protease inhibitors were discovered.

The same principles could be applied to metalloprotease inhibitors, where knowledge obtained from the development of angiotensin-converting enzyme (ACE) inhibitors was applied to the design of various other metalloprotease inhibitors. In the class of serine protease inhibitors, trypsin inhibitors paved the way to thrombin and factor Xa inhibitors. Structural elements of elastase (another serine protease) inhibitors could be successfully inserted into thrombin inhibitors. Some examples of chemogenomics approaches in different target classes in the following sections will illustrate certain aspects of this strategy.

3.5.1
GPCR Ligands

The phenethylamines are a class of GPCR ligands that bind more or less specifically to α- and β-adrenergic receptors, as well as to some other targets with phenethylamine binding sites (Tab. 3.1). Nevertheless, they must be considered as relatively specific drugs, whereas tricyclic neuroleptics and some other antipsychotic agents are promiscuous ligands, binding to various different receptors [27]. An extreme in this respect seems to be the atypical neuroleptic drug olanzapine **30**, which binds to many different GPCRs, including 5-HT$_{2A}$, 5-HT$_{2B}$, 5-HT$_{2C}$, dopaminergic D$_1$, D$_2$, D$_4$, muscarinic M$_1$, M$_2$, M$_3$, M$_4$, M$_5$, adrenergic α_1, and histaminic H$_1$ receptors, as well as to the 5-HT$_3$ ion channel (Tab. 3.2) [28,29]. Clearly, such multipotent drugs are better suited to the therapy of schizophrenia than would be a highly selective agent.

An example, where different receptor subtype selectivities have been observed for chemically related compounds, is presented in Fig. 3.7. The 5-HT$_3$ receptor is a serotonin-controlled ion channel, whereas the 5-HT$_4$ receptor is, like all other serotonin receptor subtypes, a GPCR. Despite a close chemical similarity, compound **31** is highly specific for the 5-HT$_3$ receptor, having a more than 300-fold higher affinity to the 5-HT$_3$ ion channel than to the 5-HT$_4$ receptor (K_i 5-HT$_3$ = 3.7 nM versus K_i 5-HT$_4$ >1000 nM; selectivity >250), while compound **32** binds almost exclusively to the 5-HT$_4$ receptor (K_i 5-HT$_3$ >10 000 nM versus K_i 5-HT$_4$ = 13.7 nM; selectivity >700) [30,31].

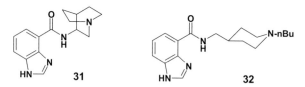

Fig. 3.7 The serotonin receptor ligands **31** and **32** differ in their subtype selectivity. Whereas **31** is a highly selective 5-HT$_3$ receptor ligand, compound **32** binds preferentially to the 5-HT$_4$ receptor.

Tab. 3.1 Biological activities of phenethylamines.

Compound (prototype)	Biological activity
Amphetamine	Stimulant
MDMA (Ecstasy)	Hallucinogen
Dopamine	Dopaminergic agonist
Norepinephrine	a-adrenergic agonist
Epinephrine	Mixed a,β-adrenergic agonist
Isoprenaline	β-adrenergic agonist
Dobutamine	β_1-specific adrenergic agonist
Salbutamol	β_2-specific adrenergic agonist
Dichloroisoprenaline	β-adrenergic antagonist
Metoprolol	β_1-specific antagonist
Xamoterol	β_2-specific partial agonist
Ephedrine	Indirect sympathomimetic agent
Norpseudoephedrine	Appetite suppressant

Sources: Textbooks of Medicinal Chemistry, *Annual Reports of Medicinal Chemistry*.

Tab. 3.2 G protein-coupled receptor (GPCR) and 5-HT$_3$ binding affinities of olanzapine **30** in different *in-vitro* models [28,29].

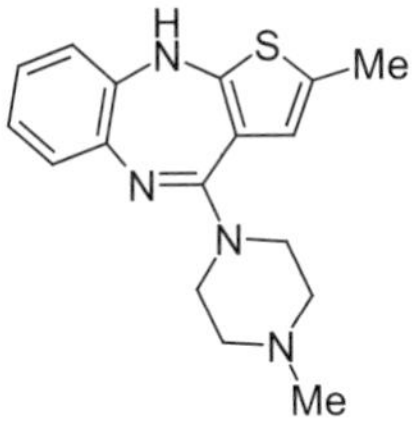

30

Olanzapine

Receptor	K_i values [nM]
5-HT$_{2A}$	2.5; 4
5-HT$_{2B}$	12
5-HT$_{2C}$	11; 29
5-HT$_3$	57
Dopaminergic D$_1$	31; 119
Dopaminergic D$_2$	11
Dopaminergic D$_4$	27
Muscarinic M$_1$	1.9; 2.5
Muscarinic M$_2$	18
Muscarinic M$_3$	13; 25
Muscarinic M$_4$	10; 13
Muscarinic M$_5$	6
Adrenergic a_1	19
Adrenergic a_2	230
Histaminic H$_1$	7

3.5.2
Nuclear Receptor Ligands

The steroid ring system is most probably the best-investigated privileged structure. Nature has already created estrogenic, progestogenic and androgenic hormones, mineralo- and glucocorticoids, cholesterol and the bile acids, as well as cardiac glycosides in plants. By systematic chemical variation, a large group of nuclear receptor ligands with different selectivities and analogues with other mechanisms of action resulted (Tab. 3.3), providing a broad variety of specific drugs that were effective against a wide variety of diseases.

3.5.3
Integrin Ligands

Integrins are a group of cell surface receptors, and several of them recognize the RGD (R = arg, G = gly, D = asp) motif of different proteins. On binding of such proteins, various specific cell–cell contacts are mediated. Interesting therapeutic applications should result from the inhibition of fibrinogen binding to the so-called fibrinogen receptor (integrin $a_{2b}\beta_{3a}$, GP IIb/IIIa), to prevent platelet aggregation, and inhibition of the vitronectin receptor (integrin $a_v\beta_3$), to suppress angiogenesis. The benzodiazepines **33** (lotrafiban; SmithKline Beecham; clinical investigation terminated in Phase III) and **34** mimic the RGD motif in different confor-

Tab. 3.3 Biological activities of steroids against nuclear receptors and other targets.

Compound (prototype)	Biological activity
Estradiol	Estrogen
Clometherone	Anti-estrogen
Danazole	Gonadotropin-releasing hormone (GRH) antagonist (indirect estrogen antagonist)
Formestane	Aromatase inhibitor
Progesterone	Progestogen
Mifepristone	Antiprogestogen
Testosterone	Androgen
Cyproterone acetate	Anti-androgen
Nandrolone decanoate	Anabolic
Finasteride	5a-Reductase inhibitor (testosterone biosynthesis inhibitor)
Cortisol	Nonselective gluco-/mineralocorticoid
Dexamethasone	Glucocorticoid
Aldosterone	Mineralocorticoid
Spironolactone	Aldosterone antagonist
Cholesterol	Membrane-modulating agent
Cholic acid	Choleretic
Chenodiol	Anticholelithogenic
Digoxin	Na^+/K^+-ATPase inhibitor (cardiac glycoside)
Cholecalciferol (vitamin D_3)	Antirachitic
Pancuronium bromide	Muscle relaxant
Hydroxydione	Anesthetic
Allo-pregnanolone	Neuromodulator
Tirilazad	Lipid peroxidation inhibitor
Edifolone	Cardiac depressant/anti-arrhythmic
Conessin	Anti-amebic
a-Ecdysone	Insect juvenile hormone

Sources: Textbooks of Medicinal Chemistry, *Annual Reports of Medicinal Chemistry*.

mations; correspondingly they are potent and highly selective inhibitors of these two integrins (Fig. 3.8) [32,33].

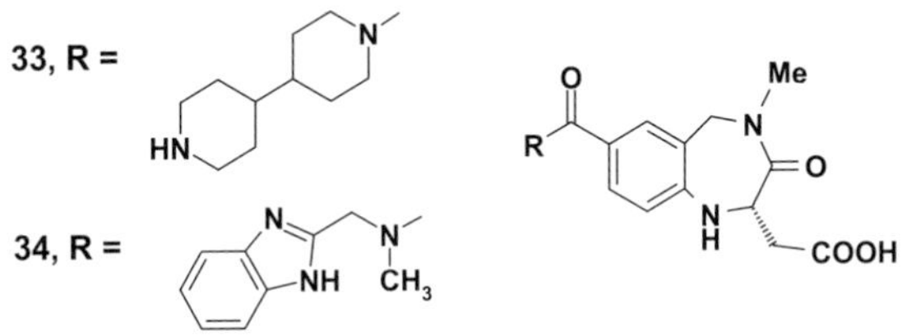

33, R =

34, R =

Fig. 3.8 Despite their common scaffold, compound **33** is a selective fibrinogen (GP IIb/IIIa) receptor ligand, whereas **34** is a selective vitronectin ($\alpha_v\beta_3$) receptor ligand; their selectivities differ by more than seven orders of magnitude.

3.5.4
Kinase Inhibitors

Kinases are phosphate group-transferring enzymes. Within our body, they constitute the largest enzyme family, with more than 500 different members. They control signal transduction pathways, in this manner mediating cellular processes such as metabolism, transcription, cell cycle progression, apoptosis, and differentiation. The catalytic domains of all kinases have as a common feature an ATP binding site to transfer the γ-phosphate of ATP to a serine, threonine or tyrosine hydroxyl group of a protein or peptide, or to a lipid. Whereas the ATP binding site is very similar in all kinases, the different members have special hydrophobic pockets in the close vicinity of these binding sites, in this manner enabling medicinal chemists to design specific ligands [34].

The discovery of imatinib (STI571, Gleevec®, Glivec®; Novartis) is a striking example of a successful chemogenomics approach, the selective optimization of a drug side activity. Most cases of chronic myelogenous leukemia result from a cross-over between chromosomes 9 and 22, by which a short chromosome 22-, the so-called Philadelphia chromosome, is formed. The DNA sequence at the fusion site codes for the bcr-abl kinase, a new protein with constitutionally enhanced tyrosine protein kinase activity [35]. At Novartis, the general lead structure **35** (Fig. 3.9), a protein kinase C (PKC) inhibitor, was the starting point of systematic structural variation. The introduction of amide residues into **36** produced an optimized PKC inhibitor **37**, for which also bcr-abl inhibition was observed. A methyl group in position R1 (compound **38**) surprisingly eliminated the undesired PKC activity. The result of the optimization was imatinib **39**, a bcr-abl kinase inhibitor [35], which was introduced into human therapy in May 2002.

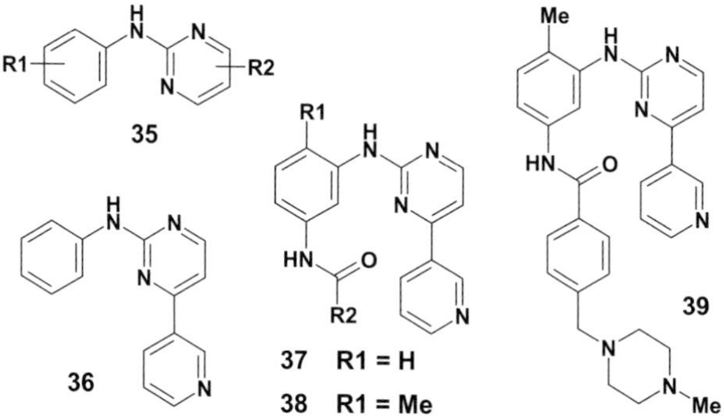

Fig. 3.9 Compounds of the general structure **35** and the optimized structure **36** are protein kinase C (PKC) inhibitors. The introduction of an amide residue led to mixed PKC and bcr-abl kinase inhibitors **37**, which lost their PKC-inhibitory activity by introduction of a methyl group in the para position to the amide group (compound **38**). The result of further optimization was the specific bcr-abl kinase inhibitor imatinib **39** (Gleevec®; Novartis).

3.5.5
Phosphodiesterase Inhibitors

The family of phosphodiesterases cleaves the cyclic nucleotides cGMP and cAMP to their hydrolysis products 5'-GMP and 5'-AMP, in this manner terminating the signals of these "second messengers" [36]. Eleven isoenzymes (PDE1 to PDE11) have been described in the literature. Following the discovery of sildenafil (Viagra®; Pfizer) – the first drug to be effective in erectile dysfunction ED – PDE5 became a target of great interest [36,37].

More than 30 years ago, the company May & Baker began research on anti-allergic xanthine derivatives [38]. Their first leads, **40** and **41** (Fig. 3.10), were about 40 times [38,39] and 1000 times [40] more active than cromoglycate, the standard drug at this time, and were already structurally closely related to sildenafil. The orally active drug candidate zaprinast **40** (M&B 22,948; Fig. 3.10) has, in addition to its "mast cell-stabilizing" activity against histamine- and exercise-induced asthma also vasodilatory and antihypertensive activities. Later, Pfizer started a search for a new antihypertensive principle [41], attempting initially to enhance the biological activity of the atrial natriuretic peptide (ANP) by prolonging the action of the second messenger of the corresponding receptor response. As zaprinast was one of the very few cGMP PDE inhibitors known at that time, its structure was modified to the PDE5-selective inhibitor sildenafil **42** (Fig. 3.10). Clinical tests showed the drug to be safe, but its clinical activity as an anti-anginal drug was disappointing. In 1992, a 10-day tolerance study in healthy volunteers led to the serendipitous observation of penile erections. Further clinical profiling went into this direction, and Viagra was approved for the treatment of ED in March 1998 [41]. As an undesired side effect, sildenafil **42** and vardenafil **43** inhibit PDE6 in a double-digit nanomolar range, whereas the structurally unrelated indoline derivative tadalafil **44** (Fig. 3.10) acts only at considerably higher concentrations.

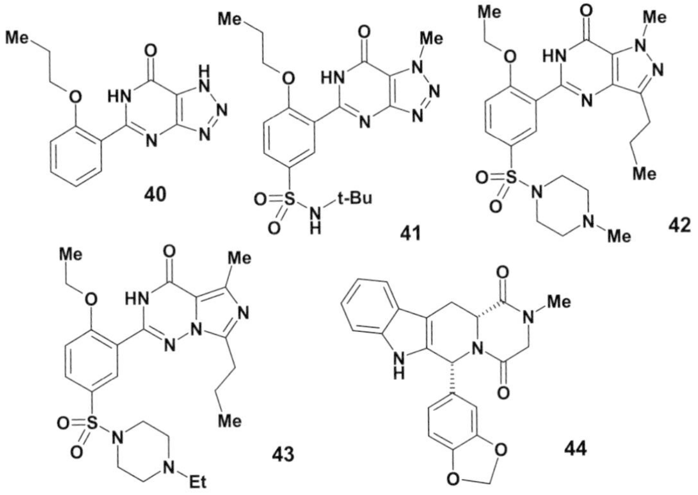

Fig. 3.10 Structural optimization of the anti-allergic xanthines **40** and **41** led to the PDE5 inhibitor sildenafil. Whereas vardenafil **43** is a close analogue of sildenafil, tadalafil **44** belongs to a different class of compounds; both, **43** and **44**, are highly potent PDE5 inhibitors.

There is now an ongoing interest to develop highly selective inhibitors also against certain other phosphodiesterases, because general structure–activity relationships can be transferred from one PDE isoenzyme to others [37].

3.5.6
Neurotransmitter Uptake Inhibitors

Talopram **45** and citalopram **46** (Fig. 3.11) are closely related analogues, but talopram is a norepinephrine uptake blocker with a selectivity factor of about 550 against serotonin uptake inhibition, whereas citalopram is a serotonin uptake blocker, with a selectivity of 3400 against norepinephrine uptake inhibition. A similar selectivity difference is observed for the even more closely related analogues nisoxetine **47** (norepinephrine uptake selectivity ca. 180) and fluoxetine **48** (serotonin uptake selectivity 54) (Fig. 3.11) [42].

3.6
Summary and Conclusions

This chapter has limited the discussion to only certain examples of chemogenomics in drug discovery, though further examples and systematic reviews are available in Ref. 4 and in recent volumes of the *Annual Reports in Medicinal Chemistry*. Nevertheless, these examples show the great potential of searching for new

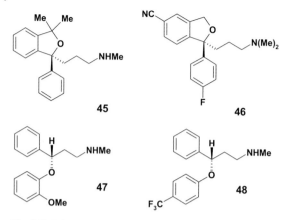

Fig. 3.11 Talopram **45** and nisoxetine **47** are norepinephrine-specific uptake inhibitors, whereas the closely related analogues citalopram **46** and fluoxetine **48** are specific serotonin uptake inhibitors.

drug candidates within target classes and of a chemistry focus on privileged scaffolds (albeit with the risk of patent problems). Wermuth [17,18] recommends optimization of the side effects of drugs, which might be a time- and cost-saving strategy in the search for new therapies. A recent editorial [43] recommended the investigation of every possible "alternative in therapeutic space" of existing drugs. Chemogenomics goes beyond these approaches – it starts from any active ligands within a series of structurally related biological targets and attempts to convert them into subtype-specific agonists, antagonists or inhibitors, by systematic modification, following the expertise of classical medicinal chemistry.

References

1 (a) Frantz S. *Nature Rev. Drug Discov.*, **2004**, 3, 103–104; (b) Frantz S, Smith A. *Nature Rev. Drug Discov.*, **2003**, 2, 95–96.
2 Caron PR, Mullican MD, Mashal RD, Wilson KP, Su MS, Murcko MA. *Curr. Opin. Chem. Biol.*, **2001**, 5, 464–470.
3 Bleicher KH. *Curr. Med. Chem.*, **2002**, 9, 2077–2084.
4 Jacoby E, Schuffenhauer A, Floersheim P. *Drug News Perspect.*, **2003**, 16, 93–102.
5 Kubinyi H, Müller G (Eds.). *Chemogenomics in Drug Discovery – A Medicinal Chemistry Perspective. Vol. 22. Methods and Principles in Medicinal Chemistry.*

Mannhold R, Kubinyi H, Folkers G (Eds.). Wiley-VCH, Weinheim, **2004**.
6 Evans BS, Rittle KE, Bock MG, DiPardo RM, Freidinger RM, Whitter WL, Lundell GF, Veber DF, Anderson PS, Chang RSL, Lotti VJ, Cerino DJ, Chen TB, Kling PJ, Kunkel KA, Springer JP, Hirshfield J. *J. Med. Chem.*, **1988**, 31, 2235–2246.
7 Patchett AA, Nargund RP. *Annu. Rep. Med. Chem.*, **2000**, 35, 289–298.
8 Fischer E. *Ber. Dtsch. Chem. Ges.*, **1894**, 27, 2985–2993.
9 Müller G. *Drug Discovery Today*, **2003**, 8, 681–691.

10 Müller G. In: Kubinyi H, Müller G (Eds.). *Chemogenomics in Drug Discovery – A Medicinal Chemistry Perspective. Vol. 22. Methods and Principles in Medicinal Chemistry.* Mannhold R, Kubinyi H, Folkers G (Eds.). Wiley-VCH, Weinheim, **2004**, pp. 7–41.

11 Burger A. *A Guide to the Chemical Basis of Drug Design.* John Wiley & Sons, New York, **1983**.

12 de Stevens G. *Prog. Drug Res.*, **1986**, 30, 189–203.

13 Roberts RM. *Serendipity. Accidental Discoveries in Science.* John Wiley & Sons, New York, **1989**.

14 Sneader W. *Drug Prototypes and their Exploitation.* John Wiley & Sons, Chichester, **1996**.

15 Kubinyi H. *J. Receptor Signal Transduct. Res.*, **1999**, 19, 15–39.

16 Kubinyi H. In: Kubinyi H, Müller G (Eds.). *Chemogenomics in Drug Discovery – A Medicinal Chemistry Perspective. Vol. 22. Methods and Principles in Medicinal Chemistry.* Mannhold R, Kubinyi H, Folkers G (Eds.). Wiley-VCH, Weinheim, **2004**, pp. 43–67.

17 Wermuth CG. *Med. Chem. Res.*, **2001**, 10, 431–439.

18 Wermuth CG. *J. Med. Chem.*, **2004**, 47, 1303–1314.

19 Underwood DJ, Strader CD, Rivero R, Patchett AA, Greenlee W, Prendergast K. *Chem. Biol.*, **1994**, 1, 211–221.

20 Perlman S, Costa-Neto CM, Miyakawa AA, Schambye HT, Hjorth SA, Paiva ACM, Rivero RA, Greenlee WJ, Schwartz TW. *Mol. Pharmacol.*, **1997**, 51, 301–311.

21 Beeley NRA. *Drug Discovery Today*, **2000**, 5, 354–363.

22 Ooms F. *Curr. Med. Chem.*, **2000**, 7, 141–158.

23 Dukat M, El-Zahabi M, Ferretti G, Damaj MI, Martin BR, Young R, Glennon RA. *Bioorg. Med. Chem. Lett.*, **2002**, 12, 3005–3007.

24 Römer D, Büscher HH, Hill RC, Maurer R, Petcher TJ, Zeugner H, Benson W, Finner E, Milkowski W, Thies PW. *Nature*, **1982**, 298, 759–760.

25 Chang RSL, Lotti VJ, Chen TB, Keegan ME. *Neurosci. Lett.*, **1982**, 72, 211–214.

26 Pernerstorfer J. In: *Handbook of Combinatorial Chemistry. Drugs, Catalysts, Materials, Volume II.* Nicolaou KC, Hanko R, Hartwig W (Eds.). Wiley-VCH, Weinheim, **2002**, pp. 725–742.

27 Schaus JM, Bymaster FP. *Annu. Rep. Med. Chem.*, **1998**, 33, 1–10.

28 Bymaster FP, Calligaro DO, Falcone JF, Marsh RD, Moore NA, Tye NC, Seeman P, Wong DT. *Neuropsychopharmacology*, **1996**, 14, 87–96.

29 Bymaster FP, Nelson DL, DeLapp NW, Falcone JF, Eckols K, Truex LL, Foreman MM, Lucaites VL, Calligaro DO. *Schizophr. Res.*, **1999**, 37, 107–122.

30 Ursini A, Capelli AM, Carr RA, Cassara P, Corsi M, Curcuruto O, Curotto G, Dal Cin M, Davalli S, Donati D, Feriani A, Finch H, Finizia G, Gaviraghi G, Marien M, Pentassuglia G, Polinelli S, Ratti E, Reggiani AM, Tarzia G, Tedesco G, Tranquillini ME, Trist DG, Van Amsterdam FT, Reggiani A. *J. Med. Chem.*, **2000**, 43, 3596–3613; erratum in *J. Med. Chem.*, **2000**, 43, 5057.

31 Lopez-Rodriguez ML, Morcillo MJ, Benhamu B, Rosado ML. *J. Comput.-Aided Mol. Design*, **1997**, 11, 589–599.

32 Samanen JM, Ali FE, Barton LS, Bondinell WE, Burgess JL, Callahan JF, Calvo RR, Chen W, Chen L, Erhard K, Feuerstein G, Heys R, Hwang SM, Jakas DR, Keenan RM, Ku TW, Kwon C, Lee C-P, Miller WH, Newlander KA, Nichols A, Parker M, Peishoff CE, Rhodes G, Ross S, Shu A, Simpson R, Takata D, Yellin TO, Uzsinskas I, Venslavsky JW, Yuan CK, Huffman WF. *J. Med. Chem.*, **1996**, 39, 4867–4870.

33 Keenan RM, Miller WH, Kwon C, Ali FE, Callahan JF, Calvo RR, Hwang SM, Kopple KD, Peishoff CE, Samanen JM, Wong AS, Yuan CK, Huffman WF. *J. Med. Chem.*, **1997**, 40, 2289–2292.

34 (a) Klebl BM, Daub H, Keri G. In: Kubinyi H, Müller G (Eds.). *Chemogenomics in Drug Discovery – A Medicinal Chemistry Perspective. Vol. 22. Methods and Principles in Medicinal Chemistry.* Mannhold R, Kubinyi H, Folkers G

(Eds.). Wiley-VCH, Weinheim, **2004**, pp. 167–190; (b) Buijsman R. In: Kubinyi H, Müller G (Eds.). *Chemogenomics in Drug Discovery – A Medicinal Chemistry Perspective. Vol. 22. Methods and Principles in Medicinal Chemistry.* Mannhold R, Kubinyi H, Folkers G (Eds.). Wiley-VCH, Weinheim, **2004**, pp. 191–219.

35 Capdeville R, Buchdunger E, Zimmermann J, Matter A. *Nature Rev. Drug Discov.*, **2002**, 1, 493–502.

36 Hendrix M. Kallus C. In: In: Kubinyi H, Müller G (Eds.). *Chemogenomics in Drug Discovery – A Medicinal Chemistry Perspective. Vol. 22. Methods and Principles in Medicinal Chemistry.* Mannhold R, Kubinyi H, Folkers G (Eds.). Wiley-VCH, Weinheim, **2004**, pp. 243–288.

37 Stamford AW. Annu. Rep. Med. Chem., **2002**, 37, 53–64.

38 Broughton BJ, Chaplen P, Knowles P, Lunt E, Marshall SM, Pain DL, Wooldridge KRH. *J. Med. Chem.*, **1975**, 18, 1117–1122.

39 Fujita T. In: *Quantitative Drug Design.* Ramsden CA (Ed.). Vol. 4 of *Comprehensive Medicinal Chemistry. The Rational Design, Mechanistic Study & Therapeutic Application of Chemical Compounds.* Hansch C, Sammes PG, Taylor JB (Eds.). Pergamon Press, Oxford, **1990**, pp. 497–560.

40 Wooldridge KRH. Personal communication, **1976**.

41 Kling J. *Modern Drug Discovery,* November/December **1998**, pp. 31–38.

42 Gundertofte K, Bogeso KP, Liljefors T. In: Computer-Assisted Lead Finding and Optimization. Proceedings, 11th European Symposium on Quantitative Structure–Activity Relationships, Lausanne, 1996. van de Waterbeemd H, Testa B, Folkers G (Eds.). Verlag Helvetica Chimica Acta and VCH, Basel, Weinheim, **1997**; pp. 445–459.

43 Anonymous. *Nature Rev. Drug Discov.*, **2004**, 3, 101.

Part II
Selected Examples Analogue-Based Drug Discoveries

Analogue-based Drug Discovery. IUPAC, János Fischer, and C. Robin Ganellin (Eds.)
Copyright © 2006 WILEY-VCH Verlag GmbH & Co. KGaA, Weinheim
ISBN: 3-527-31257-9

1
Development of Anti-Ulcer H$_2$-Receptor Histamine Antagonists

C. Robin Ganellin

1.1
Introduction

For many years, the main medical treatment for peptic ulcer disease relied on the use of antacids to neutralize the gastric acid; alternatively, patients had to resort to undergoing surgery. Thus, most major pharmaceutical companies had research programs aimed at the discovery of anti-ulcer agents. For example, Smith Kline & French Laboratories in Philadelphia, USA had an anti-ulcer program, but their subsidiary laboratories in Welwyn Garden City, UK initiated an unconventional approach in 1964 under James Black to seek a novel type of agent, namely to block the action of histamine as a stimulant of gastric acid secretion. It was hypothesized that histamine as a secretagogue acted at a hitherto uncharacterized histamine receptor (later designated as H$_2$).

1.2
The Prototype Drug, Burimamide, Defined Histamine H$_2$-Receptors

So started the investigations that led to the new class of anti-ulcer drugs, the H$_2$-receptor histamine antagonists. Histamine served as the chemical starting point for drug design and, like histamine, the early antagonists were all imidazole derivatives. The first compound used to define the pharmacological class of histamine H$_2$ receptors was burimamide (**1**) (Fig. 1.1) a thiourea derivative of a histamine homologue. This was also the first compound to be taken into human studies. It was a rather weakly active antagonist, but was used to verify that H$_2$ receptors did indeed mediate the gastric acid secretion in man as in the rat, cat and dog (reported in 1972) [1]. Burimamide was administered intravenously, but was insufficiently active to be administered orally and not considered worthwhile for further development.

Analogue-based Drug Discovery. IUPAC, János Fischer, and C. Robin Ganellin (Eds.)
Copyright © 2006 WILEY-VCH Verlag GmbH & Co. KGaA, Weinheim
ISBN: 3-527-31257-9

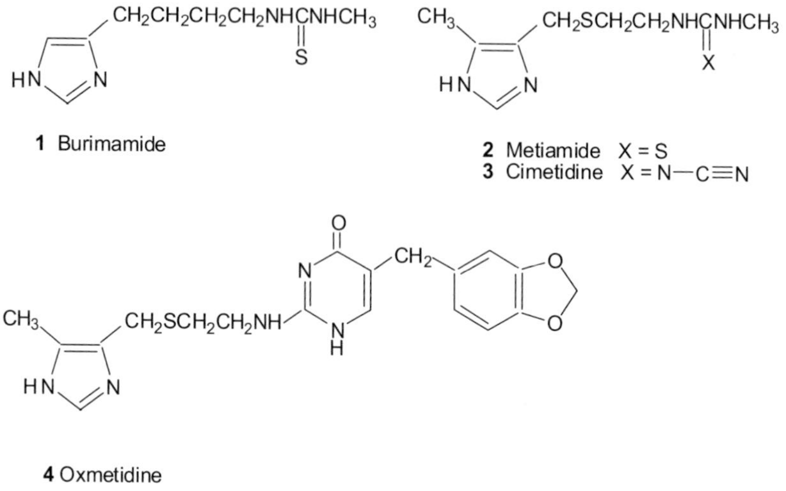

1 Burimamide

2 Metiamide X = S
3 Cimetidine X = N—C≡N

4 Oxmetidine

Fig. 1.1 The pioneer drugs.

1.3

The Pioneer Drug, Cimetidine: A Breakthrough for Treating Peptic Ulcer Disease

Burimamide was followed by metiamide (**2**) which was designed to alter the electronic properties of the imidazole ring in burimamide by possessing an electron-releasing methyl group in the imidazole 4(5)-position and an electron-withdrawing isosteric S replacement for CH_2 in the side chain [2]. Metiamide was orally active, and was the first H_2-receptor histamine antagonist shown to be clinically effective in inhibiting gastric acid secretion and assisting duodenal ulcers to heal [3]. However, it was suspended from clinical trials following the detection of a low incidence (ca. 1%) of granulocytopenia. The hypothesis that the granulocytopenia might have derived from a toxic side effect caused by the thioureido group in metiamide led to bioisosteric replacements for thiourea. These investigations provided the cyanoguanidine analogue [3], cimetidine (**3**) (Fig. 1.1), in which cyanoimino (=N–CN) replaces thione sulfur (=S).

Cimetidine closely resembled metiamide but did not exhibit the granulocytopenia side effect. It was marketed as Tagamet® at the end of 1976, as the pioneer drug which revolutionized the medical treatment of peptic ulcer disease [4]. Indeed, in many countries it became the best-selling prescription medicine and was the first of the "block-buster" products (billion dollar annual sales).

Although cimetidine was very effective for symptomatic treatment and permitted ulcers to heal, there were some side effects associated with its use, albeit at a very low level. In particular, there was a low incidence of gynaecomastia in men; this was believed to be due to an anti-androgen effect of cimetidine seen at high doses.

This anti-androgen activity was attributed to the cyanoguanidine group. Hence, a follow-up analogue was developed in which the cyanoguanidine was replaced by a 5-benzyl-isocytosine moiety, giving oxmetidine [5a] (SK&F 92994) (4). This compound was the first putative analogue product, but was withdrawn during the later stages of clinical development (within months of the intended marketing date) after liver-related side effects had been observed [5b].

1.4
Ranitidine: The First Successful Analogue of H$_2$ Antagonists

At Allen and Hanburys Ltd, which was part of the Glaxo Group, a research team had been established during the late 1960s to identify improved drugs for the treatment of peptic ulcer. Among various approaches the team had investigated was the activity of thioamides [e.g., 2-phenyl-2-(2-pyridyl)thioacetamide; SC 15396], which had been reported by Searle scientists to block the secretion of gastrin [6]. In particular, they had synthesized various tetrazole derivatives. When the report by Black et al. (1972) [1] was published defining H$_2$ receptors, the Glaxo chemists turned their attention to the synthesis of tetrazole analogues of burimamide (the first H$_2$-antagonist). The compound where 5-aminotetrazole replaced the imidazole ring (AH 15475; Tab. 1.1) was found to be equipotent with burimamide. Subsequently, many other 5-substituted analogues were made, but none showed any improvement.

Following the SK&F discovery of active thiourea replacements in H$_2$-antagonists which were published initially in the SK&F patent applications (see above), the Glaxo chemists introduced groups such as cyanoguanidine into the aminotetrazole structure. As for the SK&F compounds, these replacement groups did not increase potency [7]. By 1976, the research studies at Glaxo had still not been able to improve on AH 15475, and the management argued strongly to abandon H$_2$-antagonists in favor of effective anticholinergics as a target for inhibiting acid secretion. Furthermore, the number of chemists left working on the project was reduced to three [7]. The importance that personal experience can make to research cannot be overemphasized, however, and in this respect the experience of the senior chemist, Dr. John Clitherow, was to make a crucial contribution. Many years earlier, Dr. Clitherow had included in his Ph.D. research the synthesis of furan derivatives as intermediate in making analogues of the cholinergic agent muscarine [8]. Since aminotetrazole is only weakly basic and yet AH 15475 (in which aminotetrazole functions as an imidazole replacement) was active, the chemists were led to construct a furan analogue – that is, a nonnitrogen heterocycle which was also nonbasic. The butyl chain was replaced by $CH_2SCH_2CH_2$ for ease of synthesis, following the lead given by the SK&F studies on metiamide; the resulting compound (AH 18166) showed some activity but was poorly water-soluble (Tab. 1.1). In order to improve solubility, a Mannich base of furan (from dimethylamine and formaldehyde) was synthesized. This was the breakthrough that the researchers had been waiting for, as the product, AH 18665, was almost as active [9] as metiamide (Tab. 1.1).

Tab. 1.1 Some key steps in the discovery of ranitidine.

An aminotetrazole analogue of burimamide is synthesized and found to have activity equipotent with burimamide:

AH 15475

$$N\text{—}N\diagup(CH_2)_4NHCNHCH_3$$

with $\|$ S below, and the tetrazole ring N, N, N, N bearing NH_2.

Replacing aminotetrazole by furan and incorporating an -S- link in the chain provides an active compound, but this has poor water-solubility:

AH 18166 $CH_2SCH_2CH_2NHCNHCH_3$ (furan ring with O), with $\|$ S below.

A dimethylaminomethyl group (Mannich base) is introduced to confer solubility, and the compound is nearly as potent as metiamide:

AH 18665 $(CH_3)_2NCH_2$—(furan, O)—$CH_2SCH_2CH_2NHCNHCH_3$, with $\|$ S below.

Replacing the thiourea group by a cyanoguanidine group gives a compound equipotent with cimetidine, but it has poor crystallinity for pharmaceutical development:

AH 18801 $(CH_3)_2NCH_2$—(furan, O)—$CH_2SCH_2CH_2NHCNHCH_3$, with $\|$ NCN below.

Replacing the cyanoguanidine group by a diaminonitroethene group increases potency 4- to 10-fold to give ranitidine:

AH 19065
Ranitidine $(CH_3)_2NCH_2$—(furan, O)—$CH_2SCH_2CH_2NHCNHCH_3$, with $\|$ CHNO$_2$ below.

The chemists also prepared the cyanoguanidine (AH 18801; Tab. 1.1), and this proved to be as active [9] as cimetidine. However, two problems facing the development chemists were the poor crystallinity and low melting point of this product.

The research team then followed up another "end group" which had been designed by the chemists at SK&F, who had continued to explore other polar end group structures in place of cyanoguanidine. One moiety which had not been used previously in a drug was that of 1,1-diamino-2-nitroethene. This was designed in an attempt to reduce hydrophilicity, since an inverse correlation had been obtained between potency and hydrophilicity [10], as indicated by octanol:-water partition studies. One aim had been to replace the N of cyanoguanidine by CH, but this led to complications as the resulting diamino-2-cyanoethene was in tautomeric equilibrium with the cyanoacetamidine.

Consideration of CH versus NH acidities [10] focused attention onto the nitro compound, as the 1,1-diamino-2-nitroethene tautomer predominates, and this led to the synthesis of 1,1-diamino-2-nitroethene analogue (Tab. 1.2; X = $CHNO_2$) of cimetidine. This compound was found to be similarly active in animals, and was subsequently studied in human volunteers to confirm this finding. Since this compound showed no apparent improvement over cimetidine, it was not developed further at SK&F, although the compound was patented. This discovery was taken up by chemists at Glaxo Research who synthesized the nitrovinyl analogue of their compound in the expectation that it would be crystalline. Although this compound (AH 19065) turned out to be an oil, fortuitously, it proved to be ten times more potent in the rat than the cyanoguanidine (Tab. 1.2) and consequently was selected for development and given the name *ranitidine*.

Tab. 1.2 Comparison between imidazole and furan series.

$$HetCH_2SCH_2CH_2NHCNHCH_3$$
$$\overset{\|}{X}$$

Het =	CH₃ imidazole	(CH₃)₂NCH₂ furan
X	**Imidazole** ED_{50} [mg kg^{-1}]	**Furan** ED_{50} [mg kg^{-1}]
S	0.52	2.32
NCN	1.12[a]	1.39
$CHNO_2$	1.75	0.18[b]

Inhibition of histamine-stimulated gastric acid secretion in the perfused stomach preparation in the anesthetized rat [9]. Compounds were administered intravenously.
a. Cimetidine.
b. Ranitidine.

Ranitidine is about four times more potent than cimetidine in man as an inhibitor of gastric acid secretion [11], and in 1981 was marketed as Zantac®. Ranitidine is not only more potent than cimetidine but also proved to be more selective. Cimetidine also inhibits cytochrome P-450, an important drug-metabolizing enzyme. This interaction has the effect of inhibiting the metabolism of certain drugs such as propranolol, warfarin, diazepam and theophylline, thus producing effects equivalent to an overdose of these medicines. It is therefore advisable to avoid co-administration. This effect and binding to androgen receptors, are avoided by ranitidine.

Clinically, therefore, ranitidine offered some advantages, and eventually built up its own safety record for treating peptic ulcer disease. Indeed, by 1987 (i.e., 10 years after the launch of cimetidine), its world-wide sales had overtaken those of cimetidine and it then became the best-selling prescription product.

An analogue of ranitidine is nizatidine (Fig. 1.2), developed by the Eli Lilly Company and marketed as Axid®. In nizatidine, the 5-dimethylaminomethyl-furan ring is replaced by the 2-dimethylaminomethylthiazole ring [12]. It is interesting that Eli Lilly were able to enter this market very quickly, and this was probably due to their previous efforts during the 1950s to identify an inhibitor of histamine-induced gastric acid secretion, and for which they collaborated with the leading gastroenterologist, Mort Grossman [13a]. Indeed it is poetic justice because Reuben Jones [13b], an outstanding medicinal chemist at Eli Lilly, laid the foundation for histamine analogues which was greatly appreciated and built upon by the SK&F medicinal chemists in their pioneering studies to discover burimamide.

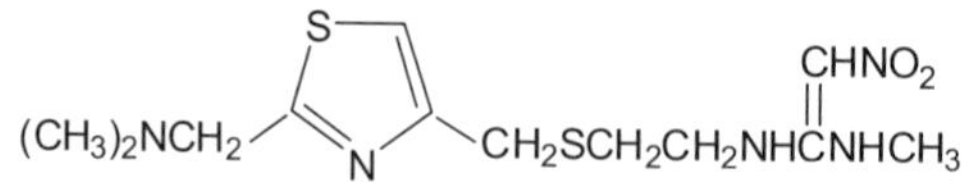

$$(CH_3)_2NCH_2 \overset{S}{\underset{N}{\diagup\!\!\diagdown}} CH_2SCH_2CH_2NH\overset{\overset{CHNO_2}{\|}}{C}NHCH_3$$

Fig. 1.2 Nizatidine: an analogue of an analogue (replacing the furan ring of ranitidine).

1.5
The Discovery of Tiotidine and Famotidine

At ICI Pharmaceuticals, anti-ulcer research had been aimed at peptide analogues of gastrin as potential gastrin antagonists, but no effective agents were obtained. Following the publication of Black et al. [1], the ICI team turned to histamine as a target and began to screen compounds which they had already had on file. From this emerged 2-guanidino-4-methylthiazole (Fig. 1.3) which was shown [14a] to be an H_2-receptor histamine antagonist on the isolated guinea-pig right atrium, with an apparent dissociation constant (K_B) of 3.6×10^{-6} M; this compound served as a new lead which was outside the scope of the SK&F patents, as SK&F had not covered the guanidine as a substituent in the heterocyclic part of their molecules. This was followed up by the synthesis of compounds in which the appropriate side chain replaced the methyl group. The cimetidine analogue, ICI 125211, was approximately 40 times more potent than cimetidine as an H_2-receptor antagonist in vitro [14b]. This compound, given the name tiotidine, was selected for drug development. Tiotidine was eight times more potent than cimetidine in man, with a relatively slow onset and long duration of action. It looked to be a very promising compound as a potential therapeutic agent but, unfortunately, after prolonged high-dose safety studies in rats a problem was discovered: tiotidine was found to have a toxic effect on the cells lining the stomach, and so clinical investigations had to be suspended.

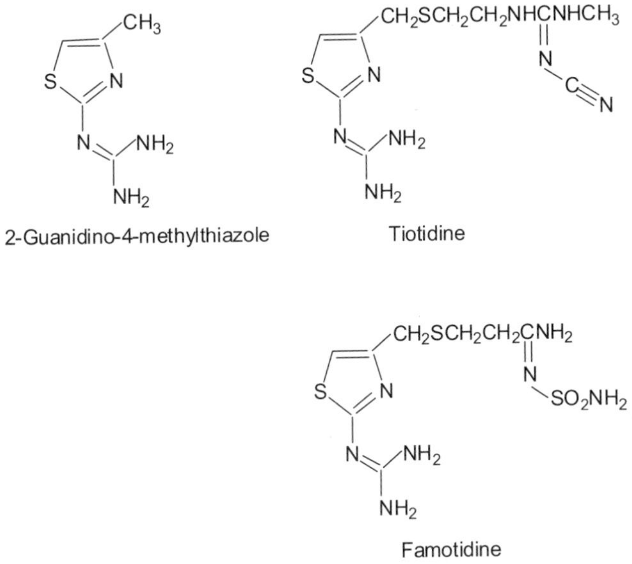

2-Guanidino-4-methylthiazole   Tiotidine

Famotidine

Fig. 1.3 Guanidinothiazoles (replacing imidazole) as analogues.

The guanidinothiazole group appears to confer high affinity at the H_2-receptor, and many active derivatives are now known. Only one of these, however, has become a therapeutically useful drug; this is a sulfamidoamidine analogue discovered in the laboratories of the Japanese Company, Yamanouchi. This compound, famotidine (YM 11170; Fig. 1.2) is about 30 times more potent [15] than cimetidine in man, and is marketed in Japan as Gaster® and in Western countries by Merck Sharp and Dohme (under license from Yamanouchi), under the trademark Pepcid®.

1.6
Other Compounds

A further development from the Glaxo chemists was to replace the furan ring by phenoxy, in effect moving the oxygen atom from within the ring to outside the ring. Activity is also dependent upon the amine substituent, and a *meta*-piperidinomethyl group was found to be particularly favorable. The use of *m*-piperidinomethylphenoxypropyl attached to the "polar" group considerably expanded the scope for active structures (Fig. 1.4), for example lamtidine (AH 22216) [16] and loxtidine (AH 23844) [17], both of which were very long-lasting inhibitors of gastric acid secretion *in vivo* and appeared to induce an almost irreversible blockade of receptor function *in vitro*. This finding was applied by chemists in many other pharmaceutical companies, including Teikoku, Wakamoto, Bristol Laboratories, Merck Sharp and Dohme, and Wyeth. It appears that *m*-piperidinomethylphenoxypropyl confers high affinity at the H_2 receptor, and many active derivatives have been synthesized. Particularly noteworthy was the finding by researchers at the Teikoku Hormone Manufacturing Company that the "polar" group need contain only one N atom, and these researchers found that a simple derivative such as -NHCOCH$_3$ had a K_B value of 0.22 μM determined *in vitro* on guinea-pig atrium. They then developed the -NHCOCH$_2$OCOCH$_3$ analogue, TZU 0460, roxatidine acetate [18], which was marketed in Japan as Altat® and in the West by Hoechst-Roussel under the name of Rotane®.

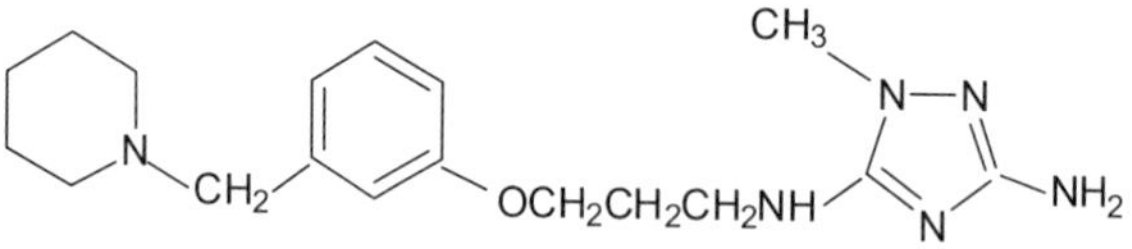

Lamtidine

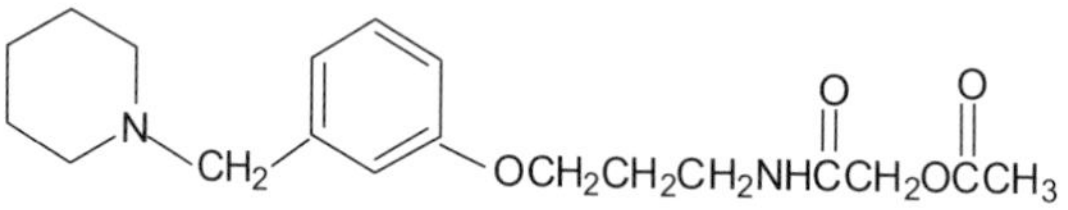

Roxatidine

Fig. 1.4 Piperidinomethylphenoxypropyl analogues (exocyclic O).

The discovery and development of cimetidine and ranitidine provided a revolution in the medical treatment and management of peptic ulcer disease. Subsequently, many pharmaceutical companies became involved in research programs to discover additional compounds as H_2-receptor histamine antagonists. As a result, a very wide range of chemical structures now exists for this class of drug (for a review, see Cooper et al. [19]). Many of these compounds have been investigated in human studies, but only the above-mentioned five drugs – cimetidine, ranitidine, nizatidine, famotidine, and roxatidine – are marketed as medicines.

1.7
The Use of H_2-Receptor Histamine Antagonists as Medicines

As stated above, cimetidine revolutionized the medical treatment of peptic ulcer disease, and other products which followed continued the same type of treatment. Thus, H_2-receptor histamine antagonists in general were given daily for 4–6 weeks to heal duodenal ulcers. They were also used to treat benign gastric ulceration (i.e., where this is not due to gastric cancer) and other conditions involving the hypersecretion of gastric acid, including gastroesophageal reflux disease, dyspepsia, and Zollinger–Ellison syndrome. They are also used for prophylaxis of gastrointestinal hemorrhage from stress ulceration in seriously ill patients, before general anesthesia in patients thought to be at risk of acid aspiration (Mendelson's syndrome), and for the peptic ulceration associated with concomitant use of non-steroidal anti-inflammatory drugs (NSAIDs).

All of the H_2 antagonists have similar clinical applications, the main difference being that the products which came after cimetidine avoided cimetidine's associated problems of anti-androgen effects and the inhibition of P450 enzymes. These drugs are also more potent than cimetidine. It is difficult to determine a precise quantitative comparison of clinically effective doses, although there is no doubt that cimetidine is the least potent and famotidine is the most potent. Approximately equi-potent daily doses for clinical treatment are listed in Tab. 1.3. All of these are normally given orally as tablets, the absorption is rapid with good bioavailability, low protein binding, and a half-life of ca. 2 hours.

Tab. 1.3 Comparison of approximately equivalent daily doses of H_2 antagonists (administered orally as tablets) for treatment of hyperacidity.

Drug	Daily dose [mg]	Molecular weight [Da]
Cimetidine	800	252
Nizatidine	300	331
Ranitidine	300	314
Roxatidine	150	385[a]
Famotidine	40	337

a. Administered as acetate hydrochloride.

The treatment of diseases is under continuous development, and the past 40 years has seen considerable changes in gastroenterology. Until cimetidine (Tagamet®) was made available, the medicinal managements of peptic ulcer disease and related problems of acid hypersecretion (e.g., dyspepsia) were not really effective, and surgery had become the mainstay for treatment. However, cimetidine dramatically changed this situation, by offering physicians a medicine which acted pharmacologically to turn off the "tap" of acid secretion. Histamine H_2-receptor blockade also provided physiologists with a tool to study gastric acid secretion, and this in turn led to the development of omeprazole, the first of the H^+/K^+ ATPase inhibitors – a new class of anti-ulcer product, the "proton-pump" inhibitors (PPIs).

There remained, however, a mystery problem – why did patients relapse after treatment had successfully permitted their ulcers to heal? There followed another breakthrough – the discovery that ulcers were caused by infection of the stomach with the bacterium, *Helicobacter pylori*. Thus, treatment for ulcers has developed into a regimen which aims first to reduce acidity by administering an H_2-antagonist or PPI to relieve patients of pain; and second, to treat them with a combination of antibiotics and antibacterial agents to eradicate the infection.

References

1 Black JW, Duncan WAM, Durant GJ, Ganellin CR, Parsons ME. *Nature*, **1972**, 236, 385.

2 Black JW, Durant GJ, Emmett JC, Ganellin CR. *Nature*, **1974**, 248, 65.

3 (a) Durant GJ, Emmett JC, Ganellin CR, Miles PD, Prain HD, Parsons ME, White GR. *J. Med. Chem.*, **1977**, 20, 901; (b) Brimblecombe RW, Duncan WAM, Durant GJ, Emmett JC, Ganellin CR, Parsons ME. *J. Int. Med. Res.*, **1975**, 3, 86; (c) Brimblecombe RW, Duncan WAM, Durant GJ, Emmett JC, Ganellin CR, Leslie GB, Parsons ME. *Gastroenterology*, **1978**, 74, 339.

4 Freston JW. *Ann. Intern. Med.*, **1982**, 97, 573.

5 (a) Blakemore RC, Brown TH, Durant GJ, Emmett JC, Ganellin CR, Parsons ME, Rasmussen AC. *Br. J. Pharmacol.*, **1980**, 70, 105P; (b) Helfrich HM, Evers PW, Shriver RC, Jacob LS. *Am. J. Gastroenterol.*, **1985**, 80, 959.

6 Cook DL, Bianchi RG. *Life Sci.*, **1967**, 6, 1381.

7 Bradshaw J. In: *Chronicles of Drug Design*. Lednicer D (Ed.). American Chemical Society, New York, **1992**.

8 Beckett AH, Harper NJ, Clitherow JW. *J. Pharm. Pharmacol.*, **1963**, 15, 362.

9 Bradshaw J, Butcher ME, Clitherow JW, Dowie MD, Hayes R, Judd DB, McKinnon JM, Price BJ. In: The Chemical Regulation of Biological Mechanisms. Creighton AM, Turner S (Eds.). Special publications No. 42. Royal Society of Chemistry, London, **1982**, pp. 45–57.

10 Ganellin CR. *J. Med. Chem.*, **1981**, 24, 913.

11 Ireland A, Colin-Jones OG, Gear P, Golding PL, Ramage JK, Williams JG, Leicester RJ, Smith CL, Ross G, Bamforth J, De Gara CJ, Gledhill T, Hunt RH. *Lancet*, **1984**, ii, 274.

12 Lin TM, Evans DC, Warrick MW, Pioch RP, Ruffalo RR. *Gastroenterology*, **1983**, 84, 1231.

13 (a) Grossman MI, Robertson C, Rosiere CE. *J. Pharmacol. Exp. Ther.*, **1952**, 104, 277; (b) Jones RG. In: *Handbook of Experimental Pharmacology, XVIII/I Histamine and Antihistaminics*. Rocha e Silva M (Ed.). Springer-Verlag, Berlin, **1966**, pp. 1–43.

14 (a) Gilman DJ, Jones DF, Oldham K, Wardleworth JM, Yellin TO. In: The Chemical Regulation of Biological Mechanisms. Creighton AM, Turner S (Eds.). Special publications No. 42. Royal Society of Chemistry, London, **1982**, pp. 58–76. (b) Yellin TO, Buck SH, Gilman DJ, Jones DF, Wardleworth JM. *Life Sci.*, **1979**, 25, 2001.

15 (a) Takeda M, Tagaki T, Yashima Y, Maeno H. *Arzneim.-Forsch.*, **1982**, 32 (ii), 734; (b) Yanagisawa I, Hirata Y, Ishii Y. *J. Med. Chem.*, **1987**, 30, 1787.

16 Brittain RT, Daly MJ, Humphray JM, Stables R. *Br. J. Pharmacol.*, **1982**, 76, 195P.

17 Brittain RT, Jack D, Reeves JJ, Stables R. *Br. J. Pharmacol.*, **1982**, 85, 843.

18 Tarutani M, Sakuma H, Shiratsuchi K, Mieda M. *Arzneim.-Forsch.*, **1985**, 35, 703.

19 Cooper DG, Young RC, Durant GJ, Ganellin CR. In: *Comprehensive Medicinal Chemistry*. Emmett JC (Ed.). Vol. 3. Pergamon Press, Oxford, **1990**, pp. 323–421

2

Esomeprazole in the Framework of Proton-Pump Inhibitor Development

Per Lindberg and Enar Carlsson

2.1
Towards Omeprazole: The First Proton-Pump Inhibitor

In January 1979, Ylva Örtengren, a chemist working at Hässle in Mölndal (part of the Swedish drug company Astra, now AstraZeneca), first described the synthesis of H 168/68. In this code number, H stood for Hässle, 168 was the laboratory book number, and 68 the page number (Fig. 2.1). Ylva should have used page 40, but the senior chemist Ulf Ljunggren wanted a code number that was easy to remember, and had a hunch that this was going to be "the compound". Ulf was correct, and today H 168/68 is better known as omeprazole. It was the culmination of an intensive research program which began at Hässle some 10 years earlier, and aimed at developing a safe and effective inhibitor of gastric acid secretion.

Omeprazole was the first of a new class of drugs that control acid in the stomach – the proton pump inhibitors (PPIs) [1–6]. This introduced a new approach for the effective inhibition of gastric acid secretion and the treatment of acid-related diseases. Omeprazole was quite quickly shown to be clinically superior to the histamine type-2 (H₂)-receptor antagonists, and was launched in 1988 as Losec® in Europe, and in 1990 as Prilosec® in the United States. In 1996, Losec® became the world's biggest ever selling pharmaceutical, and by 2004 over 800 million patients had been treated with the drug worldwide. In the same year, Nexium®, the follow-up drug of omeprazole, had been used to treat over 200 million patients.

2.2
The Treatment of Acid-Related Disorders Before Losec®

Almost a century ago, in 1910, the Croatian physician Karl Schwarz had formulated the dictum, "No acid – no ulcer", yet 50 years later doctors still lacked the means of effectively inhibiting acid secretion. Inappropriate levels of gastric acid underlie several widespread pathological conditions, including gastroesophageal

Analogue-based Drug Discovery. IUPAC, János Fischer, and C. Robin Ganellin (Eds.)
Copyright © 2006 WILEY-VCH Verlag GmbH & Co. KGaA, Weinheim
ISBN: 3-527-31257-9

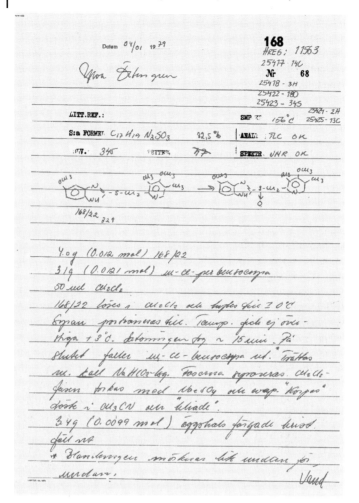

Fig. 2.1 The laboratory notebook of Ylva Örtengren describing the synthesis of H 168/68 – omeprazole.

reflux disease (GERD), for which heartburn is the most common symptom. Additionally, peptic ulcers cause pain and suffering in millions of people, and only 30 years ago were considered to be life-threatening if untreated. The treatment options of these conditions were limited, however. For example, for peptic ulcers the main treatment was to administer antacids to neutralize excess gastric acid (which in turn promoted ulcer formation and prevented healing), though this provided only temporary relief. The alternative was an operation: perhaps gastrectomy, in which part of the stomach is removed; and/or vagotomy, in which the nerves to the stomach are sectioned. Such surgery could, however, have serious side effects, and consequently a pharmacological control of the complex mechanism of gastric acid secretion had long been desirable.

A breakthrough came during the late 1970s with the introduction of the antise-cretory drug cimetidine. This was an antagonist of the H_2-receptor, which plays a key role in some of the pathways leading to gastric acid secretion. Cimetidine (Tagamet®) – and later comparable compounds with the same mechanism of action – have a marked gastric acid inhibitory effect, and considerably improved the lives of millions of people, as well as reducing the need for surgery. However, H_2-receptor antagonists have limited efficacy and a relatively short duration of action.

2.3
Pioneer Research at Hässle during the 1960s and 1970s

During the 1960s, Astra was one of several pharmaceutical companies initiating research projects on gastric acid inhibition. The local anesthetic lidocaine had been reported to decrease acid secretion in dogs. The hypothesis was that lido-caine inhibited the secretion of the acid-stimulating hormone, gastrin, locally in the stomach. Thus, the "Gastrin Project" was initiated, aimed at identifying a com-pound that inhibited gastrin release from the antral G-cells based on this mecha-nism. Whilst numerous compounds were synthesized and tested, the first of many setbacks in the project occurred in 1972 when a compound with the target activity was tested in man, and no effect whatsoever was found on acid secretion. Consequently, other ideas were sought.

Meanwhile, Searle in the USA and Servier in France had both attempted to develop 'blockers' of the gastrin receptor and had used the structure of the C-ter-minal of gastrin as the basis for synthesizing new compounds [7,8]. Although neither company was successful, Searle described antigastrin and Servier pub-lished data on a related compound, CMN 131, which had some inhibitory effect on acid secretion in animals. However, neither compound was further developed due to either toxicity or low potency (Scheme 2.1). It was believed that CMN 131 was toxic because of its thioamide function, and suggestions were made as to how the structure might be changed to avoid toxicity. Hence, CMN 131 was chosen as a new lead compound in the project.

Parenthetically, during investigations with gastrin C-terminal-related com-pounds at Searle, one chemist made a serendipitous observation that led to the development of the sweetener, aspartame.

Within a couple of years, the Hässle team had identified several compounds that inhibited acid secretion more effectively than CMN 131. At the same time, a new animal model had been established, namely the conscious gastric fistula dog, in col-laboration with the consultant Lars Olbe, a surgeon at the local Sahlgrenska Hospi-tal in Göteborg. The gastric fistula is a tube which passes from the stomach through the skin and allows the collection and analysis of secreted gastric juice. This model was much more predictable than the rat model used previously, and was a vital tech-nique for the screening and evaluation of compounds in this project.

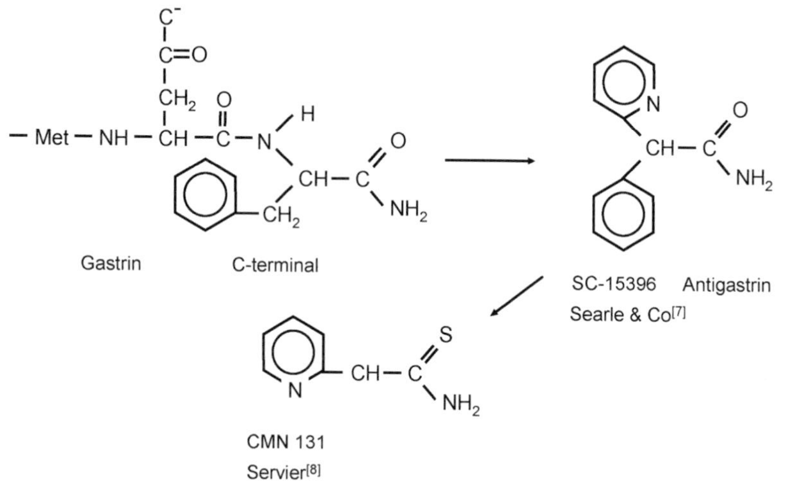

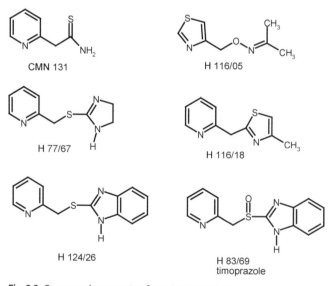

Scheme 2.1 The development of gastrin receptor blockers:
SC-15396 antigastrin (Searle & Co) and CMN 131 (Servier) [7,8].

The chemical synthesis followed two main pathways emanating from CMN 131. The first led to sulfur-containing heterocyclics (e.g., H 116/05 and H 116/18), and the second to sulfides with an imidazoline ring (e.g., H 77/67), and to compounds such as benzimidazole derivatives. The first two compounds to undergo extensive testing were the sulfide H 124/26 and its corresponding sulfoxide, H 83/69. These were both significantly more potent than CMN 131, and were selected as candidate drugs, H 124/26 in 1973, and H 83/69 in 1974 (Fig. 2.2).

Fig. 2.2 Compounds emanating from CMN 131 that were synthesized in the search for a candidate drug.

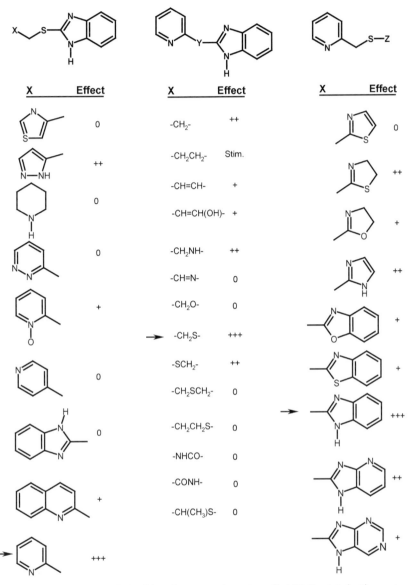

Fig. 2.3 Structure–activity relationships of analogues of H 124/26. Biological testing was conducted in the gastric fistula dog. 0 = lacking antisecretory effect; +++ = maximal antisecretory effect [3]. Reprinted with permission of Wiley-Liss Inc., a subsidiary of John Wiley & Sons Inc. *Omeprazole, the first proton pump inhibitor*, P. Lindberg, **1990**.

By the time that H 124/26 had been prepared, it was considered to be the most powerful inhibitor of gastric acid secretion, and it became the new lead compound. In the initial chemical synthesis program (in collaboration with Abbott Laboratories in Chicago), this lead compound was treated as a molecular template consisting of two heterocyclic ring systems connected by a chain – that is, three elements. Each element was systematically varied in the synthesis program during 1974, but the original three elements of H 124/26 turned out to be optimal (Fig. 2.3) [3,9].

While the sulfone was inactive as an inhibitor of gastric acid secretion, the sulfoxide, H 83/69, was found to be a very potent inhibitor of acid secretion. Pharmacokinetic studies of H 124/26 conducted at Abbott Laboratories indicated a rapid formation of two main metabolites, the sulfoxide H 83/69 and the corresponding sulfone [9]. However, there followed further setbacks, initially when H 124/26 turned out to be known from a Hungarian patent aimed at tuberculosis treatment. Moreover, both compounds (H 124/26 and H 83/69) were found to be toxic, particularly affecting the thyroid gland. Despite this, H 83/69 – which became known as timoprazole – proved to be a breakthrough, mainly because of a number of key observations that were made about the compound.

2.3.1
Toxicological Challenges

Long-term toxicological studies of timoprazole revealed that it caused enlargement of the thyroid gland (this was later shown to be due to inhibition of iodine uptake), as well as atrophy of the thymus gland. Thiourea compounds are well-known inhibitors of iodine uptake in the thyroid. A literature search of the chemistry of thiourea compounds showed that some substituted mercapto-benzimidazoles (which may as such be considered as thioureas), when substituted, had no effect on iodine uptake. Consequently, the introduction of such substituents into timoprazole resulted in an elimination of the toxic effects, without reducing the antisecretory effect. Tests on several substituted benzimidazoles showed that separation of the inhibition of acid secretion from the inhibition of iodine uptake was obtained in compounds within a specific range of lipophilicity. The most potent antisecretory compound without thyroid/thymus effects was H 149/94 or picoprazole, which was synthesized and selected as a candidate drug in 1976.

However, during the initial repeat-dose toxicological study of picoprazole, a few of the treated beagle dogs developed necrotizing vasculitis. Fortunately, from the perspective of the project, one control dog also developed this condition. It was subsequently found that all dogs with vasculitis had a male dog called Fabian in their pedigree, and all had antibodies against intestinal worms, which were probably obtained after de-worming. A new toxicological study in another beagle strain, and in nonparasitized dogs, was completely clear of necrotizing vasculitis. Today, it is known that Fabian and his offspring carried the genes for Beagle Pain Syndrome, a rare, latent hereditary beagle dog disease. Vasculitis can be triggered in these dogs by different factors such as stress and exposure to chemicals. Picoprazole was later used in a concept study in human volunteers, and showed a potent antisecretory action of very long duration. Indeed, it seemed to be the most powerful inhibitor of acid secretion tested in humans to that date.

2.3.2
Discovery of H⁺, K⁺-ATPase: The Gastric Proton Pump

Intracellular cyclic adenosine monophosphate (cAMP)-stimulated acid secretion in the isolated guinea-pig gastric mucosa was not inhibited by administration of an H_2-receptor antagonist, as expected, although H 83/69 (timoprazole) induced a dose-dependent inhibition. This was the first experimental evidence for a site of inhibitory action beyond the panel of stimulatory cell membrane receptors. Interestingly, it was found that the initial lead compound (CMN 131), had no inhibitory effect on dibutyryl-cAMP-stimulated acid secretion, nor was it an H_2-receptor antagonist [9].

By the end of the 1970s, however, evidence was emerging that the activation of a newly discovered proton pump (an H^+, K^+-ATPase) in the secretory membranes of the parietal cell was the final step in acid secretion [10,11]. Immunohistological data obtained using antibodies against a crude preparation from the secretory membranes of parietal cells revealed strong immunoreactivity in the parietal cell region of the stomach, as well as some activity in the thyroid gland [12]. Coupled with knowledge of the side effects of timoprazole on the thyroid (as discussed earlier), these findings raised the intriguing possibility that benzimidazoles such as timoprazole could be inhibitors of the H^+, K^+-ATPase. Research was initiated in this area in parallel with the further development of benzimidazoles, and it was indeed subsequently shown – in collaboration with Professor George Sachs – that substituted benzimidazoles inhibit gastric acid secretion by blocking the H^+, K^+-ATPase [1,13]. The mode of action of substituted benzimidazoles, and the implications of this for their clinical benefits, is discussed further in the sections below.

2.3.3
Analogue Optimization

While the gastric fistula dog remained the most important testing technique for these compounds, simpler *in-vitro* techniques were developed and used to test a large number of different substituted benzimidazoles for the optimal inhibition of gastric acid secretion. The isolated gastric-acid secreting mucosa of the guinea-pig was introduced as an appropriate *in-vitro* model, though at a later time isolated rabbit acid-secreting gastric glands were used, and a micromethod for isolating acid-secreting parietal cells from human gastric biopsies was developed. These techniques each allowed the testing of a large number of compounds, including tests on the human target tissue.

The problem persisted, however, of how the antisecretory effect of substituted benzimidazoles could be optimized. As weak bases accumulate in the acidic compartment of the parietal cell close to the proton pump, substituents were now added to the pyridine ring of timoprazole to obtain a higher pK_a value, thereby facilitating accumulation within the parietal cell (Fig. 2.4). Compound H 159/69, with the same benzimidazole substitution as picoprazole, was found to be 10 times more potent than this compound in the dog. However, the final result of the optimization was compound H 168/68 (omeprazole), synthesized in 1979. It was later shown in a thorough mechanistic investigation that the higher pK_a value of the omeprazole pyridine ring (~1 pK_a unit higher than that in timoprazole) also increased the rate of acid-mediated conversion to the active species (the sulfenamide; see Section 2.5 on the mechanism of action), which turned out to be a prerequisite for the acid-inhibitory activity. The 5-methoxy-substitution in the benzimidazole moiety of omeprazole also made the compound much more stable to conversion at neutral pH compared with, for example, the ester substitution in the benzimidazole moiety of picoprazole and H 159/69.

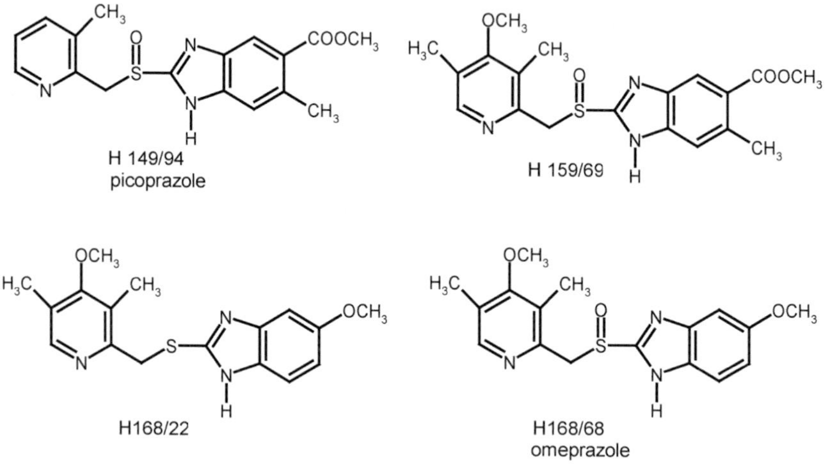

H 149/94
picoprazole

H 159/69

H168/22

H168/68
omeprazole

Fig. 2.4 Substituted benzimidazoles tested in the search for the optimal inhibition of gastric acid secretion. Substituents added to the pyridine ring resulted in the synthesis of H 168/88 (omeprazole).

Omeprazole was found to be the most efficient inhibitor of stimulated gastric acid secretion in rats and dogs *in vivo*, thereby signaling a new era in the treatment of acid-related disorders of the gastrointestinal tract [2]. Moreover, omeprazole had no effects on iodine uptake, did not induce thymus atrophy or cause necrotizing vasculitis, and showed no other signs of toxicity. An Investigational New Drug (IND) application was filed in 1980, and omeprazole was taken into Phase III human trials in 1982. Although the initial pharmacological and clinical results in man were highly encouraging however (Fig. 2.5) [14–16], there remained further challenges to address (see below).

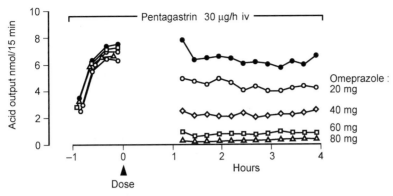

Fig. 2.5 Acid output during 4 h post-dose with omeprazole 20 mg, 40 mg, 60 mg, and 80 mg orally in healthy subjects (n = 6) [14]. Reproduced *Gut*, **1983**, 24, 270–276, with permission from the British Medical Journal Publishing Group.

2.4
The Development of Omeprazole

In 1980, after more than 10 years of research, the research team had a candidate drug which was thought to have optimal properties with regard to efficacy, and to have no signs of serious toxicity in initial safety studies in animals. The pressure from Astra management was to develop the compound in record time, and consequently the New Drug Application submission was scheduled for 1985. Most of the investigations were focused on preclinical and clinical documentation, large-scale synthesis, analytical methods, and the development of suitable formulation and stability tests.

The chemical instability and decomposition presented special challenges, both with regard to shelf stability as well to working on a formulation that would protect the compound in the acidic environment of the stomach, before absorption in the small intestine. Although chemical instability was a major problem for the formulation studies, it was a prerequisite for the mechanism of action. For some time, the quest was for chemically stable prodrugs of omeprazole, as there were suspicions that a formulation would not simultaneously solve both the problem of shelf stability and protection of the compound before absorption. Along this line, the acid-stable omeprazole sulfide (H 168/22) was developed as a back-up compound, and documented in Phase I human studies. A significant step towards higher stability was the discovery that alkaline salts of omeprazole showed a great increase in stability.

2.4.1
Further Toxicological Challenges and the Halt of the Clinical Program

Lifelong toxicological studies of high-dose omeprazole in rats revealed the development of endocrine tumors (carcinoids) in the stomach, and this led to the halting of all clinical studies in 1984. The carcinoids originated from enterochromaffin-like (ECL) cells, a type of endocrine cell in the gastric mucosa that synthesizes and secretes histamine in response to stimulation by the gastric hormone gastrin. However, longer-term stimulation by gastrin also has a potent trophic action on ECL cells. By combining this with the fact that gastrin was known to be released in increasing amounts from the antrum of the stomach as the amount of acid secretion decreases, a possible explanation was suggested for the observed effects of lifelong, very high-dose omeprazole in rats – namely, the elimination of gastric acid secretion, resulting in massive hypergastrinemia. This was shown to be the cause of ECL cell hyperplasia in omeprazole-treated rats, as the condition did not occur in omeprazole-treated rats subjected to resection of the gastric antrum, which is the source of gastrin. Furthermore, the ECL cell carcinoids were also produced in lifelong studies of rats receiving an H_2-receptor antagonist (ranitidine) at high doses, as well as by a surgical procedure that created massive hypergastrinemia. Fortunately, these data allowed the clinical studies with omeprazole to be restarted [6].

2.4.2
Resumption of Clinical Studies

In Phase III clinical trials, omeprazole was found to be significantly superior to previous treatment regimens of H_2-receptor antagonists in patients with duodenal and gastric ulcers [17]. A particularly notable advantage of omeprazole over ranitidine was found in GERD patients, in whom healing rates were about twice as high with omeprazole [18]. Clinical doses of omeprazole produce a modest hypergastrinemia in the same range as the surgical procedure vagotomy, and neither treatment has produced ECL cell carcinoids over the long-term (i.e., >10 years) follow-up. Massive hypergastrinemia in man is observed in patients with gastrin-producing tumors, who develop hyperplasia of the ECL cells but not ECL cell carcinoids. Clearly, the response of the ECL cells to hypergastrinemia is different in man and rat [6].

2.4.3
Omeprazole Reaches the Market and Supersedes H_2-Receptor Antagonists

In 1988, after extensive documentation and careful evaluation, omeprazole was registered in Sweden for the treatment of duodenal ulcer and reflux esophagitis, and was launched in Europe as Losec® in the same year. The clinical program continued to document the value of omeprazole in further acid-related disorders, though its full therapeutic potential became apparent only during the 1990s, when it superseded the H_2-receptor antagonists in the treatment of all these conditions.

2.5
The Unique Action of Omeprazole

2.5.1
Inhibition of the Final Step

The success of omeprazole in the clinic can be ascribed to the very effective inhibition of gastric acid secretion, that is achieved through specific inhibition of the gastric H^+, K^+-ATPase, which constitutes the final step of acid secretion. Thus, blocking this pump results in a more specific inhibition of acid secretion compared with blocking the more widely distributed H_2 and cholinergic receptors. Furthermore, as omeprazole interacts with the final step of acid production, the inhibition of gastric acid secretion is independent of how acid secretion is stimulated – an important advantage over other pharmacological approaches to inhibiting acid secretion [3]. For example, the inhibition of acid secretion by H_2-receptor antagonists can be overcome by food-induced stimulation of acid secretion via release of gastrin and acetylcholine and stimulation of the corresponding receptors.

2.5.2
Omeprazole Binds Strongly to the H^+, K^+-ATPase

At this point, the abiding question is how omeprazole inhibits the H^+, K^+-ATPase. In whole-body autoradiography studies in mice, the radiolabel originating from ^{14}C-labeled omeprazole was found to be confined to the gastric mucosa (Fig. 2.6). With a similar technique, and using both light and electron microscopic evaluation, omeprazole was found to label only the tubulovesicles and secretory membranes of the parietal cell, which contain the H^+, K^+-ATPase [3]. Electrophoretic analyses of the proteins from such membranes, purified after systemic administration of

After 1 minute

After 16 hours

Fig. 2.6 Distribution of ^{14}C-labeled omeprazole in the mouse at 16 h after intravenous administration.

labeled omeprazole to rats showed that the radiolabel was associated specifically with the 92kDa protein known to hold the catalytic subunit of H^+, K^+-ATPase [3]. Based on these data, it was concluded that omeprazole binds only to the H^+, K^+-ATPase in the gastric mucosa (Fig. 2.7).

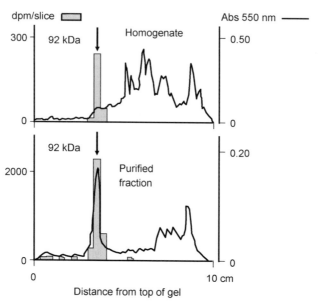

Fig. 2.7 Specific binding of radiolabeled omeprazole to the H^+, K-ATPase in the gastric mucosa [19]. Reprinted from Fryklund J, Gedda K and Wallmark B, *Biochem. Pharmacol.*, **1988**, 37, 2543–2549. Specific labeling of gastric H^+, K^+-ATPase by omeprazole, with permission from Elsevier.

2.5.3
Inhibition of Acid Secretion and H^+, K^+-ATPase Activity

The observations described above offered a possibility to investigate the relationship between acid secretion and the activity of the gastric H^+, K^+-ATPase. In a set of experiments, omeprazole was given to rats in submaximal doses, with the effect on acid secretion and gastric mucosal H^+, K^+-ATPase activity being determined in separate groups of animals. It was shown that omeprazole, dose-dependently and in a parallel manner, inhibited both maximally stimulated acid secretion and H^+, K^+-ATPase activity. The ED_{50} was similar (10 μmol kg^{-1}) for the inhibition of acid secretion and for H^+, K^+-ATPase activity, which strongly suggested that acid secretion is highly correlated to the activity of the gastric H^+, K^+-ATPase. The data given above provide direct evidence that omeprazole inhibits acid secretion by blockade of the gastric H^+, K^+-ATPase [3].

2.5.4
Omeprazole Concentrates and Transforms in Acid

Omeprazole is a lipophilic weak base (pK_a 4.0), which means that it easily penetrates cell membranes and concentrates in a protonated form in acid compartments such as the secretory canaliculi of the parietal cell. It is unstable in acid, with a half-life of 2 min at pH 1–3, but is significantly more stable at pH 7 (half-life ca. 20 h). All these characteristics are important for omeprazole's unique mechanism of action and selective action on acid secretion.

Omeprazole itself is not the active inhibitor of the H^+, K^+-ATPase, however. Rather, the transformation of omeprazole in acid is required to inhibit the H^+, K^+-ATPase *in vitro* and *in vivo*, whereas intact omeprazole is devoid of inhibitory action. Isolated H^+, K^+-ATPase is blocked by omeprazole only after the pretreatment of omeprazole with acid. Conversely, neutralization of the acid-secretory canaliculi of isolated gastric gland and parietal cell preparations by permeable buffers, which blocks the acid-catalyzed transformation of omeprazole, prevents inhibition of the enzyme. Furthermore, *in-vivo* blockade of acid secretion using an H_2-receptor antagonist prior to omeprazole administration decreases the inhibitory potency of omeprazole, whereas the stimulation of acid secretion (e.g., by food intake) increases the potency.

Thus, in conclusion, omeprazole has a highly specific action on the H^+, K^+-ATPase. Three factors of primary importance for this specificity are: (a) the unique location of the H^+, K^+-ATPase in the parietal cell and the steep proton gradient generated by the H^+, K^+-ATPase; (b) the concentration of the protonated form of omeprazole in the acidic canaliculus; and (c) the conversion of omeprazole in the acidic compartments close to its target enzyme, the H^+, K^+-ATPase [3].

2.5.5
Disulfide Enzyme–Inhibitor Complex on the Luminal Side

As indicated above, binding of the inhibitor generated from omeprazole takes place in the luminal sector of the H^+, K^+-ATPase. The binding levels obtained *in vivo* and *in vitro* were shown to be very similar, with 2 mol inhibitor bound per 1 mol active site H^+, K^+-ATPase. The bound inhibitor could be displaced from enzyme prepared from either *in-vivo* or *in-vitro* experiments by the addition of β-mercaptoethanol, which suggests that the bond between the omeprazole-derived inhibitor and the H^+, K^+-ATPase is a disulfide linkage. This proposal was supported by earlier observations which showed a linear relationship between inhibitor binding and the modification of SH groups in the H^+, K^+-ATPase preparation [3].

2.5.6
Short Half-Life in Plasma and Long Half-Life for Enzyme–Inhibitor Complex

Whilst the half-life of omeprazole in the plasma of humans is rather short (1–2 h), that of the inhibition complex is long, thus providing a long duration of action. Based on the duration of action in humans, the half-life at the site of action is estimated to be approximately 24 h. Dissociation of the inhibitory complex is probably due to the effect of the endogenous antioxidant glutathione; this leads to the release of omeprazole sulfide and reactivation of the enzyme (see below). The fact that the sulfide is found in gastric juice is consistent with this proposal. Reactivation of the acid-producing capacity may also in part be due to the *de-novo* synthesis of enzyme molecules [3,5].

2.5.7
Mechanism at the Molecular Level

In order to obtain information about the mechanism of action at the molecular level, an extensive study of the acid degradation of omeprazole was undertaken [3,20]. Degradation studies on omeprazole and analogues have also been performed by others [3]. Of crucial importance to this work was the great simplification of decomposition by adding β-mercaptoethanol to the acid medium before the addition of omeprazole (A) (Scheme 2.2). This caused only two compounds to form: the sulfide S and an adduct with β-mercaptoethanol (ESSR). At high dilution (10^{-5} M) in dilute hydrochloric acid (0.001–0.5 M), the conversion of omeprazole (A) into an intermediate D (taking 5–15 min to completion at 37 °C) could be followed kinetically using UV absorption. When a slight excess of β-mercaptoethanol was added to the resulting solution, a very rapid conversion of D into the above-mentioned adduct (ESSR) could also be followed kinetically using UV. It was also possible to obtain crystals for X-ray analysis of the β-mercaptoethanol adduct (ClO_4^- salt from methanol) generated from an omeprazole analogue 2-[[(4-methoxy-3,5-dimethyl-2-pyridinyl)methyl]sulfinyl]-1H-benzimidazole, resulting in the disulfide structure (Fig. 2.8). On decomposition of omeprazole (A) in methanol instead of water, the intermediate D was significantly more stable, even at moderate concentrations.

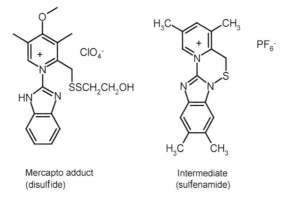

Mercapto adduct
(disulfide)

Intermediate
(sulfenamide)

Fig. 2.8 Two important structures elucidated by X-ray.

The addition of 70% HPF_6 led to the production of crystals of this intermediate. Although too unstable to recrystallize, it was possible to obtain crystals for an X-ray investigation of a PF_6^- salt of an omeprazole analogue, revealing the cyclic sulfenamide structure of the intermediate (Fig. 2.8) [20].

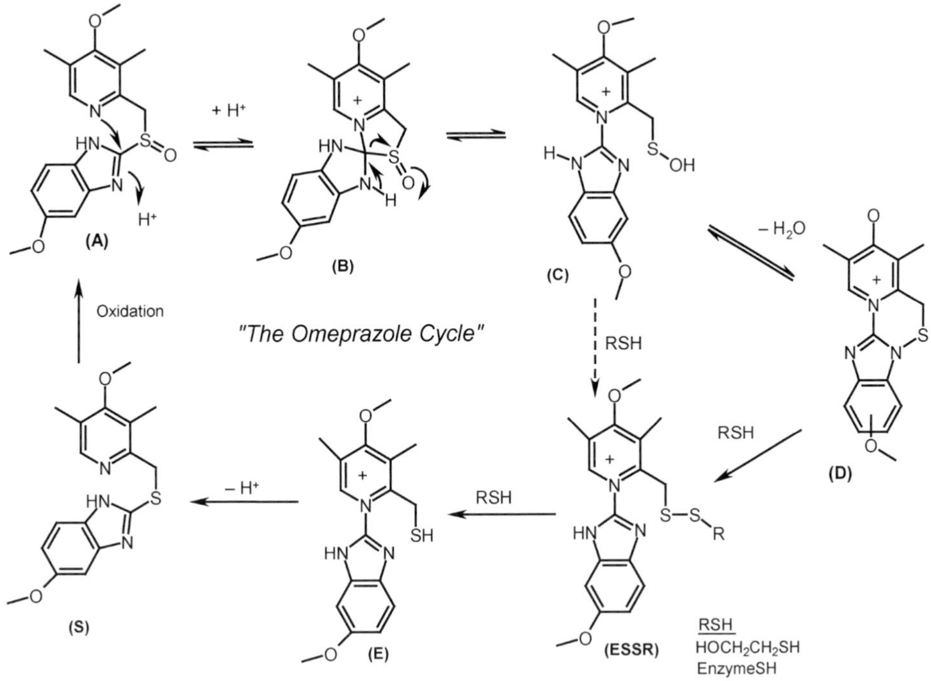

Scheme 2.2 Mechanism for the transformation of omeprazole in acid. Formation of the active sulfenamide (or, alternatively, the sulfenic acid), the inhibitory reaction with the enzyme and cleavage of the enzyme–inhibitor complex by mercaptan (the "omeprazole cycle").

The reaction mechanism proposed for the acid transformation of omeprazole (A) to the sulfenamide isomers D is outlined in Scheme 2.2 [3,20]. The reaction is reversible, and proceeds via the spiro intermediate B and the sulfenic acid C. The reversibility was formally proved by kinetic measurements in both directions – that is, starting from A or D. The formation of the spiro intermediate B in the rate-limiting step is supported by kinetic measurements.

As the H^+, K^+-ATPase inhibition is associated with the modification of mercapto groups in the enzyme, the disulfide adduct (ESSR) can be considered as a model of the enzyme–inhibitor complex, and the sulfenamide, or possibly the sulfenic acid (C), formed from omeprazole can be considered to be the active inhibitor, which binds covalently to cysteine residues of the H^+, K^+-ATPase.

The last step in the "omeprazole cycle" (Scheme 1.2) finds support in the fact that the sulfide S is formed from omeprazole A in isolated gastric glands and in the gastric mucosa *in vivo*. It is also known that the sulfide S is oxidized *in vivo* (by the liver and not in the mucosa) to the sulfoxide omeprazole, thus closing the cyclic process shown in Scheme 1.2. However, due to simultaneous metabolism, and an observed short plasma half-life (1–2 h in man), probably only a fraction of the omeprazole dose completes the cyclic process, and therefore it does not contribute to the long duration of action. [3]

The "omeprazole cycle" (Scheme 1.2) can also be considered as a "prodrug cycle". The sulfenamide D, which is active when tested on a purified H^+, K^+-ATPase preparation, is inactive when tested in the gastric gland preparation. This can be explained by a very efficient conversion of the sulfenamide D into the sulfide S as a result of its reaction with glutathione during its passage through the neutral part of the parietal cell. Interestingly enough, the sulfenamide is very active when tested *in vivo*; intraduodenal administration in the dog of 4 μmol kg^{-1} of the omeprazole sulfenamide D led to 40% inhibition of histamine-stimulated acid secretion. This can be explained by the expected conversion of the sulfenamide D into the sulfide S in the intestine due to reaction with mercapto groups, leading to sulfide formation. The sulfide is then absorbed, oxidized in the liver to the sulfoxide A, and transported to the acid compartments of the parietal cell where it is converted back into the sulfenamide D, which can then exert its inhibitory effect on the proton pump. This implies that the sulfenamide D is acting as a prodrug to a prodrug to a prodrug, and so forth to itself (!).This is also the background for selection of the sulfide as a back-up compound. Therefore, the "prodrug cycle" is of great theoretical interest but, as mentioned above, it has no relation to the long duration of action of omeprazole [3].

2.5.8
The "Targeted Prodrug" Omeprazole means a Highly Specific Action

Several factors make H^+, K^+-ATPase a unique target for enzyme inhibition and provide a potential for selective drug targeting. Thus, the following factors most likely contribute to the highly specific action of omeprazole [3,4]:

- The limited distribution of the H^+, K^+-ATPase, which is found mainly in the tubulovesicular and canalicular membranes of the gastric parietal cell. This is in contrast to, for example, the more widely distributed H_2 and cholinergic receptors.
- The tubulovesicular and canalicular structures of the parietal cell are the only compartments in the body with an extremely low pH (1–2).
- Omeprazole is a lipophilic weak base, which rapidly penetrates the cell membrane of the parietal cell, and therefore concentrates in acidic regions (with a pH <4). Omeprazole will thus be concentrated predominantly in the parietal cell.
- Omeprazole is rapidly transformed, via an acid-catalyzed reaction, into an active inhibitor close to its target enzyme.

- The active inhibitor reacts very rapidly, via an acid-catalyzed mechanism, with mercapto groups on the luminal side of the target enzyme forming a covalently bound inhibitor complex.
- Omeprazole is a compound, which is inactive *per se*, and it is essentially stable at neutral pH. Furthermore, it is rapidly eliminated from the circulation via metabolism and excretion.
- The active inhibitor is a permanent cation, which cannot easily penetrate back through the secretory membrane.

2.6
pH-Stability Profile

An understanding of the pH-dependence of the reactions of omeprazole and its analogues requires knowledge of the protolytic nature of these compounds. Omeprazole and its analogues are protolytically active primarily in two reactions. These molecules can accept a proton on the pyridine nitrogen atom, and the release of this proton is described by the dissociation constant [pK_a (1); the value for omeprazole is [4.0], or a proton can be released from the NH group of the benzimidazole [pK_a (2); the value for omeprazole is 8.7] (Scheme 1.3) [3].

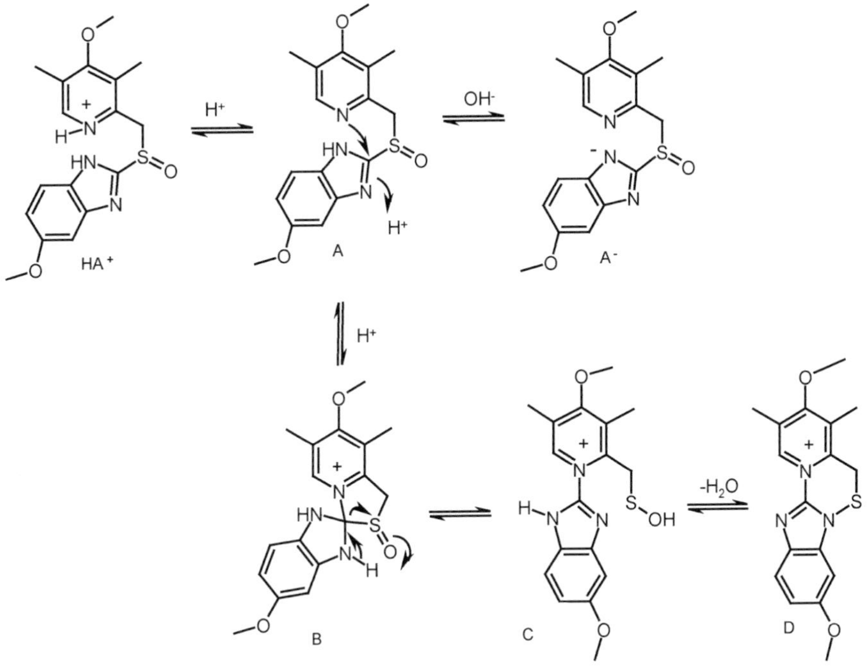

Scheme 2.3 The acid-catalyzed reaction of omeprazole to the sulfenamide. The neutral form of omeprazole is the only reactive species.

At pH values below pK_a (1), omeprazole and its analogues are predominantly protonated on the pyridine nitrogen, at pH values above pK_a (2) these molecules are predominantly deprotonated on the benzimidazole, and at pH values between the two pK_as, the neutral form will predominate. Only the neutral form is a reactive species [3]. The protonated form of the drug is unreactive, since there is no electron pair available on the pyridine nitrogen atom for an attack of the electron-deficient 2-carbon atom of the benzimidazole ring. However, in the negatively charged, depronated form, this 2-carbon atom is no longer electron-deficient, and thus this form is also unreactive (Scheme 2.3).

2.7
Omeprazole Analogues Synthesized by Other Companies

More than 40 pharmaceutical companies have been investigating the development of H$^+$, K$^+$-ATPase inhibitors, frequently called PPIs. The majority of these compounds were built on the same basic structure as that of omeprazole, the 2-(2-pyridinylmethylsulfinyl)-benzimidazole framework – that is, the timoprazole molecule. In some analogues, however, one of the heterocyclic rings has been exchanged for another group, for example, the pyridine for a dialkyl-anilino group or the benzimidazole for the thieno-imidazole group. However, it appears that all compounds can be expected to have the same or a very similar mechanism of action as that of omeprazole [3].

Early on, Hoffmann-La Roche & Co. Ltd (Switzerland) initiated analogue-based chemical studies, probably based on Astra's earlier patent application including timoprazole and analogues, and modified the benzimidazole moiety by introducing a further carbocyclic ring, aliphatic (patented in 1977) or aromatic, the latter modification giving naphthimidazole analogues (patented in 1980). The naphthimidazole compounds (in their sulfoxide forms) were probably too unstable under both acid and neutral conditions to be useful as drugs. HoffmannLa Roche continued their efforts with compounds containing a 4-alkoxy group in the pyridine moiety and an extra aliphatic ring on the benzimidazole. Their candidate drug Ro 18-5364 [21] was later discontinued (Fig. 2.9).

In 1981, Byk Gulden Research Laboratories (Germany) identified analogues of timoprazole that contained the important alkoxy group in the 4-position of the

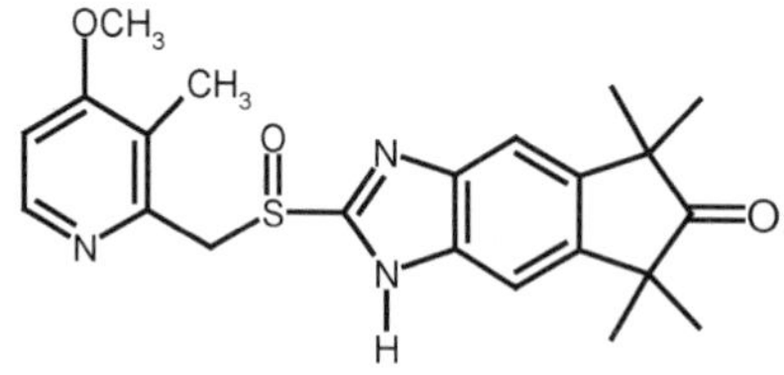

Ro 18-5364

Fig. 2.9 Hoffmann-LaRoche candidate drug Ro 185364.

pyridine ring, as described in Hässle's omeprazole patent (EP 5129). Substitution of the benzimidazole moiety with a CF_3 group resulted in the candidate drugs BY 319 and BY 308 (Fig. 2.10) [10]. The latter drug was a sulfide, since its corresponding sulfoxide form was unstable in neutral solution and too difficult to formulate. The solution instability of the sulfoxide forms was due to the strongly electron-withdrawing effect of the CF_3 group [3]. The sulfide BY308, albeit very potent in rat animal models, was not sufficiently potent in man to warrant further development.

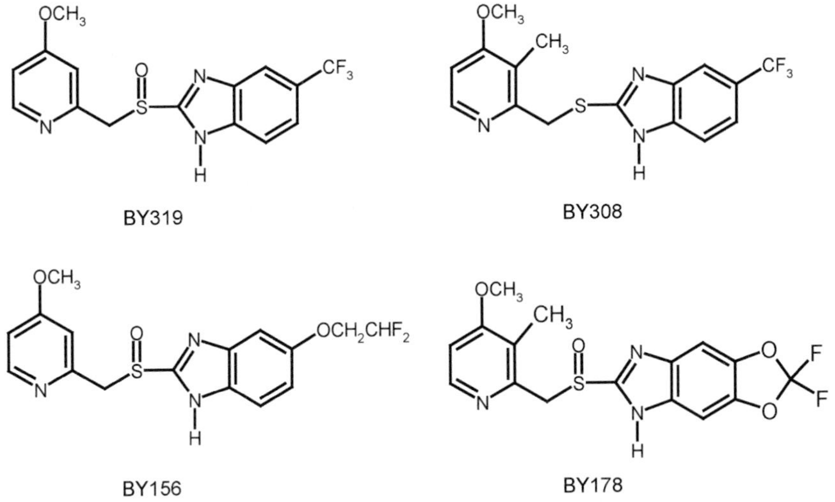

Fig. 2.10 Byk Gulden candidate drugs BY319, BY308, BY156, and BY178.

Apparently, Byk Gulden then changed their strategy and synthesized analogues that had straight and cyclic fluoroalkoxy substituents in the benzimidazole, of which BY 156 [23] and BY 178 are examples (see Fig. 2.10). Although less electron-withdrawing than CF_3, the electron-withdrawing capacity of the fluoroalkoxy substituents in the benzimidazole 5-position required compensation for by appropriate substitution in other positions to achieve solution stability. Solution stability became the primary objective, and this was finally achieved by utilizing a second alkoxy group in the 3- or 5-position of the pyridine moiety. One of those analogues with this substitution pattern was BY 1023 [24] – later named pantoprazole and synthesized in 1985 (Fig. 2.11). In the mid-1980s, Byk Gulden and Smith Kline & French (UK) collaborated in the area of PPIs. The UK Company synthesized the candidate compound SK&F 95601 [25] with a strongly electron-releasing 4-morpholino group in the pyridine moiety, which needed the strongly electron-withdrawing adjacent 3-Cl substituent to obtain sufficient solution stability and selectivity (Fig. 2.11). Among these compounds, only pantoprazole eventually reached the market [3] (see also Chapter II-3).

The UpJohn Company (USA) showed an early interest in the area of PPIs, and by 1980 they had filed a use patent application on all of the Astra compounds as

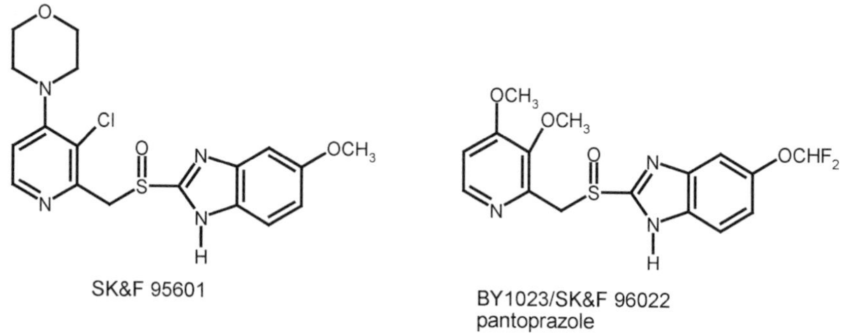

Fig. 2.11 SK&F and Byk Gulden candidate drugs SK&F 95601 and BY1023. BY1023 (pantoprazole) was later launched by Byk Gulden.

cytoprotectives for prophylactic use against gastric ulcers. They also made an early attempt to elucidate the mechanism of action of omeprazole and its analogues for inhibition of the proton pump [3]. UpJohn later also synthesized many new compounds and selected at least one candidate drug, a 4-ethylthio-3-methylpyridine analogue, disuprazole, [26] with no substituent on the benzimidazole ring (Fig. 2.12), but this compound never reached the market.

Disuprazole

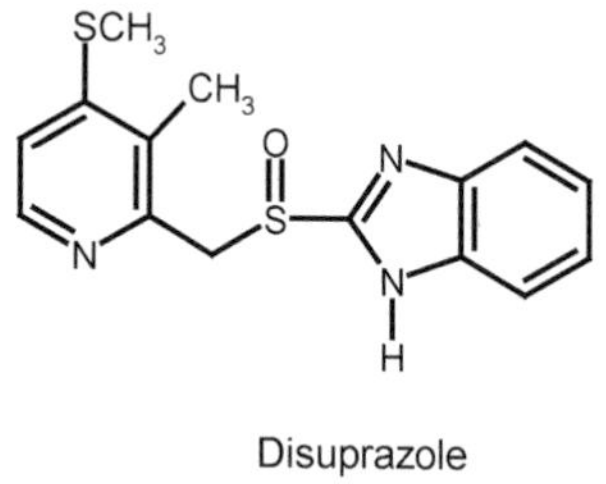

Fig. 2.12 A candidate drug, disuprazole, synthesized by the UpJohn Company.

Tokyo Tanabe KK synthesized omeprazole analogues with an additional nitrogen in the six-membered ring of the benzimidazole moiety, but with the same substitution as omeprazole in the pyridine ring. The analogue, TU-199 (later named tenatoprazole), also contained a methoxy substituent in the "benzimidazole" unit, and was selected for development (Fig. 2.13). Currently, the compound has been in-licensed by the French company Negma GILD [27].

Fisons PLC (UK) also had an early interest in the PPI area. Apparently, by having an early understanding of the chemical mechanism for the acid rearrangement reaction, they devised two novel ideas: (a) to exchange the pyridine in omeprazole for an N-alkyl- and N,N-dialkyl-substituted aniline moiety; and (b) to exchange the methylene as a connecting chain for a phenylene unit. At least one compound of this latter type, FPL 65372 [28] was quite active in early human testing, but no compound of this type ever reached the market (Fig. 2.14).

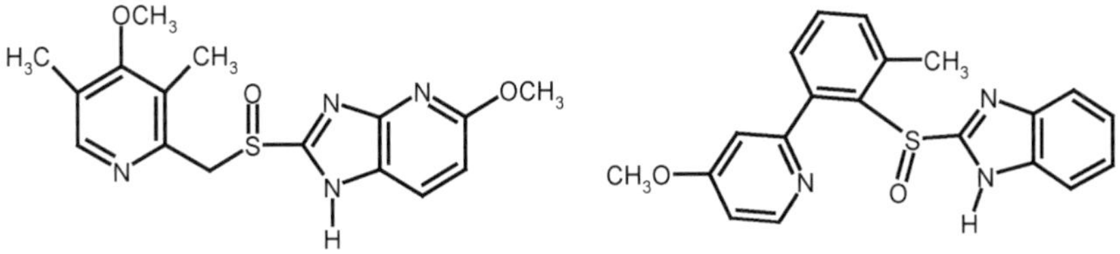

TU 199
tenatoprazole

FPL 65372

Fig. 2.13 Tokyo Tanabe synthesized the analogue TU-199, later named tenatoprazole. Fisons Plc synthesized FPL 65372.

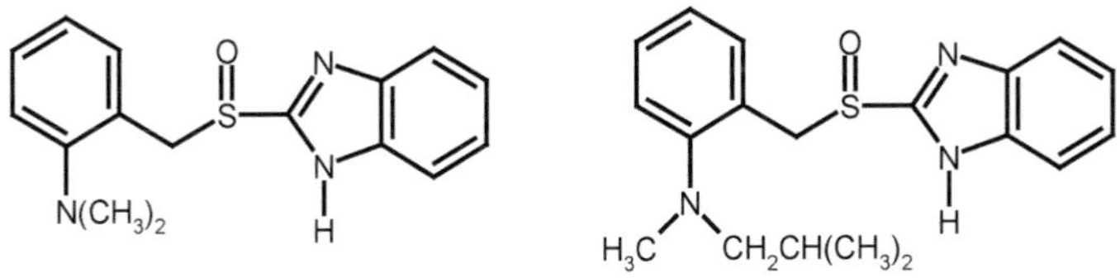

NC-1300

NC-1300-0-3
leminoprazole

Fig. 2.14 Compounds NC-1300 and NC-1300-0-3 (leminoprazole) synthesized by Nippon Chemiphar.

Nippon Chemiphar, more-or-less simultaneously with Fisons PLC, synthesized the *N,N*-dimethylamino analogue NC-1300 [26]. A derivative of this compound, leminoprazole [30], was later selected for development, but to date has not reached the market.

Takeda Pharm KK synthesized omeprazole analogues having a 4-fluoroalkoxy-substitution in the pyridine ring. AG 1749 [31], with a 4-trifluoroethoxy group in the pyridine ring (later named lansoprazole) was one of their candidate drugs (Fig. 2.15). By avoiding the 5-alkyl group in the pyridine ring (compared with that of the omeprazole pyridine), the resulting pK_a of this moiety became approximately 4, despite the electron-withdrawing capacity of the CF_3 group. This is because the trifluoroethoxy group becomes more electron-donating as it is not twisted out of the plane, which is the case in omeprazole due to hindered rotation of the methoxy group in the pyridine [3]. Lansoprazole became the second compound (after omeprazole) to reach the PPI market.

Toa Eiyo Ltd. (Japan) synthesized omeprazole analogues with an aliphatic sevenmembered ring condensed to the omeprazole skeleton so as to include the 2- and 3-positions of the pyridine ring and the methylene carbon in the connecting chain, for example, TY-11345 [32]. To date, no compound of this type has reached the market.

Chemie Linz AG (Austria) and Hoechst AG (Germany) synthesized thienoimidazole analogues of omeprazole, such as the "Chemie Linz" compound [33] and

HOE-731 (Fig. 2.16) [35]. Hoechst also made aniline analogues of omeprazole. However, none of the above compounds has yet been launched.

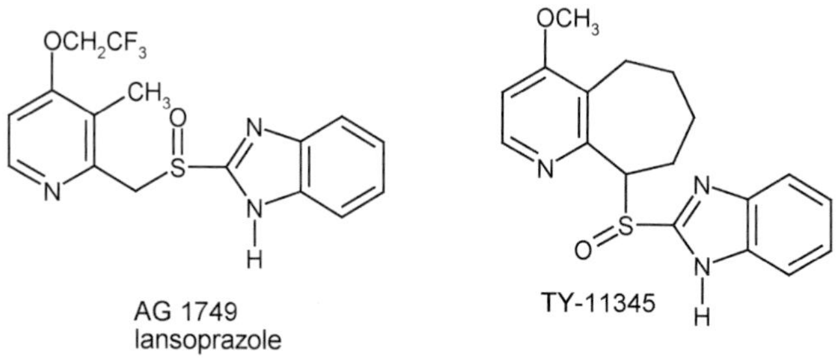

AG 1749
lansoprazole

TY-11345

Fig. 2.15 Takeda Pharm KK synthesized a candidate dug AG 1749 (lansoprazole) later launched. Toa Eiyo Ltd synthesized TY-11345.

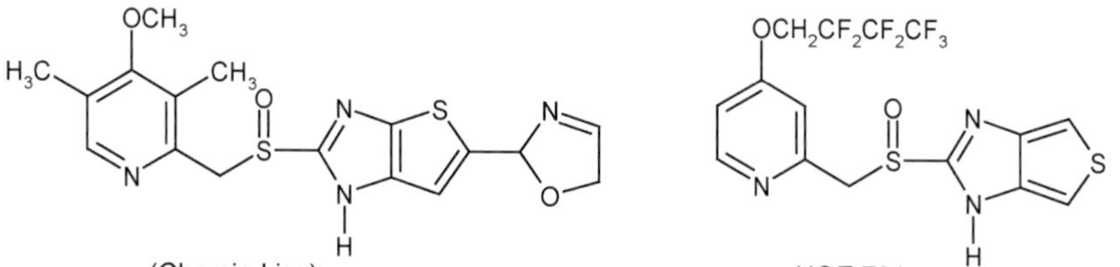

(Chemie Linz)

HOE 731

Fig. 2.16 Chemie Linz AG and Hoechst AG synthesized the compounds Chemie Linz and HOE-731.

Eisai Co Ltd. (Japan) synthesized 4-alkoxyalkoxy-pyridine analogues of omeprazole and selected E-3810 [35], rabeprazole, as their candidate drug (Fig. 2.17). Rabeprazole, which has the 4-methoxypropoxy-3-methyl-pyridinyl moiety in its chemical structure, is largely similar to omeprazole (as are pantoprazole and lansoprazole) and has now also reached the market.

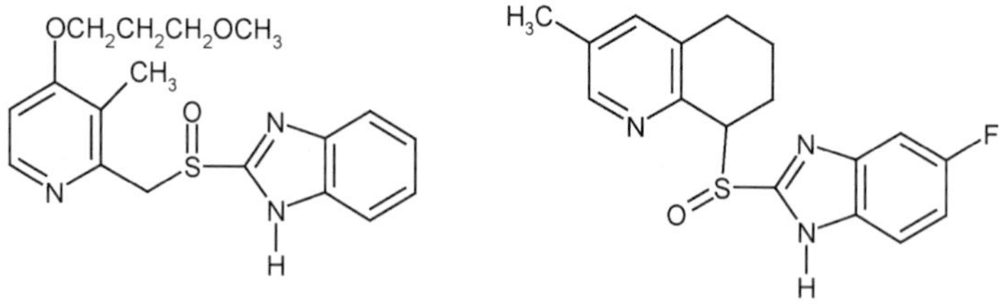

E-3810
rabeprazole

OPC-22321

Fig. 2.17 Eisai Co. Ltd synthesized E-3810 (rabeprazole) that was later launched. Otsuka Pharm KK synthesized the compound OPC-22321.

Other companies that were quite active in the PPI area during the 1980s include Beecham Group PLC, Banyu Pharm KK, Otsuka Pharm KK (with, for example, OPC-22321; [36]), Shionogi Seiyaku KK, Yoshitomi Pharm Ind KK, and Yamanouchi.

2.8
Omeprazole: A Need for Improvement?

Although omeprazole provided more effective control of acid secretion than previous therapies [37–39], it was not equally effective in all patients. A significant number of patients with acid-related disorders required higher or multiple doses to achieve symptom relief and healing of esophagitis [40,41]. Therefore, there was a need to identify a therapy that provided even more effective control of acid secretion than that achieved with omeprazole, and which reduced inter-patient variation in acid inhibition.

2.8.1
The Omeprazole Follow-Up Program

Although the synthesis and screening of new analogues of omeprazole continued even after 1979 (primarily to find a back-up compound), a more focused research program was initiated in 1987. The aim of this was to identify a new PPI with improved pharmacokinetic and metabolic properties, for example, with reduced inter-individual variability. One idea to achieve this was to reduce clearance by the liver – that is, to increase the bioavailability. Additionally, the compound could not induce its own hepatic metabolism, as did omeprazole. An earlier compound in the back-up program, when administered to dogs, had been found to induce its own metabolism, to the extent that plasma levels were immeasurable after five days of repeated administration of an estimated therapeutic dose.

2.8.2
No Good Alternative to the Omeprazole Structural Template

Based on chemical and mechanistic knowledge, a variety of alternative molecular approaches were investigated. Although simple sulfenamides and disulfides, as the closest functional analogues to the sulfenamide (and sulfenic acid), are highly effective inhibitors on the enzymic level, these types of compound cannot survive and reach the target SH group on the enzyme after systemic administration. In order to be active, a compound must fulfill the important criterion of acting as a "masked" sulfenic acid, with release of the sulfenic acid moiety only in an acidic environment. A good alternative to the omeprazole analogues with the same basic structural template as that of timoprazole (or omeprazole) could not be identified, despite a detailed mechanistic understanding.

2.8.3
Chemical Approach

In an effort to increase bioavailability relative to that of omeprazole, chemical approaches were used to change the substitution pattern on the pyridine and benzimidazole rings by, for example, avoiding the methyl group in the 5-position of the pyridine ring and introducing halogen atoms or other electron-withdrawing substituents in the benzimidazole moiety. The idea of avoiding the 5-methyl group came from previous studies of the metabolism of omeprazole made by Renberg et al. [42], who showed omeprazole to be metabolized primarily by oxidation of the 5-methyl group on the pyridine ring. The idea of introducing halogens was based on a general view that these types of atom often slow down metabolism. A rat model was used for testing bioavailability. The more important metabolic pathways are shown in Scheme 2.4.

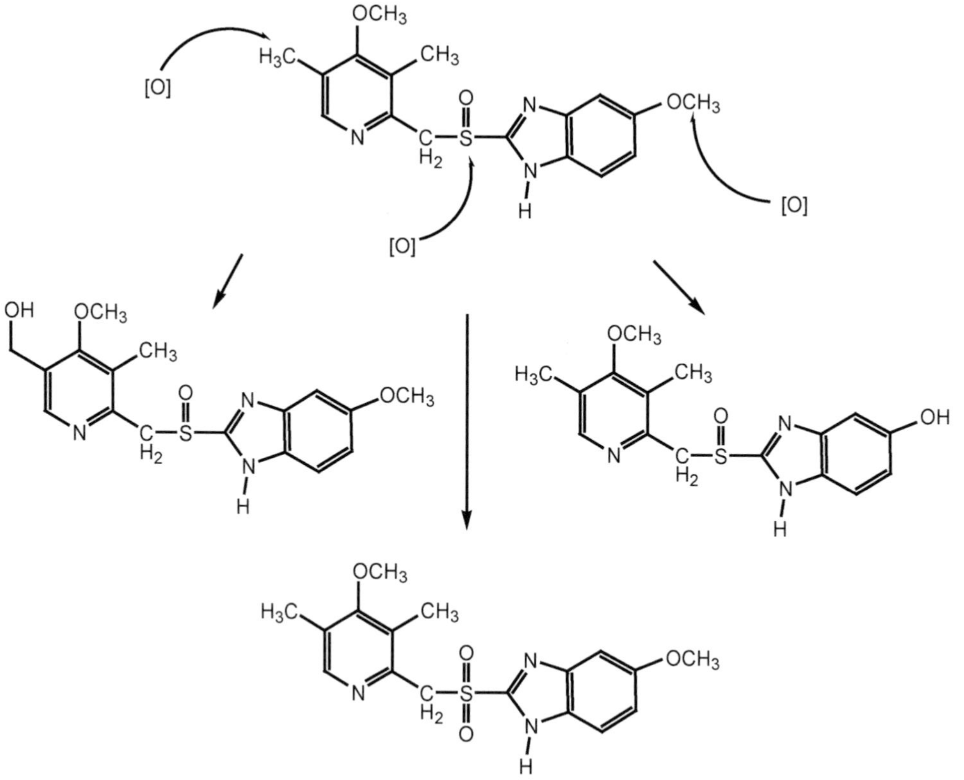

Scheme 2.4 Important metabolic pathways of omeprazole.

2.8.4
Synthesis and Screening

During the following years, more than 30 scientists synthesized and screened several hundred compounds in the search for a compound which fulfilled the screening goal, namely to produce a drug with a better bioavailability than omeprazole.

Only a few omeprazole congeners were discovered that satisfied the preclinical goals. In this respect, H 259/31 and H 326/07 were two of the most promising candidates (Fig. 2.18). Between 1989 and 1994, a total of seven compounds were selected for further tests beyond screening, but gradually it was realized exactly how difficult an act omeprazole was to follow. However, four of the compounds – including H 259/31 and H 326/07 – passed the rigorous preclinical tests and were subsequently tested in humans. When each of the key target parameters – pharmacokinetic properties, acid inhibitory effect, and safety – was assessed, only one compound proved superior to omeprazole. This was H 199/18 (see Fig. 2.18), the S-(–)-enantiomer of omeprazole, which was referred to as esomeprazole (originally perprazole).

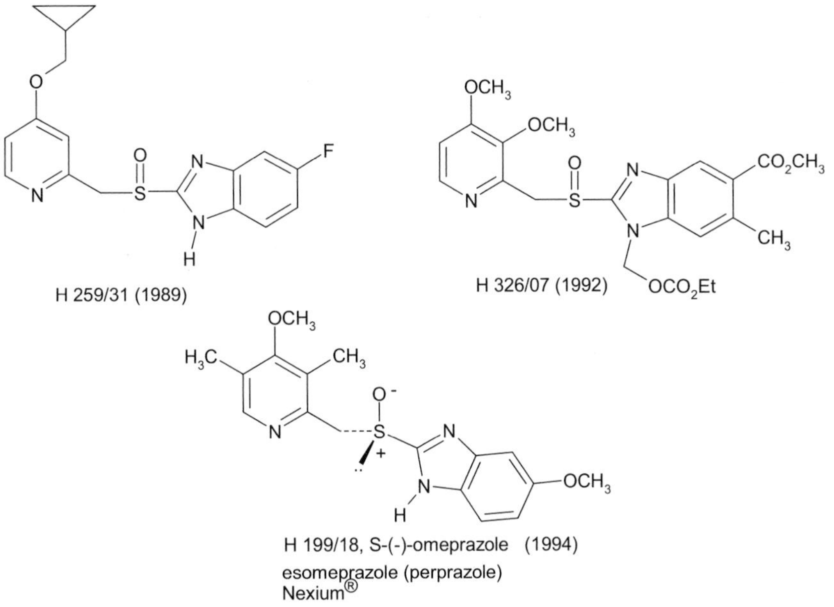

H 259/31 (1989)

H 326/07 (1992)

H 199/18, S-(-)-omeprazole (1994)
esomeprazole (perprazole)
Nexium®

Fig. 2.18 Three candidate drugs synthesized by Astra that passed preclinical tests and were tested in humans. H 199/18 (esomeprazole) was later launched in 2000 in Europe, and 2001 in the USA as Nexium®.

2.8.5
Isomers Seemed Unattractive

The discovery of esomeprazole was surprising. Knowledge of the mechanism of acid inhibition had led to the prediction that both isomers of omeprazole would have exactly the same effect. The acid-catalyzed conversion of either isomer to the same active non-chiral sulfenamide species should occur at the same rate for both isomers. In accordance with this, and at an early stage, it had been found that the two isomers showed identical dose–response curves for the inhibition of acid production in isolated gastric glands [43] when tested *in vitro*. Thus, whilst it was known that the isomers of chiral drugs often have different activities, the logical conclusion was that this was not the case here. Moreover, before 1990, it had been possible to prepare only milligram quantities of the single isomers, and this was not enough for *in vivo* testing. Lastly, it was believed that the single enantiomers would racemize too easily, especially as a racemization of the isomers had, in fact, been observed previously (*in vitro*).

However, even if there was no difference between the isomers at the target cell level, the idea of a possible difference in metabolism between the two isomers could not be dismissed. To test this idea, larger quantities of the pure isomers were required. Initially, isomer-selective production using microbial and enzymatic systems [44a,b] were only partly successful, but more promising was the concept of separating isomers via chromatography of diastereomers. As a spin-off from previous studies with prodrugs of omeprazole, a technique was identified to use a temporary covalent diastereomeric complex with mandelic acid for chromatographic separation. Not only were hundreds of milligrams of the single isomers obtained, but it was also found that alkaline salts of the isomers were stable against racemization and, moreover, that they, as opposed to the neutral forms, were crystalline. The prerequisites to begin *in vivo* testing were now in position.

2.8.6
Isomer Pharmacokinetics and Pharmacodynamics in Animals

Using the first few hundred milligrams of the pure isomers available, the first *in vivo* experiments in rats were carried out, and the plasma concentrations and inhibition of stimulated acid secretion following identical oral doses of the two isomers and the racemate were compared. In the rat, the R-isomer showed a higher bioavailability than the S-isomer, although in this first study there was no significant difference in the inhibition of acid secretion. Subsequent studies, however, gave dose-response curves. In later studies in the dog no significant differences could be detected between the two isomers. If the initial *in vivo* experiments had been performed solely in dogs, it is most likely that further studies with the isomers would have been stopped. However, on the basis of the initial findings in rats, the decision was taken to continue the project and to compare the effects of the isomers in man.

2.8.7
The Key Experiment in Man

After intensive procedures to produce sufficient quantities of the pure isomers (separated chromatographically), the very first experiments in man were initiated during autumn 1994.

The racemate and the two isomers (15 mg each, all as sodium salts) were given orally and in a random order to each of four healthy subjects, who were defined genetically as extensive metabolizers (Fig. 2.19a and b). Drug plasma concentrations and the inhibition of pentagastrin-stimulated acid secretion were monitored.

As can be seen from Fig. 2.19a and b, the most significant differences in plasma levels and effect on acid inhibition were found between the isomers and the racemate. Surprisingly, however, the opposite order to that found in rats was

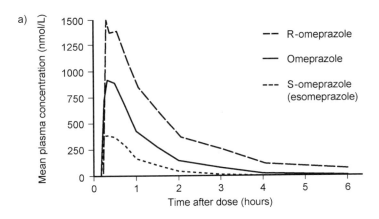

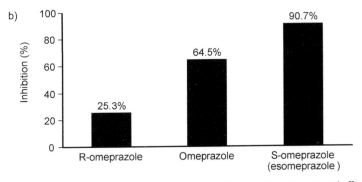

Fig. 2.19 (a,b) AUCs and degree of acid inhibition after repeated administration (7 days) of omeprazole 15 mg and its *S*- and *R*-isomers (all as sodium salts) in extensive metabolizers [46]. Reprinted from Andersson T, Bredberg E, Sunzel, M, et al., *Gastroenterology*, **2000**, 118 (4 pt. II), A1210, 5551. Pharmacokinetics (PK) and effect on pentagastrin-stimulated peak acid output (PAO) of omeprazole (O) and its two optical isomers, *S*-omeprazole/esomeprazole (E) and *R*-omeprazole (R-O). Reproduced with permission from the American Gastroenterological Association.

observed. In man, the S-isomer produced the highest area under the curve (AUC), four to five times that of the R-isomer, and about double that of the racemate (Fig. 2.19a). These differences were reflected in the effect pattern (Fig. 2.19b), with 15 mg of the S-isomer inhibiting stimulated acid secretion by more than 90%, while 15 mg of the R-isomer inhibited the pentagastrin response by only ca. 20%. The corresponding value for the racemate was 60%. Thus, the S-isomer fulfilled the project's aim of identifying a compound with significantly higher bioavailability together with a higher oral potency than that of omeprazole [45–47].

In the second part of the experiment, five poor metabolizers (who normally exhibit several-fold higher AUC values after dosing with omeprazole compared with extensive metabolizers) each received 60 mg oral daily doses of omeprazole, the S-isomer and R-isomer, for one week (Fig. 2.20) [45]. The average AUC of the S-isomer was 30% less than that of omeprazole, whereas plasma levels of the R-isomer were more than 20% higher. The pattern in poor metabolizers was thus reversed compared with that seen in extensive metabolizers; exposure to the drug in poor metabolizers, which with omeprazole was higher than was needed to obtain optimal inhibition of gastric acid secretion, was decreased with the S-isomer. Moreover, the exposure in extensive metabolizers, which with omeprazole was insufficient to achieve optimal acid inhibition in all subjects, was increased with the S-isomer. This resulted in less overall variability in the pharmacokinetics of the S-isomer than that of the R-isomer and the racemate, omeprazole. These data showed that the impact of polymorphic metabolism was less pronounced for the S-isomer than for the racemate omeprazole, and especially for the R-isomer [48] – a finding that was totally unexpected when the isomers were selected as candidate drugs.

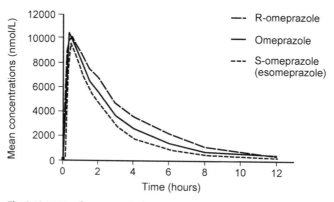

Fig. 2.20 AUCs after repeated administration (7 days) of omeprazole 15 mg and its S- and R-isomers (all as sodium salts) in poor metabolizers [45]. Reproduced from Andersson T, Hassan-Alin M, Hasselgren G, et al., *Clin. Pharmacokinet.*, **2001**, 40(6), 411–426, with permission from Wolters Kluwer.

Based on these exciting findings, the S-isomer of omeprazole (in its alkaline salt form) was chosen for development. Documenting esomeprazole – the generic name – for routine clinical use, including pivotal studies that demonstrated for the first time that a new antisecretory agent provided a significant clinical advance over omeprazole and other PPIs in direct head-to-head clinical studies [49–55]. Esomeprazole magnesium salt (or Nexium®) was launched in Sweden in August 2000, and in the rest of the Europe during the autumn that year. Early in 2001, Nexium was also approved and launched in the United States.

2.8.8
Production of Esomeprazole (Mg Salt)

The large-scale production of esomeprazole is now successfully achieved by asymmetric oxidation of the same sulfide intermediate as is used in the production of omeprazole (Scheme 2.5). Using the titanium-based catalyst originally developed by K. Barry Sharpless for allyl alcohol oxidation [56] and by H.B. Kagan for certain sulfide oxidations [57], a process was developed that could achieve initial enantiomeric excesses of about 94% [53]. During the production process, the optical purity is further enhanced by the preparation of esomeprazole magnesium salt, with subsequent re-crystallization.

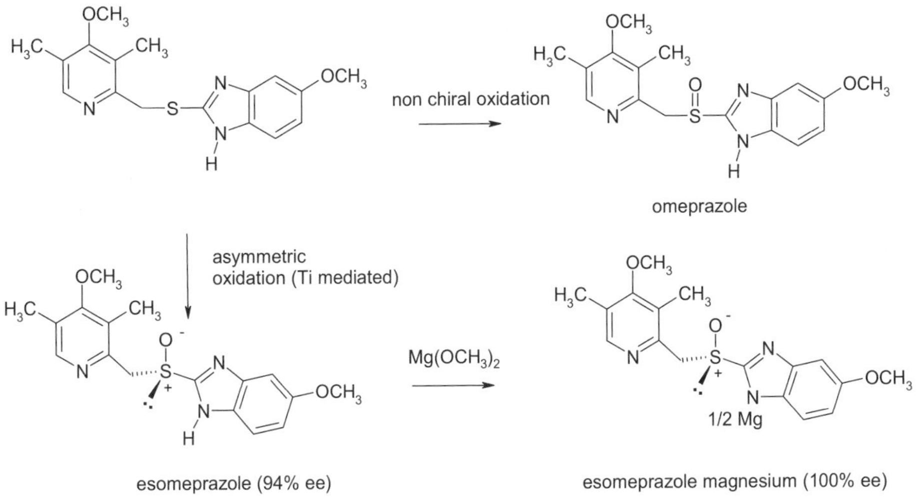

Scheme 2.5 Synthesis of omeprazole and esomeprazole magnesium.

2.8.9
Omeprazole Isomers: Differences in Clearance and Metabolic Pattern

Omeprazole is mainly metabolized by cytochrome P450 (CYP 2C19). This enzyme is expressed polymorphically, which means that some individuals do not express a functioning CYP 2C19 enzyme. In these individuals, another CYP isoform (CYP 3A4), which normally plays a less dominant role in the metabolism of

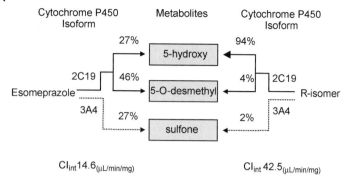

Fig. 2.21 Elimination pathways for the *S*- and the *R*-isomers of omeprazole via cytochrome P450 (CYP) enzymes, based on data from an *in vitro* study in human liver microsomes [45]. CL_{tot} = total clearance. Reproduced from Andersson T, Hassan-Alin M, Hasselgren G, et al., *Clin. Pharmacokinet.*, **2001**, 40(6), 411–426, with permission from Wolters Kluwer.

omeprazole, takes over. The metabolism of omeprazole by CYP 3A4 is slower than that by CYP 2C19 and, therefore, these individuals may be considered poor metabolizers (3% of Caucasians and 15–20% of Asians have this phenotype) compared to the majority of the population, the so-called extensive metabolizers. Studies *in vitro* have shown that the drug-metabolizing CYP enzymes in the liver exhibit a stereoselective metabolism of omeprazole. Although the same types of metabolites are formed from the two isomers, there are marked quantitative differences, and the intrinsic clearance of esomeprazole (*S*-isomer) is three times lower than that of the *R*-isomer in human liver microsomes (Fig. 2.21) [48,58].

The differences in pharmacokinetic properties of esomeprazole and the *R*-isomer in man can be summarized as follows:
- A threefold difference in metabolic clearance as assessed *in vitro*.
- A significant difference in relative dependency on CYP 2C19 and CYP 3A4 as assessed *in vitro*.
- As a consequence, less impact of polymorphic metabolism *in vivo* for esomeprazole, resulting in overall less variability in pharmacokinetics and pharmacodynamics (inhibitory effect) [48].

These different properties of the two omeprazole isomers, demonstrated both *in vitro* and *in vivo*, result in the following pharmacokinetic differences in the clinical situation when comparing esomeprazole and the racemate omeprazole:
- A given oral dose of esomeprazole appears to result in an approximately two-fold higher AUC than the same dose of omeprazole.
- Less difference in AUC between poor and extensive metabolizers, due to a lower AUC in poor metabolizers and a higher AUC in extensive metabolizers than with omeprazole at similar doses.
- The inter-individual variation in AUC is significantly less for esomeprazole than for the same dose of omeprazole [48].

Hence, due to a two-fold increase in AUC and less inter-patient variability compared with omeprazole and concomitant superior acid inhibition and higher predictability, esomeprazole, the *S*-isomer of omeprazole (in alkaline salt form), provides clinical benefit in a wide range of acid-related disorders.

2.9
Summary

Omeprazole was developed by screening in a conscious dog model. Starting from CMN 131, a known acid-secretion inhibitor from Servier as a lead and with an idea of how to overcome its toxic effects, the significantly more potent compound H 83/69 (timoprazole) was synthesized in 1974. This first pyridinyl methyl sulfinylbenzimidazole, however, caused toxicity problems and made human testing impossible. By 1976, the analogue H 149/94 (picoprazole), had been identified and shown not only to be devoid of toxicity problems but also to be more potent than timoprazole. Furthermore, it was revealed that picoprazole inhibited the H^+,K^+-ATPase (the proton pump) in the stomach. The first human testing confirmed picoprazole to be the most powerful inhibitor of acid secretion identified so far. By then (1979), several compounds had been identified by Astra as being superior to picoprazole in the dog, and one of these (omeprazole) was chosen for development. Omeprazole – the first PPI used in clinical practice – was launched in 1988 as Losec® in Sweden, and in 1990 as Prilosec® in the United States.

During the 1980s, about 40 other companies entered the PPI area, but few achieved market success: Takeda with lansoprazole, Byk Gulden with pantoprazole, and Eisai with rabeprazole, all of which were analogues of omeprazole.

Since omeprazole showed an inter-individual variability, and a significant number of patients with acid-related disorders required higher or multiple doses to achieve symptom relief and healing, Astra started a new research program in 1987. One goal was to identify a new analogue to omeprazole with less inter-patient variability, and during the following years several compounds were selected for further study. Eventually, only one compound proved superior to omeprazole, this being its *S*-isomer – esomeprazole – and this was developed as the magnesium salt. Esomeprazole magnesium (Nexium®), which received its first approval in 2000, provided a more pronounced inhibition of acid secretion and less inter-patient variation compared to omeprazole. Subsequently, this has translated into clinical superiority, as has been shown clearly in terms of predictability of response and efficacy relative to omeprazole.

References

1 Fellenius E, Berglindh T, Sachs G, et al. *Nature*, **1981**, 290, 159–161.

2 Larsson H, Carlsson E, Junggren U, et al. *Gastroenterology*, **1983**, 85, 900–907.

3 Lindberg P, Brändström A, Wallmark B, et al. *Med. Res. Rev.*, **1990**, 10, 1–54.

4 Lindberg P, Brändström A, Wallmark B, *Trends Pharmacol. Sci.*, **1987**, 8, 399–402.

5 Carlsson E, Lindberg P, von Unge S. *Chem. Brit.*, **2002**, 38, 42–45.

6 Olbe L, Carlsson E, Lindberg P. *Nature Rev.*, **2003**, 2, 132–139.

7 Cook DL, Bianchi RG. *Life Sci.*, **1967**, 6, 1381–1387.

8 Malen CE, Danree BH. *J. Med. Chem.*, **1971**, 14, 244–246.

9 Sjöstrand SE, Olbe L, Fellenius E. The discovery and development of the proton pump inhibitor. In: *Proton Pump Inhibitors*. Olbe L. (Ed.), Birkhauser, Switzerland, **1999**.

10 Forte JG, Lee HC. *Gastroenterology*, **1977**, 73, 921–926.

11 Sachs G, Chang H, Rabon E, et al. *Gastroenterology*, **1977**, 73, 931–940.

12 Saccomani G. *Acta Physiol. Scand.*, **1978** (Suppl.), 293–305.

13 Wallmark B, Sachs G, March S, et al. *Biochim. Biophys. Acta*, **1983**, 728, 31–38.

14 Lind T, Cederberg C, Ekenved G, et al. *Gut*, **1983**, 24, 270–276.

15 Bonnevie O, et al. Gastric acid secretion and duodenal ulcer healing during treatment with omeprazole. *Scand. J. Gastroenterol.*, **1984**, 19, 882–884.

16 Lauritsen K, Rune SJ, Bytzer P, et al. *N. Engl. J. Med.*, **1985**, 312, 958–961.

17 Walan A, Bader JP, Classen M, et al. *N. Engl. J. Med.*, **1989**, 320, 69–75.

18 Sandmark S, Carlsson R, Fausa O, et al. *Scand. J. Gastroenterol.*, **1988**, 23, 625–632.

19 Fryklund J, Gedda K, Wallmark B. *Biochem Pharmacol.*, **1988**, 37, 2543–2549.

20 Lindberg P, Nordberg P, Alminger T, et al. *J. Med. Chem.*, **1986**, 29, 1327–1329.

21 Sigrist-Nelson K, Müller RKM, Fischli AE. *FEBS Lett.*, **1986**, 197, 187–191.

22 Bohnenkamp W, Eltze M, Heintze K, et al. *Pharmacology*, **1987**, 34, 269–278.

23 Bohnenkamp W, Eltze M, Heintze K, et al. 8th World Congress of Gastroenterology, Sao Paulo, Brazil, September 7–12, **1986**. Abstract 1377.

24 Simon B, Müller P, Marinis E, et al. *Aliment. Pharmacol. Ther.*, **1990**, 4, 373–379.

25 Ife RJ, Dyke CA, Keeling DJ, et al. *J. Med. Chem.*, **1989**, 32, 1970–1977.

26 Sih JC, Im WB, Robert A, et al. *J. Med. Chem.*, **1991**, 34, 1049–1062.

27 Galmiche JP, Bruley des Varannes S, Ducrotte P, et al. *Aliment. Pharmacol. Ther.*, **2004**, 19, 655–662.

28 Cox D, Dowlatshahi NP, Gensmantel AH, et al. 4th SCI-RSC Medicinal Chemistry Symposium, Cambridge, England, September 6–9, **1987**. Abstract P5; and private communication with Fisons PLC.

29. S. Okabe, Y. Akimoto, S. Yamasaki, et al., *Digest. Dis. Sci.*, **1988**, 33, 1425–1434.

30 Igata H, Takagi K, Okabe S, *Digestion*, **1991**, 49, 54.

31 Takemoto T, Okazaki Y, Tada M, et al. *Clin. Adult Dis.*, **1991**, 21, 769–783.

32 Yamada S, Narita S, *Chem. Pharm. Bull.*, **1994**, 42, 1679–1681.

33 Binder D, Rovenszky F, Ferber H, et al. Patent EP 261478.

34 Herling AW, Bickel M, Land H-J, et al. International Conference on Gastroenteric Biology, Oxnard, California, October 25–28, **1988**, Abstract P27.

35 Fujisaki H, Shibata H, Oketani K, et al. *Biochem. Pharmacol.*, **1991**, 42, 321–328.

36 Uchida M, Morita S, Chihiro M, et al. *Chem. Pharm. Bull.*, **1989**, 37, 1517–1523.

37 Brunner G, Creutzfeldt W. *Scand. J. Gastroenterol. Suppl.*, **1989**, 166, 111–113.

38 Koop H, Hotz J, Pommer G, et al. *Aliment. Pharmacol. Ther.*, **1990**, 4, 593–599.

39 Lind T, Cederberg C, Idstrom JP, et al. *Scand. J. Gastroenterol.*, **1991**, 26, 620–626.

40 Klinkenberg-Knol EC, Meuwissen SG. *Digestion*, **1992**, 51(Suppl. 1), 44–48.

41 Lundell L, Backman L, Ekstrom P, et al. *Aliment. Pharmacol. Ther.*, **1990**, 4, 145–155.

42 Renberg L, Simonsson R, Hoffmann K-J. *Drug Metab. Dispos.*, **1989**, 17, 69–76.

43 Erlandsson P, Isaksson R, Lorentzon P, et al. *J. Chromotogr.*, **1990**, 532, 305–319.

44 (a) Holt R, Lindberg P, Reeve Ch, et al. US Patent 5,840,552, **1998**; (b) Graham D, Holt H, Lindberg P, et al. US Patent 5,776,765, **1998**.

45 Andersson T, Hassan-Alin M, Hasselgren G, et al. *Clin. Pharmacokinet.*, **2001**, 40, 411–426.

46 Andersson T, Bredberg E, Sunzel M, et al. *Gastroenterology*, **2000**, 118 (4 pt. II), A 1210, 5551.

47 Andersson T, Röhss K, Bredberg E, et al. *Aliment. Pharmacol. Ther.*, **2000**, 14, 861–867.

48 Andersson T. *Clin. Pharmacokinet.*, **2004**, 43, 279–285.

49 Kahrilas PJ, Falk GW, Johnson DA, et al. *Aliment. Pharmacol. Ther.*, **2000**, 14, 1249–1258.

50 Richter JE, Kahrilas PJ, Johansson J, et al. *Am. J. Gastroenterol.*, **2001**, 96(3), 656–665.

51 Castell D, Kahrilas P, Richter J, et al. *Am. J. Gastroenterol.*, **2002**, 97, 575–583.

52 Lauritsen K, Deviére J, Bigard MA, et al. *Aliment. Pharmacol. Ther.*, **2003**, 17, 333–341.

53 Cotton H, Elebring T, Larsson M, et al. *Tetrahedron: Asymmetry*, **2000**, 11, 3819–3825.

54 Labenz J, Armstrong D, Katelaris PH, et al. *Gut*, **2004**, 53 (Suppl. VI), A110.

55 Labenz J, Armstrong D, Katelaris PH, et al. *Gut*, **2004**, 53 (Suppl. VI), A108.

56 Katsuki T, Sharpless KB. *J. Am. Chem. Soc.*, **1980**, 102, 5974–5976.

57 Pitchen P, Dunach E, Desmukh MN, et al. *J. Am. Chem. Soc.*, **1984**, 106, 8188–8193.

58 Äbelö A, Andersson TB, Antonsson MA, et al. *Drug. Metab. Dispos.*, **2000**, 28, 966–972.

3

The Development of a New Proton-Pump Inhibitor:
The Case History of Pantoprazole

Jörg Senn-Bilfinger and Ernst Sturm

3.1
Introduction

For over 100 years, medicine has known that the presence of acid is essential for the formation of duodenal and gastric ulcers and for the development of gastroesophageal reflux disease (no acid – no ulcer) [1]. It is, however, only over the past 30 years that drugs have been developed which could control acid secretion adequately enough to bring these three major acid-related diseases under control. During the early part of the last century and beyond, effective treatment of acid-related diseases centered on surgery which set out to reduce acid secretion either by partial or total gastrectomy (removal of sections of the stomach wall), or by various types of vagotomy (interfering with the nervous pathways which stimulate the acid secretion) (1).

Antacids have long been used to counteract symptoms and, when given in high and frequent doses, they remain effective in milder forms of acid-related diseases. By the first quarter of the twentieth century, the three major stimuli of acid secretion were recognized as gastrin, acetylcholine, and histamine (Fig. 3.1), but it took another 50 years to develop antagonists for acetylcholine and histamine. The first histamine-2 receptor antagonist validated in human studies in 1972 was buriamide, discovered in the UK by James Black and colleagues. This was followed by cimetidine, which reached the market in 1976 [2,3]. For the first time, effective healing was possible without the expense of surgery. Further analogues, such as ranitidine and famotidine, followed during the early 1980s [4]. However, because of the multiple pathways for the activation of acid secretion (gastrin, acetylcholine, and histamine), these drugs could not completely suppress the hypersecretion of acid. In addition, acid rebound on discontinuation of therapy and development of tolerance to the therapy made histamine-2 blocker therapy ineffective in a large proportion of patients.

Analogue-based Drug Discovery. IUPAC, János Fischer, and C. Robin Ganellin (Eds.)
Copyright © 2006 WILEY-VCH Verlag GmbH & Co. KGaA, Weinheim
ISBN: 3-527-31257-9

Resting Parietal Cell   Secreting Parietal Cell

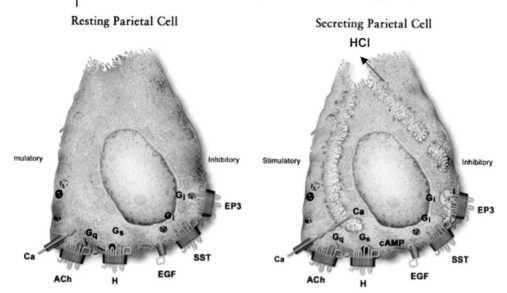

Fig. 3.1 Parietal cell in its resting and secreting state with the receptors for stimulation (ACh, H) and inhibition (EGF,SST,EP3) of acid secretion on its basolateral surface.
(ACh acetylcholine, H histamine-2, EGF epidermal growth factor, SST somatostatin, EP3 prostaglandin subtype).

During the 1970s, with the discovery of the gastric acid pump [5,6] (Fig. 3.1) and its mechanism and means of regulation, the stage was set for the development of a newer class of anti-secretory inhibitors [7]. These were the substituted benzimidazoles (Fig. 3.2), also known as proton-pump inhibitors (PPIs), because they specifically target the acid (proton) pump itself. By blocking the very last step of acid secretion, these PPIs provide a more effective and reliable reduction of gastric acid secretion, and hence healing of acid-related diseases. The unique environment of the proton pump in the apical membrane of the parietal cell separating the neutral cytosol from the acidic lumen is distinct from other H-translocating ATPases in the body, thus providing an excellent conceptual basis for the identification of selective drugs [8]. The substituted benzimidazoles harness this unique pH gradient in a twofold way because they accumulate in the acidic secretory canaliculus of the parietal cell (luminal surface of the gastric ATPase) due to protonation of basic functionalities present. Further, in this compartment they then undergo an unique acid-catalyzed chemical rearrangement that is a prerequisite for their activity [9,10].

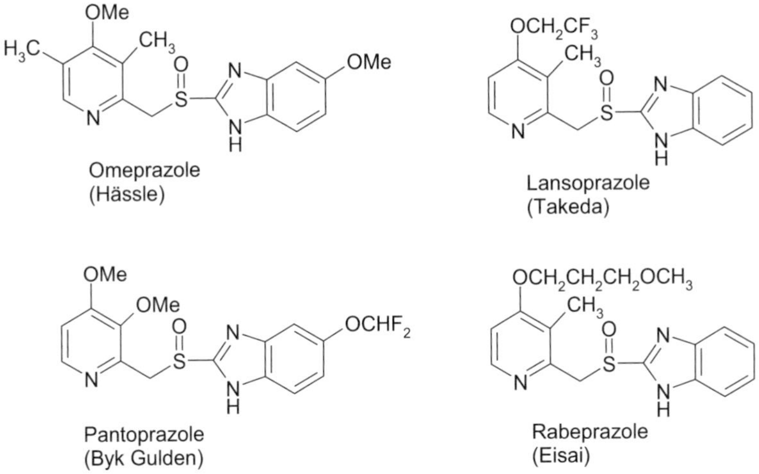

Fig. 3.2 The four currently marketed racemic proton-pump inhibitors (PPIs).

These mechanistic consequences on the design of new PPIs were not known at the outset of the synthesis program in late 1980 at Byk Gulden. Initial efforts were directed towards the preparation of acid-stable compounds, which turned out to be completely devoid of any anti-secretory activity. Later, efforts focused on the synthesis of analogues, which had high inhibitory activity in the stimulated gastric glands possessing acidic pH [11], but had low reactivity (high chemical stability) at neutral pH. This was achieved by varying the substituents in the benzimidazole and pyridine parts of the timoprazole-like backbone, resulting in balanced nucleophilic/electrophilic properties of the molecule. The identification of pantoprazole [12,13] as a valuable drug with high anti-secretory activity was finally achieved in 1985 after the synthesis and evaluation of more than 450 analogues. Here, on the basis of the longstanding gastrointestinal research history, the investigations at Byk Gulden into PPIs over a period of almost eight years is summarized, and the process how pantoprazole sodium was identified and eventually developed to a clinically valuable drug is described.

3.2
History of Gastrointestinal Research at Byk Gulden

3.2.1
The Antacids and Cytoprotectives Projects and the Set-Up of In-Vivo Ulcer Models

Initially, Byk Gulden's anti-ulcer investigations concentrated on the development of antacids. In 1957 [14,15], Byk Gulden invented the antacid magaldrate (Riopan®). This compound consists of a hydrotalcite-like layered structure, which shows a rapid partial neutralization of the acid to pH 4.0–4.5 and a strong additional binding of bile acids and lysolecithin. Bile salts, in particular, play a detrimental role in reflux esophagitis (inflammation of the lower part of the esophagus). The drug was licensed-out to Wyeth (formerly Wyeth-Ayerst) for the US market; it has been sold in Europe since 1961 by Tosse, a subsidiary of Byk Gulden.

During the 1970s, Byk Gulden moved on from the development of antacids to concentrate on gastroprotective drugs. The main product developed during this period was clanobutin. This was successfully developed for veterinarian use, and no drug for human use resulted. The clanobutin research studies provided the company with significant experience with animal models in the gastrointestinal field, and this proved to be very important later during the development of suitable PPI compounds.

In this regard, the Shay rat model was first used at Byk Gulden in 1977, whilst the modified Shay rat model, in which, following ligation of the pylorus, aspirin is administered into the rat's stomach, was used from 1978. Byk Gulden also used the Ghosh Schild rat model since 1974. Several advantages of the Shay rat model [16] are recognized. First, with the Shay rat model, the duration of the test compound's activity can be assessed. Second, in Shay rats acid secretion is at a basal level, and this reflects the situation in man more closely than other animal models which require constant stimulation of acid secretion. Third, due to biological variations it is essential to include a large number of animals in the tests; for this reason, rats are preferable to dogs as they are easier to handle, require less test substance, and are less expensive to maintain.

3.2.2
Decision to Concentrate on Anti-Secretory Treatments and the Study of Compounds with an Unknown Mechanism of Action

By 1978, there was a move at Byk Gulden to concentrate on a smaller number of clinical indications and to move away from the more diverse approach. The company consulted widely – both internally and externally – on what direction to take. There were several areas in which the company could specialize, including anti-ulcer treatments. These treatments were appealing because of the company's experience with antacids and gastroprotectives, and this was due in particular to the great experience with relevant animal models. These had been developed not only during the company's studies on gastroprotectives but also on anti-inflammatory compounds, when the ulcer-inducing adverse action of nonsteroidal anti-inflammatory drugs (NSAIDs) had been assessed.

Externally, most investigations in the anti-ulcer field during the 1970s was directed towards new anti-secretory treatments. These included H_2-blockers, anti-muscarinics, and compounds with unknown mechanisms of action; Byk Gulden considered compounds in each of these classes. In 1976, Smith Kline & French (now Glaxo SmithKline, GSK) – Byk Gulden's later partner in the identification and development of pantoprazole-sodium – had launched cimetidine (marketed as Tagamet®), a histamine receptor antagonist (H_2-blocker), whilst in 1981 Glaxo (now GSK) launched ranitidine (marketed as Zantac®), an improved H_2-blocker

[17]. H_2-blockers are anti-secretory drugs because they inhibit histamine-stimulated acid secretion. The multitude of the synthesized H_2-blockers and patents relating to H_2-blockers led Byk Gulden scientists to the conclusion that this field was already – or would in the near future – be fully optimized, and that there was little or no room for further improvement. Another relatively new area was that of anti-muscarinic treatments [18,19]. The only relevant anti-muscarinic drug then on the market was pirenzepine (made by Boehringer Ingelheim). The level of optimization of anti-muscarinic drugs was, according to Byk Gulden, comparatively low as known compounds with anti-muscarinic activity had some unwelcome side effects, such as dry mouth and blurred vision. Based on its broad experience with similar tricyclic chemical structures studied in previous research on the central nervous system, Byk Gulden wished to be able to separate the unwanted centrally acting effects of those chemical analogues from the desired anti-secretory activity. After a great deal of work and the investigation of some 190 new compounds, two possible candidate drugs were developed, zolenzepine and telenzepine (Fig. 3.3), which were first tested in animals in 1979 and 1980, respectively.

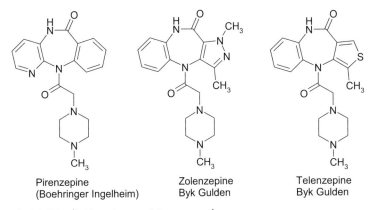

Pirenzepine
(Boehringer Ingelheim)

Zolenzepine
Byk Gulden

Telenzepine
Byk Gulden

Fig. 3.3 Tricyclic M_3-anti-muscarinic compounds.

Although these drugs showed good activity also in man at a low dose, they had still some side effects such as dry mouth and blurred vision. They were therefore inferior to the H_2-blockers and were not pursued; indeed, no other company has managed to avoid these side effects to date. However, these drugs provided Byk Gulden with very useful experience in conducting clinical trials and established important contacts to the clinicians in the field throughout Europe. All of this know-how was invaluable later during the clinical development of pantoprazole.

During early 1979, a literature search for new anti-secretory compounds with unknown mechanisms of action had been carried out to identify interesting compounds and to analyze the competitive situation in the anti-ulcer field. New anti-secretory compounds of unknown mechanism of action (Fig. 3.4) appeared in the scientific or patent literature, and included CMN-131 (Science-Union), tiquinamide (Wyeth-Ayerst), trithiozine (ISF Lab. Biomed. Res.), nolinium bromide (Norwich), and timoprazole (Hässle) [20].

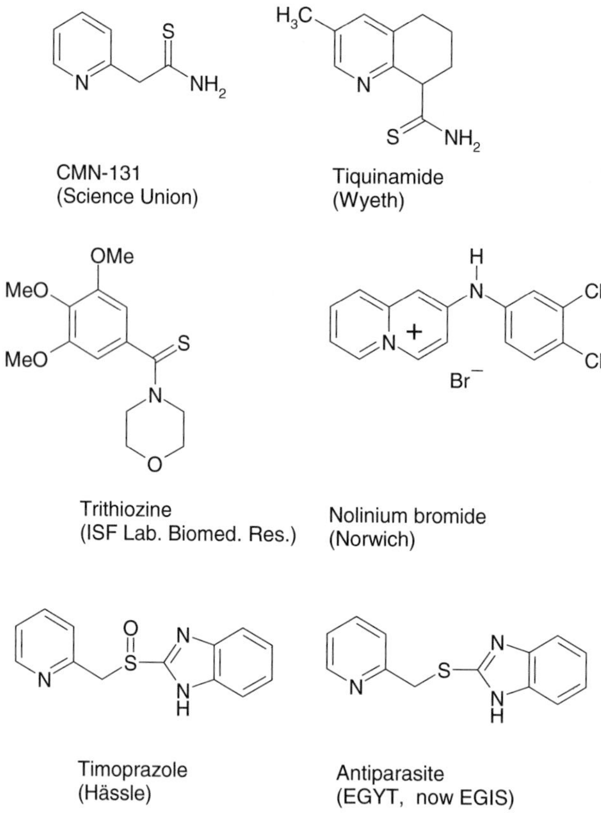

CMN-131
(Science Union)

Tiquinamide
(Wyeth)

Trithiozine
(ISF Lab. Biomed. Res.)

Nolinium bromide
(Norwich)

Timoprazole
(Hässle)

Antiparasite
(EGYT, now EGIS)

Fig. 3.4 Anti-secretory agents with unknown mechanism of action as of 1979, without H_2-blockers and anti-muscarinics. (The producing companies are indicated in brackets. The EGYT (now EGIS) compound is a prodrug form of timoprazole, identified by Hässle.)

The first three of these compounds were thioamides with suspected toxicological problems, noliniumbromide had a quaternary salt structure with inherent low oral bioavailability, and timoprazole appealed due to its simple chemical structure and its surprisingly high level of anti-secretory activity seen in the animal models established at Byk Gulden. The EGYT (now EGIS)-compound could be regarded as a prodrug of timoprazole, the first compound reported that contained the appropriate molecular backbone for anti-secretory activity.

In late 1977, Hässle published their finding, showing that the substituted benzimidazole timoprazole might act as a specific inhibitor of acid secretion by the proton pump present in the stomach wall [21]. The experts were confused by the very steep dose–response curve of timoprazole, and its propensity to decompose to colored products in solution. It quickly became apparent that this compound needed to be assessed carefully by measuring its solution stability at various pH values. The experts also considered whether it would be possible to separate this instability from activity.

In early 1981, Byk Gulden decided gradually to redirect more of the research resources away from anti-muscarinics programme [22,23] to PPIs. At this time, picoprazole became the new state of the art (Fig. 3.5) [24,25]. Some data on pico-prazole had been published previously in *Nature* [26]. Furthermore, in spring 1981 Byk Gulden became aware of Hässle's patent EP 5129, which focused on the pyridine part of the timoprazole skeleton [27].

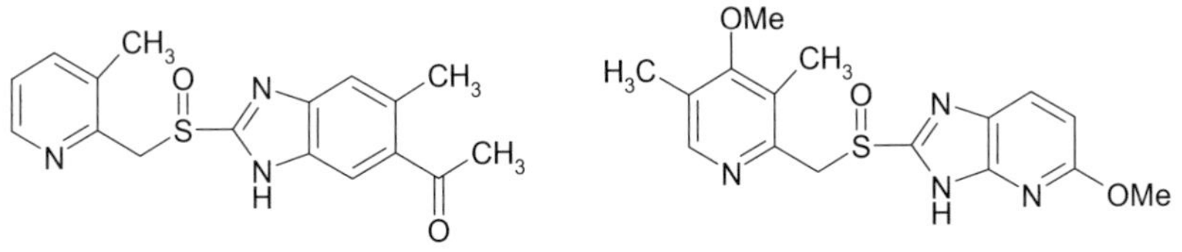

Picoprazole
(Hässle)

TU-199,Tenatoprazole
(Tokyo Tanabe)

Fig. 3.5 Experimental proton-pump inhibitors. Tenatoprazole is currently under development.

From the patent, and from a general knowledge of chemistry, it became evident that the introduction of a 4-alkoxy-group in the pyridine moiety would increase the basicity of the compounds. However, the exact influence of the adjacent alkyl-groups on the 4-alkoxy-group and hence the resulting pyridine basicity could not be derived from the given data. With Byk Gulden's expertise with the established *in-vivo* models, the belief was that by experimental work it could be determined whether such changes were beneficial. The patent contained no indication that the level of stability (in other words, whether it decomposed in neutral or mildly acidic solution) was important, and that activity was not the sole selection criter-ion. Therefore, the chemists had to synthesize a variety of compounds referred to in the EP 5129 patent, and to submit them for biological testing. The patent also included Hässle's later development compound, omeprazole, for which also no explicit information on its synthesis was given. When Byk Gulden focused on their own compounds, it was only then realized that careful experimentation with different substituents in both parts of the molecules was required to obtain accept-able solution-stable and selective compounds with the required level of activity.

3.3
Identification of the First PPI Project Candidates

3.3.1
Optimizing the Benzimidazole Moiety

Initially, Byk Gulden's efforts were mainly devoted to modification of the benzimi-dazole part of the lead structure timoprazole, and linking it to various substituted pyridines previously described in the patent literature. Thereby, the medicinal chemists were able to study preliminary structure–activity relationships (SARs) of this new type of compound. Adding an additional N-atom to the benzo part of the benzimidazole led to imidazopyridines (see also the most recent development compound, tenatoprazole; Fig. 3.5). These new inhibitors turned out to be active, but even more unstable than timoprazole, and very difficult to prepare. On the other hand, modifications of the linkage to the pyridine moiety to the 3- or 4-position instead of the 2-position (as in timoprazole) led to stable but completely inactive compounds (Fig. 3.6).

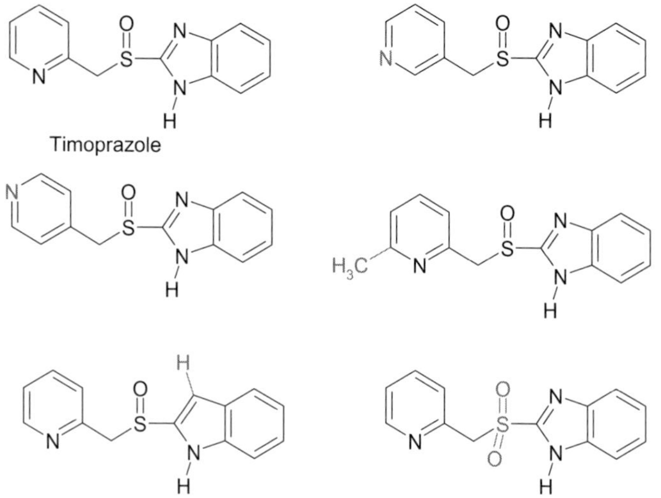

Fig. 3.6 Timoprazole and acid-stable but inactive analogues.

Further modifications of the timoprazole skeleton, such as using indoles instead of benzimidazoles or sulfones in place of sulfoxides, resulted surprisingly in solution-stable analogues with no activity. In another approach, the attachment of a trifluoromethyl group to the benzimidazole moiety of timoprazole led to a series of very active compounds with varying solution-stability (Fig. 3.7). The most interesting compound from this series was compound BY319, which later became the first project compound. Its 3-methyl pyridine-substituted congener was even more potent, but its solution instability prevented its further use. Surprisingly, its corresponding sulfide, BY308, turned out to be very potent in the rat models, presumably by metabolic oxidation to the corresponding sulfoxide form. The sulfides were generally stable but highly lipophilic. A patent application was filed in November 1981 which encompassed both compounds [28]. By the end of October 1982, it was apparent that BY308 and BY319 were suitable candidates for further study, although these were the only compounds to show promise among some 160 that had been synthesized.

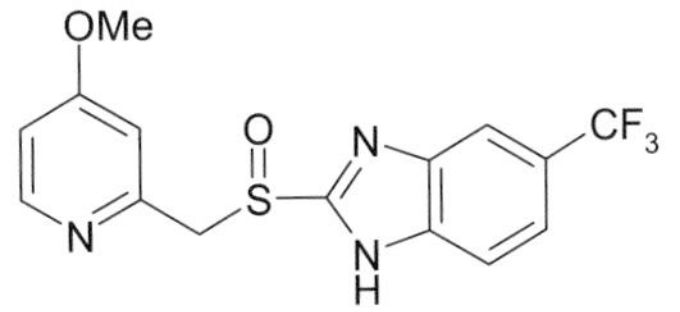

BY319

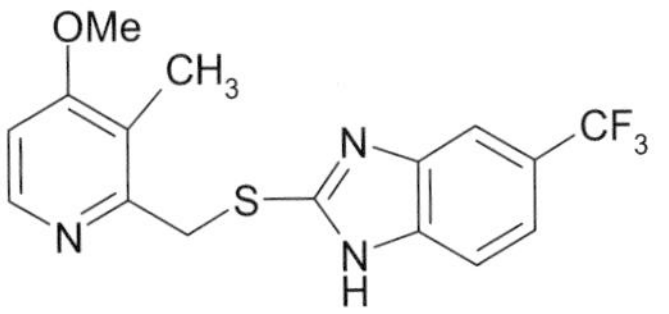

BY308

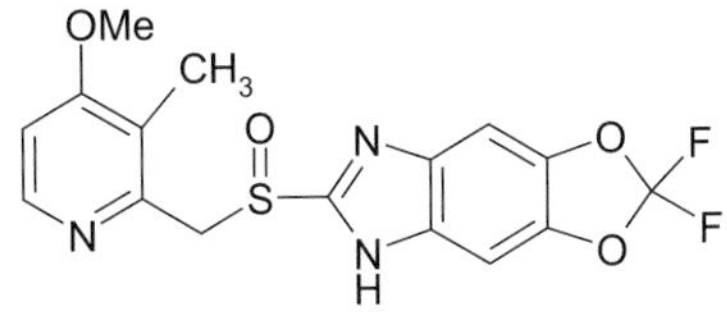

BY178

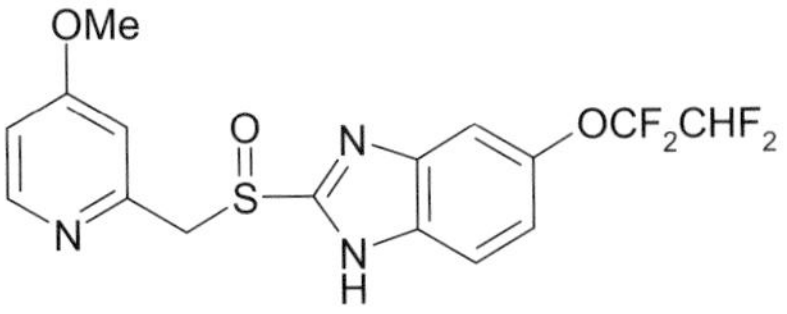

BY156

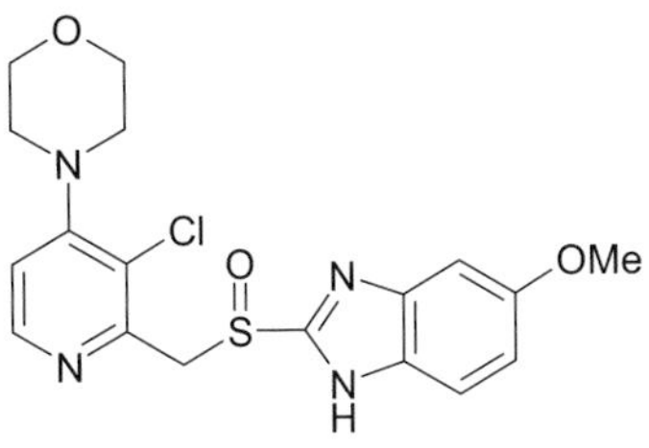

SK&F 95601

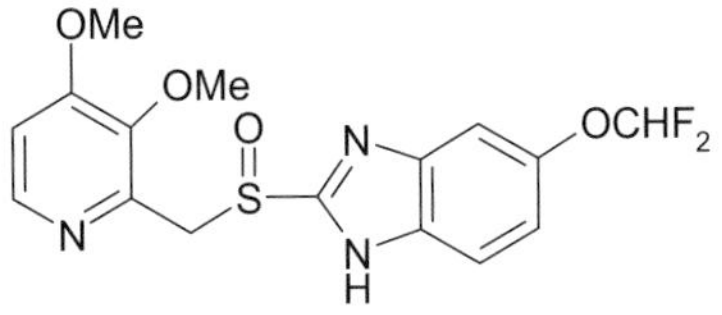

Pantoprazole
(BY1029/SK&F96022)

Fig. 3.7 Byk Gulden's PPI project compounds.

Toxicity studies on BY308 and BY319 began in August 1983 in both rats and dogs (as well as studies relating to pharmaceutics, pharmacology and pharmacokinetics) with a view to fulfilling the regulatory requirements to conduct clinical trials with these compounds. It was proposed that BY319 would be tested in humans from September 1984. There were concerns, however, with regard to BY319 adversely affecting the thyroid gland (enlargement and atrophy), and unfortunately the compound did not reach the clinical trials stage.

3.3.2
Impact of the First PPI Project Compounds

Although BY319 was far from being optimized, its research had a far-reaching impact on the further course of Byk Gulden's investigations. It was, for example, very helpful for the pantoprazole project, in that the chemists were able to overcome the instability problems of BY319 by making alkaline salts and cyclodextrine inclusion complexes of the molecule.

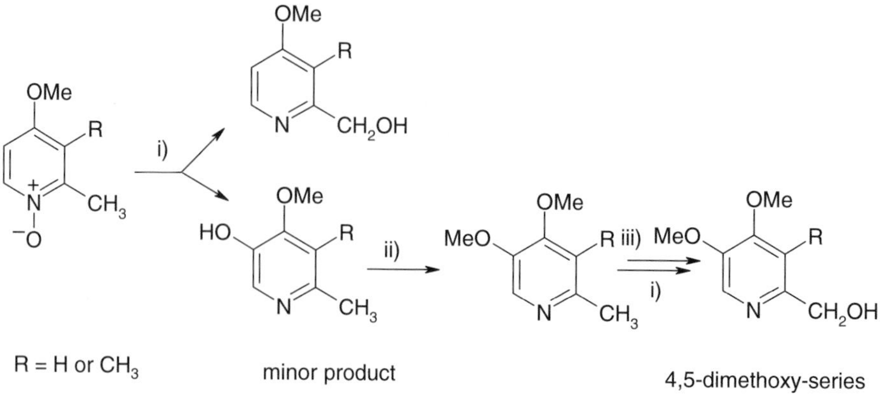

R = H or CH$_3$ minor product 4,5-dimethoxy-series

i) Ac$_2$O, ii) dimethylsulphate , iii) H$_2$O$_2$

Fig. 3.8 Initial synthesis of 4,5-dimethoxy-2-hydroxymethylpyridines.

The instability of BY319 also led pharmacists to develop a form of the compound for administration to man by using an enteric coating. In addition, the problems associated with the final step of the synthesis – oxidation to the sulfoxide – were overcome by working under alkaline conditions and using inexpensive and readily available hypochlorite solution. Thus, the scale-up of this oxidation step was no longer a problem issue. Most importantly, during the scale-up of BY308 and BY319, byproducts in the pyridine-N-oxide rearrangement steps of each precursor could be isolated, and this paved the way to the invention of pantoprazole sodium (Fig. 3.8). Finally, it was of great importance that both project compounds drew the attention of Smith Kline & French to Byk Gulden's proton pump studies.

In mid-1984, Byk Gulden entered into collaboration with the UK pharmaceutical company, Smith Kline & French. The latter company's main anti-ulcer product, the H$_2$-blocker cimetidine, was approaching the end of its patent life and the company was seeking a new product, specifically a PPI. Their initial interest was BY308, and they wanted to know the activity of the compound, both by itself and in comparison with BY319 and omeprazole, which had been regarded as the state of the art since December 1982 [29]. Smith Kline & French's interest and involvement was of considerable encouragement to Byk Gulden, and their expertise in gastrointestinal research was of great assistance.

BY308 finally went into clinical trial, but its anti-secretory activity was regarded as noncompetitive. However, Byk Gulden continued to seek out other possible compounds. Fluoro substituents in the benzimidazole part were investigated in more depth. In general, fluoro substituents block metabolism at the point were they are attached. The more balanced fluoroalkoxy substituent instead of the highly lipophilic and strongly electron-withdrawing trifluoromethyl substituent in the benzimidazole moiety, as used in BY319 and BY308, led to highly active compounds with supposed longer half-lives and higher solution-stability. These studies, in turn, led to the candidate compounds BY156 and BY178 (see Fig. 3.7), both of which were first submitted for testing in animals in April 1983 [30].

In early 1986, the decision was taken to postpone further development of BY178 and to concentrate on BY156, because this compound had several advantages over BY178. For example, it had greater solubility and bioavailability of its tablet formulation, as well as relatively high pharmacological potency and lack of toxicity in subchronic toxicology studies. Possible back-up compounds, which were chemically different from BY156 and BY178, were also identified. However, it was ultimately decided not to proceed with BY156, largely because of toxic effects found in dogs, the most significant problem being that several dogs developed necrotizing vasculitis [31].

3.4
Elucidation of the Mechanism of Action of PPIs

3.4.1
A Surprising Interrelationship Between Stability and Activity

During the initial SAR studies in which 25 new molecular skeletons of timoprazole-like compounds were synthesized, the chemists were puzzled by the findings that there was an inherent solution-stability activity relationship; the most active compounds were the least stable in solution. Even very small changes in the timoprazole backbone resulted in major decreases in anti-secretory activity, and this was inversely proportional to an increased solution stability. For example, as mentioned above, the 3-position-linked pyridine derivatives (see Fig. 3.6) were completely inactive, whereas its 2-position-linked equivalents were active, but unstable. Other major changes such as the use of an indole instead of a benzimidazole moiety in the timoprazole backbone, and the use of a sulfonyl instead of a sulfinyl group in the bridge linking the benzimidazolyl and the 2-pyridinyl functionalities, led to stable but inactive compounds. It was then realized that activity was somehow linked to instability in solution, and thus a program was initiated to gain insight into stability–structure relationships, and to elucidate the underlying chemistry of the decomposition reaction.

3.4.2
Isolation and Identification of the Active Principle of the PPIs

The most salient products in the decomposition reaction were red-colored products of tetracyclic structure, which showed some inhibitory activity *in vitro* and were suggested by two other leading groups as the active species exerting acid inhibition. In biochemical experiments it was found that pretreatment of methanolic solutions of timoprazole-like compounds with acids enhanced their inhibitory activity *in vitro*. Surprisingly, on closer examination of these acidic methanolic pretreatment mixtures at preparative scale and at concentrations used in prepara-

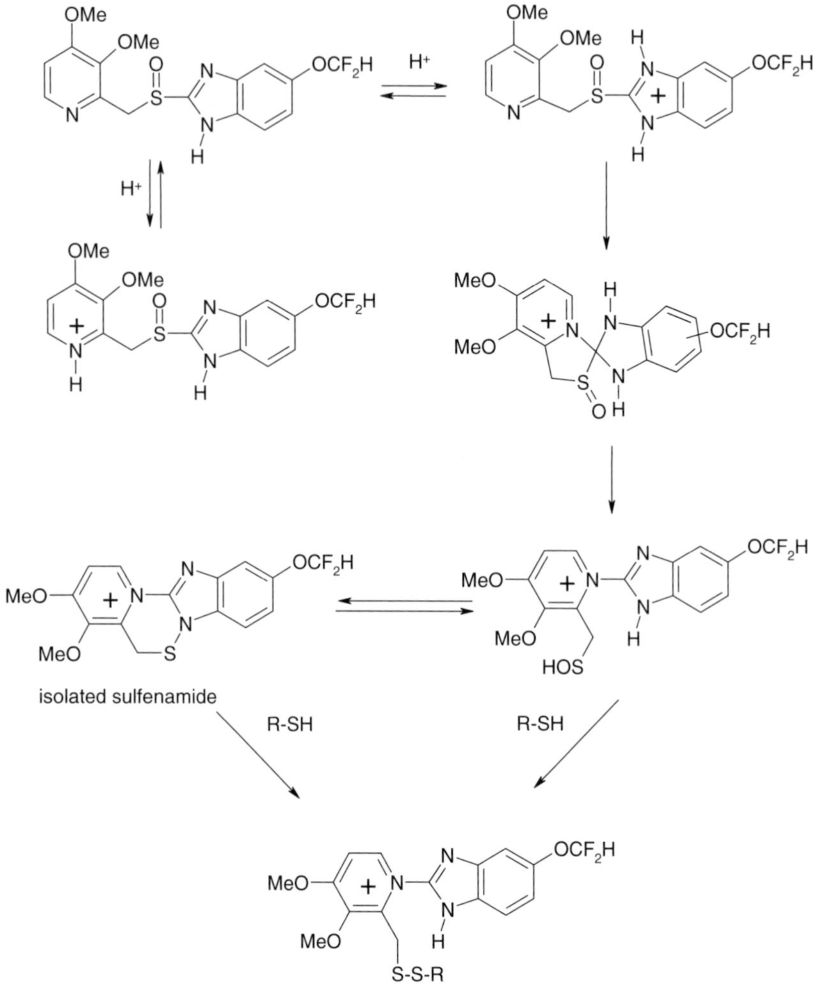

Fig. 3.9 The acid-catalyzed rearrangement of the substituted benzimidazole pantoprazole (prodrug) to the cyclic sulfena-mide (active principle) and its reaction with thiols.

tive organic chemistry, precipitations of in solution very instable, but in solid state very stable, salts were observed and isolated (Fig. 3.9). The unambiguous structural assignment to the novel cyclic sulfenamide was supported by X-ray data [32].

The elusive nature of the cyclic sulfenamide is exemplified by its instability in aqueous weakly acidic or neutral solution. However, under conditions which resemble those of the parietal cell (e.g., 0.1 M HCl), the cyclic sulfenamide was fairly stable if, in addition, the concentrations were in the micromolar range. With isolated cyclic sulfenamides in hand, the study of the reaction of this species with thiols (e.g., mercaptoethanol) in different solvents and in different acidic solutions could easily be performed, and compared with similar reactions starting with the original PPIs with stopped flow techniques [33]. In these experiments, UV-detection techniques could be used since the coplanar arrangements in the cyclic sulfenamides give rise to bathochromic UV-absorbance shifts. The isolation and characterization of identical disulfides both, from *in-situ*-reacted PPIs as well as starting from isolated sulfenamides, led to the conclusion that the cyclic sulfenamides were the active principle of the PPIs. In 1984, investigations conducted by the Hässle group on the mode of action of the PPIs revealed evidence that, *in vitro*, the PPIs are activated and transformed in acidic media into an intermediate which exerts its inhibitory activity by reaction with one or more essential SH-groups on the gastric acid pump. Later, it was found *in vitro* that, using radiolabeled PPIs and thiol reagents such as glutathione, the inhibition of the gastric acid pump is paralleled by the covalently incorporated labeled inhibitor.

Despite the complicated chemistry involved and the erroneous results published by two prominent groups, the Byk Gulden research group was confident that the correct mechanism of action had been identified [34]. The results of these studies were first presented in September 1985 at the Third SCI-RSC Medicinal Chemistry Symposium in Cambridge, at which Hässle's group also published their version of the mechanism of action. This was essentially identical to Byk Gulden's results [35]. Later, other groups also published similar results with their compounds.

The complete elucidation of the mechanism of action of the PPIs and detailed studies of the reaction pathway of the involved far-reaching rearrangement provided partial answers to the previously described interrelationship of acid instability (or chemical reactivity) and anti-secretory activity. As a consequence, it was finally understood that seemingly small alterations in the backbone of timoprazole led nowhere, and focus had to be centered on the substituents on the backbone. It also became clear that a certain inherent instability (or reactivity) at low pH was a prerequisite for activity, and that the optimal compounds would be those that were stable at neutral pH but were quickly activated at low pH. In this respect, a crucial aspect of the mechanism of action is that, before interacting with the active site in the body, the substituted benzimidazole undergoes an intramolecular rearrangement. However, such a rearrangement, in being intramolecular, poses severe geometric constraints.

Therefore, the seemingly small or simple changes to the backbone or the substituents on that backbone, such as the introduction of a sterically demanding 6-methyl group in the pyridine moiety (see Fig. 3.6), can have a dramatic effect on the ease of rearrangement and the conditions under which this could occur. However, even with detailed knowledge of the mechanism involved in the chemical reaction cascade, a clear-cut design of active inhibitors was still not possible because in the complex multi-step chemistry the influence of a substituent on each step in the cascade could be different, and therefore not predictable for the overall rate of the prerequisite acid activation.

Furthermore, the decisive electrophilicity/nucleophilicity (basicity) balance of the involved 2-benzimidazolyl-center and the pyridine nitrogen, respectively, cannot be predicted. The basicity of the latter can only be used as a surrogate parameter. Therefore, Byk Gulden chemists had to scrutinize all available experimental data for the already synthesized timoprazole-like analogues, such as the activity, selectivity, stability in solution, solubility, and pharmacokinetics. Another aspect of the mechanism of action was the loss of the stereochemical information of the sulfinyl-group during the reaction cascade because the thiophilic cyclic sulfenamides are devoid of chirality [36].

3.5
Identification of Pantoprazole as a Candidate for Development

3.5.1
Optimizing the Pyridine Moiety and the First Synthesis of Pantoprazole

During the scale-up process of the first project candidate compound BY319, the 5-hydroxy-4-methoxypicoline (see Fig. 3.8) was isolated as a by-product in the rearrangement reaction of the corresponding pyridine-N-oxide, along with the required 2-hydroxymethyl-4-methoxy-pyridine. This by-product was O-methylated and further processed in the usual mode and finally linked to substituted 2-mercaptobenzimidazoles, especially to those substituted with the metabolically inert fluoroalkoxy groups to produce, after sulfoxidation, the first series of 4,5-dimethoxypyridinyl-substituted inhibitors (Fig. 3.10).

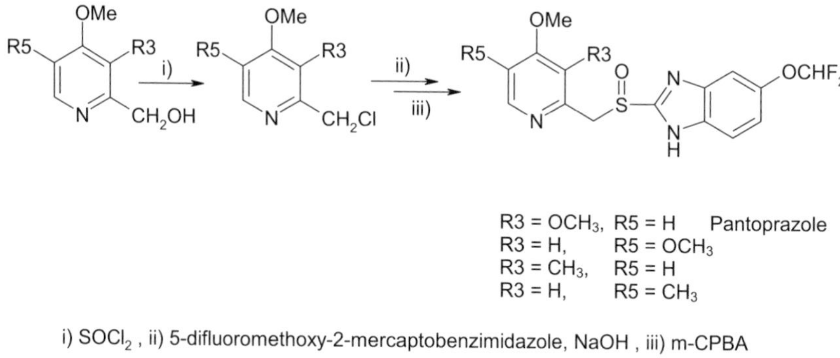

R3 = OCH$_3$, R5 = H   Pantoprazole
R3 = H,   R5 = OCH$_3$
R3 = CH$_3$,   R5 = H
R3 = H,   R5 = CH$_3$

i) SOCl$_2$, ii) 5-difluoromethoxy-2-mercaptobenzimidazole, NaOH , iii) m-CPBA

Fig. 3.10 Synthesis of pantoprazole and congeners.

The compounds showed mediocre activity but very high stability in neutral solution [12]. Despite these unsatisfactory results, efforts were made to prepare the corresponding 3,4-dimethoxy-isomers. The synthesis of the required precursor pyridines is outlined in Figs. 3.11 and 3.12.

The basicity of both series of compounds was almost identical but, surprisingly, these congeners displayed high anti-secretory activity along with good neutral solution stability and good selectivity against the sodium pump. The most interesting compound in the latter series was pantoprazole, which was prepared in April

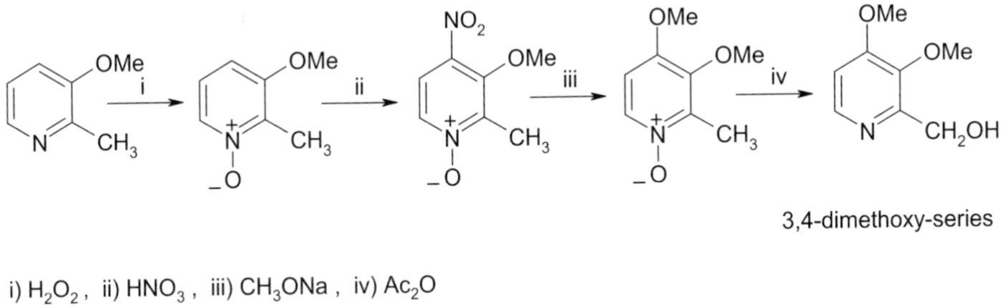

i) H$_2$O$_2$, ii) HNO$_3$, iii) CH$_3$ONa, iv) Ac$_2$O

Fig. 3.11 Initial synthesis of 3,4-dimethoxy-2-hydroxymethyl-pyridine as a building block for pantoprazole.

i) POCl$_3$, ii) H$_2$O$_2$, iii) CH$_3$ONa, iv) Ac$_2$O

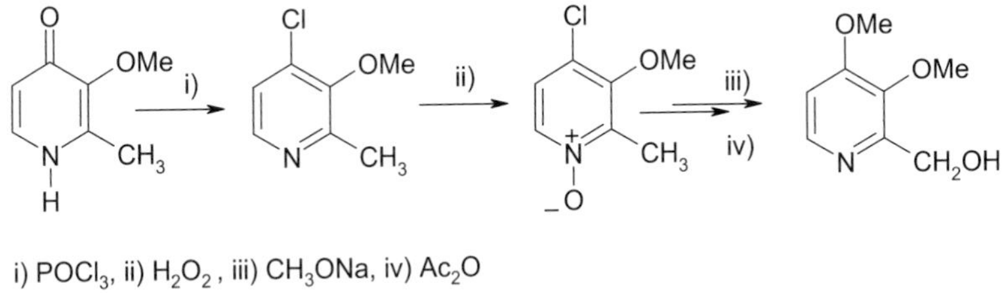

Fig. 3.12 Current synthesis of 3,4-dimethoxy-2-hydroxymethyl-pyridine as a building block for pantoprazole.

1985 by a small group of scale-up chemists not directly involved in the medicinal chemistry program. They prepared a series of compounds with dimethoxy-substituted pyridines, and these were filed in a patent application in June 1984 [37a]. These valuable compounds drew full attention when the previous four project compounds failed due to unwanted toxicity findings in animals high-dose toxicity studies; the unwanted toxicity findings were believed to be associated with the solution instability and insufficient selectivity.

3.5.2
Selection Criteria

As experience was gained with PPIs, it became clear that activity, though important, was not the only criterion for the selection of a development compound. The correct level of stability was also important. In order that the drug was converted to the active principle in the parietal cell, it was important that any compound was inherently unstable at low pH. However, if it was too instable (i.e., unstable at neutral pH), it could be converted to the active principle in other areas of the body and potentially cause unwanted side effects [13]. Furthermore, elucidation of the mechanism of action showed that there were other key criteria. These included good selectivity, namely inhibiting only the proton pump. Also of importance were solubility, lipophilicity, bioavailability, interaction with other drugs and the effect on the liver enzymes. In this respect, Smith Kline & French greatly assisted in determining these criteria.

From late 1985, the aim was to identify a compound with good stability at neutral pH, sustaining this higher level of stability down to pH 5 but being rapidly activateable at lower pHs, combined with a high level of (H^+/K^+)-ATPase inhibition. It was believed that by this activation profile, interactions with weakly acidic lysosomes (particles found in most cells containing digestive enzymes) could be avoided and that the depletion of protective glutathione in other cells could be circumvented [12].

Since the beginning of Byk Gulden's work in this area, the stability characteristics of numerous compounds (expressed as the half-life of decomposition at pH 7.4, 5, and 3) had been determined. Once these criteria were established, and having elucidated the mechanism of action, the researchers were able to identify potential candidate compounds, which had already been synthesized fulfilling these criteria. The most promising of these candidates was pantoprazole and its salt, pantoprazole sodium.

3.5.3
The Selection of Pantoprazole and Internal Competition with SK&F95601

A toxicity screening in late 1985 compared six compounds, including BY 319 and pantoprazole. The results for pantoprazole were very promising. In a four-week toxicity study in early 1986, pantoprazole displayed the most promising data from the series of dimethoxy pyridine compounds. Based on the various animal models, pantoprazole was found to be approximately equipotent to omeprazole [37b], much more stable at neutral and weakly acidic pH, and displayed higher selectivity in enzyme models (pantoprazole is more selective to the H^+/K^+- as opposed to the Na^+/K^+- ATPase than omeprazole). Pantoprazole was not the only possible candidate compound at this time. Smith Kline & French also had a potential candidate compound, SK&F95601 (see Fig. 3.7), which was a 4-amino-3-chloropyridine compound [38].

The development work on SK&F95601 was about six months behind that of pantoprazole. Pantoprazole had higher hydrophilicity and higher solubility. Smith Kline & French wanted the compound also to be administered parenterally, and its solubility was therefore very important. Ultimately, the key factor in Smith Kline & French's recognition that pantoprazole was a more suitable compound for the treatment was that pantoprazole had less effect on cytochrome P450 and less potential for interaction with other drugs [39–41]. Pantoprazole was given full support of both companies and went into a fully fledged development program with the highest priorities.

In 1986, pantoprazole sodium sesquihydrate was synthesized and from 1987 onwards the development was switched to the sodium salt (BY1023). This was more soluble and more stable, and had better compatibility with other excipients used in the drug formulation. Pantoprazole was identified after nearly seven years of research and registered for clinical use after a further seven years of development. Finally, it reached its first market in 1994 in Germany. During the course of the studies on pantoprazole, more than 650 PPIs had been synthesized and evaluated.

3.5.4
Toxicological Problems: Project Development at Risk

At the start of the development of pantoprazole, serious toxicity findings in dogs after the high-dose application for four weeks had to be overcome, and this halted the development for several weeks. Several hypotheses were developed to explain these findings which, fortunately, were not seen in other species. At this point, the toxicological expertise of Smith Kline & French was of critical importance to the fate of the project. In their pioneering development of H_2-blockers, Smith Kline & French had seen a similar pattern of toxicity in dogs treated with the early development candidate metiamide (Fig. 3.13), which had a suspect thioamide group [42].

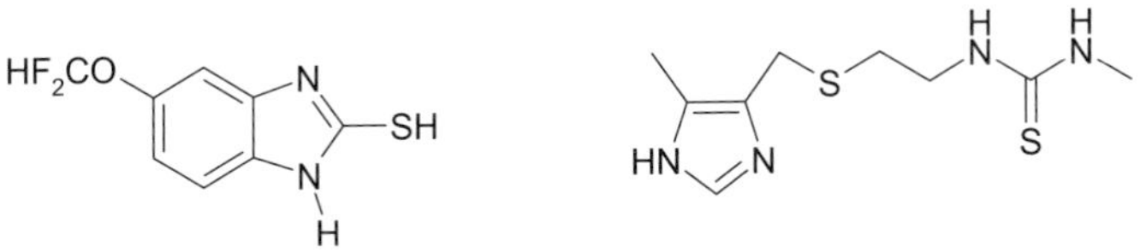

5-difluoromethox-2-mercaptobezimidazole   metiamide (SK&F)

Fig. 3.13 Thioamides.

The similarity of both unwanted findings in dogs led Smith Kline & French to hypothesize that the disguised thioamide group in pantoprazole was uncovered by metabolism to the corresponding 2-mercapto-benzimidazole derivative of pantoprazole (Fig. 3.13). Indeed, this metabolite was formed in dogs in large quantities in a dose-related manner, and its serum concentrations correlated clearly with the degree of toxic effects. A speedy one-week confirmation study with the separately synthesized metabolite was initiated. A threshold blood AUC-level of 700 h μg/mL was observed for the toxicity when increasing doses were used. Fortunately, the formation of this metabolite was species-specific, and it was never observed in humans. Even with sophisticated analytical methods, the detection level was never reached, even when higher doses were applied at a later date. Following this period of serious problems, when the development of pantoprazole was at risk, the compound successfully completed all further required high-dose animal studies and was submitted for registration in 1993.

It was always of great interest whether the very high selection criteria described above – especially the favorable low potential for interaction with other drugs – would translate into human use. Indeed, this was confirmed later in extended drug interaction studies. Pantoprazole's good solubility and very high stability in solution allowed it to become the first marketed PPI for intravenous use in critical care patients [39,40,43,44].

3.5.5
Benefits of Pantoprazole for the Patient

Pantoprazole-sodium inhibits gastric acid secretion by selectively binding to the gastric enzyme (H^+/K^+)-ATPase (gastric proton pump). Pantoprazole-sodium has a clear clinical application in the treatment of gastric acid-related diseases, including peptic ulcers, gastroesophageal reflux disease (GERD), Zollinger–Ellison syndrome, as a prophylactic drug against peptic mucosa damage in high-risk patients requiring regular use of NSAIDs, and in combination with antibiotics for the eradication of *Helicobacter pylori* infection in patients with peptic ulcers [39,45].

In humans, pantoprazole-sodium has linear, highly predictable pharmacokinetics, and minimal potential for metabolic drug–drug interactions [40]. Pantoprazole-sodium is available as gastroresistant tablets containing 20 mg or 40 mg of the active moiety pantoprazole, and as lyophilized powder containing 40 mg pantoprazole for intravenous use. The 40 mg oral and intravenous formulations have equivalent potency in raising the intragastric pH; this allows easy change from one administration route to the other, as required by the clinical condition of the patient.

Short-term as well as long-term studies with pantoprazole-sodium have proven its excellent efficacy in healing acid-related diseases, preventing relapse, and providing rapid relief from disease-related symptoms. Direct comparison with other anti-secretory drugs showed that pantoprazole-sodium is significantly more effective than H_2-receptor blockers and is either equivalent to or better than other clinically used PPIs [39].

Pantoprazole-sodium has a lower variability in pharmacokinetics compared with omeprazole, particularly with respect to bioavailability. The pharmacokinetics of pantoprazole-sodium are almost the same in patients with gastrointestinal diseases and those with renal failure, and in the elderly, so that no dose adjustment is required. In addition, pantoprazole-sodium has a low potential for drug–drug interaction, which is a considerable benefit because many patients who require pantoprazole are elderly and are receiving co-medications [13].

Since its introduction for therapeutic use in 1994, safety data on pantoprazole-sodium are based on clinical experience and safety documentation from about 65 000 patients who have received the drug in controlled clinical studies, about 629 000 therapy courses documented in observational studies, and an estimated number of 300 million patients treated in daily practice. Information derived from these studies has firmly established the therapeutic efficacy of pantoprazole-sodium and its excellent safety and tolerability profile [39,45].

3.5.6
Summary

Over the past three decades, a number of major advances have been made in the development of acid-suppressive agents, with PPIs now being considered as the treatment of choice in most countries. The PPIs are targeted to the gastric $(H^+/$

K^+)-ATPase and the gastric acid pump. Because this is the final common step of acid secretion, this class of drug showed superior acid suppression as compared with the H_2-receptor antagonists (H_2-RAs) and anti-muscarinic drugs. Especially in GERD, symptom severity and esophageal mucosal damage are correlated with the degree of acid exposure, both in terms of the absolute intragastric pH level and the proportion of time during which the intragastric pH is maintained above a certain level. Therefore, the PPIs have become the standard of therapy in this disease.

There are currently four racemic PPIs available on the market: omeprazole, lansoprazole, pantoprazole, and rabeprazole. (More recently, enantiomerically pure versions have also been studied and developed, e.g., S-omeprazole, marketed by AstraZeneca as esomeprazole; see Chapter II-2.) Proton pump inhibitors share the same core structure, the substituted pyridylmethyl-sulfinyl-benzimidazole, but differ in terms of substituents on this core structure. The absolute requirements of the core structure for the activity of PPIs was not understood until it became clear that the active PPIs are derived from inactive prodrugs; the prodrugs are transformed, in the acid-secreting parietal cells, by a unique cascade of chemical structural transformations leading to the active principle, a cyclic sulfenamide species. Inhibition of acid secretion in turn is then achieved by formation of covalent disulfide bonds with key cysteines of the (H^+/K^+)-ATPase.

3.6
Outlook on Further Developments

Despite the efficacy of the PPIs, there is still potential for clinical improvement in GERD pharmacotherapy [46]. A faster onset of complete acid inhibition, a lower inter-patient variability in the inhibition of acid secretion, and an improved duration of efficacy are potential areas for improvement [47]. Concomitant antacid use among PPI users (37%) suggests that current therapies may not always achieve sufficient acid control when used as monotherapy for GERD [48].

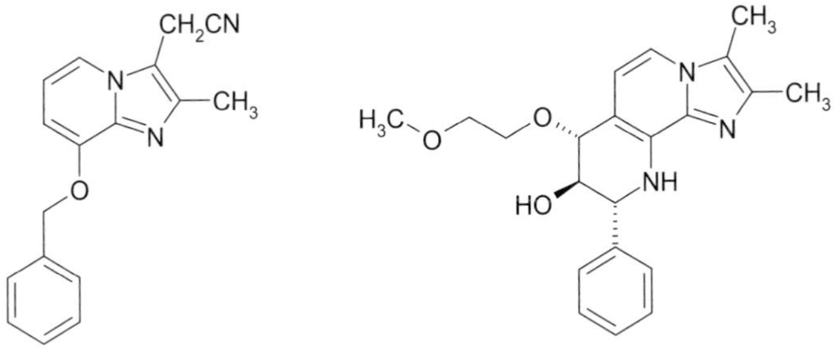

SCH28080
(Schering Co)

Soraprazan
(ALTANA Pharma)

Fig. 3.14 Potassium-competitive acid blockers.

During the early 1980s, research on the imidazopyridine compound SCH28080 (Fig. 3.14) signaled the development of a new class of (H^+/K^+)-ATPase inhibitors, which block the (H^+/K^+)-ATPase independently of acid secretion and do not require conversion to an active drug but are active in their own. The blockade of acid secretion is achieved via reversible potassium-competitive inhibition of the (H^+/K^+)-ATPase [49]. However, development of the lead compound SCH28080 was dropped because hepatotoxicity precluded its clinical use. Several analogues with improved bioavailability and better toxicological profiles have been studied at ALTANA Pharma, and Soraprazan (Fig. 3.14) is currently being investigated in Phase II trials. Soraprazan [50] produces a more rapid and more profound inhibition of acid secretion in man than the currently used PPIs, and thus is potentially useful in accelerating symptom relief as well as being suitable for on-demand therapy. Several companies are developing such potassium-competitive acid blockers, and studies are ongoing to further characterize these newer compounds. It remains to be seen whether they can achieve the desired improvements in clinical efficacy, duration of acid control, and variability.

References

1 Modlin IM, Sachs G. *Acid related diseases: Biology and treatment.* 2nd edn. Lippincot, Williams & Wilkins, Philadelphia, **2004**.

2 Black JW, Duncan WA, Durant CJ, Ganellin CR, Parsons EM. Definition and antagonism of histamine H_2-receptors. *Nature*, **1972**; 236, 385–390.

3 Cooper DG, Young RC, Durant GJ, Ganellin CR. *Comprehensive Medicinal Chemistry.* Pergamon Press, Oxford, **1990**, pp. 323–421.

4 Colin-Jones DG. The role and limitations of H_2-receptor antagonists in the treatment of gastro-oesophageal reflux disease. *Aliment. Pharmacol. Ther.*, **1995**, 9(Suppl. 1), 9–14.

5 Forte JG, Lee HC. Gastric adenosine triphosphatases: a review of their possible role in HCl secretion. *Gastroenterology*, **1977**, 73(4 Pt. 2), 921–926.

6 Sachs G, Chang H, Rabon E, Shackman R, Sarau HM, Saccomani G. Metabolic and membrane aspects of gastric H^+ transport. *Gastroenterology*, **1977**, 73(4 Pt. 2), 931–940.

7 Sjostrand SE, Ryberg B, Olbe L. Analysis of the actions of cimetidine and

metiamide on gastric acid secretion in the isolated guinea pig gastric mucosa. *Naunyn Schmiedebergs Arch. Pharmacol.*, **1977**, 296(2), 139–142.

8 Saccomani G. *Acta Physiol. Scand. Suppl.*, **1978**, 293–305.

9 Figala V, Klemm K, Kohl B, Krüger U, Rainer G, Schaefer H, et al. Acid activation of (H^+-K^+)-ATPase inhibiting 2-(2-pyridylmethyl-sulphinyl) benzimidazoles: Isolation and characterization of the thiophilic 'active principle' and its reactions. *J. Chem. Soc., Chem. Commun.*, **1986**, 125–127.

10 Lindberg P, Nordberg P, Alminger T, Brandstrom A, Wallmark B. The mechanism of action of the gastric acid secretion inhibitor omeprazole. *J. Med. Chem.*, **1986**, 29(8), 1327–1329.

11 Berglindh T, Obrink KJ. A method for preparing isolated glands from the rabbit gastric mucosa. *Acta Physiol. Scand.*, **1976**, 96(2), 150–159.

12 Kohl B, Sturm E, Senn-Bilfinger J, Simon WA, Krüger U, Schaefer H et al. (H^+-K^+)-ATPase inhibiting 2-[(2-pyridylmethyl)sulfinyl]benzimidazoles. 4. A novel series of dimethoxypyridyl-

substituted inhibitors with enhanced selectivity. The selection of pantoprazole as a clinical candidate. *J. Med. Chem.*, **1992**, 35(6), 1049–1057.

13 Huber R, Kohl B, Sachs G, Senn-Bilfinger J, Simon WA, Sturm E. Review article: the continuing development of proton pump inhibitors with particular reference to pantoprazole. *Aliment. Pharmacol. Ther.*, **1995**, 9(4), 363–378.

14 Kurtz W. [A layered lattice antacid with long-term effectiveness. Profile of action and safety exemplified by magaldrate]. *Fortschr. Med.*, **1993**, 111(6), 93–96.

15 Magaldrate and almasilate – complex buffering antacids. *Drug Ther. Bull.*, **1985**, 23(10), 39–40.

16 Riedel R, Bohnenkamp W, Eltze M, Heintze K, Prinz W, Kromer W. Comparison of the gastric antisecretory and antiulcer potencies of telenzepine, pirenzepine, ranitidine and cimetidine in the rat. *Digestion*, **1988**, 40(1), 25–32.

17 Bell NJ, Burget D, Howden CW, Wilkinson J, Hunt RH. Appropriate acid suppression for the management of gastro-oesophageal reflux disease. *Digestion*, **1992**, 51 (Suppl. 1), 59–67.

18 Bertaccini G, Coruzzi G. Control of gastric acid secretion by histamine H_2 receptor antagonists and anticholinergics. *Pharmacol. Res.*, **1989**, 21(4), 339–352.

19 Feldman M. Inhibition of gastric acid secretion by selective and nonselective anticholinergics. *Gastroenterology*, **1984**, 86(2), 361–366.

20 Sjostrand SE, Olbe L, Fellenius E. The discovery and development of the proton pump inhibitor. In: Olbe L (Ed.). *Proton Pump Inhibitors*. Birkhäuser, Basel, **1999**.

21 Sundell G, Sjostrand SE, Olbe L. Gastric antisecretory effects of H83/69, a benzimidazolyl-pyridyl-methyl-sulfoxide. *Acta Pharmacol. Toxicol.*, **1977**, Suppl. 4, 77.

22 Kilian U, Beume R, Eltze M, Häfner D, Hanauer G, Schudt C. Telenzepine and its enantiomers, M1-selective antimuscarinics, in guinea pig lung function tests. *Agents Actions Suppl.*, **1991**, 34, 131–147.

23 Stockbrügger RW. Antimuscarinic drugs. *Methods Fund. Exp. Clin. Pharmacol.*, **1989**, 11 (Suppl. 1), 79–86.

24 Olbe L, Haglund U, Leth R, Lind T, Cederberg C, Ekenved G, et al. Effects of substituted benzimidazole (H 149/94) on gastric acid secretion in humans. *Gastroenterology*, **1982**, 83(1 Pt. 2), 193–198.

25 Wallmark B, Sachs G, Mardh S, Fellenius E. Inhibition of gastric (H^+/K^+)-ATPase by the substituted benzimidazole, picoprazole. *Biochim. Biophys. Acta*, **1983**, 728(1), 31–38.

26 Fellenius E, Berglindh T, Sachs G, Olbe L, Elander B, Sjostrand SE, et al. Substituted benzimidazoles inhibit gastric acid secretion by blocking (H^+/K^+)-ATPase. *Nature*, **1981**, 290, 159–161.

27 Junggren UK, Sjöstrand SE, Eur Pat 005129, **1979**.

28 Senn-Bilfinger J, Schaefer H, Figala V, Klemm K, Rainer G, Riedel R. European Patent 080602, **1981**.

29 Larsson H, Carlsson E, Junggren U, Olbe L, Sjostrand SE, Skanberg I, et al. Inhibition of gastric acid secretion by omeprazole in the dog and rat. *Gastroenterology*, **1983**, 85(4), 900–907.

30 Rainer G, Riedel R, Senn-Bilfinger J, Klemm K, Schaefer H, Figala V. European Patent 134400, **1984**.

31 Olbe L, Carlsson E, Lindberg P. A proton-pump inhibitor expedition: the case histories of omeprazole and esomeprazole. *Nat. Rev. Drug Discov.*, **2003**, 2(2), 132–139.

32 Senn-Bilfinger J, Krüger U, Sturm E, Figala V, Klemm K, Kohl B et al. (H^+-K^+)-ATPase inhibiting 2-[(2-pyridyl-methyl)sulfinyl]benzimidazoles. 2. The reaction cascade induced by treatment with acids. Formation of 5H-pyriodo [1′,2′:4,5] [1,2,4] thiadiazino benzimidazole-13-ium salts and their reactions with thiols. *J. Org. Chem.*, **1987**, 52(20), 4582–4592.

33 Sturm E, Krüger U, Senn-Bilfinger J, Figala V, Klemm K, Kohl B, et al. (H^+-K^+)-ATPase inhibiting 2-[(2-pyridyl-methyl)sulfinyl]benzimidazoles. 1. Their reaction with thiols under acidic conditions. Disulfide containing 2-pyridinio-benzimidazolides as mimics for the

inhibited enzyme. *J. Org. Chem.*, **1987**, 52(20), 4573–4581.

34 Krüger U, Senn-Bilfinger J, Sturm E, Figala V, Klemm K, Kohl B, et al. (H⁺-K⁺)-ATPase inhibiting 2-[(2-pyridyl-methyl)sulfinyl]benzimidazoles. 3. Evidence for the involvement of a sulfenic acid in their reactions. *J. Org. Chem.*, **1990**, 55(13), 4163–4168.

35 Brandstrom A, Lindberg P, Bergman N-A, Alminger T, Ankner K, Jungren U. Chemical reactions of omeprazole and omeprazole analogues. 1. A survey of the chemical transformations of omeprazole and its analogues. *Acta Chim. Scand.*, **1989**, 43, 536–548.

36 Lindberg P, Keeling D, Fryklund J, Andersson T, Lundborg P, Carlsson E. Review article: Esomeprazole – enhanced bio-availability, specificity for the proton pump and inhibition of acid secretion. *Aliment. Pharmacol. Ther.*, **2003**, 17(4), 481–488.

37a Kohl B, Klemm K, Riedel R, Rainer G, Schaefer H, Senn-Bilfinger J. Eur Pat 166287, **1985**.

37b Kromer W, Postius S, Riedel R, Simon WA, Hanauer G, Brand U. et al. BY 1023/SK&F 96022 INN pantoprazole, a novel gastric proton pump inhibitor, potently inhibits acid secretion but lacks relevant cytochrome P450 interactions. *J. Pharmacol. Exp. Ther.*, **1990**, 254(1), 129–135.

38 Ife RJ, Dyke CA, Keeling DJ, Meenan E, Meeson ML, Parsons ME, et al. 2-[[(4-Amino-2-pyridyl)methyl]sulfinyl]-benzimidazole H⁺/K⁺-ATPase inhibitors. The relationship between pyridine basicity, stability, and activity. *J. Med. Chem.*, **1989**, 32(8), 1970–1977.

39 Cheer SM, Prakash A, Faulds D, Lamb HM. Pantoprazole: an update of its pharmacological properties and therapeutic use in the management of acid-related disorders. *Drugs*, **2003**, 63(1), 101–132.

40 Steinijans VW, Huber R, Hartmann M, Zech K, Bliesath H, Wurst W, et al. Lack of pantoprazole drug interactions in man: an updated review. *Int. J. Clin. Pharmacol. Ther.*, **1996**, 34(Suppl. 1), S31–S50.

41 Kliem V, Bahlmann J, Hartmann M, Huber R, Lühmann R, Wurst W. Pharmacokinetics of pantoprazole in patients with end-stage renal failure. *Nephrol. Dial. Transplant.*, **1998**, 13, 1189–1193.

42 Malen CE, Danree BH. New thiocarboxamide derivatives with specific gastric antisecretory properties. *J. Med. Chem.*, **1971**, 14(3), 244–246.

43 Wurzer H, Hofbauer R, Worm HC, Frass M, Kaye K, Kaye AD. Intravenous administration of pantoprazole. An Austrian multicenter study. *Clin. Drug Invest.*, **2002**, 22(8), 1173–2563.

44 Pisegna JR. Switching between intravenous and oral pantoprazole. *J. Clin. Gastroenterol.*, **2001**, 32(1), 27–32.

45 Bardhan KD. Pantoprazole: a new proton pump inhibitor in the management of upper gastrointestinal disease. *Drugs Today*, **1999**, 35(10), 773–808.

46 Vakil N. Review article: new pharmacological agents for the treatment of gastro-oesophageal reflux disease. *Aliment Pharacol Ther* **2004**, 19, 1041–1049.

47 Kenneth R, McQuaid and Loren Laine. Early heartburn relief with proton pump inhibitors: A systematic review and meta-analysis of clinical trails. *Clinical Gastroenterology and Hepatology* **2005**, 3, 553–563.

48 (48) Crawley J A, Schmitt C M. How satisfied are chronic heartburn suffers with their prescription medications? Results of the Patient Unmet Needs Survey. *J Clin Outcomes Management* **2000**, 7, 29–34.

49 Keeling D J, Laing S M, Senn-Bifinger J. SCH 28080 is a lumenally acting, K+-site inhibitor of the gastric (H+; K+)-ATPase. *Biochem Pharmacol* **1988**, 37, 2231–6.

50 Senn-Bilfinger J, Postius S, Simon W, Grundler G, Hanauer G, Huber R, Kromer W, Sturm E. WO 200017200-A1.

4

Optimizing the Clinical Pharmacologic Properties of the HMG-CoA Reductase Inhibitors

Sándor Kerpel-Fronius and János Fischer

4.1
Introduction

The statins, which are specific inhibitors of hydroxymethyl-glutaryl Coenzyme A (HMG-CoA) reductase, were introduced into clinical practice during the late 1980s. The aim of their introduction was to reduce plasma concentrations of low-density lipoprotein-cholesterol (LDL-C) which is known to play an active role in the development of atherosclerotic cardiovascular diseases. In low doses, the statins exert a more or less selective effect on hepatic cholesterol synthesis, since most of the statins are taken up by the liver. In this dose range, the drugs were considered practically harmless. However, at a later time when the beneficial cardiovascular effects were shown to increase proportionally to the decreasing LDL-C level, statin therapy became more aggressive, with the aim of reducing LDL-C concentrations further, especially in high-risk groups. As a consequence, in more patients – and especially in those treated with high-dose statin therapy – severe myopathies became more frequently observed which, in some unfortunate cases, progressed to rhabdomyolysis with fatal outcome. An outstandingly high number of rhabdomyolysis cases were seen in patients who received statins in combination with gemfibrozil for lowering both LDL-C and triglyceride levels. A similar trend was described in subjects who received concomitantly drugs that inhibit different cytochrome P-450 (CYP) isoforms involved in oxidative metabolism of the statin analogues. These drug interactions, which all might lead to severe myopathies, are due to increased plasma levels of the various statins.

The weight of evidence supports the conclusion that the more expressed inhibition of HMG-CoA reductase by a higher statin blood level reduces the concentrations of other essential products, primarily of isoprenylated proteins and possibly ubiquinone, synthetized downstream from mevalonic acid within the peripheral cells. In parallel, it was also recognized that statins exert pleiotropic effects in various cells far beyond the originally described inhibition of hepatic cholesterol synthesis. All of these effects are considered to be class-specific for the statins. It is important to emphasize that the frequency of untoward side effects observed with the various statins can be related to their potency, the number of metabolic inter-

Analogue-based Drug Discovery. IUPAC, János Fischer, and C. Robin Ganellin (Eds.)
Copyright © 2006 WILEY-VCH Verlag GmbH & Co. KGaA, Weinheim
ISBN: 3-527-31257-9

actions that they have with other drugs, and the relative importance of the affected metabolic pathway for the elimination of statins. Additionally, their lipid solubility – which determines the extent of statin distribution to the peripheral tissues – is of crucial importance. The treating physician must be aware of the fact that the higher the statin dose, the more side effects will develop when using statins either alone or with the addition even of small doses of interacting agents. Therefore, in such situations concomitant medication must be thoroughly checked for possible metabolic interactions with the statin applied. In the following sections, the properties of the various statin analogues will be reviewed, the aim being to emphasize those characteristics which are crucial for their effective and safe use in medical practice.

4.2
Medicinal chemistry of the Statins

HMG-CoA reductase catalyzes the rate-limiting conversion of 3-hydroxy-3-methyl-glutaryl coenzyme A to mevalonic acid which is a key intermediate in biosynthesis of cholesterol (Fig. 4.1)

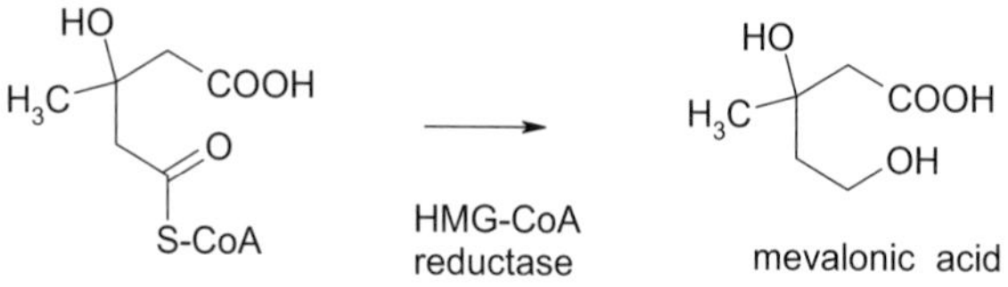

Fig. 4.1 Formation of mevalonic acid.

Research groups in Japan [1] have screened over 8000 microbial extracts for their ability to produce an inhibitor of sterol synthesis *in vitro*. These studies led to the isolation of mevastatin from cultures of *Penicillium citrinum*. Mevastatin (Fig. 4.2) served as the lead-molecule for seven statins which currently are among the best-selling drugs.

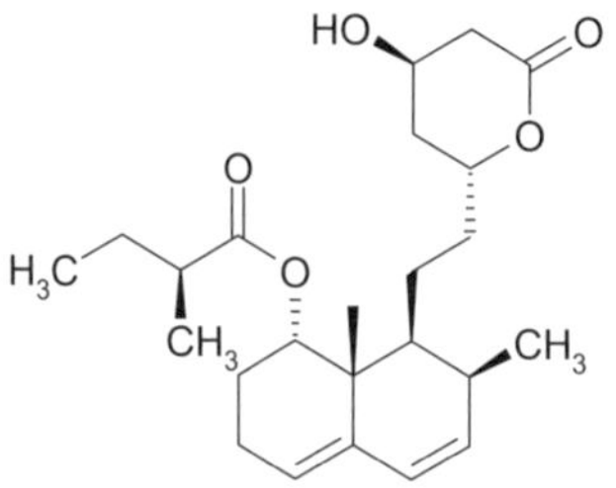

Fig. 4.2 Mevastatin.

Lovastatin was the first statin to be developed, and was isolated from the fungus *Aspergillus terreus*. It was launched in 1987 by Merck, Sharpe and Dohme (USA), and is a two-fold more potent methyl-homologue of mevastatin (Fig. 4.3).

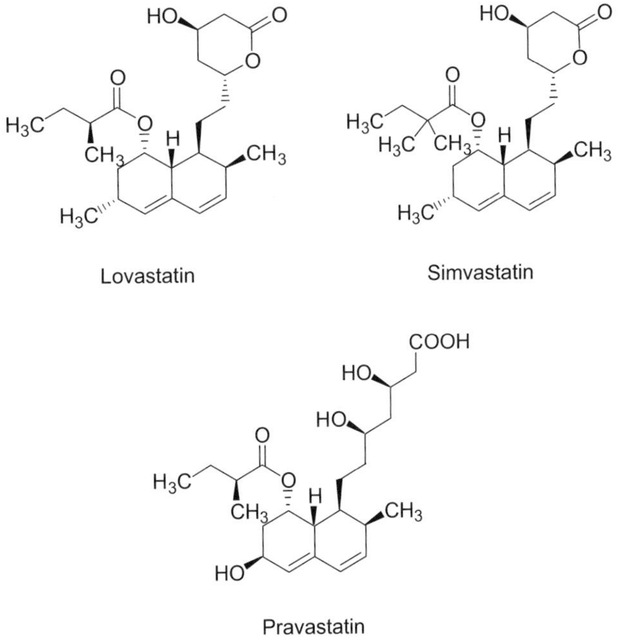

Lovastatin

Simvastatin

Pravastatin

Fig. 4.3 First generation statins having substituted decalin rings.

Simvastatin has been prepared from lovastatin and was launched also by Merck (USA) in 1988. Both lovastatin and simvastatin are prodrugs; their lactone-rings need to be hydrolyzed to the 3,5-dihydroxy heptanoic acid derivatives which are the active metabolites.

Pravastatin was obtained by the microbiological hydroxylation of mevastatin. It is a ring-opened derivative which is much more hydrophilic than lovastatin and simvastatin. Lovastatin, simvastatin and pravastatin all share the decalin (hexahydronaphthalene) ring of natural origin. These drugs represent the first generation of statins.

The research teams of Merck, Sharpe and Dohme (USA) were the first to succeed in finding a synthetic statin analogue that contained a biphenyl moiety [2] instead of a decalin ring (Fig. 4.4).

At approximately the same time, *fluvastatin* was discovered as having an indole ring. This was the first synthetic statin launched in 1994.

In the case of *atorvastatin*, a 1H-pyrrole ring system was selected [3]. The synthetic 2-(4-fluoro-phenyl)-5-isopropyl derivative (Fig. 4.5) inhibited [^{14}C]-acetate conversion to cholesterol in a crude rat liver homogenate. A optimization of its 3,4-disubstituted analogues resulted in atorvastatin.

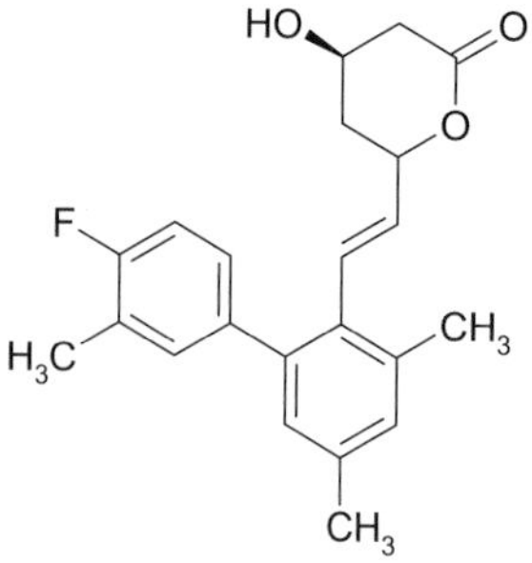

Fig. 4.4 Merck's compound having a biphenyl ring system.

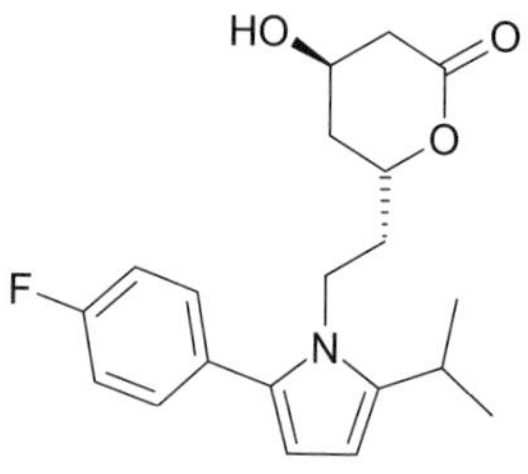

Fig. 4.5 An active compound based on a substituted pyrrole.

In 2003, two further statins have been launched, namely rosuvastatin and pitavastatin. Fluvastatin, atorvastatin, rosuvastatin, and pitavastatin represent the second subtype of statins having heterocyclic ring systems.

Analysis of the statin structures reveals that they possess a common 3,5-dihydroxy-haptanoic acid or 3,5-dihydroxy-6-heptenoic acid or their lactonic form as the characteristic pharmacophore moiety bound to a ring system R (Fig. 4.6).

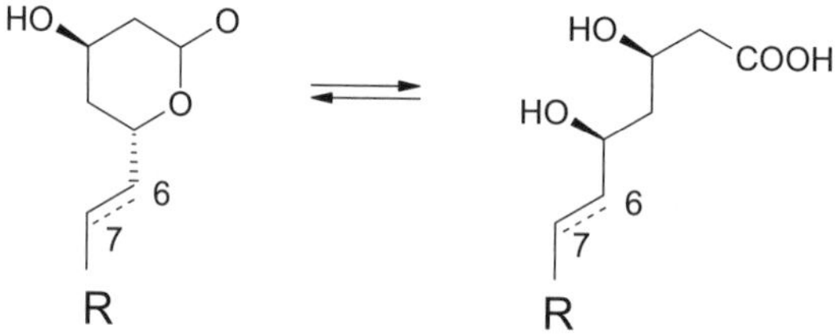

Fig. 4.6 General structure of statins. The characteristic moiety is shown in its lactone ring and open acidic structure.

The variable moiety is also an essential part of a drug molecule as highlighted above. The ring systems of the HMG CoA reductase inhibitors bind to the HMGR enzyme in the same general area where the coenzyme A component of the endogenous HMG CoA substrate binds. To date, two generations of statins have been created according to the variable moieties: in the first generation (lovastatin, simvastatin and pravastatin), R is a decalin group with only minor differences (Fig. 4.7).

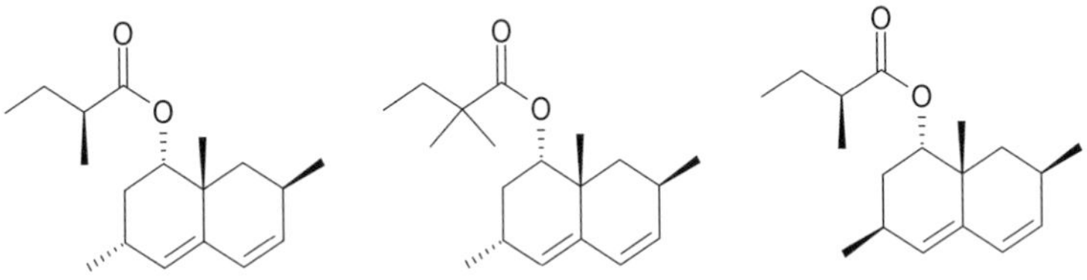

Fig. 4.7 Substituted decalin systems in first-generation statins.

The second generation are characterized by nitrogen-containing heteroaromatic ring-systems (Fig. 4.8).

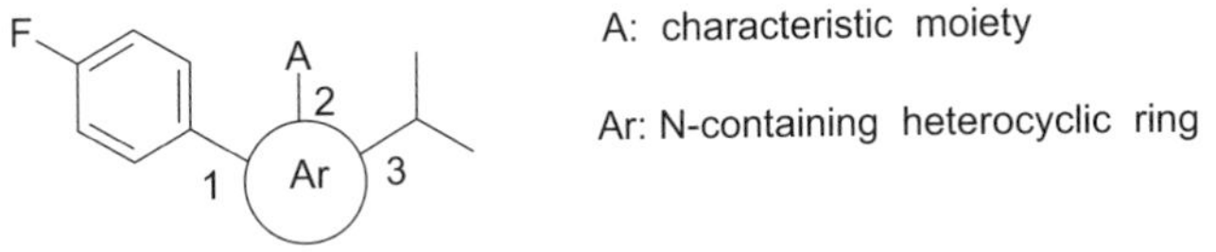

A: characteristic moiety

Ar: N-containing heterocyclic ring

Fig. 4.8 Characteristic arrangement in second-generation statins.

The rings of the second-generation statins show greater differences, despite the same ortho-substituents (4-fluoro-phenyl and isopropyl) being present in four cases. However, the N-containing heterocyclic rings are different: fluvastatin is an indole-, atorvastatin a pyrrole-, rosuvastatin a pyrimidine-, and pitavastatin a quinoline-derivative (Fig. 4.9). Fluvastatin, the single representative of the second generation of statins, is a racemic compound, while the members of the others are single enantiomers. They are the most active statins known. Pitavastatin is still undergoing broad clinical evaluation, and therefore an in-depth comparative evaluation of its clinical pharmacologic properties cannot be undertaken at the time of writing this chapter.

In summary, the statins can be considered to be *direct* analogues as they share an identical pharmacophore interacting with the target. Within the group, two subtypes can be differentiated: the first-generation statins having the decalin ring, while those of the second-generation possessing nitrogen-containing heterocyclic ring-systems.

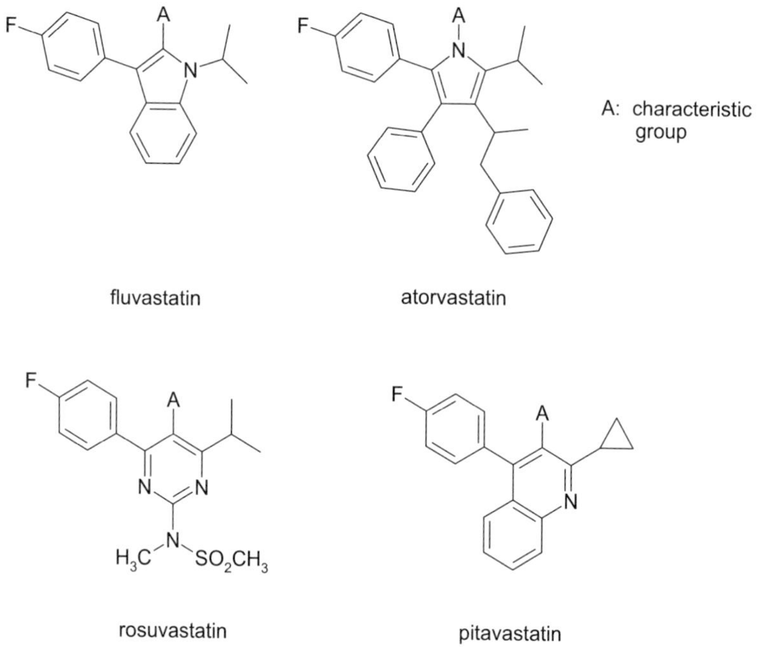

A: characteristic group

fluvastatin   atorvastatin

rosuvastatin   pitavastatin

Fig. 4.9 Ring systems in second-generation statins.

4.3
Clinical and Pharmacologic Properties of the Statin Analogues

For the inhibition of HMG-CoA reductase, the hydrophilic pharmacophore (the dihydroxy 3,5-dihydroxy-6-heptenoic acid side chain) is responsible and must fit into the HMG binding site of the enzyme. The hydrophobic groups of the statins occupy a binding site which is exposed by the movement of flexible helices within the enzyme catalytic domain. The hydrophobic bonds ensure firm anchorage of the molecule into the active site of the HMG-CoA reductase. The "superstatins" – atorvastatin and rosuvastatin – exhibit additional hydrogen bonds that are absent from other types of statins. In rosuvastatin, the sulfonamide group forms additional bonds that are unique to the molecule [4]. The differences in bonding affinity are related to the ring systems of the various statins, and are responsible for the differences in potency between the drugs. *In vitro* testing in microsomal preparations has shown the most potent statin to be rosuvastatin, followed by atorvastatin and simvastatin. The IC_{50} values of fluvastatin and pravastatin are manyfold higher (Tab. 4.1) [5,6].

Tab. 4.1 Inhibitory effects of various statins *in vitro* (data from Refs. [5,6]).

Parameter	Atorvastatin	Fluvastatin	Pravastatin	Rosuvastatin	Simvastatin
Inhibition of HMG-CoA reductase IC_{50} [nM]	8.2	27.6	44.1	5.4	11.2
Inhibition of cholesterol synthesis in rat hepatocytes IC_{50} [nM]	1.15	3.78	6.93	0.16	2.74
Inhibition of cholesterol synthesis in rat fibroblasts IC_{50} [nM]	193	3.43	21500	331	7.07

The lipid solubility of the statins is the key physico-chemical property which influences their activity in cell cultures and *in vivo*. High lipophilicity is also considered to be essential for the rapid uptake of the drugs into the liver and the cells of the other tissues. Lipophilicity of the statins is much higher if the dihydroxy-heptenoic side chain is present in its lactone ring form, as it is the case of lovastatin and simvastatin (see Figs. 4.3 and 4.6; Tab. 4.2) [7,8]. As a consequence, most of the lactone prodrug is eliminated from the circulation by hepatic uptake during the first-pass. Nonetheless, considerable amounts of drug penetrate into the cells of other tissues by passive diffusion. Indeed, the concentrations of the lactone prodrug simvastatin and of the similarly highly lipid-soluble fluvastatin required to inhibit cholesterol synthesis by 50%, both in the hepatic and fibroblast cells, are similar (see Tab. 4.1) [5,6].

In pravastatin, the pharmacophore is present in the open hydroxyl acid form. In addition, the molecule carries a hydroxyl moiety at position 6 of the ring, whereas the other two derivatives have methyl groups. Although these structural features make pravastatin the most hydrophilic statin (see Tab. 4.2), it is nevertheless an active substance which also shows a remarkable specificity for uptake into the liver. This finding was unexpected at the time when only the lactone-containing prodrugs were known, and it was believed that the hepatic specificity of drug action was primarily the consequence of first-pass uptake into the liver due to the very high lipid solubility. This contradiction was resolved by the identification of an active statin uptake mechanism which is linked to the organic acid transporter polypeptides (OATP) [7,9]. Since this uptake – and especially the OATP2 isoform-linked process – is not present, or its activity is much less in the other tissues than in the liver, the preferential uptake of statins into hepatic cells is guaranteed. This

Tab. 4.2 Pharmacokinetic properties of statin analogues in clinical use.

Parameter	Atorva-statin	Fluva-statin	Lova-statin	Pitava-statin	Prava-statin	Rosuva-statin	Simva-statin
Lipophilicity	+	+	++	++	Hydro-philic	Hydro-philic	++
(log P)	1.11	1.27	1.7	0.59	−0.84	−0.13	1.6
Prodrug	No	No	Yes	No	No	No	Yes
Absorption (%)	30	98	30	(>80)	34	50	60–80
Oral bioavailability [%]	14	19–29	5	80	17	20	5
t_{max} [h]	1–2	0.5–1.5	2–2.4	0.8	1.0–1.5	3–5	0.5–1.5
$t_{1/2}$ [h]	11.5	1.0–3.0	1.0–1.7	11	2.6–3.2	19	2.0–3.0
Protein binding [%]	>98	>98	>95	96	43–55	90	95
Renal excretion [%]	1–2	6	10	2	20	5–10	13
Excretion Feces	++	++	++	++	++	++	++
Urine [%]	<2	5	10		20	10	13
Metabolism by CYP3A4	++	+	++				++
CYP2C9		++		+		+	
CYP2D6		+					
Glucuronidation	++	++	++	++	Sulfation ++		++

conclusion is supported by the much higher IC_{50} value of the hydrophilic pravastatin needed to inhibit cholesterol synthesis in fibroblasts than in hepatocyte cultures (see Tab. 4.1) [5,6].

The first entirely synthetic derivative, fluvastatin (see Ref. [19]), and its subsequent derivative, atorvastatin, have a lipid solubility that is intermediate between pravastatin and the lactone prodrugs (see Tab. 4.2), and consequently their liver specificities are less expressed. In order to benefit from selective statin uptake mechanism into the liver cells and thereby decrease passive diffusion into other cell types, the recently introduced rosuvastatin molecule was purposefully made more hydrophilic by the introduction of a sulfonamide group. Indeed, the ratio of IC_{50} values measured in fibroblasts and hepatocyte cultures became considerable higher than that of atorvastatin (see Tab. 4.1). The pronounced differences of inhibitory potency, lipophilicity and the extent of active OATP-linked transport jointly define the rank order of cholesterol synthesis blocking activity of the various statins in hepatocytes as follows (see Tab. 4.1) [5,6]:

rosuvastatin > atorvastatin > simvastatin > fluvastatin > pravastatin.

The rank order of *in-vivo* lipid-lowering effects of the various statins are very similar, although not entirely overlapping with that obtained in hepatocyte cultures (Fig. 4.11a–c) [10,11]. As a consequence of their low potency, the least pharmacological effect is produced by pravastatin and fluvastatin, respectively. Simvastatin, which is much more potent and also highly lipid-soluble, exerts a considerably higher activity. The most expressed lipid-lowering effect is seen after rosuvastatin and atorvastatin administration, and consequently these drugs deserve the frequently applied designation as "superstatins". According to the few data available, pitavastatin might also belong to this group [12]. It is interesting to note that the beneficial high-density lipoprotein-cholesterol (HDL-C) -increasing effects of the analogues are much more different and are not dose-dependent. Fluvastatin exhibits the least effect, while simvastatin and especially rosuvastatin elevate the HDL-C level by the most significant degree, although the effect levels off with higher doses. Surprisingly, atorvastatin produces only a modest effect, and with higher doses even this modest effect is decreased.

Although the majority of the statins are well absorbed from the gut, their bioavailabilities are low due to the first-pass metabolism in the liver (see Tab. 4.2). The compounds are highly protein-bound both in their lactone and hydroxy-acid forms, with the only exemption being pravastatin. Strong protein binding is important to minimize statin uptake into the peripheral tissues (see Tab. 4.2) [7]. Differences in the metabolism of statins underlie the various types and different severities of their interactions with other drugs. In the case of the two prodrugs simvastatin and lovastatin, the lactone ring must be hydrolyzed in the cells to the active, open hydroxy-acid form by carboxylesterases or the recently identified paroxonases. The main inactivation proceeds through cytochrome P-450-dependent oxidative metabolism and glucuronidation of the parent compounds and their various metabolites, respectively. β-Oxidation of the dihydroxy heptenoic chain produces inactive metabolites, though this pathway has a lesser role in man than in rodents [14]. Various isoforms of the UDP-glucuronosyltransferases (UGT1A1 and UGT1A3) are involved in the glucuronidation of the statins. Following glucuronidation, the open acid side chain of all statins undergoes extensive, spontaneous cyclization to the corresponding inactive lactones. This inactivation process was only recently recognized as being one of the most important mechanisms for reducing the level of the active hydroxy-acid form. Conversion of the open acid form of the parent compounds and their respective metabolites to lactones might also proceed via a CoA-dependent pathway, while their reconversion to the open acid form is catalyzed by esterases and paroxonases [13–15]. As a result, both the glucuronidated open acid and lactone forms of the parent compounds and their metabolites can be demonstrated in the bile. An active enterohepatic recirculation also occurs. The major route of elimination is via the feces, with only a small fraction of the statins and their metabolites appearing in the urine. Most of the statins

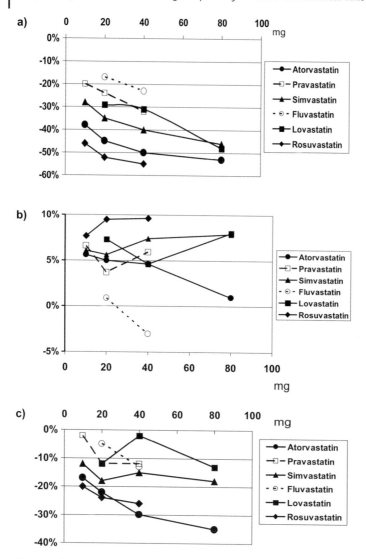

Fig. 4.11 Plasma levels of (a) LDL-C, (b) HDL-C and (c) triglycerides after statin treatment in human subjects.

are eliminated rapidly from the circulation, with only the "superstatins" exhibiting terminal half-lives of more than 10 h (see Tab. 4.2). It is likely that this also contributes to their increased lipid-lowering potential.

Oxidative metabolism occurs with all of the statins, though the extent of metabolism and importance in the inactivation of the various compounds are different. Usually, more than one cytochrome isoform is implicated in the metabolism of various statins, though many of these have only very limited functional significance. In Tabs. 4.2 and 4.3, only those cytochromes which play a significant role in the metabolism of statins are listed [16,17,19]. Lovastatin, simvastatin and atorvastatin are all extensively metabolized by CYP3A4, and consequently any inhibition of this cytochrome will increase substantially the plasma level of the parent compounds; conversely, the induction of CYP3A4 will reduce the plasma concentration. The former situation might result in severe toxicity, whilst on the other hand the more rapid elimination of statins will reduce the effectiveness of such therapy. Serious interactions leading to severe rhabdomyolysis have been described following the coadministration of some of the inhibitory compounds listed in Tab. 4.3, whilst other known inhibitors have the potential to cause similar toxicity [16,17]. The metabolism of fluvastatin is primarily related to CYP2C9, with less-pronounced contributions from CYP2C19 and CYP2D6; consequently, only those interactions with CYP2C9 are of clinical relevance [19]. CYP2C9-dependent metabolism contributes only minimally to the inactivation of rosuvastatin, and therefore inhibition of this pathway is not considered to be functionally important in clinical practice [20]. Finally, it should be pointed out that drug interaction with statins might also alter the metabolism of the interfering compound, and lead in turn to side effects related to them.

Tab. 4.3 Effects of metabolic interactions on the plasma level of statin analogues in clinical use.

	Drugs causing interaction	Increase plasma level of	Decrease plasma level of
CYP3A4 Induction	Barbiturates, carbamazepine, griseofulvin, nafcillin, phenytoin, rifampin, troglitazone		Atorvastatin Lovastatin Simvastatin
CYP3A4 Inhibition	Amiodarone, clarithromycin, erythromycin, cimetidine, cyclosporine, fluoxetine fluvoxamine, itraconazole, ketoconazole, nefazodone, verapamil, diltiazem HIV antivirals: delaviridine, indanavire, nelfinavire, ritonavire, sequinavire	Atorvastatin Lovastatin Simvastatin	
CYP2C9 Induction	Barbiturates, carbamazepine, phenytoin, primidone, rifampin		Fluvastatin
CYP2C9 Inhibition	Amiodarone, cimetidine, trimethoprimsulfamethoxazole, fluoxetine, fluvoxamine, isoniazid, itraconazole, ketoconazole, fluconazole, metronidasole, sulfinpyrazone, ticlopidine	Fluvastatin Rosuvastatin (±)	

4.3.1

Fibrate Coadministration

Fibrates are frequently coadministered with statins to enhance their lipid-lowering effect, and especially for the simultaneous reduction of both cholesterol and triglyceride levels. In particular, the simultaneous use of gemfibrozil was seen to lead more frequently to muscle impairment. It transpired that gemfibrozil significantly interferes with glucuronidation of the statins by competitively inhibiting the UGT1A1 and UGT1A3 isoforms. As a result, lactonization and inactivation of the glucuronide conjugate of the open acid form cannot take place. Furthermore, the level of the hydroxy-acid form increases. This seems to be the case with all statins that are glucuronidated. The other fibrates have a much lesser interaction with statins because they affect other isoforms of the UDP-glucuronosyltransferase, examples being fenofibrate affecting UGT1A9 and UGT2B7 [13–15]. Due to a better understanding of statin metabolism, the tragic increase in numbers of fatal rhabdomyolysis caused by the coadministration of gemfibrozil and cerivastatin can now be conclusively related to several simultaneous interactions at the level of glucuronidation and CYP2C8 which is responsible for the oxidative metabolism of cerivastatin. Due to the inhibited hepatic metabolism of cerivastatin being the plasma level increased significantly. To make the situation worse, this statin derivative is the most lipophilic compound in the group, and the level of active compound easily reached a sufficiently high concentration in striated muscle to cause significant inhibition of HMG-CoA reductase activity. Inhibition also occurred in the formation of mevalonate, a primary intermediate in the formation of farnesyl pyrophosphate, a key intermediate to several functionally very important products, including cholesterol and ubiquinone. Farnesyl pyrophosphate and its product, geranylgeranyl pyrophosphate, activate GTP-binding regulatory proteins by prenylation, and this promotes cell maintenance and attenuates apoptosis. Indeed, the results of several studies have indicated that statins induce apoptosis by inhibiting prenylation, for example in pulmonary vein endothelial cells or vascular smooth muscle cells *in vitro* [21,22]. The advantageous reduction of atherosclerotic plaques after statin therapy might be partly due to the increased apoptosis of such cells within the lesions. The same mechanism is most probably responsible for the statin-induced myotoxicity which affects striated muscles [23]. Finally, it should be pointed out that despite most severe cases of rhabdomyolysis being observed after combination treatments, muscle injury can also occur after monotherapy with practically all statins, especially if used at high dosage. Other interactions, such as those at the level of OATP-related transport or with P-glycoprotein which actively extrudes statins from the cells might occur, though they have not been proven to cause severe side effects.

4.4
Clinical Efficacy of the Statins

During the past decade, the great clinical benefit of lipid lowering for reducing death due to coronary artery disease has been amply demonstrated. Statins are effective in both secondary prevention trials in patients with previous major cardiac events, and in primary prevention in asymptomatic subjects with increased risk for coronary disease [24–29]. The results indicated a close relationship between the degree of plasma level decrease in LDL-C and clinical outcome [30,31]. Direct proof of such a close relationship was provided in a landmark study in patients with acute coronary syndrome (ACS). In this trial, which was reported in 2000, 80 mg atorvastatin was compared with 40 mg pravastatin, the broadly accepted standard treatment [32]. The LDL-C levels achieved during treatment were 2.46 mmol L^{-1} and 1.60 mmol L^{-1}, respectively. At the end of two years, the composite end-point of death due to any cause, myocardial infarction, recurring unstable angina, revascularization and stroke occurred in 22.4% and 26.3% of the cases, which corresponds to a 16% risk reduction in favor of atorvastatin (p = 0.005). In an accompanying editorial, it was predicted by Topol [33] that the treatment strategy level of ACS will be fundamentally changed, and especially the desired lipid target level will be significantly lower on the basis of this result. In another study carried out in a similar, but slightly poorer risk, patient population, a 4-month placebo treatment followed by 20 mg per day simvastatin was compared with 40 mg per day simvastatin started shortly after the event and followed after 1 month by an increased dose of 80 mg per day. This study failed to prove any overall advantage of the more aggressive statin therapy, although the survival curve showed an advantage in favor of higher-dose therapy if survival was calculated for the period between 4 and 24 months [34]. The reasons for this discrepancy between the two studies remains unknown, but the authors speculated that although the final LDL-C levels reached after high-dose simvastatin and atorvastatin in the two studies were similar, the difference achieved between the low- and high-dose simvastatin treatments in the LDL-C concentration was much less (0.36 mmol L^{-1}; 18% relative difference) than was observed in the pravastatin versus atorvastatin comparison (0.85 mmol L^{-1}; 33% relative difference). In addition, high-dose simvastatin caused more serious side effects, especially an increased rate of severe myositis (0.4%) (creatinine kinase level > 10-fold the upper limit of normal value, ULN), whereas no corresponding increase was observed following atorvastatin treatment.

The results of these studies provided direct proof for the importance of the extent of lipid lowering in order to improve cardiac outcome. In addition, they clearly indicated that, in order to achieve a large reduction in lipid levels and the corresponding cardiovascular event rates, the statins are not equally suitable. The correlation between the degree of lipid lowering and therapeutic benefit, as well as the different effectiveness of the statins to achieve this goal, were corroborated by a meta-analysis of the available clinical data [35]. In 164 short-term studies, the lipid-lowering potency of all statins was shown to correlate more or less linearly with increasing doses between 5 and 80 mg per day. The least effective agents were pravastatin and fluvastatin, which decrease LDL-C levels by no more than 27–33%, even at the highest dose of 80 mg per day. The maximal effect of 80 mg per day lovastatin and simvastatin was 42–45%. The clinically accepted highest doses of 80 and 40 mg per day of atorvastatin and rosuvastatin caused decreases of 53–55% in LDL-C levels. Rosuvastatin, which is clearly the most potent agent available, can lead to even higher (58%) reduction if the dose is increased to 80 mg per day. However, this dose is not registered because an increased rate of rhabdomyolysis was observed. This rank order of efficacy derived from many trials corroborates the direct comparisons of lipid levels within individual studies by Jones (Fig. 4.11a) [10,11]. The percentage reduction in LDL-C level was independent of the baseline value. In 58 trials, each of at least 24 months' duration, in which both cholesterol levels and clinical outcomes were measured, any correlation between the degree of lipid lowering and therapeutic results could be evaluated. The absolute reduction of LDL-C concentration achieved in these trials could be clustered into three groups: 0.2–0.7, 0.8– 1.4 and ≥1.5 mmol L^{-1}. There was a significant correlation between the LDL-C reduction and therapeutic benefit, and this increased in line with the duration of treatment. At between three and five years the ischemic heart disease events were reduced in the three dosage groups by 19%, 31%, and 50%, respectively. The HDL-C concentration was increased by 0.06–0.08 mmol L^{-1} on average, with no detectable dose–effect relationship. The distinct difference between efficacy of the statins was also apparent in this meta-analysis, as simvastatin and rosuvastatin each increased HDL-C levels much more than atorvastatin, though with increasing doses a plateau level was reached [11,36].

Considering these excellent clinical results of statin therapy, the treatment goals defined by cardiologists became increasingly ambitious, especially in patient groups with increased risks. The evidence-based National Cholesterol Education Program, Adult Treatment Panel III (NCEP- ATPIII) defined an LDL-C level of <2.6 mmol L^{-1} as optimal for adults [37]. Target LDL-C goals for patients with cardiovascular disease are set according to the risk factors of those patients. In subjects who have coronary heart disease (CHD) or a CHD risk-equivalent condition (non-coronary atherosclerosis, diabetes and >20% risk in 10 years to develop CHD), the aim is to reduce the level below 2.6 mmol L^{-1}. In patients with two or more risk factors (<10% and 10–20% risk of CHD in 10 years) and finally in subjects with less than two CHD risk factors, the goals are 3.4 mmol L^{-1} and 4.1 mmol L^{-1}, respectively. In patients with a triglyceride level ≥2.3 mmol L^{-1}, a non-HDL-C level 0.8 mmol L^{-1} higher than the LDL-C goal is specified. Essentially, very similar guidelines were developed in Europe by the Third Joint Task Force [38]. In general the LDL-C level should be <3 mmol L^{-1}. For patients with clinically established CHD, peripheral vascular disease, cerebrovascular syndrome or having diabetes and/or metabolic syndrome, the targets are for LDL-C 2.5 mmol L^{-1}, for triglyceride 1.7 mmol L^{-1} and finally for HDL-C >1.0 and 1.2 mmol L^{-1} in males and females, respectively. These goals are difficult to achieve. Lifestyle changes are essential, but the more ambitious end-points for high-risk cases cannot be reached without using aggressive, high-dose statin treatment either alone or in combination with fibrates and/or niacin. These latter drugs are especially required when the joint decrease in triglyceride level and/or increase in HDL-C level is desired.

The recent STELLAR trial, which was an open, parallel group randomized design, compared 10 to 40 mg per day doses of pravastatin with 10 to 80 mg per day of simvastatin, atorvastatin, and rosuvastatin. (The 80 mg per day rosuvastatin treatment arm was later stopped due to severe muscular side effects [39]). The baseline LDL-C levels were between 4.1 and 6.5 mmol L^{-1}, and the triglyceride level was <4.5 mmol L^{-1}. It was shown that even the <2.6 mmol L^{-1} LDL-C concentration could be reached with the lowest dose of rosuvastatin in some patients. The potency of rosuvastatin was followed (in rank order) by atorvastatin, simvastatin, and pravastatin. However, it should be emphasized that the goal could not be reached even with the highest clinically applicable doses in 20%, 30%, 47%, and 92% of the cases using rosuvastatin, atorvastatin, simvastatin, and pravastatin, respectively. Rosuvastatin efficacy was found to be statistically superior to that of atorvastatin in the dose range between 10 and 40 mg per day, but not for 80 mg per day atorvastatin. In addition, 20 mg rosuvastatin produced close to the maximal result. In order to reach the triglyceride goal and the corresponding HDL-C goal, rosuvastatin was found also to be superior compared to atorvastatin, though statin monotherapy was not sufficient in approximately 20% of the patients. In translating the results for clinical practice, it is clear that for moderate goals in patients with a relatively low risk, the weaker statins are adequate, or the stronger statins can be used at low dose. For poor-risk patients, however, statin therapy must be pushed to the highest clinically acceptable dose, which is close to the level where the frequency of serious adverse events increases rapidly. In this dangerous dose range, the clinician must monitor the patients extremely carefully and scrutinize meticulously all other drugs that might be needed for the treatment of any other concomitant disease.

As a class, the statins can be considered to be a very safe group of drugs, and serious adverse reactions occur rarely [36]. However, with increased doses hepatic and especially muscle injury became more frequent and increased in severity. The patient-reported symptom of myalgia also occurred relatively frequently in patients treated with placebo. An elevated creatine kinase (CK) level on the other hand, indicates muscle injury. In the clinical setting, an asymptomatic elevation of CK less than five times ULN can be considered to be benign, but above this level a thorough evaluation is required. Myopathy is defined as a CK level 10-times ULN, with symptoms such as generalized myalgia, fatigue, and weakness. Rhabdomyolysis is a clinical syndrome that is characterized by severe and widespread skeletal muscle injury, accompanied by subsequent accumulation of muscle breakdown products in the blood and urine. These products can block the renal tubules, and in turn cause irreversible kidney failure and death. If diagnosed early, the progression from myopathy to rhabdomyolysis can be prevented with hydration and cessation of the eliciting drug therapy [23,36,40]. By using prescription rate as denominator, the overall reporting rate per 100 000 prescriptions for rhabdomyolysis was 0.18, 0.02, 0.11, 0.0, 0.03, and 4.29 for lovastatin, pravastatin, simvastatin, fluvastatin, atorvastatin, and finally cerivastatin, respectively. For the newly introduced rosuvastatin, a preliminary judgment made on the basis of almost 10 000 000 prescriptions showed the occurrence of myopathy-rhabdomyolysis cases, evaluated together in one group, to be about 0.3–0.4 per 100 000 prescriptions. According to the critical evaluation of all reported side effects, the FDA concludes that the risk of this very active agent is well within the range of the entire statin group. The slightly higher number of reported cases is partly due to the increased awareness of myopathy and the consequent higher reporting rate by physicians [41]. The occurrence of statin-related rhabdomyolysis is related to the dose, and especially to drug interactions. Rhabdomyolysis was observed 15 to 30 times more frequently if the statins were used in combination with gemfibrozil [17,23,40,42,43] (for details of this interaction, see Section 4.3). In the case of cerivastatin, which has multiple interactions with gemfibrozil, the frequency of rhabdomyolysis increased from 1.81 in monotherapy to 1248 per 100 000 prescriptions, and as a consequence the drug was withdrawn from the market [40]. Bezafibrate or fenofibrate cause less muscle damage if administered jointly with statins, and therefore they are preferred for combination therapy [13,15,44–46].

Other frequent precipitators of side effects are drugs that interact with atorvastatin, lovastatin, and simvastatin at CYP3A4, for example macrolide antibiotics, antifungals, HIV antivirals, amiodarone, and cyclosporine (see Tab. 4.3). The latter interaction is clinically especially important, since following organ transplantation many patients develop dyslipidemia which must be treated with statins concomitantly with cyclosporine used to inhibit transplant rejection [17]. The fact that rosuvastatin is not metabolized by CYP3A4 makes the drug very useful in this situation. However, rosuvastatin might be susceptible to drugs interacting with its metabolism at CYP2C9. Although in a healthy volunteer study, fluconazole (a well-known substrate of this cytochrome) caused only minor pharmacokinetic interaction [47], caution is required if higher doses of rosuvastatin are intended to be coadministered with drugs metabolized by CYP2C9 (see Tab. 4.3). Statins also cause a more than three-fold increase over the ULN of alanine and aspartate aminotransferases in about 3% of patients. Since all other lipid-lowering agents cause similar toxicity, it might be suspected that the liver injury is in some way related to the inhibition of lipid metabolism This elevation of enzyme levels is accompanied only very rarely by hepatocellular injury and jaundice [48]. Finally, all statins affect the kidney tubular cells, leading to proteinuria probably as a result of impaired tubular reabsorption, and hematuria in 1–3% of patients, though without any increase in creatinine levels. No serious kidney damage could be definitely related to statin therapy.

When examining the history of lipid-lowering therapy, the development of statin analogues is a clear example of the interaction between clinical needs and the efforts of drug designers to meet this need. The analogues produced became increasingly effective during the years and, encouraged by these excellent results, the treatment goals became correspondingly bolder. Unfortunately, it appears that the limits of statin therapy were approached very closely, although the therapeutic window of atorvastatin and rosuvastatin became larger compared to the older statins. The result is that muscle and liver toxicities appear at drug concentrations which produce a much greater LDL-C reduction than was observed with earlier statins [49]. Since the other products derived from mevalonic acid play an essential role in many other cellular functions, the miracle drug without the above toxicities will most likely remain forever an utopia. Therefore, the call by the Public Citizen's Health Research Group [50] to withdraw rosuvastatin – the most effective drug – from the market, cannot be easily understood from the clinical pharmacologic point of view. Based on the critical review of all data obtained before and after marketing of rosuvastatin, the FDA recently concluded that neither the rate nor the severity of the muscle and renal side effects indicated a real risk for the population, and hence the request to withdraw this very powerful agent was denied. However, the Agency pointed out again that the highest permitted dose of 40 mg should only be used if the target lipid value could not be reached with lower doses [41]. Moreover, the close observation of the drug will be continued since the real benefit:risk ratio of a new drug can be characterized adequately only during many years of experience in the broad clinical practice. In order to guarantee the effectiveness and safety of high-dose treatment with statins, medical practitioners and patients must share the responsibilities. The therapy should be performed in close cooperation between the treating physicians and the patients in order to detect possibly dangerous side effects at a very early stage, when the process can still be reversed. In this context it is surprising that, according to a Dutch survey, 23% of patients who received rosuvastatin soon after its introduction to the market had risk factors for developing rhabdomyolysis and were, therefore, only conditionally eligible for intensive rosuvastatin treatment. The risk factors were properly stated in the drug information leaflet [51]. Similar findings were reported by the FDA [41]. It is important that both the medical and patient communities should understand that further improvement of the therapeutic results cannot be expected without more effective statins with slightly increased toxicity. However, these drugs must be used very prudently, especially if the therapy must be performed with higher doses. Moreover, in order to meet the still unmet need to reduce lipid levels further in severe cardiovascular conditions, agents with other mechanisms of action must also be developed. In the case of LDL-C, a new drug ezetimibe – which blocks cholesterol uptake from the gut – is a promising innovation [52–54]. For the adequate treatment of abnormal triglyceride and HDL-C levels, drugs with new mechanisms of action must be developed by drug designers.

References

1 Endo A, Kuroda M, Tsujita Y, *J. Antibiotics*, **1976**, 29, 1346.

2 Willard AK, Novello FC, Hoffman, WF, et al. (Merck and Co.) US Patent **1982**, 4,459,422.

3 Roth BC. *Progress in Medicinal Chemistry* (Eds. King FD, Osford AW). Elsevier, Amsterdam, **2002**, 40, 1–22.

4 Istvan E. Statin inhibition of HMG-CoA reductase: a 3-dimensional view. *Atherosclerosis*, **2003** (Suppl. 4, Issue 1), 3–8.

5 Stein EA. New statins and new doses of older statins. *Curr. Sci.*, **2001**, 3, 14–18.

6 Department of Health and Human Services, Food and Drug Administration, Center for Drug Evaluation and Research. Rosuvastatin (Crestor). Pharmacology review(s). http://www.fda.gov/cder/fai/nda/2003/21-366_Crestor_Pharmr.P1.pdf.

7 Hamelin BA, Turgeon J. Hydrophilicity/lipophilicity: relevance for the pharmacology and clinical effects of HMG-CoA reductase inhibitors. *Trends Pharm. Sci.*, **1998**, 19, 26–37.

8 Klotz U. Pharmacologic comparison of the statins. *Arzneim.-Forsch./Drug Res.*, **2003**, 53, 605–611.

9 Hsiang B, Zhu Y, Wang Z, et al. A novel human hepatic organic anion transporting polypeptide (OATP2). Identification of a liver-specific human organic anion transporting polypeptide and identification of rat and human hydroxymethylglutaryl-CoA reductase inhibitor. *J. Biol. Chem.*, **1999**, 274, 37161–37168.

10 Jones PH, Kafonek S, Laurora I, et al. Comparative dose efficacy study of atorvastatin versus simvastatin, pravastatin, lovastatin and fluvastatin in patients with hypercholesterolemia. (The CURVES study). *Am. J. Cardiol.*, **1998**, 81, 582–587.

11 Jones PH, Davidson MH, Stein EA, et al. Comparison of the efficacy and safety of rosuvastatin, versus atorvastatin, simvastatin and pravastatin across doses (STELLAR trial). *Am. J. Cardiol.*, **2003**, 92, 152–160.

12 Hegde S, Carter J. Pitavastatin (Hypercholesterolemic). In: *Annual Reports in Medicinal Chemistry* (Ed. Doherty AM). **2004**, 39, 380.

13 Prueksaritanont T, Zhao JJ, Ma B, et al. Mechanistic studies on metabolic interactions between gemfibrozil and statins. *J. Pharmacol. Exp. Ther.*, **2002**, 301, 1042–1051.

14 Prueksaritanont T, Tang C, Qiu Y, et al. Effects of fibrates on metabolism of statins in human hepatocytes. *Drug Metab. Dispos.*, **2002**, 30, 1280–1287.

15 Prueksaritanont T, Subramanian R, Fang X, et al. Glucuronidation of statins in animals and humans: a novel mechanism of statin lactonization. *Drug Metab. Dispos.*, **2002**, 30, 505–512.

16 Knopp RH. Drug treatment of lipid disorders. *N. Engl. J. Med.*, **1999**, 341, 498–511.

17 Ballantyne CM, Corsini A, Davidson MH, et al. Risk of myopathy with statin therapy in high-risk patients. *Arch. Intern. Med.*, **2003**, 163, 553–564.

18 P-450 Drug Interaction Table; http://www.medicine.iupui.edu./flockhart/clinlist.htm

19 Langtry HD, Markham A. Fluvastatin. *Drugs*, **1999**, 57, 583–606.

20 Cooper KJ, Martin PD, Dane AL, et al. The effect of fluconazole on the pharmacokinetics of rosuvastatin. *Eur. J. Clin. Pharmacol.*, **2002**, 58, 527–531.

21 Guijjaro C, Blanco-Colio LM, Ortego Monica, et al. 3-Hydroxy-3-methylglutaryl coenzyme A reductase and isoprenylation inhibitors induce apoptosis of vascular smooth muscle cells in culture. *Circ. Res.*, **1998**, 83, 490–500.

22 Kaneta S, Satoh K Kano S, et al. All hydrophobic HMG-CoA reductase inhibitors induce apoptotic death in rat pulmonary vein endothelial cells. *Atherosclerosis*, **2003**, 170, 237–243.

23 Thompson PD, Clarkson P, Karas RH. Statin-associated myopathy. *JAMA*, **2003**, 289, 1681–1690.

24 4S Investigators. Randomized trial of cholesterol lowering in 4444 patients with coronary heart disease: The Scandi-

navian Simvastatin Survival Study (4S). *Lancet*, **1994**, 344, 1383–1389.

25 Shepherd J, Cobbe SM, Ford I, et al. Prevention of coronary heart disease with pravastatin in men with hypercholesterolemia. West of Scotland Coronary Prevention Study Group. *N. Engl. J. Med.*, **1995**, 333, 1301–1307.

26 Sacks FM, Pfeffer MA, Moye LA, et al. The effect of pravastatin on coronary events after myocardial infarction in patients with average cholesterol levels. Cholesterol and Recurrent Events Trial Investigators. *N. Engl. J. Med.*, **1996**, 335, 1001–1009.

27 LIPID Study Group. Prevention of cardiovascular events and death with pravastatin in patients with coronary heart disease and a broad range of initial cholesterol levels. The Long-term intervention with Pravastatin in Ischemic Disease (LIPID) Study Group. *N. Engl. J. Med.*, **1998**, 339, 1349–1357.

28 Downs JR, Clearfield M, Weis S, et al. Primary prevention of acute coronary events with lovastatin in men and women with average cholesterol levels: results of AFCAPS/TexCAPS. Air Force/ Texas Coronary Atherosclerosis Prevention Study. *JAMA*, **1998**, 279, 1615–1622.

29 MRC Investigators. MRC/BHF Heart Protection Study of cholesterol lowering with simvastatin in 20 536 high-risk individuals: a randomized-placebo contolled trial. *Lancet*, **2002**, 360, 7–22.

30 Sheperd J. The statin era: in search of the ideal lipid regulating agent. *Heart*, **2001**, 85, 259–264.

31 Jones PH. Lower is better: LDL-C goal achievement and statin efficacy in coronary prevention. *Medscape Cardiology*, **2004**, 8(1), 1–8. (http://www. Medscape.com/viewarticle/472518)

32 Cannon CP, Braunwald E, McCabe CH, et al. Comparison of intensive and moderate lipid lowering with statins after acute coronary syndromes. *N. Engl. J. Med.*, **2004**, 350, 1495–1504.

33 Topol EJ. Intensive statin therapy. A sea change in cardiovascular prevention. *N. Engl. J. Med.*, **2004**, 350, 15–17.

34 De Lemos JA, Blaying MA, Wiviott SD, et al. Early intensive vs. a delayed conservative simvastatin strategy in patients with acute coronary syndromes. *JAMA*, **2004**, 292, 1307–1316.

35 Law MR, Wald NJ, Rudnicka AR. Quantifying effect of statins on low density lipoprotein cholesterol, ischaemic heart disease and stroke: systemic review and meta-analysis. *Br. Med. J.*, **2003**, 326, 1423–1430.

36 Ballantine CM, Blazing MA, Hunninghake DB, et al. Effect on high-density lipoprotein cholesterol of maximum dose simvastatin and atorvastatin in patients with hypercholesterolemia: results of the comparative HDL efficacy and safety study. *Am. Heart J.*, **2003**, 146, 862–869.

37 National Institute of Health. Third Report of the National Cholesterol Education Program Expert Panel on Detection, Evaluation and Treatment of High Blood Cholesterol in Adults (Adult Treatment Panel III). Final Report, Washington, DC: National Institute of Health, US Department of Health and Human Services, September, **2002**, II-27. NIH Publication 02–5215.

38 De Backer G, Ambrosioni E, Borch-Johnsen K, Brotons C, et al. European guidelines on cardiovascular disease prevention in clinical practice. *Eur. Heart J.*, **2003**, 24, 1601–1610.

39 McKenney JM, Jones PH, Adamczyk MA, et al. Comparison of the efficacy of rosuvastatin versus atorvastatin, simvastatin and pravastatin in achieving lipid goals: results from the STELLAR trial. *Curr. Med. Res. Opin.*, **2003**, 19, 689–698.

40 Chang JT, Staffa JA, Parks MD, et al. Rhabdomyolysis with HMG-CoA reductase inhibitors and gemfibrozil combination therapy. *Pharmacoepidemiol. Drug Safety*, **2004**, 13, 417–426.

41 Food and Drug Administration. Docket No. 2004–0113/CPI. http://www.fda.gov.

42 Backmann JT, Kyrklund C, Kivistö KT, et al. Plasma concentrations of active simvastatin acid are increased by gemfibrozil. *Clin. Pharmacol. Ther.*, **2000**, 68, 122–129.

43 Schneck DW, Birmingham BK, Zalikowski JA, et al. The effect of gemfi-

brozil on the pharmacokinetics of rosu-vastatin *Clin. Pharmacol. Ther.*, **2004**, 75, 455–463.

44 Ballantine CM, Davidson MH. Possible differences between fibrates in pharma-cokinetic interactions with statins. *Arch. Intern. Med.*, **2003**, 163, 2394.

45 Alseikh-Ali A, Kuvin JT, Karas HR. Risk of adverse events with fibrates. *Am. J. Cardiol.*, **2004**, 94, 935–938.

46 Kyrklund C, Backmann JT, Kivistö KT, et al. Plasma concentrations of active lovastatin acid are markedly increased by gemfibrozil but not by benzafibrate. *Clin. Pharmacol. Ther.*, **2001**, 69, 340–345.

47 Cooper KJ, Martin PD, Dane AL, et al. The effect of fluconazole on the phar-macokinetics of rosuvastatin. *Eur. J. Clin. Pharmacol.*, **2002**, 58, 527–531.

48 Gotto AM. Safety of statin therapy. *Arch. Intern. Med.*, **2003**, 163, 637–639.

49 Brewer HB. Benefit-risk assessment of rosuvastatin 10 to 40 milligrams. *Am. J. Cardiol.*, **2003**, 92, 23K–29K.

50 Wolfe SM. Dangers of rosuvastatin iden-tified before and after FDA approval. *Lancet*, **2004**, 363, 2189–2190.

51 Florentinus SR, Heerdink ER, Klungel OH. Should rosuvastatin be withdrawn from the market? *Lancet*, **2004**, 364, 1577.

52 Anonymous. Ezetimibe – a new choles-terol-lowering drug. *Drug Ther. Bull.*, **2004**, 42, 65–67.

53 Feldman T, Koren M, Insull W, et al. Treatment of high-risk patients with ezetimibe plus simvastatin co-adminis-tration versus simvastatin alone to attain National Cholesterol Education Program Adult Treatment Panel III low-density lipoprotein cholesterol goals. *Am. J. Cardiol.*, **2004**, 93, 1481–1486.

54 Ballantyne CM, Blazing MA, King TR, et al. Efficacy and safety of ezetimibe co-administration with simvastatin com-pared with atorvastatin in adults with hypercholesterolemia. *Am. J. Cardiol.*, **2002**, 93, 1487–1494.

5

Optimizing Antihypertensive Therapy by Angiotensin Receptor Blockers

Csaba Farsang and János Fischer

5.1
Medicinal Chemistry

The angiotensin II antagonists were investigated several years before the angiotensin-converting enzyme (ACE) inhibitors. For example, during the early 1970s, saralasin – an octapeptide analogue of angiotensin II – was the first angiotensin II antagonist (Fig. 5.1), where Asp_1, Ile_5, and Phe_8 had been replaced with Sar (sarcosine, N-methylglycine), Val, and Ala, respectively. Unfortunately, saralasin – similar to teprotide – could not be used clinically due to a lack of oral activity.

Asp - Arg - Val - Tyr - Ile - His - Pro - Phe

Angiotensin II

Sar - Arg - Val - Tyr - Val - His - Pro - Ala

Saralasin

Fig. 5.1 The structures of angiotensin II and saralasin.

In 1982, research investigators at Takeda discovered weak nonpeptide angiotensin II antagonists among a group of imidazole-5-acetic acid derivatives (S-8307, S-8308) (Fig. 5.2) [1]. These molecules were weak, but selective, antagonists of angiotensin II.

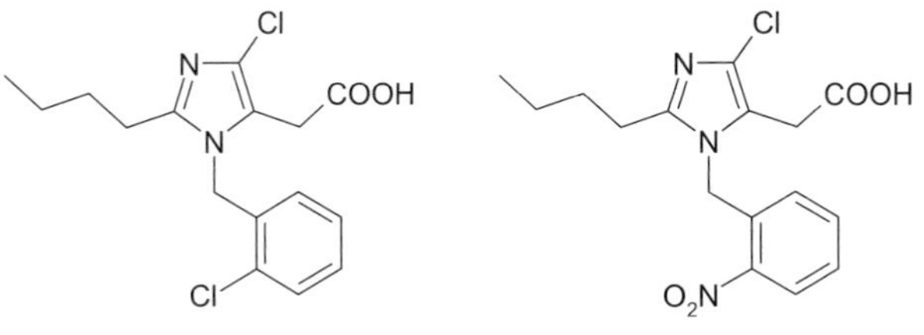

S-8307

$IC_{50} = 40\ \mu M$

S-8308

$IC_{50} = 15\ \mu M$

Fig. 5.2 Lead compounds of Takeda.

Analogue-based Drug Discovery. IUPAC, János Fischer, and C. Robin Ganellin (Eds.)
Copyright © 2006 WILEY-VCH Verlag GmbH & Co. KGaA, Weinheim
ISBN: 3-527-31257-9

Later, a group at Du Pont Merck postulated that both angiotensin II and the Takeda-leads were bound at the same receptor site. The suggestion was to obtain analogues by modifying the Takeda-leads, as these would better mimic angiotensin II. Initially, computer modeling was used to compare angiotensin II with S-8307 and S-8308, but because angiotensin II contains two acidic residues near the NH_2 terminus (namely Asp_1 and Tyr_4) that were not mimicked by the Takeda-leads, it was hypothesized that the latter compounds were missing an additional acidic functionality. Indeed, it was subsequently found that the 4-carboxy-derivative (Fig. 5.3) had a binding activity which was 10-fold greater than that of S-8308 [2].

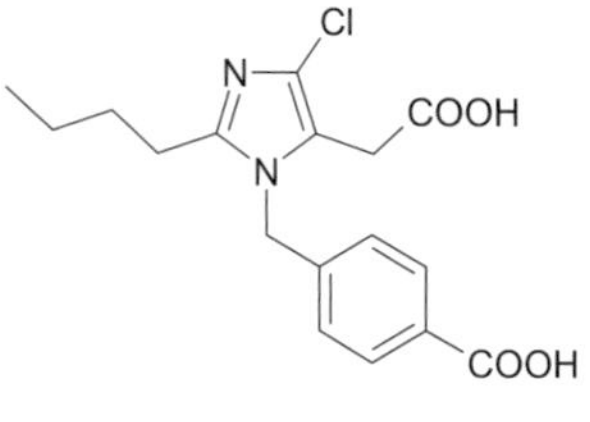

EXP-6155

$IC_{50} = 1.2 \ \mu M$

Fig. 5.3 Lead compound of Du Pont Merck.

Replacement of the 4-carboxy-group by a 2-carboxy-benzamido-moiety (EXP-6803) increased the binding affinity by yet another order of magnitude, but the compound was only active when administered intravenously. Replacement of the 2-carboxy-benzamido-group by a 2-carboxy-phenyl-group afforded a lipophilic biphenyl-containing EXP-7711 (Fig. 5.4), which for the first time exhibited good oral activity ($ED_{30} = 11$ mg kg^{-1}). In order to increase the oral activity further, the polar carboxy-group was replaced with a more lipophilic tetrazole group, affording losartan (Fig. 5.5), and this became the first successful angiotensin II antagonist drug.

Losartan (1986) was followed in 1990 by valsartan, candesartan, and irbesartan, while telmisartan was developed in 1991 and olmesartan in 2002. All of these molecules were based on the same lead-molecule and contained the characteristic biphenyl-methyl-group (Fig. 5.6).

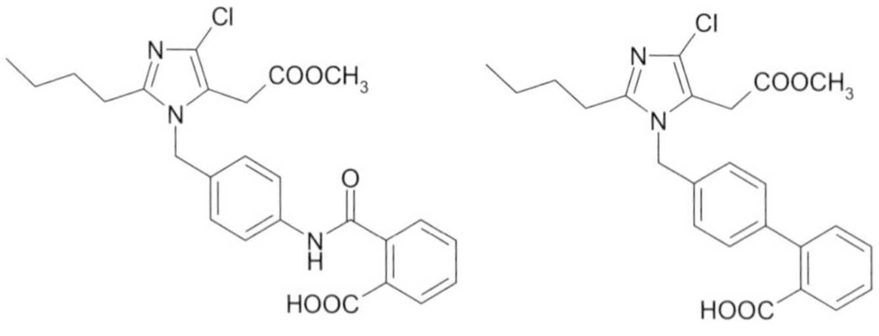

EXP-6803

EXP-7711

Fig. 5.4 Compounds of the lead optimization.

Eprosartan was developed in 1997 using a different lead-optimization from S-8308. The main structural change was not the extension of the N-benzyl group, but in order to mimic the C-terminal end of angiotensin II the 5-acetic acid group was replaced with an α-thienylacrylic acid and a 4-carboxy-substituent was introduced, as in the case of the above optimization (Fig. 5.7).

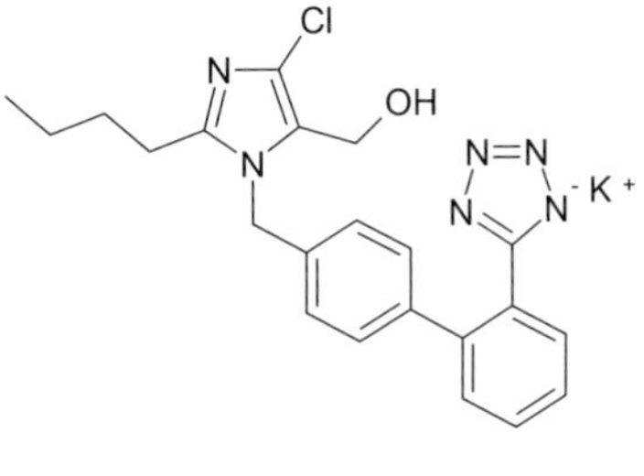

IC$_{50}$ 1,5 nM

Fig. 5.5 Losartan kalium.

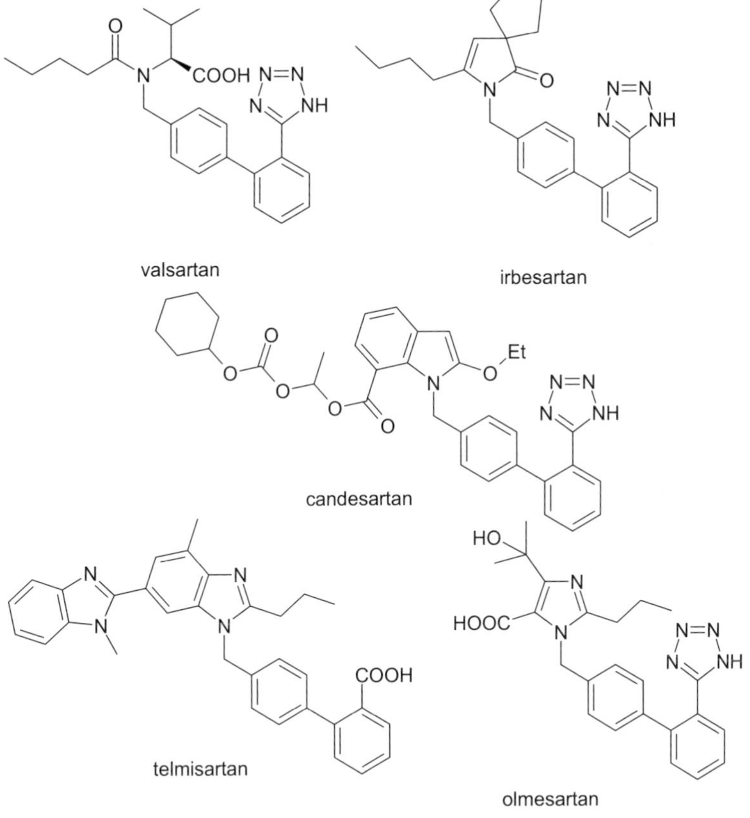

valsartan

irbesartan

candesartan

telmisartan

olmesartan

Fig. 5.6 Analogues: valsartan, irbesartan, candesartan, telmisartan, and olmesartan.

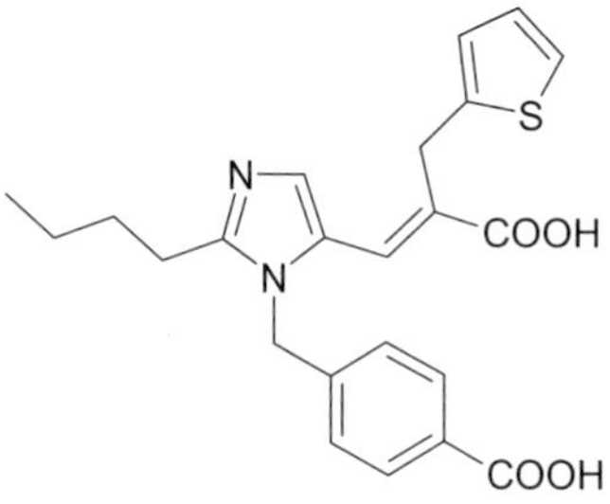

Fig. 5.7 Eprosartan.

5.2
Clinical Results with Angiotensin II Antagonists

5.2.1
Mechanisms of Action

Angiotensin II receptor blockers (ARBs) inhibit angiotensin II activity by selectively binding to and blocking the angiotensin II type 1 (AT_1) receptor. Virtually all known deleterious responses to angiotensin II, which is the main effector peptide of the renin–angiotensin–aldosterone system (RAAS), are mediated via stimulation of the AT_1 receptor. This is followed by vasoconstriction, aldosterone and vasopressin secretion, sympathetic nervous system activation, renal tubular sodium reabsorption, and decreased renal blood flow [3]. Chronic AT_1 receptor stimulation also contributes independently to diseases such as vascular smooth-muscle cell growth and proliferation, left ventricular hypertrophy (LVH), glomerulosclerosis, vascular media hypertrophy, endothelial dysfunction, neointima formation, atherosclerosis, stroke, and dementia [3,4]. Studies conducted in animals have shown that an effective blockade of brain AT_1 receptors has neuroprotective effects and improves the outcome in ischemic neuronal tissue.

A second key angiotensin II receptor (AT_2) has largely favorable effects on tissue growth and repair, and also causes vasodilation. Furthermore, it is involved in the control of cell proliferation, cell differentiation, angiogenesis, wound healing, tissue regeneration, and apoptosis. Thus, the actions of the AT_2 receptor tend to counteract those resulting from AT_1 stimulation. Selective AT_1 blockade with an ARB inhibits the negative cardiovascular consequences of AT_1 receptor stimulation [5]. Circulating angiotensin II levels show a compensatory rise during ARB therapy, thereby activating only the unopposed AT_2 receptors. This should preserve or even augment the favorable effects of angiotensin II, potentially producing benefits above and beyond those due to blood pressure control [3]. Because they act at the final step of the RAAS, ARBs block the effects of angiotensin II, regardless of whether it is generated systemically by ACE or within tissues by ACE-independent pathways (e.g., tonin, cathepsin, chimase).

Several studies have confirmed the antihypertensive efficacy of ARBs, although blood pressure reduction is only a surrogate end-point. In contrast, during the past few years clinical outcome trials have confirmed the value of these agents in "hard end-points", for example target organ protection.

None of the ARBs block degradation of the vasodilator bradykinin, and are therefore unlikely to cause the cough associated with the ACE inhibitor treatment. A differentiating class feature of ARBs is their excellent tolerability. In controlled clinical trials, the side-effect profiles of these agents are indistinguishable from those of placebo. The incidence of cough with ARBs is less than with ACE inhibitors [6]. The main differences among ARBs are apparent in their pharmacokinetic profiles (see below). An important consideration in the selection of an antihypertensive agent is its duration of action, as calculated from its half-life. Long-acting drugs that provide 24-h efficacy are preferred over short-acting agents for several reasons: compliance is better with once-daily dosing, fewer tablets may mean lower cost to the patient, and blood pressure control is smooth and sustained rather than intermittent.

5.2.1.1 Other Effects of ARBs

Losartan was shown to increase uric acid excretion by inhibiting its tubular reabsorption. This effect does not lie within the spectrum of the blockade of AT_1 receptors, but rather is a genuine action of the drug on renal tubuli. Losartan was also shown to decrease ocular pressure in normotensive as well as in hypertensive patients with or without glaucoma. Losartan and eprosartan also reduce central sympathetic tone. A metabolite of losartan, EXP-3179 (Fig. 5.8), does not block AT_1 receptors but may inhibit the atherosclerotic process by suppressing intercellular cell adhesion molecules (ICAM-1), cyclooxygenase-2 (COX-2) and thromboxane-A_2 (TXA-2). These effects of losartan may lead to potential indications other than blood pressure reduction in hypertensive patients [7].

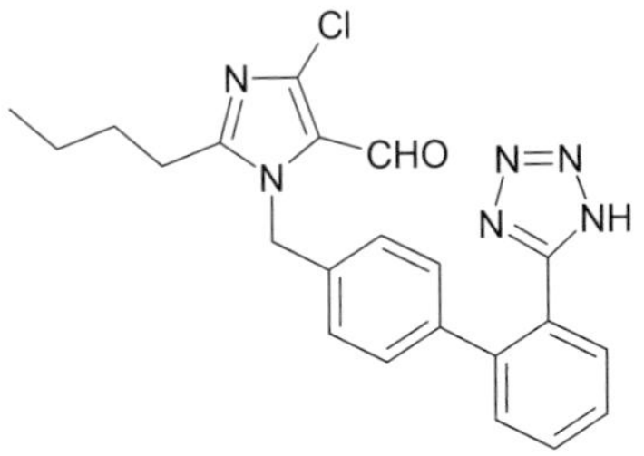

Fig. 5.8 A metabolite of losartan, EXP-3179.

5.2.2
Target Organ Protection

5.2.2.1 Left Ventricular Hypertrophy

Almost 40% of hypertensive patients have LVH which, after age, is the strongest predictor of cardiovascular disease, such as congestive heart failure (CHF), stroke, and coronary artery disease (CAD). Cardiovascular events occur in relation to left ventricular mass. Blood pressure reduction causes LVH regression, and therefore decreases the risk of all-cause, cardiovascular, and CAD mortality. ARBs have been shown to reduce LVH in a number of trials, including losartan in the LIFE [8] and irbesartan in the SILVHIA study [9].

5.2.2.2 Diabetic and Nondiabetic Nephropathy

Diabetes is the most frequent cause of end-stage renal disease. Hypertension, which is common among patients with type 2 diabetes, accelerates the development and progression of renal disease. Early and tight blood pressure control in diabetic patients, preferably with antihypertensive agents that have proven renoprotective properties, is therefore essential to minimize loss of kidney function. Several controlled clinical trials have investigated and proved the beneficial effects of ARBs on type 2 diabetic nephropathy [10–14].

5.2.2.3 Diabetes Prevention

Hypertension and type 2 diabetes mellitus are frequent co-morbid conditions resulting in a very high risk population. ARBs – losartan as compared to the beta-blocker atenolol in the LIFE Study [8], candesartan as compared to the control therapies in the SCOPE [16] and CHARM [33] Studies, valsartan as compared to amlodipine in the VALUE Study [34] have been found to have beneficial effects in preventing the development of type 2 diabetes mellitus.

5.2.2.4 Coronary Heart Disease (CHD)

Acute mycocardinal infarction (MI) is still associated with significant mortality. Effects of losartan, in the OPTIMAAL Study, and that of valsartan, in the VALI-ANT study, were proved to be equivalent or superior (less side effects) to ACE inhibitors after MI [15,35].

5.2.2.5 Congestive Heart Failure

The effects of losartan on morbidity/mortality outcomes were not different statistically from those of ACE-inhibitor captopril in a head-to-head comparative trial in the ELITE II Study [36], while the beneficial effects of valsartan was reported in the ValHeFT [37] Study, furthermore, candesartan in the CHARM Study [33] was significant and additional to the concomitant conventional heart failure therapy.

5.2.2.6 Stroke Prevention and Other CNS Effects

Losartan was proved to be more effective for stroke prevention in patients with hypertension and LVH than the beta adrenoceptor blocker atenolol in the LIFE Study [8]. Furthermore, losartan in a comparative trial with hydrochlorothiazide/atenolol [38], valsartan [39] were shown to inhibit the cognitive decline associated with hypertension in the elderly patients.

5.3
Differences Among Angiotensin AT$_1$ Receptor Blockers

Although the mechanisms of action of ARBs are similar – but not identical (see below) – these drugs differ in their pharmacokinetic features, as well as in their other actions (pleiotropic effects), not related to AT$_1$ receptor blockade. Some of these effects are listed in Tab. 5.1 and are detailed below.

Tab. 5.1 Some pharmacokinetic properties of ARBs.

Drug	Half-life [h]	Bioavailability [%]	Volume of distribution [L kg^{-1}]	Renal/hepatic clearance [%]
Candesartan	9	15	0.13	60/40
Eprosartan	5	13	13	30/70
Irbesartan	11–15	60–80	53–93	1/99
Losartan	2	33	34	10/90
EXP-3174[a]	6–9	–	12	50/50
Telmisartan	24	42–58	500	1199
Valsartan	6	~25	7	30/70

a. Active metabolite of losartan.

5.3.1
Structural Differences

Structural differences of ARBs influence the variation in pharmacological (pharmacokinetic /pharmacodynamic) properties of these drugs. Affinity for the AT$_1$ receptor is highest with candesartan, followed by irbesartan, losartan, valsartan, and telmisartan. Differences and affinities at the AT$_1$ receptor are also involved in cognitive function. Candesartan is the most effective ARB in crossing the blood–brain barrier.

5.3.2
AT$_1$ Receptor Antagonism

An interesting difference among the ARBs is the classification of insurmountable or surmountable angiotensin AT$_1$ receptor-blocking capacity. Insurmountable antagonism indicates suppression of agonist response, despite increasing agonist concentration. Insurmountable antagonists (e.g., candesartan, telmisartan, irbesartan, valsartan) are semi-irreversibly bound to their receptors and thereby may provide a greater antihypertensive effect [17–20]. Losartan, and its active metabolite (EXP-3174) (Fig. 5.9) is a classical competitive antagonist of AT$_1$ receptor.

The duration of receptor antagonism also varies, with candesartan having the longest duration, irbesartan having intermediate duration, and losartan having the shortest duration of antagonism, but its active metabolite (EXP-3174) has a much longer duration of action.

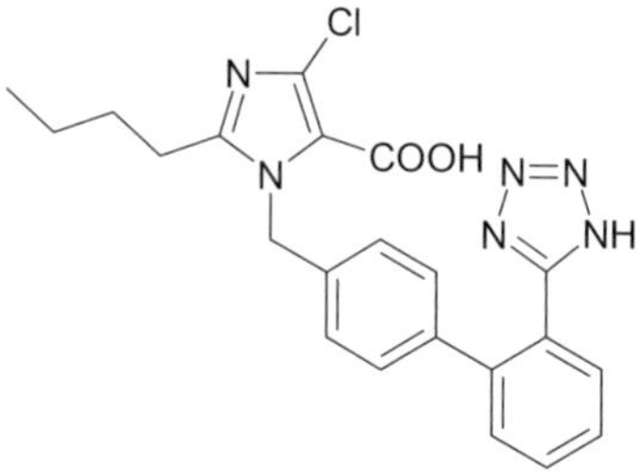

Fig. 5.9 The active metabolite of losartan, EXP-3174.

5.3.3
Pharmacokinetics/Dosing Considerations

Oral bioavailability does not appear to be of clinical significance due to the large therapeutic index of these drugs, and it is generally unaffected by food.

Candesartan cilexetil (converted to candesartan), olmesartan medoxomil (converted to olmesartan) and losartan (converted to EXP-3174 which is 10- to 40-times more potent than losartan) are prodrugs. While losartan is a reversible, surmountable competitive inhibitor, EXP-3174 is a reversible noncompetitive inhibitor of the AT$_1$ receptor; therefore, losartan may compete for the same receptor. All other ARBs are pharmacologically active substances. Losartan requires a 50% initial dose reduction in patients with impaired hepatic function [21]. All agents can be dosed once daily, and telmisartan has the longest plasma half-life.

5.3.4
Drug Interactions/Adverse Effects

Losartan and irbesartan have the highest, while candesartan, valsartan, and eprosartan have variable but modest affinity for cytochrome P-450 (CYP-450) isoenzymes. Fluconazole may inhibit the metabolism of losartan, causing increased antihypertensive effects. Indomethacin (and possibly other nonsteroidal antiinflammatory drugs), phenobarbital, and rifampicin may decrease the antihypertensive effect of losartan [22]. Telmisartan has no affinity for any of the CYP isoenzymes, but may significantly increase peak and trough concentrations of digoxin (49% and 20%, respectively). Thus, serum concentrations of digoxin should be monitored when initiating telmisartan therapy [21,23]. Telmisartan was found to activate PPAReceptor-γ [40] resulting in enhanced insulin sensitivity, an effect which may help preventing diabetes mellitus. This is to be proved by clinical trials.

5.3.5
Efficacy in Hypertension

There are some differences in blood-pressure lowering efficacy among ARBs [20]. The dose of ARB used appears critically important, confounding the results of many clinical trials. For example, a meta-analysis of five large multicenter trials revealed that telmisartan 80 mg was significantly more effective than losartan 50 mg and valsartan 80 mg in reducing 24-h mean blood pressure. The ARBs did not have equal duration of action. Towards the end of the dosing interval, a decrease in blood pressure control was seen with losartan 50 mg and valsartan 80 mg, while telmisartan (40 mg and 80 mg) maintained efficacy comparable to amlodipine throughout the 24-h dosing period, controlling the early morning surge in blood pressure [24]. It is important to note that this meta-analysis compared the maximum daily dose of telmisartan (80 mg once daily) to the initial (and not the maximal) recommended doses of losartan and valsartan, prohibiting a thoroughly objective comparison, as other studies revealed that the most effective dose for losartan is 100 mg, while that for valsartan is 320 mg.

The Antihypertensive Efficacy of Candesartan in Comparison to Losartan (CLAIM) study demonstrated a higher efficacy of candesartan 32 mg over losartan 100 mg in lowering both trough (by 3.5/2.2 mmHg) and peak (by 2.6/1.5 mmHg) systolic/diastolic blood pressures when both drugs are administered once daily [18]. However, if a diastolic blood pressure (DBP) reduction of 3 mmHg or more is selected as a meaningful clinical difference among ARBs (as in other hypertensive trials), the CLAIM study provides little evidence to show clear superiority of any ARB. Other clinical studies have demonstrated a greater antihypertensive effect of irbesartan over losartan [25]. The weighted average DBP reduction of all six ARBs from 51 prospective, double-blind, randomized, placebo-controlled trials including more than 12 000 patients showed the largest effect with irbesartan, and the least effect with eprosartan. The largest effect on systolic blood pressure

(SBP) was seen with candesartan, and the smallest with eprosartan. The control of SBP attracts more attention from the fact that SBP is more closely related to target organ damage than DBP. The beneficial effects of ARBs in patients with *isolated systolic hypertension* have been well-documented by the LIFE (losartan), Val-Syst (valsartan), and SCOPE (candesartan) studies [26–28].

With the newest ARB, olmesartan, a comparison with starting doses of losartan 50 mg, irbesartan 150 mg, and valsartan 80 mg, in a prospective randomized trial, showed a significantly greater reduction in DBP (11.5 mmHg) with olmesartan, followed by the other three ARBs (8.2, 9.9, 7.9 mmHg, respectively). Olmesartan appeared to have a greater dose-dependent blood pressure-lowering effect than irbesartan or candesartan, followed by losartan, telmisartan, and valsartan [29].

In addition to its antihypertensive effects, the ARB losartan demonstrated cardiovascular benefits beyond lowering blood pressure in the Losartan Intervention for Endpoint (LIFE) Reduction in Hypertension Study [8]. Similarly, results from the Valsartan in Acute Myocardial Infarction Trial showed that valsartan improves survival after acute MI in high-risk patients [15]. Due to a lack of clinical studies involving populations with the same cardiovascular risk profile, it is not possible to extrapolate these findings to all ARBs, particularly since the dose of ARB used appears critically important [7]. Several ARBs carry additional indications, including treatment of hypertension with LVH, stroke prevention, CHF, and nephropathy in typ. 2 diabetes mellitus.

Losartan appears to be unique among the ARBs because it inhibits the urate anion transport in renal proximal tubuli; hence, it increases uric acid excretion and decreases plasma levels of uric acid in hypertensive patients. This effect is not the consequence of AT_1 receptor blockade. Since high uric acid levels have been associated with cardiovascular morbidity/mortality, losartan may be the best ARB for patients with gout [7,30,31]. Moreover, losartan was found to decrease ocular pressure in normotensive as well as in hypertensive patients [32].

The question as to whether the differences in efficacy (a few mmHg) among the ARBs translates to meaningful end-points (e.g., reduced cardiovascular morbidity and mortality) remains to be answered by further clinical pharmacological trials. It is possible, however, that ARBs may have further potential indications in the prevention of cognitive decline.

5.4
Summary

The ability of ARBs to reduce blood pressure in hypertensive patients is unequivocal, and several large-scale, prospective, controlled trials with ARBs have provided evidence that blocking the actions of angiotensin II at the AT_1 receptor also improves the prognosis above and beyond the effect on blood pressure in a broad range of patients at high cardiovascular risk. The results of some of these trials have already expanded the indications for ARBs. For example, losartan is now indicated for reducing the risk of stroke in patients with hypertension and LVH;

irbesartan, losartan and valsartan are indicated for the treatment of diabetic nephropathy in patients with typ. 2 diabetes and hypertension; and losartan and valsartan are indicated for the treatment of heart failure.

Although variations have been identified in the antihypertensive effects (efficacy and duration of action) of different ARBs, the results are conflicting, and hence no definitive conclusion can be drawn on the superiority of one ARB over any other. Special features of the drug (e.g., the uricosuric effect of losartan) may be important in selected patients. Differences in safety and side-effect profile are nil or minimal among the different agents, with all ARBs being extremely well tolerated as compared to other classes of antihypertensive medications.

References

1 Furukawa Y, Kishimoto S, Nishikawa K. **1982**. US Patents 4,340,598 and 4,355,040.

2 Duncia JV, Chiu AT, Carini DJ, Gregory GB, Johnson AL, Price WA, Wells GJ, Wong PC, Calabrese JC, Timmermans PBMWM. *J. Med. Chem.*, **1990**, 33, 1312.

3 Unger T. *Am. J. Cardiol.*, **2002**, 89 (2A), 3–9A.

4 Kaschina E, Unger T. *Blood Press.*, **2003**, 12(2), 70–88.

5 McClellan KJ, Balfour JA. Eprosartan. *Drugs*, **1998**, 55(5), 713–718.

6 Pylypchuk GB. *Ann. Pharmacother.*, **1998**, 32(10), 1060–1066.

7 Goa KL, Wagstaff AJ. *Drugs*, **1996**, 51, 820–845.

8 Dahlöf B, Devereux RB, Kjeldsen SE, Julius S, Beevers G, de Faire U, Fyhrquist F, Ibsen H, Kristiansson K, Lederballe-Pedersen O, Lindholm LH, Nieminen MS, Omvik P, Oparil S. *Lancet*, **2002**, 359, 995–1003.

9 Malmqvist K, Kahan T, Edner M, Held C, Hagg A, Lind L, Muller-Brunotte R, Nystrom F, Ohman KP, Osbakken MD, Ostergern J. *J. Hypertens.*, **2001**, 19(6), 1167–1176.

10 Brenner BM, Cooper ME, de Zeeuw D, Keane WF, Mitch WE, Parving HH, Remuzzi G, Snapinn SM, Zhang Z, Shahinfar S. *N. Engl. J. Med.*, **2001**, 345(12), 861–869.

11 Parving HH, Lehnert H, Brochner-Mortensen J, Gomis R, Andersen S, Arner P. *N. Engl. J. Med.*, **2001**, 345, 870–878.

12 Viberti G, Wheeldon NM. *Circulation*, **2002**, 106, 672–678.

13 Mogensen CE, Neldam S, Tikkanen I, Oren S, Viskoper R, Watts RW, Cooper ME. *Br. Med. J.*, **2000**, 321, 1440–1444.

14 Nakao N, Yoshimura A, Morita H, Takada M, Kayano T, Ideura T. *Lancet*, **2003**, 361(9352), 117–124.

15 Pfeffer MA, McMurray JJV, Velazquez EJ, Rouleau JL, Køber L, Maggioni AP, Solomon SD, Swedberg K, Van de Werf F, White H, Leimberger JD, Henis M, Edwards S, Zelenkofske S, Sellers MA, Califf RM. *N. Engl. J. Med.*, **2003**, 349, 1893–1906.

16 Lithell H, Hansson L, Skoog I, Elmfeldt D, Hofman A, Olofsson B, Trenkwalder P, Zanchetti A. *J. Hypertens.*, **2003**, 21, 875–886.

17 Morsing P, Adler G, Brandt-Eliasson U, Karp L, Ohlson K, Renberg L, Sjöquist P-O. *Hypertension*, **1999**, 33, 1406–1413.

18 Bakris G, Gradman A, Reif M, Wofford M, Munger M, Harris S, Vendetti J, Michelson EL, Wang R. *J. Clin. Hypertens.*, **2001**, 3, 16–21.

19 Maillard MP, Perregaux C, Centeno C, Stangier J, Wienen W, Brunner H-R, Burnier MJ. *J. Pharmacol. Exp. Ther.*, **2002**, 302, 1089–1095.

20 Sica DA. *J. Clin. Hypertens.*, **2001**, 3, 45–49.

21 Kastrup EK. (Ed.) *Drug Facts and Comparisons.* 56th edn. St. Louis: Wolters Kluwer Health, **2002**: 427–613.

22 Conlin PR, Elkins M, Liss C, Vrecenak AJ, Barr E, Edekman J. *J. Hum. Hypertens.*, **1998**, 12, 693–700.

23 Unger T., Kaschina E. *Drug Safety,* **2003**, 26, 707–720.

24 Neutel JM, Smith DH. *J. Clin. Hypertens.*, **2003**, 5, 58–63.

25 Oparil S, Guthrie R, Lewin AJ, Marbury T, Reilly K, Triscari J, Witcher JA. *Clin. Ther.*, **1998**, 20, 398–409.

26 Kjeldsen SE, Dahlöf B, Devereux RB, Julius S, Aurup P, Edelman J, Beevers G, de Faire U, Fyhrquist F, Ibsen H, Kristianson K, Lederballe-Pedersen O, Lindholm LH, Nieminen MS, Omvik P, Oparil S, Snapinn S, Wedel H. *JAMA,* **2002**, 288(12), 1491–1498.

27 Malacco E, Vari N, Capuano V, Spagnuolo V, Borgnino C, Palatini PA. *Clin. Ther.*, **2003**, 25(11), 2765–2780.

28 Papademetriou MD, Farsang C. *J. Am. Coll. Cardiol.*, **2004**, 44, 1175–1180.

29 Greathouse M. *Congest. Heart Fail.*, **2002**, 8, 313–320.

30 Puig JG. *J. Hypertens.*, **2002**, 20 (Suppl. 5), S29–S31.

31 Weir CJ. *Stroke,* **2003**, 34, 1956–1957.

32 Costagliola C. *Clin. Drug Invest.*, **1999**, 17(4), 329–332.

33 Pfeffer MA, et al. *Lancet,* **2003**, 362, 759–66.

34 Julius,S. et al. *Lancet,* **2004**, 363, 2022–2031.

35 Dickstein K, et al. *Lancet,* **2002**, 360, 752–760.

36 Pitt B, et al. *Lancet,* **2000**, 355, 1582–7.

37 Cohn JN, et al. *N Engl J Med.*, **2001**, 345, 1667–75.

38 Fogari R, et al. *J Hum Hypertens,* **2003**, 17, 781–785.

39 Fogari R,et al. *Eur J Clin Pharmacol.*, **2004**, 59, 863–868.

40 Benson SC. et al. *Hypertension,* **2004**, 43, 993–1002.

6

Optimizing Antihypertensive Therapy by Angiotensin-Converting Enzyme Inhibitors

Sándor Alföldi and János Fischer

6.1
Medicinal Chemistry of ACE-inhibitors

The renin–angiotensin system (RAS) controls blood pressure, blood volume and electrolytic balance. Two main enzymes influence the release of angiotensin II from its endogenous precursor, angiotensinogen: renin and angiotensin-converting enzyme (ACE) (Fig. 6.1).

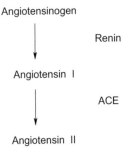

Fig. 6.1 The renin–angiotensin system.

Angiotensinogen, an a_2-globulin, consists of 452 amino acids and is synthesized by the liver. Renin cleaves this protein, affording angiotensin I, an inactive decapeptide. In turn, this decapeptide is cleaved by ACE to give angiotensin II, a potent vasoconstrictor. ACE is not a specific enzyme, but also cleaves bradykinin, a potent vasodilator. Both activities are involved in raising blood pressure.

One of the most successful stories of drug research is in connection with the ACE-inhibitor drugs. This research was initiated by the discovery of a natural product, a nonapeptide, which was isolated from snake venom and called teprotide (Fig. 6.2).

pyro-Glu - Trp - Pro - Arg - Pro - Gln - Ile - Pro - Pro - OH

Fig. 6.2 Teprotide.

Analogue-based Drug Discovery. IUPAC, János Fischer, and C. Robin Ganellin (Eds.)
Copyright © 2006 WILEY-VCH Verlag GmbH & Co. KGaA, Weinheim
ISBN: 3-527-31257-9

Teprotide was used as an active antihypertensive drug in patients with essential hypertension, but due to a lack of oral activity it was administered parenterally. Teprotide has been the first lead for novel orally active ACE-inhibitors, with three important lead-molecules having been discovered as follows: succinyl-L-proline, (2S)-methylsuccinyl-L-proline and 3-mercaptopropanoyl-L-proline affording at the end the first successful drug in this class: captopril, as discovered by Ondetti, Rubin and Cushman [1] (Fig. 6.3).

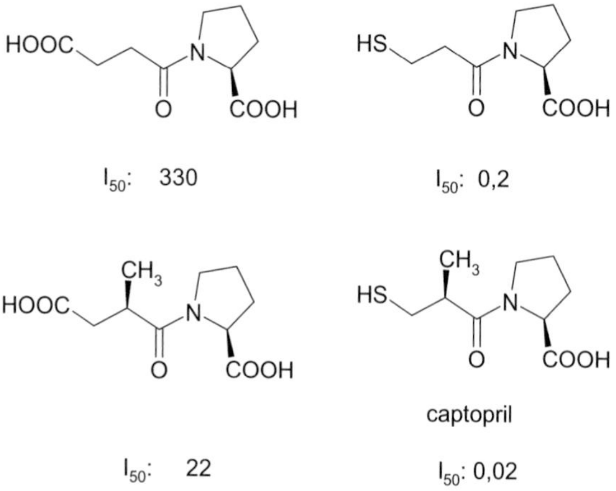

I_{50}: 330

I_{50}: 0,2

captopril

I_{50}: 22

I_{50}: 0,02

Fig. 6.3 The design of captopril: the lead molecules.
I_{50}: concentration (μM) required to inhibit 50% of rabbit lung ACE activity.

Captopril was the first orally active ACE-inhibitor to reach the market-place. Captopril represents the sulfhydryl-containing subtype of ACE-inhibitors and, when given orally it is rapidly absorbed, with a bioavailability of ca. 75%. Peak drug concentrations in plasma occur within 1 h of dosing, and the drug is then cleared rapidly ($t_{1/2}$ ca. 2 h); therefore, a dosage regimen of two- to three-times daily is necessary. Since food reduces the oral bioavailability of captopril, the drug should be given 1 h before meals. The only direct analogue of captopril is zofenopril (Fig. 6.4), which has a somewhat longer elimination half-life (4.5 h) (see Tab. 6.1).

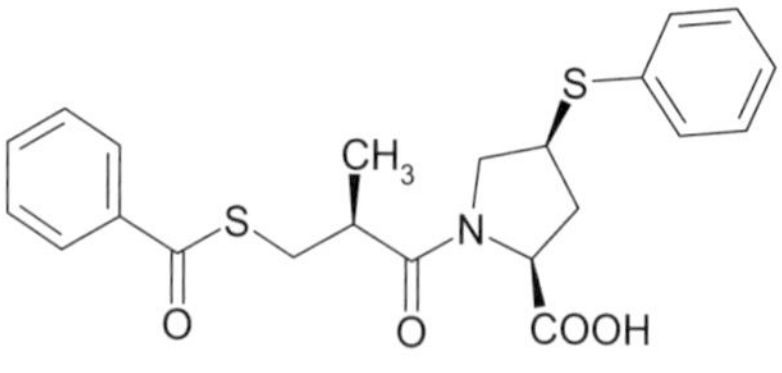

Fig. 6.4 Zofenopril.

Researchers at Merck sought the possibility of replacing the sulfhydryl-group of captopril, and enalapril, as an early-phase analogue was subsequently discovered by this group [2]. Enalapril is the prototype of several dicarboxylate-containing ACE-inhibitor-analogues, and Patchett and colleagues used the same lead-molecule, namely succinyl-L-proline, to follow the route: a homologue, a (2S)-methyl-derivative, its NH-isostere and ultimately its substituent-optimization by a (S)-phenylethyl-moiety, which resulted in the discovery of enalaprilat [3] (Fig. 6.5).

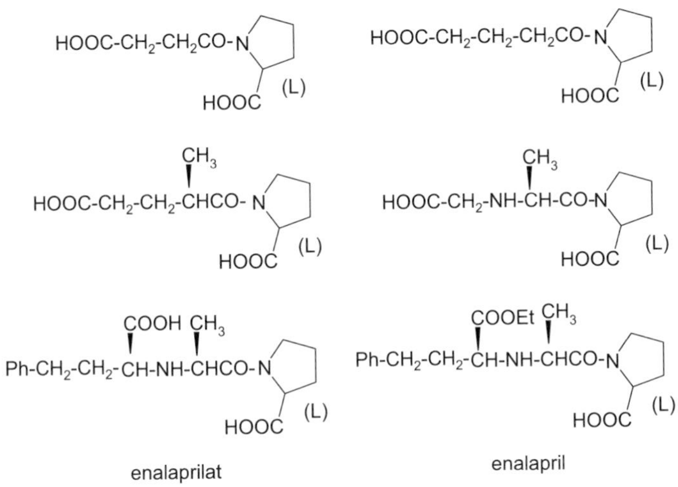

Fig. 6.5 The design of enalapril: the lead molecules.

Enalaprilat proved to be significantly more active than captopril, although its oral bioavailability was low. The monoethyl-ester pro-drug, enalapril, was used to solve this problem to provide an ACE-inhibitor with a longer duration of action because of the longer plasma half-life of enalaprilat (11 h). Enalapril has one further advantage over captopril, in that its oral bioavailability is not reduced by the presence of food.

Today, there are several enalapril analogues available on the market. Lisinopril is a lysine analogue of enalaprilat but, unlike enalaprilat, is itself active. The *in-vitro* potency of lisinopril is slightly higher than that of enalaprilat. Moreover, lisinopril is cleared unchanged by the kidney and its half-life in plasma is about 12 h, which is slightly longer than that of enalapril. The enalapril-like dicarboxylate-containing inhibitors (benazepril, cilazapril, imidapril, lisinopril, moexipril, perindopril, quinapril, ramipril, spirapril, temocapril, and trandolapril) (Figs. 6.6–6.16) have rather similar therapeutic applications.

The lone phosphinate-containing inhibitor that is currently available is fosinopril (Fig. 6.17), which is also a prodrug. This molecule is a third subtype of ACE-inhibitors, and can also be used in patients with renal impairment, since it is eliminated by both renal and hepatic pathways. This contrasts with the other ACE-inhibitors (with the exception of spirapril and trandolapril), all of which are eliminated only via the renal route.

Fig. 6.6 Benazepril.

Fig. 6.7 Cilazapril.

Fig. 6.8 Imidapril.

Fig. 6.9 Lisinopril.

Fig. 6.10 Moexipril.

Fig. 6.11 Perindopril.

Fig. 6.12 Quinapril.

Fig. 6.13 Ramipril.

Fig. 6.14 Spirapril.

Fig. 6.15 Temocapril.

Fig. 6.16 Trandolapril.

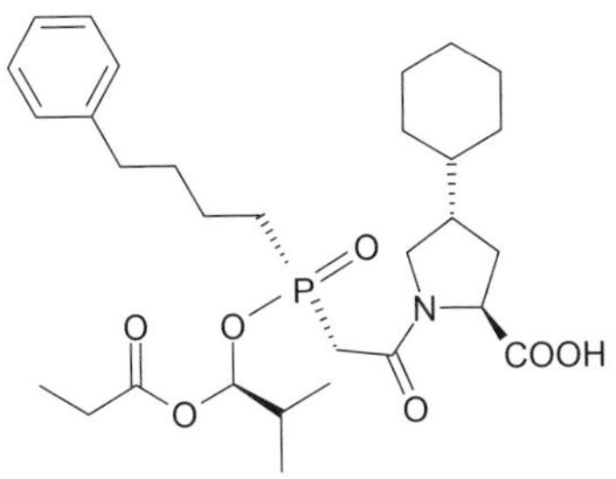

Fig. 6.17 Fosinopril.

6.2
Clinical Results with ACE-Inhibitors

6.2.1
Hemodynamic Effects

ACE-inhibitors reduce total peripheral vascular resistance and promote natriuresis, but cause little change in heart rate. In contrast to other vasodilators, no reflex tachycardia is observed, possibly due to an effect on baroreceptor sensitivity, vagal stimulation and/or reduced stimulation of sympathetic nerve activity. ACE-inhibitors reverse cardiac hypertrophy in hypertensive patients and reduce endothelial dysfunction in normotensive patients with coronary artery disease, hypertension, non-insulin-dependent diabetes mellitus and heart failure. An improvement in endothelial function is related to the attenuation of vasoconstriction and to the increased bradykinin-induced production of endothelium-derived nitric oxide (NO). In patients with congestive heart failure (CHF), ACE-inhibitors induce venous and arterial vasodilatation. Venous vasodilatation leads to increased peripheral venous capacitance, and to reductions in right atrial pressure, pulmonary arterial pressure, and capillary wedge pressures. The left ventricular filling volume and pressure are also reduced, producing a rapid relief of pulmonary congestion.

All components of the RAS can be found in the brain, heart, vasculature, adipose tissue, gonads, pancreas, placenta, and kidney, among others. Biochemical measurements of ACE activity show that the enzyme is tissue-based. Indeed, <10% of ACE is found circulating in the plasma [4]. The potential importance of the tissue RAS is supported by observations that the beneficial effects of RAS blockers cannot reliably be predicted by measurements of the activity of the circulating RAS. The antihypertensive actions of ACE-inhibitors are better correlated with inhibition of tissue ACE rather than plasma ACE, and hypertensive patients with normal or even low levels of systemic RAS activity can be effectively treated with inhibitors of the RAS. The intrarenal RAS is hypothesized to regulate systemic blood pressure and aspects of renal function such as blood flow and sodium reabsorption. In the brain, the RAS may facilitate neurotransmission and stimulate vasopressin release and sympathetic outflow, whereas the cardiovascular RAS is probably not required for the maintenance of normal cardiac and vascular function, though it may have a role in promoting atherosclerosis and heart failure. The degree of inhibition of tissue ACE produced by an ACE-inhibitor is directly dependent on the binding affinity of the inhibitor and the concentration of the free inhibitor in the tissue (depending mainly on tissue retention). The rank order of potency of several different ACE-inhibitors has been determined by investigators using competition analyses and by direct binding of tritium-labeled ACE-inhibitors to tissue ACE. The ranked potency determined was: quinaprilat = benazeprilat > ramiprilat > perindoprilat > lisinopril > enalaprilat > fosinoprilat > captopril [5].

6.2.2
Effects of ACE-Inhibitors

In most patients, ACE-inhibitors are well tolerated, although several adverse reactions may occur.

6.2.2.1   Hypotension
Symptomatic hypotension due to the withdrawal of angiotensin II-mediated vasoconstrictor tone can occur, especially after the first dose of an ACE-inhibitor, and particularly in patients with high plasma renin activity (e.g., patients with salt depletion due to high doses of diuretics, or with CHF).

6.2.2.2   Dry Cough
This appears in 5–10% of patients, and is not always easy to distinguish from that caused by pulmonary congestion or concomitant diseases (e.g., respiratory disease). The etiology is unknown, but it may be related to increased levels of bradykinin and/or substance P in the lungs. The cough is not dose-dependent, and is more frequent among women and in Asian populations. It usually develops between 1 week and a few months of treatment, and sometimes requires treatment discontinuation, even if some patients may tolerate reinstitution of the ACE-inhibitor after a drug-free period. Once the therapy is stopped, the cough usually disappears within 3–5 days. There are no differences in the propensity of cough among the different ACE-inhibitors.

6.2.2.3   Hyperkalemia
Hyperkalemia due to a decrease in aldosterone secretion is rarely found in patients with normal renal function, but it is relatively common in those with CHF and in the elderly. Hyperkalemia is more frequent in patients with renal impairment, diabetes, and in those receiving either K^+ or potassium K^+-sparing diuretics, heparin or non-steroidal anti-inflammatory drugs (NSAIDs).

6.2.2.4 Acute Renal Failure

ACE-inhibitors can increase blood urea nitrogen or creatinine levels. In most patients, creatinine levels either will remain stable or decrease towards pre-treatment values during continued therapy. Acute renal failure is more frequent in patients with volume depletion due to high doses of diuretics, hyponatremia, bilateral renal artery stenosis, stenosis of the dominant renal artery or a single kidney and renal transplant recipients. Under these circumstances, renin release is increased, and this in turn leads to an increase in angiotensin II levels that produces a selective efferent arteriolar constriction and helps to maintain the glomerular filtration rate. ACE-inhibitors reduce angiotensin II levels, produce efferent arteriolar vasodilatation, and reduce glomerular filtration, leading to an increase in creatinine levels. Older patients with CHF are particularly susceptible to ACE-inhibitor-induced acute renal failure, although in almost all patients a recovery of renal function occurs after discontinuation of ACE-inhibitor therapy.

6.2.2.5 Proteinuria

Although ACE-inhibitor therapy can cause proteinuria, pre-existing proteinuria is not a contraindication for ACE-inhibitor treatment, as they have been found to exert nephroprotective effects in renal diseases associated with proteinuria (i.e., diabetic nephropathy).

6.2.2.6 Angioedema

Angioedema is a rare, but potentially life-threatening, side effect of ACE-inhibitor therapy. Symptoms range from mild gastrointestinal disturbances (nausea, vomiting, diarrhea, colic) to severe dyspnea due to laryngeal edema, and even death. Angioedema is more frequent within the first month of therapy, and among black patients, but disappears within hours after cessation of the ACE-inhibitor treatment. The mechanism appears to involve an accumulation of bradykinin and its metabolite des-arginine-bradykinin, together with inhibition of the complement-1 esterase inactivator.

6.2.2.7 Teratogenic Effects

When administered during the second or third trimester of pregnancy, ACE-inhibitors can cause fetal abnormalities, including oligohydramnios, pulmonary hypoplasia, fetal growth retardation, renal dysgenesis neonatal anuria and neonatal death.

6.2.2.8 Other Side Effects

Other side effects, not related to ACE inhibition, include ageusia and other taste disturbances (especially in the elderly), neutropenia, and maculopapular rash.

Neutropenia is rare, and occurs more frequently in patients with renal or collagen vascular disease. These side effects are dose-related and might derive from toxic effects caused by the sulfhydryl group (captopril).

6.2.3
Contraindications

A history of angioneurotic edema and bilateral renal artery stenosis are absolute contraindications for the initiation of ACE-inhibitor treatment. Although ACE-inhibitors are not contraindicated in women of reproductive age, they should be discontinued as soon as pregnancy is suspected or diagnosed. Low blood pressures (systolic blood pressure ≥ 90 mmHg) during ACE-inhibitor treatment are acceptable if the patient is asymptomatic. However, if the serum potassium level rises to ≥ 6.0 mmol L^{-1}, or the serum creatinine level increases by $\geq 50\%$ or to >3 mg dL^{-1} (265 mmol L^{-1}), the administration of ACE-inhibitors should be stopped. Moderate renal insufficiency (serum creatinine 3 mg dL^{-1} or up to 265 μmol L^{-1}), mild hyperkalemia (<6.0 mmol L^{-1}) and relatively low blood pressure (systolic blood pressure as low as 90 mmHg) are not contraindications to ACE-inhibitor treatment, but therapy should be maintained with renal function being carefully monitored. The risk of hypotension and renal dysfunction increases with high doses, in elderly patients or in patients with severe heart failure, those treated with high doses of diuretics, with renal dysfunction or hyponatremia. ACE-inhibitors, as well as other vasodilators, should also be avoided in patients with dynamic left ventricular outflow tract obstruction.

6.2.4
Drug Interactions

Antacids may reduce the availability of ACE-inhibitors, while NSAIDs may reduce their vasodilatory effects. K^+-sparing diuretics, K^+ supplements or low-salt substitutes with a high K^+ content may exacerbate ACE-inhibitor-induced hyperkalemia and thus, these combinations should be avoided. However, with careful monitoring, the combination of an ACE-inhibitor and spironolactone may be advantageous. If serum urea or creatinine levels rise excessively, then discontinuation of concomitant nephrotoxic drugs (e.g., NSAIDs, cyclosporine) should be considered. ACE-inhibitors may increase plasma levels of digoxin and lithium, whilst patients taking diuretics may be particularly sensitive to the vasodilatory effects of ACE-inhibitors. In some studies, the concomitant administration of salicylate reduced the effectiveness of ACE-inhibitors in patients with CHF. However, in a large study which included over 12 000 patients, there was little evidence for reducing the benefit of ACE-inhibition in the presence of aspirin, as long as low-dose aspirin (≥ 100 mg) was used [6].

6.3
Differences Among ACE-Inhibitors

There appear to be significant differences between the pioneer drug, captopril, and the enalapril analogues. Differences in these characteristics influence mainly the onset and duration of action. Captopril has a short onset time (1 h) and a relatively short duration of action, and therefore is administered three times daily. Captopril has potential drug–food interactions, and is the only agent that should be spaced from meals. Therefore, captopril is no longer used as a first-choice ACE-inhibitor in clinical practice, except for hypertensive emergencies, acute myocardial ischemia, and acute CHF.

For patients with severe liver disease, captopril and lisinopril (i.e. not prodrugs, and not requiring hepatic activation and having almost solely renal elimination) are recommended. Enalaprilat (Fig. 6.18), the active metabolite of enalapril, is the only available ACE-inhibitor which is given intravenously, and can be used in patients with severe liver dysfunction.

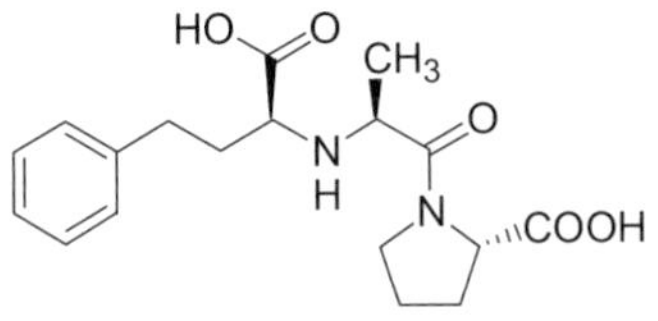

Fig. 6.18 Enalaprilat.

Fosinopril and spirapril – the only drugs with compensatory dual routes of elimination – do not require dosage adjustment in patients with reduced renal function, as do other ACE-inhibitors.

The ACE-inhibitors also differ in their dialyzability, half-life, lipophilicity, trough-to-peak ratios, approved indications, and therapeutic information available for many indications. ACE-inhibitors are generally characterized by their flat dose–response curves; lisinopril is the only such drug that exhibits a linear dose–response curve. Only trandolapril, lisinopril, ramipril and fosinopril have trough-to-peak effect ratios in excess of 50%. The importance of long-acting (>24 h) once-daily dosing has recently been underlined [9]. The practical benefits observed with a once-daily antihypertensive drug include improved patient compliance, which is of particular interest in long-term treatments, and the fact that patients need to take fewer tablets. The clinical advantage of full 24-h control of blood pressure also becomes clear when considering the incidence of end-organ damage and most adverse cardiovascular events (e.g., MI and stroke). These seem to follow a circadian rhythm, reaching a peak in the early morning hours after patients wake and get up [10,11].

ACE-inhibitors differ in their lipid solubility and in their affinity for tissue-bound ACE. It has been hypothesized that tissue ACE affinity might be responsible for some of the beneficial cardiovascular properties of ACE-inhibitors. However, an examination of this point by a meta-analysis did not identify any correlation between tissue ACE affinity and risk of first nonfatal MI in patients with hypertension [7].

The pharmacokinetic properties and doses of the ACE-inhibitors are summarized in Tab. 6.1.

Tab 6.1 Pharmacokinetic properties and doses of ACE inhibitors.

Drug	Elimination half-life [h]	Renal elimination [%]	Dose [mg] (i) Standard dose in hypertension (ii) Starting dose (iii) Target dose in heart failure
Pioneer drug			
Captopril	2	95	(i) 25–100 t.i.d. (ii) 6.25 t.i.d (iii) 50–100 t.i.d.
Analogue drugs			
Benazepril	11	85	(i) 2.5–20 b.i.d.
Cilazapril	10	80	(i) 1.25–5 daily
Enalapril	11	88	(i) 2.5–20 b.i.d. (ii) 2.5 b.i.d. (iii) 10–20 daily
Fosinopril	12	50[a]	(i) 10–40 daily
Lisinopril	12	70	(i) 2.5–10 daily (ii) 2.5–5 daily (iii) 30–35 daily
Perindopril	>24	75	(i) 4–8 daily
Quinapril	2–4	75	(i) 10–40 daily
Ramipril	8–14	85	(i) 2.5–10 daily (ii) 2.5 daily (iii) 10 daily
Spirapril	1.6	50[a]	(i) 3–6 daily
Trandolapril	16–24	15[a]	(i) 1–4 daily (ii) 1 daily (iii) 4 daily
Zofenopril	4.5	60[a]	(i) 7.5–30 b.i.d.

a. Significant hepatic elimination.
b.i.d.: Twice daily; t.i.d.: three times daily.

6.4
Summary and Outlook

ACE-inhibitors may be considered as first-choice therapy in patients with all forms of primary hypertension, but they are preferred in hypertension associated with heart failure, reduced systolic left ventricular ejection fraction or diabetic nephropathy, previous MI or stroke, chronic kidney disease and patients with high coronary disease risk, based on the compelling evidence of the efficacy of these drugs in such patient populations [8].

The pioneer drug, captopril, had a relatively short duration of action. Nonetheless, with sublingual administration it is used to elicit beneficial hemodynamic and clinical effects in hypertensive crises, acute myocardial ischemia, and acute CHF. Several long-acting analogue drugs are used nowadays as a first-choice therapy in cardiovascular diseases, and their minor differences have been summarized in this chapter.

Recent trials have suggested that inhibitors of the RAS, such as ACE-inhibitors and angiotensin II receptor blockers, may reduce the incidence of new-onset diabetes in patients with or without hypertension and at high risk of developing diabetes. This has been explained by hemodynamic effects, such as improved delivery of insulin and glucose to the peripheral skeletal muscle, and nonhemodynamic effects, including direct effects on glucose transport and insulin signaling pathways, all of which decrease insulin resistance. Improved beta-cell survival and preservation of islet function are also important pathways. Prospective clinical studies with the primary end-point of diabetes prevention are now indicated to explore further whether inhibitors of the RAS are superior to other antihypertensive agents. As the prevalence of type 2 diabetes is increasing worldwide, and diabetes is a major risk factor for cardiovascular mortality and morbidity, the development of new ACE-inhibitors with a particular antidiabetic effect would be a very promising target.

References

1 Ondetti MA, Rubin B, Cushman DW. *Science*, **1977**, 196, 441–444.

2 Patchett AA, Harris E, Tristram EW, et al. *Nature*, **1980**, 288, 280–283.

3 Patchett AA. In: *Chronicles of Drug Design*. Lednicer D (Ed.). American Chemical Society, New York, **1993**.

4 Cushman DW, Cheung HS. *Biochim. Biophys. Acta*, **1971**, 250, 261–265.

5 Dzau VJ, Bernstein K, Celermajer D, et al. *Am. J. Cardiol.*, **2001**, 88, 1L–20L.

6 Flather MD, Yusuf S, Kober L, et al. *Lancet*, **2000**, 355, 1575–1581.

7 Sauer WH, Baer JT, Berlin JA, et al. *Am. J. Cardiol.*, **2004**, 94(9), 1171–1173.

8 The Task Force on ACE-inhibitors of the European Society of Cardiology. *Eur. Heart J.*, **2004**, 25, 1454–1470.

9 Weber MA. *Am. J. Cardiol.*, **2002**, 89 (Suppl.), 27A–33A.

10 Muller JE, Stone PH, Turi ZG, et al. *N. Engl. J. Med.*, **1985**, 313, 1315–1322.

11 Lago A, Geffner D, Tembl J, et al. *Stroke*, **1998**, 29, 1873–1875.

7
Case Study of Lacidipine in the Research of New Calcium Antagonists

Giovanni Gaviraghi

7.1
Introduction

The calcium channel-blocking agents are a chemically, pharmacologically, and therapeutically heterogeneous group of drugs exemplified by three prototypes: nifedipine (Fig. 7.1), verapamil (Fig. 7.2), and diltiazem (Fig. 7.3). These drugs owe their pharmacological action to their ability to interfere with the L class of voltage-gated calcium channels which can be considered as pharmacological receptors with specific sites of interaction for agonists and antagonists. Calcium antagonists are widely used in cardiovascular therapy, for example in the treatment of essential hypertension and coronary vascular disease in which increased intracellular calcium concentration provokes vascular smooth muscle hyperreactivity and contractility. Calcium antagonists are able to block calcium inward movements into the cell and therefore to relax arterial smooth muscles, while they have little effect on venous beds.

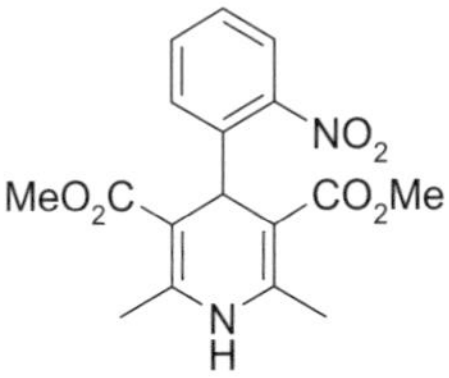

Fig. 7.1 Nifedipine.

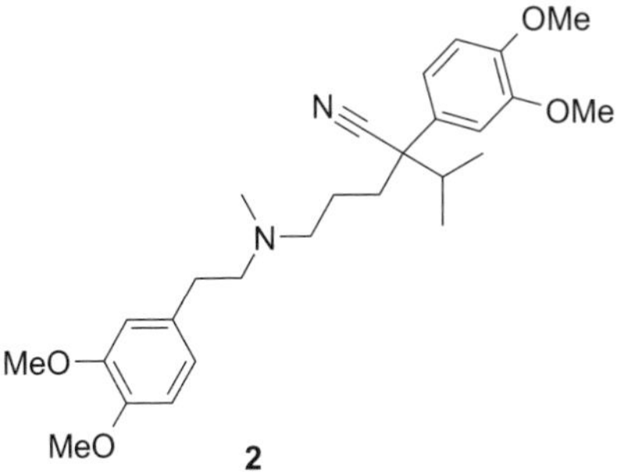

Fig. 7.2 Verapamil.

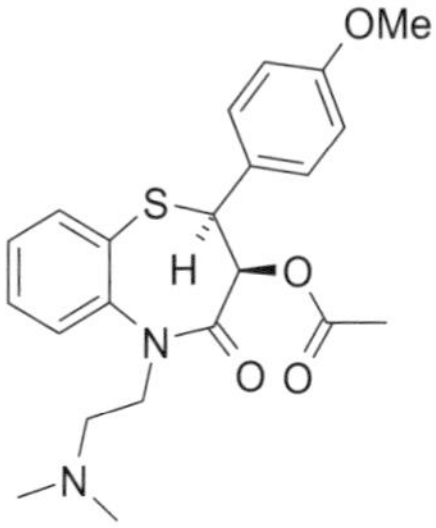

Fig. 7.3 Diltiazem.

Analogue-based Drug Discovery. IUPAC, János Fischer, and C. Robin Ganellin (Eds.)
Copyright © 2006 WILEY-VCH Verlag GmbH & Co. KGaA, Weinheim
ISBN: 3-527-31257-9

Therapeutic uses of calcium channel-blocking agents include:

- Variant angina: in this condition the reduction of coronary flow is due to a vasospasm; calcium antagonists are very effective in the control of this disease.
- Exertional angina: calcium antagonists are useful in the treatment of exercise-induced angina, through their ability to increase coronary dilatation and blood flow.
- Unstable angina: although the therapy of choice for unstable angina rests on the use of beta-blockers and aspirin, calcium antagonists may add further value due to their spasmolytic effects.
- Essential hypertension: all calcium antagonists, particular the dihydropyridine type, are widely used for the treatment of hypertension due to not only their strong effects on the reduction of peripheral resistances, but also their excellent safety profile.
- Atherosclerosis: most calcium antagonists, particularly the lipophilic dihydropyridines, possess experimental and clinical anti-atherosclerotic effects.

7.2
Dihydropyridine Calcium Channel-Blocking Agents

1,4-Dihydropyridines possess significant selectivity towards vascular versus myocardial cells, and therefore have a greater vasodilatory effect with respect to other calcium antagonists. They also show a minor negative inotropic effect with respect to verapamil (Fig. 7.2) and diltiazem (Fig. 7.3). Therefore, they do not have any significant direct effect on the heart as tachycardia, which is sometimes present in the patient after acute administration, is due to reflex sympathetic activation. The *in-vivo* metabolism of these agents is due to oxidation of the 1,4-dihydropyridine ring by the cytochrome P450 3A4 isoform to an aromatic pyridine ring, and to oxidative cleavage of the carboxylic esters. Dihydropyridines are insoluble in water and are light-sensitive; thus, they must be kept in dark environment during their preparation and testing. A number of molecules belong to this important class, some of which are described below.

7.2.1
Nifedipine

Nifedipine (Fig. 7.1) is the lead compound which was first introduced for the treatment of coronary angina. However, its use in the treatment of hypertension was blunted by a short plasma half-life (in the range of 1.5–2 h); this led to the need for multiple daily administration, and consequently blood pressure control and patient compliance were not fully achieved. However, slow-release formulations – for example, the once-daily Nifedipine Oros – made possible the wide use of nifedipine in cardiovascular therapy.

7.2.2
Felodipine

Felodipine (Fig. 7.4) is an analogue of nifedipine in which the *ortho*-nitro group in the aryl moiety has been replaced by 1,2-dichloro group. Felodipine possesses a very potent calcium antagonist activity which is exerted through a specific binding to the dihydropyridine site. It is completely absorbed in the gut, but is also extensively metabolized by the liver; thus, the duration of action of the immediate-release formulation is limited. Slow-release formulations of felodipine can provide a more sustained duration of action. Felodipine is able to reduce both systolic and diastolic blood pressures in a dose-related manner. Felodipine also increases the secretion of potassium, calcium, and magnesium. Patients taking felodipine should avoid cytochrome P450 3A4 and 1A2 isoform inhibitors, such as grapefruit juice, which can interfere with the hepatic metabolism of felodipine.

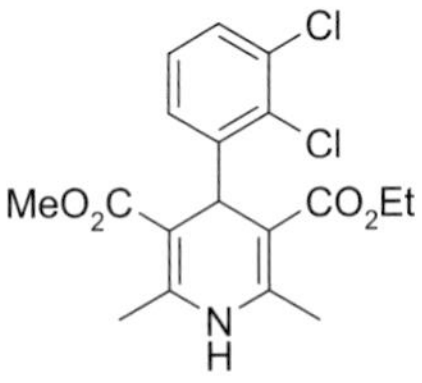

4

Fig. 7.4 Felodipine.

7.2.3
Isradipine

Isradipine (Fig. 7.5) is a further nifedipine analogue, which has been used in the treatment of hypertension either alone or in combination with diuretics. Isradipine possesses high calcium antagonist activity, which results in a peripheral vasodilatation without any detrimental effect on cardiac conduction. Its oral bioavailability is 90–95% of the administered dose, but the hepatic first-pass effect reduces this to 15–24%. Isradipine was shown to possess anti-atherosclerotic activity in a number of experimental models, but clinical studies (MIDAS) failed to confirm these findings in patients.

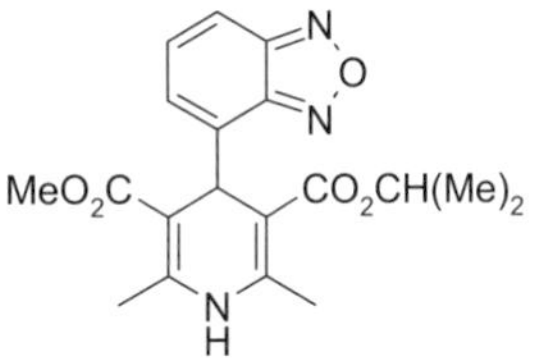

5

Fig. 7.5 Isradipine.

7.2.4
Nimodipine

Nimodipine (Fig. 7.6) is an asymmetric structural analogue of nifedipine, and calcium channel-blocking activity resides in the S-enantiomer. This compound exerts its pharmacological activity through an ability to bind to the same site of the other dihydropyridines. It is interesting, though, to note that nimodipine shows a preferential binding activity to the cerebral vessels, thus indicating a more pronounced effect at the cerebral level. Nimodipine has been widely used for the treatment of neurological diseases such as stroke of the subarachnoid-hemorrhagic type.

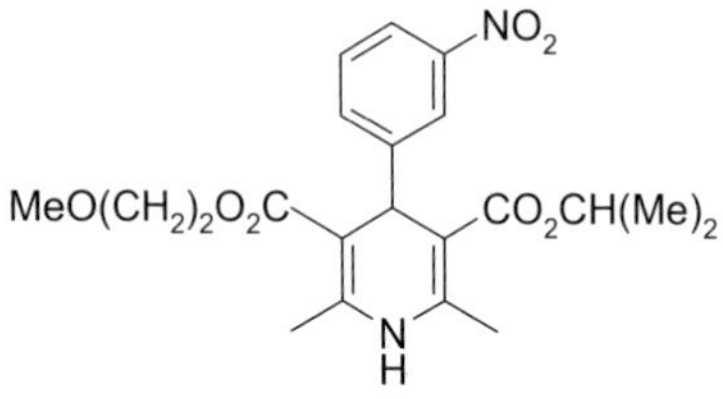

6

Fig. 7.6 Nimodipine.

7.2.5
Nisoldipine

Nisoldipine (Fig. 7.7) is an analogue of nifedipine, but it is at least five to ten times more potent than nifedipine on arterial smooth muscle, without affecting myocardial contractility. It also possesses a very potent and selective relaxation activity on small arteries, and therefore it can be used for the treatment of hypertension. Since its duration is limited by a massive hepatic first-pass effect, it is currently used as an extended-release formulation for once-daily administration.

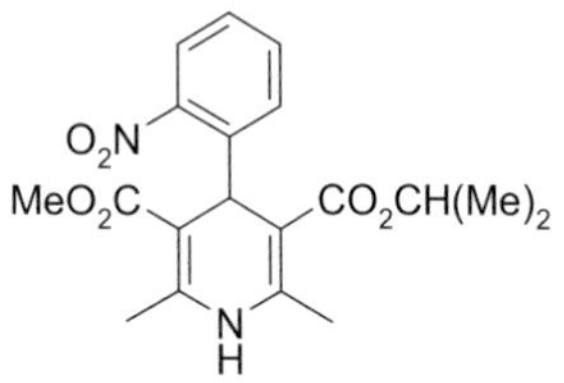

7

Fig. 7.7 Nisoldipine.

7.2.6
Amlodipine

Amlodipine besylate (Fig. 7.8), is a long-acting calcium antagonist, the long duration of action being achieved through a long kinetic half-life. Its bioavailability is
in the range of 60%, and the hepatic first-pass effect is less than that of other compounds of the same family. Amlodipine is widely used either alone or in combination to treat hypertension and angina. It also possesses anti-atherosclerotic properties which can be further enhanced by combination with statins.

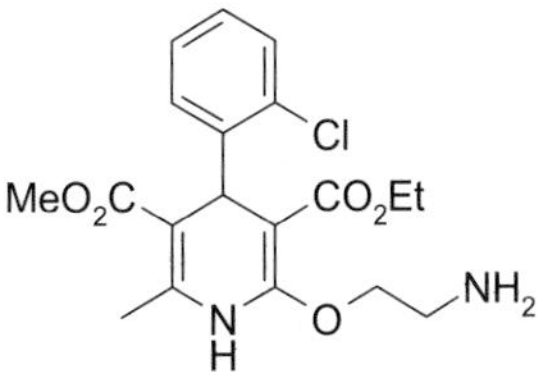

8

Fig. 7.8 Amlodipine.

7.2.7
Lacidipine

Lacidipine (Fig. 7.9) is extensively discussed in Section 7.3.

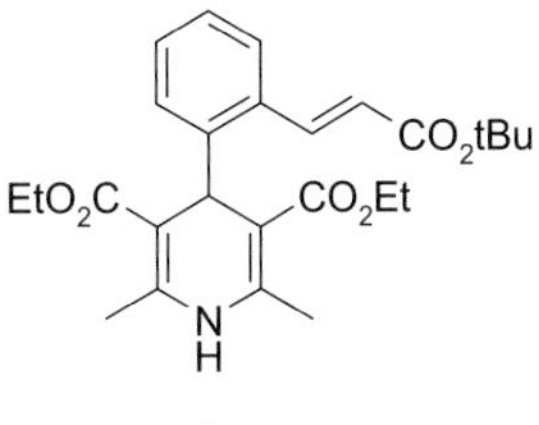

9

Fig. 7.9 Lacidipine.

7.2.8
Lercanidipine

Lercanidipine (Fig. 7.10) is one of the last dihydropyridines introduced into the
market. Its high lipophilicity explains its long duration of action. Lercanidipine is
extensively metabolized in the liver after oral administration, and is then retained
by vascular cells to achieve a sustained duration of action. It is used in the treatment of hypertension and angina. Like lacidipine, its smooth onset of action

reduces the vasodilatory adverse effects such as ankle edema, and it is indicated for the treatment of isolated systolic hypertension in elderly patients.

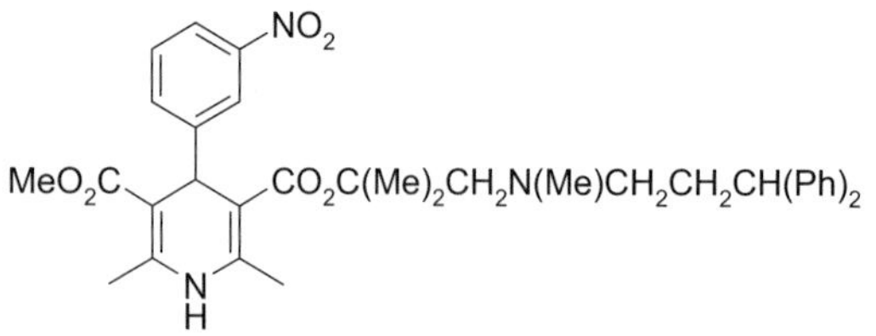

10

Fig. 7.10 Lercanidipine.

7.2.9
Manidipine

Manidipine (Fig. 7.11) is a nifedipine analogue which shows a potent and selective calcium antagonist activity that translates into a potent antihypertensive and coronary vasodilatory activity. Manidipine shows some selectivity towards the renal vasculature, making it particularly useful for the treatment of renovascular hypertension.

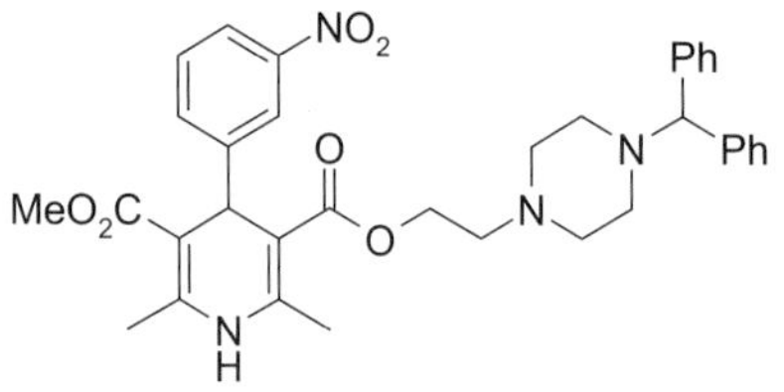

11

Fig. 7.11 Manidipine.

In conclusion, all dihydropyridines show a potent calcium channel antagonist activity, which is in turn translated into direct arteriolar spasmolytic effect that results in a beneficial vasodilatory activity. This is useful in some cardiovascular diseases, such as hypertension and angina, in which the peripheral resistances are raised due to increased calcium entry into the cells. Many analogues of nifedipine have been synthesized and introduced into the market, and each of these presents some common features and some peculiar differences. In particular:

• All dihydropyridines share the same mechanism of action,
  through selective binding to the L-type calcium channel; they are
  able to inhibit calcium entry into the cell.
• All are competitive antagonists.
• All show a preferential selectivity on arterial smooth muscle;
  some seem selective for specific vascular beds: nisoldipine for

small arteries, nimodipine for cerebral smooth muscles, manidipine for kidney vasculature.
- All show significant hepatic oxidative degradation after oral
  administration, with the exception of amlodipine.
- Oral bioavailability is low, with the exception of amlodipine.
- The duration of action is short, except for lacidipine and lercanidipine (high lipophilicity) and amlodipine (long kinetic half-life);
  extended-release formulations have been developed to ensure
  once- daily oral administration for many of these agents.
- All are effective in the clinical situation, and are well tolerated.
- Side effects such as ankle edema seem less pronounced for lipophilic compounds (lacidipine, lercanidipine), most likely due to
  their slow onset of action.

7.3
Lacidipine: A Long-Lasting Calcium Channel-Blocking Drug: Case Study

Following the introduction of nifedipine into the market, other dihydropyridines
were prepared starting from its structure. Among these, nitrendipine, isradipine,
and felodipine have emerged as novel therapeutic agents [1,2]. All of these belong
to the first-generation dihydropyridines which are characterized by a short duration of pharmacological action.

Structure–activity relationship (SAR) studies have shown that the presence of
certain chemical features are essential to maintain the calcium channel-blocking
activity, including:
- the 1,4-dihydropyridine ring;
- the secondary nitrogen in the dihydropyridine ring; and
- the aromatic or heteroaromatic substituent in the 4 position.

The substitution pattern of the 4-aromatic residue is also important for the activity,
the *ortho*-substitution being the best one in terms of potency and selectivity. A
Hansch analysis on a series of *ortho*-derivatives has shown a significant correlation
between calcium antagonist activity and steric hindrance of the substituent, while
no relationship was found for either electronic or lipophilic parameters [3]. The
best SAR correlation was obtained when the B1 steric parameter (the Verloop
parameter) was introduced into the analysis [4]. The calcium channel-blocking
activity increases as B1 increases, which probably indicates that steric hindrance
in the *ortho*-position is required to fix the dihydropyridine structure into a favorable conformation in which the aromatic group is approximately perpendicular to
the dihydropyridine ring (Fig. 7.12).

This predicted conformation is supported by SAR studies performed in a series
of compounds in which the torsion angle of the bond between the aryl and the
1,4-dihydropyridine ring is fixed by lactone bridges of different chain lengths [5],
and also by X-ray analysis [6].

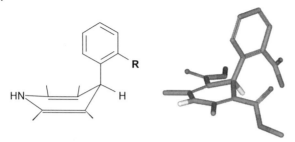

Fig. 7.12 Conformation of the dihydropyridines.

7.3.1
The Lacidipine Project

The main objective of the project was to identify a new potent and selective dihy-dropyridine calcium antagonist which was endowed with a sustained duration of action and was amenable to become a once-daily antihypertensive drug.

The first attempt to obtain a long duration of action whilst maintaining high potency and selectivity was to prepare and test a set of simple *ortho*-substituted derivatives of general formula shown in Tab. 7.1, in which the lipophilicity and the bulk of the side chain were varied.

Tab 7.1 *In- vitro* and *in- vivo* properties of the first set of compounds.

Compound	X	IC_{50}[a]	Dw[b]	ED_{25}[c]	Duration[d]
12	OCH_2	3	40	0.5	2
13	CH_2CH_2	4	60	1	1
14	$CH=CH$[e]	0.5	120	0.04	4

a. Concentration (nM) that inhibits 50% of K^+-induced contraction in isolated rabbit ear artery (REA).
b. Duration of action (min) *in vitro* determined by wash-out experiments on REA.
c. Intravenous dose (mg kg^{-1}) which reduces 25% of blood pressure in spontaneously hypertensive rats (SHR).
d. Duration of antihypertensive action expressed as $t^1/_2$ (min) in SHR.
e. E-isomer.

Calcium channel antagonist activity was determined in isolated rabbit ear artery rings depolarized by a high K^+ concentration [7]. The duration of action *in vitro* (Dw) was determined by means of wash-out experiments. Rabbit ear artery rings were contracted by a submaximal concentration of calcium, and incubated with the tested compounds for 60 min. Subsequently, they were washed with saline solution until the initial contraction was restored. Dw was defined as the time required by the tissue to recover the basal effect elicited by a submaximal concentration of calcium [8]. Antihypertensive activity and Dw were determined in chronically implanted spontaneously hypertensive rats (SHR) [9].

In the first set of compounds 12–14 (Tab. 7.1), replacement of the nitro group of nifedipine with ethoxycarbonylalkyl groups retained the calcium channel potency *in vitro*, while the metabolic instability of the ethyl ester *in vivo* did not allow the antihypertensive effect to materialize [9].

A lead optimization program was started to improve the duration of activity *in vivo*. A new series of compounds starting from 14 (Tab. 7.1) was prepared and tested [9], as shown in Tab. 7.2.

Tab 7.2 *In-vitro* and *in-vivo* properties of the second series of compounds.

Compound	X	IC_{50}[a]	ED_{25}[b]	Duration[c]	Log P	$v\ OR$[d]
15	C_2H_5	0.5	0.04	4	4.9	0.48
16	$(CH_2)_7CH_3$	45	0.021	62	8.1	0.61
17	$CH(CH_3)_2$	1	0.013	27	5.2	0.75
18	$CH_2CH(CH_3)_2$	3	0.01	32	5.8	0.62
19	C_6H_{11}	10	0.009	52	6.4	0.81
9	$C(CH_3)_3$	7	0.006	130	5.6	1.22

a. Concentration (nM) that inhibits 50% of K^+-induced contraction in isolated rabbit ear artery (REA).
b. Intravenous dose (mg kg^{-1}) which reduces 25% of blood pressure in spontaneously hypertensive rats (SHR).
c. Duration of antihypertensive action expressed as $t^{1/2}$ (min) in SHR.
d. Steric hindrance of substituent OR.

The compound with the longest-lasting duration of action was lacidipine (Fig. 7.9), in which the ester group was the bulky *tert*-butyl moiety [9].

7.3.2
Synthesis

The nucleus of lacidipine was constructed following standard methods used for dihydropyridines. The *ortho*-phenyl side chain was then introduced either through the Wittig reaction of an aldehyde with the required ylide, or by a Heck reaction of a corresponding bromide (Scheme 7.1)

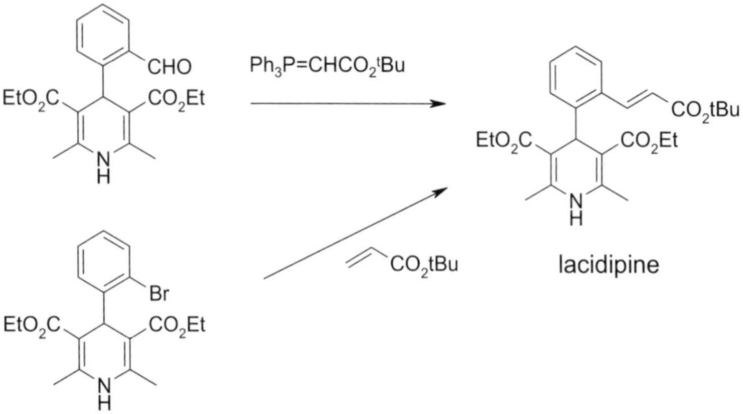

Scheme 7.1 Functionalization of dihydropyridine nuclei to produce lacidipine.

7.3.3
The Pharmacological Profile of Lacidipine

The pharmacological profile of lacidipine has been studied extensively both *in vitro* [10] and *in vivo* [11]. In vascular smooth muscle preparations, lacidipine caused an equipotent concentration-dependent shift to the right of the calcium cumulative dose–response curves, without any significant depression of the maximum response, with a pA$_2$ of 9.4, indicating that it is a competitive antagonist. In extravascular smooth muscle, the spasmolytic activity was at least 100 times less potent. The duration of action *in vitro*, determined by wash-out experiments in isolated smooth muscle preparations, was found to be longer than 10 h, indicating persistency of the molecule in the tissue, even after repeated wash-out. This effect can be explained by the drug's high lipophilicity (logP = 5), which allows the molecule to be concentrated in lipid storage close to the receptor, from where it is released to cause continuous blockade of the channel [9].

In vivo, lacidipine –when administered orally – was found to be an effective and potent antihypertensive agent in different animal models, including SHR (ED$_{25}$ = 0.5 mg kg^{-1}, per os) and renally hypertensive dogs (ED$_{25}$ = 0.004 mg kg^{-1}, i.v.) with a sustained duration of action [11,12]. Lacidipine was also found to possess coronary and cerebral vasodilatory activity [13].

Furthermore, lacidipine was found to possess antioxidant activity at the same level of that of vitamin E in many tests, including hydrogen peroxide oxidation of rat neuronal cells [14]. The antioxidant property of lacidipine was further confirmed by experiments *in vivo*, in which low-density lipoprotein (LDL) oxidation was completely abolished [15]. Lacidipine also shows a direct protective effect on the vasculature at non-antihypertensive doses, indicating that its high lipophilicity combined with antioxidant potential can exert an additional therapeutic benefit [16–18].

In clinical trials, lacidipine was found to be a very potent and long-lasting antihypertensive agent; a dose of 4 mg emerged as the recommended daily dose to control mild to moderate hypertension. The drug was well tolerated, and the side effects – including headache – were mild in nature and common to those of the other dihydropyridines.

The peculiar vascular protective and anti-atherosclerotic properties of lacidipine were further confirmed in the European Lacidipine Study on Atherosclerosis (ELSA), where the carotid atherosclerotic process was slowed down by lacidipine in comparison with atenolol [19].

7.4
Conclusion

Lacidipine, a second-generation analogue of nifedipine, has fulfilled the concept of optimization of the prototype in medicinal chemistry. Although structurally related to other dihydropyridines, lacidipine possesses improved physico-chemical properties (both lipophilic and antioxidant), together with an excellent pharmacological profile, including selectivity, potency, and a sustained duration of action. This achievement has been completely confirmed in the clinical setting, in which the improved profile is translated into a long-lasting and effective antihypertensive agent, in combination with a significant anti-atherosclerotic action that is beneficial for patients, notably those who are elderly.

References

1 Triggle DJ. In: *Calcium Antagonists in Clinical Medicine*, 2nd edn. Epstein M (Ed.). Hanley & Belfus, Inc., **1997**: 1–26.

2 Joahi GS, et al. In: *Burger's Medicinal Chemistry & Drug Discovery*. Vol. 3. Abraham DJ (Ed.). J. Wiley & Sons, **2003**: 1–54.

3 Massoud M, et al., *J. Pharm. Pharmacol.*, **1986**, 38(4), 272–276.

4 Gaudio AC, et al., *J. Pharm. Sci.*, **1994**, 83, 1110–1115.

5 Meyer H, et al. Bayer-Symposium, 9th Cardiovascular Eff. Dihydropyridine-Type Calcium Antagonists Agonists, Rigid Calcium Agonists of the Nifedipine Type: Geometric Requirements for the Dihydropyridine Receptor 1985: 90–103.

6 Rovnyak G, et al., *J. Med. Chem.*, **1988**, 31, 936–944.

7 Giacometti A, et al., *J. Pharmacol. Exp. Ther.*, **1992**, 263, 1241–1247.

8 Micheli D, et al., *J. Cardiovasc. Pharmacol.*, **1990**, 15, 666–675.

9 Gaviraghi G. *Trends Medicinal Chem.*, **1989**, 12, 675–690.

10 Kawada T, et al., *Jpn. J. Pharmacol.*, **0000**, 62, 289–296.

11 Micheli D, et al., *J. Cardiovasc. Pharmacol.*, **1991**, 17, S1–S8.

12 Feron O, et al., *J. Cardiovasc. Pharmacol.*, **1995**, 26, S459–S461.

13 Napoli C, et al., *Stroke*, **1999**, 30, 1907–1915.

14 Van Amsterdam FTM, et al., *Free Radical Biol. Med.*, **1992**, 12, 183–187.

15 Cominacini L, et al., *J. Hypertens.*, **1999**, 17, 1837–1841.

16 Massart PE, et al., *Hypertension*, **1999**, 34, 1197–1201.

17 Lupo E, et al., *Biochem. Biophys. Research Commun.*, **1994**, 203, 1803–1808.

18 Berkels R, et al., *Vasc. Pharmacol.*, **2005**, 42, 145–152.

19 Zanchetti A, et al., *Circulation*, **2002**, 106(19), 2422–2427.

8

Selective Beta-Adrenergic Receptor-Blocking Agents

Paul W. Erhardt and Lajos Matos

8.1
Introduction

The designation "analogue-based drug discovery" (ABDD) [1] is used within this monograph to denote a special case of "ligand-based drug design" [2] in which the parent ligands are specifically represented by clinically useful drugs. Thus, a significant advantage of ABDD lies in the fact that its structural starting points have already proven themselves to be "drug-like" [2]. Although apparent when deploying ABDD in pursuit of a "new use" for an existing therapeutic agent, this advantage is also retained when the parent drug is thought to have a significant shortcoming in one or more of its "ADMET" [3] -related properties such that an enhancement of the overall pharmacokinetic or "PK" [4] profile is deemed to be worthwhile for a "next-generation" agent having the same clinical indication. Indeed, it is precisely an interplay between ABDD, certain key pharmacological observations, and the theme of further tailoring the ADMET/PK profile of established drug-like molecules for both the same and new indications, that so clearly characterizes the evolution of the beta-adrenergic receptor blockers ('beta-blockers'). This situation becomes evident in the following account that provides a historical introduction to the beta-blocker field in general, while also serving to identify what can be considered to be its specific "pioneer drug" [1]. The remaining portion of this chapter will then focus upon the further development of selected analogues that demonstrate a preference for beta-1 adrenergic receptors.

Norepinephrine (noradrenaline) and epinephrine (adrenaline), **1** and **2**, are the endogenous receptor ligands that respectively serve as the neurotransmitter- and endocrine-derived agonists for both alpha- and beta-adrenergic receptors associated with the mammalian sympathetic nervous system. Isoproterenol (isoprenaline) is a long-standing, racemic derivative of these endogenous ligands for which its (R)-enantiomer, **3**, demonstrates potent agonist properties at beta-receptors while exhibiting only minimal activity at alpha-receptors, the latter owing to the a-receptor's intolerance of steric bulk near the vicinity of the requisite amino-group. Still used as a bronchodilator, isoproterenol was the first synthetic catecholamine derivative to achieve clinical prominence [5]. It should be noted that while

Analogue-based Drug Discovery. IUPAC, János Fischer, and C. Robin Ganellin (Eds.)
Copyright © 2006 WILEY-VCH Verlag GmbH & Co. KGaA, Weinheim
ISBN: 3-527-31257-9

the distinct enantiomeric spatial orientation depicted in **1** to **3** is required for the hydroxy-group to bind with beta-receptors, the vast majority of ligands that have been studied within this field have been prepared as their racemates. Thus, for convenience, the remaining compounds to be discussed within this chapter should be considered to be racemic unless stipulated to be otherwise. Similarly, unless a specific point of interest is to be made, compounds will generally be drawn as their free base amines whether or not their generic names may actually include some type of acid salt.

1	R = H
2	R = CH$_3$
3	R = CH(CH$_3$)$_2$

As part of initial efforts to define structure–activity relationships (SAR) for these agonist compounds and, in particular, to identify more stable replacements for the air-sensitive catechol moiety, the dichloro-containing derivative **4** was prepared. Although this modification did not prove to be a useful "bioisosteric substitution" in that agonist activity was largely lost, **4** was instead found to be a weak partial agonist that could be used to block the effects of the sympathomimetic amines **1–3** on bronchodilation, uterine relaxation and cardiac stimulation [6]. This key, medicinal chemistry observation – coupled with Black's earlier pharmacological suggestion that beta-adrenergic receptor antagonists might protect the ischemic myocardium from sympathetic drive during angina pectoris [7] – served to usher in the era of beta-blocker research. Subsequent replacement of the dichloro substituents with an unsaturated carbon bridge afforded a naphthyl moiety that, in terms of seeking beta-blockade, did prove to be an effective bioisosteric replacement. The resulting beta-blocker **5**, pronethalol, was taken into the clinic where it was indeed shown to be effective in the treatment of angina pectoris [8]. However, after having clearly "validated" beta-blockade as a useful therapeutic "target" by establishing "clinical proof of principle" in line with today's phraseology, within a year pronethalol had to be withdrawn due to inherent carcinogenicity [9].

The insertion of an oxymethylene bridge, -OCH$_2$-, into the ethanolamine C2-substituent of pronethalol so as to form an oxypropanolamine, was found to retain beta-blocking activity, particularly when the entire oxypropanolamine appendage was also moved to the C1-position of the naphthyl ring. The resulting compound **6**, propranolol, does not exhibit carcinogenicity and it soon became the first sustained, clinically successful beta-blocker. Thus, subsequent to its own history of ligand-based drug design, within the context of the present monograph, propranolol should be considered to be the "pioneer drug" that, as we shall see, has gone on to inspire numerous other scenarios of ABDD. For example, during its initial evaluation as a treatment for angina pectoris [10], propranolol was also found to have useful anti-arrhythmic [11] and antihypertensive [12] properties, both of which represented new indications that needed somewhat differently tailored overall pharmacologic profiles and, thereby, represented ideal targets for further ABDD. In addition to serving as the pioneer drug, propranolol's classical synthetic scheme remains as the standard method for the assembly of aryloxypropanolamine systems. This general methodology is shown in Scheme 8.1.

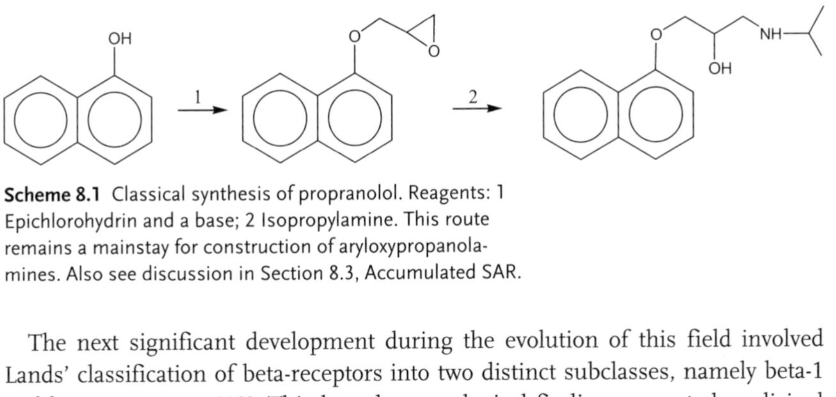

Scheme 8.1 Classical synthesis of propranolol. Reagents: 1 Epichlorohydrin and a base; 2 Isopropylamine. This route remains a mainstay for construction of aryloxypropanolamines. Also see discussion in Section 8.3, Accumulated SAR.

The next significant development during the evolution of this field involved Lands' classification of beta-receptors into two distinct subclasses, namely beta-1 and beta-2 receptors [13]. This key pharmacological finding prompted medicinal chemistry's ABDD activities to turn toward the pursuit of compounds having beta-1 selectivity for the various cardiac-related indications. Such selectivity was regarded as being particularly important for an anti-anginal drug to be used by patients prone to bronchospasm because of the latter tissue's preponderance of beta-2 receptors. In addition, ADMET-related properties to be generally enhanced included lowering lipophilicity as it specifically pertained to removing side-effect "membrane-stabilizing activity" (MSA) [14–16], and prolonging the *in-vivo* half-life as it specifically pertained to achieving convenient "once a day dosing". Residual partial agonist properties called "intrinsic sympathomimetic activity" (ISA) [17], and mixed patterns of selectivity between alpha- and beta-receptors, as well as among the two known beta-receptor subtypes, represented indication-dependent features that could also be further manipulated. Thus, as the era of beta-blocker research unfolded, numerous compounds became available which afforded a wide range of profiles when compared across the types of properties mentioned above. Several representative beta-blockers which were either in the marketplace or were undergoing preclinical development during the mid to late 1970s are listed in Tab. 8.1 [17,18]. Many of these compounds can still be found within the monograph's master listing of presently marketed drugs. All of their structures are shown within Fig. 8.1. That all but six of the 40 compounds shown in Fig. 8.1 are aryloxypropanolamines which can thus be traced back to the pioneer drug 6, clearly underscores the important role that ABDD can play during the overall course of drug discovery for a given pharmacological area or a distinct biochemical mechanism.

Tab 8.1 Representative beta-blockers in the market or undergoing development during the mid to late 1970s [17,18].

Compound #	Generic name[a]	B$_1$-selectivity[b]	MSA[b]	ISA[b]	Reference
7	Acebutolol	+	+	–	19
8	Alprenolol	–	+	+	20
9	Atenolol	+	–	–	21
10	Bevantolol	+	+	–	22
11	Bucumolol	–	+	–	23
12	Bufetolol	–	+	+	24
13	Bufuralol	–	+	+	25
14	Bunitrolol	–[c]	(+)	(–)	26
15	Bunolol	–	+	–	27
16	Bupranolol	(–)	(+)	(–)	28
17	Butidrine	–	+	(–)	29
18	Butocrolol	(–)	+	–	30
19	Butoxamine	–[c]	(+)	–	31
20	Carazolol	(–)	(–)	–	32
21	Carteolol	(–)	(–)	+	33
22	Exaprolol	–	+	–	34
23	Indenolol	(–)	(+)	(–)	35

Tab 8.1 Continued.

Compound #	Generic name[a]	B$_1$-selectivity[b]	MSA[b]	ISA[b]	Reference
24	Iprocrolol	–	(+)	–	36
25	Labetolol	–	+	–	37
26	Mepindolol	–	+	+	38
27	Metipranolol	–	–	–	39
28	Metoprolol	+	(+)	–	40
29	Moprolol	(–)	+	(–)	41
30	Nadolol	–	–	–	42
31	Nifenalol	–	–	+	43
32	Oxprenolol	–	+	+	44
33	Pamatolol	+	(–)	(–)	45
34	Pargolol	–	+	(–)	46
35	Penbutolol	–	+	(–)	47
36	Pindolol	–	+	+	48
37	Practolol	+	–	+	49
38	Procindolol	–	–	(–)	50
5	Pronethalol	–	+	–	51
6	Propranolol	–	+	–	52
39	Sotalol	–	–	–	53
40	Tazolol	–	–	–	54
41	Timolol	–	+	–	55
42	Tiprenolol	–	(+)	(–)	56
43	Tolamolol	+	–	–	57
44	Toliprolol	–	(+)	(–)	58

a. Chemical structures are shown in Fig. 8.1.
b. A '+' sign indicates that the compound exhibits the property within the specified column, while a '–' sign indicates that it does not. Entries in parentheses indicate that either definitive data were not provided in Refs. [17,18], or that there was conflicting data between these two references. In such cases, an estimate has been suggested by the present authors after considering the structure relative to general SAR (see Section 8.3). MSA: membrane-stabilizing activity or local anesthetic activity; ISA: intrinsic sympathomimetic (partial agonist) activity.
c. Exhibits selectivity for beta-2-adrenergic receptors.

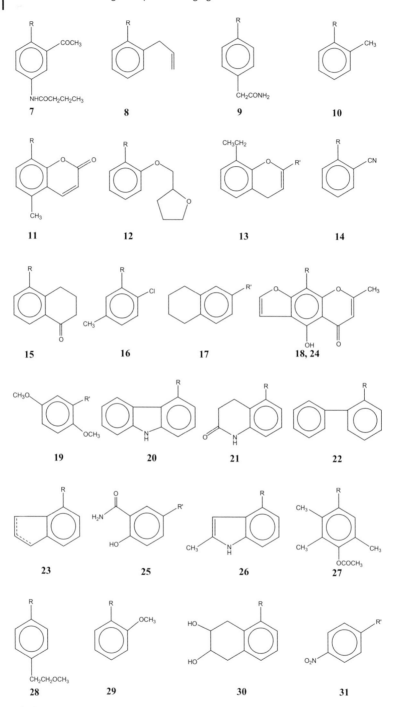

Fig. 8.1 Continues on next page.

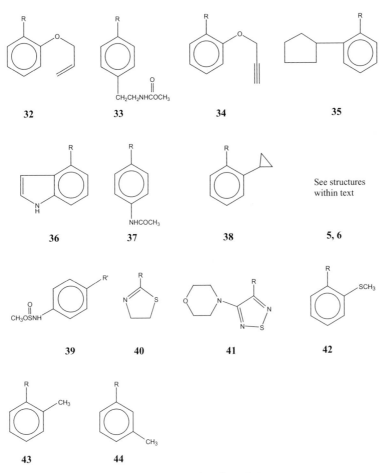

Fig. 8.1 Chemical structures for compounds listed in Tab. 8.1.
R = OCH$_2$CH(OH)CH$_2$NHR″ with R″ = CH(CH$_3$)$_2$ for compounds
7, 8, 9, 20, 22, 23,[a] **24, 26, 27, 28, 29, 32, 33, 36, 37, 38, 40,** and **44**;
R″ = C(CH$_3$)$_3$ for compounds **11, 12, 14, 15, 16, 18, 21, 30,**[b] **34, 35, 41,**
and **42**; R″ = CH$_2$CH$_2$Ph(3,4-di-OCH$_3$) for compound **10**; and
R″ = CH$_2$CH$_2$OPh(4-CONH$_2$) for compound **43**. R′ = CH(OH)CH$_2$NHR‴
with R‴ = CH(CH$_3$)$_2$ for compounds **31** and **39**; R‴ = C(CH$_3$)$_3$
for compounds **13** and **19**[c]; R‴ = CH(CH$_3$)CH$_2$CH$_3$ for compound **17**;
and, R‴ = CH(CH$_3$)CH$_2$CH$_2$Ph for compound **25**. [a]Tautomeric mixture
of the 7- and 4- indenyloxy isomers in a 2:1 ratio, respectively.
[b]Dihydroxy groups are cis. [c]Also contains an alpha-methyl group
on the beta-hydroxyethylamino side chain.

Although selective for beta- over alpha-receptors, propranolol is classified as a nonselective beta-blocker in that it has nearly equipotent actions at both beta-1 and beta-2 receptors. In addition, propranolol can cause cardiac depression by MSA-related mechanisms that are independent from its actions at beta-receptors. Thus, despite the significant pioneering role that 6 has played in the overall field of beta-blockers, in terms of treating angina there was an immediate need for more selective beta-1 blockers that could be used in patients with obstructive airway diseases so as to avoid blockade of bronchial beta-2 receptors and concomitant bronchoconstriction. It was at this stage of the beta-blocker field's overall development that a second, important medicinal chemistry observation relating to a key SAR would occur, namely that the para-substituted phenyl derivative 37, practolol, was found to block cardiac beta-1 receptors at lower doses than those at which it blocked beta-2 receptors located within the bronchi and vascular beds [49,59]. Thus, 37 became the first selective beta-1 adrenergic receptor blocking agent.

37

Clinically, 37 initially proved to be a valuable anti-anginal agent that was much better tolerated by patients also suffering from obstructive airway disease [60]. Unfortunately, as with the earlier experience with pronethalol, 5, practolol also had to be quickly withdrawn from the clinic due to side effects that this time were related to an oculomucocutaneous syndrome [60]. The remaining sections of this chapter will now focus specifically upon the successful clinical development of several compounds that have sustained themselves as selective, beta-1 adrenergic receptor blocking agents. In concluding this introduction, it should be appreciated that beyond the beta-1 selective compounds which are primarily directed toward cardiac-related indications such as angina and myocardial infarction [61] as well as hypertension, the overall family of beta-blockers has found a much wider range of therapeutic applications including hypertension using nonselective beta-blockers [62], arrhythmias [63], glaucoma [64], migraine [65], anxiety [66], tremor [67], drug dependence [68,69], thyrotoxicosis [70], schizophrenia [71], and mania [72]. Likewise, although numerous companies eventually came to be involved with beta-blocker research, the early efforts and tenacity of Imperial Chemical Industries (ICI) should be especially noted at this point because it was their introduction (and later withdrawal) of pronethalol (Alderlin) that eventually led them to propranolol (Inderal) along with everyone else to the aryloxypropanolamine, pioneer beta-blocker pharmacophore. Then it was again their introduction (and later withdrawal) of practolol (Eraldin) that eventually led them to atenolol (Tenormin; discussed below) along with everyone else down the pathway towards beta-1 selectivity via para-substitution.

8.2
Beta-1 Selective Blockers

Continuing in historical chronology, representative compounds to be highlighted in this section have been selected from the monograph's master table under the subheading for selective beta-blocking agents and will include: atenolol, betaxolol, celiprolol, and nebivolol. Although esmolol also represents a very distinct type of beta-1 blocker, it is not included herein because it is covered in a separate case study (see Chapter 9 in Part II). Each of these four compounds will be considered in terms of their initial ABDD and SAR, interesting synthetic chemistry, and characteristic clinical profile, including a brief summary of their relevant ADMET properties.

8.2.1
Atenolol

8.2.1.1 Discovery

As mentioned in Section 8.1, it was recognized that it would be beneficial to further tailor the overall profile of propranolol, **6**, for use as an anti-anginal agent within the setting of obstructive airway disease. Toward this end, ICI immediately embarked on a program that sought to identify a compound which would have the following profile [60]:
- Possess a potency equivalent to **6** at cardiac, beta-1 adrenoceptor sites;
- Possess a greater affinity for cardiac beta-1 adrenoceptors than for beta-2 adrenoceptors in the bronchial tree, peripheral vasculature and coronary circulation, or for those involved in sugar metabolism where an additional receptor subclass at the time had not yet been defined [73];
- Be devoid of partial agonist activity or ISA; and,
- Be free of membrane-depressant or membrane-stabilizing activity (MSA).

Taking advantage of the beta-1-selectivity afforded by practolol, **37**, two structural features of **37**'s acetylamino functionality were considered for further exploration on propranolol's aryloxypropanolamine template [60]:
- That the electron-withdrawing effect of the carbonyl on imparting acidity to the proton of the -NH-CO-group might be mimicked by a -CH$_2$-CO- group; and,

- That the hydrogen bonding capability might be retained by incorporating it into the above, even if in an extended manner, involving a -CH$_2$CONHR arrangement.

Approximately 5000 compounds were prepared and tested in the screening tree depicted in Scheme 8.2 [60]. From this effort, atenolol, **9**, was ultimately identified as the compound that came closest to the desired pharmacological profile.

Anesthetized Rats
(beta-blocking activity) ⟶ Anesthetized Cats
(beta-blocking selectivity: cardiac versus vascular) ⟶ Guinea Pigs (beta-blocking selectivity: cardiac versus bronchial)

⟶ Anesthetized Rats (ISA) ⟶ Frog Sciatic Nerve
(Membrane-Stabilizing Activity)

Scheme 8.2 Biological screening sequence deployed by ICI during the discovery of atenolol [60].

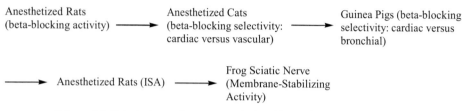

The production of a rich and useful SAR dataset also accompanied the selection of **9**. For example, the *para*-acetamido moiety gives rise to beta-1 selectivity with the amide -NH- group apparently playing an important role in this activity since the tertiary amide versions were found to be much less active. Likewise, while the methylene group does not appear to play a major role in activity or selectivity, it does have an important impact upon ISA, namely to reduce ISA to negligible levels [60,74]. Thus, while *para*-amides with either the nitrogen or carbonyl group attached to the aromatic nucleus have varying levels of ISA, the insertion of a methylene group between the phenyl-ring and the amide substituent significantly reduces this property. Interestingly, extension of such a spacer beyond one methylene group, leads to a fall-off in inherent activity. Finally, while additional substituents can be present on the phenyl-ring, these are best tolerated when placed in an *ortho*-relationship to the ether rather than when in a *meta*-relationship, the latter perhaps reflecting steric interference with the hydrogen-bonding role that the amide may be playing.

8.2.1.2 Synthesis

The synthetic route used to prepare atenolol is shown in Scheme 8.3 [60]. One of the notable features of the synthesis involved simply stirring a suspension of the methyl or ethyl ester intermediate in aqueous ammonia at ambient temperature whereupon quantitative yields of the primary amides were obtained.

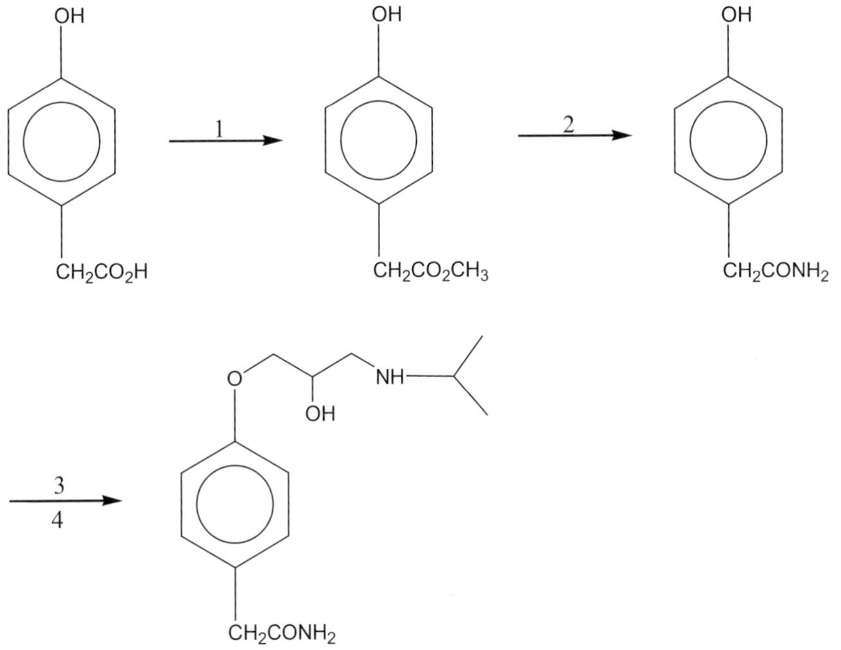

Scheme 8.3 Chemical synthesis of atenolol [60]. Reagents:
1 MeOH, H_2SO_4; 2 Aq. NH_3 (see text); 3 and 4 analogous to propranolol (see Scheme 8.1).

8.2.1.3 Clinical Pharmacology

In terms of ADMET, following oral administration about half of the atenolol dose is absorbed. Plasma-protein binding is minimal (3–5%). Peak plasma concentrations, as well as peak action, are reached in 2–4 h. Atenolol has low lipid solubility, and only small amounts cross the blood–brain barrier. Thus, atenolol's CNS side effects are less than with other beta-blockers [75]. Atenolol is excreted mainly by the kidneys, with little or no hepatic metabolism. It crosses the placenta, and concentrations in breast milk can be similar or even higher than those in maternal blood [76]. Atenolol is not recommended in asthma, even though its high beta-1 selectivity makes it safer in obstructive pulmonary disease than nonselective beta-blocking agents. Atenolol's important ADMET characteristics are listed in Tab. 8.2.

The major clinical uses of atenolol include: hypertension; angina pectoris; acute myocardial infarction (MI); heart rhythm disturbances; and migraine prophylaxis. In hypertonia disease, various studies have addressed the efficacy of atenolol in different groups of patients and in different combinations. In the 'STOP-Hypertension trial' [77], elderly men and women (aged 70–84 years) were treated with either atenolol, metoprolol, pindolol, a combination of amiloride and hydrochlorothiazide, or placebo with follow-up at 25 months (mean). The study was terminated prematurely due to the positive outcome on active treatments. In "STOP-Hypertension-2" [78], the effectiveness of "old" antihypertensives including beta-blockers such as atenolol and diuretics, and those of "new" drugs such as calcium antagonists and ACE inhibitors, were compared in hypertensive patients. There were no significant differences in the risk of cardiovascular events between patients receiving conventional therapy and those receiving newer therapies.

The "MRC study" [79] compared the effectiveness of atenolol, amiloride plus hydrochlorothiazide, or placebo as antihypertensives. Both active treatments reduced blood pressure below that for the placebo group, and the diuretics significantly reduced the risk for serious cardiovascular events. The atenolol group showed no significant reduction in the latter end-points. The "UKPDS/HDS trial" [80] addressed hypertensive patients who also suffered from type 2 diabetes. Atenolol and the ACE-inhibitor captopril were equally effective in reducing the risk of macrovascular end-points. In the "ELSA study" [81], the effects of atenolol and the calcium antagonist lacidipine were compared relative to the development and progression of atherosclerosis in hypertensive patients. Lacidipine was found to be more effective (p <0.0001). Similarly, the aim of the "LAARS study" [82] was to compare the effects of atenolol and the angiotensin-receptor blocker losartan on the atherosclerotic process. In this case, there were no significant differences between the two treatments. In the "LIFE study" [83], hypertensive patents with left ventricular hypertrophy (LVH) were treated with either atenolol or losartan. A greater reduction in LVH was seen with losartan than with atenolol, and the incidence of new-onset diabetes mellitus was lower with losartan than with the beta-blocker. "REASON" [84] compared the effects of atenolol versus a combination of the ACE-inhibitor perindopril plus the diuretic indapamide on blood pressure and pulse-wave velocity. The reduction in pulse-wave velocity was similar with both drugs, but aortic wave reflections were reduced to a higher degree with the ACE-inhibitor plus diuretic combination than with atenolol. In a recent meta-analysis [85], the effect of atenolol on cardiovascular morbidity and mortality in hypertensive patients was reviewed. Despite major differences in antihypertensive efficacy, there were no outcome differences between atenolol and placebo in four studies comprising 6825 patients on all-cause mortality, cardiovascular mortality or MI. The risk of stroke tended to be lower in the atenolol group than in the placebo group. However, lethal cardiovascular events, as well as stroke, were more frequent with atenolol treatment than with other antihypertensive therapy. These results cast doubts on atenolol as a suitable drug for hypertensive patients in general.

In coronary heart disease with chronic stable angina pectoris, the purpose of the "CAPE II trial" [86] was to compare the efficacy of the calcium-antagonist amlodipine plus other agents including atenolol, in the management of the circadian pattern of transient myocardial ischemia. The amlodipine and atenolol combination was significantly superior in lowering nitroglycerine consumption and in maintaining ischemia reduction. The aim of "TIBET" [87] was to investigate whether the total ischemic burden of the heart could be influenced either by atenolol, the calcium antagonist nifedipine, or their fixed combination within the setting of stable angina pectoris. All three treatments showed a comparably favorable influence on myocardial ischemia when compared to placebo.

In acute MI, the "ISIS-1 study" [88] was organized to assess the effects of early beta-blockade with atenolol on cardiovascular mortality during the first week following infarction and after long-term (mean 20 months) follow-up. There was a 15% reduction in vascular deaths, especially in the early phase of MI. The difference in early mortality was mainly due to a reduction in electromechanical dissociation in the presence of atenolol. In this regard, atenolol is usually given by intravenous injection or infusion to treat cardiac arrhythmias, and it should be noted that atenolol induced atrial fibrillation in half of the so-predisposed patients [89].

Migraine prophylaxis is also possible with atenolol, where it has compared favorably against placebo [90] and has been found to have efficacy similar to that of propranolol [91].

Tab 8.2 Clinical ADMET (Absorption, Distribution, Metabolism, Elimination and Toxicity) characteristics of atenolol, betaxolol, celiprolol, and nebivolol. For details and reference sources, see the text.

ADMET characteristic	Atenolol	Betaxolol	Celiprolol	Nebivolol
Absorption from a dose by oral administration [%]	~50	~98	20–30	~95
Time to onset of action [h]	~1	~2	0.5–1	0.5–2
Time to peak action [h]	2–4	3–4	2–3	2–3
Volume of distribution [L kg^{-1}]	0.6–1.2	4–7.3	6.5	10–40
Protein binding [%]	3–10	About 50	25–30	~98
Metabolism	Minimal	Extensive	Minimal	Extensive (fast or slow metabolizers)
Excretion	Renal	Renal and some hepatic	Renal and intestinal	Renal and intestinal
Plasma half-life [h]	~6	16–20	5–6	~10[a]
Tolerability side effects[b, c] [%]	8.0 (0–12)	3.8 (0.8–8.1)	12 (0.8–20)	5.6 (0.2–6.0)

a. In slow metabolizers this could be increased up to five-fold.
b. Using recommended doses.
c. Lowest and highest values in brackets from the available literature.

8.2.2
Betaxolol

8.2.2.1 Discovery

Although a clinically relevant alteration in pharmacologic profile (e.g., pertaining to a significant enhancement of selectivity or PK properties) generally serves as the criterion to establish demarcations between the first and second generations (etc.) of various members in an extended family of related therapeutic agents, this type of lineage can also be traced very conveniently within the context of ABDD. For example, betaxolol and seemingly bisoprolol (**45** and **46**, respectively), which were launched in the mid-1980s after using metoprolol as a clinically successful template during ABDD, can be considered to be fourth-generation compounds (structures shown in Fig. 8.2).

Metoprolol, **28**, which was launched in the mid-1970s, can be considered to be a third-generation ABDD compound that, while appending a distinctly different structural motif than either practolol or atenolol (**37** or **9**), still took advantage of practolol's *para*-substituted phenyl-ring SAR to bestow selectivity for beta-1 receptors. As mentioned in Section 8.1, practolol – significantly endowed with cardioselectivity – should be recognized as a key, second-generation compound, even though it ultimately was withdrawn from the clinic. Finally, practolol took advantage of the aryloxypropanolamine system present in propranolol, **6**, during an ABDD program that allowed its market launch in the late 1960s to occur very shortly after that of propranolol which was launched in the early 1960s and represents the field's pioneer drug and first-generation agent. This complete family tree, along with the lineage for atenolol and several of the other beta-1-selective compounds, is depicted in Scheme 8.4.

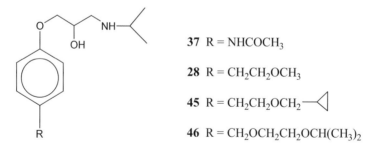

37 R = NHCOCH$_3$

28 R = CH$_2$CH$_2$OCH$_3$

45 R = CH$_2$CH$_2$OCH$_2$—◁

46 R = CH$_2$OCH$_2$CH$_2$OCH(CH$_3$)$_2$

Fig. 8.2 Structures of betaxolol **45** and bisoprolol **46**, wherein an apparent lineage from metoprolol **28** seems reasonable given their approximately ten-year later launch dates, as well as their obvious similarities in structure. Likewise, that the *para*-substitution pattern utilized in **28** was able to draw from the clinical relevance established for such a relationship nearly ten years in advance by practolol **37**, further suggests an even earlier lineage between **28** and **37**.

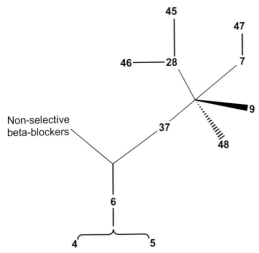

Scheme 8.4 Beta-blocker family tree showing major branching for cardioselective members based upon ABDD criteria. **4** and **5** are dichloroisoproterenol and pronethalol respectively which, by ligand-based drug design served as the roots eventually to sprout propranolol, **6**, the first clinically successful pioneer drug that proceeded to grow into a trunk for this entire family tree of all beta-blocker compounds. **37** is practolol, and even though it was not sustained clinically, its initial success as a cardioselective agent served as a key branch point for ABDD leading immedi- ately to atenolol, **9**, acebutolol, **7**, and metoprolol, **28**. Acebutolol was, in turn, soon followed by celiprolol, **47** (structure shown in next section), while metoprolol further branched into betaxolol, **45**, and seemingly bisoprolol, **46**. **48** (structure shown in later section) is nebivolol which, even though a much more recent member, further draws from the *para*-substituted arrangement as initially established by practolol, along with the N-substituted aralkyl relationship, to bestow cardioselectivity.

Betaxolol's ABDD program began with Synthelabo's (later Senofi-Synthelabo) desire to find a beta-adrenergic receptor blocking agent having:
- a high degree of cardioselectivity;
- a long duration of action allowing for once-a-day dosing; and
- a very high bioavailability upon oral administration [92].

Existing SAR data suggested that either the presence of selected substituents in the *para*-position of the aromatic ring analogous to acebutolol **7** [19,93], atenolol **9** [94] and metoprolol **28** [95], or replacement of the N-isopropyl group by an aralkyl system [96], could be used to confer cardioselectivity to beta-blockers. Considering **28** as a clinically successful parent drug from which to make analogues "... *it was reasoned that introduction of fairly bulky, stable substituents at the end of the chain might result in compounds which would be slowly metabolized*" [92], as well as being cardioselective. The metabolism of **28** is shown in Scheme 8.5, wherein with **28a** representing the major metabolite, the rationale for increasing the bulk of the terminal methyl group becomes clear.

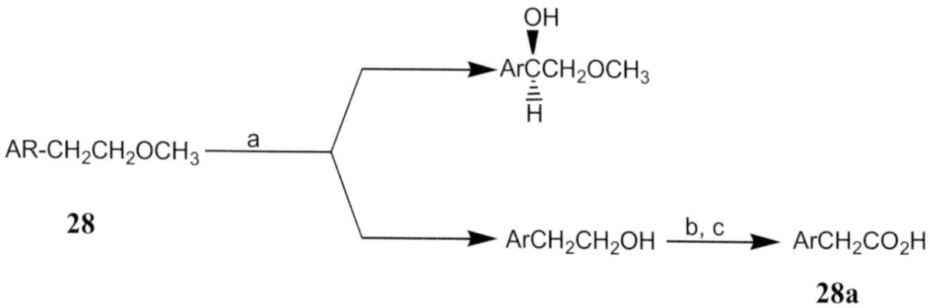

OH
↓
AR-CH₂CH₂OCH₃ —a—→ $ArCCH_2OCH_3$
‖ $\underline{\underline{H}}$

28

→ $ArCH_2CH_2OH$ —b, c—→ $ArCH_2CO_2H$

28a

Scheme 8.5 Metabolism of metoprolol, **28**, is thought to involve "...*rapid cleavage of the methoxyethyl chain at the methoxy group to afford a 2-phenylethanol derivative*" [92] which is subsequently oxidized to the corresponding phenylacetic acid derivative, **28a** [97]. Another metabolite is produced by oxidation of the benzylic carbon, a process recently shown to be stereoselective in its formation of only the (*R*)-alcohol as depicted [98]. Cytochrome P450 (CYP) 2D6 is thought to mediate both transformations. Steps b and c are likely to be catalyzed by alcohol dehydrogenase and aldehyde dehydrogenase, respectively.

Several analogues having the general structure shown in **49** were prepared and tested in *in-vitro* screens to assess their beta-1 blocking potency (rat atria) versus effects on beta-2 receptors (trachea), and as local anesthetics (frog sciatic nerve) [92]. Compounds exhibiting good potency as blockers and high selectivity for only beta-1 receptors were then tested in spontaneously hypertensive rats in which drug was given orally twice a day over a two-day period.

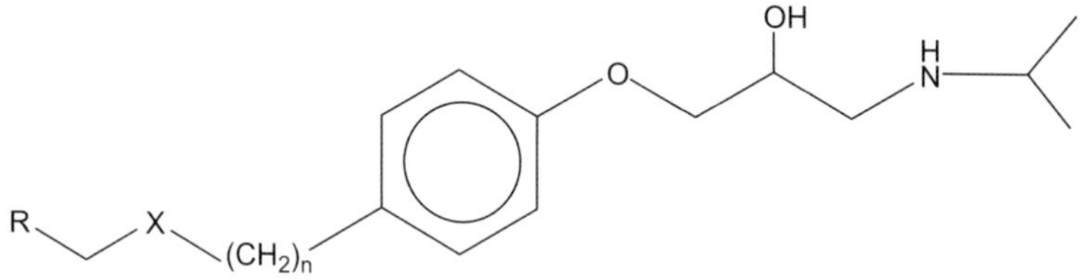

49

SAR data revealed that beta-blocker "... *activity depends simultaneously on the size of R, the nature of X, and the length of the alkyl chain between the aromatic ring and X*" [92]. Specifically, potency rose when R was initially increased in size from H (metoprolol) to cyclopropyl (betaxolol) and finally to cyclobutyl, but was then found to fall-off very sharply for cyclopentyl and cyclohexyl. Much better potency was observed when X was oxygen than when X was sulfur (thioether), although a sulfonyl group (SO₂ as linker) was not nearly as detrimental. Analogues having chains shorter (n = 1) or longer (n = 3) than the ethylene system, as present in metoprolol and betaxolol, were much less active than betaxolol. Eventually, betaxolol **45** was selected in that its overall profile was regarded as making it "... *simultaneously a more potent, more cardioselective, more bioavailable and longer-acting compound*" [92] than atenolol [94], acebutolol [93] or metoprolol [95].

The greater beta-1 blocking activity of betaxolol compared to metoprolol is thought to be due to its approximately seven-fold higher lipophilicity [99], an assessment deriving from the higher potency observed during *in-vitro* studies, as well as during *in-vivo* studies where its high lipophilicity also contributes to excellent oral absorption. The increased half-life compared to that of metoprolol is thought to be due "... *to a slower rate of cleavage of the* [terminal] *ether function* (see Scheme 8.5) *due to the steric hindrance conferred by the cyclopropane*" [92], a property that likely also explains betaxolol's greatly reduced hepatic first-pass metabolism [92].

8.2.2.2 Synthesis

The synthetic route used to prepare betaxolol is shown in Scheme 8.6 [92]. Interestingly, the yields for each of the eight steps are all above 80%, except for the last step (73%), which would generally be considered by drug discovery chemists to be the most straightforward in that it simply involves formation of the hydrochloride

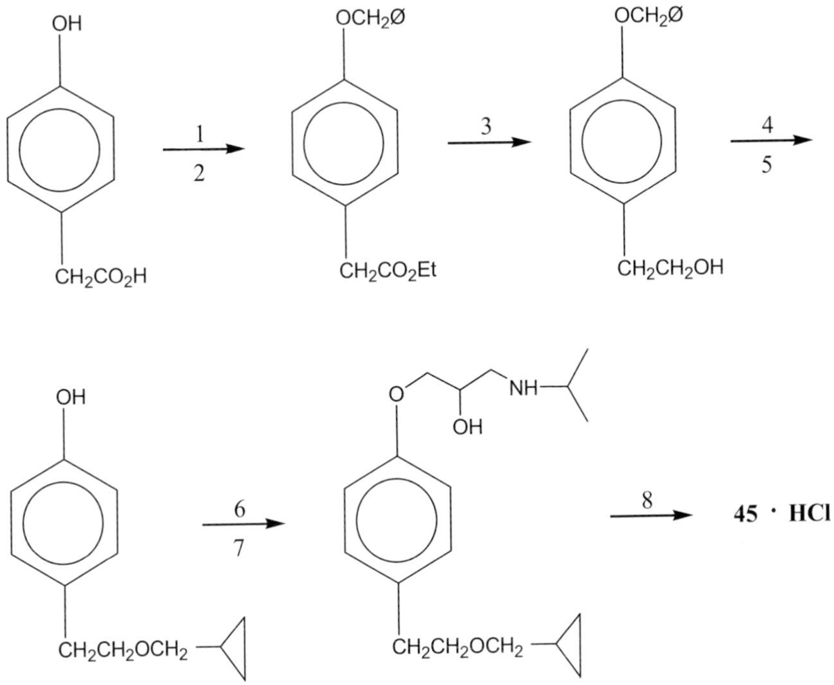

45

Scheme 8.6 Synthetic route to betaxolol, **45** [92]. Reagents: 1 Ethanol with catalytic HCl; 2 Benzyl chloride; 3 Potassium borohydride and lithium chloride; 4 Cyclopropylmethyl bromide with NaH as base; 5 Catalytic hydrogenolysis with Pd/C; 6 and 7 analogous to propranolol (Scheme 8.1); 8 HCl.

salt from the free base form of the secondary aliphatic amine. This lower yield makes the point, however, that even a salt-forming step is something that cannot be taken for granted as an inherently efficient procedure during the rigors of serious process chemistry, particularly when the latter is being conducted amidst a competitive market already having several related compounds.

8.2.2.3 Clinical Pharmacology

In terms of ADMET, following oral administration the absorption of betaxolol is rapid and almost complete, with bioavailability ranging from 80 to 89% [100]. Plasma protein binding is about 50%. The peak plasma concentration is achieved in 3–4 h, and is proportional to the dose. Plasma half-life is about 16 h and there are no active metabolites. The PK profile of racemic betaxolol accurately reflects the behavior of the betaxolol enantiomers in healthy male subjects [101]. The volume of distribution (4.0–7.3 L kg^{-1}) is larger than that of other beta-blockers. In pregnant women there is substantial placental transfer of the drug, and concentrations in breast milk can become very high – perhaps three-times those in blood [100]. The first-pass effect is minimal, and the elimination of betaxolol is mainly by hepatic biotransformation with the remaining portion being excreted unchanged by the kidneys [102]. The most common side effects are bradycardia, fatigue, and headache. The incidence of these and of CNS side effects are similar to those seen with atenolol [103]. The most important ADMET characteristics of betaxolol are shown in Table 8.2.

Major clinical uses of betaxolol include: hypertension; angina pectoris; chronic congestive heart failure (CHF); heart rhythm disturbances; and glaucoma. For hypertonia disease, efficacy and tolerability of betaxolol were studied in a large group (n = 4685) of ambulatory patients [104]. A significant reduction in blood pressure and heart rate were observed after 3 and 6 months of treatment. Clinical tolerance and compliance with once-daily treatment were high, and elderly patients responded well to this therapy. In comparative clinical trials, betaxolol was found to be superior to long-acting propranolol [105], and may be more effective than once-daily atenolol in patients with mild to moderate hypertension [103]. In studying the antihypertensive efficacy of betaxolol at three dose levels, the absolute magnitude of blood pressure reduction was greater with increasing dose [106]. Headache occurred significantly more often in the placebo group than in the betaxolol group, while the frequency of other adverse events was not significantly different between the two groups. In obese hypertensive patients, betaxolol was found to be a safe and effective long-term antihypertensive agent, without adverse effects on lipoproteins [107]. Because noncompliance is one of the major problems of chronic medical treatment in general, the effects of missing daily doses were studied in a double-blind trial, comparing once-daily oral betaxolol and atenolol [108]. Although the two beta-blocking agents were equally effective in controlling blood pressure when taken consistently, both arterial pressure and heart rate response with betaxolol were significantly superior for at least one day after missing a dose. The more reliable antihypertensive efficacy of once-a-day administration of betaxolol as monotherapy was also shown by 24-h, ambulatory blood pressure monitoring [109].

In patients with angina pectoris, betaxolol had the same anti-anginal effect and a comparable anti-ischemic effect as atenolol at 3 h. The beta-blocking, anti-anginal and anti-ischemic effects of atenolol were significantly decreased after 24 h of drug administration, while betaxolol maintained its beta-blocking and anti-anginal activities at 24 h, and compared favorably with atenolol in anti-ischemic activity [110]. During treadmill exercise, the heart rate remained lower with betaxolol than with atenolol, while exercise time was significantly prolonged only by betaxolol [111]. These results were confirmed by other studies, wherein it was also observed that the combination of betaxolol with long-acting nitrates can lead to further potentiated effects [112].

The ongoing 'BETACAR' study is comparing the safety and efficacy of betaxolol compared to carvedilol in heart failure [113]. The two beta-blockers appear to have similar effects on left ventricular ejection fraction [New York Heart Association (NYHA) class], exercise capacity and quality of life over 6 months [114]. However, final results are not yet available.

In atrial fibrillation, it was found that combination therapy with low-dose betaxolol plus digoxin was superior to low-dose diltiazem plus digoxin in controlling ventricular rate and reducing rate-pressure products [115]. In monotherapy, betaxolol effectively controlled the ventricular rate of fibrillating patients at rest and during exercise, but also caused considerable reductions in maximal oxygen uptake and cardiac output during exercise [116].

By decreasing the production of aqueous humor, betaxolol can reduce intraocular pressure, and thus it is widely used to treat glaucoma. The "Early Manifest Glaucoma Trial" was the first, adequately powered randomized trial in patients with open-angle glaucoma [117]. Laser trabeculoplasty plus topical betaxolol reduced intraocular pressure, and this effect was maintained throughout the follow-up period. Whereas progression varied across patient categories, treatment effects were present in both older and younger patients, high- and normal-tension glaucoma, and eyes with less and greater visual field loss.

8.2.3
Celiprolol

8.2.3.1   Discovery

Tracking the ABDD pathway leading to celiprolol, **47**, reveals that its discovery and development appears to be very similar to that of betaxolol. Referring to Scheme 8.4, one can again trace a lineage from propranolol, **6**, to practolol, **37**, and then to acebutolol, **7**, the latter occurring at about the same time as that for metoprolol, **28**. While **28** then leads to betaxolol, **45**, acebutolol leads to celiprolol, **47**, with both of these offspring again appearing about 10 years later and thus representing fourth-generation agents. Furthermore, the rationale for designing **47** from **7** appears to be driven by the very same PK-related need that prompted the design

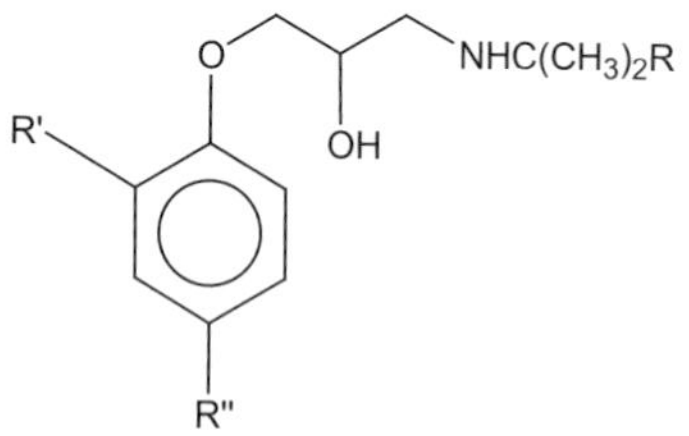

37 R = R' = H; R" = NHCOCH$_3$

7 R = H; R' = COCH$_3$; R" = NHCOCH$_2$CH$_2$CH$_3$

47 R = CH$_3$; R' = COCH$_3$; R" = NHCON(CH$_2$CH$_3$)$_2$

Fig. 8.3 Structure of celiprolol **47**. An apparent ABDD lineage from acebutolol **7** seems reasonable, given its approximately ten-year later launch date, as well as its obvious similarity in structure. Likewise, the even closer resemblance of **7**'s *para*-substituent to that established for cardioselectivity nearly ten years in advance by practolol **37**, further conveys the earlier ABDD-related lineage between these two compounds as well.

of **45** from **28**, namely that of avoiding the rapid metabolism to which **7**, like **28**, is subject. The structures for practolol, acebutolol and celiprolol are provided in Fig. 8.3 for direct comparison. The structural similarities along this lineage are evident.

Celiprolol was first synthesized by Chemie Linz (later known as Hafslund Nycomed Pharma) [118], and eventually inherited by what is now Aventis Pharma. With a desire to attenuate the significant hepatic first-pass metabolism [119] and prolong the approximate 3-h half-life [120] observed for acebutolol, **7**, analogues were directed toward protecting the latter's amide function from metabolic attack. The metabolism of **7** is shown in Scheme 8.7, wherein the rationale for protecting this functionality becomes clear.

Following the preparation and initial testing of several analogues, celiprolol **47** was selected for preclinical development. Compound **47** "... *is not affected by significant first pass metabolism*" [118,125], and its half-life is approximately 5 h [125–128] compared to 3 h for its immediate predecessor, acebutolol, **7**. Furthermore, celiprolol is only minimally metabolized, with about 84% and 11% being excreted unchanged via the liver and kidney, respectively, after oral dosing in humans [125,129,182]. With its *ortho*-substituent contributing to potency (pA$_2$ value in gui-

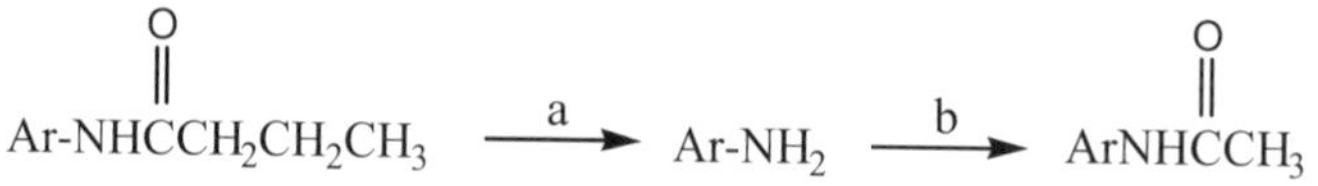

Scheme 8.7 Metabolism of acebutolol, **7** [119]. After oral administration to man, **7** is extensively metabolized upon first-pass through the liver to first produce **7a** or 'acetolol' [121], which then undergoes N-acetylation to form **7b** or 'diacetolol' [122]. While the N-acetylation is thought to be independent of acetylator status [123] and **7b** retains similar beta-blocking potency [119,124], these biotransformations were regarded as complicating factors for deploying **7** clinically, particularly when a longer, overall half-life was desirable.

nea-pig atrium of 8.05) and its *para*-substituent contributing to selectivity for beta-1 receptors (ca. 20-fold less active on tracheal tissue), **47** was thus a potent, cardio-selective beta-blocker with an improved PK profile compared to acebutolol [130]. In addition, whilst **47** retained some degree of ISA, it was completely devoid of any MSA [130].

A few points pertaining to SAR are noteworthy. First, that the acetyl function as a second substituent on the aryl-ring in both **7** and **47**, resides in a position that is ortho to the oxypropanolamine side chain is in line with accumulated SAR. The suggestion is that this can enhance overall beta-blocker potency, even if not contributing toward beta-1 selectivity. Likewise, this same arrangement is in line with the aforementioned SAR associated with atenolol, **9**, wherein such "... *additional substituents ... are best tolerated when placed in an ortho-relationship to the ether rather than when in a meta-relationship, the latter perhaps reflecting steric interference with the hydrogen-bonding role that the amide* (or in **47**'s case the urea) *may be playing.*" Upon further tracking the origin of the *ortho*-acetyl function which appears first in acebutolol, **7**, it becomes interesting to note from the latter's related literature that not only were medicinal chemists simply mapping multiple aryl-substitutions within the context of practolol, **37**, [e.g., 131], they were also exploring "immobilized" versions [132] of propranolol's oxypropanolamine system wherein the latter had been folded-back into a cyclic arrangement involving the *ortho*-position of the aryl-ring. These types of classical, "semi-rigid analogues" are shown below as structures **50** to **52** [132].

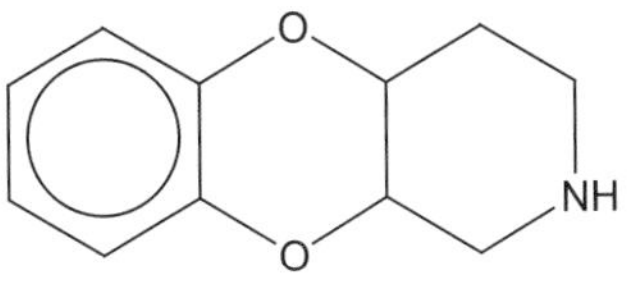

50 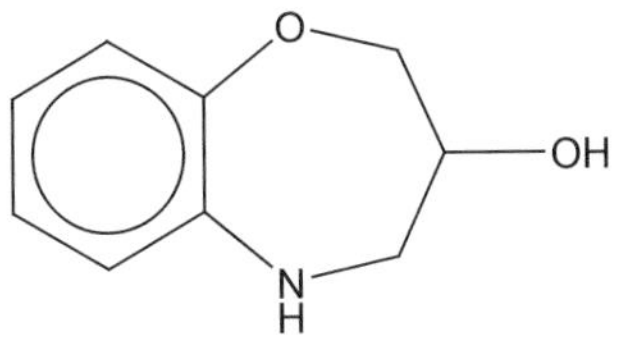   **51**

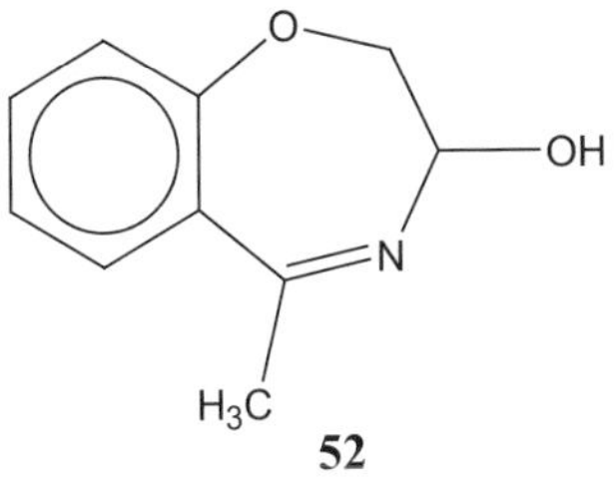

52

As part of such investigations, *ortho*-substituted oxime intermediates were found to be active, and this led to an examination of their precursor *ortho*-substituted ketones, again within the context of simple aryl systems like that present in practolol and its *para*-substituent [132]. This gradual accumulation of SAR data further demonstrates propranolol's key role as a pioneer drug during ABDD, as well as practolol's key role as an early offspring that prompted so much interest in beta-1 selectivity as manifest via "*para*-substitution." Another interesting SAR point is that the diethyl urea functionality of 47 clearly represents both a sterically hindered and an inherently less susceptible system toward metabolic degradation when compared to the simple amide that is present in acebutolol, 7. Particularly noteworthy for today's medicinal chemists as they continue to struggle with optimization of ADMET issues that most often involve attempts to enhance PK properties, is the repeated use of steric hindrance to decrease (betaxolol over metoprolol) or completely avoid (celiprolol over acebutolol) metabolic degradation for which in both cases, the results clearly proved to be clinically successful. Indeed, given today's apparent fondness for simple but firmly anchored, numeric-related guideposts that can be confidently utilized during drug design, the aforementioned steric strategy might be referred to as "metabolism's rule of one" because no other approach – short of completely removing the culprit functionality itself – comes even close to being a distant second in terms of the reliability for actually accomplishing such an objective within the context of attenuating drug metabolism [133]. A final structural point of interest to the adrenergics arena in general, as well as to beta-blockers specifically, involves the change from an *N*-isopropyl to an *N-t*-butyl group, for which SAR at that time [e.g. 134] was suggesting somewhat higher potency and an even further stability toward monoamine oxidase (MAO) (see the discussions about this fundamental metabolic pathway and this key SAR feature in Section 8.3, Accumulated SAR).

8.2.3.2 Synthesis

The synthetic route used to prepare celiprolol is shown in Scheme 8.8 [118,135,136]. It is perhaps interesting to first note that while the early syntheses of acebutolol, 7, took advantage of a Fries rearrangement from an acetylated phenol already bearing the desired amide in the *para*-position so as to also establish the *ortho*-acetyl functionality [119], for the case of celiprolol this convenient, one-step process has been replaced by a Friedel-Crafts acylation reaction followed by a rather harsh acid treatment to liberate the ether-protected phenol (steps 2 and 3, respectively). Although probably no more problematic than for the syntheses of most other beta-blockers, side products 47a and 47b, as distinctly pointed out within the pertinent literature [135,136], are also displayed in Scheme 8.8. Clearly, 47a is limited to 47's unique urea substituent, and simply reflects a displacement of the diethylamine by *tertiary*-butylamine during the final step. Alternatively, 47b deserves additional comment because over-alkylation of the amine system being used either to open a substituted 1,2-cyclo-oxypropane system (as shown for propranolol in step 2 of Scheme 8.1) or to displace a halide from its beta-hydroxy halide equivalent (as shown in step 6 of Scheme 8.8) can be quite problematic for beta-blockers in general. However, the magnitude of this problem is inversely proportional to the degree of steric bulk present on the amine system. Thus, the present case of having a very bulky, *tertiary*-butylamine system should actually present minimal problems and even an isopropylamine can be regarded as being reason-

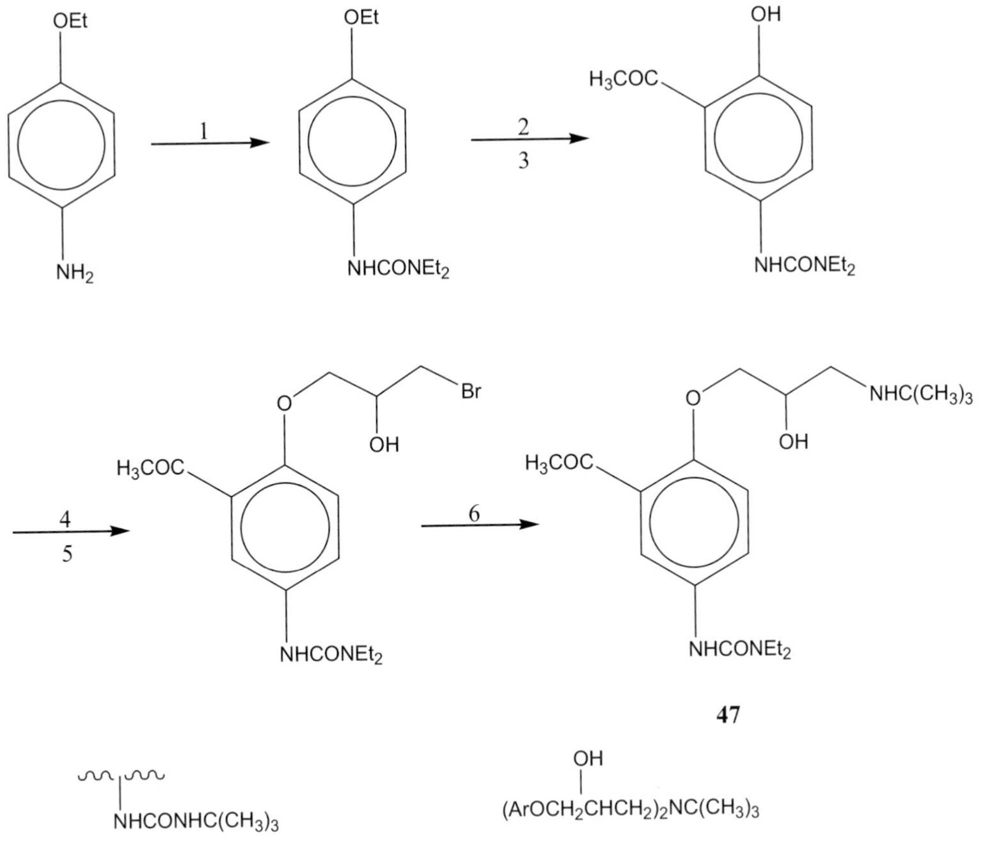

Scheme 8.8 Synthesis of celiprolol **47** [118,135,136].
Reagents: 1 ClCONEt₂; 2 ClCOCH₃, AlCl₃; 3 HCl;
4 Epichlorohydrin; 5 HBr; 6 H₂NC(CH₃)₃, Et₃N. **47a** and **47b**
are minor side-products.

ably well-behaved during such chemistry. Alternatively, for primary aliphatic amines bearing only a single substituent on the alpha-carbon atom, this particular chemical conversion does indeed become troublesome [137].

8.2.3.3 Clinical Pharmacology

In terms of ADMET, the absorption of celiprolol from the gastrointestinal tract is rapid, and its time to onset of action is only 30–60 min, with a peak effect in about 2–3 h. The presence of food in the gut can reduce absorption by about 30%. Itraconazole increases, while grapefruit juice significantly decreases, plasma concentrations of celiprolol [138]. Likewise, orange juice substantially reduces the bioavailability of celiprolol, and because it is consumed widely, this interaction is likely to have significant clinical importance [139]. Bioavailability ranges from 30 to 70% and is dose-dependent. Plasma protein binding is about 25–30%; the plasma half-life is 5–6 h. The volume of distribution is similar to that of betaxolol (i.e., a mean value of 6.5 L kg^{-1} [140]). Celiprolol has no clinically relevant active metabolites. Excretion occurs mainly as the unchanged form in urine and bile. Celiprolol crosses the placenta, and its use during lactation is contraindicated. When administered at the recommended doses, celiprolol has no negative inotropic effect and appears to be hemodynamically advantageous compared to metoprolol in patients with coronary artery disease and hypertension [141]. It is of special importance that, in a recent study in patients with mild-to-moderate chronic obstructive pulmonary disease, celiprolol was the only beta-blocking agent having no detrimental pulmonary effects [142]. The most important ADMET characteristics of celiprolol are shown in Tab. 8.2.

Celiprolol is used clinically to treat: hypertension; angina pectoris; and CHF. There have not been any large, multicenter and properly controlled international trials for the use of celiprolol in treating hypertension. One acronym-marked trial, the "CELIMENE study" [143] assessed whether celiprolol or the ACE-inhibitor enalapril is able to reduce carotid artery wall hypertrophy through a reduction in carotid pulse pressure rather than by lowering mean blood pressure, and whether the influence of local pulse pressure reduction can be detected at the site of the radial artery. The reduction in carotid pulse pressure, but not in mean blood pressure, was a major determinant of the reduction in carotid intima-media thickness. Radial artery intima-media thickness and pulse pressure decreased significantly with both treatments. The reduction in radial artery intima-media thickness was not related to the changes in radial artery pulse pressure. Unlike patients with either hypertension or angina pectoris alone, patients with both pathologic conditions usually have a reduced left ventricular compliance and may, therefore, have an impaired capability to cope with acute hemodynamic changes generated by standard beta-blockers. In an open-label, sequential comparison of standard monotherapy with other beta-blockers and calcium antagonists versus celiprolol [144], celiprolol produced fewer occurrences of fatigue, dizziness, and edema. Celiprolol controlled blood pressure and anginal attacks to the same extent as did standard monotherapy. The results suggested that patients with hypertension and angina pectoris are prone to adverse effects of standard drug treatments, and celiprolol, whilst equally effective, is largely devoid of the adverse effects typically seen with standard therapy. In a randomized, 21-month cross-over trial, it was found that celiprolol improves insulin sensitivity of hypertensive patients with dyslipidemia [145]. A comparative study with propranolol, atenolol, bisoprolol, or celiprolol showed that in hypercholesterolemic hypertensive patients, selective beta-blockers are less likely to adversely effect plasma lipids than nonselective ones, and celiprolol was actually able to improve the lipid profile [147]. These results have also been confirmed in elderly hypertensive patients [147].

In patients with coronary heart disease and hypertension, the acute effects of intravenously administered celiprolol appeared to be hemodynamically advantageous [141]. Celiprolol and atenolol were given once-a-day to patients with stable angina pectoris in a double-blind, placebo-controlled study [148]. Both drugs were equally effective in reducing the frequency of anginal attacks and delaying the onset of ischemia during exercise. However, only atenolol lowered cardiac output at rest and during exercise. Thus, the ancillary properties of celiprolol, including partial beta-2 agonist activity and direct vasodilating effects have detectable influences on cardiac function. According to data from animal studies [149], celiprolol mediates coronary vasodilatation and improves myocardial ischemia through nitric oxide-dependent mechanisms. For the treatment of severe heart failure (NYHA class IV), the initiation of intravenous beta-blocker therapy with low doses of a beta-1 blocker with vasodilating effects (i.e., celiprolol) may have hemodynamic advantages over conventional beta-blockade [150].

8.2.4
Nebivolol

8.2.4.1 Discovery

Although an ABDD-derived lineage leading to nebivolol, **48**, is not as readily discernable as it is for betaxolol, **45**, or for that of celiprolol, **47**, nebivolol's market launch by Menarini in the late 1990s is, again, about ten years after the launch of these same ABDD-defined fourth-generation beta-blockers, both of which were discussed in preceding sections (also see Scheme 8.4). Thus, it is clear that the discovery of **48** was able to draw from all of the prior beta-1 blockers' SAR data along with the valuable SAR likewise provided by their predecessors. Nebivolol **48** will be discussed in such a context.

An examination of nebivolol's structure (Fig. 8.4) reveals first that it contains two propranolol-like oxypropanolamine systems in a bis-like arrangement that takes advantage of a single amino-group by disubstituting the latter in a near-symmetrical manner. One of these near-bis-arranged beta-blocking pharmacophores has been highlighted in Fig. 8.4 by emboldening the relevant atoms.

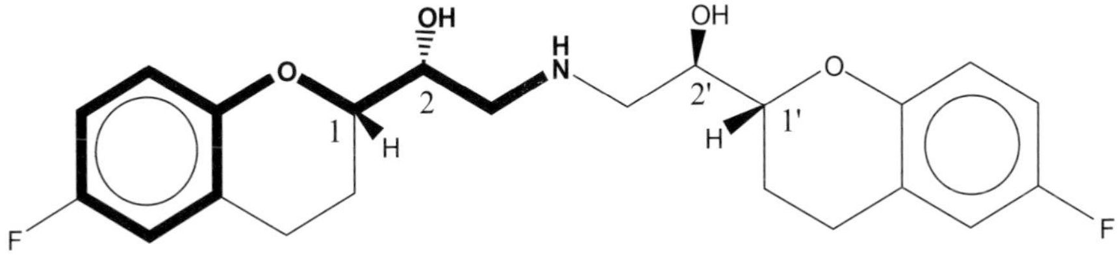

48

Fig. 8.4 Structure of (S, R, R, R) nebivolol [151,152] which exhibits nearly 200 times higher beta-1 adrenergic binding affinity than its enantiomer. The marketed product is a racemate [151]. The bold-faced atoms depict one of the two, nearly perfect bis-arranged, beta-blocking pharmacophores that are present in nebivolol (see text). Note that the indicated numbering has been arbitrarily assigned for convenient use in the text discussion.

Upon closer examination, however, it can be seen from the depicted stereo-chemistry that the two partners are not quite perfectly bis or symmetrical in that while the emboldened portion has 1-(S), 2-(R) stereochemistry, the other partner has 1'-(R), 2'-(R) stereochemistry, which makes them diastereomeric rather than being either identical (perfectly bis) or enantiomeric (perfectly symmetrical). Note that the indicated numbering has been arbitrarily assigned for discussion pur-poses and that nebivolol is actually marketed as its racemic mixture of the (S,R, R,R) and (R,S,S,S) enantiomers. Only the former is depicted in Fig. 8.4 because it is thought to be the active isomer at beta-1 adrenergic receptors [151]. Further scrutiny of these stereocenters reveals one final but extremely interesting SAR detail, namely that the 2- and 2'-position hydroxy groups, as present in the pur-portedly more active (S, R, R, R)-enantiomer, are oriented in a manner that does not appear to allow either one of them to occupy the same three-dimensional space that is required for optimal binding with beta-adrenergic receptors (i.e., see structures **1** to **3** and the accompanying text in Section 8.1). This finding does indeed stand-out as an anomaly amongst beta-blockers in general (see Section 8.3, Accumulated Structure–Activity Relationships) where, as has also been noted by others, while (S,R,R,R) nebi-volol with a 2-(R)-hydroxy configuration "*carries beta-adrenoreceptor blocking properties, in other* [aryloxypropanolamine-derived] *beta-adrenoreceptor antagonists these properties are carried by the* [opposite] *enantiomer*" [152,153] wherein the analogous hydroxyl-groups are consistently found to bear an (S) configuration (also see futher discussion in Section 8.3).

Returning to the two "pseudosymmetrical" [151] portions of **48**, we see that these scaffolds are quite similar to the semi-rigid, conformational analogues **50** to **52** that were previously traversed as part of the ABDD explorations ultimately leading to celiprolol, **47**. In this regard, it follows that the *ortho*-connection to the *para*-fluoro-substituted ring then becomes reminiscent of the enhanced potency and beta-1 selectivity that can be derived in a respective manner upon establishing this type of di-substituted pattern on the phenoxy-ring. Likewise, if one or the other of the pseudosymmetrical partners binds to the beta-adrenergic receptor pocket, then the other partner becomes an N-aralkyl type of substituent that is also thought to be capable of endowing selectivity for beta-1 adrenergic receptors by associating with an auxiliary binding site (this is further discussed in Section 8.3, 'Accumulated Structure–Activity Relationships). Finally, that each of the two partners is distinct and must therefore be assembled in a step-wise manner, clearly shows that nebivolol did not simply result from fortuitous testing of syn-thetic byproducts resulting from the over-alkylation problems mentioned in Sec-tion 8.2.3 (celiprolol) because that would result in true bis-products. An enantiose-lective, total synthesis of nebivolol is illustrated below.

8.2.4.2 Synthesis

The reported [152] synthetic route depicted in Schemes 8.9 and 8.10 was selected herein to further showcase the intriguing stereochemistry that is uniquely present in nebivolol, **48**, when compared to the vast majority of other beta-blockers.

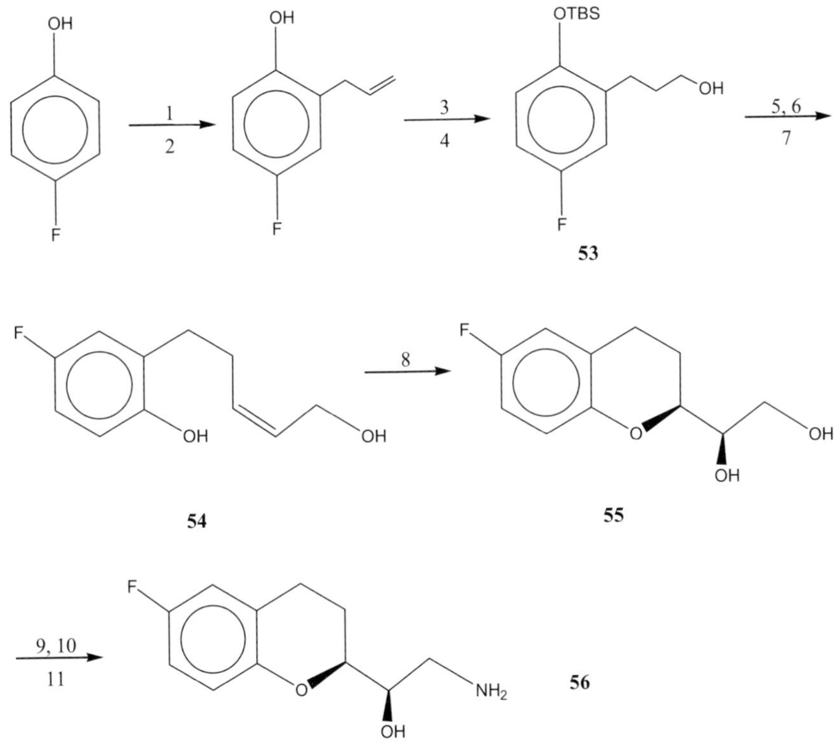

Scheme 8.9 Synthesis [152] of **48**'s bold-faced (Western) partner as shown in Fig. 8.4. Reagents: 1 Allylbromide, K_2CO_3; 2 Heat; 3 TBDMSCl, Imidazole; 4 $BH_3.Me_2S$, H_2O_2, Base; 5 Dess-Martin Periodinane, PPh_3=$CHCO_2Et$; 6 DIBAL; 7 TBAF; 8 (–)-DET, Ti (isopropoxide)$_4$, TBHP, Base; 9 TsCl, Pyr.; 10 NaN_3; 11 Pd/C.

Unlike the other synthetic schemes shown in this chapter, this synthesis is not the same process that is used in large-scale manufacturing which produces the racemic form of nebivolol for the market place.

Briefly, as shown in Scheme 8.9, a Claisen rearrangement of the *para*-fluorophenol's O-allyl ether followed by TBS-protection, hydroboration and treatment with hydrogen peroxide, produced the propyl alcohol intermediate **53**. The latter was subjected to a combination of a Dess–Martin oxidation and Wittig homologation reactions, followed by DIBAL reduction of the ester and removal of TBS with TBAF to provide the extended alkene-alcohol intermediate **54**. Sharpless asymmetric epoxidation using a sodium hydroxide work-up provided either intermediate **55** on treatment with (–)-DET, or intermediate **57** (Scheme 8.10) on treatment with (+)-DET. Tosylation of **55** followed by displacement with sodium azide and subsequent reduction provided **56** as the stereochemically defined, amine-containing, Western partner of **48** having an (S, R) designation. Alternatively, inversion of the secondary hydroxyl groups stereochemistry in **57** was accomplished by a Mitsunobu reaction followed by removal of the *para*-nitro-benzoate (PNB) and

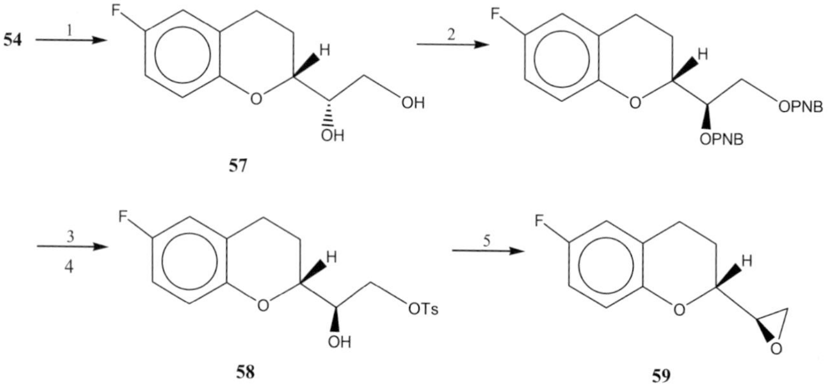

Scheme 8.10 Synthesis [152] of **48**'s non-bold-faced (Eastern) partner as shown in Fig. 8.4. Reagents: 1 (+)-DET, Ti(isoprop-oxide)$_4$, TBHP, Base; 2 p-NO$_2$C$_6$H$_4$CO$_2$H, DEAD, TPP; 3 NaOMe; 4 TsCl, Pyr.; 5 NaOMe.

tosylation to give intermediate **58** which was, in turn, readily converted to epoxide **59**. The latter compound represents the Eastern partner in **48** and has defined (R, R) stereochemistry (Fig. 8.4). East was finally joined with West by nucleophilic opening of epoxide **59** using the hydroxylamine version of **56** produced *in situ* (BF$_3$O(Et)$_2$, t-BuOH) in a manner ultimately reminiscent of step 2 for propranolol (see Scheme 8.1).

Although the synthetic route shown in Schemes 8.9 and 8.10 was chosen for its overall elegance and absolute control of stereochemistry while producing the active enantiomeric form of racemic **48**, it should be pointed out that nebivolol, along with many of its other distinct stereoisomers, were all initially prepared by Janssen Pharmaceutica [151].

8.2.4.3 Clinical Pharmacology

In terms of ADMET, nebivolol is rapidly absorbed following oral administration, and even the presence of food does not appear to alter the rate or extent of absorption. Peak plasma concentrations are reached within 2–6 h after dosing. The volume of distribution ranges from 10.1 to 39.4 L kg^{-1}. Plasma protein binding is about 98%. Nebivolol is extensively metabolized in the liver, but the metabolism is rather complex: the main metabolic pathway is aromatic hydroxylation which is subject to a debrisoquine-type metabolic genetic polymorphism [154]. This means that humans are divided into two types of metabolizers, namely subjects with fast metabolism and others with slow metabolism. In fast metabolizers, the elimination half-life is about 10 h, and that of the still active (hydroxy-) metabolites is about 24 h. Peak plasma concentrations of unchanged nebivolol plus active metabolites are 1.3- to 1.4-fold higher in slow metabolizers. However, a substantial hepatic first-pass effect of active metabolites in fast metabolizers seems to compensate for differences in unchanged drug between the two phenotypes. Therefore, nebivolol ultimately has similar therapeutic effects in the two types of meta-bolizers. It is excreted in the urine and feces, almost entirely as metabolites. In a study with healthy volunteers, there was no interaction between the H_2-receptor antagonist ranitidine and nebivolol. Although cimetidine inhibited nebivolol metabolism, it did not have any significant effect on the pharmacodynamics of the beta-blocker [155]. The most common side effects, having an incidence of 1 to 10 %, are headache, dizziness, tiredness, and paraesthesia. The most important ADMET characteristics of nebivolol are listed in Tab. 8.2.

Nebivolol's clinical uses include: hypertension; angina pectoris, and CHF. Nebivolol induced a dose-dependent reduction in blood pressure within hyperten-sive patients, with no significant differences between peak-to-trough levels [156]. The drug was well tolerated and was equi-effective in Afro-American patients. In a double-blind, placebo-controlled study, nebivolol not only decreased elevated blood pressure, but also suppressed plasma renin and aldosterone levels, and stimulated plasma atrial natriuretic peptide [157]. Comparing antihypertensive actions in hypertension across general practice settings, both atenolol and nebivolol caused similar reductions in systolic and diastolic pressure without orthostatic hypoten-sion or significant falls in heart rate [158]. Nebivolol achieved more pronounced falls in blood pressure than enalapril, wherein the trough-to-peak, sitting diastolic ratios also favored nebivolol (84% versus 60%, p = 0.002) and the incidence of coughing was much higher with enalapril than with nebivolol [159]. A pilot study has suggested that the use of nebivolol in hypertensive patients with mild-to-mod-erate chronic obstructive pulmonary disease (COPD) was safe during a 2-week treatment [160]. In the "NEBIS trial" [161], the antihypertensive efficacy of nebivo-lol and that of bisoprolol were compared. High proportions of responders (92% versus 89.6%) were observed in both groups, and there was no significant differ-ence between the two treatments.

The blood pressure-lowering effects of nebivolol are at least partially due to direct vasodilation as a result of nitric oxide (NO) release from endothelial cells [162]. Nebivolol appears to interact with the endothelial NO pathway in two com-plementary ways: (a) it increases NO synthase (NOS) activity; and (b) it reduces the NO-scavenging radical superoxide anion by redirecting deranged NO activity from superoxide to NO production. In addition, it appears also to have a comple-mentary anti-oxidant activity. A recent study [163] indicated that nebivolol, through its anti-oxidant properties, increases NO by decreasing its oxidative inacti-vation. Finally, it has also been suggested that the nebivolol-elicited vasorelaxation is partially triggered by estrogen-dependent pathways in the endothelium [164].

Unlike classic beta-blockers, nebivolol maintains or even improves left ventricular function. In a single-blind study, the anti-anginal and anti-ischemic activities of nebi-volol were assessed in patients with stable angina pectoris [165]. Nebivolol increased the time to onset of angina during exercise testing when compared to placebo, and reduced nitroglycerine consumption. In patients with angina pectoris, nebivolol prolonged exercise time to anginal pain and improved exercise tolerance [166].

Because of its unusual multiple modes of action, nebivolol may improve toler-ability in elderly patients with heart failure, where endothelial vasodilator reserve is typically limited. Data were also reported that nebivolol may have advantages in patients with diastolic heart failure and arterial hypertension [167]. The "SENIORS trial" was undertaken to determine the effects of nebivolol on morbid-ity and mortality in elderly patients with CHF, regardless of ejection fraction [168]. The primary outcome was the composite of all causes of mortality or cardio-vascular hospital admission (time to first event) and, when compared to the pla-cebo group, such markers occurred less frequently in the patient group treated with nebivolol (p = 0.039, hazard ratio 0.86, 95% confidence interval 0.74–0.99). There was no significant influence of age, gender or ejection fraction on the advantageous effect of nebivolol. The drug was found to be an effective and well-tolerated treatment for heart failure in the elderly.

8.3
Accumulated Structure–Activity Relationships

Although X-ray diffraction and/or detailed nuclear magnetic resonance (NMR) studies are still not available to allow for explicit "structure-based drug design" [133] or for precise "topographical maps" [169,170] from which to orient SAR data, conceptual models based upon the known amino acid sequences for these well-established G protein-coupled receptors are available. In this regard, the beta-2 receptor has received the most attention [171–173], and its models can be conveni-ently used to assist in categorizing SAR that pertain to beta-receptors in general. Taking this approach, three principal binding domains need to be considered, namely an aromatic region, a specific interaction with the alkyl-hydroxy group, and a likely ionic bond area since the amino group is thought to be protonated during receptor interaction.

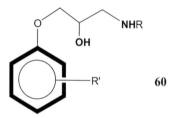

These three pharmacophoric elements are shown as emboldened atoms within the generalized beta-blocker structure **60**. The key atoms in **60** should be consid-ered to be discretely displayed in three-dimensional (3D) space via the oxypropyl-connecting chain, so as to be able to dynamically occupy a specified residence within the beta-receptor pocket, rather than to be merely residing in a static man-ner on the rigid plane that becomes defined by three points. For example, reconsi-dering the previously noted historical development wherein a surprising enhancement in beta-blocking potency occurred when an oxymethylene unit was inserted into the ethyl connecting chain that characterizes the endogenous agonists, two salient SAR features should be emphasized by also referring to model structures **61** and **62** shown side-by-side below.

61   **62**

First, as suggested by computational studies [174], the oxypropanolamine system present in **62** can adopt a conformation that places its hydroxy and amino groups into nearly the same 3D space relative to the aryl moiety, as that occupied by the ethanolamine system's analogous groups relative to **61**'s aryl moiety. Second, it is clear that in both cases, the hydroxyl-group resides in the same spatial orientation relative to the asymmetric carbon present in each system, namely (*R*)-**61** and (*S*)-**62** wherein the designated switch from (*R*) to (*S*) reflects only a switch in the nomenclature-associated group priorities made during such assignments. That is, the C–O ether group present in **62** gains priority over the C–N amino group that is present in both structures, even though all these groups are still displayed in identical 3D space. Coupled with these chemical findings is the biological observation that while the (*R*)-arylethanolamine and (*S*)-aryloxypropanolamine systems produce active beta-receptor ligands, their enantiomeric counterparts are essentially inactive, being comparable to when no hydroxyl-group is present. Likewise, attempts to substitute the secondary hydroxyl group with other functionality that can form hydrogen bonds either as a donor, acceptor or both, have proven to be highly detrimental for activity [e.g., 175]. The one notable exception to this key 3D relationship is that of nebivolol, **48**, which as was mentioned earlier, clearly stands-out as an anomaly among the numerous other beta-blockers. As a final point to be made about stereochemistry it should also be appreciated that the "*superficially simple reaction*" [176] that can be used so conveniently to establish the stereochemical disposition of the hydroxy-group, namely that of the epichlorohydrin reaction step 1 in the initial propranolol synthesis as shown in Scheme 8.1, actually " *... proceeds via a Payne [like] rearrangement, such that the carbon atom initially bonded to chlorine in epichlorohydrin becomes the terminal atom of the* [aryloxy-substituted] *epoxide*" [176] intermediate [177,178]. However, in this case it can be noted that, unlike a true Payne rearrangement, the C-2 position of epichlorohydrin is never itself attacked (Scheme 8.11). Thus, as shown in Scheme 8.11, stereochemistry at this position is retained throughout while nomenclature flips once and, in the end, it is (*R*)-epichlorohydrin that should be deployed when seeking an (*S*)-aryloxypropanolamine having beta-receptor blocking activity.

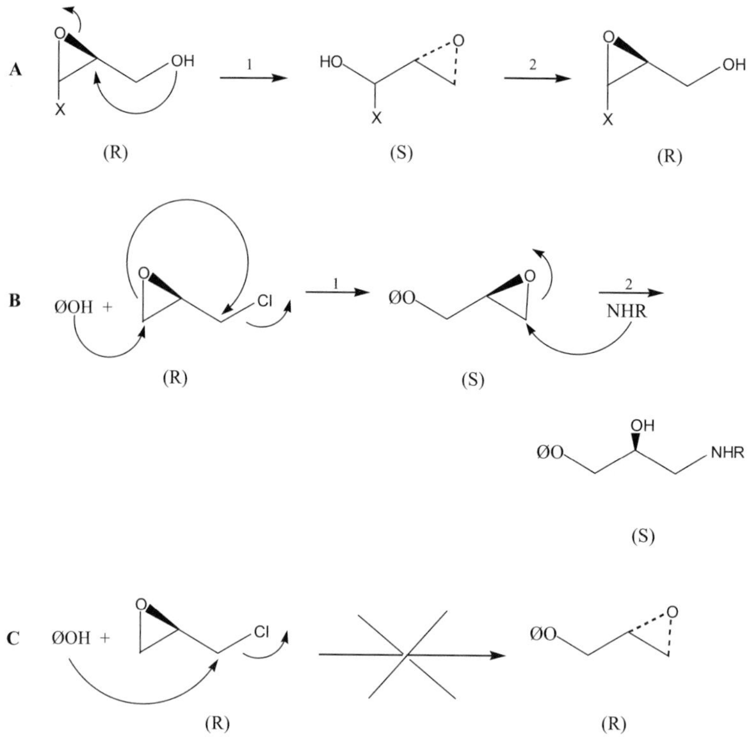

Scheme 8.11 Payne-like opening of epichloro-hydrin during syntheses of aryloxypropanola-mine beta-blockers. Pathway A shows a true Payne rearrangement [178] as it pertains to how a 2,3-epoxy alcohol becomes isomerized when treated with aqueous base. Step 1 is meant to show how the initial intermediate alkoxide first attacks the 2-position to cause an inversion of stereochemistry. Since the same process can then be repeated from the other direction (step 2), an equilibrium is eventually obtained where the preponderance of one isomer over the other is dictated by whatever other substituents may be present. In the present case, the other substituent, X, has a higher nomenclature ranking than oxy-gen and there are no changes in group priori-ties. Pathway B shows a Payne-like isomeriza-tion of the epoxide when epichlorohydrin is attacked by a phenolate anion. Step 1 can be conveniently conducted by using K_2CO_3 in ac-etone, thus exploiting the acidity of a phenolic hydroxy group while also operating under the mildly basic and nonaqueous conditions often deployed to prepare epoxides from alpha-halohydrins. Step 2 reflects opening of the epoxide from the least sterically hindered site by attack of a primary amine. When the latter is isopropylamine, as is commonly found among beta-blockers, step 2 can be conveniently conducted by refluxing a mixture of isopropylamine and methanol deployed as cosolvents. Note that in this pathway, the 2-position of epichlorohydrin is never attacked, although there is a flip in absolute assign-ment based upon a change in the nomencla-ture-related priority assignments after step 1. Pathway C depicts an alternate reaction that can also lead to an arylether intermediate wherein the epoxide of epichlorohydrin would similarly assume a "terminal" [176] position. However, because attack by the phenolate occurs from the opposite side of epichlorohy-drin, in this case an inversion of stereochem-istry relative to the position of the aryl system does occur even though the 2-position is again untouched. This reaction does not gen-erally compete in a significant manner with the epoxide-opening pathway shown in B, as it is typically undertaken during the synthesis of aryloxypropanolamines. However, as has been noted by others, pathway C can become problematic when other solvents are deployed (e.g., dimethylformamide) [179–181].

Referring again to **60**, it is interesting to note that while these secondary amines show good activity as beta-blockers, neither primary nor tertiary amines exhibit much activity, the primary amines likely being subject to rapid metabolism, and the tertiary amines being outright unacceptable to the receptor. In terms of selectivity, although the oxypropanolamine system itself serves to reduce binding with alpha-receptors, it can be further noted that in general, as the steric bulk of R is increased, interaction with alpha-receptors is decreased while interaction with beta-receptors is increased. The latter is presumably due to association with an auxiliary lipophilic binding pocket lying adjacent to an ionized Asp (e.g., Asp 113 in the beta-2 receptor) residue that is thought to form an ionic bond with the protonated amine. Steric bulk near the amine also avoids the rapid metabolic degradation alluded to above, namely by MAO. Likewise, steric bulk in this area helps to retard cytochrome P450-mediated oxidative dealkylations. The metabolic pathways commonly encountered by aryloxypropanolamines are shown in Scheme 8.12. The importance of having a bulky R group to minimize these pathways is apparent.

Increased interaction with beta-1 receptors over beta-2 receptors can also be achieved by manipulation of the nitrogen's R group. In this regard, two types of substitutions have been found to provide for such selectivity, namely a 3,4-dimethoxyphenethyl (DMPE) group as present in bevantolol, **10**, [22] and an alkyl or aralkyl extended amide system as present in tolamolol, **43** [57].

For the last pharmacophoric element, a simple phenyl ring is all that is needed to serve as the aromatic system. However, when such a simple system is not further substituted (e.g., R′ = H in **60**), such compounds also demonstrate significant partial agonist effects or "intrinsic activity". The latter is probably due to the uptake of such drugs into the presynaptic neuron, followed by release of norepinephrine (an indirect rather than a direct receptor-related mechanism), although this has never been established in a definitive manner. Interestingly, substitution with just about anything in any location on the ring appears to eliminate intrinsic activity, with the exceptions, of course, being the placement of an alcohol in the 3′ or 4′ position, or of functionality that mimics one or the other of the endogenous

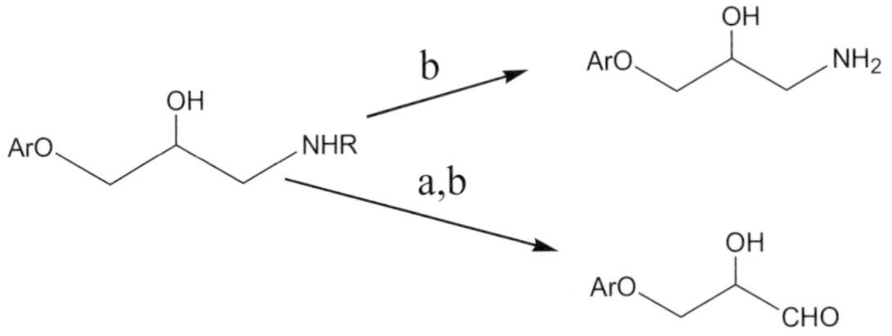

Scheme 8.12 Typical amine-related metabolism reactions that aryloxypropanolamines can be subject to [182]. (a) Monoamine oxidase (MAO), particularly when R = H. (b) Cytochrome P450 oxidation, particularly when R is not alpha-substituted. Note that for both cases, the initial aldehydes produced from the parent alkyl system quickly undergo further metabolism, namely continued oxidation to the carboxylic acid by aldehyde dehydrogenase [183].

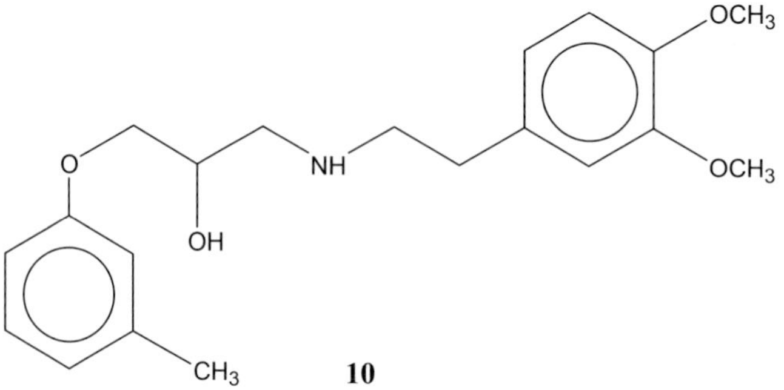

10

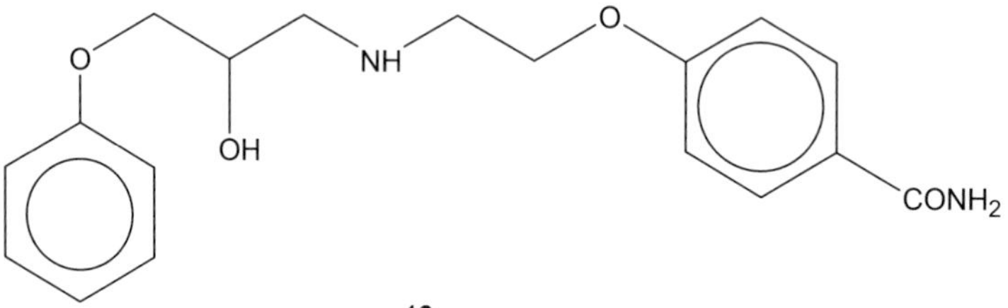

43

sympathomimetic amines' two phenolic hydroxyl groups that constitute the catechol moiety. Once R′ groups are placed on the phenyl-ring, the following distinct
trend is repeatedly observed as one moves from the *ortho–* to *meta–* to the *para-*
positions: decreased potency overall, but significantly more so on beta-2 receptors.
Thus, *para*-substituted systems generally yield weaker but more beta-1-selective
beta-blockers [184–186]. Multiple substitution on the ring can lead to variable
effects, and in some cases this has proven to be quite useful; for example, as we
have seen for the combination of *ortho–* and *para*-placements in celiprolol, **47**, providing for both enhanced potency and beta-1 selectivity, respectively.

8.4
Summary

ABDD has clearly dominated the strategic approaches that have been deployed to
discover new beta-adrenergic receptor blocking agents. This has been the case during efforts to develop novel therapeutic paradigms for new indications, as well as
during efforts to enhance the overall clinical profile of established drugs within
the context of their same clinical indications. That ABDD has been fruitful in this
regard is underscored by both the numerous beta-blockers that have, in turn,

become available to treat several different therapeutic indications, as well as by those for which the properties have been finely tailored for a given use. From a clinical perspective, the various beta-blockers exhibit a wide array of differing ancillary actions that physicians can utilize selectively to provide optimal medical therapy for their various patients. That the new and rapidly evolving field of pharmacogenomics [187–189] is already prompting the additional need for exactly such drug arrays across all types of therapeutic categories, further emphasizes the critical value of deploying ABDD to construct the appropriate medical armamentarium.

Importantly, it should be just as clear that ABDD also generates new, fundamental knowledge wherein the latter can then be applied toward future drug design in a general manner. In this regard, several lessons can be learned from the success stories associated with the examples conveyed herein, such as those where steric bulk was effectively deployed to avoid metabolism. The latter represents a particularly relevant scenario for today's medicinal chemists, who must frequently engage in further optimization of the PK profiles for lead compounds so as to better position them for success in the clinic.

References

1 Fischer J. *Med. Chem. Res.*, **2004**, 13, 218.

2 Erhardt PW. *J. Pure Appl. Chem.*, **2002**, 74, 703.

3 *ADMET* is a common abbreviation that pertains to the absorption, distribution, metabolism, excretion and toxicity of xenobiotics by humans or animals [2].

4 *PK* is a common abbreviation that pertains to the timeline associated with a xenobiotic's presence in the body of humans or animals and, as such, encompasses all of the ADMET parameters except toxicity [2].

5 Konzett H. *Arch. Exp. Path. Pharmacol.*, **1940**, 197, 27.

6 Moran NC. *Ann. N.Y. Acad. Sci.*, **1967**, 139, 545.

7 Black, JW, Stephenson JS. *Lancet*, **1962**, 2, 311.

8 Dornhorst AC, Robinson BF. *Lancet*, **1962**, 2, 314.

9 Paget GE. *Br. Med. J.*, **1963**, 5367, 1266.

10 Srivastava SC, Dewar HA, Newell DJ. *Br. Med. J.*, **1964**, 5411, 724.

11 Rowlands DJ, Howitt G, Markman P. *Br. Med. J.*, **1965**, 5439, 891.

12 Prichard BN, Gillman PM. *Br. Med. J.*, **1964**, 5411, 725.

13 Lands AM, Arnold A, McAuliff JP, Luduena FP, Brown TG, Jr. *Nature*, **1967**, 214, 597.

14 Sood NK, Harvard CW. *Thorax*, **1973**, 28, 331.

15 Nordenfelt I, Olsson L, Persson S. *Eur. J. Clin. Pharmacol.*, **1974**, 7, 157.

16 As opposed to the numerous research groups seeking to remove or reduce membrane depressant properties, it should additionally be noted that at this same point in time there were other groups operating under the impression that there was also a considerable amount of *"evidence that the non-specific depressant properties have no clinical significance"* [17].

17 Clarkson R, Tucker H, Wale J. *Annu. Rep. Med. Chem.*, **1975**, 10, 51.

18 Evans DB, Fox R, Hauck FP. *Annu. Rep. Med. Chem.*, **1979**, 14, 81.

19 Basil B, Jordan R, Loveless AH, Maxwell DR. *Br. J. Pharmacol.*, **1973**, 48, 198.

20 Ablad B, Brogard M, Ek L. *Acta Pharmacol. Toxicol.*, **1967**, Suppl., 25, 9.

21 Barrett AM, Carter J, Fitzgerald JD, Hull R, LeCount D. *Br. J. Pharmacol.*, **1973**, 48, 340P.

22 Hastings SG, Smith RD, Corey RM, Essenburg AD, Pettway CE, Tessman DK. *Arch. Int. Pharmacodyn. Ther.*, **1977**, 226, 81.

23 Sato Y, Kobayashi Y, Nagasaki T, Oshima T, Kumakura S, Nakayama K, Koike H, Takagi H. *Chem. Pharm. Bull.*, **1972**, 20, 905.

24 *Drugs of Today*, **1974**, 10, 332.

25 Hamilton TC, Parkes NW. *Arzneim.-Forsch.*, **1977**, 27, 1410.

26 Baum T, Rowles G, Shropshire AT, Gluckman MI. *J. Pharmacol. Exp. Ther.*, **1971**, 176, 339.

27 Robson RD, Kaplan HR. *J. Pharmacol. Exp. Ther.*, **1970**, 175, 157.

28 *WHO Chronicle*, **1972**, 26, 1125.

29 Ferrini R, Miragoli G, Croce G. *Arzneim.-Forsch.*, **1968**, 18, 829.

30 Martin M, Cautain M, Sado M, Zuckerkandl F, Fourneau JP, Linee P, Lacroix P, Quiniou P, Van den Driessche J. *Eur. J. Med. Chem.*, **1974**, 9, 563.

31 Levy B. *J. Pharmacol. Exp. Ther.*, **1966**, 151, 413.

32 Bartsch W, Dietmann K, Leinert H, Sponer G. *Arzneim.-Forsch.*, **1977**, 27, 1022.

33 Yabuochi Y, Kinoshita D. *Jap. J. Pharmacol.*, **1974**, 24, 853.

34 Carissimi M, Gentili P, Grumelli E, Milla E, Picciola G, Ravenna F. *Arzneim.-Forsch.*, **1976**, 26, 506.

35 Tachikawa S, Takenaka T. *Arch. Int. Pharmacodyn.*, **1973**, 202, 79.

36 Van den Driessche J. *Therapie*, **1977**, 32, 111.

37 Farmer JB, Kennedy I, Levy GP, Marshall RJ. *Br. J. Pharmacol.*, **1972**, 45, 660.

38 Gugler R, Kreis L, Dengler HJ. *Arzneim.-Forsch.*, **1975**, 25, 1067.

39 Zakhari S. *Eur. J. Pharmacol.*, **1974**, 29, 22.

40 Ablad B, Carlsson E, Ek L. *Life Sci.*, **1973**, 12, 107.

41 Ferrini R, Miragoli G, Groce G. *Arzneim.-Forsch.*, **1970**, 20, 1074.

42 Lee RJ, Evans DB, Baky SH, Laffan RJ. *Eur. J. Pharmacol.*, **1975**, 33, 371.

43 Fitzgerald JD. *Clin. Pharmacol. Ther.*, **1969**, 10, 191.

44 Raper C, Wale J. *Eur. J. Pharmacol.*, **1968**, 4, 1.

45 Carruthers SG, Hosler JP, Pentikainen P, Azarnoff DL. *Clin. Pharmacol. Ther.*, **1978**, 24, 168.

46 Matsubara I, Hashimoto K, Katano Y, Tsukada T, Matsuda H. *Folia Pharmacol. Jap.*, **1976**, 72, 557.

47 Hartfelder G, Lessenich H, Schmitt K. *Arzneim.-Forsch.*, **1972**, 22, 930.

48 Guidicelli JF, Schmidt H, Boissier JR. *J. Pharmacol. Exp. Ther.*, **1969**, 168, 116.

49 Dunlop D, Shanks RG. *Br. J. Pharmacol.*, **1968**, 32, 201.

50 Boissier JR, Guidicelli JF, Viars P, Advenier C, Mouille P, Larno S. *Eur. J. Pharmacol.*, **1971**, 15, 151.

51 Barrett AM. In: *Recent Advances in Cardiology*, 6th ed. Hamer J (Ed.) Churchill Livingstone, Edinburgh, **1973**, p. 289.

52 Black JW, Crowther AF, Shanks RG, Smith LH Dornhorst, A.C. *Lancet*, **1964**, 303, 1.

53 Lish PM, Weikel JH, Dungan KW. *J. Pharmacol. Exp. Ther.*, **1965**, 149, 161.

54 Vauguelin G, Lacombe ML, Guellaen G, Strosberg D, Hanoune J. *Biochem. Pharmacol.*, **1976**, 25, 2605.

55 Scriabine A, Torchiana ML, Stavorski JM, Ludden CT, Minsker DH, Stone CA. *Arch. Int. Pharmacodyn.*, **1973**, 205, 76.

56 Allen JD, Shanks RG. *Br. J. Pharmacol.*, **1974**, 51, 179.

57 Augstein J, Cox DA, Ham AL, Leming PR, Snarey M. *J. Med. Chem.*, **1973**, 16, 1245.

58 Stock K, Westermann E. *Biochem. Pharmacol.*, **1965**, 14, 227.

59 Barrett AM, Crowther AF, Dunlop D, Shanks RG, Smith LH. *Arch. Pharmacol. Exp. Pathol.*, **1968**, 259, 152.

60 Le Count D. *Chronicles of Drug Discovery*, **1982**, 1, 113.

61 Lee RJ. *Life Sci.* **1978**, 23, 2539.

62 Waal-Manning HJ. *Drugs*, **1979**, 17, 129.

63 Goldberg LI. *Proc. Royal Soc. Med.*, **1977**, 70, 7.

64 Katz IM. *Ann. Ophthalmol.*, **1978**, 10, 847.

65 Nanda R. *Headache*, **1978**, 18, 20.

66 Granville-Grossman K. *Br. J. Clin. Pharmacol.*, **1974**, 1, 361.

67 Sevitt I. *Practitioner*, **1974**, 213, 91.

68 Jefferson JW. *Arch. Gen. Psychiatr.*, **1974**, 31, 681.

69 Resnick RB, Kestenbaum RS, Schwartz LK, Smith A. *Arch. Gen. Psychiatr.*, **1976**, 33, 993.

70 Rubenfeld S, Silverman VE, Welch KM, Mallette LE, Kohler PO. *N. Engl. J. Med.*, **1979**, 300, 353.

71 Yorkston NJ, Zaki SA, Malik MK, Morrison RC, Havard CW. *Br. Med. J.*, **1974**, 4, 633.

72 Yorkston N, Zaki SA, Themen JF, Harvard CW. *Postgrad. Med. J.*, **1976**, 52 175.

73 This subclass of receptors has more recently been termed beta-3. Since its ligands appear to have a rather divergent SAR, it is not clear that propranolol will again represent a useful pioneer drug to allow for advantageous ABDD. As yet, no beta-3-selective compounds have arrived in the marketplace.

74 Smith LH. *J. Med. Chem.*, **1977**, 20, 1254.

75 McAninsh H, Cruickshank JM. *Pharmacol. Ther.*, **1990**; 46, 163.

76 White WB, Andreoli JW, Wang SH, Cohn RD. *Obstet. Gynecol.*, **1984**, 63, 42.

77 Dahlöf B, Lindholm LH, Hansson L, Scherstéin B, Ekbom T, Wester P-O. *Lancet*, **1991**, 338, 1281.

78 Hansson L, Lindholm LH, Ekbom T, Dahlöf B, Lanke J, Scherstéin B, Wester P-O, Hedner T, de Faire U. *Lancet*, **1999**, 354, 1751.

79 MRC Working Party *Br. Med. J.*, **1992**, 304, 405.

80 UK Prospective Diabetes Study Group *Br. Med. J.*, **1998**, 317, 713.

81 Zanchetti A, Bond MG, Hnnig M, Neiss A, Mancia G, Dal Palú C, Hansson L, Magnani B, Rahn K-H, Reid JL, Rocicio J, Safar M, Eckes L, Rizzine P. *Circulation*, **2002**, 106, 2422.

82 Ludwig M, Stapff M, Ribeiro A, Frischka E, Tholl U, Smith RD, Stumpe KO. *Clin. Ther.*, **2002**, 24, 1175.

83 Dahlöf B, Devereux RB, Kjeldsen SE, Julius S, Beevers G, de Faire W, Fyhrquist F, Ibsen H, Kristiansson K, Lederballe-Pedersen O, LIndholm LH, Nieminen MS, Omvik P, Oparil S, Wedel H. *Lancet*, **2002**, 359, 995.

84 Pannier B, Guérin A, London G, Asmar R, Safar M. *Arch. Mal. Coeur Vaiss.*, **2002**, 95, 11.

85 Carlberg B, Samuelsson O, Lindholm LH. *Lancet*, **2004**, 364, 1684.

86 Deanfield JE, Detry J-M, Sellier P, Lichtlen PR, Thaulow E, Bultas J, Brennan C, Young ST, Beckerman B. *J. Am. Coll. Cardiol.*, **1992**, 40, 917.

87 Dargie HJ, Ford I, Fox KM. *Eur. Heart J.*, **1996**, 17, 104.

88 ISIS-1 (First International Study of Infarct Survival) Collaborative Group. *Lancet.* **1988**, i, 921.

89 Rassmussen K, Anderson K, Wang H. *Eur. Heart J.*, **1982**, 3, 276.

90 Forssman B, Lindblad CJ, Zbornikova V. *Headache*, **1983**, 23, 188.

91 Stensrud P, Sjaastad O. *Headache*, **1980**, 20, 204.

92 Manoury P. In: *Betaxolol and Other β_1-Adrenoceptor Antagonists.* Morselli PL, Kiborn JR, Cavero I, Harrison DC, Langer SZ (Eds.) L.E.R.S. Monograph Series, Vol. 1. Raven Press, New York, **1983**, p. 13.

93 Cuthbert OA, *Br. J. Pharmacol.*, **1971**, 43, 639.

94 Barrett AM, Carter J, Fitzgerald JD, Hull R, LeCount D. *Br. J. Pharmacol.*, **1973**,48, 340P.

95 Ablad B, Carlsson E, Ek L. *Life Sci.*, **1973**, 12, 107.

96 Smith LH, Tucker H. *J. Med. Chem.*, **1977**, 20, 1653.

97 Arfwidsson A, Borg KO, Hoffman KJ, Skanberg I. *Xenobiotica*, **1976**, 6, 691.

98 Cerqueira PM, Cesarino EJ, Bertucci C, Bonato PS, Lanchote VL. *Chirality*, **2003**, 15, 542.

99 Sada H, Ban T. *Experientia*, **1981**, 37, 171.

100 Beresford R, Heel RC. *Drugs*, **1986**, 31, 6.

101 Stagni G, Davis PJ, Ludden TM. *J. Pharm. Sci.*, **1991**, 80, 321.

102 Ferrandes B, Durand A, André-Fraisse J, Thénot J, Hermann P. In: *Betaxolol and Other β_1-Adrenoceptor Antagonists.* Morselli PL, Kiborn JR, Cavero I, Harrison DC, Langer SZ (Eds.) L.E.R.S. Monograph Series, Vol. 1. Raven Press, New York, **1983**, p. 51.

103 Mroczek WJ, Burris JF, Hogan LB, Citron DC, Barker AH, McDonald RH. *Am. J. Cardiol.*, **1988**, 61, 807.

104 Djian J. *Br. J. Clin. Pract.*, **1985**, 39, 188.

105 Coupez JM, Bachy C, Coupez-Lopinot R. In: *Betaxolol and Other â₁-Adrenoceptor Antagonists.* Morselli PL, Kiborn JR, Cavero I, Harrison DC, Langer SZ (Eds.) L.E.R.S. Monograph Series, Vol. 1. Raven Press, New York, **1983**, p. 315.

106 Williams RL, Goyle KK, Herman TS, Rofman BA, Ruoff GE, Hogan LB. *J. Clin. Pharmacol.*, **1992**, 32, 360.

107 van Os JS, van Brummelen P, Woittiez AJJ. *Neth. J. Med.*, **1992**, 40, 227.

108 Johnson BF, Whelton A. *Am. J. Ther.*, **1994**, 1, 260.

109 Hwang Y-S, Yen H-W, Wu J-C, Lin C-C. *Curr. Ther. Res.*, **1998**, 59, 307.

110 de Backer G, Derese A. In: *Betaxolol and Other β_1-Adrenoceptor Antagonists.* Morselli PL, Kiborn JR, Cavero I, Harrison DC, Langer SZ (Eds.) L.E.R.S. Monograph Series, Vol. 1. Raven Press, New York, **1983**, p. 261.

111 McLenachan JM, Findlay IN, Wilson JT, Dargie HJ. *J. Cardiovasc. Pharmacol.*, **1992**, 20, 311.

112 Chrysant SG, Bittar N. *Cardiology*, **1994**, 84, 316.

113 Böhler S, Saubadu S, Scheldewaert R, Figulla H-R. *Arzneim.-Forsch./Drug Res.*, **1999**, 49(I), 311.

114 Coletta AP, Cleland JGF, Freemantle N, Clark AL. *Eur. J. Heart Fail.*, **2004**, 6, 673.

115 Koh KK, Song JH, Kwon KS, Park HB, Baik SH, Park YS, In HH, Moon TH, Park GS, Cho SK, Kim SS. *Int. J. Cardiol.*, **1995**, 52, 167.

116 Atwood JE, Myers J, Quaglietti S, Grumet J, Gianrossi R, Umman T. *Chest*, **1999**, 115, 1175.

117 Heijl A, Leske MC, Bengtsson B, Hyman L, Bengtsson B, Hussein M, for the Early Manifest Glaucoma Trial Group. *Arch. Ophthalmol.*, **2002**, 120, 1268.

118 Mazzo DJ, Obeetz CL, Shuster JE. *Anal. Profiles of Drug Substances*, **1991**, 20, 237.

119 Foster RT, Carr RA. *Anal. Profiles of Drug Substances*, **1990**, 19, 1.

120 Kaye CM, Kumana CR, Leighton M, Hamer J, Turner P. *Clin. Pharmacol. Ther.*, **1976**, 19, 416.

121 Zaman R, Jack DB, Wilkins MR, Kendall MJ. *Biopharm. Drug Dispos.*, **1985**, 6, 131.

122 DeBono G, Kaye CM, Roland E, Summers AJH. *Am. Heart J.*, **1985**, 109, 1211.

123 Gulaid A, James IM, Kaye CM, Lewellen ORW, Roberts E, Snakey M, Smith J, Templeton R, Thomas RJ. *Br. J. Clin. Pharmacol.*, **1978**, 5, 261.

124 Kirch W, Kohler H, Berggren G, Braun W. *Clin. Nephrol.*, **1982**, 18, 88.

125 Riddell JG, Harron DWG, Shanks RG. *Clin. Pharmacokinet.*, **1987**, 12, 305.

126 Norris RJ, Lee EH, Muirhead D, Sanders SW. *J. Cardiovasc. Pharmacol.*, **1986**, 8, S91.

127 Doshan HD, Berger BM, Costello R, Applin W, Caruso FS, Neiss ES. *Clin. Pharmacol. Ther.*, **1985**, 37, 192.

128 Hitzenberger G, Takacs F, Pittner H. *Arzneim.-Forsch.*, **1983**, 33 50.

129 Riddell JG, Shanks RG, Brogden RN. *Drugs*, **1987**, 34, 438.

130 Pittner H. *Arzneim.-Forsch.*, **1983**, 33, 13.

131 Smith LH. *J. Med. Chem.*, **1976**, 19, 1119.

132 Basil B, Clark JR, Coffee ECJ, Jordan R, Loveless AH, Pain DL, Wooldridge KRH. *J. Med. Chem.*, **1976**, 19, 399.

133 Erhardt PW, Proudfoot JR. In: *Comprehensive Medicinal Chemistry II*, Vol. 1 Kennewell P (Ed.); Taylor J, Triggle D (Series Eds.). Elsevier Ltd, Oxford, UK (in press).

134 Hieble JP, Ellis S. *Pharmacologist*, **1975**, 17, 223.

135 Zolss G. *Arzneim.-Forsch.*, **1983**, 33, 2.

136 Joshi RA, Gurjar MK, Tripathy NK, Chorghade MS. *Org. Proc. Res. Dev.*, **2001**, 5, 176.

137 Erhardt PW. *Synth. Commun.*, **1983**, 13, 103. (Also see discussion of this topic in Chapter 9, Esmolol Case Study.)

138 Lilja JJ, Backman JT, Laitila J, Luurila H, Neuvonen PJ. *Clin. Pharmacol. Ther.*, **2003**, 73, 192.

139 Lilja JJ, Juntti-Patinen L, Neuvonen PJ. *Clin. Pharmacol. Ther.*, **2004**, 75, 184.

140 Borchard U. *J. Clin. Bas. Cardiol.*, **1998**, 1, 5.

141 Heublein H, Modersohn D, Franz N, Panzner B. *Eur. Heart J.*, **1991**, 12, 617.

142 van der Woude HJ, Zaagsma J, Postma DS, Winter TH, van Hulst M, Aalbers R. *Chest*, **2005**, 127, 818.

143 Boutouyrie P, Bussy C, Tropeano A-I, Hayoz D, Hengstler J, Dartois N, Laloux B, Brunner H, Laurent S. *Arch. Mal. Coeur Vaiss.*, **2000**, 93, 911.

144 Cleophas TJM, Niemeyer MG, Bernink PJLM, Zwinderman KH, Wijk AV, Wall EEVD. *Curr. Ther. Res.*, **1996**, 57, 614.

145 Malminiemi K, Lahtela J, Malminiemi O, Ala-Kaila K, Huupponen R. *J. Cardiovasc. Pharmacol.*, **1998**, 31, 140.

146 Fogari R, Zoppi A, Corradi L, Preti P, Mugellini A, Lusardi P. *J. Cardiovasc. Pharmacol.* **1999**, 33, 534.

147 Takeda K, Mitsunami K, Muso E, Kinoshita M, Sasayama S, Nakagawa M. *Curr. Ther. Res.*, **2000**, 61, 49.

148 McLenachan JM, Wilson JT, Dargie HJ. *Br. Heart J.*, **1988**, 59, 685.

149 Asanuma H, Node K, Minamino T, Sanada S, Takashima S, Ueda Y, Sakata Y, Asakura M, Kim J, Ogita H, Tada M, Hori M, Kitakaze M. *J. Cardiovasc. Pharmacol.*, **2003**, 41, 499.

150 Felix SB, Stangl V, Kieback A, Doerffel W, Staudt A, Wernecke KD, Baumann G, Stangl K. *J. Cardiovasc. Pharmacol.*, **2001**, 38, 66.

151 Pauwels PJ, Gommeren W, van Lommen G, Janssen PAJ, Leysen JE. *Mol. Pharmacol.*, **1988**, 34, 843.

152 Chandrasekhar S, Reddy MV. *Tetrahedron*, **2000**, 56, 6339.

153 Van de Water A, Xhonneux RS, Reneman RS, Janssen PAJ. *Eur. J. Pharmacol.*, **1988**, 156, 95.

154 Van Peer A, Snoeck E, Woestenborghs R, Van de Velde V, Mannens G, Meuldermans W, Heykants J. *Drug Invest.*, **1991**, 3(Suppl.1), 25.

155 Kamali F, Howes A, Thomas SHL, Ford GA, Snoeck E, Kamall F. *Br. J. Clin. Pharmacol.*, **1997**, 43, 201.

156 De Cree J, Van Rooy P, Genkens H, Haeveransk K, Verhaegen H. *Angiology*, **2000**, 43, 369.

157 Chan TYK, Woo KS, Nicholls MG. *Int. J. Cardiol.*, **1992**, 35, 387.

158 Van Nueten L, Taylor FR, Robertson JIS. *J. Hum. Hypertens.*, **1998**, 12, 135.

159 Van Nueten L, Schelling A, Vertommen C, Dupont AG, Robertson JIS. *J. Hum. Hypertens.*, **1997**, 11, 813.

160 Cazzola M, Matera MG, Ruggeri P, Sanduzzi A, Spicuzza L, Vatrella A, Girbino G. *Respiration*, **2004**, 71, 159.

161 Czuriga I, Riecansky I, Bodnar J, Fulop T, Krusicz V, Kristof E. *Cardiovasc. Drug Ther.*, **2003**, 17, 257.

162 Ignarro LJ. *Blood Press.*, **2004**, Suppl. 13, 2.

163 Pasini AF, Garbin U, Nava MC, Stranieri C, Davoli A, Sawamura T, Cascio V, Cominacini L. *J. Hypertens.*, **2005**, 23, 589.

164 Garban HJ, Buga GM, Ignarro LJ. *J. Cardiovasc. Pharmacol.*, **2004**,43, 638.

165 Ulvenstam G. *Drug Invest.*, **1991**, 3 (Suppl. 1), 199.

166 Ruf G, Trenk D, Jahnchen E, Roskamm H. *Int. J. Cardiol.*, **1994**, 43, 279.

167 Nodari S, Metra M, Dei Cas L. *Eur. J. Heart Failure*, **2003**, 5, 621.

168 Flather MD, Shibata MC, Coats AJS, Van Veldhuisen DJ, Parkhomenko A, Borbola J, Cohen-Solal A, Dumitrascu D, Ferrari R, Lechat P, Soler-Soler J, Tavazzi L, Spinarova L, Toman J, Böhm M, Anker SD, Thompson SG, Poole-Wilson PA. *Eur. Heart J.*, **2005**, 26, 215.

169 Erhardt PW. *J. Pharm. Sci.*, **1980**, 69, 1059.

170 Erhardt PW, Chou Y-L. *Life Sci.*, **1991**, 49, 553.

171 Ostrowski J, Kjelsberg MA, Caron MG, Lefkowitz RJ. *Annu. Rev. Pharmacol. Toxicol.*, **1992**, 32, 167.

172 Strader CD, Candelore MR, Hill WS, Dixon RA, Sigal IS. *J. Biol. Chem.*, **1989**, 264, 16470.

173 Kobilka B. *Annu. Rev. Neurosci.*, **1992**, 15, 87.

174 Jen T, Frazee JS, Schwartz MS, Erhard KF, Kaiser C. *J. Med. Chem.*, **1977**, 20, 1263.

175 Goehring R, Lumma WC, Jr., Erhardt PW, Topiol S, Sabio M, Wiggins J, Wong S, Greenberg S, Pang D, Cantor E. *Eur. J. Med. Chem.*, **1987**, 22, 165.

176 Akisanya J, Parkins AW, Steed JW. *Org. Proc. Res. Dev.*, **1998**, 2, 274.

177 Kleidernigg OP, Maier NM, Uray G, Lindner W. *Chirality*, **1994**, 6, 411.

178 Payne GB. *J. Org. Chem.*, **1962**, 27, 3819.

179 Baldwin JJ, Raab AW, Mensler K, Arison BH, McClure DE. *J. Org. Chem.*, **1978**, 43, 4676.

180 McClure DE, Arison BH, Baldwin JJ. *J. Am. Chem. Soc.*, **1979**, 101, 3666.

181 Baldwin JJ, McClure DE, Gross DM, Williams M. *J. Med. Chem.*, **1982**, 25, 931.

182 Benedetti MS, Dostert P. *Drug Metab. Rev.*, **1994**, 26, 507.

183 Goedde HW, Agarwal DP. In: *Pharmacogenetics of Drug Metabolism*. Kalow W (Ed.). New York, Pergamon Press, **1992**, p. 281.

184 Erhardt PW, Woo CM, Gorczynski RJ, Anderson WG. *J. Med. Chem.*, **1982**, 25, 1402.

185 Erhardt PW, Woo CM, Anderson WG, Gorczynski RJ. *J. Med. Chem.*, **1982**, 25, 1408.

186 Erhardt PW, Woo CM, Matier WL, Gorczynski RJ, Anderson WG. *J. Med. Chem.*, **1983**, 26, 1109.

187 Bailey DS, Dean PM. *Annu. Rep. Med. Chem.*, **1999**, 34, 339.

188 Lau KF, Sakul H. *Annu. Rep. Med. Chem.*, **2000**, 35, 261.

189 Murphy MP. *Pharmacogenomics*, **2000**, 1, 115.

9
Case Study: "Esmolol Stat" [1]

Paul W. Erhardt

9.1
Introduction [2]

As conveyed in Chapter II-8, the clinical success of propranolol, **1**, served to establish it as the "pioneer drug" from which numerous companies deployed "analogue-based drug discovery" (ABDD) [3] to pursue additional beta-advenergic receptor blocking agents having enhanced pharmacological profiles. For most cardiac indications, the preferred profile typically sought was:

- selectivity for beta-1 receptors over alpha- and beta-2 receptors;
- low partial agonist or intrinsic sympathomimetic activity (ISA);
- low membrane-depressant or -stabilizing activity (MSA); and
- a long duration of action conducive to once-a-day dosing.

Within the setting of acute myocardial infarction (MI), however, the seemingly beneficial use of such compounds to preserve ischemic cardiac tissue by lowering the latter's energy and oxygen requirements, was accompanied by considerable reservation. This was because, in this particular setting, cardiac muscle often has to rely upon beta-1 adrenergic drive to sustain its otherwise inherent contractility, and the early use of beta-1 blockers had occasionally resulted in heart block and death [4]. Thus, in this case the clinical challenge became the need for an effective level of drug-induced beta-receptor antagonism to be finely balanced against the need for a critical level of inherent, endogenously driven beta-agonism, with the

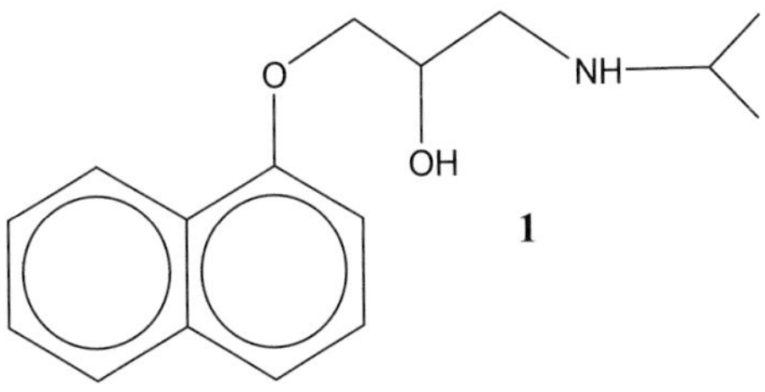

1

Analogue-based Drug Discovery. IUPAC, János Fischer, and C. Robin Ganellin (Eds.)
Copyright © 2006 WILEY-VCH Verlag GmbH & Co. KGaA, Weinheim
ISBN: 3-527-31257-9

combination in a dynamic state of flux associated with an emergency situation wherein the precise level of adrenergic tone across the heart was very difficult to ascertain. From this backdrop, during the late 1970s Arnar-Stone Laboratories, whose niche, strategic focus was directed toward emergency medications [5], set out to discover and develop a beta-blocker compound that could be used safely in the critical setting of MI [4].

9.2
Pharmacological Target

It was clear that the desired balance between inherent adrenergic tone and its antagonism could be best achieved and continually adjusted by titrating the beta-blocker's effects against some readily measurable parameter of cardiac function such as heart rate. Thus, an ultra-short-acting compound (e.g., with half-life of ca. 10 min) that could be administered by the intravenous (iv) route was called for. Not only could the level of blockade be adjusted for such a compound, if it was given in too large an amount for a given patient its ultra-short duration would also allow for the antagonistic effects to be quickly removed by simply stopping the iv infusion [4].

9.3
Chemical Target

9.3.1
Internal Esters

Interestingly, at this point in the esmolol story it can be seen that another pioneer drug served as an additional starting-point for ABDD, this time to pursue an ultra-short-acting compound. As shown in Scheme 9.1, procaine is an ultra-short-acting anesthetic agent that is rapidly deactivated by esterase-mediated hydrolysis because its aryl and amino pharmacophoric elements become severed from their requisite contiguous arrangement [6].

Combining these two pioneer drugs, namely **1** for its efficacy and **2** for its ultra-short duration, resulted in the "double ABDD" target compound **3**. By analogy to procaine, it was anticipated that metabolic hydrolysis of the ester within **3** would fragment the requisite beta-blocker pharmacophore (see Chapter II-8) and thus rapidly deactivate such compounds. This unique, ABDD-related situation is depicted in Scheme 9.2.

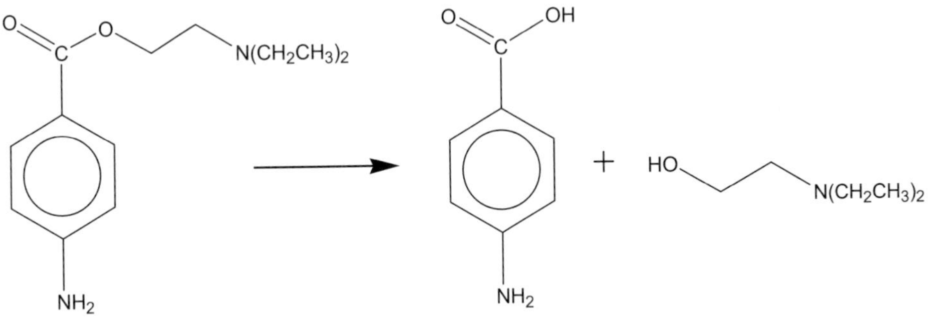

Scheme 9.1 Metabolic hydrolysis of procaine, **2**. Esterases convert **2** into two inactive fragments **2a** and **2b**. Since this process is very rapidly accomplished in the body by these ubiquitous enzymes, procaine exhibits an ultra-short duration of action (USA) after intravenous administration.

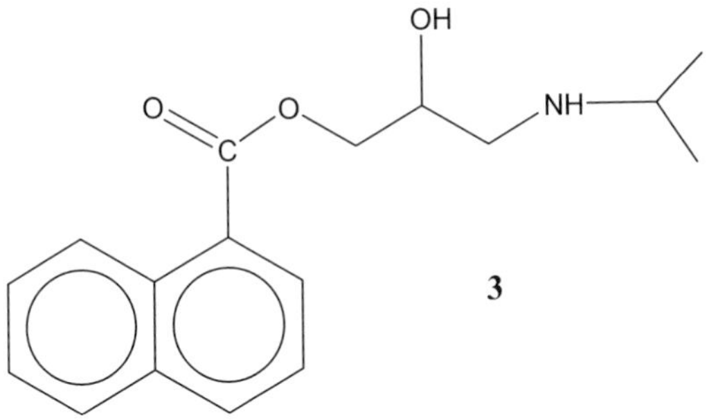

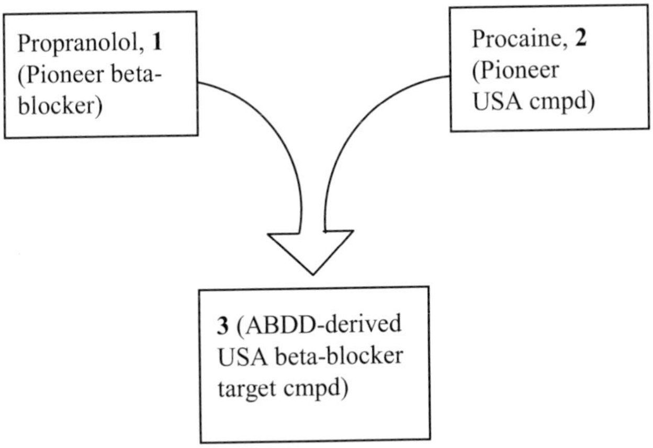

Scheme 9.2 Unique ABDD-related starting-point for the design of beta-blockers containing an internal ester moiety. USA: ultra-short-acting; cmpd: compound.

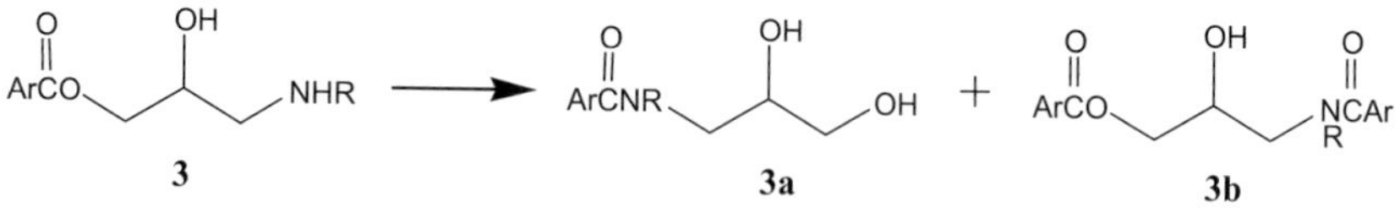

Scheme 9.3 Problematic side reactions encountered during initial attempts to prepare **3**. **3a** represents an intramolecular rearrangement and **3b** represents an intermolecular reaction, the latter also being capable of further reaction (only the Ar-containing side products are shown for the intermolecular reaction).

Because an ester linkage had been incorporated within the oxypropanol connecting chain, these types of target molecules were called "internal esters" [2]. Structure–activity relationships (SARs) in the literature suggested that an increase in the oxypropyl chain's length by one carbon did not dramatically alter the interaction with beta-receptors. Unfortunately, however, these types of structures proved difficult to synthesize, owing in part to their propensity to undergo rapid side-product reactions that become problematic whenever the amine is not protonated (Scheme 9.3).

9.3.2
External Esters

Although these side-product problems would eventually be solved [7], at this early point in esmolol's history an alternate chemical strategy was proposed. By virtue of their lipophilic nature, it was assumed that the aryl and typical N-aliphatic substituent groups interacted with the beta-receptor in a hydrophobic manner, or within a pocket that at least preferred lipophilic groups. Thus, while the presence of an ester placed in either of these areas of the parent drug molecule might still be tolerated, the acid resulting from hydrolysis, upon further ionization at physiological pH accompanied by solvation with water molecules, should not be well tolerated. The new target molecules are shown below as **4** and **5**. Since their ester moieties had now been placed outside of the requisite beta-blocking pharmacophore, these compounds were called "aryl-" and "N-external esters", respectively. Because later reviews have not always conveyed this key conceptual design step in a clear manner [8,9], it is probably important to emphasize at this juncture that this critical aspect of the overall strategy pertaining to the discovery of esmolol was based exactly upon the stated "lipophilic versus hydrophilic hypothesis", and that it had nothing to do with the metabolism observed by others for metoprolol (an altogether different beta-1 blocker that instead has a long duration of action) or with any resemblance to "external" amide appendages as present in practolol (which although found by others to bestow cardioselectivity to beta-blockers, again were known to have half-lives of several hours' duration).

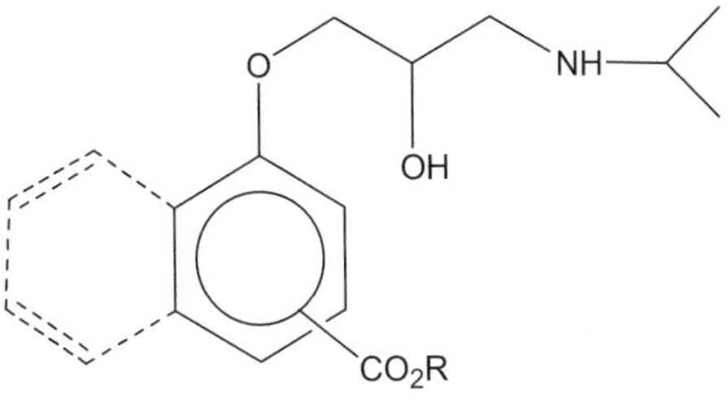

4

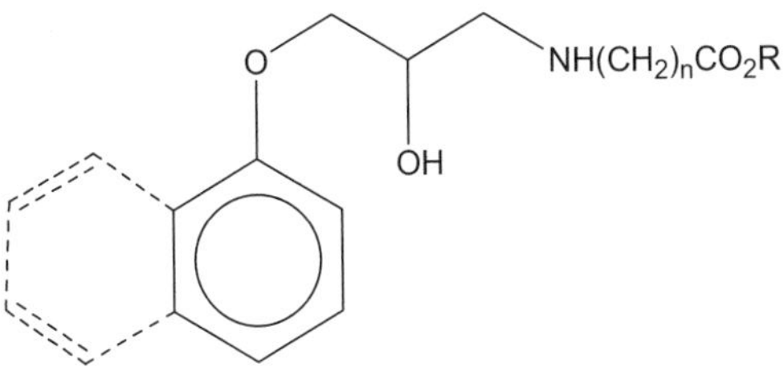

5

9.3.3
Structure–Activity Relationships

While **5** provided certain synthetic challenges in terms of controlling the degree of N-alkylation (see Scheme 9.4 and its caption) [10], several representatives for both series were prepared reasonably quickly [11–19]. Interestingly, the resulting increase in the overall lipophilicity for these types of esters when first placed on a naphthyl system analogous to **1**, proved to lower the aqueous solubility to the point that biological assessment became problematic. Thus, the single, phenyl-ring system was immediately deployed as an alternative that provided an aryl pharmacophoric element with much less of an effect toward lowering overall aqueous solubility. However, when a single-ring system was unsubstituted, significant ISA became problematic. Given the earlier synthetic difficulties encountered with the internal esters ("strike one"), at this point the poor behavior toward testing encountered with the first, low-aqueous-solubility series immediately followed by the second, ISA-problem series of external esters ("strikes two and three") almost caused the entire ultra-short-acting beta-blocker program to be dropped (rather than just returning it to the "dugout" to contemplate a better fate upon making yet another attempt).

Finally, after quickly exploring a series of substituted phenyl rings (Fig. 9.1) to identify a suitable scaffold for use with the N-external ester targets, an ortho-methyl-phenyl ring was deployed for series **5** (**4**'s ring being inherently substituted via the ester placement was not showing problems with ISA).

Importantly, in both **4** and **5**, these types of esters were found to retain their potency as beta-blockers. Even more important for the hypothesis and new design strategy, however, the carboxylic acid versions were found to be essentially inactive – just what was needed. As mentioned for **5**, a substituted phenyl-ring was re-

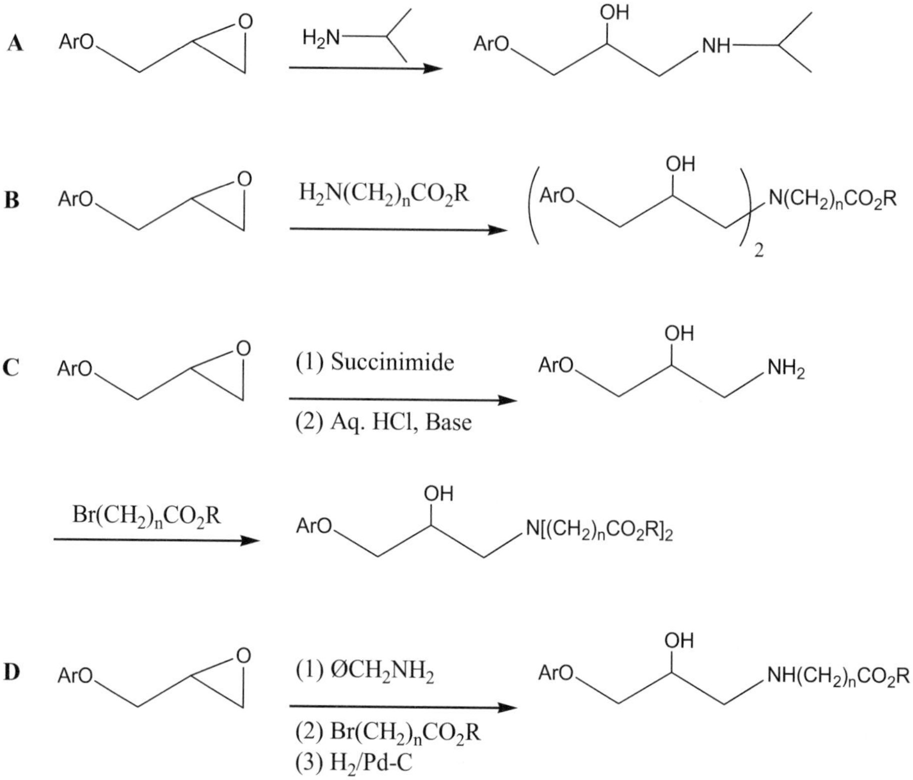

A ArO [epoxide] $\xrightarrow{H_2N-}$ ArO [structure with OH] $NH-$

B ArO [epoxide] $\xrightarrow{H_2N(CH_2)_nCO_2R}$ $\left(ArO \text{ [structure with OH]} \right)_2 N(CH_2)_nCO_2R$

C ArO [epoxide] $\xrightarrow[\text{(2) Aq. HCl, Base}]{\text{(1) Succinimide}}$ ArO [structure with OH] NH_2

$\xrightarrow{Br(CH_2)_nCO_2R}$ ArO [structure with OH] $N[(CH_2)_nCO_2R]_2$

D ArO [epoxide] $\xrightarrow[\substack{\text{(2) } Br(CH_2)_nCO_2R \\ \text{(3) } H_2/Pd\text{-}C}]{\text{(1) } \varnothing CH_2NH_2}$ ArO [structure with OH] $NH(CH_2)_nCO_2R$

Scheme 9.4 Problems encountered during opening of typical epoxide intermediate (see also Section 9.4, Chemical Synthesis) when intended N-external ester substituents do not bear alpha-substituents [10]. A: Normal, well-behaved reaction as typically deployed for beta-blockers wherein the N-alkyl group bears an alpha-substituent as most commonly represented by isopropylamine. B: The over-alkylation problem that occurs significantly when the N-alkyl group is not alpha-substituted. Also note that when n = 3 or 4, the first alkylation is instead followed immediately by an intramolecular ring-closure reaction to form the five- or six-membered lactam. These unwanted side products, although readily isolable, are not depicted above. C: The over-alkylation problem still persists when the amine substitution sequence is reversed. D: The route eventually deployed in a very convenient manner to successfully overcome these types of side-product reactions so as to finally prepare the desired N-external ester target series [10]. Note that the steric hindrance afforded by the benzyl group causes reactions (1) and (2) to be so clean that they can actually be conducted "back-to-back" within the same reaction flask (one-pot fashion). Likewise, catalytic debenzylation is equally facile, as would be expected when going from tertiary to secondary amines.

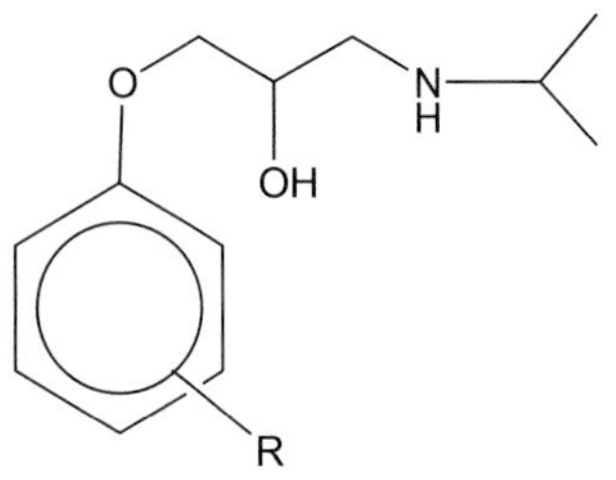

Fig. 9.1 Exploration of scaffolds containing a beta-blocker pharmacophore in order to gain overall properties suitable for use with lipophilic N-external ester types of target compounds (see text). Approximately a dozen compounds were prepared wherein R was a methyl, methoxy, or halide. Based on a combination of *in-vitro* beta-blocking potency, lack of ISA, low logP value and high aqueous solubility, R = *ortho*-methyl was selected as the most suitable scaffold for deployment within the N-external ester target series.

quired to eliminate ISA for the N-external esters, and in both **4** and **5** the effects of aromatic substitution followed the pattern where *ortho*-substituents enhanced potency greater than *meta*-substituents which, in turn, were more active than those placed in a *para*-position. Furthermore, the fall-off in potency was much greater at beta-2 receptors than at beta-1 receptors. Thus, while the *para*-substituted series had lower potency overall, they were also much more beta-1 selective in their actions.

Another important medicinal chemistry-derived hypothesis stemming from these early observations was associated with a comparison of some of the initial aryl external ester compounds, namely **6** and **7** as shown below.

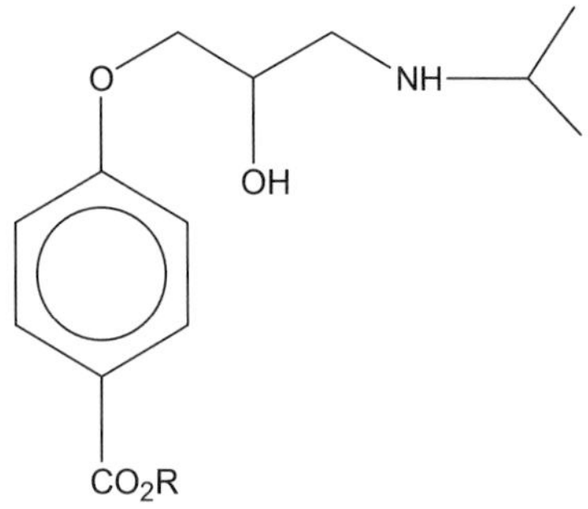

6 R = CH$_3$

7 R = CH$_2$CH$_3$

For **6** and **7**, the half-lives for beta-blocking activity in an *in-vivo* dog model were found to be 30 min and 60 min, respectively. Assuming that this difference reflected simple steric effects, it was speculated that hydrolysis might be even quicker for esters placed further from the bulky phenyl ring. To address this hypothesis, compound series **8** was prepared. These compounds and their corresponding half-life data are shown in Fig. 9.2 [12]. The "Goldilocks" nature [20] of **8** where n = 2, likely reflects the removal of the steric impediment caused by the

aryl-ring and thus supports the hypothesis behind the design of this series. Likewise, the seeming paradox observed for **8** when n = 3 probably reflects conformational folding of the *para*-substituent such that its ester group tends to again reside closer to the overall bulk displayed by the entire aryloxy propanolamine system. Ultimately, the desired 10-min half-life was reproduced in humans and **8** where n = 2 became esmolol. Further illustrating the importance of the exquisite, inverse relationship that was uncovered between enzymatic hydrolysis rates and steric bulk, is the fact that all of our attempts to increase metabolic hydrolysis by the construction of "good leaving group" esters had only a negligible influence upon this process [11,12]. Presumably, once Mother Nature is freed from steric constraints and gains access to a given site, she needs no further assistance toward lowering the transition state energy for effecting an enzymatic conversion.

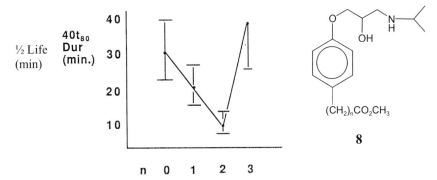

Fig. 9.2 Correlation between biological half-life and added methylene units assessed by measuring 80% recovery from 50% beta-blockade of isoproterenol-induced tachycardia after iv administration to anesthetized dogs (n = 5) [12]. Error bars indicate 1 SD from the mean. Note that when n = 0, then **8** = **6**.

9.4
Chemical Synthesis

The synthesis of esmolol was adopted from the classical procedures that had been developed for propranolol, **1** (see Chapter II-8). The synthetic route is depicted in Scheme 9.5 [12].

Since esmolol is marketed as its racemic mixture, there were no stereochemical issues associated with the Payne-like [21] formation of the intermediate aryloxyepoxide, particularly when the coupling reaction was carried out under the K_2CO_3/ acetone conditions conducive to the formation of epoxides from alpha-halohydrins (see discussion in Chapter II-8). Likewise, even though over-alkylation had been a major problem when preparing the N-external ester targets because the N-alkyl-

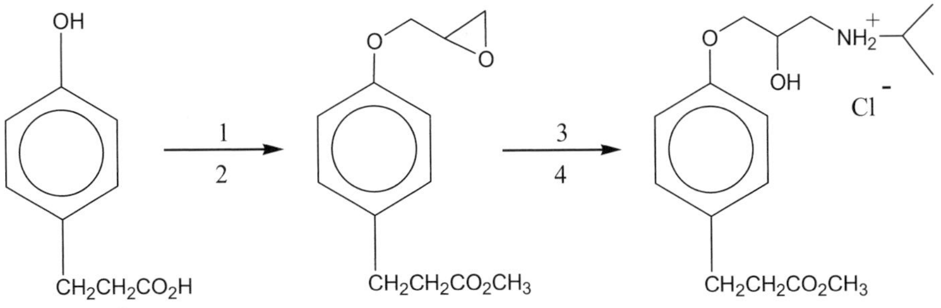

Scheme 9.5 Synthesis of esmolol **8** (n = 2 in Fig. 9.2) [12].
Reagents: Step 1 Methanol, Acid cat.; 2 Epichlorohydrin,
K$_2$CO$_3$, Acetone; 3 Isopropylamine, MeOH; 4 HCl.

ester groups had intentionally lacked alpha-substituents so as to make them more accessible to esterases [10,11], such problems were not encountered when isopropylamine was deployed to open the intermediate aryloxyepoxide. Furthermore, amide side products, which were prepared intentionally from the starting material, were not observed during step 3, indicating that the ester was not subject to transamidation under these conditions (see also the discussion on atenolol chemistry in Chapter II-8, wherein the opposite result is actually used to advantage).

9.5
Pharmacology and Clinical Profile

The required pharmacological profile as ultimately obtained by esmolol is depicted in Fig. 9.3, which also clearly shows its difference from propranolol [12]. Upon intravenous infusion, esmolol quickly produces a pseudo-steady state of beta-blockade and when the infusion is stopped, the effects of esmolol rapidly disappear with a half-life of about 10 min. Alternatively, propranolol requires almost 2 h simply to reach a pseudo-steady state of beta-blockade, and then dissipates only very slowly with a half-life of almost 4 h.

In the clinic, esmolol's distribution half-life is 2 min and its elimination half-life is 9 min. "*Esmolol hydrochloride is rapidly metabolized by hydrolysis of the ester linkage, chiefly by esterases in the cytosol of red blood cells and not by plasma cholinesterases or red cell membrane acetylcholinesterase*" [22]. Its volume of distribution is 3.4 L kg^{-1}, and its total clearance is 285 mL kg^{-1} min^{-1}, "*... which is greater than cardiac output; thus the metabolism of esmolol is not limited by the rate of blood flow to metabolizing tissues such as the liver or affected by hepatic or renal blood flow*" [22]. As expected from such a "*... high rate of blood-based metabolism, less than 2% of the drug is excreted unchanged in the urine*" [22]. Within 24 h after infusion, approximately

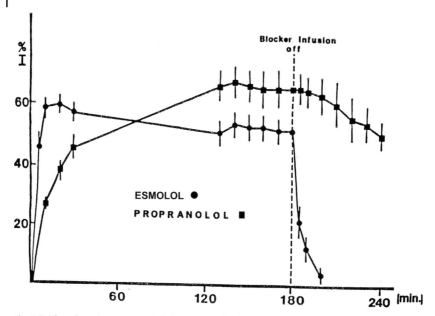

Fig. 9.3 Three-hour intravenous infusion of esmolol versus propranolol in dogs (n = 5) wherein percentage inhibition (% I) of isoproterenol-induced tachycardia was measured against time (min) [12]. Error bars indicate 1 SD from the mean.

73–88% of the total dose can be accounted for in the urine as the acid metabolite. Esmolol becomes about 55% bound to plasma proteins, while the acid metabolite is only 10% bound. "*The acid metabolite has been shown in animals to have about 1/1500th the activity of esmolol, and in normal volunteers its blood levels do not correspond to* [any significant] *level of beta blockade. The acid metabolite has an elimination half-life of about 3.7 hours and is excreted in the urine with a clearance approximately equivalent to the glomerular filtration rate*" [22]. The other metabolic byproduct is methanol. Monitoring the latter in subjects receiving esmolol for up to 6 h at 300 μg kg^{-1} min^{-1} and 24 h at 150 μg kg^{-1} min^{-1}, approximated endogenous levels and these were still less than 2% of those usually associated with methanol toxicity [22].

Esmolol is used to control adrenergic tone across the heart in emergency situations during critical care medicine of very young, as well as adult, patients [1]. It also finds use for alleviating the adrenergic burst-prompted trauma that can be experienced during intubation prior to surgery.

During the later stages of esmolol's clinical development, these types of drugs (i.e., drugs having a deliberately appended functional group that is constructed to program a specified metabolic deactivation pathway and rate) were come to be known as "soft drugs" [23]. Thus, ultimately esmolol itself has become the "pioneer soft drug" for which the template portrayed below in bold-faced atoms can now serve as an effective starting point for "soft ABDD" [24].

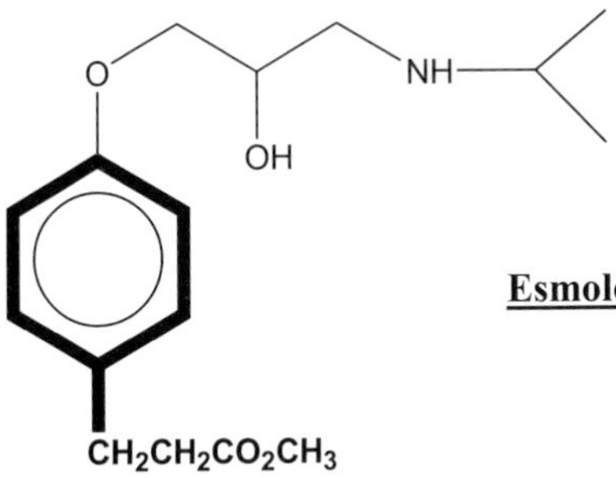

Esmolol

CH₂CH₂CO₂CH₃

9.6
Summary and Some Lessons for Today

The discovery of esmolol took advantage of a combination of ABDD from two different types of pioneer drugs, namely propranolol for efficacy and procaine for an ultra-short duration of action. However, for the latter, a novel, medicinal chemistry hypothesis led to a significant departure from procaine's internal ester structure and ultimately prompted the external ester arrangement. The latter eventually led to esmolol upon subsequent exploration of the relationship between enzymatic ester hydrolysis rates and steric factors. Some of the hurdles traversed during the discovery of esmolol provide useful lessons for today's medicinal chemists, who often become immersed in problems associated with a lead compound's overall pharmacokinetic profile [25]. These final comments are listed below in bullet fashion under headings common to the jargon of today's drug discovery process [25]. Taken together, these comments convey the important role that ABDD can play toward contributing to fundamental medicinal chemistry principles, as well as toward the practical improvement or specified tailoring of selected therapeutic entities.

9.6.1
Compound Libraries

- While the attributes of a "drug-like" profile are in no way to be minimized, the first and foremost properties for a proposed test series during early drug discovery is certainly their "assay-likable" profile – that is, if it cannot be tested you will not get a result, or even worse, one may perceive that one has an inactive compound and thus erroneously derive a negative SAR data point. Indeed, for the case of esmolol, the entire discovery process was almost halted when its first two series of target compounds were either too insoluble or too plagued by ISA, to be able to be effectively screened.

- Likewise, while "preferred scaffolds" and "privileged structures" generally refer to the attributes of such molecules relative to their potential clinical performance as drug candidates, this concept need not be so restrictive. For example, it may also be useful to think in such terms relative to the various preclinical testing steps so as to then exploit such attributes in that particular setting for the better or easier examination of a specific hypothesis that is to ultimately be applied to the actual lead series. Note that this was done for esmolol wherein the N-external esters served as important testing surrogates to help confirm the merits of the overall "external esters hypothesis". Alternatively, it would also be useful to be forewarned about what might otherwise appear to be attributes for the clinic when the structural domains of such pre-ferred testing templates happen to overlap with the structural domains of the actual lead series during evaluation of the latter's screening results.
- While the redundancy of homologues may indeed detract from "molecular diversity", their pursuit should not be forgotten alto-gether and certainly should not be regarded as a mundane opera-tion once a lead has been selected (see related bullet in SAR cate-gory).

9.6.2
Biological Testing

- While "high-throughput testing" (HTS) can certainly work its way through huge libraries of test compounds in short-order, it is also worthwhile to periodically check template representatives in an *in-vivo* setting as early as possible during the screening process. For example, the discovery of esmolol immediately relied upon a dog model to ascertain overall metabolic stability in a manner that allowed the entire *in-vivo* hydrolytic complement to be dis-cerned, rather than conducting a biochemical battery of discrete enzymatic assays. Had the latter approach been undertaken, SAR would have been very likely to have crossed back-and-forth from one test system to another for various molecular series, perhaps even to the extent of being noninterpretable relative to the key hypotheses being explored.

9.6.3
SAR

- The efficacy and/or toxicity "positive hit" data typically derived from HTS methods is simply not enough for creative, problem-solving medicinal chemistry. Note that the verification of the key hypothesis leading to pursuit of the external esters was actually gleaned from the novel, negative SAR data found for the proposed carboxylic acid metabolites, with the anticipated neutral SAR data associated with the esters themselves also falling nicely into place as fully anticipated.

- While hydrophilic/hydrophobic and/or substituent electronic properties may be important for various, selected interactions with a given biological surface, steric effects are by far the most reliable and universally applicable parameter when trying to monitor and exploit such SAR. Recall that while the esmolol ester family was essentially nonresponsive to the electronics of the "leaving group," it was exquisitely sensitive to localized steric parameters in terms of enzymatic hydrolysis rates. Indeed, the inverse relationship between the degree of steric bulk and the rate of drug metabolism within such localized vicinities, is so much more generally reliable compared to any other attempted physico-chemical correlation, that it probably deserves to be considered as "metabolism's rule of one."

- The importance of pursuing methyl, ethyl, propyl, etc., as one fine-tunes a lead is worth mentioning again in this category – that is, the "Goldilocks" nature of esmolol with its distinct 10-min half-life was found to reside only at n = 2 within series 8 (what more can be said about the merits of carefully conducting such molecular scrutiny in a patient and deliberately methodical manner?).

References and Notes

1 Esmolol enjoys a certain level of name recognition among the lay-public in that its common use within emergency room settings has prompted the long-running television series entitled "ER" often to open its first scene with a physician shouting: "*Esmolol, STAT!*", i.e. "stat" means immediately.

2 It has been a pleasure once again to provide a review about the discovery of esmolol, particularly at a time when it has just recently gone-off its initial U.S. patent. An earlier and differently focused discourse can be found in Erhardt PW, *Chronicles of Drug Discovery*, **1993**, 3, 191–206.

3 Fischer J. *Med. Chem. Res.*, 2004, 13, 218.

4 Zaroslinski J, Borgman RJ, O'Donnell JP, Anderson WG, Erhardt PW, Kam S-T, Reynolds RD, Lee RJ, Gorczynski RJ. *Life Sci.*, **1982**, 31, 899.

5 Arnar-Stone Laboratories (ASL), was staffed by four practicing, drug discovery-related scientists. Originally part of American Hospital Supply Corporation (AHSC), the latter soon renamed ASL to American Critical Care (ACC). As esmolol began to undergo serious development, AHSC was purchased by Baxter-Travenol who, in turn, sold ACC to Dupont in the mid-1980s. Interestingly, *"the purchase of ACC by Dupont was made for $425 million, $190 million of which was dependent upon the progression of esmolol into the clinic and marketplace;* (*C&E News*, p. 13, January 12, 1987) and

as a result, accounts of esmolol's discovery by such a small company are often referred to as a "Cinderella Story." Eventually, Dupont merged with Merck and among these giants the original, ASL niche strategy of critical care therapeutics would no longer be pursued. Esmolol, already doing well in the clinic, was eventually sold-off for others to market and, after a series of such transactions, ultimately wound-up back in the hands of Baxter and its affiliates.

6 Seifen AB. *Anesth. Analg.*, **1979**, 58, 382.

7 Kam ST, Martier WL, Maj KX, Barcelon-Yang C, Borgman RJ, O'Donnell JP, Stampfli HF, Sum CY, Anderson WG, Gorczynski RJ, Lee RJ. *J. Med. Chem.*, **1984**, 27, 1007.

8 Bodor N, Buchwald P. *Medicinal Res. Rev.*, **2000**, 20, 58.

9 Gringauz A. *Introduction to Medicinal Chemistry. How Drugs Act and Why.* Wiley, New York, 1997, p. 435.

10 Erhardt PW. *Synth. Commun.*, **1983**, 13, 103.

11 Erhardt PW, Woo CM, Gorczynski RJ, Anderson WG *J. Med. Chem.*, **1982**, 25, 1402.

12 Erhardt PW, Woo CM, Anderson WG, Gorczynski RJ. *J. Med. Chem.*, **1982**, 25, 1408.

13 Erhardt PW, Woo CM, Matier WL, Gorczynski RJ, Anderson WG. *J. Med. Chem.*, **1983**, 26, 1109.

14 Erhardt PW, Borgman RJ, O'Donnell JP. **1983**. U.S. Patent 4,387,103.

15 Erhardt PW, Borgman RJ, O'Donnell JP. **1986**. U.S. Patent 4,593,119.

16 Erhardt PW, Borgman RJ. **1984**. U.S. Patent 4,450,173.

17 Borgman RJ, Erhardt PW, Kam ST, O'Donnell JP. **1986**. U.S. Patent 4,604,481.

18 Erhardt PW, Matier WL. **1986**. U.S. Patent 4,623,652.

19 Matier WL, Erhardt PW, Patil G. **1985**. U.S. Patent 4,508,725.

20 Given the unique half-life for esmolol (8 n = 2; 10 min) relative to either the higher or lower homologues (n = 3 or n = 1 as being too long or too short), this result has been referred to as the *"Goldilocks effect"* when Dr. Erhardt has lectured on this topic. "Goldilocks" from an Anglo-Saxon fairy-tale, in this context means "just right".

21 Payne GB. *J. Org. Chem.*, **1962**, 27, 3819.

22 Quote and details taken directly from package insert as provided by Bedford Laboratories™ for "Esmolol Hydrochloride Injection ready-to-use 10 mL vials."

23 Bodor N. In: *Strategy in Drug Research.* Buisman JAK (Ed.). Elsevier, Amsterdam, **1982**, p. 137.

24 For example, Erhardt PW. **2004**. U.S. Patent 6,750,238 B1.

25 Erhardt PW. *Pure Appl. Chem.*, **2002**, 74, 703.

10
Development of Organic Nitrates for Coronary Heart Disease

László Dézsi

10.1
Introduction

The primary symptom of coronary heart disease (CHD) is angina pectoris, caused by transient episodes of myocardial ischemia. This is due to an imbalance in the myocardial oxygen supply and demand caused either by an increase in myocardial oxygen demand, or by a decrease in myocardial oxygen supply. Approved anti-anginal agents improve the myocardial oxygen supply to demand ratio either by increasing supply by dilating coronary vessels, or by decreasing demand by reducing cardiac work, or both. Organic nitrates meet these dual requirements to a great extent. The therapeutic use of organic nitrates is best indicated by the fact that this class of compounds represents the most frequently prescribed drugs for the treatment of myocardial ischemia [1].

Organic nitrates (group C01DA according to the ATC nomenclature) comprise nitroglycerin, isosorbide dinitrate, and isosorbide-5-mononitrate. However, in a broad sense classic nitrovasodilators, organic nitrate and nitrite esters also include the pioneer drug isoamyl nitrite, as well as nicorandil [2]. Although the latter is listed by ATC in group C01DX "Other vasodilators used in cardiac diseases", as part of a wide therapeutic area nicorandil has been found useful for the treatment of CHD [3], and therefore it is also discussed here.

10.2
Empirical Observations Leading to the Therapeutic Use of Classic Nitrovasodilators

The application of organic nitrates as anti-anginal agents has a long history, since they have been used for more then 100 years in the management of myocardial ischemia [1,4,5], without proper knowledge of their mechanism of action. It had been observed as early as the 19th century that people working with nitroglycerin in the production of explosives suffered from a syndrome consisting of severe headache and syncopal tendency but that, after a few days, these symptoms disappeared. These subjects were clearly developing tolerance to nitrate exposure [6,7] (see below).

Analogue-based Drug Discovery. IUPAC, János Fischer, and C. Robin Ganellin (Eds.)
Copyright © 2006 WILEY-VCH Verlag GmbH & Co. KGaA, Weinheim
ISBN: 3-527-31257-9

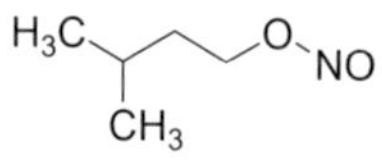

Fig. 10.1 Chemical structure of isoamyl nitrite.

10.3
Isoamyl Nitrite: The Pioneer Drug

Isoamyl nitrite (Fig. 10.1) was first synthesized in 1844 by Balard [9], who reported that its vapor had given him a severe headache. Further experiments revealed [10] that the chemical produced throbbing of the carotid artery, flushing of the face, and an increase in heart rate. The first report on the clinical use of isoamyl nitrite in the treatment of angina pectoris was published by Brunton [10], who thought that its pain-relieving therapeutic effect was due to lowering of the blood pressure. The route of administration of isoamyl nitrite was by inhalation. During the following years, conflicting results emerged on the efficacy of organic nitrites and nitrates, and in an attempt to clarify this confusion Murell [12] compared the action of isoamyl nitrite and nitroglycerin on volunteers. He established that although the effects of isoamyl nitrite appeared in 10 s, it took 2–3 min for nitroglycerin to produce any changes. Moreover, the effects of the former agent wore off after 5 min, as opposed to about 30 min for nitroglycerin. These observations led Murell to consider nitroglycerin as being superior to amyl nitrite in the treatment of angina pectoris. It also transpired that much higher inhaled doses of amyl nitrite (100–300 mg kg^{-1}) were necessary to produce effects similar to those caused by nitroglycerin (0.5–1 mg kg^{-1}) applied in tablet form [5]. Currently, isoamyl nitrite and some other organic nitrate analogues (e.g., pentaerythritol tetranitrate) have limited clinical relevance [13].

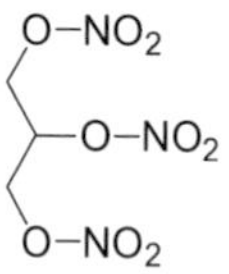

Fig. 10.2 Chemical structure of glyceryl trinitrate.

10.4
Nitroglycerin (Glyceryl Trinitrate): The Most Successful Analogue

Nitroglycerin (correctly termed glyceryl trinitrate) is the most successful analogue of organic nitrates (Fig. 10.2). The compound is a triester formed by the nitration of glycerol (1,2,3-propanetriol), and was discovered by Sobrero in 1847 [14]. In pure form, nitroglycerin is an explosive, as was discovered by Alfred Nobel, but when combined with an inert carrier (such as lactose) it is stable enough to be used as a medication. Its beneficial effects in CHD were revealed by empirical observations [1]. Sobrero reported that upon tasting a drop of it he experienced an intense headache [14]. A severe headache had also been reported by Field [15], but he also claimed the alleviation of toothache by applying an alcoholic solution of nitroglycerin to the tongue. Murell, in his aforementioned investigations, had also noted that the characteristic headache was due to an overdose, whereas smaller sublingual doses were effective treatments for the relief of pain resulting from acute anginal attacks [12]. These pioneering studies led to the adoption of organic nitrates into general clinical practice [5].

A wide variety of formulations and routes of administration have been utilized for the treatment and prevention of anginal attacks, including sublingual tablets and spray (in a dose of 0.3–0.8 mg) as well as buccal tablets (1–3 mg). Intravenous infusion or transdermal patches are used for longer-lasting nitroglycerin therapy [16]. At lower infusion rates (<50 μg min^{-1}), nitroglycerin is principally a venodilator, whereas at higher infusion rates more balanced venous and arterial dilating effects are seen [2]. In a prospective randomized study, the number of spontaneous ischemic episodes was reduced to one-third in 24 h [17]. The transdermal therapeutic system (releasing 10 mg in 24 h) ensured constant nitroglycerin levels over 24 h with once daily administration [18]. However, adverse effects included headache [19], and in many patients tolerance developed within 24 h, with loss of anti-anginal effects. There is also a sustained-release oral formulation available. Nitroglycerin has a plasma half life of 1–4 min, and undergoes extensive hepatic first-pass metabolism, as well as intravascular metabolism. The resulting dinitrate metabolites (1,2-glyceryl dinitrate and 1,3-glyceryl dinitrate) are also biologically active, with half lives of ca. 40 min [16].

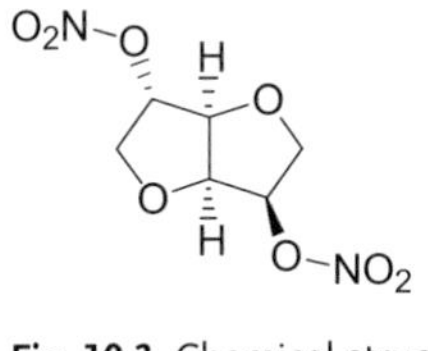

Fig. 10.3 Chemical structure of isosorbide dinitrate.

10.5
Isosorbide Dinitrate: A Viable Analogue with Prolonged Action

Several attempts have been made to synthesize longer-acting analogues for prophylactic use by the nitration of simple sugars. The first analogue, pentaerythritol tetranitrate, was reported in 1901 [20], but its clinical application is rather limited [13]. Furthermore, it is controversial whether these nitrate esters reach the circulation in sufficiently high concentrations because of their high rate of metabolism in the liver.

A new series of organic nitrates was discovered during the late 1930s. Among these, nitrates of sugar alcohols and their anhydrides (e.g., isosorbide dinitrate;

Fig. 10.3) turned out to be useful depressor agents [21]. The latter compound had an advantage compared with nitroglycerin, presumably due to the presence of an ether linkage between the nitrates and the sugar alcohol in the molecule. Although its potency was diminished, the duration of the depressor response was prolonged, which made it a useful therapeutic agent. It was first marketed in the form of tablets (2.5–10 mg) for sublingual administration [22], which has a slightly longer onset of effects (5 min) than that of glyceryl trinitrate (1–4 min), although it is longer acting (4 h or more) [23]. Later on, both standard formulations (10–45 mg) and sustained-release (20–80 mg) tablets were produced for oral administration.

Orally administered isosorbide dinitrate shows clear hemodynamic and anti-anginal effects, but it undergoes a rapid hepatic first-pass metabolism, with a plasma half-life of ca. 40 min. Its major metabolites are isosorbide-2-mononitrate and isosorbide-5-mononitrate, with half-lives of 2 and 4 h, respectively [16]. A transdermal spray of isosorbide dinitrate can be administered in order to avoid the first-pass effect, and the unchanged drug penetrates through the skin within 20 min so that it is slowly released into the capillaries in therapeutically effective amounts [24]. A very recent study showed intravenously infused isosorbide dinitrate to be ideal anti-anginal agent, especially in acute conditions [25].

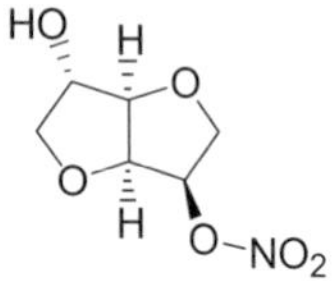

Fig. 10.4 Chemical structure of isosorbide mononitrate.

10.6
Isosorbide Mononitrate: The Metabolite of Isosorbide Dinitrate

The next chapter of the organic nitrates is connected to the patent of the pharmaceutical company American Home Products in 1971 [26], which introduced O-substituted mononitrate esters of isosorbide dinitrate. One of these O-mononitrate esters, isosorbide-5-mononitrate (Fig. 10.4), is a metabolite of the previously mentioned isosorbide dinitrate (see Fig. 10.3). The 5-mononitrate possesses the characteristics of lowering systemic blood pressure and coronary resistance, thereby increasing coronary blood flow in standard pharmacological tests, and indicating utility as an anti-anginal agent. A major advantage of this analogue over earlier molecules is that it does not undergo hepatic first-pass metabolism, and is completely bioavailable. Therefore, both standard formulations (10–20 mg) and sustained-release (10–240 mg) oral tablets are available [16]. After oral administration of isosorbide mononitrate tablet (20 mg), there is a rapid onset of action (5–10 min) that is maintained for up to 10–12 h [25]. In combination with beta-blockade therapy, isosorbide mononitrate was also found to be effective in prolonging exercise duration [27].

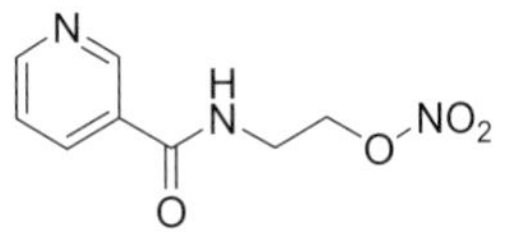

Fig. 10.5 Chemical structure of nicorandil.

10.7
Nicorandil: The Potassium Channel Opener Analogue with a Broad Cardiovascular Spectrum

In addition to the classic organic nitrates, nicorandil (Fig. 10.5), the nitrate ester of *N*-(2-hydroxyethyl)nicotinamide, discovered by Chugai Pharmaceutical Co. of Japan in 1980 [3], is listed by ATC in another group (C01DX) referred to as "Other vasodilators used in cardiac diseases". This novel pyridine derivative has been shown to stimulate the circulatory system in the case of CHD. It exhibited a wide therapeutic efficacy compared with classical organic nitrates for treating circulatory disease, having coronary vasodilating action, antihypertensive action, anticoagulative action, and peripheral vasodilating action. Thus, it was found to be useful when treating CHD, as an antihypertensive, anticoagulant, anti-arrhythmic, and as a peripheral vasodilator, including cerebral and renal vasodilation.

Among its actions, nicorandil has been shown to activate ATP-dependent potassium (K^+_{ATP}) channels, and this action is thought to play a significant role in its cardioprotective effects. This argument is supported by the observation that K^+_{ATP} blockers such as glibenclamide are known to exacerbate myocardial ischemia, and that this effect can be reversed by K^+_{ATP} openers such as pinacidil and cromakalim [28]. Patients receiving 40 mg kg^{-1} nicorandil either sublingually or orally showed central and peripheral hemodynamic changes within 15 min [29]. Following oral administration of 5–40 mg nicorandil, peak plasma levels are attained within 0.3–1 h after dosing, although due to the drug's rapid plasma clearance the duration of action is merely 3–4 h [30]. In addition to commercially available tablets, dry emulsion-loaded fast-disintegrating tablets were utilized to improve the sustained release property of the drug [31]. Nitroglycerin compared with nicorandil in 24 h intravenous treatment of congestive heart failure indicated the development of significantly less hemodynamic tolerance with the latter compound (60% versus 15%, respectively) [32].

The basic pharmacokinetic parameters of organic nitrovasodilators, including the time of onset (with favored routes of administration) and duration of action are summarized in Tab. 10.1.

Tab 10.1 Basic pharmacokinetic parameters of organic nitrovasodilators.

Organic nitrate	Time of onset	Duration of action
Isoamyl-NO$_2$	10 s[a]	5 min
NTG	1–4 min[b]	0.5 h
ISDN	5 min[b]	4 h
ISMN	5–10 min[c]	10–12 h
Nicorandil	15 min[b,c]	3–4 h

Isoamyl-NO$_2$, amyl nitrite; NTG, nitroglycerin; ISDN, isosorbide dinitrate; ISMN; isosorbide mononitrate.
Routes of administration: a, inhalation; b, sublingual; c, oral.

10.8
Cardiovascular Efficacy of Organic Nitrates

Vast amounts of clinical and experimental data are available in support of the therapeutic use of organic nitrates. In addition to angina pectoris, clinical syndromes such as acute myocardial infarction, congestive heart failure, blood pressure control, adjunctive therapy for interventional cardiac procedures also benefit from organic nitrate therapy [13]. Most of these states involve ischemic syndromes related to CHD, for the management of which organic nitrates represent a safe and effective choice [2]. Organic nitrates are potent vasodilators that dilate both arterial and venous smooth muscle [4,13]. At low doses, they produce dilatation of the venous capacitance system, which results in decreased right and left ventricular filling pressures and a subsequent reduction in cardiac work, but little change in systemic vascular resistance. A slight fall in systemic arterial pressure and a reflex increase in heart rate may be seen, while pulmonary vascular resistance and cardiac output are both slightly reduced [1,13].

At higher doses, further venous pooling occurs. There are also more marked arterial effects, and decreased systemic arteriolar resistance is accompanied by a reduction in systolic and diastolic blood pressure and cardiac output. Activation of compensatory sympathetic reflexes results in tachycardia and peripheral arteriolar vasoconstriction tending to restore systemic vascular resistance. Venodilating and arterial dilating effects are often termed "indirect or peripheral effects" since they relieve ischemia by reducing myocardial oxygen demand [1,13].

The "direct or central" effects of organic nitrates relieve ischemia by direct action on coronary vessels, improving myocardial oxygen supply. Direct vasodilator effects are exerted mostly on large conductance, rather than small resistance coronary vessels including intercoronary collateral vessels. Coronary stenosis dilatation is potentially a major component of the beneficial response. These mechanisms preferentially increase perfusion to the ischemic areas in patients with acute myocardial infarction [13,33].

10.9
Mechanism of Action of Organic Nitrates

The principal action of organic nitrates is vasorelaxation. However, the mechanism of this effect was for a long time unknown, and some of its exact details are still to be elucidated. For many years the vasodilatory action was attributed to the formation of NO_2^- in the bloodstream, until Krantz [34] showed that the total conversion of a therapeutic nitroglycerin dose into NO_2^- would not result in a sufficient vasodilatation. Starting in the 1970s, several mechanisms for the vasodilating effects of organic nitrates were proposed [13], including: interaction with sulfhydryl (SH) group-dependent nitrate receptors [8,35]; lowering of intracellular calcium levels; stimulation of soluble guanylate cyclase and subsequent synthesis of cyclic guanosine monophosphate (cGMP); or simulation of vasodilating prostaglandins, for example prostacyclin (PGI_2). While the latter suggestion turned out to be of minor importance, a dose-dependent increase in cGMP levels by organic nitrates was demonstrated [36,37]. From the 1980s onwards, the discovery of endothelium-derived relaxing factor (EDRF) by Furchgott [38] had a major impact on the elucidation of the mechanism of action of organic nitrates. The humoral agent EDRF was shown to mediate vasodilation in vivo [39] and was later identified as nitric oxide (NO) by Moncada and coworkers, and termed the "Endogenous Nitrovasodilator" [40].

NO is synthesized in the vascular endothelium and also in other cells (e.g., neurons and macrophages) from L-arginine, catalyzed by a family of nitric oxide synthases (NOSs) [2]. NO diffuses into the target cells (e.g., smooth muscle cells) and, according to its classical pathway, activates soluble guanylate cyclase to produce cGMP. Subsequent actions, such as vasorelaxation are mediated by the cGMP-dependent protein kinase through the phosphorylation of different proteins [41].

Organic nitrates exert their vasodilatory activity following conversion to NO in vascular smooth muscle cells [40]. Endogenous NO and nitrovasodilators apparently compete for the activation of guanylate cyclase [36]. After inhibition of endogenous NO, the development of supersensitivity to exogenous nitrovasodilators can be demonstrated *in vivo*, which is attributed to the up-regulation of guanylate cyclase that serves as a "receptor" for NO. This developing supersensitivity contributes to the efficacy of organic nitrates in pathological conditions involving endothelial dysfunction (e.g., myocardial ischemia and atherosclerotic coronary arteries).

The mechanisms of NO release from organic nitrates are complex, and still not fully explored. Figure 10.6, based on the scheme of Van de Voorde [7], dates back to the early 1990s but can be considered still valid after slight modification based on additional data [42]. It summarizes the proposed mechanisms by which NO-

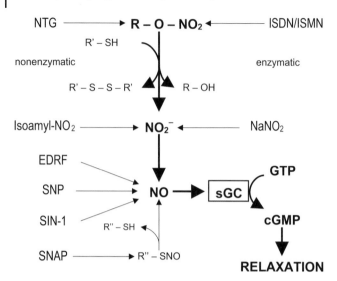

Fig. 10.6 Schematic diagram of proposed mechanisms by which NO-containing molecules relax vascular smooth muscle (redrawn and modified after Van de Voorde 1991 [7]). NTG, nitroglycerin; R-O-NO$_2$, organic nitrate; ISDN, isosorbide dinitrate; ISMN, isosorbide mononitrate; R′-SH and R″-SH, thiol-containing molecules; R′-S-S-R′ disulfides; R-OH, denitrated organic nitrite; NO$_2^-$, nitrite; Isoamyl-NO$_2$, isoamyl nitrite; NaNO$_2$, sodium nitrite; EDRF, endothelium-derived relaxing factor; SNP, sodium nitroprusside; SIN-1; sydnonimine; SNAP, S-nitroso-N-acetyl-penicillamine; NO, nitric oxide; sGC, soluble guanylate cyclase; GTP, guanosine triphosphate; cGMP, cyclic guanosine monophosphate.

containing substances relax vascular smooth muscle. Organic nitrates (nitroglycerin, isosorbide dinitrate and isosorbide mononitrate) do not spontaneously release NO, unlike some other NO-containing substances, but they are subject to biotransformation in smooth muscle cells [7,8]. Therefore, organic nitrates can be considered as prodrugs [16]. The nitrovasodilators can presumably be metabolized both via enzymatic and nonenzymatic processes. Nonenzymatic metabolism has been shown to require thiol-containing groups (most probably cysteine) to form NO$_2^-$ [43], which subsequently reacts with H$^+$, yielding NO. According to Ignarro, at near-neutral pH the NO derived from nitroglycerin and the excess cysteine forms S-nitrosocysteine, which together with other S-nitrosothiols might serve as intermediates to activate soluble guanylate cyclase [8]. Alternatively, it was proposed that free radicals, including NO, activate soluble guanylate cyclase [44]. A very recent study confirmed this mechanism, but also revealed that other reaction products (magnesium/cGMP/pyrophosphate) must be present for guanylate cyclase to reach an active state [45].

The enzymatic mechanisms of nitrate bioactivation have long been the subject of debate. The glutathione-S-transferase and cytochrome P-450 systems were thought to be involved [8], while recently a nitrate reductase that specifically catalyzes the formation of 1,2-glyceryl dinitrate from glyceryl trinitrate was purified and identified as a mitochondrial aldehyde dehydrogenase [46]. Interestingly, this enzyme requires the presence of SH groups in the molecule for activity, while di-sulfide bonds inactivate it. These findings suggest that organic nitrate biotransfor-mation takes place in the mitochondrion. Nicorandil is also metabolized in myo-cardial mitochondria, but it seems that glutathione-S-transferases are not primar-ily involved in the conversion of nicorandil [47].

10.10
Tolerance to Organic Nitrates

As mentioned above, headache and other symptoms observed by people working with explosives disappeared upon maintained exposure [6]. The development of tolerance was first reported by Stewart in 1888 [48], soon after nitroglycerin was introduced into clinical practice. Tolerance is defined as the loss of hemodynamic and anti-anginal effects during continued therapy [16]. Flaherty has proposed the term "attenuation" instead of tolerance, since complete loss of anti-ischemic effects has rarely been observed and patients continue to respond to elevated doses [13]; however, the term tolerance is generally accepted.

The mechanism of tolerance, despite intensive research, remains poorly under-stood. The most important hypotheses about nitrate tolerance are: sulfhydryl depletion, neurohormonal (counter-regulatory) changes, plasma volume expan-sion, and free radical generation; while further hypotheses include decreased sen-sitivity of guanylate cyclase [16]. Mülsch and colleagues have proposed that impaired biotransformation and not desensitization of guanylate cyclase is the cause of the nitrate tolerance [49]. The sulfhydryl depletion hypothesis [35] is based on the notion that loss of efficacy of nitrates during continued treatment is due to the depletion of SH groups which are thought to be necessary for the bio-transformation of organic nitrates to NO. However, experimental data do not always support this idea. For example, thiols such as N-acetylcysteine may improve vasodilatation, independently of the reversal of nitrate tolerance [50]. The free radical theory is based on the observation that high mitochondrial concentra-tions of NO inhibit the respiratory chain, thus inducing the generation of super-oxide anions ($.O_2^-$) [51]. By this mechanism, the biotransformation of nitrogly-cerin generates $.O_2^-$ [52], which in turn leads to the formation of peroxynitrite. Superoxide anion and peroxynitrite generation may result in a chain of events, including NO quenching, NOS uncoupling, sympathetic activation, as well as superoxide dismutase inhibition. According to a recent "unifying hypothesis" by Gori and Parker [41], such reactions might form the basis of nitrate tolerance.

The clinical relevance of nitrate tolerance is extensively demonstrated (for reviews, see Refs. [13,16]). In stable angina pectoris, treadmill walking time dur-ing isosorbide dinitrate treatment is significantly increased, although after sus-tained therapy for two weeks the walking time to angina was diminished [53]. Similar to angina, hemodynamic attenuation (both in pressure-lowering and in arterial and venous dilating effects) was observed in patients receiving chronic nitrate therapy (glyceryl trinitrate or isosorbide dinitrate) for congestive heart failure, but this was less common with oral than with transdermal nitrate administration [13].

The prevention of nitrate tolerance is also an unresolved issue. Although several pharmacological interventions based on the above hypotheses have been tested, the only accepted method so far is the use of nitrate-free intervals in the dosing strategy during each day. The basis of this intermittent therapy or "eccentric" (unevenly timed) dosing regimen is that although tolerance develops rapidly, it is also rapidly reversed [16]. The mechanisms underlying this rapid reversal remain to be elucidated, however.

10.11
Concluding Remarks

Organic nitrovasodilators are successfully used for the treatment of CHD manifested in angina pectoris and related symptoms. The therapeutic use of organic nitrates began in the 19th century and was based on empirical observations, although the precise mechanism of action was not known until the end of the 20th century. Nevertheless, analogues of nitroglycerin afforded longer-acting drugs. Meanwhile, the metabolic pathways of organic nitrates within the vascular smooth muscle cells were elucidated, showing that biotransformation of these compounds releases NO, which activates guanylate cyclase, and in turn causes the vascular smooth muscle to relax. Thus, organic nitrates function as a replacement for the endogenous nitrovasodilator NO, which is unavailable when the vascular endothelium becomes dysfunctional [4]. In line with this discovery, future avenues of analogue-based drug research in the field of nitrovasodilators are two-fold. First, the development of analogues or combined molecules for broader therapeutic use (see the example of nicorandil). Second, the development of pharmacological analogues, the so-called "nitric oxide donors" or "nitric oxide-releasing drugs", which comprise various classes of compounds with very heterogeneous chemical structures (e.g., *S*-nitrosothiols, sydnonimines, NONOates, sodium nitroprusside; for reviews, see Refs. [2,54,55]). These compounds are either not dependent on biotransformation, as organic nitrates are, or are metabolized by different pathways and thus are free of tolerance-related problems. Therefore, they may represent the next generation of NO-releasing therapeutic agents in the treatment of cardiovascular diseases.

References

1 Kerins DM, Robertson RM, Robertson D. In: *Goodman & Gilman's The Pharmacological Basis of Therapeutics*. 10th edn. Hardman JG, Limbird LE, Goodman Gilman A (Eds.). McGraw-Hill, USA, **2002**, pp. 843–870.

2 Napoli C, Ignarro LJ. *Annu. Rev. Pharmacol. Toxicol.*, **2003**, 43, 97–123.

3 Chugai Pharm. Co. Patent No. US4200640, **1980**.

4 Mehta J. *Am. Heart J.*, **1995**, 129, 382–391.

5 Sneader W. *Drug Prototypes and their Exploitation*. John Wiley & Sons, Chichester, **1996**.

6 Swartz AM. *N. Engl. J. Med.*, **1946**, 235, 241–244.

7 Van de Voorde J. *J. Cardiovasc. Pharmacol.*, **1991**, 17 (Suppl. 3), S304–S308.

8 Ignarro L, Lippton H, Edwards JC, Baricos WH, Hyman AL, Kadowitz PJ, Gruetter CA. *J. Pharmacol. Exp. Ther.*, **1981**, 218, 739–749.

9 Balard M. *C. R. Acad. Sci.*, **1844**, 19, 634.

10 Guthrie F. *J. Chem Soc.*, **1859**, 11, 245–252.

11 Brunton TL. *Lancet*, **1867**, 90, 225–227.

12 Murell W. *Lancet*, **1879**, 113, 80–81.

13 Flaherty JT. *Drugs*, **1989**, 37, 523–550.

14 Sobrero A. *Annalen*, **1847**, 64, 398.

15 Field A. *Med. Times Gaz.*, **1858**, 37, 291.

16 Parker JD, Parker JO. *N. Engl. J. Med.*, **1998**, 338, 520–531.

17 Curfman GD, Heinsimer JA, Lozner EC, Fung HL. *Circulation*, **1983**, 67, 276–282.

18 Colombo G, Favini G, Aglieri S, Pollavini G, de Vita C. *Int. J. Clin. Pharmacol. Toxicol.*, **1985**, 23, 211–214.

19 Santoro A, Rovati LC, Setnikar I. *Arzneim.-Forsch.*, **2001**, 51, 29–37.

20 Vignon P, Gerinm F. *C.R. Acad. Sci.*, **1901**, 133, 590.

21 Krantz JC, Carr CJ, Forman SE. *J. Pharmacol. Exp. Ther.*, **1939**, 67, 187.

22 Goldberg L. *Acta Physiol. Scand.*, **1948**, 15, 173.

23 Reichek N. *Am. J. Med.*, **1983**, 74 (Suppl.), 33–39.

24 Wildfeuer A, Laufen H, Leitold M. *Arzneim.-Forsch.*, **1985**, 35, 1289–1291.

25 Chen J, Jiang XG, Cai L, Lu W, Gao KP, Shi ZQ, Zhang QZ. *Arzneim.-Forsch.*, **2004**, 54, 203–206.

26 American Home Products. Patent No. 1356374, **1971**.

27 Akhras F, Jackson G. *Lancet*, **1991**, 338, 1036–1039.

28 D'Alonzo AJ, Darbenzio RB, Hess TA, Sewter JC, Sleph PG, Grover GJ. *Cardiovasc. Res.*, **1994**, 28, 881–887.

29 Coltart DJ, Signy M. *Am. J. Cardiol.*, **1989**, 63, J34–J39.

30 Frydman AM, Chapelle P, Diekmann H, Bruno R, Thebault JJ, Bouthier J, Capain H, Ungethuem W, Gaillard C, Le Liboux A, Renard A, Gaillot J. *Am. J. Cardiol.*, **1989**, 63, J25–J33.

31 Jin Y, Ohkuma H, Wang H, Natsume H, Sugibayashi K, Morimoto Y. *Yakugaku-Zasshi*, **2002**, 122, 989–94.

32 Larsen AI, Goransson L, Aarsland T, Tamby JF, Dickstein K. *Am. Heart J.*, **1997**, 134, 435–41.

33 Brown GB, Bolson E, Petersen RB, Pierce C, Dodge HT. *Circulation*, **1981**, 64, 1089–1097.

34 Krantz JC, Carr CJ, Forman SE, Cone N. *J. Pharmacol. Exp. Ther.*, **1940**, 70, 323–327.

35 Needleman P, Jakschik B, Johnson EM. *J. Pharmacol. Exp. Ther.*, **1973**, 187, 324–331.

36 Murad F, Mittal CK, Arnold WP, Katsuki S, Kimura H. *Adv. Cyclic Nucleotide Res.*, **1978**, 9, 145–158.

37 Kukovetz WR, Holtmannn S, Wurm A, Poch B. *Naunyn-Schmiederberg's Arch. Pharmacol.*, **1979**, 310, 129–138.

38 Furchgott RF, Zawadzki JV, *Nature*, **1980**, 288, 373–6.

39 Pohl U, Dezsi L, Simon B, Busse R. *Am. J. Physiol.*, **1987**, 253, H234–H249

40 Palmer RM, Ferrige AG, Moncada S, *Nature*, **1987**, 327, S24–6.

41 Gori T, Parker DJ. *Circulation*, **2002**, 106, 2510–2513.

42 Noack E, Feelisch M. *Basic Res. Cardiol.*, **1991**, 86 (Suppl. 2), 37–50.

43 Heppel L, Hilmoe RJ. *J. Biol. Chem.*, **1950**, 183, 129–138.

44 Katsuki S, Arnold W, Mittal C, Murad F. *J. Cyclic Nucleotide Res.*, **1977**, 3, 23–25.

45 Russwurm M, Koesling D. *EMBO J.*, **2004**, 23, 4443–4450.

46 Chen Z, Zhang J, Stamler JS. *Proc. Natl. Acad. Sci. USA*, **2004**, 99, 8306–8311.

47 Sakai K, Akima M, Saito K, Saitoh M, Matsubara SJ. *J. Cardiovasc. Pharmacol.*, **2000**, 35, 723–728.

48 Stewart DD. *Policlinic*, **1888**, 6, 171–172.

49 Mülsch A, Busse R, Bassenge E. *Z. Kardiol.*, **1989**, 78 (Suppl. 2), 22–25.

50 Fung HL, Chong S, Kowaluk E, Hough K, Kakemi M. *J. Pharmacol. Exp. Thes.*, **1988**, 245, 524–530.

51 Brown GC. *Biochim. Biophys. Acta*, **1999**, 1411, 351–369.

52 Fink B, Dikalow S, Bassenge E. *Free Radic. Biol. Med.*, **2000**, 28, 121–128.

53 Thadani U, Fung HL, Drake AC, Parker JO. *Am J. Cardiol.*, **1982**, 49, 411–419.

54 Felisch M. *Naunyn-Schmiederberg's Arch. Pharmacol.*, **1998**, 358, 113–122.

55 Hou YC, Janczuk A, Wang PG. *Curr. Pharmaceut. Design*, **1999**, 5, 417–441.

11
Development of Opioid Receptor Ligands

Christopher R. McCurdy

11.1
Introduction

The term opioid has come to represent the definition of any compound (exogenous or endogenous) that interacts with opioid receptors. Other terms have been used in the past like, opiate and narcotic, to describe these ligands. Opioids have been used since the first written records of history, and probably well before history was recorded. The milky white exudates or latex of the opium poppy has been found to contain several alkaloids that posses analgesic properties. The German chemist, Friedrich Wilhelm Serturner, first identified the chemical structure responsible for the main analgesic and euphoriant effects of the opium latex in 1805 [1]. This compound, known as morphine, named after the Greek god of dreams, *Morpheus*, is still widely prescribed today. It has led the field as the structural prototype for analgesic drug design and development, and has remained as the foundation for the most widely prescribed class of analgesics today. Initially, it was postulated that morphine functioned as an analgesic compound through ingestion, and no specific receptor was known. However, as early as the 1954, Beckett and Casy hypothesized that morphine had a unique and distinct receptor in the human body [2]. In 1973, Candace Pert and Solomon Snyder played a key role in discovering that morphine worked on specific receptors, termed opioid receptors [3]. In 1975, Hans Kosterlitz and his colleagues [4] isolated an endogenous opioid receptor ligand, enkephalin, in the human brain. At this time, research in the field of opiates exploded. Philip S. Portoghese postulated that multiple opioid receptor subtypes existed, based on his studies with morphine-skeleton analogues to produce antagonist ligands that demonstrated multiple pharmacological effects *in vivo* [5]. This idea was expanded upon by medicinal chemistry and the subsequent development of pharmacological tools that led to the postulation of at least three main opioid receptors. Indeed, more endogenous peptides were discovered that had differing pharmacological activities at these purported receptor types. Pharmacology led to the discovery of these three distinct subtypes of opioid receptors, known as mu, delta, and kappa. Many years later, during the early 1990s, the cDNA was identified that encoded the existence of the previously

Analogue-based Drug Discovery. IUPAC, János Fischer, and C. Robin Ganellin (Eds.)
Copyright © 2006 WILEY-VCH Verlag GmbH & Co. KGaA, Weinheim
ISBN: 3-527-31257-9

postulated, mu [6], delta [7], and kappa [8] opioid receptors. Along with the three receptors that had been, to this point, putative due to the pharmacological actions of selective antagonists, a fourth receptor with homology close enough to classify it as a member of the opioid receptor family was cloned and designated as the ORL-1 receptor [9]. The ORL-1 receptor was distinct in recognition characteristics as well as pharmacological characteristics from the traditional three opioid receptors. This receptor, although homologous to the classical opioid receptors, does not share in the recognition of classical opioid ligands such as the endogenous peptides or the majority of the morphine-based analogues, and therefore the discovery of the ORL-1 was suspect. However, one year later in 1995, the endogenous ligand for the ORL-1 receptor was elucidated. This ligand was termed as nociceptin [10] or orphaninan FQ [11], since it was discovered by two independent groups simultaneously. Since then, the role of the ORL-1 receptor has been difficult to define. Many publications have demonstrated differing effects of the endogenous peptide. Depending on the route of administration or concentration used, the peptide has produced either analgesic or hyperanalgesic responses. Thus, it is unclear if an agonist or antagonist would be therapeutically beneficial. Further research is needed to elucidate fully the role of this new opioid receptor and to define its pharmacological role in the opioid receptor system. The ORL1 receptor will not be discussed further at this point, as the true meaning of its existence is still unclear and it will require medicinal chemists to develop selective probes to define fully the role of this receptor. At this point, we will focus our attention on current knowledge and explore the role that morphine – the prototype opioid ligand – has played as the template for analogue design and in the synthesis of new ligands for the classical opioid receptors.

The classical opioid receptors, mu, delta, and kappa, have led to an even further definition of the relative pharmacology previously associated with them [12]. Moreover, it has more recently been reported that opioid receptors can exist as heterodimers [13]. This remarkable discovery opens the future of ligand design to target these dimeric and potentially oligomeric forms of the opioid receptor complexes.

Morphine has affinities for all three receptors, and acts as an agonist at each of them. From the structure of morphine, several semisynthetic and synthetic ligands have been developed and marketed. This chapter will focus on the development of agonists and antagonists from the morphine lead compound. Some of these compounds have been developed to be more selective at certain opioid receptor subtypes, while others have helped to define the roles of the endogenous opioid system. It will be clear, from a structure-based analogue approach, exactly what are the necessary requirements for opioid receptor affinities according to current knowledge of these ligands.

11.2
Pharmacology Related to the Classic Opioid Receptors.

Cloning of the opioid receptors led to a further definition of their classification and naming [12]. The delta (δ) opioid receptor (DOP) was the first to be cloned, and is known as the OP_1 receptor [7]. The mu (μ) opioid receptor (MOP) was cloned secondly, and carries the designation of OP_2 [6]. Finally, the kappa (κ) opioid receptor (KOP) was cloned, and is referred to as the OP_3 receptor [8]. Any one of the above designations is acceptable to define the opioid receptors. Analgesia can be produced through activation of each of the three classic opioid receptor subtypes [14]. Coupled with the benefits of pain relief are several side effects that are produced with opioid receptor activation, and have limited the clinical use of opioids [15]. However, most ligands designed from the morphine template are MOP agonists. The MOP therefore produces an analgesia that is referred to as "morphine-like". In addition, activation of the MOP produces euphoria, or the "high" associated with these ligands. The side effects or negative effects produced by MOP activation include: increased gastrointestinal transit time (constipation); respiratory depression; suppression of the immune system (opioid receptors are present on immune cells); emesis; along with tolerance and physical dependence. The selective activation of KOP produces analgesia with sedation and dysphoria. In addition, these agents cause diuresis and miosis. In contrast to the MOP, activation of the DOP produces stimulation of the immune system. However, DOP activation also results in analgesia and decreased respiration rates.

11.3
Alkaloids from the Latex of *Papaver somniferum* Initiate Research

Isolation of the alkaloids from the latex of the opium poppy, *Papaver somniferum*, led to the identification of over 20 compounds (Fig. 11.1). The most abundant alkaloids were morphine (**1**), codeine (**2**), thebaine (**3**), noscapine (**4**), and papaverine (**5**).

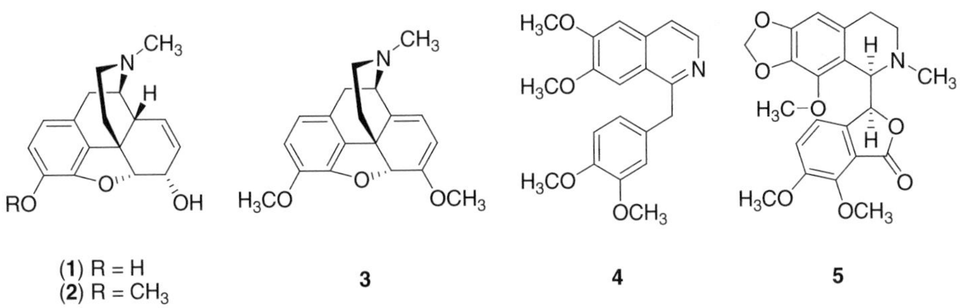

(1) R = H
(2) R = CH$_3$

3

4

5

Fig. 11.1 The major alkaloids from *Papaver somniferum*.

It has long been known that only morphine and codeine possess analgesic properties, whereas thebaine acts as a convulsant, similar to strychnine [16]. However, analogues produced from thebaine have provided some of the more interesting and clinically useful drugs such as buprenorphine (see Section 11.5). Indeed, an early structure–activity relationship was realized between morphine and its methylated phenolic relative, codeine. Codeine (see below) is known to have about one-tenth the analgesic activity of morphine, and was also less likely to produce addiction. Additionally, an early diacetylated analogue of morphine, heroin (6) (Fig. 11.2), demonstrated an increased potency but also an increased addictive potential.

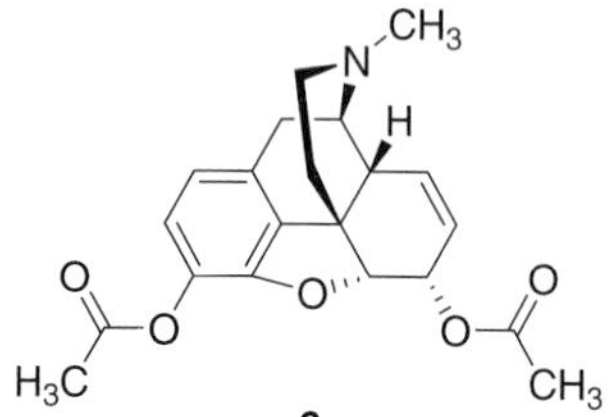

Fig. 11.2 The structure of heroin.

Based on an ability to modify the activity of morphine by these slight chemical changes to the molecule, a large-scale experimental program of analogue synthesis was instituted in 1929 by the National Research Council in the United States [17]. This program involved chemists from the University of Virginia (Small, Mosettig, Burger, and coworkers) and pharmacologists from the University of Michigan (Eddy and coworkers). Over the years, these groups investigated over 450 compounds for analgesic activity. About 25% of these compounds were derived from the morphine, codeine, or thebaine skeletons. The remainder were synthetic derivatives prepared to elucidate the pharmacophore of morphine.

11.4
Morphine: The Prototype Opioid Ligand

Opium was first utilized as an extract tar from the poppy, and later found to be enhanced by the addition of brandy to the extract (laudanum) to further enhance the solubility of the compounds. Opium quickly became utilized not only as a treatment for pain but also as a recreational concoction for abuse. Purification of the extract provided morphine (1), which became the first studied naturally occurring alkaloid from *Papaver somniferum*. Morphine quickly made a mark in clinical use and abuse. Currently, morphine is marketed as an injection, an oral solution, and in oral tablet formulations. Morphine is not well absorbed from the gastrointestinal tract when given orally, and what is absorbed is glucuronidated in the liver in both the 3- and the 6-positions. The 3-glucuronide of morphine undergoes extensive enterohepatic recycling, and this leads to a decreased bioavailability [18]. Moreover, the 3-glucuronide has been reported to have antagonist activity [19]. Morphine 6-glucuronide, on the other hand, is about 650 times more potent as an analgesic agent than morphine itself [20], and many argue that most of the analgesic activity may come from this metabolite. Due to differences in patient metabolism, the individual dosages vary greatly when morphine is administered orally. In contrast, when morphine is administered by parenteral routes it bypasses the hepatic first-pass metabolism and has rapid and more predictable analgesic effects.

11.4.1
Initial Studies of Morphine Analogues

Codeine (**2**), another naturally occurring alkaloid from *Papaver somniferum*, was first isolated by Robiquet in 1832 [21], and provided another analgesic substance to the market. By the mid-19th century, the use of pure alkaloids became more prevalent than the use of crude opium preparations. Codeine is 3-methoxy morphine, and derived its name from the Greek term describing the head of the poppy. Today, most codeine is produced semisynthetically from morphine. Codeine also suffers from hepatic first-pass metabolism when given orally, with the majority being metabolized to the glucuronidated form and the remainder (ca. 10% of the dose) being O-demethylated to morphine [18]. It has been established that less than 10% of the population has a genetic deficiency of the cytochrome P450 2D6 (CYP2D6) enzyme responsible for the conversion of codeine to morphine [22]. In this patient population, codeine has been reported to have no analgesic or addictive properties. However, a recent report indicates that persons deficient in CYP2D6 may obtain analgesia from codeine 6-glucuronide [23]. Regardless of individual patient metabolic capability, codeine is more effective as an antitussive agent than morphine, although the exact mechanism of this antitussive action remains the subject of debate. Nonetheless, codeine has enjoyed tremendous clinical success as an antitussive agent in combination with other products for the common cold.

Heroin (**6**, diacetyl morphine) was perhaps the first semisynthetic opioid ligand to be created [24]. Interestingly, when heroin was first marketed by Bayer in 1898, it was promoted as a nonaddicting opiate agonist for the treatment of severe pain. However, it was not long before Dreser published a paper on the clinical use and abuse of heroin [25]. Soon after this, heroin was removed from the market and placed into restricted use and finally Schedule I of the Controlled Substance Act in the United States, making it illegal with no defined medical use. However, it is still prescribed in the most of Europe and Canada for very severe pain. Due to the ease of synthesis from morphine, heroin is still a widely abused illicit substance. Chemically, heroin is more lipophilic and is believed to penetrate the blood–brain barrier very quickly, giving it a much faster onset of action. Heroin (hydrochloride) is about two-fold more potent on a weight basis than morphine and is more soluble in water. Heroin is actually a prodrug and is converted to 6-acetylmorphine, which is much more potent [26]. Further hydrolysis produces morphine, and the effects are seen from both of these metabolites.

Dihydromorphine (**7**) [27] and dihydrocodeine (**8**) [28] were some of the first compounds produced by saturation of the 7,8-olefin of the respective scaffolds (Fig. 11.3), although this modification led to the production of compounds with less potent analgesic effects.

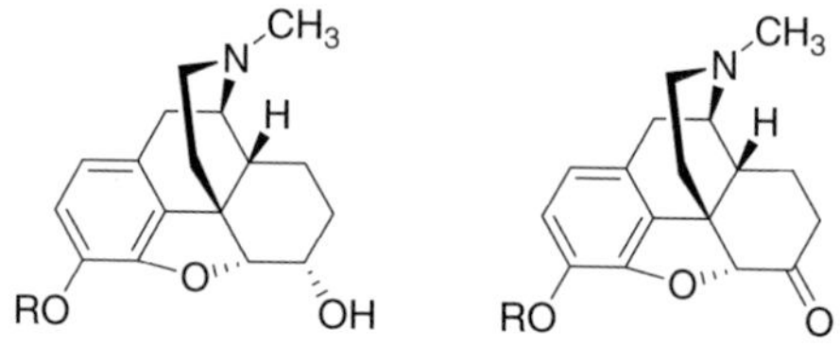

(**7**) R = H
(**8**) R = CH$_3$

(**9**) R = H
(**10**) R = CH$_3$

Fig. 11.3 Structures of dihydromorphine, dihydrocodeine, hydromorphone, and hydrocodone.

However, these compounds led in turn to the discovery of hydromorphone (**9**; Dilaudid) [29] and hydrocodone (**10**) [30], which are oxidized 6-keto derivatives of dihydromorphine and dihydrocodeine, respectively. Hydromorphone is about seven-fold more potent than morphine, and has seen an increased use among cancer patients. In particular, the hydrochloride salt is more water-soluble than morphine and can therefore be injected in a smaller volume, causing less aggravation to the patient. Hydrocodone is about three times more potent than codeine in relieving cough [15]. Like codeine, it is primarily marketed in combination products with acetaminophen (paracetamol) (as Vicodin®, Lortab®), or with homatropine (Hycodan®) for cough. Hydrocodone is also produced semisynthetically from morphine. The first step, hydrogenation of the 7,8-olefin, produces dihydromorphine, which is then oxidized to convert the 6-hydroxy to the ketone (hydromorphone). Finally, methylation of the phenolic hydroxy group produces hydrocodone. These modifications produce compounds that are more potent than the parent compounds and are marketed for either moderate pain or cough control.

Nalorphine (**11**; N-allylnormorphine; Fig. 11.4) [31] was found to antagonize the properties of morphine, and was utilized as an antidote for morphine poisoning,

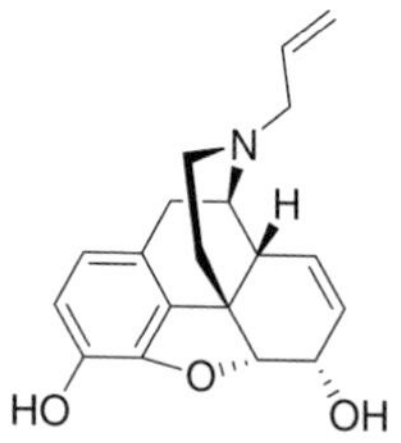

11

Fig. 11.4 Structure of nalorphine (N-allylnormorhpine).

until Lasagna and Beecher [32] reported that nalorphine had analgesic actions in postoperative patients. However, nalorphine use as an analgesic agent was limited due to its dysphoric actions through activation of the KOP. Nalorphine was recognized as being one of the first mixed agonist-antagonist derivatives of morphine. Nonetheless, the structure of nalorphine led to the development of new drugs such as the pure opioid antagonist naloxone and compounds with mixed agonist-antagonist actions (see Section 11.5).

11.5
Structure–Activity Relationships of Morphine Derivatives

Although structure–activity relationships (SAR) have been derived from morphine derivatives over the years, it is imperative to remember that small changes to the core structure of this molecule can impart large changes in receptor selectivity and pharmacological activity. Therefore, profiles for each derivative must be established, and the SAR are ever-changing. Nonetheless, some generalizations can be realized [18]. First and foremost in importance is the presence of a basic nitrogen (almost always protonated at physiological pH) for opioid receptor affinity. This basic nitrogen must be two carbons away from a central quaternary carbon that is attached to a phenol or bioisosteric aromatic ring. The A-ring and basic nitrogen are essential components to every known MOP agonist to date. These two features are not sufficient for activity alone however, and must be accompanied by other essential groups to define the pharmacophore. The SAR for clinically important morphine derivatives are illustrated in Fig. 11.5.

The basic nitrogen is most often tertiary for optimal activity. Substituents on this nitrogen produce a wide array of pharmacological profiles. The native methyl moiety can be easily removed using the von Braun [33] reaction to produce normorphine or norcodeine. Other alkyl groups can be added, and increasing the chain length produces compounds with mixed agonist-antagonist activity [18].

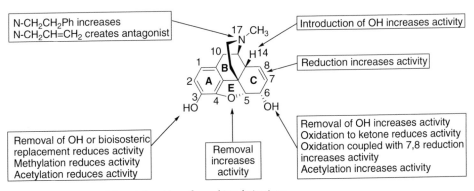

Fig. 11.5 Structure–activity relationships of morphine derivatives.

The substitution pattern begins to change in relation to the size of the group on the basic nitrogen, and upon reaching larger substituents (e.g., *N*-phenethyl) the analogues impart greater agonist activity at opioid receptors than morphine itself [18]. However, these compounds have not yet reached the prescription market.

The 4,5-epoxide bridge defines the morphine-based analogues. Removal of the epoxide is not trivial, and analogues lacking it are derived from synthetic methods. These analogues are discussed in greater detail in Section 11.5.

The 3-hydroxy moiety on the A-ring has proven to provide an important pharmacophoric element [18]. Replacement by hydrogen results in a decrease of activity, whilst methylation (as in codeine) also decreases the receptor affinity and the analgesic activity. Bioisosteric replacements of the 3-phenolic hydroxy have yielded little success [34], and therefore it is believed that the 3-hydroxy moiety is important for high affinity and intrinsic activity through MOP. In fact, much of the analgesic activity of codeine can be attributed to the O-demethylation to morphine. This occurs for about 10% of the dose, and is accomplished through cytochrome P450 2D6 (CYP2D6) [22]. Codeine itself has potent antitussive activity and has been utilized clinically for this purpose for many years.

Modifications to the C-ring of the morphinan structure can produce compounds with increased activity, as well as changes in the receptor selectivity profiles [18]. Most pronounced is oxidation of the 6-hydroxy to the 6-keto derivative. With the 7,8-olefin in place, the activity is reduced when compared to morphine. However, reduction of the 7,8-olefin produces compounds with increased activity. Combining the 7,8-dihydro and 6-keto modifications to the morphinan skeleton produces compounds with greatly increased activities (10-fold) over morphine.

11.6
Synthetic Analogues of Thebaine Further Define Morphinan SAR

Synthetic analogues of *thebaine* (3) that introduce a β-hydroxy moiety into the 14-position and a ketone into the 6-position of the C-ring provide compounds with increased analgesic activity, while reducing the antitussive properties [18]. This is accomplished by 1,4-addition to thebaine with hydrogen peroxide [35] or *m*-chloroperbenzoic acid [36]. The prototype drugs from this class are oxycodone [37] (12) and oxymorphone [38] (13) (Fig. 11.6). These drugs are some of the most potent marketed opioid agonists created to date. Oxycodone (OxyContin®, Roxicodone®) has similar efficacy to morphine but improves on bioavailability, making it more predictable in its actions. It is also marketed in combination with acetaminophen (paracetamol) (as Percocet®) or with aspirin (as Percodan®, Tylox®). Oxycodone is also available in sustained-release dosage forms. The metabolism is similar to that of codeine [18]. Oxymorphone (Numorphan®), which is synthesized from oxycodone by cleavage of the methyl ether, is 10-fold more potent than morphine but lacks the antitussive properties thought due to the 14-hydroxy moiety.

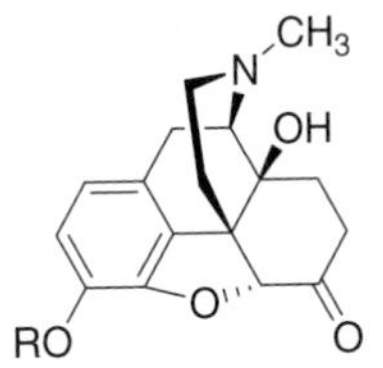

(12) R = CH₃
(13) R = H

Fig. 11.6 Structures of oxymorphone and oxycodone.

In addition, compounds prepared from thebaine (**3**) utilizing the Diels–Alder reaction produce the oripavine class of derivatives [39]. These are defined by buprenorphine (**14**; Fig. 11.7) [40], which is up to 50 times more potent than morphine. Buprenorphine is a mixed agonist-antagonist with agonist activities at the MOP and KOP, and antagonist activity at the DOP. It also produces fewer side effects than the typical morphinan derivatives, and has been reported as being incapable of producing tolerance and addiction comparable to other MOP agonists [41]. Interestingly, buprenorphine has very little potential for producing withdrawal symptoms in patients unless they are highly addicted to other opioids [42]. However, it suppresses withdrawal symptoms in individuals undergoing withdrawal from opioids. It has also been reported to effectively block the effect of high heroin doses. For these reasons, in 2001, buprenorphine received approval from the United States Food and Drug Administration (FDA) for use as an opioid withdrawal agent.

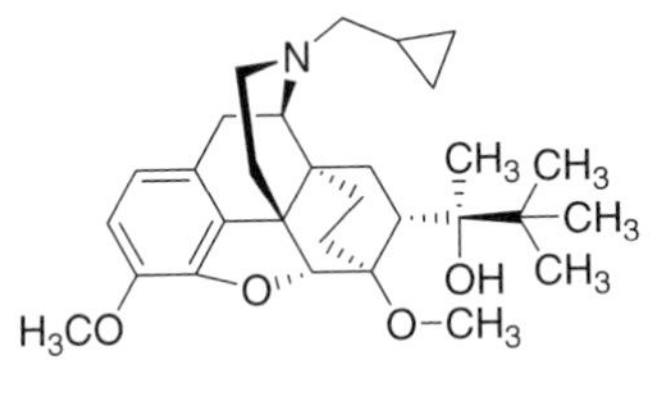

14

Fig. 11.7 Structure of the oripavine, buprenorphine.

As seen in the morphine analogues, placing larger alkyl chains on the basic nitrogen imparts opioid antagonist activity [18], and on this scaffold, with a 14-hydroxy, true opioid antagonists have been realized. When the nitrogen is substituted with an allyl group or a cyclopropylmethyl group, the resultant compounds impart antagonist activity at all three opioid receptors [18]. *Naloxone* (**15**; Narcan®; Fig. 11.8) [43], which contains the *N*-allyl moiety, acts rapidly as an opioid antagonist and has the distinction of being the drug of choice for the reversal of respiratory depression in cases of opioid overdose. Naloxone must be administered parenterally in order to avoid hepatic first-pass metabolism as it is almost completely degraded [18]; consequently, naloxone is ineffective when administered orally. This property has been utilized in oral formulations of opioid analgesics that can be dissolved and injected by drug abusers to achieve a more rapid effect. The incorporation of naloxone into a tablet makes it ineffective when it is used in this way. However, the incorporation does not affect the oral administration of agents for analgesic control, since naloxone is so readily degraded by hepatic first-pass metabolism. *Naltrexone* (16; Trexan®; Fig. 11.8) [44], which contains the N-cyclo-propylmethyl moiety, is also a nonspecific opioid antagonist, but has the distinction of being readily bioavailable by the oral route of administration, with long-lasting effects. The actions last for up to 72 h, making naltrexone a good prophy-lactic drug in abusers [45]. Both naloxone and naltrexone can precipitate opioid withdrawal and may be troublesome in some patients. Nonetheless, there is a large body of literature indicating that naltrexone has no effect in individuals with no opioid use. More recently, naltrexone has been studied for use in the treatment of addiction to alcoholism [46] and gambling [47].

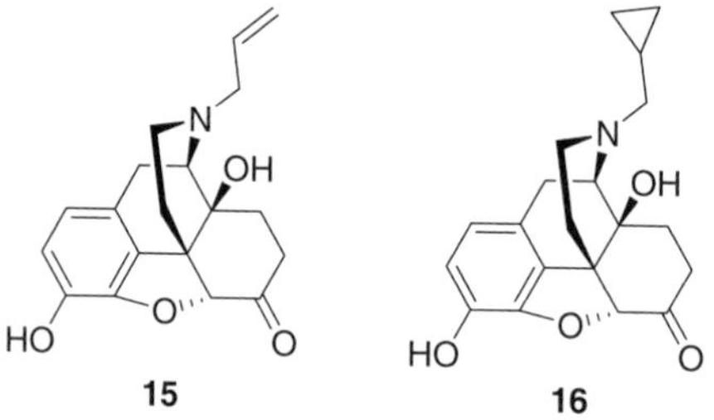

Fig. 11.8 Structures of the opioid antagonists, naloxone and naltrexone.

Interestingly, cyclobutyl methyl groups on this scaffold also create mixed ago-nist-antagonist compounds [18]. For example, nalbuphine (17; Nubain®; Fig. 11.9) [48], which contains a cyclobutyl methyl substituent on the nitrogen, is an antago-nist of MOP and an agonist of KOP. The analgesic potency is similar to that of morphine, and the side-effect profile of nalbuphine was improved. Sedation is the most common side effect, and the dysphoria usually seen with KOP agonists is reduced. This agent led to the great hope that compounds could be produced and

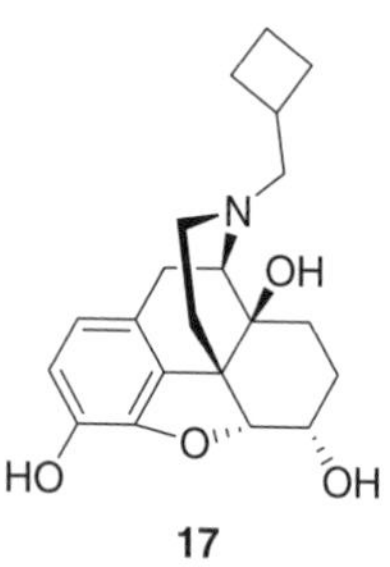

Fig. 11.9 Structure of nalbuphine.

marketed with mixed agonist-antagonist activity and potentially begin to reduce addiction to opioids. Nalbuphine has a low abuse potential, and remains unscheduled in the United States. However, the excitement in the clinical arena for these types of compounds has not met with increased usage or study of these mixed agents.

11.7
Compounds of the Morphinan Skeleton Produce New Agents

As mentioned previously, several synthetic derivatives were investigated that contained portions of the morphine skeleton. Compounds lacking the 4,5-epoxide bridge were also potent analgesics, and classified as morphinans (see Fig. 11.10). Levorphanol (18) [49], which represents the same chirality as morphine, showed enhanced analgesic activity over morphine. However, like morphine, it also had a side-effect profile that limits its clinical usage somewhat. Interestingly, the opposite chirality or enantiomer of levorphanol completely lacks opioid activity. This compound is known as dextrorphan (19) [50], and it demonstrated potent antitussive properties [51]. Methylation of the phenolic hydroxy of dextrorphan produced dextromethorphan (DM) (20) [52], which became a more potent antitussive agent [53] and subsequently was marketed over-the-counter as such. More importantly, it was discovered that the (+) stereochemistry of morphine-like structures was completely devoid of opioid effects [18]. DM, although lacking opioid abuse potential, has not been devoid of abuse [54], and large doses produce psychic effects similar to those of phencyclidine (PCP). Indeed, it was found that DM is first metabolized by CYP2D6 in a similar fashion as codeine to dextrorphan, and then has actions on the NMDA (*N*-methyl-D-aspartate) receptor at the same site to which PCP binds [55]. Many high-school-aged persons abuse DM to achieve this "high". Despite the ingestion of an entire 237 mL (8-oz) bottle of DM being sufficient to produce the effects, DM remains available over-the-counter in almost all countries.

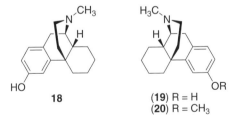

HO   **18**   OR
                            (19) R = H
                            (20) R = CH$_3$

Fig. 11.10 Structures of the morphinans; levorphanol, dextrorphan, and dextromethorphan.

Again, substitution on the nitrogen with a cyclobutylmethyl group on the morphinan scaffold produces mixed agonist-antagonist activity [18]. The resultant compound is known as butorphanol (**21**; Stadol®; Fig. 11.11) [56], and this has increased potency when compared to morphine. It is mainly utilized in the form of a nasal spray for the treatment of moderate to severe pain. However, very severe pain must still be treated with traditional opioids. Butorphanol is a potent KOP agonist and an antagonist at MOP [18]. Due to its hepatic first-pass metabolism, butorphanol has not been used as an oral agent. Butorphanol remains unscheduled in the United States.

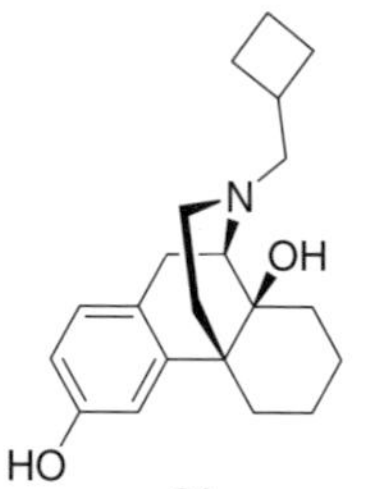

21

Fig. 11.11 Structure of butorphanol.

11.8
Further Reduction of the Morphinan Skeleton Produced the Benzomorphans

Removal of the C-ring of the morphinan scaffold produced the benzomorphan class of opioid ligands, as shown in Fig. 11.12. These compounds have a greater affinity for KOP, but still interact weakly with MOP. The central quaternary carbon atom must remain intact and therefore, these compounds are really truncated C-ring analogues [18]. Pentazocine (**22**; Talwin®) [57] is the only marketed drug in this class, and is about one-sixth as potent as morphine [18]. Pentazocine, which is used to treat moderate pain, is a potent KOP agonist and a weak MOP antagonist. The side effects of pentazocine are similar to those of other KOP agonists. Pentazocine was felt to have no abuse liability due to a similar pharmacological profile to the other mixed agonist-antagonists. However, it was quickly recognized as an abused substance where addicts would dissolve the tablet and inject it for

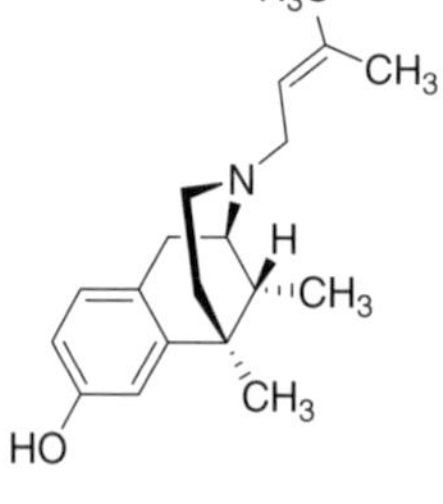

22

Fig. 11.12 Structure of the benzomorphan, pentazocine.

euphoric effects (or dysphoric as the case may be). Indeed, the addicts found a way of achieving a pleasurable high by injecting it along with the antihistamine, tripelennamine [58]. For this reason, the manufacture of pentazocine incorporated naloxone in the oral dosage formulation to prevent abuse, as discussed previously. Pentazocine also suffers from hepatic first-pass metabolism, which reduces the bioavailability by up to 80%. All metabolites of pentazocine are inactive [18].

11.9
Another Simplified Version of Morphine Creates a New Class of Opioid Ligand

Tramadol hydrochloride (23; Ultram®) [59] was originally touted as being structurally unrelated to morphine. However, upon closer examination the relationship can easily be seen (Fig. 11.13).

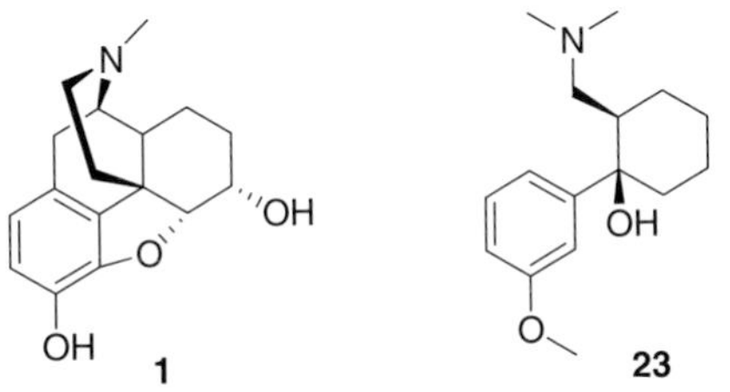

Fig. 11.13 Comparison of the structures of morphine and tramadol.

Tramadol, however, is thousands of times less potent than morphine as an analgesic agent [18]. It is marketed in the racemic form, and each enantiomer has distinct pharmacological actions. The (+)-isomer is a weak MOP agonist, while the (−)-isomer inhibits neurotransmitter reuptake (norepinephrine and serotonin). The O-demethylated metabolite has improved opioid receptor affinity but is still much less potent (35-fold) than morphine. The ability for this metabolite to ameliorate the analgesic effects of tramadol has not been well studied and remains questionable. The drug has been used for decades in Europe, but was only recently introduced in the United States. It has a greater safety profile than morphine, and produces no respiratory depression or constipation. It is also claimed to be nonaddictive, but remains unscheduled.

11.10
A Breakthrough in the Structural Design of Opioid Ligands

A significant discovery came about in 1939 at the laboratories of Farbenindustrie (Germany), during the investigation of antispasmodic medications related to atropine. At this time, Schaumann [60] examined a compound synthesized by Eisleb for *in-vivo* antispasmodic activity, and observed a Straub tail reaction which was usually associated with the centrally exciting component of opioid activity. This observation led to the performance of a standard battery of tests for opioid analgesic activity. Indeed, the compound known as meperidine (**24**; Demerol®) or pethidine (as it is known in Europe) introduced a new structural class of opioid receptor ligands. Upon investigation of the relationship of meperidine to morphine, it can be noticed that the removal of another ring from the benzomorphan class of ligands can produce the phenylpiperidine class, to which meperidine belongs.

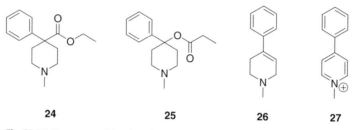

24 **25** **26** **27**

Fig. 11.14 Structures of the phenylpiperidine, meperidine; reverse meperidine (MPPP), MTP, and MPP⁺.

Meperidine is widely used today as it is both fast-acting and rapidly metabolized, allowing for greater control of administration. The major metabolite produced is normeperidine, which has the liability of producing seizures if it accumulates [61]. Normally, this is not a problem in controlled prescribing, but it can be a liability in abusers or those with impaired kidney function (the main organ of elimination). Meperidine is one of the simplest structures endowed with opioid activity. In fact, its structure is so simple that clandestine chemists were able to produce the reverse ester derivative of meperidine, 1-methyl-4-phenyl-4-propionoxypiperidine (MPPP; **25**; Fig. 11.14). This often contained a trace amount of unpropionylated precursor material that, upon prolonged heating under acidic conditions, dehydrated to afford 1-methyl-4-phenyl-1,2,5,6-tetrahydropyridine (MPTP; **26**; Fig. 11.14). When MPTP is administered and subsequently metabolized by monoamine oxidase B in the brain, it produces a neurotoxin known as MPP⁺ (1-methyl-4-phenylpyridinium; **27**) [62]. As MPP⁺ is produced in the brain, it becomes "trapped" inside the tissues, cannot cross the blood–brain barrier, and this results in an accumulation and finally neurotoxicity. Indeed, this careless street chemistry led to several emergency room cases of young, mostly male, individuals presenting in a catatonic state resembling parkinsonism. These individuals became known as "frozen addicts" and had suffered irreversible brain damage due to the neurotoxic MPP⁺ that was formed from MPTP. It has been recognized that MPTP destroys dopaminergic neurons in the substantia nigra [63], and MPTP is now utilized in research as a chemical means of producing Parkinson's disease in some species-specific animal models.

11.11
Discovery of the 4-Anilidopiperidines

A modification of the phenylpiperidine class by Janssen Pharmaceuticals led to the 4-anilidopiperidine class of opioid ligands. This class, which is exemplified by fentanyl (28; Fig. 11.15) [64], is perhaps the most potent class of opioid ligands known. Because of the introduction of a second nitrogen into the structure of the anilidopiperidines, they no longer follow the traditional SAR for opioid receptor ligands. Fentanyl is reported to be 100-times more potent than morphine [18], and analogues of fentanyl can approach 10 000-times the potency of morphine. Fentanyl and its analogues have a more rapid onset of action and a shorter duration of action than traditional opioid ligands based on morphine. This effect has allowed for the use of fentanyl and its derivatives for general anesthesia as well as analgesia. Fentanyl is dosed in microgram quantities for analgesic activity, and is principally delivered to the patient in the form of topical patches for transdermal use. Fentanyl was quickly adopted by clandestine laboratories, and several analogues are available on the street. China white, which referred in the past to high-purity heroin, became the street name for clandestine fentanyl derivatives. Due to the much higher potency (in some cases 6000-times that of morphine), many deaths have been attributed to overdoses.

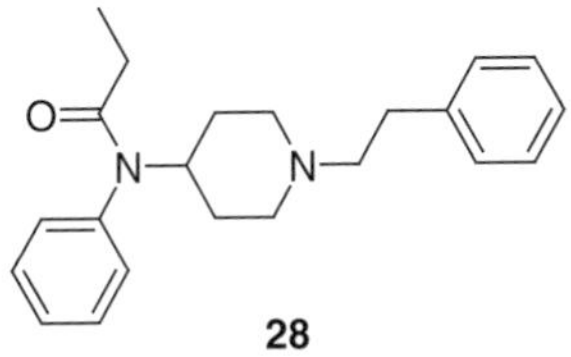

28

Fig. 11.15 Structure of the anilidopiperidine, fentanyl.

11.12
Phenylpropylamines: The Most Stripped-Down Opioids Still Related to Morphine

Removal of one of the ethylene portions of the piperidine ring leaves the phenylpropylamine skeleton of the phenyl piperidines. These are the most simplified structures still to have opioid receptor activity. The phenylpropylamines still follow the SAR of morphine on the basis that they possess a basic nitrogen which is two carbons away from a central quaternary carbon atom having a benzene ring attached [18]. In the case of methadone (29; Fig. 11.16) [65], the prototype phenylpropylamine, the quaternary carbon contains another benzene ring and a propionyl moiety. The basic nitrogen is still tertiary with two methyl groups present.

This molecule was actually prepared at the same time as meperidine as an antispasmodic agent. Methadone is marketed as a racemic mixture, and again the

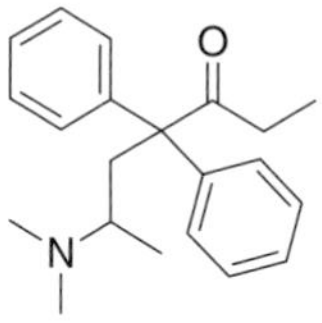

29

Fig. 11.16 Structure of the phenylpropylamine, methadone.

analgesic activity resides in the (–)-form. The (+)-form does have antitussive activity, as seen in other (+)-isomers related to morphine. When methadone is given by parenteral routes it has a more pronounced effect than morphine. However, methadone is well absorbed orally as compared to morphine, and has a longer-lasting and more potent effect. Methadone has many of morphine's side effects, including addictive capabilities. However, it is interesting that the withdrawal symptoms are less severe and have been reported to be more gradual as compared to morphine. This may be, in part, due to the extensive metabolism of methadone, as all but one of the metabolites retain activity. This profile has led to the widespread use of methadone in opioid-withdrawal programs.

11.13
The Use of Opioid Analgesics in Clinical Practice: Hope of the Future

Opioid analgesics still constitute the largest class of prescription medications used for the treatment of pain, although in this respect they are still underutilized – most likely due to their legendary addictive profiles. It is well documented that the compounds are safe and effective when used in the correct manner, and if a patient is experiencing severe chronic pain there should be little concern of addiction. Nonetheless, the fear of addiction and the frequency of malpractice law suits continue to curb the needed use. Medicinal chemists have toiled for many years to find a "holy grail" of opioid analgesic. Such an ideal opioid would have superior analgesic effects, without tolerance or side effects that would liken it to currently available opioids. Recently, derivatives of morphine – either semisynthetic or completely synthetic – have shed light on the notion of one day creating this reality, and because the field still lacks the ideal opioid analgesic there remains great justification to continue the search. The future holds great promise for novel agents acting through opioid receptors. These may either act through a single receptor interaction or by the targeting of compounds to oligomeric receptor types. In addition, novel structures have begun to appear in the literature that interact with opioid receptors but meet none of the previously known pharmacophores. This has been exemplified by the KOP-selective, naturally occurring, non-nitrogenous agonist, salvinorin A (**30**; Fig. 11.17). The future of opioid receptor ligand research is very much open and exciting. Moreover, the hope that a nonaddictive, side-effect free analgesic is yet to be discovered will continue the aggressive investigations of medicinal chemists for years to come.

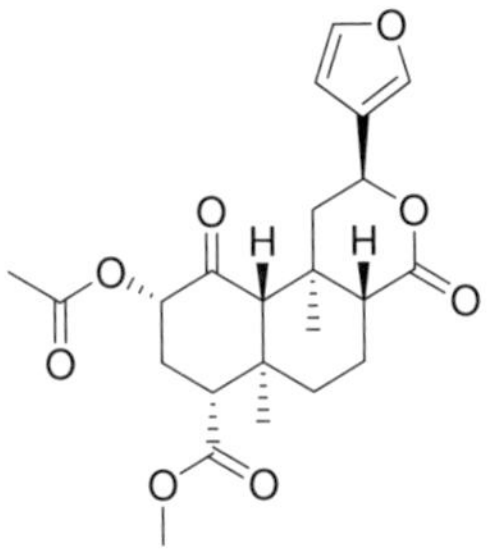

30

Fig. 11.17 Structure of the non-nitrogenous kappa opioid receptor agonist, salvinorin A.

References

1 Serturner FW. *Trommsdorff's J. Pharm.*, **1805**, 13, 234.
2 Beckett AH, Casy AF. *J. Pharm. Pharmacol.*, **1954**, 6, 986–1001.
3 Pert CB, Snyder SH. *Science*, **1973**, 179, 1011–1014.
4 Hughes J, Smith TW, Kosterlitz HW, Fothergill LA, Morgan BA, Morris HR. *Nature*, **1975**, 258, 577–580.
5 Portoghese PS. *J. Med. Chem.*, **1965**, 8, 609–616.
6 Wang JB, Johnson PS, Persico AM, Hawkins AL, Griffin CA, Uhl GR. *FEBS Lett.*, **1994**, 338, 217–222.
7 Evans CJ, Keith DE, Jr., Morrison H, Magendzo K, Edwards RH. *Science*, **1992**, 258, 1952–1955.
8 Zhu J, Chen C, Xue JC, Kunapuli S, DeRiel JK, Liu-Chen LY. *Life Sci.*, **1995**, 56, PL201–PL207.
9 Mollereau C, Parmentier M, Mailleux P, Butour JL, Moisand C, Chalon P, Caput D, Vassart G, Meunier JC. *FEBS Lett.*, **1994**, 341, 33–38.
10 Meunier JC, Mollereau C, Toll L, Suaudeau C, Moisand C, Alvinerie P, Butour JL, Gillemot JC, Ferrara P, Monserrat B, Mazarguil H, Vassart G, Parmentier M, Costentin J. *Nature*, **1995**, 337, 532–535.
11 Reinscheid RK, Nothacker HP, Bourson A, Ardati A, Henningsen RA, Bunzow JR, Grandy DK, Langen H, Monsma FJ, Jr., Civelli O. *Science*, **1995**, 270, 792–794.
12 Dhawan BN, Cesselin F, Raghubir R, Reisine T, Bradley PB, Portoghese PS, Hamon M. *Pharmacol. Rev.*, **1995**, 48, 567–592.
13 Rios CD, Jordan BA, Gomes I, Devi LA. *Pharmacol. Ther.*, **2001**, 92, 71–87.
14 Kanjhan R. *Clin. Exp. Pharmacol. Physiol.*, **1995**, 22, 397–403.
15 Gutstein HB, Akil H. In: *Goodman and Gilman's The Pharmacological Basis of Therapeutics*. 10th edn. Hardman J, Limbird L, Gilman A (Eds.). McGraw-Hill, New York, **2001**, pp. 569–619.
16 Hall JA, Nelson RB, Edlin AI. *J. Pharm. Sci.*, **1967**, 56, 298–299.
17 Burger A. In: *Analgetics*. deStevens G (Ed.). Academic Press, New York, **1965**, pp. 1–8.
18 Fries DS. In: *Foye's Principles of Medicinal Chemistry*. Williams DA, Lemke TL (Eds.). Lippincott Williams & Wilkins, Baltimore, **2002**, pp. 453–496.
19 Smith MT, Watt JA, Cramond T. *Life Sci.*, **1990**, 47, 579–585.
20 Pasternak GW, Bodnar RJ, Clark JA, Inturrisi CE. *Life Sci.*, **1987**, 41, 2845–2849.
21 Robiquet PJ. *Ann. Chim.*, **1832**, 51, 259.
22 Cascorbi I. *Eur. J. Clin. Invest.*, **2003**, 33, 17–22.
23 Vree TB, van Dongen RT, Koopman-Kimenai PM. *Int. J. Clin. Pract.*, **2000**, 54, 395–398.
24 Wright CRA. *J. Chem. Soc.*, **1874**, 27, 1031.

25 Dreser H.. *Dtsch. Med. Wochenschr.*, **1898**, 24, 185.

26 Sawynok J. *Can. J. Physiol. Pharmacol.*, **1986**, 64, 1–6.

27 Oldenberg L. *Berichte*, **1911**, 44, 1829–1831.

28 Skita A, Franck HH. *Berichte*, **1912**, 44, 2862–2867.

29 DRP 365,683. Knoll, **1922**.

30 DE 415,097. E. Merck, **1923**.

31 McCawley EL, Hart ER, Marsh DF. **J. Am. Chem. Soc.**, **1941**, 63, 314.

32 Lasagna L, Beecher HK. *J. Pharmacol. Exp. Ther.*, **1954**, 112, 356–363.

33 von Braun J. Berichte, **1907**, 40, 3914.

34 McCurdy CR, Jones RM, Portoghese PS. *Org. Lett.*, **2000**, 2, 819–821.

35 Freund M, Speyer E. J. *Prakt. Chem.*, **1916**, 94, 135.

36 Hauser FM, Chen T-K, Carroll FI. *J. Med. Chem.*, **1974**, 17, 1117.

37 DRP 411,530. E. Merck AG, **1925**.

38 Lewenstein MJ, Weiss U. US Patent 2,806,033, **1955**.

39 Bentley KW, Horsewood P, Kirby GW, Singh S. *J. Chem. Soc. D.*, **1969**, 23, 1411.

40 DE 1,620,206. Reckitt and Sons Ltd., **1965**.

41 Heel RC, Brogden RN, Speight TM, Avery GS. *Drugs*, **1979**, 17, 81–110.

42 Robinson SE. *CNS Drug Rev.*, **2002**, 8, 377–390.

43 Lewenstein MJ. US Patent 3,254,088, **1961**.

44 Lewenstein MJ, Fishman J. US Patent 3,320,262, **1967**.

45 Crabtree BL. *Clin. Pharm.*, **1984**, 3, 273–280.

46 Rohsenow DJ. CNS *Drugs*, **2004**, 18, 547–560.

47 Toneatto T, Ladoceur R. *Psychol. Addict. Behav.*, **2003**, 17, 284–292.

48 US Patent 3,332,950. Endo, **1963**.

49 CH Patent 252,755. F. Hoffmann-La Roche & Co., A.-G, **1948**.

50 GB Patent 725,763. Roche Products Ltd., **1955**.

51 Fossati A, Vimercati MG, Caputo R, Valenti M. *Arzneim.-Forsch.*, **1995**, 45, 1188–1193.

52 Schnider O, Grussner A. *Helv. Chim. Acta*, **1951**, 34, 2211–2217.

53 Gravenstein JS, Beecher H. *Arzneim.- Forsch.*, **1955**, 5, 364–367.

54 Boyer EW. *Pediatr. Emerg. Care*, **2004**, 20, 858–863.

55 Murray TF, Leid ME. *Life Sci.*, **1984**, 34, 1899–1911.

56 US Patent 3,775,414. Bristol-Myers, **1972**.

57 BE 611,000. Sterling Drug, **1960**.

58 Butch AJ, Yokel RA, Sigell LT, Hanenson IB, Nelson ED. *Clin. Toxicol.*, **1979**, 14, 301–306.

59 GB Patent 997,399. Gruenenthal, **1963**.

60 Eisleb O, Schaumann O. *Dtsch. Med. Wochenschr.*, **1939**, 65, 967.

61 Umans JG, Inturrisi CE. *J. Pharmacol. Exp. Ther.*, **1982**, 223, 203–206.

62 Glover V, Gibb C, Sandler M. *J. Neural Transm. Suppl.*, **1986**, 20, 65–76.

63 Javitch JA, Snyder SH. *Eur. J. Pharmacol.*, **1984**, 106, 455–456.

64 US Patent 3,141,823. Janssen, **1961**.

65 DE Patent 865,314. Farbw. Hoechst, **1941**.

12
Stigmines

Zeev Tashma

The biological activity of "carbamates", meaning mainly aryl carbamates, was discovered during the second half of the 19th century, whilst many of the important investigations in the field were conducted during the first half of the 20th century. The structure–activity relationship (SAR) studies performed by the medicinal chemists of the period were naturally rather basic. Nonetheless, we can admire their achievements by acknowledging that some of the drugs developed at that time (e.g., pyridostigmine) still occupy pharmacy shelves today. What changed considerably during those decades is that the original wide range of carbamates, which activated both the central and peripheral nervous systems, evolved during the 1930s into compounds that function only in the peripheral system (mainly against myasthenia gravis), in which the central activities became annoying untoward effects. This swing changed direction when carbamates were developed as agricultural insecticides and, more extremely, during the 1980s when centrally acting carbamates were developed for the treatment of Alzheimer's disease (AD), and the peripheral activities were looked upon as annoying, untoward effects.

12.1
Historical Background

Physostigmine, also known as eserine, is the main alkaloid of the Calabar bean (*Physostigma venenosum*, Balf), and is responsible for the plant's typical biological activity and toxicity (for a historical review, see Ref. [1]). The plant, a twining vine with pinkish flowers, is a native of West Africa. The fruit is a seedpot that carries two or three of these beans. The native population of the area used the beans as a kind of "truth drug": people who denied committing a crime were made to consume a preparation containing the bean. It was expected that the innocent would survive the test, while the sinner would die. Thus, the drug served both as a proof and as an executioner. Interestingly, the cholinergic system, which is so dramatically activated by physostigmine, is considered also today to be related to the ability to tell a lie, but in an opposite way: anticholinergics drugs such as scopolamine were sometimes used as "truth drugs".

Analogue-based Drug Discovery. IUPAC, János Fischer, and C. Robin Ganellin (Eds.)
Copyright © 2006 WILEY-VCH Verlag GmbH & Co. KGaA, Weinheim
ISBN: 3-527-31257-9

The modern era of Calabar bean research began in the mid-19th century, mainly in Edinburgh. The opposing pharmacological effects of the drug and of atropine were discovered and investigated. From a chemical aspect, the alkaloid was isolated in a pure state in 1864 and its rather complicated chemical structure (Fig. 12.1), comprising a tricyclic aminophenol, eseroline, esterified ("carbamylated") by N-methyl carbamic acid, was gradually revealed. It was not, however, until 1925 that the fully correct structure was determined by Stedman [2]. Much later, the chemistry and stereochemistry of physostigmine were proven by X-ray analysis of the crystals [3].

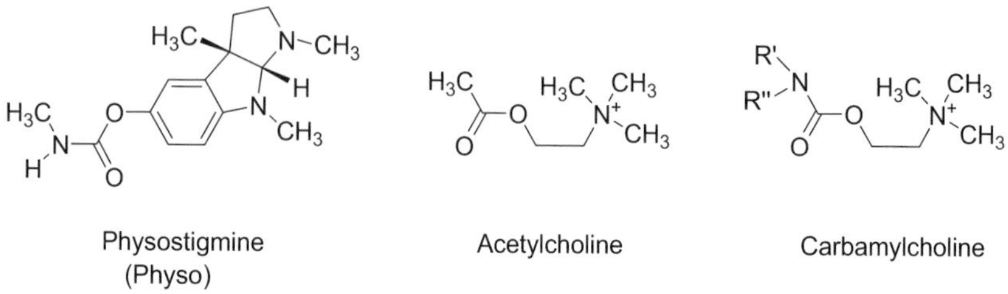

Physostigmine
(Physo)

Acetylcholine

Carbamylcholine

Fig. 12.1 The chemical structure of physostigmine, acetylcholine and carbamylcholine.

The understanding of physostigmine's pharmacology was also much improved at the time following the discovery of acetylcholine (ACh; Fig. 12.1) as the neurotransmitter of the newly defined cholinergic system, as well as the enzyme acetylcholinesterase (AChE) which proved responsible for the rapid degradation of ACh. In about 1930, Stedman discovered that physostigmine and related carbamates strongly inhibit AChE, so that the level of ACh in the cholinergic synapses is much enhanced. In 1940, a second hydrolytic enzyme for ACh was discovered in blood [4]. This proved to be somewhat less efficient in degrading ACh, but to be capable of hydrolyzing a much wider range of esters of aminoalcohols. This enzyme, which today is known as butyrylcholinesterase (BChE), is also inhibited by physostigmine.

Each of these two enzymes is transcribed from its own (single) gene, although it should be mentioned that AChE has a number of different aggregational, and even molecular, forms which, though not differing in their active sites, behave somewhat differently towards inhibitors [5]. This is probably also true for BChE.

12.2
Pharmacological Activities of Physostigmine

As described previously, the physiological activities of physostigmine and the related carbamates are mainly a reflection of the physiological activity of ACh itself. Moreover, the spectrum of physiological activities of ACh is much better demonstrated by physostigmine than by applying ACh itself, since ACh is quickly hydrolyzed in the blood. The biological activity is divided into two groups [6], which differ in the cholinergic receptor that they employ: (a) the muscarinic group (of activities and of receptors); and (b) the nicotinic group. In the peripheral nervous system (PNS), the muscarinic activities include: contraction of the smooth muscles such as the bowel, bladder and the iris; the activation of glands, which manifests as enhanced salivation, sweating and accumulation of fluid in the lungs; and slowing down of the heart rate (bradycardia). Nicotinic activity in the PNS leads to contraction of the skeletal muscles which, when in excess, can cause severe twitching that can lead further to severe weakness and paralysis. Besides these so-called peripheral activities, ACh also mediates transmissions between nerves in the central nervous system (CNS), using receptors of both muscarinic and nicotinic types. It is believed that the brain nicotinic system is mainly active in regulating the work of other neurotransmitters [7].

While normal cholinergic activity in the brain is essential for its normal functioning, strong activation of the cholinergic system in the CNS leads to loss of consciousness, to epilepsy-type convulsions, to muscular fasciculations (CNS-caused) and to down-regulation of the breathing center.

Since physostigmine causes these peripheral and central activities, and is also able to activate nicotinic receptors in the neuromuscular junction (NMJ) [8], it is no wonder that the alkaloid is relatively toxic. Indeed, the LD_{50} of physostigmine in mice is about 0.6 mg kg^{-1} [9].

Today, physostigmine is used only rarely in medicine, as a miotic, which helps to reduce intra-ocular pressure in the glaucomic eye, and as an antidote following the intake of anticholinergic drugs (e.g., atropine and scopolamine).

12.3
Chemistry and Biochemistry of Physostigmine

Of the four possible configurations of the two chiral centers at the pyrrolidine-indole interface, only that depicted in Fig. 12.1 (of cis– and S-configuration and levo rotation) is found in nature. Its enantiomer is devoid of significant AChE-inhibitory activity. As an alkaloid, it reacts with acids to produce salts, but its basicity is rather low. Assuming a pK_a value of 8.2 [8], it is only about 80% ionized in the body fluids at p. 7.4. This simplifies passage of the drug through the lipophilic body membranes. It can be assumed that the anilinic nitrogen is the weaker base, so that a far larger percentage of the protons is located on the outer nitrogen.

As in related esters, nucleophilic attack by water on the carbonyl group will lead to expulsion of the phenolic eseroline, and to the unstable carbamic acid, $CH_3NH(C=O)–OH$, which decomposes further to carbon dioxide and methylamine. The rate of such a reaction is considerably lower for carbamates than for similar esters [10], due to a lower positive charge on the carbonyl carbon in the carbamate group. This reduced activity of carbamates with water is reflected in the interaction of carbamates with AChE, as discussed in the following section.

Monomethyl carbamates hydrolyze more rapidly than dimethyl carbamates, because of the lesser steric congestion near the carbonyl. In addition, NH-containing arylcarbamates may also be hydrolyzed to phenols via an E1cB decomposition of the ionized (anionic) carbamate group [11], but at neutral pH conditions the contribution of this mechanism is probably low (Fig. 12.2).

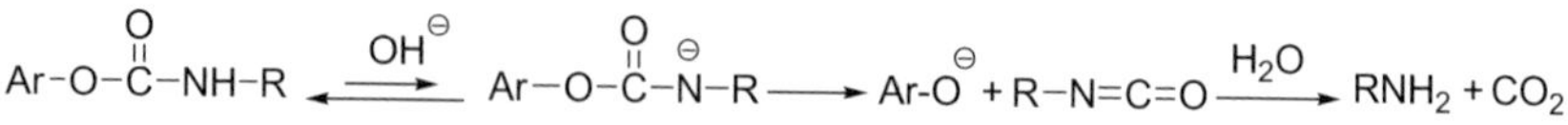

$$Ar-O-\overset{\overset{O}{\|}}{C}-NH-R \xrightleftharpoons{OH^{\ominus}} Ar-O-\overset{\overset{O}{\|}}{C}-\overset{\ominus}{N}-R \longrightarrow Ar-O^{\ominus} + R-N=C=O \xrightarrow{H_2O} RNH_2 + CO_2$$

Fig. 12.2 E1cB mechanism for hydrolysis of arylcarbamates.

12.4
Interaction of Acetylcholinesterase with Carbamates

The detailed mechanism by which AChE and BChE hydrolyze ACh has been the subject of much research, especially since the crystal structure of the *Torpedo californica* AChE was elucidated by Sussman et al. in 1991 [12]. (Reviews of these enzymes and their interactions can be found in Refs. [5,13]). This mechanism will be described here only briefly, as an introduction to the reaction of the enzyme with carbamates. The active site of AChE is located at the bottom of a 20 Å-deep gorge, where acetylcholine fits in by attachment of the quaternary ammonium group to the so-called "anionic site" (mainly through cation interaction with the π electrons of Trp84), and by dipole interactions between the ester group and Ser200 at the "esteratic site".

As shown in the upper part of Fig. 12.3, the enzymatic hydrolysis consists of two discrete reactions. The nucleophilicity of Ser200 hydroxyl is much enhanced by hydrogen bonding with His440 (and further by Glu327; not shown in the scheme), so that the hydroxyl group attacks the carbonyl of ACh to form a tetrahedral moiety. This breaks down immediately with the release of choline to form an acetylated inactive enzyme. In the second reaction, a water molecule, which is helped by acidic catalysis, attacks the acyl group, releasing the acetate and freeing the enzyme. The fact that both steps proceed very quickly is evidenced by the turnover rate of the enzyme, which is close to one million molecules per minute. In fact, AChE is considered to be one of the most efficient enzymes known.

As explained above, carbamates are expected to react with AChE at a slower pace than esters. *N*-methylcarbamylcholine (Fig. 12.1, R′ = Me, R″ = H) , with all its similarity to ACh, is hydrolyzed very slowly by the enzyme [14]. If the choline moiety is replaced by a much better leaving group X, for which the HX has a pK_a value <9, the carbamate becomes a rather good substrate for the first enzymatic reaction, leading to a carbamylated enzyme, much like the acetylated enzyme mentioned above, as in the lower part of Fig. 12.3. However, even in this case the second, deacylation step, is slow in the case of carbamylated enzyme, since the free enzyme, which must act as the leaving group at this stage, is again not good enough as a leaving group for a rapid reaction. The enzyme is therefore blocked for a considerable length of time. This rather unusual case of reversible inhibition by forming a covalent bond, which is nevertheless reversible within (usually) about an hour or two, is the basis for the roles that carbamates have as drugs.

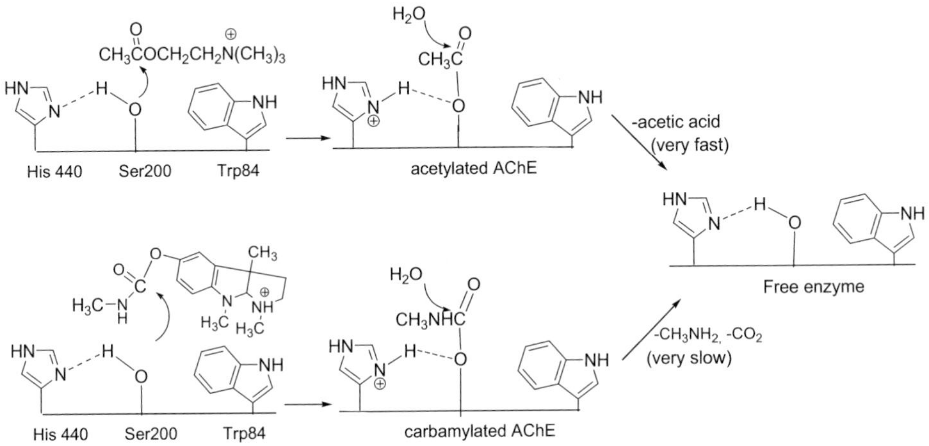

Fig. 12.3 Schematic representation of the interactions of acetylcholine esterase (AChE) with acetylcholine (ACh) (upper part) and with physostigmine (lower part). Left: AChE complexed with the substrate. Middle: acylated AChE (inactive). Deacylation by hydrolysis or otherwise leads to the regenerated active enzyme (Right).

In numerical terms, we can compare the carbamylation rates (the on-rate) and the decarbamylation rates (off-rate) as determined [15] for the reaction at 37 °C between bovine erythrocyte AChE and 10^{-3} M monomethyl or dimethyl carbamylcholine (Fig. 12.1, right, R′ = Me, R″ = H, and R′ = R″ = Me). The carbamylation rate was found to proceed at 2.3×10^{-2} s^{-1} (t½ = 30 s) and 6.2×10^{-3} s^{-1} (t½ = 112 s), respectively, while for the decarbamylation stage the rate values are 2.6×10^{-3} s^{-1} (t½ = 267 s) and 9.3×10^{-4} s^{-1} (t½ = 745 s), respectively. The on-rate of physostigmine, under the same conditions, is 55 s^{-1}, which is over 2300-fold higher than that of *N*-methyl carbamylcholine, demonstrating the effect of improving the leaving group. The off-rate is similar to that of the methyl carbamylcholine. Compilations of additional kinetic data can be found elsewhere [16,17].

The half-life for the off-rate does not in any way represent the useful *in-vivo* life of the drug, since the decarbamylated enzyme is re-carbamylated repeatedly. Soldiers take the carbamate pyridostigmine for "nerve gas" protection (see below) every 8 h – many times the molecular off-rate.

Carbamylated enzymes that carry a much longer (C=O)-N-substituent, have a considerably longer off-rate. The *N*-heptyl analogue of physostigmine (see below) has a t½ of 11 days, but this was attributed to a hydrophobic interaction between the chain and the enzyme.

12.5
SAR of the Eseroline Moiety, and the Development of Miotine

During the 1920s, as the chemical structure of physostigmine was being resolved by Stedman and coworkers, the SAR of the drug was also studied. The carbamate moiety, including its methyl group, proved essential for meaningful activity. When the methyl was replaced with an ethyl group in the benzylamine series of derivatives (see below), no miotic activity was detected. On the other hand, the pharmacological activity was preserved even when the two heterocyclic rings in the eseroline part were removed, on condition that a substituted amine is kept not far from the carbamylated phenolic ring. When evaluating the analogues for their miotic activity in the eye, these workers tried to locate the optimal position on the aromatic ring for the amino-containing group. This proved to be complicated, and after checking the o-, m-, and p-dimethylamino-substituted phenyl carbamates, and a similar series of compounds in which the dimethylamino moiety was moved to a benzylic position (Fig. 12.4), the order of activity was found to be o > p > m in both series. In the quaternary nitrogen analogues of the benzylic series, the activity hierarchy was o > m > p [14]. The preference for the *ortho*-position was later interpreted as if the *ortho*-isomer holds the optimal distance between the nitrogen cation and the carbamate C=O oxygen (about 4.7 Å according to Long, as quoted in Ref. [18]).

For closer examination, Stedman wished to determine whether different spatial conformations of the dimethylamino group on the benzylic carbon would result in different activities. Therefore, a synthesis of the α-methyl chiral analogues of

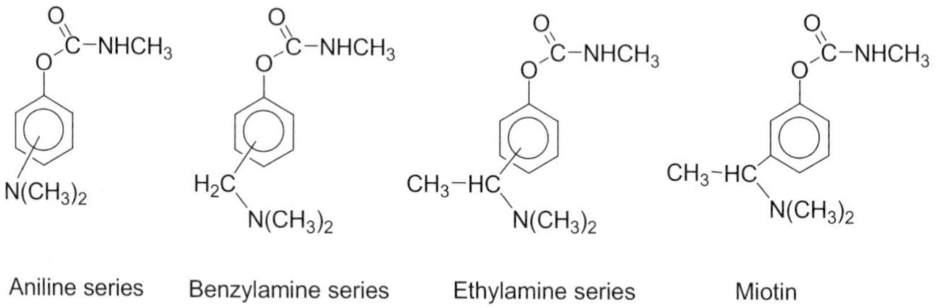

Aniline series   Benzylamine series   Ethylamine series   Miotin

Fig. 12.4 Stedman's series of physostigmine analogues, including miotin (racemic).

the benzyl series was undertaken (ethylamine series in Fig. 12.4), This new group (as racemates) revealed a different trait: the order of activity was now m >> p = o. The *meta* derivative, details of which were published in 1929 [14], proved to be the most active miotic among all Stedman's simplified physostigmine derivatives, and was appropriately termed miotin (Fig. 12.4). Its activity was described as near to that of physostigmine (see also Tab. 12.1). However, the question about the biological difference between the enantiomers could not be solved at that time because of problems encountered in resolving the racemates.

Much later, it was shown that substituting the indolic nitrogen in physostigmine with a methylene did not change the level of activity [19], thus proving clearly that there is no unique optimal distance between the positive nitrogen and the carbonyl. It seems that while the ester moiety must be located precisely near Ser200, there is more flexibility in accommodating the cationic nitrogen, and this is much influenced by the way that the phenyl ring and the other groups close by fit into the active site.

12.6
The Development of Quaternary Carbamates for Myasthenia Gravis

Despite its strong activity, miotine's resemblance to physostigmine offered only minimal advantage in its clinical use, since it also suffered from a wide range of untoward health effects. Medically improved carbamates had to wait for another couple of years, for the development of the two compounds neostigmine and pyridostigmine (Fig. 12.5), both of which were strongly active, more water-stable, and quaternary in nature.

Both neostigmine and pyridostigmine are dimethyl carbamates that differ in their phenolic moieties: in neostigmine, it is the *m-N,N,N*-trimethylammoniophenol, while in pyridostigmine the same "meta" orientation of the quaternary nitrogen is basically kept, but in a surprisingly different chemical setting. As is common with quaternary compounds, these two drugs have a limited access to the brain since their entry is blocked by the lipophilic blood–brain barrier. Glaucoma patients, and others who require a peripheral neuromuscular boost (e.g., enhancing bowel movement) are thus spared the unneeded and annoying CNS effects of the drug.

At about the same time, it was realized (this was credited mainly to the physician Mary Walker) that this family of drugs is very useful in the treatment of the crippling and paralyzing disease, myasthenia gravis. In this condition, an autoimmune antibody against a component of the cholinergic-nicotinic receptor blocks cholinergic transmission in the NMJ, thereby diminishing skeletal muscle activity. Since the nicotinic receptors of the brain differ from those in the NMJ, this disease of the nervous system is confined to the periphery. Enhancing the level of ACh in the NMJ by blocking AChE affords the patient significant symptomatic relief, and with time pyridostigmine has become established as the mainstay treatment for multiple sclerosis. The daily dose required can exceed 1 g, but this quantity can be tolerated by patients only because of the drug's low penetration into the CNS.

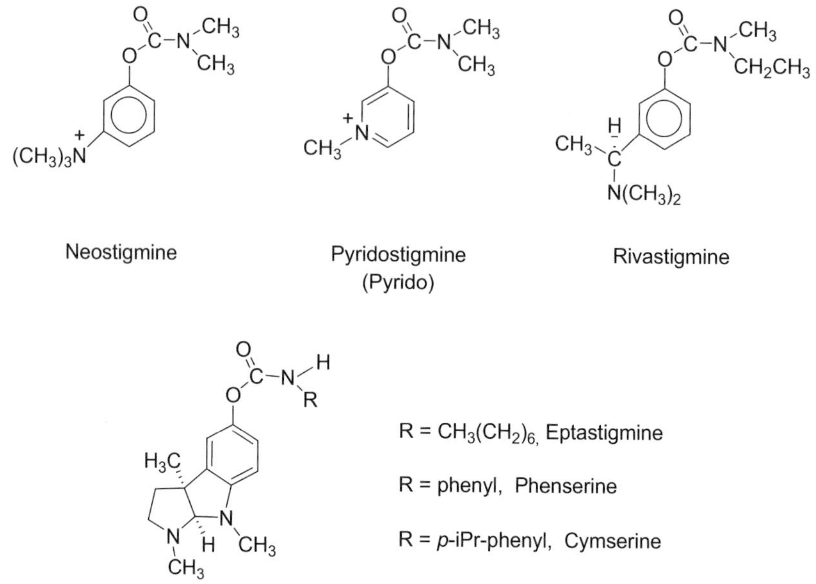

Neostigmine   Pyridostigmine   Rivastigmine
                        (Pyrido)

R = CH₃(CH₂)₆, Eptastigmine

R = phenyl, Phenserine

R = p-iPr-phenyl, Cymserine

Fig. 12.5 Carbamates which are currently in clinical use, or which have been suggested for such use.

12.7
Carbamates as Pre-Exposure Treatment against Organophosphate Intoxication

Organophosphate compounds that block AChE are highly toxic, and are used in agriculture as insecticides, as well as being a constant threat as chemical warfare agents (CWA), commonly known as "nerve gases". The biological and enzyme inhibition activities of the organophosphates have been the subject of very detailed investigations that cannot be summarized here, except in a very general terms (for reviews, see Refs. [20,21]). In short, their chemical structure are in most cases as in Fig. 12.6(a) for agricultural insecticides, and as in Fig. 12.6(b) for CWA.

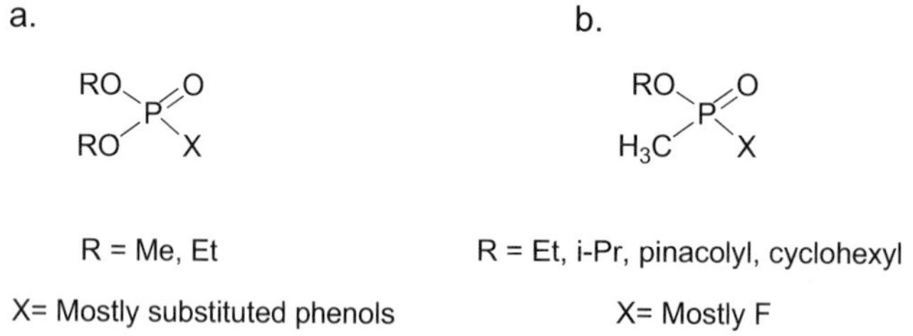

a.   b.

R = Me, Et   R = Et, i-Pr, pinacolyl, cyclohexyl

X= Mostly substituted phenols   X= Mostly F

Fig. 12.6 Basic structures of organophosphate AChE inhibitors. X represents the leaving group. (a) Typical agricultural insecticides; (b) typical chemical warfare agents.

In Fig. 12.6, the X represents the leaving group which is expelled when an organophosphate reacts with Ser200 of AChE to form a phosphorylated, inhibited, enzyme. Thus, this series of events is remarkably similar to what occurs in the reaction of AChE with carbamates (as depicted in Fig. 12.3). The most important difference is that the rate of release of the phosphate moiety from the enzyme is so slow that phosphorylation is practically irreversible. It should be mentioned, however, that during the first few hours the organophosphate-blocked enzymes can in most cases be de-phosphorylated and reactivated to a significant degree with the help of an antidotal class of drugs known as "oximes", such as pralidoxime and toxogonin [22].

Since it is impossible to measure the inhibition inside the synapses, it is customary in cases of carbamate and organophosphate intoxication to measure the residual enzymatic activity in the blood of the victim and to calculate the percentage of inhibition. Normally, light symptoms appear not before 30% of inhibition, becoming an imminent life-threatening condition when this figure is above the 90–95% level.

When the issue of organophosphates as CWA was declassified after World War II, a surprising finding was published by the British [23]: giving a modest dose (that causes 25–30% inhibition) of physostigmine, or of some other carbamates, before intoxication with an organophosphate, not only fails to aggravate the clinical symptoms of the organophosphate intoxication but also actually increases the survival rate of the (fully treated) victims. Whilst it is not yet clear why there is no amplification of intoxication at the height of the exposure, when the presence of the carbamylated enzyme must reduce the level of the remaining active enzyme, the smoother recovery can be adequately explained by the continuous supply of a small but a steady stream of free enzyme (supplied by the decarbamylation step) throughout the critical period of inhibition.

The interest in this clinical use of carbamates was minimal during the 1950s and 1960s, when much effort was directed to developing better oximes. However, with the realization that the inadequacies in the use of oximes (e.g., no oxime is active against all organophosphates) could not be rectified, the interest in pretreatment with carbamate resurfaced during the early 1970s [24], and pyridostigmine use has since been adopted by armies almost worldwide, in the form of 3×30 mg tablets per day. The treatment has usually been considered to be safe, except for minor inconveniences, mainly in the gastrointestinal tract. However, the large-scale use of pyridostigmine during the Gulf War raised the suspicion that a minor proportion of the population reacts to pyridostigmine by developing (with time) muscle weakness and fatigue. This became part of the "Gulf War Syndrome" that, despite much research, remains ambiguous, with many questions still to be answered [25].

It can be seen that one of the quaternary carbamates has been chosen for this use, mainly because they are safer and also have fewer untoward effects, and since in the early years organophosphate intoxication was considered to be mainly a peripheral breathing problem (paralysis of the diaphragm muscles and edema of the lungs). The realization during the 1980s and 1990s that inhibition of the brain

AChE significantly contributes to morbidity and mortality, led to a number of trials which clearly showed that physostigmine, either alone or in combinations, is indeed superior to pyridostigmine as a pretreatment agent [26,27]. However, since by definition pretreatment is given to a healthy person not yet exposed to a CWA, and who might never be exposed, it seems that the highly (and centrally) active physostigmine is not going to be used commonly as pretreatment. In this respect, other more moderate carbamates with central activity are more of a possibility.

12.8
Carbamates as Insecticides

It is worth noting that carbamates serve in agriculture not only as insecticides but also as fungicides and herbicides [28]. The noninsecticidal compounds do not inhibit AChE, and consequently will not be discussed at this point.

Strangely, the idea of using carbamates as insecticides arose neither from the high toxicity of physostigmine, nor from its already known inhibitory activity on AChE. Instead, some cycloaliphatic carbamates were developed during 1950s as insect-repellents, and it was through their investigation that the insecticidal activity of phenolic (and some other) carbamates surfaced [1], and eventually developed into a major market. Typically, the toxicity is based on inhibition of AChE, and since this enzyme is rather similar in insects and mammals, it has not been possible to construct a molecule that will be insecticidal and at the same time totally safe for mammals. Many carbamates have indeed a relatively moderate toxicity to mammals

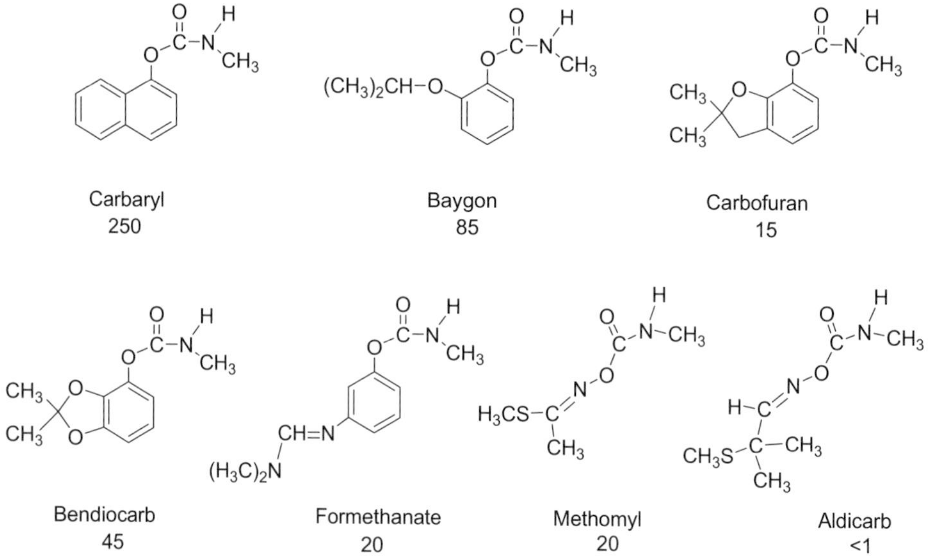

Fig. 12.7 Insecticidal carbamates. The numbers below each structure indicate approximate oral LD_{50} values (mg kg^{-1}) in mice or rats. Values taken from Ref. [28].

(oral LD_{50} values in the range 100–300 mg kg^{-1}), notably carbaryl. However, some agents (e.g., aldicarb) are also highly toxic to mammals, as will be discussed below. The structures of some representative insecticidal carbamates are shown in Fig. 12.7.

12.8.1
Structural Features

With the aim of increasing biological permeability into the insects, it is only natural that this group of carbamates does not consist of quaternary salts. What is slightly more surprising is that in most cases the amino group, which is present in physostigmine and is so common in the therapeutic carbamates, is missing – perhaps to reduce the mammalian toxicity. Usually, the leaving groups are substituted phenols (1-naphthol in carbaryl). In many of these insecticides the benzene is further oxidized, and in some the other oxygen forms part of a benzofuran (and related) rings (this may be an historic relic from the early insecticidal carbamates, which were based on resorcinol). Another type of leaving group is found in carbamylated oximes such as oxamyl and aldicarb. Here, the rather acidic nature of oximes (pK_a 8–9) is used to construct a nonaromatic good leaving group, which is more in the shape of choline. This leads to a higher affinity to the enzyme and to a higher toxicity. When the oxime moiety is shaped closely along the structure of choline, the mammalian toxicity is very high, culminating in the oral LD_{50} of <1 mg kg^{-1} for aldicarb. An amino group at the *meta*-position of the aromatic ring also heightens the affinity and toxicity: the oral mammalian toxicity (LD_{50}) for formetanate is about 20 mg kg^{-1}.

Carbamate intoxication is treated with atropine (or scopolamine). Despite the fact that oximes, such as pralidoxime or toxogonin, given after intoxication by most carbamates, quicken the decarbamylation step and reactivation of the enzyme, the medical practice is to avoid the use of oximes. Two reasons are commonly given for this:

- It has been shown that people and animals poisoned by carbaryl are adversely affected by the treatment. Since in many cases of carbamate exposure the physician cannot rule out the possibility that carbaryl is involved, it is better to avoid the use of oximes.
- Even without the oxime, the enzymatic activity will recover within a few hours. As discussed above, the second argument is not always valid, but it is also not proven that in those cases the oximes will be of much assistance.

12.9
Carbamates in the Treatment of Alzheimer's Disease

Alzheimer's disease (AD) is one of a series of neurodegenerative disorders that can affect people, mainly during the later part of their lives. A basic feature of these disorders is the heavy loss of neurons from specific areas of the brain. In

AD, it is the hippocampus and some cortex areas that suffer the main loss; this results in memory impairment that proceeds during the course of a few years to cognitive deficiency and to dementia. The same areas are characterized also by an abundance of intercellular β-amyloid plaques, which are known to be toxic to neurons. It has been shown that cholinergic neurons – especially in the forebrain – are more affected, giving rise to the notion that cholinergic underactivity is the most important immediate cause of the mental deficit. This, together with the knowledge that cholinergic agonists boost mental activity, led in the early 1980s to an experimental use of physostigmine in AD patients. The encouraging results opened the door to a wave of experiments with AChE-blockers and with centrally acting cholinergics, and to a new era in the treatment of AD.

There are three limitations to this line of thought:

1. Many AD patients have damage to other neurotransmitter pathways as much as to their cholinergic pathways; thus, they would not be helped much by manipulating only ACh levels in their brains.
2. The unknown underlying causes of the disease are not changed, so progression of the illness is expected to continue.
3. Advanced AD patients, in whom too few cholinergic neurons remain, are not expected to be significantly helped by blocking brain AChE.

With the experience gathered, we now know that whilst the third limitation is correct, the situation with the first two is less clear. The drugs might also, indirectly, favorably affect the levels of other transmitters (as well as in other possible ways, such as slowing down amyloid plaque accumulation). There is also evidence that in those patients who respond symptomatically, the rate of decline seems to be reduced [29,30].

In the following sections we will discuss the development of carbamates as AD drugs. A full list of these can be found elsewhere [31], but in the following section we will describe only some of the close derivatives of physostigmine, with special focus on rivastigmine.

12.9.2
Close Derivatives of Physostigmine

Contrary to the need for periphery-selective drugs that avoid the brain for the treatment of myasthenia gravis, AD treatment demands carbamates which easily penetrate the blood–brain barrier. Moreover, since any activity in the periphery will cause undesirable "side effects", it is preferable to develop and use a CNS-selective drug. Since physostigmine had already been shown to be of value in AD patients, it is only natural that the new carbamates were modeled on its structure.

Some effort was directed toward developing long-acting physostigmine preparations, and the transdermal delivery patches were especially promising [32]. The idea was to secure a slow but continuous supply of this otherwise toxic agent. However, physostigmine appears to have too strong central and peripheral activities, with too low a therapeutic index, to be regarded as a safe drug for AD.

Eptastigmine (also called heptylphysostigmine; see Fig. 12.5), a close analogue of physostigmine, was developed by the Italian company Mediolanum Farmaceutici and almost made it to market. Its chemical structure differs from that of physostigmine only in the replacement of the carbamatic CH_3NH with n-$C_7H_{15}NH$. The resulting compound has certain advantages over the original alkaloid. Being a strongly lipophilic compound, it is not only capable of good penetration of the blood–brain barrier, but may also accumulate to some extent in the lipophilic brain. Moreover, unlike the very short life of the physostigmine-blocked AChE, which decarbamylates back to the active enzyme in about 1 h, the eptastigmine-blocked enzyme decarbamylates very slowly, with a $t_{1/2}$ of at least 11 days [33]. In addition, it is an order of magnitude lower in toxicity and in activity compared to physostigmine.

Clinical trials with eptastigmine were successful up to a 6-month Phase III study, and to experimental administration of the drug for two years. Unfortunately, among more than 1500 patients who took the drug a few serious cases of aplastic anemia caused withdrawal of eptastigmine from clinical use [34]. While there is no known physiological connection between the carbamylated AChE and bone marrow dysfunction, and aplastic anemia has never been found in users of physostigmine, pyridostigmine, or rivastigmine, it is interesting to note that cases of leukemia arising from exposure to agricultural AChE organophosphate inhibitors have been reported, and amplification of the AChE gene has been identified in leukemic cells [35]. Other compounds in this series, some of which contain weakly basic nitrogens in the carbamic N-substituents (this is said to enhance affinity to the brain), are currently under investigation [34].

Beginning during the 1990s, a related group of carbamates has been investigated by scientists at the National Institutes of Health and the University of North Carolina. In this series, the carbamoyl N-CH_3 group is replaced by a substituted aniline [36]. The main interest in these compounds stems from the fact that some of them (e.g., cymserine; see Fig. 12.5) are strong and selective BChE inhibitors, some (e.g., phenserine) are selective AChE inhibitors, while others are nearly equipotent. Recently, it has been reported that so far phenserine has failed to be better than placebo in a Phase III clinical trial. These compounds are valuable for investigating a still-disputed hypothesis that the selective inhibition of BChE in the ailing AD brain will result in a better AD treatment with fewer (peripheral) side effects, since BChE becomes progressively the dominant cholinesterase in the brain of AD patients.

12.9.3
Rivastigmine

In rivastigmine (see Fig. 12.5), which was modeled in miotin's skeleton, the tricyclic eseroline of physostigmine is simplified into a benzylamine system. As detailed above, Stedman found that the m-Me_2N-$CH(CH_3)$-aryl derivative (miotin;

see Fig. 12.4) was the most active of his tertiary phenylcarbamate series, with activity close to that of physostigmine. Stedman did not investigate modifications in the carbamoyl N-substituent, after being disappointed with the N-Et derivative. During the early 1980s, Weinstock and colleagues at the Hebrew University in Jerusalem re-evaluated this type of carbamate, but this time as candidate AD drugs [37]. The pathway to a longer-acting and more centrally selective compound seemed to be through lengthening or branching of the carbamoyl N-substituent(s), so that the blocked enzyme will not be hydrolyzed quickly, and the molecule as a whole will become more lipophilic. At that stage, the phenolic leaving group of miotine was not modified, as it had already been optimized by Stedman, so that changes were likely to reduce AChE blocking activity. As will be described below, recent modifications of rivastigmine include changes in the phenolic side of the molecule, with the aim of widening the range of their pharmacological activity.

Tab 12.1 Toxicity values in mice, and *in-vitro* and *in-vivo* values for inhibition of mouse brain AChE, for some rivastigmine analogues [37].

		ED_{50} [a] $[\mu mol\ kg^{-1}]$	IC_{50} [b] [M]	LD_{50} [c] $[\mu mol\ kg^{-1}]$	Inhibition by ED_{50} dose of AChE activity [d] [%]
	R', R''				
1	Physostigmine	0.91	1.1×10^{-8}	3.0	Approaching 0
2	H, Me miotin	0.92	1.3×10^{-8}	4.50	Approaching 0
3	Me, Me	1.14	2.7×10^{-8}	12.4	Approaching 0
4	H, Et	8.50	4.0×10^{-7}	95.7	47
5	Et,Et	56.8	3.5×10^{-5}	>560	31
6	Me, Et rivastigmine	4.20	3.0×10^{-6}	46.0	41
7	H, *n*-Pr	2.80	1.1×10^{-7}	30.5	26
8	H, cyclohexyl	7.24	9.3×10^{-8}	41.5	

a. Subcutaneous dose for *in vivo* inhibition of AChE activity in mouse brain, measured *in vitro*.
b. *In-vitro* inhibition of AChE activity, in mouse brain homogenate.
c. Subcutaneous in mice.
d. In mouse brain, measured 3 h after s.c. administration.

The inhibitory activity of the some representative compounds toward mouse brain AChE is shown in Tab. 12.1. From the data in the table, it can be deduced that the monomethyl and dimethyl derivatives (Entries 2 and 3) are the most active. The monoethyl analogue (Entry 4) has about 1% of the monomethyl activity, which might explain Stedman's inability to detect miosis when using its dilute solutions. Lengthening the second substituent from methyl to *n*-hexyl gradually restores the activity to the dimethyl level [38,39]. From the SAR performed with the eptastigmine series, it is probable that further lengthening will decrease activity. It is clear that there is no need for long hydrophobic chains for a useful CNS activity, though this is definitely an option.

Based on the *in-vitro* data in the table, the *N*-methyl-*N*-ethyl compound (Entry 6) does not seem to be the best candidate. In fact, it seems to be positioned on the line between the *N,N*-dimethyl and the barely active *N,N*-diethyl compounds. It was the *in-vivo* activities, including the better oral absorption and the much longer duration of action, that first placed it at the top of the group. The enantiomeric mixture, that first attracted Stedman to miotine, was resolved for the methylethyl compound, and the more active enantiomer, *S*, was developed by Novartis into the AD drug rivastigmine (Exelon®) [40].

Beside the better oral absorption, rivastigmine has some features that probably contributed to its clinical outcome. The CNS activity of the drug is much more prominent than in the periphery, and this probably accounts for its remarkably lower toxicity. Also, in the brain itself the drug has been shown to be rather selective for the cerebral cortex and hippocampus regions, which are the areas most affected by AD [39,40].

Rivastigmine is equipotent against both AChE and BChE, but is four to six times more potent against a globular monomeric form of AChE known as G1, than against another, tetrameric, aggregation form of the enzyme, termed G4 (physostigmine has the same potency towards both forms). In AD patients, the G4 level is progressively reduced in the brain with the advancement of the disease, while the G1 gains prominence among the declining AChE forms. Clearly, selectivity towards this specific form can lead to a more focused treatment with fewer side effects [40].

The LD_{50} value of rivastigmine can be raised by the muscarinic blocker atropine almost 11-fold, while the value for the other carbamates in this series is usually about 3 (except for 5.8 for the dimethyl derivative). The quaternary drug atropine methylnitrate, which cannot penetrate the brain, raises the value about two-fold in all cases. This not only adds a safety feature to the use of rivastigmine, but also indicates that rivastigmine's activity – even at high and toxic doses – is more focused than the others in central mechanisms that either directly or indirectly stem from activation of muscarinic receptors. This might contribute to the clinical results if it is assumed that those unexplored nonmuscarinic effects are mostly detrimental to the AD treatment.

Whether these explanations are the full reason for the pharmacological advantage of rivastigmine in this series, or there is more to be discovered, rivastigmine can serve as an example to the notion that because of the complexity of biological systems *in vitro* and even *in vivo*, gross biochemical inhibition models cannot fully predict the outcome of clinical-type testing in whole animals and eventually in humans.

As explained above, N,N-dimethyl carbamates keep the mammalian enzyme blocked for about 25–30 min ($t_{1/2}$). Lengthening the N-alkyl to N-n-heptyl (in eptastigmine) increased this $t_{1/2}$ value to 11 days. With rivastigmine, the $t_{1/2}$ for the inhibition of human brain AChE has been found to be about 10 h [41], which is seemingly a reasonable time for its N-methyl N-ethyl substituents. That this is not the case has been revealed recently by Silman's group [42], who found that BChE and recombinant human AChE, carbamylated by MeEtN-(C=O)-Cl, recover their activity at 25 °C with half-lives of just 34 and 4 min, respectively. The same enzymes were inhibited by rivastigmine decarbamylate at 37 °C with $t_{1/2}$ values of about 24 and 44 h, respectively. The reason for the very slow decarbamylation must clearly lie with the influence of the phenolic leaving group of the drug (which in itself has a negligible inhibitory activity). This indeed was found to be the case. X-ray crystallography of the inhibited torpedo enzyme unexpectedly revealed that the enzyme "traps" the free phenolic compound inside its active site. This disturbs the catalytic machinery, and thus slows down the pathway to decarbamylation, contributing greatly to the clinical usefulness of the drug. It is not known if other long-acting members of the series behave in a similar manner.

Among the several modifications that were tried in the rivastigmine structure, one group of compounds deserves special note [43,44].

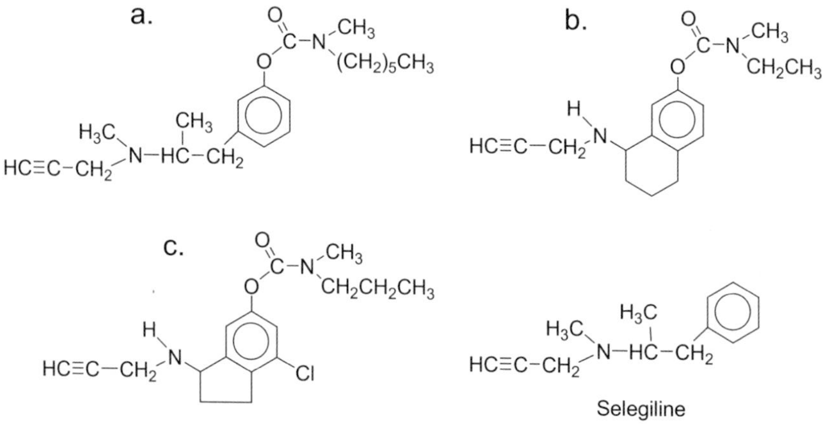

Fig. 12.8 Carbamates (a), (b), and (c) are related to rivastigmine and to selegiline; these are both AChE and MAO-B inhibitors.

Molecules a, b, c, in Fig. 12.8 manifest the same structure of the carbamic moiety as rivastigmine, but their phenolic part is modified, with regard to both the distance of the amine from the aromatic ring, and the exact N-substituents. This led to compounds that, beside their central AChE-blocking activity remindful of rivastigmine (albeit at a higher dose), also possess a new pharmacological feature.

The *free* phenols that are produced in the brain (when they break off during the reaction with AChE), are related to the drug selegiline (Fig. 12.8), which is a monoamine oxidase B (MAO-B) inhibitor, so that they also exhibit such inhibitory activity. MAO-B inhibitors are helpful in Parkinson's disease, mainly because they cause an increase in the level of dopamine, and are also antidepressants. These compounds are currently under investigation as treatment for Alzheimer's disease complicated by other cognitive deficits.

References

1 Holmstedt B. Cholinesterase inhibitors: an introduction. In: *Cholinesterase and cholinesterase inhibitors*. Giacobini E (Ed.). Martin Dunitz, London, **2000**, pp. 1–8.

2 Stedman E, Banger G. Physostigmine (eserine). *J. Chem. Soc.*, **1925**, 127, 247–258.

3 Petcher TJ, Pauling P. Cholinesterase inhibitors: structure of eserine. *Nature*, **1973**, 241(5387), 277.

4 Alles GA, Hawes RC. Cholinesterases in the blood of man. *J. Biol. Chem.*, **1940**, 133, 374–390.

5 Soreq H, Zakut H. *Human Cholinesterase and Anticholinesterase*. Academic Press, New York, **1993**.

6 Lefkowitz RJ, Hoffman BB, Taylor P. Neurotransmission. In: *Goodman and Gilman's the Pharmacological Basis of Therapeutics*. Hardman JGL (Ed.). McGraw-Hill, New York, **1996**.

7 Dajas-Bailador F, Wonnacott S. Nicotinic acetylcholine receptors and the regulation of neuronal signalling. *Trends Pharmacol. Sci.*, **2004**, 25(6), 317–324.

8 Pagala MK, Sandow A. Physostigmine-induced contractures in frog skeletal muscle. *Pfluger's Arch.*, **1976**, 1976(363), 223–229.

9 Sofia RD, Knobloch LC. Influence of acute pretreatment with delta 9-tetrahydrocannabinol on the LD_{50} of various substances that alter neurohumoral transmission. *Toxicol. Appl. Pharmacol.*, **1974**, 28(2), 227–234.

10 Hegarty AG. Derivatives of carbon dioxide. In: Comprehensive Organic Chemistry. Ollis D (Ed.). Pergamon Press, Oxford, **1979**, pp. 1067–1099.

11 Hegarty AG, Frost LN. Isocyanate intermediate in E1Cb mechanism of carbamate hydrolysis. *J. Chem. Soc. Chem. Commun.*, **1972**, pp. 500–501.

12 Sussman JL, et al. Atomic structure of acetylcholinesterase from *Torpedo californica*: a prototypic acetylcholine-binding protein. *Science*, **1991**, 253(5022), 872–879.

13 Silman I, Sussman JL. Cholinesterase inhibitors: an introduction. In: Cholinesterase and cholinesterase inhibitors. Giacobini E (Ed.). Martin Dunitz, London, **2000**, pp. 9–25.

14 Stedman E. Studies on the relationship between chemical constitution and physiological action. *Biochem. J.*, **1929**, 23, 17–24.

15 Watts P, Wilkinson RG. The interaction of carbamates with acetylcholinesterase. *Biochem. Pharmacol.*, **1977**, 26(8), 757–761.

16 Reiner E. Spontaneous reactivation of phosphorylated and carbamylated cholinesterases. *Bull. World Health Org.*, **1971**, 44(1), 109–112.

17 Wetherell JR, French MC. A comparison of the decarbamoylation rates of physostigmine-inhibited plasma and red cell cholinesterases of man with other species. *Biochem. Pharmacol.*, **1991**, 42(3), 515–520.

18 Cannon J. Anticholinergic drugs. In: *Burger's Medicinal Chemistry and Drug Discovery.* Wolff M. (Ed.), **1990**, p. 47.

19 Chen YL, et al. Syntheses, resolution, and structure–activity relationships of potent acetylcholinesterase inhibitors: 8-carbaphysostigmine analogues. *J. Med. Chem.*, **1992**, 35(8), 1429–1434.

20 Ballantyne B, Marrs TC. *Clinical and Experimental Toxicology of Organophosphates and Carbamates*. Butterworth-Heinemann Ltd., Oxford, **1992**.

21 Wiener SW, Hoffman RS. Nerve agents: a comprehensive review. *J. Intensive Care Med.*, **2004**, 19(1), 22–37.

22 Worek F, et al. Reappraisal of indications and limitations of oxime therapy in organophosphate poisoning. *Hum. Exp. Toxicol.*, **1997**, 16(8), 466–472.

23 Koster R. Synergism and antagonism between physostigmine and diisopropyl fluorophosphate in cats. *J. Pharmacol. Exp. Ther.*, **1946**, 88, 39–46.

24 Berry WK, Davies DR, Gordon JJ. Protection of animals against soman (1,2,2-trimethylpropyl methylphosphono-fluoridate) by pretreatment with some other organophosphorus compounds, followed by oxime and atropine. *Biochem. Pharmacol.*, **1971**, 20(1), 125–134.

25 Golomb BA. *Pyridostigmine bromide*. Titles in Gulf War Illness series. Vol. 2., Rand, Santa Monica, **1999**.

26 Jenner J, Saleem A, Swanston D. Transdermal delivery of physostigmine. A pretreatment against organophosphate poisoning. *J. Pharm. Pharmacol.*, **1995**, 47(3), 206–212.

27 Meshulam Y, et al. Prophylaxis against organophosphate poisoning by sustained release of scopolamine and physostigmine. *J. Appl. Toxicol.*, **2001**, 21 (Suppl. 1), S75–S78.

28 The Pesticide Manual. 10th edn. Tomlin C (Ed.). Crop Protection Publication, Farnham, **1994**.

29 Giacobini E. Do cholinesterase inhibitors have disease-modifying effects in Alzheimer's disease? *CNS Drugs*, **2001**, 15(2), 85–91.

30 Svensson A-L, Giacobini E. Cholinesterase inhibitors do more than inhibit cholinesterase. In: *Cholinesterase and Cholinesterase Inhibitors*. Giacobini E (Ed.). Martin Dunitz Ltd., London, **2000**, pp. 227–236.

31 Camps P, Munoz-Torrero D. Cholinergic drugs in pharmacotherapy of Alzheimer's disease. *Mini Rev. Med. Chem.*, **2002**, 2(1), 11–25.

32 Levy A, et al. Transdermal physostigmine in the treatment of Alzheimer's disease. *Alzheimer Dis. Assoc. Disord.*, **1994**, 8(1), 15–21.

33 Perola E, et al. Long-chain analogs of physostigmine as potential drugs for Alzheimer's disease: new insights into the mechanism of action in the inhibition of acetylcholinesterase. *Biochim. Biophys. Acta*, **1997**, 1343(1), 41–50.

34 Brufani M, Filocamo L. Rational design of cholinesterase inhibitors. In: *Cholinesterase and Cholinesterase Inhibitors*. Giacobini E (Ed.). Martin Dunitz Ltd., London, **2000**, pp. 27–46.

35 Lapidot-Lifson Y, et al. Coamplification of human acetylcholinesterase and butyrylcholinesterase genes in blood cells: correlation with various leukemias and abnormal megakaryocytopoiesis. *Proc. Natl. Acad. Sci. USA*, **1989**, 86(12), 4715–4719.

36 Yu Q, et al. Synthesis of novel phenserine-based-selective inhibitors of butyrylcholinesterase for Alzheimer's disease. *J. Med. Chem.*, **1999**, 42(10), 1855–1861.

37 Weinstock M, et al. Pharmacological activity of novel acetylcholinesterase agents of potential use in the treatment of Alzheimer's disease. In: *Advances in Behavioral Biology*. Lachman C (Ed.). Plenum Press, New York, **1986**, pp. 539–551.

38 Weinstock M, et al. Acetylcholinesterase inhibition by novel carbamates: a kinetic and nuclear magnetic resonance study. In: *Multidisciplinary Approaches to Cholinesterase Functions*. Velan B (Ed.). Plenum Press, New York, **1992**, pp. 251–259.

39 Weinstock M, et al. Pharmacological evaluation of phenyl-carbamates as CNS-selective acetylcholinesterase inhibitors. *J. Neural Transm. Suppl.*, **1994**, 43, 219–225.

40 Enz A, Chappius A, Probst A. Different influence of inhibitors of acetylcholinesterase molecular forms G1 and G4 isolated from Alzheimer's disease and control brains. In: *Multidisciplinary Approaches to Cholinesterase Functions*. Velan B (Ed.). Plenum Press, New York, **1992**, 243–249.

41 Jann MW, Shirley KL, Small GW. Clinical pharmacokinetics and pharmacodynamics of cholinesterase inhibitors. *Clin. Pharmacokinet.*, **2002**, 41(10), 719–739.

42 Bar-On P, et al. Kinetic and structural studies on the interaction of cholinesterases with the anti-Alzheimer drug rivastigmine. *Biochemistry*, **2002**, 41(11), 3555–3564.

43 Sterling J, et al. Novel dual inhibitors of AChE and MAO derived from hydroxy aminoindan and phenethylamine as potential treatment for Alzheimer's disease. *J. Med. Chem.*, **2002**, 45(24), 5260–5279.

44 Weinstock M, et al. A novel cholinesterase and brain-selective monoamine oxidase inhibitor for the treatment of dementia comorbid with depression and Parkinson's disease. *Prog. Neuropsychopharmacol. Biol. Psychiatry*, **2003**, 27(4), 555–561.

13
Structural Analogues of Clozapine

Béla Kiss and István Bitter

13.1
Introduction

Schizophrenia is one of most serious, debilitating mental illness as its strikes at some of the most advanced functions of the human brain. It affects about 1% of the population, regardless of economical and cultural differences. Psychotic symptoms usually begin in late adolescence or early adulthood. Because of its chronic nature, severity and early onset, schizophrenia is an expensive illness, especially in industrialized nations where the symptoms are incompatible with the highly structured nature of the workplace and where social expectations are high.

The introduction of the phenothiazine derivative, chlorpromazine, has started a dramatic improvement in the clinical treatment of schizophrenia. During the past four decades, beside phenothiazines, various antipsychotics having different chemical structures have been identified and introduced into clinical practice (e.g., butyrophenones and benzamides). These drugs ("typical antipsychotics") decreased the duration of hospitalizations and, with maintenance treatment, reduced the risk of relapse and re-hospitalization. However, they had significant adverse side effects such as tardive dyskinesia, orthostatic hypotension, prolactin increase, and QT prolongation.

The synthesis and discovery of the antipsychotic effects of the piperazinyl-dibenzoazepine, clozapine (Fig. 13.1) and its launch in 1972 was an important turning point in the drug treatment of schizophrenia [1–3]. Clozapine was called an "atypical antipsychotic" as it did not produce side effects characteristic for compounds of the chlorpromazine- or haloperidol-type (i.e., extrapyramidal symptoms) either in animal models or in the clinic. Its use, however, became very limited when it was recognized that clozapine might cause a severe, and sometimes fatal, form of agranulocytosis.

In 1989, Meltzer and coworkers, when comparing the pK_i values of several antipsychotics for dopamine D_1, D_2 and 5-HT$_{2A}$ receptors, came to the conclusion that the affinity of an antipsychotic compound to D_2 and 5-HT$_{2A}$ receptors is an important factor for the atypicality [4]. It was observed that some antipsychotics possess significantly lower affinity to dopamine D_2 than to serotonin 5-HT$_{2A}$ receptors,

Analogue-based Drug Discovery. IUPAC, János Fischer, and C. Robin Ganellin (Eds.)
Copyright © 2006 WILEY-VCH Verlag GmbH & Co. KGaA, Weinheim
ISBN: 3-527-31257-9

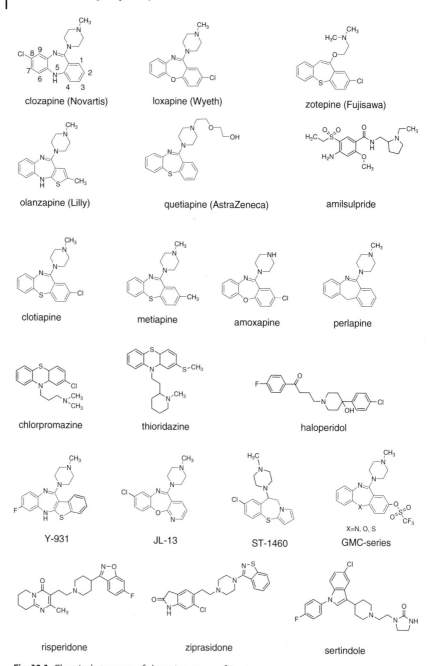

Fig. 13.1 Chemical structure of clozapine, some of its structural analogues, typical (chlorpromazine, haloperidol) and atypical antipsychotics (risperidone, ziprasidone, sertindole and amisulpride) unrelated to clozapine.

and that compounds having a pK_i ratio for 5-HT$_{2A}$ to D$_2$ (i.e., 5-HT$_{2A}$/D$_2$ affinity) higher than 1.2 fell into the atypical category. This recognition led to the "5-HT$_{2A}$/D$_2$ hypothesis" which provided a relatively "simple" receptor-based approach for antipsychotic drug research.

These important points – that is, the need to overcome the severe side effects of clozapine and the "D$_2$/5-HT$_{2A}$ hypothesis" (i.e., maintaining the atypicality) – became the major elements in the screening programs initiated by pharmaceutical companies during the 1980s and 1990s. These efforts resulted in several D$_2$/5-HT$_{2A}$-acting "atypical antipsychotics". Some of these, such as the thienobenzodiazepine, olanzapine [5] and the dibenzothiazepine, quetiapine [6], are structural and pharmacological analogues of clozapine that are the results of the D$_2$/5-HT$_{2A}$ hypothesis. Other atypicals discovered during this period, such as risperidone [7], ziprasidone [8], and sertindole [9] (Fig. 13.1) came from different chemical approaches but stemmed also from the "D$_2$/5-HT$_{2A}$ hypothesis". These compounds were introduced into the treatment of schizophrenia (sertindole was later withdrawn due to cardiovascular side-effect liability) and represented significant progress in its therapy; today, they dominate the lucrative schizophrenia therapy market.

13.2
Clozapine: The Prototype "Atypical" Antipsychotic; Some Chemical Aspects

Clozapine, the prototype atypical antipsychotic was synthesized in 1967 by Sandoz-Wander chemists (see Fig. 13.1). During the early stages of the company's chemistry program, conscious efforts were made to understand structure–activity relationships in this series of piperazinyl-dibenzodiazepines. This paved the way to the discovery of several pharmacologically active compounds. The replacement of one diazepine nitrogen atom with sulfur or oxygen, for example, resulted in thiazepines (clotiapine, metiapine, which possess typical antipsychotic actions) and dibenzooxazepines (e.g., loxapine, with typical antipsychotic and amoxapine with antidepressant activity), respectively. Omitting one diazepine nitrogen atom yielded morphanthridines (e.g., perlapine, which is a hypnotic). Substitutions on the benzene rings at positions 2 and 8 greatly modified the *in-vivo* pharmacological activity (cataleptogenic activity and apomorphine antagonism in rat). An important finding was that only derivatives having an intact di-basic piperazinyl moiety were active pharmacologically [1]. The most important discovery was, however, that the 8-Cl-dibenzodiazepine derivative (i.e., clozapine) had outstandingly interesting pharmacological features. In contrast to neuroleptics of the phenothiazine (e.g., chlorpromazine) and butyrophenone classes (e.g., haloperidol), it did not cause catalepsy, did not antagonize apomorphine's action, and elevated the dopamine metabolite dihydroxyphenylacetic acid content only at high doses [10]. Moreover, it proved to be a clinically active antipsychotic with a profile that differed greatly from that of compounds of the butyrophenone- (i.e., haloperidol) or phenothiazine-type (e.g., chlorpromazine). Interestingly, the 2-Cl-isomer of clozapine displayed classical neuroleptic properties.

Beside clozapine, only the 2-Cl-dibenzoxazepine loxapine (launched in 1975 by Wyeth) and 2-Cl-dibenzothiepine zotepine (launched in 1982 by Fujisawa) were introduced into clinical practice for the treatment of schizophrenia. Interestingly, in contrast with other dibenzodiazepines, all of which have a piperazinyl substitution at position-11, zotepine (which is a dibenzothiepine) has a 2-dimethylaminoethoxy-substitution at the same position.

The unique pharmacological and clinical profile of clozapine and the need to eliminate its most serious side effects (i.e., agranulocytosis, weight gain liability) prompted several research groups to initiate chemistry programs to identify antipsychotics with pharmacological and clinical profiles similar to clozapine, but without its clinical disadvantages. In the search for such compound(s), from a chemistry viewpoint, an obvious approach was to select the dibenzodiazepine moiety as a starting point. During the 1980s and 1990s, these synthetic efforts, combined with the "D_2/5-HT_{2A} hypothesis" [4] resulted in the discovery of several potential antipsychotic compounds. Among these analogues of clozapine, only olanzapine (LY-170053) [5], quetiapine (ICI-204,636) [6], and the benzisoxazole derivative, risperidone [7], were introduced into the clinic and soon became the leaders in the antipsychotic market.

Today, efforts to further modify the benzodiazepine moiety are continuing, although compounds produced from these attempts have not yet reached the clinic (see for example compounds Y-931 [11], JL-13 [12], ST1460 [13,14] and GMC-series [15]; Fig. 13.1), and their evaluation is beyond the scope of this chapter.

An interesting branch of "clozapine research" has evolved from the finding that clozapine had remarkable selectivity toward dopamine D_4 receptors. It was hypothesized that the atypical feature of clozapine might be related to its D_4 selectivity [16]. Selective D_4 compounds possessing chemical structures quite different from clozapine were developed by several companies, and these drugs seemed to be active in some (not conventional) animal models of schizophrenia [17]. However, the clinical trial with L-745,870, the first selective D_4 acting compound, produced disappointing results [18] and this research line greatly attenuated thereafter.

Because of the apparent structural similarities of these clozapine-analogues, the question arises of how different are these compounds in their receptorial, pharmacological, and clinical actions? Some of the major features of the clozapine analogues, olanzapine and quetiapine, are compared with clozapine in the following section.

13.3
Preclinical Aspects

13.3.1
Multireceptor Profile: *In-Vitro, In-Vivo* Similarities and Differences

Atypical antipsychotics including clozapine, olanzapine, and quetiapine (as well as those which do not belong to this chemical class, e.g., risperidone, ziprasidone, sertindole, etc.) show affinities to various degrees to an array of receptors in the

Tab 13.1 Ki values for clozapine, olanzapine, and quetiapine:

Reference(s)	Clozapine					Olanzapine				Quetiapine				
	[19]	[20]	[21,22]	[23]	[24]	[19]	[20]	[23]	[21,22]	[19]	[20]	[21,22]	[23]	[24]
D_1	53^{r}	540^{r}	85^{r}		321^{r}	10^{r}	250^{r}		31^{r}	390^{r}	4240^{r}	455^{r}		1243^{r}
D_2	36^{r}	$150^{r};190^{hr}$	125^{r}	210^{h}	132^{r}	2.1^{r}	$17^{r};31^{hr}$	20^{h}	11^{r}	69^{r}	$310^{r};700^{hr}$	160^{r}	770^{h}	329^{r}
D_3	160^{rr}	$360^{rr};280^{hr}$	9^{rr}			2.0^{rr}	$54^{r};49^{hr}$		27^{rr}	$>180^{rr}$	$650^{r};340^{hr}$			
D_4	22^{hr}	40^{hr}			330^{r}	17^{hr}	28^{hr}			$>500^{hr}$	1600^{hr}			
$5\text{-}HT_{1A}$	710^{r}	$180^{r};140^{hr}$	770^{r}	160^{h}		7100^{r}	2720^{hr}	610^{h}	$>1000^{r}$	$>830^{r}$	320^{hr}	2450^{r}	300^{h}	720^{r}
$5\text{-}HT_{1B}$		3280^{r}	1200^{r}				$2850^{r,\ddagger}$		1355^{r}		$>5000^{r}$	5400^{r}		
$5\text{-}HT_{1D}$		390^{hr}	980^{r}	130^{h}			540^{hr}	150^{h}	800^{r}		2050^{r}	6220^{r}	560^{h}	
$5\text{-}HT_{2A}$	4.0^{r}	$3.3^{r};9.6^{hr}$	12^{r}	2.6^{h}	13^{r}	$1.9^{r};11.8^{hr,*};7.9^{hr,\ddagger}$	$1.9^{r};2.5^{hr};7.9^{hr,\ddagger}$	1.5^{h}	4^{r}	$82^{r};100^{hr,\ddagger}$	$120^{r};96^{hr}$	220^{r}	31^{h}	148^{r}
$5\text{-}HT_{2B}$	$8.5^{hr,*};1.6^{hr,\ddagger}$	13^{p}	8^{b}	4.8^{h}		2.8^{r}	7.1^{p}	4.1^{h}	11^{b}					
$5\text{-}HT_{2C}$	5.5^{r}	110^{mr}	69^{r}				84^{mr}		57^{r}	1500^{r}	3820^{r}	615^{b}	3500^{h}	
$5\text{-}HT_{3}$		4.0^{rr}					2.5^{rr}				4060^{mr}	170^{r}		
$5\text{-}HT_{6}$		21^{mr}					120^{mr}							
$5\text{-}HT_{7}$		23^{r}					60^{r}		19^{r}	290^{mr}	290^{mr}			
α_1	3.7^{r}	$160^{r};50^{hr}$	7^{r}	6.8^{h}	48^{r}	7.3^{r}	470^{hr}	44^{h}	230^{r}	4.5^{r}	58^{r}	7^{r}	8.1^{h}	90^{r}
α_{2A}	51^{r}	22^{hr}	8^{r}	15^{h}	37^{r}	140^{r}	180^{hr}	280^{h}		1100^{r}	2230^{hr}	87^{r}	80^{h}	270^{r}
α_{2B}		9.1^{h}					210^{hr}				90^{hr}			
α_{2C}	6^{r}	$2.1^{g};0.23^{hr}$	6^{r}	3.1^{h}	288^{r}	5.6^{r}	3.5^{g}	0.09^{h}	7^{r}		350^{hr}			
H_1	17^{r}	34^{r}		9^{h}		2.1^{r}	26^{r}	36^{h}		$21^{r};56^{r}$	$19^{g};2.2^{hr}$	11^{r}	19^{h}	
ACh-M	0.98^{r}										1020^{r}		1400^{h}	$>10000^{r}$
M_1			1.9^{r}						1.9^{r}			120^{r}		
M_2			10^{r}						18^{r}			630^{r}		
M_3			14^{r}						25^{r}			1320^{r}		
M_4			18^{r}						13^{r}			660^{r}		

r: rat brain, native; b: beef brain, native; rr: rat recombinant; hr: human recombinant; mr: mouse recombinant; p: pig brain, native; h: human brain, native; g: guinea-pig brain, native.

*: data from Ref. [25].

†: data from Ref. [26].

‡: data taken from Ref. [27].

central nervous system (CNS). That is, they all possess a multireceptorial profile. For clozapine, olanzapine and quetiapine, this is illustrated by their *in-vitro* affinity (K_i, inhibitor constant) data listed in Tab. 13.1, which is collected from the most cited authors in this field. Based on these data, a rank order of their *in-vitro* affinity to various receptors is presented in Tab. 13.2.

Tab 13.2 Rank order of *in-vitro* affinity of clozapine, olanzapine and quetiapine to various receptors.

Compound	Rank order of *in-vitro* affinity for various receptors
Clozapine	$5\text{-}HT_{2A} \cong H1 \cong 5\text{-}HT_{2C} \cong 5\text{-}HT_6 \cong 5\text{-}HT_{2B} \cong M \geq 5\text{-}HT_7 \geq a_1 \cong a_2 \cong D_4 > D_2 > D3 > D_1 > 5\text{-}HT_{1A} = 5\text{-}HT_{1D}$
Olanzapine	$5\text{-}HT_{2A} \cong H1 \cong 5\text{-}HT_6 \cong 5\text{-}HT_{2C} \geq D_2 \cong M \geq D_3 \cong 5\text{-}HT_{2B} > D_4 \cong 5\text{-}HT_7 \geq a_1 \cong a_2 \cong > D_1 > 5\text{-}HT_{1D} > 5\text{-}HT_{1A}$
Quetiapine	$a_1 \cong H1 > 5\text{-}HT_{2A} > 5\text{-}HT_{2B} > D_2 > M \gg \text{all others}$

The affinities (Tab. 13.1) and rank order of affinities (Tab. 13.2) of "clozapine-type" compounds clearly show that their primary receptor targets are serotonin (excep. 5-HT$_1$), histamine H$_1$ and adrenergic receptors, followed by dopamine receptors. Whilst clozapine and olanzapine seem to have some similarities in their overall receptor profile, quetiapine differs from them not only in the significantly lower absolute affinity values but also in the rank order of potencies; that is, it has the highest affinity for adrenergic alpha-1 and histamine H$_1$ receptors followed by 5-HT$_{2A}$ and dopamine receptors.

Several preclinical animal models (or a battery of models) are available for predicting typical or atypical antipsychotic action(s) and side effects. Models such as apomorphine-induced climbing (APC), amphetamine (or methamphetamine)-induced hyperlocomotion (AHL), conditioned avoidance response (CAR), catalepsy induction, amphetamine-, apomorphine- or phencyclidine-induced disruption of prepulse inhibition, and phencyclidine- or dizocilpine-induced hyperlocomotion, are the most frequently used for testing dopamine antagonism and glutamatergic mechanisms, respectively. DOI (2,5-dimethoxy-4-iodoamphetamine)- or 5-hydroxytryptamine-induced head twitch (HTW) are reliable models for assessing *in-vivo* 5-HT$_{2A}$ antagonism. However, besides the above-mentioned models, the paw test, apomorphine-disrupted swimming, disruption of social interaction by phencyclidine (in rats and monkeys) tremulous jaw movement, c-fos expression in various regions of the brain, measuring the single unit activity in the A9 and A10 region of the substantia nigra, measurement of prolactin levels in plasma and several variations of cognitive or learning tests are also used to demonstrate atypicality. The use of local variations of the above methods (either in the individual tests or in the composition of test battery), different strains of animals, different routes of administration, etc., renders comparison of the *in-vivo* profile of the available antipsychotics rather difficult.

Comparison of the *in-vivo* profile of clozapine, olanzapine and quetiapine in the most frequently used *in-vivo* animal models predicting (atypical) antipsychotic properties of these compounds is provided in Tab. 13.3. These data (collected from Prous Science Integrity database) show that great variations exist in the ED_{50} values obtained in the most frequently used methods. Nevertheless, they support the *in-vitro* receptor data given in Tab. 13.1 in that all three compounds are the most active in tests predicting serotonin 5-HT_{2A} (HTW) as well as dopamine D_2 receptor antagonism (APC and AHL) with remarkable separation between antipsychotic and side effect (i.e., catalepsy induction) doses. This is further illustrated with results from one laboratory (Tab. 13.4) [28,29].

Tab 13.3 Range of ED_{50} values (mg kg^{-1}) in models predictive of antipsychotic activity and extrapyramidal side effects in animal models.[a]

	APC	AHL	CAR	HTW	CATL
Clozapine	5.7–48[po]	0.8–22[po]	4.9–24[po]	0.04–0.2[sc]	>100–200[po]
Olanzapine	0.7–4.8[po]	4.2–15.6[po]	1.4–4.7[po]	0.04–0.3[sc]	10–39.4[po]
Quetiapine	1.9[sc]	1.8–22[sc]	21–50[sc]	0.6–0.8[sc]	30–50[sc]

[a]: Data collected from Prous Science Integrity.
[po]: drugs given orally; [sc]: drugs given subcutaneously.
APC: apomorphine-induced climbing in mice; AHL: amphetamine-induced hyperlocomotion in rats; CAR: conditioned avoidance response in rats; HTW: DOI-induced head twitch; CATL: catalepsy inducing dose in rats.

Tab 13.4 Comparison of ED_{50} (mg kg^{-1}, s.c.) values for clozapine, olanzapine and quetiapine from one laboratory.[a]

	APC	AHL	CAR	HTW	CATL
Clozapine	2.1	8.2	5.2	0.041	>80[b]
Olanzapine	0.1	0.3	0.6	0.004	n.d.
Quetiapine	1.9	3.8	38.4	0.6	n.d.

[a]: Calculated from values in Ref. [28]
[b]: From Ref. [29].
For abbreviations, see Tab. 13.3. n.d.: not determined.

A more detailed evaluation or comparison of the receptor profiles of these compounds and explanation or understanding of their clinical profile (i.e., antipsychotic action and side effects) based on their receptor data is, therefore a controversial and complicated issue. Some aspects of this problem in connection with clozapine-type antipsychotics are addressed in the following sections.

13.3.2
The Availability of Data

The availability of a great number of data obtained with different radioligands, membrane preparations, different tissue sources and differences in the assay conditions makes the comparison of receptor profile of clozapine, olanzapine and quetiapine (as well as other antipsychotics) extremely difficult. This issue has become even more complicated since the availability of recombinant receptors from various species expressed in various expression systems that resulted in a continuously growing number of K_i data and further additions to the receptor profile. (For example, in May 2005, when the latest search in the database http://kidb.cwru.edu/pdsp.php was made, 464, 272, and 156 K_i values for various receptors and transporters were recorded for clozapine, olanzapine, and quetiapine, respectively, representing an average K_i "growth rate" of 11 to 16 per month over the past five months.) Different data from the same authors obtained at different times further complicate the comparisons. The receptor K_i data in Tab. 13.1 illustrate the situation to some extent.

13.3.3
Dopamine D$_2$ versus Serotonin 5-HT$_{2A}$ Affinity

There is no doubt that all three atypical antipsychotics have high-to-moderate affinity to different serotonin receptor subtypes, and that they have less affinity to dopamine D$_2$ receptors. It is, however, widely accepted that interaction (i.e., antagonism) of both typical and atypical antipsychotics with dopamine D$_2$ (and D$_3$) receptors is a "must" for the antipsychotic action (for the treatment of positive symptom) (for a review, see Refs. [30,31]). On the other hand, it is thought that 5-HT$_{2A}$ antagonism, beside D$_2$ antagonism, is – or may be – the most important element of atypicality [4,32,33] ("D$_2$/5-HT$_{2A}$ hypothesis"). In animal experiments, 5-HT$_{2A}$ antagonists do stimulate dopamine release in the nigrostriatal and mesocortical pathways which may help to overcome D$_2$ blockade. It is therefore, assumed that 5-HT$_{2A}$ antagonism reduces side effects attributed to D$_2$ antagonism (e.g., extrapyramidal symptoms, enhanced prolactin release) and improve both positive and negative symptoms. This hypothesis, however has recently been seriously challenged on the basis of results obtained from *in-vitro* receptor–ligand binding kinetic experiments and from *in-vivo* occupancy results [34,35]. These results indicate that 5-HT$_{2A}$ blockade may not be as important in the antipsychotic action (and in reducing the side effects) as was thought originally.

In the *in-vitro* kinetic experiments, the rates of association (K_{on}) and dissociation (K_{off}) of various (labeled) antipsychotic compounds to dopamine D_2 receptors were determined. Kapur and Seeman found that antipsychotics substantially differ (almost 1000-fold) in their K_{off} rate (whereas only 10-fold differences were found in the K_{on} rate), and that this value is highly correlated with their affinity to D_2 receptors. These authors also demonstrated that K_{off} for clozapine, olanzapine and quetiapine was 1.386 min^{-1}, 0.039 min^{-1}, and 3.013 min^{-1}, respectively, and assumed that the rate of how rapidly they left the receptor was an important mechanism in their atypical antipsychotic action. Indeed, this fully explained the lack of extrapyramidal symptoms (EPS) and hyperprolactinemia and the low risk for tardive dyskinesia [34–36]. In this regard, quetiapine (which has the lowest affinity to D_2 receptors) seems to be the most "atypical" among all tested antipsychotics, followed by clozapine and olanzapine (nevertheless, olanzapine's K_{off} value is close to those of raclopride and chlorpromazine).

In their pioneering positron emission tomography (PET) studies, Farde et al. demonstrated that clinical response in patients treated with different antipsychotics can be observed when D_2 occupancy in the brain reaches the threshold of about 65%, but side effects (i.e., EPS) appears if the occupancy exceeds 80% [37]. This observation has been confirmed by several others, and the common view is that 65–80% occupancy of D_2 receptors is necessary to achieve antipsychotic action without EPS. Occupancy of this degree is produced by olanzapine with a daily dose of 7.5–10 mg day^{-1} [38]. Quetiapine (300–600 mg day^{-1}) and clozapine (350 mg day^{-1}) produced 60% and 71% occupancy at 2 h after their administration, although the occupancy by these drugs declined rapidly (to ca. 20% at 12 h after quetiapine, and to 26% at 24 h after clozapine); this can be explained by the drugs' relatively rapid dissociation from the D_2 receptors [39–41]. *In-vivo* occupancy studies also revealed that atypical antipsychotics (i.e., risperidone, olanzapine) with high affinity to 5-HT$_{2A}$ receptors achieve 80–90% occupancy of 5-HT$_{2A}$ receptors at doses that do not produce antipsychotic responses. However, if their D_2 occupancy exceeds 80%, even 5-HT$_{2A}$ antagonism cannot prevent the appearance of side effects. These findings led Kapur et al. to conclude that "... *it would seem that 5-HT$_{2A}$ blockade is neither necessary ... nor sufficient for antipsychotic response*". According to the D_2/5-HT$_{2A}$ occupancy hypothesis of Kapur et al., the lack of EPS after clozapine and quetiapine may thus lie in the fact that they do not produce D_2 occupancy higher than 60% [34], although several animal (but not clinical) data seem to support the alleviating effect of 5-HT$_{2A}$ antagonism on the D_2-related side effects of antipsychotics.

The example of amisulpride (launched by Sanofi-Synthelabo in 1986) also supports the primary importance of dopamine D_2 (as well as D_3) but not 5-HT$_{2A}$ receptors in atypical antipsychotic action. This benzamide derivative displays high affinity only to D_2 and D_3 receptors (with some selectivity toward D_3) [42], and in low doses (i.e., 50–100 mg day^{-1}) it acts preferentially on negative [43] symptoms and at higher doses (400–800 mg day^{-1}) on depressive symptoms [44] and cognitive impairment [45].

13.3.4
Affinity to other Receptors

Little is known of the contribution of the affinity of clozapine, olanzapine and que-tiapine to receptors other than D_2 and 5-HT_{2A} to their antipsychotic action and side effects. For example, both clozapine and olanzapine have significant affinity to histamine H_1, adrenergic alpha-1 receptors and various muscarinic receptor subtype (M_1-M_5) *in vitro*. However, in the *in-vitro* functional studies and *in-vivo* animal experiments, antagonism by olanzapine at muscarinic receptors was modest, but the significant alpha-1 antagonism was confirmed [46]. It has been known for some time that clozapine is a potent antimuscarinic compound [47] and has high affinity to all five muscarinic receptor subtypes [45]. Moreover, it was found to be either an antagonist [48] or a partial agonist [49] depending on the subtype examined and the experimental settings (expression systems, assay methods). In fact, clozapine has been found to occupy cerebral muscarinic receptors [50], and it has been proposed that interaction with central muscarinic receptors might contribute beneficially to its antipsychotic action. Antagonism at these receptors may alleviate parkinsonian symptoms [51], whereas their stimulation may improve both positive and negative symptoms [52]. As to adrenergic alpha-1 receptors, it has long been assumed that postural hypotension caused by typical antipsychotics (e.g., chlorpromazine) is related to their adrenergic alpha-1 antagonism. However, it has been recently hypothesized that alpha-1 antagonism may contribute to the antipsychotic action of clozapine [53] by enhancing its mesocortical selectivity [54]. Clozapine, olanzapine, and quetiapine display significant affinity to histamine H_1 receptors, and it is widely accepted that some side effects of these agents (e.g., weight gain liability, sedation) are attributed to their histamine antagonism [55]. All three compounds have relatively high affinity to serotonin 5-HT_{2B} receptors, and clozapine and olanzapine also have remarkable affinity to 5-HT_6 and 5-HT_7 receptors, but almost nothing is known of the role of these interactions either in the antipsychotic action or in the side effects.

13.3.5
Inverse Agonism

Antipsychotic drugs – including clozapine, olanzapine, and quetiapine – have been assumed to antagonize dopamine receptors. However, using *in-vitro* recombinant expression systems, it has recently been shown that several antipsychotics act as inverse agonists at various dopamine receptors [31]. Clozapine and olanzapine have been shown to be inverse agonist at 5-HT_{2A} and 5-HT_{2C} receptors [53,54]. These mechanisms may be important elements in the antipsychotic action, but the clinical relevance of inverse agonism is not yet clear.

13.3.6
Receptor Affinity of Metabolites and Clinical Action

Receptor affinities of metabolites may variably contribute to the *in-vivo* or clinical action. For example, the olanzapine metabolite 2-hydroxmethyl-olanzapine has affinity to histamine H_1 and adrenergic alpha-1 receptors comparable to that of the parent compound [58]. Further, *N*-desmethylclozapine, the main metabolite of clozapine, has remarkable agonism at muscarinic M_1 receptors [59], and it is hypothesized that this action may greatly contribute to the uniquely beneficial clinical profile of clozapine [60].

13.4
Clinical Aspects

Chlorpromazine, the first modern drug to be used in the treatment of schizophrenia and other psychotic disorders, was introduced into psychiatry in 1952 [61]. It was followed by a number of other drugs for the treatment of these conditions (e.g., haloperidol, thioridazine). These were also called *neuroleptics* because of their neurological side effects, such as parkinsonian syndrome and tardive dyskinesia. Tardive dyskinesia is a movement disorder characterized by involuntary movements of the face and limbs. The antipsychotic properties of these drugs were inseparable from the extrapyramidal effects.

13.4.1
Terminology

Clozapine was introduced into psychiatry in Europe in the 1970s and in the US in the 1990s. The frequency of extrapyramidal neurological side effects of clozapine is comparable to that found with placebo. Clozapine was followed by other antipsychotics (e.g., risperidone, olanzapine) with a low frequency of neurological adverse events. The term "neuroleptic" was no longer appropriate for these new drugs; consequently the term "atypical neuroleptics", and later the term "second-generation antipsychotics" was introduced as opposed to the older "typical neuroleptics" or "first-generation antipsychotics". Subsequently, clozapine has become the "gold standard" of second-generation antipsychotics. It has practically no EPS, and indeed is useful in their treatment. Clozapine is effective in the treatment of negative symptoms and of depression in schizophrenia, and is also efficacious in preventing suicidal behavior in these patients. However, the side effects of clozapine limit its use, especially the potentially life-threatening agranulocytosis. Clozapine can be used only in patients who comply with regular blood monitoring.

13.4.2
Indications

Clozapine has been found effective in patients who did not improve during treatment with first-generation antipsychotics, and since the hematological side effects permit only its restricted use, this drug has a unique indication for "treatment-resistant" schizophrenia. Another unique indication for clozapine is the reduction in the risk of recurrent suicidal behavior in schizophrenia or schizoaffective disorders. The indications of clozapine and its two analogues, olanzapine and quetiapine, are summarized in Tab. 13.5. The US labels of these drugs served as the data source [62–64]. Clozapine and olanzapine, but not quetiapine, are available in intramuscular form, which is helpful in the treatment of acutely agitated patients with diagnoses as defined in Tab. 13.5.

Tab 13.5 Indications for clozapine, olanzapine and quetiapine tablets.

Indication/Drug	Clozapine	Olanzapine	Quetiapine
Treatment-resistant schizophrenia	x		
Risk of recurrent suicidal behavior in schizophrenia or schizoaffective disorder	x		
Schizophrenia		x	x
Bipolar disorder, acute mania, monotherapy		x	x
Bipolar disorder, acute mania, combination therapy		x	x
Bipolar disorder, maintenance monotherapy		x	

13.4.3
Dosage

In the USA, the target dose of clozapine is 300–450 mg day^{-1}, but this is significantly lower in Europe [65]. The target dose of olanzapine is 10–15 mg day^{-1}, and for quetiapine it is 300–400 mg day^{-1}, although clinicians tend to use higher doses of quetiapine [66]. Clozapine must be titrated slowly, with the target dose being achieved by the end of 2 weeks. Quetiapine must also be titrated, but the target dose can be achieved in 4–6 days. Olanzapine can be dosed without titration.

13.4.4
Clinical Efficacy in Schizophrenia

A systematic review of 31 studies comparing clozapine to first-generation antipsychotics concluded that improvement was more frequent with clozapine than with the comparators (OR: 0.4; CI: 0.2–0.6; NNT 6) and patients receiving clozapine had fewer relapses than patients receiving first-generation drugs (OR: 0.6; CI:

0.4–0.8; NNT: 20; CI: 17–38) [67]. [OR = odds ratio; the ratio of odds of an event (e.g., improvement or parkinsonian syndrome) in an exposed group to odds of an event in controls (nonexposed). CI = confidence interval; the range of values in which the true value most likely lies. NNT = number needed to treat; the number of patients that must be treated to produce one additional good outcome (or to prevent one additional bad outcome) as compared to the controls.]

Based on a systematic review [68], olanzapine was superior to placebo in 6-week treatment studies of schizophrenia, although the attrition rates in the placebo-controlled studies were very high (olanzapine –61%, placebo –73%). The NNT was 8 (CI: 5–340).

A systematic review of quetiapine included 12 studies [69]. The attrition rate in the four placebo-controlled studies was almost as high as in the olanzapine studies (quetiapine –53%, placebo –60%), the RR being 0.84 (CI: 0.7–0.9; NNT: 11; CI: 7–55); thus, it is rather difficult to interpret the data with such CI-values. [RR = relative risk; in randomized controlled trials, the ratio of rate of negative events in the exposed group to the rate in controls.]

Few studies have been conducted which directly compare second-generation antipsychotics, and even fewer comparing all three drugs discussed in this chapter. The restriction on the use of clozapine limits the inclusion criteria of randomized studies to treatment-resistant patients. A meta-analysis of randomized, controlled clinical trials that compared the effectiveness of typical and second-generation antipsychotics in the treatment of chronic refractory schizophrenia indicated that "... *clozapine exhibits superiority over typical antipsychotics in terms of both efficacy (as measured by improvement in overall psychopathology) and safety (in terms of reduced extrapyramidal side effects)*" [70]. Some studies have been reported showing noninferiority or no difference in the global ratings of efficacy between olanzapine and clozapine in treatment-resistant, partially resistant or intolerant-to-treatment subgroups of patients [71,72].

This short summary on the efficacy of the three selected drugs demonstrates the methodological difficulties of schizophrenia studies. The attrition rates are high, the effect sizes are rather small, and the evidence for superior efficacy in global psychopathology ratings is weak for olanzapine and quetiapine, which does not reflect the overall better outcome with these drugs. Measures of negative, depressive and cognitive symptoms and suicidality are important outcome measures contributing to the understanding of better functioning of schizophrenia patients receiving treatment with these drugs [72,73].

13.4.5
Clinical Efficacy in Bipolar Disorder (Especially in Mania)

Clozapine has well-documented efficacy in the treatment of mania, including "treatment-resistant" mania [74], although it has not been approved for this indication. Olanzapine and quetiapine both have been approved for the treatment of mania, though not for the treatment of bipolar depression. Olanzapine also has approval for maintenance treatment of bipolar disorder.

13.4.6
Adverse Events

The clinically most important adverse events of the three drugs are summarized in Tab. 13.6. The above-cited US labels and systematic reviews served as data sources. The differences in the side-effect profile of these drugs are clinically significant, and they are important in the selection of a drug. The magnitude of some listed side effects may be different for the three drugs; for example, orthostatic hypotension is most severe with clozapine, followed by quetiapine and olanzapine. Clozapine and quetiapine must be titrated, mainly because of this side effect. Weight gain is most prominent with clozapine (ca. 4 kg after 10 weeks), followed by olanzapine and quetiapine [75]; however, data have been presented indicating that quetiapine may cause more weight gain than olanzapine [76].

Tab 13.6 Clinically important adverse events.

Adverse event/Drug	Clozapine	Olanzapine	Quetiapine
Agranulocytosis	x		
Orthostatic hypotension	x	x	x
Myocarditis	x		
Seizures	x	(x)	(x)
Anticholinergic toxicity, incl. delirium	x		
Weight gain	x	x	x
Hyperglycemia and diabetes mellitus[a]	x	x	x
Hyperprolactinemia		(x)	(x)
Cataract			(x)
Hypothyroidism			x

a. Class effect of second-generation antipsychotics.
(x) limited clinical significance.

13.5
Summary and Conclusions

The discovery of the atypical antipsychotic clozapine opened a new era and set new standards in the drug treatment of schizophrenia. Modifications of the dibenzoazepine structure in clozapine resulted in olanzapine and quetiapine, which are among the most frequently used antipsychotic drugs. From a chemical viewpoint, clozapine, olanzapine and quetiapine can be considered as structural analogues. Although they share some common features in their molecular mechanism of action, the three compounds show significant differences in their clinical efficacy and adverse event profile.

References

1 Schmutz, J. *Arzneim.-Forsch.*, **1975**, 25, 712.

2 Berzewski H, Helmchen E, Hippius H, Hoffmann H, Kanowski S. *Arzneim.-Forsch.*, **1969**, 19, 496.

3 Angst J, Bente D, Berner P, Heimann H, Hippius H. *Pharmakopsychiatrie*, **1971**, 4, 200.

4 Meltzer H, Matsubara S, Lee J-C. *J. Pharmacol. Exp. Ther.*, **1989**, 251, 238.

5 Chakrabarti JK, Horsman L, Hotten TM, Pullar IA, Tupper DE, Wright FC. *J. Med. Chem.*, **1980**, 23, 878.

6 Warawa EJ, Migler BM, Ohnmacht CJ, Needles AL, Gatos GC, McLaren FM, Nelson CL, Kirkland KM. *J. Med. Chem.*, **2001**, 44, 372.

7 Janssen PAJ, Niemegers CJE, Awouters F, Schellekens KHL, Megens AAHP, Meert TF. *J. Pharmacol. Exp. Ther.*, **1988**, 244, 685.

8 Howard HR, Lowe JA, Seeger TS, Seymour PA, Zorn S, Maloney PR, Ewing FE, Newman ME, Schmidt AW, Furman JS, Robinson GL, Jackson E, Johnson C, Morrone J. *J. Med. Chem.*, **1996**, 39, 143.

9 Perregaard J, Arnt J, Bogem KP, Hyttel J, Sanchez C. *J. Med. Chem.*, **1992**, 35, 1092.

10 Bürki HR, Sayers AC, Ruch W, Asper H. *Arzneim.-Forsch.*, **1977**, 27, 1561.

11 MorimotoT, Hashimoto K, Yasumatsu H, Tanaka H, Fujimura M, Kuriyama M, Kimura K, Takehara S, Yamagami K. *Neuropsychopharmacology*, **2002**, 26, 456.

12 Ellenbroek BA, Liegeois J-F, Bruhwyler J, Cools AR. *J. Pharmacol. Exp. Ther.*, **2001**, 298, 386.

13 Campiani G, Butini S, Fattorusso C, Catalanotti B, Gemma S, Nacci V, Morelli E, Cagnotto A, Mereghetti I, Mennini T, Carli M, Minetti P, Di Cesare MA, Mastroianni D, Scafetta N, Galletti B, Stasi A, Castorina M, Pacifici L, Vertechy M, Di Serio S, Ghirardi O, Tinti O, Carminati P. *J. Med. Chem.*, **2004**, 47, 143.

14 Campiani G, Butini S, Gemma S, Nacci V, Fattorusso C, Catalanotti B, Giorgi G, Cagnotto A, Goegan M, Mennini T, Minetti P, Di Cesare MA, Mastroianni D, Scafetta N, Galletti B, Stasi A, Castorina M, Pacifici L, Ghirardi O, Tinti O, Carminati P. *J. Med. Chem.*, **2002**, 45, 344.

15 Liao Y, Venhuis BJ, Rodenhuis N, Timmerman V, Wikström H. *J. Med. Chem.*, **1999**, 42, 2235.

16 Seeman P, Corbet R, Van Tol HHM. *Neuropsychopharmacology*, **1997**, 16, 93.

17 Patel S, Freedman S, Chapman KL, Emms F, Fletcher AE, Knowles M, Marwood R, Mcallister G, Myers J, Patel Sh, Curtis N, Kulagowski JJ, Leeon PD, Ridgill M, Graham M, Matheson S, Rathbone D, Watt AP, Bristow LJ, Rupniak NM, Baskin E, Lynch JJ, Ragan I. *J. Pharmacol. Exp. Ther.*, **1997**, 283, 636.

18 Bristow LJ, Kramer MS, Kulagowski J, Patel S, Ragan CI, Seabrook GR. *Trends Pharmacol.*, **1997**, 18, 186.

19 Arnt J, Skarfeldt T. *Neuropsychopharmacology*, **1998**, 18, 633.

20 Schotte A, Janssen PFM, Gommeren W, Luyten HML, Van Gompel P, Lesage AS, De Loore K, Leysen JE. *Psychopharmacology*, **1996**, 124, 57.

21 Bymaster FP, Calligaro DO, Falcone JF, Marsh RD, Moore NA, Tye NC, Seeman P, Wong DT. *Neuropsychopharmacology*, **1996**, 14, 87.

22 Bymaster F, Perry KW, Nelson DL, Wong DT, Rasmussen K, Moore NA, Calligaro DO. *Br. J. Psychiatry*, **1999**, 174 (Suppl. 37), 36.

23 Richelson E, Souder T. *Life Sci.*, **2000**, 68, 29.

24 Saller CF, Salama AI. *Psychopharmacology*, **1993**, 112, 285.

25 Bonhaus DW, Winhardt KK, Taylor M, Desouza A, Mcneely PM, Szczepanski K, Fontana DJ, Trinh CL, Rocha MW, Dawson LA, Flippin LA, Egle RM. *Neuropharmacology*, **1997**, 36, 621.

26 Wainscott DB, Lucaites VL, Kursar JD, Baez M, Nelson DL. *J. Pharmacol. Exp. Ther.*, **1996**, 276, 720.

27 Millan MJ, Schreiber R, Monneyron S, Denorme B, Melon C, Queriaux S,

Dekeyne A. *J. Pharmacol. Exp. Ther.*, **1999**, 289, 427.

28 Millan MJ, Brocco M, Rivet J-M, Audinot V, Newman-Tancredi A, Maiofiss L, Queriaux S, Despaux N, Peglion J-L, Dekeyne A. *J. Pharmacol. Exp. Ther.*, **2000**, 292, 54.

29 Millan MJ, Schreiber R, Dekeye A, Rivet J-M, Bervoets K, Mavridis M, Sebban C, Maurel-Remy S, Newman-Tancredi A, Spedding M, Muller O, Lavielle G, Brocco M. *J. Pharmacol. Exp. Ther.*, **1998**, 286, 1356.

30 Seeman P. *Clin. Neurosci. Res.*, **2001**, 1, 53.

31 Strange PG. *Pharmacol. Rev.*, **2001**, 53, 119.

32 Meltzer HY. *Neuropsychopharmacology*, **1999**, 21, 106S.

33 Meltzer HY. *Prog. Neuropsychopharmacol. Biol. Psychiatry*, **2003**, 27, 1159.

34 Kapur S, Seeman P. *Am. J. Psychiatry*, **2001**, 158, 360.

35 Kapur S, Seeman P. *J. Psychiatry Neurosci.*, **2000**, 25, 181.

36 Seeman P. *Can. J. Psychiatry*, **2002**, 47, 27.

37 Farde L, Nordstrom AL, Wiesel FA, Pauli S, Halldin C, Sedvall G. *Arch. Gen. Psychiatry*, **1992**, 49, 538.

38 Kapur S, Zipursky RB, Remington G, Jones C, DaSilva J, Wilson AA, Houle S. *Am. J. Psychiatry*, **1998**, 155, 921.

39 Kapur S, Zipursky RB, Jones C, Shammi CS, Remington G, Seeman P. *Arch. Gen. Psychiatry*, **2000**, 57, 553.

40 Gefvert O, Bergstrom M, Langstrom B, Lundberg T, Lindstrom L, Yates R. *Psychopharmacology (Berl.)*, **1998**, 135, 119.

41 Jones C, Kapur S, Remington G, Zipursky RB. *Biol. Psychiatry*, **2000**, 47 (8 Suppl.), 112.

42 Schoemaker H, Claustre Y, Fage D, Rouquier DFL, Chergui K, Curet O, Oblin A, Carter C, Benavides J, Scatton B. *J. Pharmacol. Exp. Ther.*, **1997**, 280, 83.

43 Paillere-Martinot ML, Lecrubier Y, Martinot JL, Aubin F. *Am. J. Psychiatry*, **1995**, 152, 130.

44 Peuskens J, Moller HJ, Puech A. *Neuropsychopharmacology*, **2002**, 12, 305.

45 Wagner M, Quednow BB, Westheide J, Schlaepfer TE, Maier W, Kuhn KU. *Neuropsychopharmacology*, **2005**, 30, 381.

46 Bymaster FP, Nelson DL, DeLapp NW, Falcone JF, Eckols K, Truex LL, Foreman MM, Lucaites VL, Calligaro DO. *Schizophrenia Res.*, **1999**, 37, 107.

47 Miller RJ, Hiley CR. *Nature*, **1974**, 248, 546.

48 Bolden C, Cusack B, Richelson E. *J. Pharmacol. Exp. Ther.*, **1992**, 260, 576.

49 Olianas MC, Maullu C, Onali P. *Neuropsychopharmacology*, **1999**, 20, 263.

50 Readler TJ, Knable MB, Jones DW, Urbina RA, Egan MF, Weinberger DR. *Neuropsychopharmacology*, **2003**, 28, 1531.

51 Snyder S, Greenberg D, Yamamura HI. *Arch. Gen. Psychiatry*, **1974**, 31, 58.

52 Bymaster F, Felder CC, Tzarava E, Nomikos GG, Calligaro DO, Mckinzie DL. *Prog. Neuropsychopharmacol. Biol. Psychiatry*, **2002**, 27, 1125.

53 Baldessarini RJ, Huston-Lyons D, Campbell A, Marsh E, Cohen BM. *Br. J. Psychiatry*, **1992**, 160, 12.

54 Svensson TH, Mathe JM, Andersson JL, Nomikos GG, Hildebrand BE, Marcus M. *J. Clin. Psychopharmacol.*, **1995**, 15 (Suppl. 1), 11S–18S.

55 Kroeze WK, Hifeisen SJ, Popadak BA, Renock SM, Steinberg S, Ernsberger P, Jayathilake K, Meltzer HY, Roth BL. *Neuropsychopharmacology*, **2003**, 28, 519.

56 Egan C, Herrick-Davies K, Teitler M. *J. Pharmacol. Exp. Ther.*, **1998**, 286, 85.

57 Herrick-Davies K, Grinde E, Teitler M. *J. Pharmacol. Exp. Ther.*, **2000**, 295, 226.

58 Calligaro DO, Fairhurst J, Hotten TM, Moore N, Tupper DE. *Bioorg. Med. Chem. Lett.*, **1997**, 7, 25.

59 Sur C, Mallorga PJ, Wittmann M, Jacobson MA, Pascarella D, Williams JB, Brandish PE, Pettibone DJ, Scolnick EM, Conn PJ. *Proc. Natl. Acad. Sci. USA*, **2003**, 100, 13674.

60 Davies MA, Compton-Toth BA, Hufeisen SJ, Meltzer HY, Roth BL. *Psychopharmacology*, **2005**, 178, 451.

61 Shorter E. *A History of Psychiatry.* John Wiley & Sons, New York, Chichester, Brisbane, Toronto, Singapore, Weinheim, **1997**.

62 AstraZeneca Pharmaceuticals LP: Sero-quel. Rev. 01/01. http://www.fda.gov/cder/foi/label/2001/20639s11lbl.pdf (accessed February 14, 2005).

63 Lilly: Zyprexa (tablets; disintegrating tablets, intramuscular injection) Rev. Sept. 22, 2004 http://pi.lilly.com/us/zyprexa-pi.pdf (accessed February 15, 2005).

64 Novartis: Clozaril tablets. Based on official label in effect July 2004, http://www.drugs.com/xq/cfm/pageid_0/htm_57300575.htm/tgid_/bn_Clozaril%20Tablets/type_pdr/qx/index.htm (accessed February 14, 2005).

65 Fleischhacker WW, Hummer M, Kurz M, Kurzthaler I, Lieberman JA, Pollack S, Safferman AZ, Kane JM. *J. Clin. Psychiatry*, **1994**, 55 (Suppl. B), 78–81.

66 Kane JM, Leucht S, Carpenter D, Docherty JP. *J. Clin. Psychiatry*, **2003**, 64 (Suppl. 12), 1.

67 Wahlbeck K, Cheine M, Essali MA. Clozapine versus typical neuroleptic medication for schizophrenia. [Systematic Review] *Cochrane Schizophrenia Group Cochrane Database of Systematic Reviews*, 2, **2005**.

68 Duggan L, Fenton M, Rathbone J, Dardennes R, El-Dosoky A, Indran S. Olanzapine for schizophrenia. [Systematic Review] *The Cochrane Database of Systematic Reviews*, 2, **2005**.

69 Srisurapanont M, Maneeton B, Maneeton N. Quetiapine for schizo-phrenia. [Systematic Review]. *Cochrane Schizophrenia Group Cochrane Database of Systematic Reviews*, 2, **2005**.

70 Chakos M, Lieberman J, Hoffman E, Bradford D, Sheitman B. *Am. J. Psychiatry*, **2001**, 158, 518.

71 Bitter I, Dossenbach MRK, Brook S, Feldman PD, Metcalfe S, Gagiano CA, Füredi J, Bartko Gy, Janka Z, Banki MC, Kovacs G, Breier A. *Prog. Neuropsychopharmacol. Biol. Psychiatry*, **2004**, 28, 173.

72 Volavka J, Czobor P, Sheitman B, Lindenmayer J-P, Citrome L, McEvoy JP, Cooper TB, Chakos M, Lieberman JA. *Am. J. Psychiatry*, **2002**, 159, 255.

73 Bilder RM, Goldman RS, Volavka J, Czobor P, Hoptman M, Sheitman B, Lindenmayer J-P, Citrome L, McEvoy J, Kunz M, Chakos M, Cooper TB, Horowitz TL, Lieberman JA. *Am. J. Psychiatry*, **2002**, 159, 1018.

74 Calabrese JR, Kimmel SE, Woyshville MJ, Rapport DJ, Faust CJ, Thompson PA, Meltzer HY. *Am. J. Psychiatry*, **1996**, 153, 759.

75 Allison DB, Mentore JL, Heo M, Chandler LP, Cappelleri JC, Infante MC, Weiden PJ. *Am. J. Psychiatry*, **1999**, 156, 1686.

76 McIntyre RS, Trakas K, Lin D, Balshaw R, Hwang P, Robinson K, Eggleston A. *Can. J. Psychiatry*, **2003**, 48, 689.

14

Quinolone Antibiotics: The Development of Moxifloxacin

Uwe Petersen

14.1
Introduction

The Makonde sculpture shown in Fig. 14.1 impressively depicts how each generation relies on the one before. This applies to all fields of life, including science
and technology, as well as to drug research, and in particular to the history of antibiotics.

The treatment of infections with anti-infectives dates back to the beginning of
the twentieth century and the Nobel prize winner Paul Ehrlich, who introduced
Salvarsan for the treatment of syphilis. The fight against pathogenic microorgan-
isms is continuing today on a broad scale, and a wide range of antibiotics is the
most important weapon for defeating bacterial infections. Pharmaceutical companies and university research institutes worldwide are continuously searching for
new anti-infectives, and the literature abounds with infinite compounds having
new structures which are reported to be antibacterially active. Only rarely is it possible to develop such new compounds into substances suitable for the treatment
of human infections. Antibiotic research therefore still has to rely on and optimize
the properties of already existing classes of substances. This is why such well-
known classes of substances as the β-lactams (penicillins, cephalosporins), macrolides and quinolones remain the most widely marketed antibiotics/anti-infectives.

The most important classes of antibiotics, and the years in which they first
appeared on the market, are depicted in Fig. 14.2. This figure also shows how extremely difficult it is, despite intense efforts, to find new classes of antibiotics
with properties suitable for antibacterial treatment. The first quinolone, nalidixic
acid, which was synthesized by Lesher [1], appeared in 1962. It then took 38 years

This article reveals the reflections of the author on the chemical evolution of moxifloxacin. During the course of
Bayer's quinolone research project that resulted in the development of moxifloxacin, numerous compounds were
synthesized, and attempts were made to modify almost every quinolone substituent in a variety of ways in order to
meet the goals described herein. Some of these efforts were successful, while others were not. This chapter does not
attempt to describe this project in its entirety. Instead, for the sake of brevity and simplicity, many areas of this complex and difficult project have been excluded or substantially simplified in order to focus on those areas of the Bayer
project which, in retrospect, were most important to the development of moxifloxacin.

Analogue-based Drug Discovery. IUPAC, János Fischer, and C. Robin Ganellin (Eds.)
Copyright © 2006 WILEY-VCH Verlag GmbH & Co. KGaA, Weinheim
ISBN: 3-527-31257-9

Fig. 14.1 Each generation rests on the shoulders of the one before (a Makonde sculpture by Samaki Likankoa).

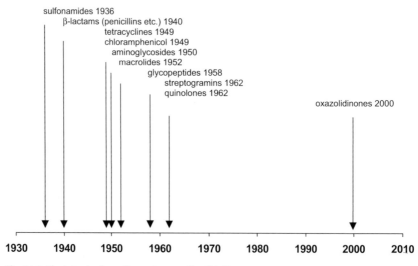

Fig. 14.2 The introduction of new classes of antibiotics.

before a new class of substances for the treatment of bacterial infections was introduced in 2000, the first oxazolidinone [2] to be marketed being linezolid.

In medicinal chemistry, the term "quinolones" or "quinolonecarboxylic acids" broadly refers to compounds having the general structures depicted in Fig. 14.3. It thus also includes bioisosteric azaquinolones, such as for example 1,8-naphthyridones, cinnolones and pyridopyrimidones and the more recently introduced pyridopyridones ("2-pyridones") [3]. The most important quinolones contain a fluorine atom in the 6-position and are therefore referred to as fluoroquinolones. In recent years, several promising quinolones have, however, been described in which the fluorine atom has been removed from the 6-position. One example of such compounds, which are referred to as 6-desfluoroquinolones, is garenoxacin (T 3811, BMS 384 756) **90** [4], which was discovered by Toyama and is presently being developed in cooperation with Schering-Plough.

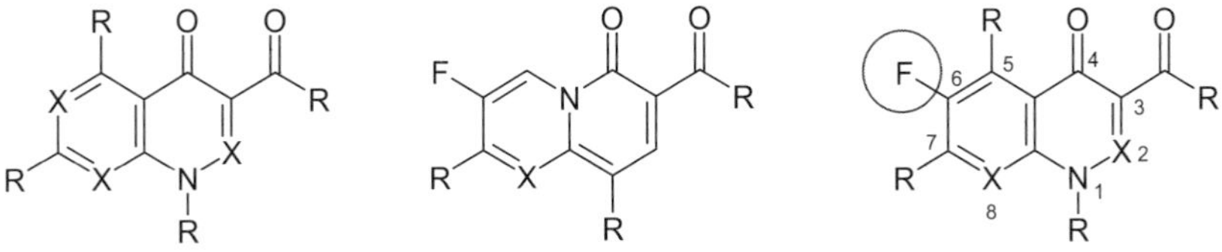

$X = N$ or C (optionally substituted)
R = various substituents; X and R can also be linked by a bridging bond

Fig. 14.3 The general structure of quinolones and fluoroquinolones.

Quinolones interfere with the replication mechanism of bacteria by inhibiting essential enzymes such as DNA gyrase (topoisomerase II) and topoisomerase IV, thereby preventing supercoiling and decatenation of the bacterial DNA and the replication of the latter [5–12].

A patent application from ICI [13] in 1954 already described the antibacterial activity of quinolonecarboxylic acids such as **1**, which were however disregarded. It was not until Lesher carried out his investigations leading to the development of nalidixic acid **2** [1] that the actual "chemical evolution of the quinolonecarboxylic acids" began (Fig. 14.4). Basically, three generations can be identified, although there is no clear-cut dividing line between them. The first-generation quinolones only had relatively poor activity against Gram-negative bacteria, as well as being relatively poorly absorbed. Resistance to these compounds also progressed at a rapid rate. They were used predominantly for the treatment of infections of the urinary tract, and included not only nalidixic acid **2** but also oxolinic acid [14], cinoxacin [15] as well as piromidic acid [16] and pipemidic acid **4** [17], in which the 7-position was substituted for the first time by cyclic amine substituents. Pipemidic acid **4**, which contained a 7-piperazinyl substituent, displayed improved activity against Gram-negative pathogens, including *Pseudomonas* spp., and was also active against some Gram-positive bacteria. Flumequine **3** [18] was the first quinolone to appear with a fluorine atom in the 6-position. In this compound, an additional ring forms a bridge between the 1- and 8-positions of the quinolone molecule.

1st Generation

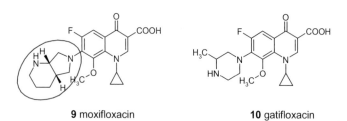

1 ICI (1954)

2 nalidixic acid

3 flumequine

4 pipemidic acid

2nd Generation

5 ofloxacin

6 norfloxacin

7 levofloxacin

8 ciprofloxacin

3rd Generation

9 moxifloxacin

10 gatifloxacin

Fig. 14.4 The chemical evolution of the quinolones.

The breakthrough in the development of quinolones came with the appearance of norfloxacin **6** [19], a second-generation quinolone which combined a 6-fluorine substituent with a piperazine ring in the 7-position of the basic compound. Additional quinolones then followed in rapid succession: pefloxacin [20], enoxacin [21] and fleroxacin [22] (Fig. 14.5). Particular mention must be made of ciprofloxacin **8** [23–25], ofloxacin **5** [26,27] and its active enantiomer levofloxacin **7** [28]. These quinolones have a broad spectrum of activity, which also includes Gram-positive bacteria and *Pseudomonas aeruginosa*, as well as favorable pharmacokinetics. The rapid absorption of these compounds from the gastrointestinal tract and their effective tissue penetration also allows them to be used for the treatment of systemic infections.

Gaps in the Gram-positive spectrum of quinolones and the increased resistance of Gram-positive cocci such as *Streptococcus pneumoniae* and *Staphylococcus aureus* to β-lactams and macrolides led to worldwide efforts to further optimize the quinolones. During the development of ciprofloxacin **8** in the mid-1980s, there was a major worldwide upsurge in research activities in the field of quinolones, as reflected by the increasing number of patent applications and publications which appeared. This was prompted by the new cycloacylation process discovered by Bayer, which allowed the introduction of special substituents into the 1-position of quinolones and therefore provided the possibility of producing compounds which had not previously been obtainable, or only with difficulty by conventional methods such as the Gould Jacobs process [29–31]. Such quinolones included for example the 1-cyclopropyl-, 1-(cis-2-fluorocyclopropyl)-, 1-(2,4-difluorophenyl)-, 1-*tert.*-butyl-, 1-(3-oxetanyl)-, 1-(1,1-dimethylpropargyl)-, 1-(bicyclo[1.1.1]pent-1-yl)-, and 1-(6-amino-3,5-difluoro-2-pyridyl)-quinolones.

Attempts to find effective structural variations included the modification of various substituents. One example of this was the search for new cyclic amines as substituents for the 7-position [32]. This research eventually resulted in third-generation quinolones, moxifloxacin **9** [33–37] and gatifloxacin **10** [38–44], which are well-tolerated compounds having considerably improved activity against Gram-positive pathogens, including those resistant to β-lactams and macrolides, while essentially retaining the same degree of activity against Gram-negative bacteria (with the exception of *P. aeruginosa*). In addition, these new quinolones are highly active against atypical pathogens (*Chlamydia*, *Mycoplasma*, *Legionella*, etc.), *M. tuberculosis* and anaerobic bacteria. Due to their activity spectrum and their rapid penetration, their use is focused on the treatment of respiratory tract infections.

Another method of classifying the quinolones based on medicinal aspects can be found in the literature [45,46].

The quinolone class of antibiotics has been investigated by many research teams in the past, and numerous reviews have reported on the chemistry [31,32,47–56], structure–activity relationships [56–66], and mode of action [7,8,56,65,67] of this class of compounds.

Some stages in the development of moxifloxacin are described in the following sections.

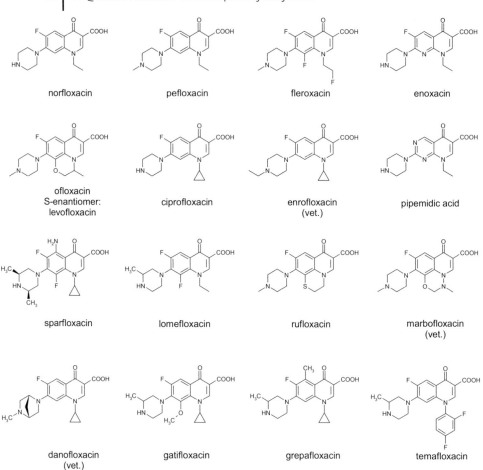

Fig. 14.5 Quinolones containing a piperazinyl residue.

14.2
Aims

Although ciprofloxacin for oral and intravenous administration already displayed a very high standard of antibacterial activity and tolerance, some properties still required optimization. These included the need for improved activity against Gram-positive pathogens (e.g., *S. pneumoniae*) and anaerobic bacteria, and kinetic properties allowing once-daily administration. The properties required by Bayer of the third-generation quinolones were therefore as follows.

The high standard provided by ciprofloxacin should be retained whilst filling in the gaps in the activity spectrum against Gram-positive bacteria. Any resulting compound should therefore:

- have excellent antibacterial activity against a broad spectrum of Gram-positive and Gram-negative bacteria, including anaerobes;
- induce a low degree of resistance;
- have excellent pharmacokinetic and pharmacodynamic properties which permit;
- "once-daily" administration;

In addition, the compound should be:
- well-tolerated and have few potential side effects and interactions with other drugs;
- suitable for oral and intravenous administration so as to provide the possibility for sequential therapy;
- clinically efficient; and
- inexpensive.

14.3
The Chemical Evolution of Moxifloxacin

At the start of and during the search for a third-generation quinolone, most of the commercially available quinolones contained an optionally substituted piperazinyl radical in the 7-position (Fig. 14.5). Of these compounds, only norfloxacin, pefloxacin, ofloxacin, levofloxacin, ciprofloxacin, enrofloxacin, marbofloxacin, danofloxacin and gatifloxacin still play a role today. The 6,8-difluoroquinolones fleroxacin, sparfloxacin and lomefloxacin, as well as grepafloxacin and temafloxacin, were of little importance or withdrawn from the market due to their adverse side effects [68,69].

The research-based pharmaceutical companies did however become interested at a very early stage in finding novel amines for the 7-position, in order to further optimize the overall properties of quinolones. After a series of extensive experiments, Bayer focused on a variety of 7-substituents, including the 3-amino- and the 3-amino-methyl-1-pyrrolidinyl substituents, which were modified in several ways.

With respect to the 1-position of the quinolones, it became clear that the cyclopropyl substituent, for which ciprofloxacin was well known, was an optimum radical, even though many attempts were made to replace this substituent [66]. 1-Cyclopropylquinolones substituted in the 8-position by chlorine or fluorine displayed the most powerful antibacterial activity. The search for suitable new amine substituents for the 7-position, which were prepared at Bayer especially for the quinolone screening program, involved first of all combining these compounds with a basic 1-cyclopropyl-8-halogenoquinolone compound. Numerous active compounds were thereby obtained, the antibacterial activity of which was tested in comparison with reference substances and for which additional biological data were gathered for the purpose of finding favorable effects.

Bayer thus developed at approximately the same time as two other companies (Kyorin and Warner-Lambert) the highly active compound 7-(3-amino-1-pyrrolidinyl)- 8-chloro-1-cyclopropyl-6-fluoro-1,4-dihydro-4-oxo-3-quinolinecarboxylic acid

(**11**; clinafloxacin, AM 1091, BAY V 3545, CI-960, PD 127 391), for which it also had patent priority [70–76]. Due to its insufficient tolerance, Bayer decided to abandon the development of BAY V 3545 while still in the preclinical phase, whereas Warner-Lambert/Parke-Davis continued to develop clinafloxacin into Phase III. It was found that the two enantiomers [77] of racemic clinafloxacin display virtually identical antibacterial activity and pharmacokinetic properties [78]. Due to the side effects involved (phototoxicity and a reduction the blood sugar levels), Parke-Davis withdrew its approval application for clinafloxacin for the intravenous treatment of hospitalized patients with severe infections [79].

11 clinafloxacin hydrochloride   **12** tosufloxacin tosylate

As with clinafloxacin, tosufloxacin **12**, which was also withdrawn from the market, has a 3-amino-1-pyrrolidinyl radical in the 7-position. This useful substituent was modified by attaching additional small radicals, thus resulting in the more recent, highly active quinolones sitafloxacin (**13**; DU 6859) [80–84], DK-507 k **14** [85,86] and olamufloxacin (**15**; HSR 903) [87–90]containing an (*S*)-3-amino-1-pyrrolidinyl radical in the 7-position, to which a cyclopropyl radical is attached spirocyclically in the 4-position. Due to the omission of the Cl substituent in sitafloxacin from DK-507 k, the latter's phototoxic potential was reduced. In 2003, Daiichi Pharmaceutical granted Pfizer the licensing rights to develop and market DK-507 k outside Asia. Pfizer did, however, abandon the development of DK-507 k in 2004 following results obtained in Phase I trials.

13 sitafloxacin   **14** DK-507 k   **15** olamufloxacin

The 3-aminomethyl-1-pyrrolidinyl radical, which was considered to be "bioisosteric" to piperazine [91], also produced highly active quinolones, which were however also found not to be tolerated. In trovafloxacin **16** [92], the 3-aminomethyl-1-pyrrolidinyl radical is integrated into a bicyclic amine structural unit with a free amino group [(1α,5α,6α)-6-amino-3-azabicyclo[3.1.0]hexane]. A correlation probably exists between the increased central nervous system (CNS) side effects obtained with this quinolone [93] and the 2,4-difluorophenyl radical in the 1-position. Severe hepatic intolerance reactions finally resulted in major restrictions on its use, which was then limited in the USA to severe clinical infections [68].

Gemifloxacin **17** contains a 3-aminomethyl-1-pyrrolidinyl radical modified by a methoxyimino group, and it has powerful antibacterial activity [94–98]. In contrast to gatifloxacin, levofloxacin and moxifloxacin, in which skin rashes occur as side effects in less than 1% of cases, this allergic reaction (which is not a photoreaction) occurs in 4.8% of gemifloxacin patients [12,99]. Gemifloxacin mesylate was originally licensed out for further development to SmithKlineBeecham by LG Life Sciences Ltd. (Seoul). Its development was, however, abandoned in the preregistration phase when the FDA expressed reservations in a "nonapprovable letter" in 2000 and upon the merger with Glaxo-Wellcome to form GSK. The rights in the product were returned in 2002 and then taken over by GeneSoft Pharmaceuticals (now Oscient), which continued with the clinical development for North America and Europe and has meanwhile obtained approval from the FDA for the treatment of (mild to moderate) community-acquired pneumonia (CAP) and acute exacerbation of chronic bronchitis (AECB). One of the analogues of gemifloxacin, which differs in an additional methyl group in the pyrrolidine ring, is the developmental product DW-286 **18** from Dong Wha. This product is characterized by a broad spectrum of activity and improved activity over gemifloxacin [100] against Gram-positive pathogens. In contrast to the racemic gemifloxacin, DW-286a is an enantiomerically pure compound [an (–)-enantiomer].

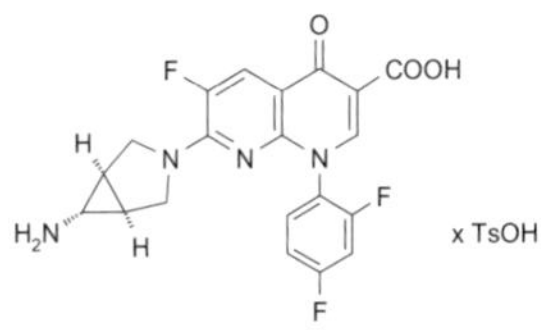

16 trovafloxacin tosylate 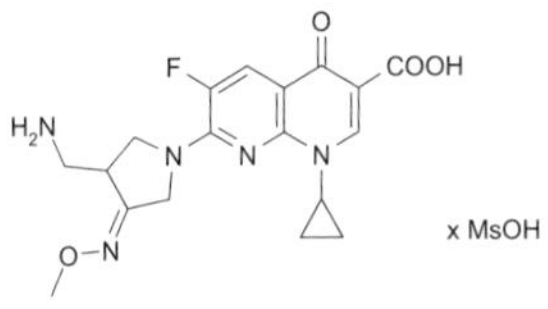   **17** gemifloxacin mesylate 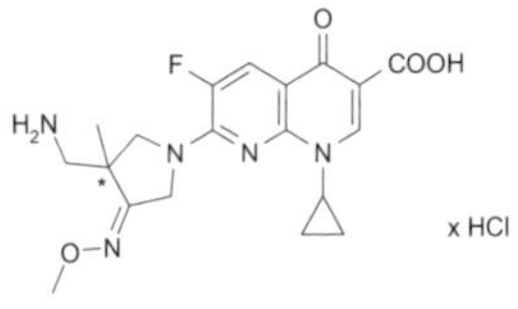   **18** DW-286

Powerful *in vitro* activity was also obtained by introducing 3-hydroxy- or 3-hydroxymethyl-pyrrolidine as a substituent, although the *in vivo* activity of such active compounds decreased considerably as a result. By combining the amino and hydroxyl groups, or by incorporating them into bridged structures, a large number of novel cyclic amines were obtained and tested for their suitability as structural units for quinolones (Fig. 14.6).

Some examples are described in the following section, which also demonstrates that successes obtained in drug research are usually preceded by a large number of disappointments as well as total or partial failures, all of which do however provide valuable information for further optimization.

Fig. 14.6 3-Amino-, 3-aminomethyl-, 3-hydroxy- and 3-hydroxymethylpyrrolidine: basic structures for novel bicyclic amines.

Thus, numerous diazabicyclo[3.3.0]octanes ("pyrrolidinopyrrolidines") were produced [101,102] and combined with various basic quinolone compounds, thereby resulting in quinolones which usually displayed powerful antibacterial activity [103,104]. However, these usually proved to be too toxic or displayed too great a reduction in their *in vivo* activity, and therefore were not investigated any further (cf. Fig. 14.7; X and R′ represent various radicals).

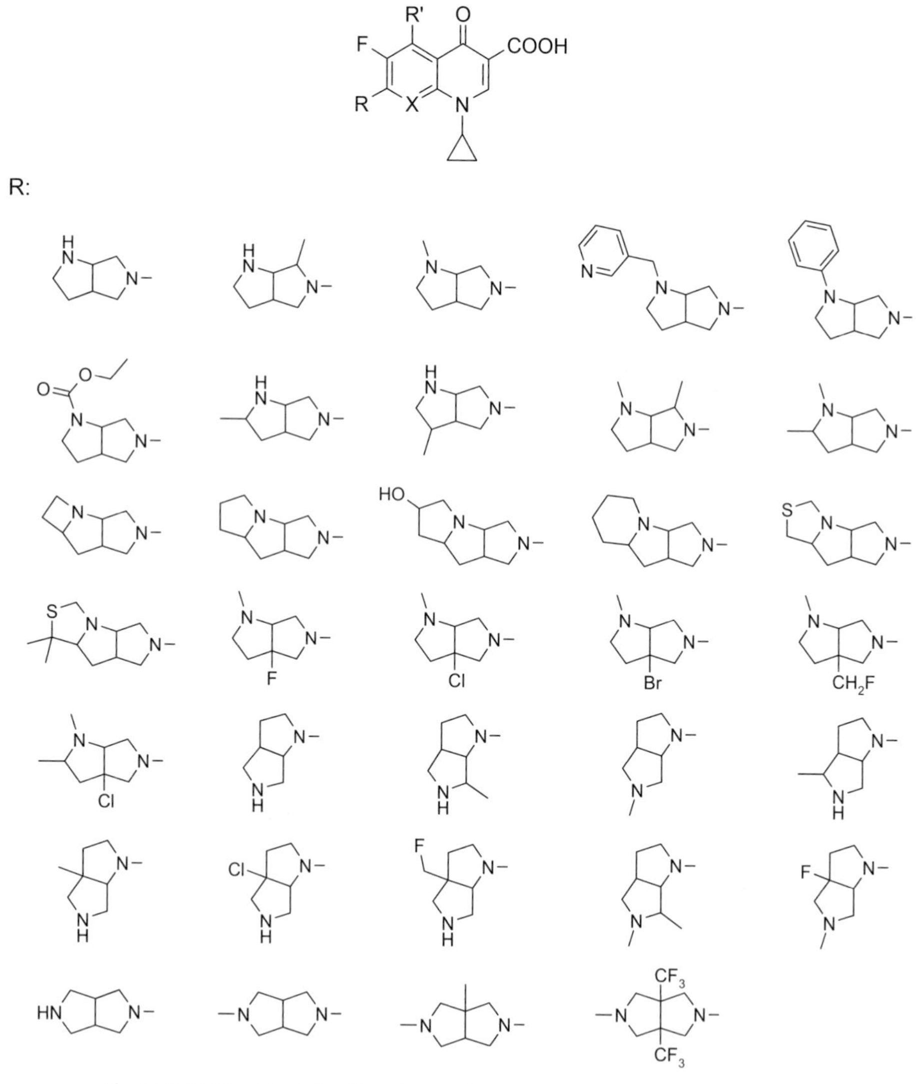

R:

Fig. 14.7 7-(Diazabicyclo[3.3.0]octyl)-quinolones.

The introduction of an additional oxygen atom into the cyclic amine resulted in 7-(2,7-diaza-3-oxabicyclo[3.3.0]oct-7-yl)-quinolones [103] which, while still displaying powerful *in vitro* activity, had only poor *in vivo* activity or none at all (Fig. 14.8).

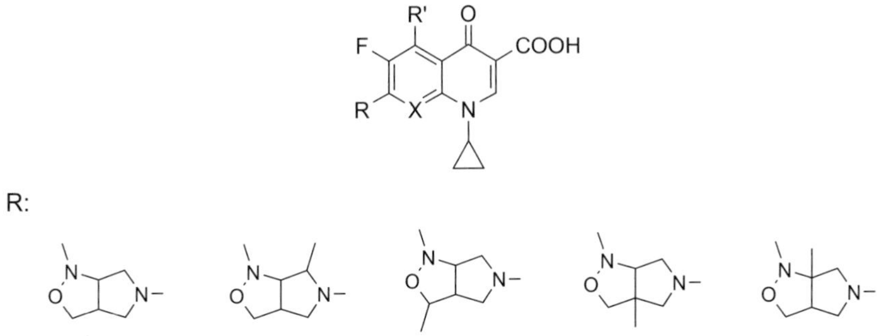

R:

Fig. 14.8 7-(2,7-Diaza-3-oxabicyclo[3.3.0]oct-7-yl)-quinolones.

The introduction of the free primary amino group of amino- or aminomethyl-pyrrolidinyl quinolones into nonfused ring systems produced active compounds which were highly active but not tolerated [105–107] (Fig. 14.9).

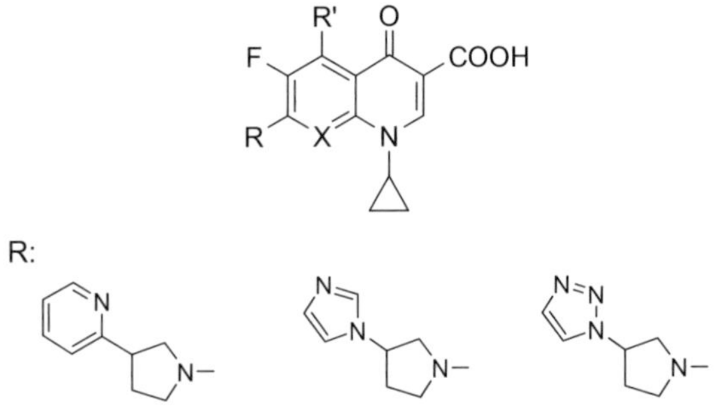

R:

Fig. 14.9 7-(3-Heterocyclyl-1-pyrrolidinyl) quinolones.

The quinolones depicted in Fig. 14.10 represent a combination of 3-amino-methyl- and 3-hydroxy-1-pyrrolidinyl quinolones [108], of which those containing less bulky radicals display improved *in vitro* activity over those containing bulky substituents. The inadequate *in vivo* properties, however, prevented any further development.

Examples of quinolones in which the 7-(3-amino-1-pyrrolidinyl) radical or the 7-(3-aminomethyl-1-pyrrolidinyl) radical is substituted by a ring fused via the 3,4-position are those listed in Fig. 14.11 (cf. also Ref. [109]) and Fig. 14.12 [109,110]. Despite their powerful *in vitro* and *in vivo* activities, most of the quinolones containing such amine structural units have problems of tolerance.

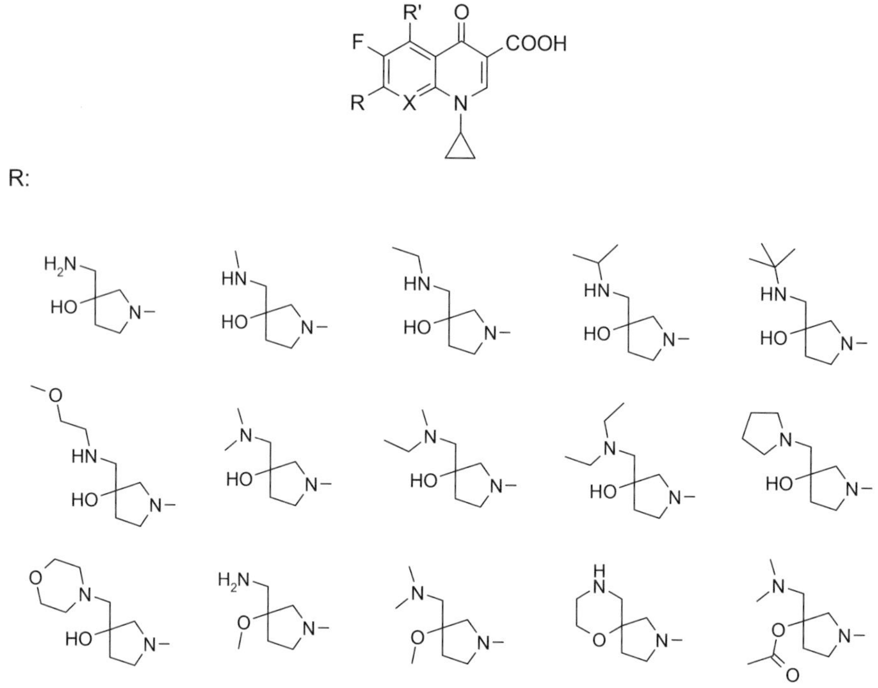

R:

Fig. 14.10 Various 7-([substituted] 3-aminomethyl-3-hydroxy-1-pyrrolidinyl) quinolones.

R:

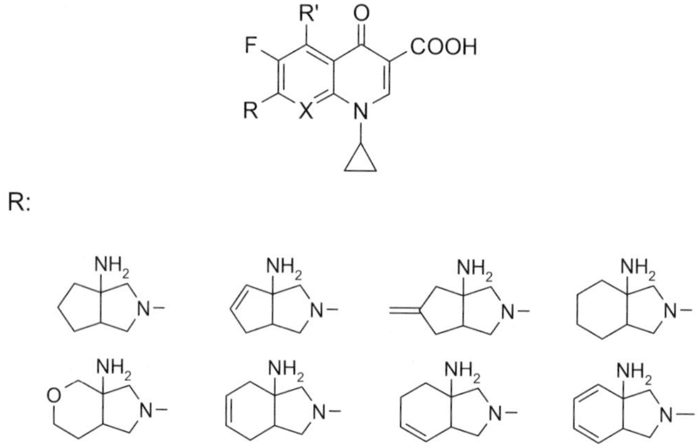

Fig. 14.11 Various 7-(3,4-bridged-3-amino-1-pyrrolidinyl)-quinolones.

The most promising quinolone from this series was the diastereomerically pure compound BAY 15–7828 **19** [111], which had an amino-tetrahydroisoindolyl substituent and was considered as a possible developmental candidate due to its excellent properties against Gram-positive and Gram-negative pathogens, including *Pseudomonas aeruginosa*, and its CHO value of 64, which shows that the sub-

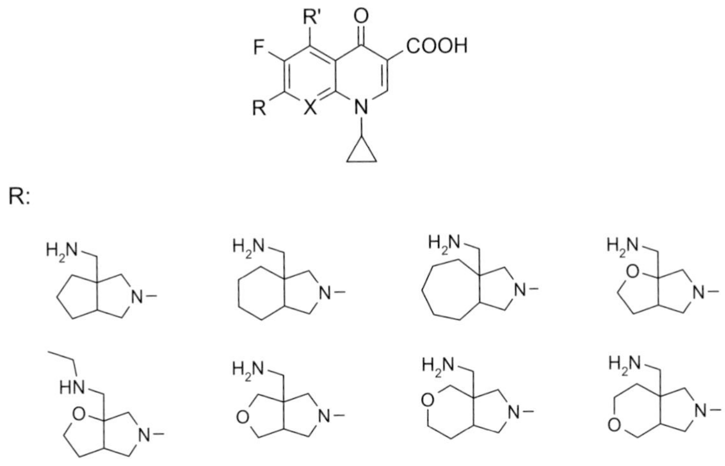

R:

Fig. 14.12 Various 7-(3,4-bridged-3-aminomethyl-1-pyrrolidinyl)-quinolones.

stance had no clastogenic effects in the Chinese hamster ovarian (CHO) cell chromosomal aberration test. Due to its neurotoxic side effects, the development of BAY 15–7828 was abandoned during the preclinical phase, however. These side effects were determined electrophysiologically in an *ex vivo* hippocampus slice model [112], in which the action potentials of brain cells were measured as a function of various active compound concentrations. In this model, BAY 15–7828 displayed CNS toxicity corresponding to that of trovafloxacin [113].

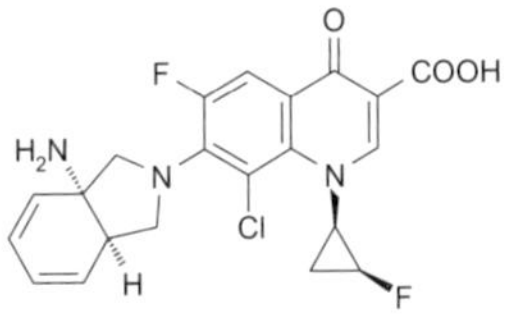

19 BAY 15-7828

Fluoroquinolones with oxygen-containing bicyclic amine radicals had the welcome effect of triggering the development of so-called "gastroquinolones" – that is, those also allowing the eradication of *Helicobacter pylori*, which colonizes the acidic gastric mucosa and is associated with a range of gastrointestinal diseases (type B gastritis, peptic ulcer disease, specific types of gastric carcinoma and MALT lymphoma). These "gastroquinolones" not only display powerful *in vitro* activity against *H. pylori* but are also, due to their slightly acidic pK_a value, capable of eradicating this bacterium *in vivo*, as shown in tests using the *H. felis* murine model and the *H. pylori* ferret model. The most effective candidates from this series of quinolones are the 7-[(+)-(1S,5R)-1-aminomethyl-2-oxa-7-aza-bicyclo[3.3.0]oct-7-yl]-8-cyano-1-cyclopropyl-6-fluoro-1,4-dihydro-4-oxo-quinolinecarboxylic acid hydrochloride **20** [114] and the (–)-8-cyano-1-cyclopropyl-6-fluoro-7-[(1S,6S)-2-oxa-5,8-diazabicyclo[4.3.0]non-8-yl]-1,4-dihydro-4-oxo-quinolinecarboxylic acid hydrochloride BAY 35–3377 **21** [115]. The clinical development of BAY 35–3377 embarked upon by Byk-Gulden in 1999 with the aim of replacing the conventional multidrug treatment of *H. pylori* infections, the compliance with which was poor, by a monodrug treatment, was abandoned.

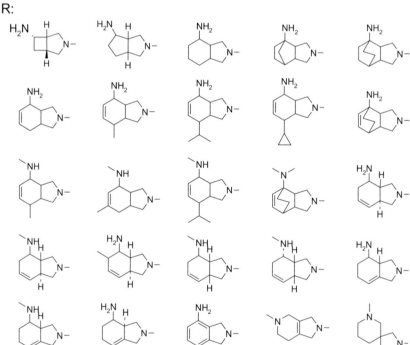

Fig. 14.13 Various quinolones substituted by 3-aminomethyl-1-pyrrolidinyl rings with the aminomethyl group inserted into a fused or spirocyclic ring system.

By introducing the aminomethyl group of 3-aminomethylpyrrolidine into a 3,4-fused ring or into a spirocycle, a large variety of bicyclic amines were synthesized which were combined with basic quinolone compounds (Fig. 14.13) (see also Ref. [109]). Quinolones with high *in vitro* and *in vivo* activity were obtained which were, however, inadequately tolerated. Quinolones with an additional double bond in the fused six-membered ring of the bicyclic amine are most effective, although their high cytotoxicity frequently makes them extremely toxic [116–125].

Among the entire range of compounds containing bridged 3-aminopyrrolidinyl radicals, the quinolonecarboxylic acids with a homologous 2,8-diazabicyclo[4.3.0]-non-8-yl radical proved to be highly useful, in contrast to the 7-(2,7-diazabicyclo[3.3.0]octyl)-quinolones shown in Fig. 14.7. As far as the 6,8-difluoroquinolones are concerned, it was for example found that the racemic BAY W 8801 **22** with a *cis*–2,8-diazabicyclo[4.3.0]non-8-yl substituent [103] had the greatest antibacterial activity compared to its C- or N-alkylation products (Tab. 14.1) and the corresponding quinolone BAY 11–4099 **29** containing the *trans*–2,8-diazabicyclo[4.3.0]non-8-yl substituent (Tab. 14.2). The isomers BAY X 8841 **30** and BAY X 8507 **31** (Tab. 14.2) were less active.

Tab 14.1 A comparison between BAY W 8801 **22** and its alkylation products. MIC values expressed as $\mu g \ mL^{-1}$.

Com-pound	R	Organism					
		E. coli Neumann	K. pneu-moniae 63	P. aeruginosa Walter	Staph. aureus 133	E. faecalis. ICB 27101	B. fragilis DSM 2151
22 W 8801		0.008	0.015	0.5	0.03	0.125	0.5
23		0.03	0.06	2	0.03	0.25	4
24		0.06	0.06	2	0.06	0.125	–

Tab 14.1 Continued.

Com-pound	R	Organism					
		E. coli Neumann	*K. pneu-moniae* 63	*P. aeruginosa* Walter	*Staph. aureus* 133	*E. faecalis.* ICB 27101	*B. fragilis* DSM 2151
25		0.06	0.125	4	0.125	0.5	2
26		0.03	0.125	4	0.03	0.25	2
27		0.06	0.25	8	0.06	0.5	1
28		1	4	2	4	128	32

Tab 14.2 A comparison between BAY W 8801 **22** and its isomers. MIC values expressed as $\mu g\ mL^{-1}$.

Com-pound	R	Organism					
		E. coli Neumann	K. pneu-moniae 63	P. aeruginosa Walter	Staph. aureus 133	E. faecalis. ICB 27101	B. fragilis ES 25
22 W 8801	cis, rac	0.008	0.015	0.5	0.03	0.125	1
29 11–4099	trans, rac	≤0.015	0.03	0.5	0.25	0.25	2
30 X 8841		≤0.015	0.03	1	0.06	0.06	4
31 X 8507		1	2	8	1	1	16
Ciprofloxacin		≤0.015	0.03	0.25	0.5	0.5	8

The replacement of one of the CH$_2$ groups of the bicyclic amine substituent of BAY W 8801 **22** by an oxygen or nitrogen atom did not produce any further increase in activity (Tab. 14.3) [103,126].

Tab 14.3 A comparison between BAY W 8801 **22** and its heteroanalogues. MIC values expressed as $\mu g\ mL^{-1}$.

Com-pound	R	Organism					
		E. coli Neumann	K. pneu-moniae 63	P. aeruginosa Walter	Staph. aureus 133	E. faecalis. ICB 27101	B. fragilis ES 25
22 W 8801	*cis, rac*	0.008	0.015	0.5	0.03	0.125	1
32	*cis, rac*	0.03	0.125	4	0.06	0.25	0.25
33*	*cis, rac*	0.03	0.125	1	0.06	0.125	16
34**	*cis, rac*	0.06	0.125	1	0.125	0.25	16
Ciprofloxacin		≤0.015	0.03	0.25	0.5	0.5	8

*, a betaine, the aminal bond is cleaved at pH 3; **, a dihydrochloride.

An increase in activity was obtained, however, by resolving the racemic BAY W 8801 **22** into its pure enantiomers. Of these, the enantiomer BAY X 8843 **36** with the *S,S* configuration proved to be two- to four-times more active *in vitro* than the corresponding *R,R* enantiomer BAY X 8842 **35** (Tab. 14.4) [127].

Tab 14.4 A comparison between BAY W 8801 **22** and its enantiomers. MIC values expressed as $\mu g\ mL^{-1}$.

Com-pound	R	Organism					
		E. coli Neumann	*K. pneu-moniae* **63**	*P. aeruginosa* Walter	*Staph. aureus* **133**	*E. faecalis.* ICB 27101	*B. fragilis* ES 25
22 W 8801	*cis, rac*	0.008	0.015	0.5	0.03	0.125	1
35 X 8842	R,R	0.015	0.03	1	0.06	0.125	2
36 X 8843	S,S	0.004	0.008	0.25	0.03	0.06	0.5

BAY X 8843 **36** appeared to be a useful developmental candidate due to its powerful *in vitro* and *in vivo* activity and insignificant neurotoxicity [113]. Preclinical development was abandoned however due to the toxicological effects obtained in monkeys.

Similar structures were also later disclosed in the literature, for example the compound described by the Korea Research Institute of Chemical Technology (KRICT): 1-cyclopropyl-6,8-difluoro-7-[(2,8-diazabicyclo[4.3.0]non-5-en)-8-yl]-1,4-dihydroquinoline-4-oxo-3-carboxylic acid with an additional double bond in the bicyclic amine **37** [128].

36 BAY X 8843   **37** quinolone described by KRICT

All of the quinolones with an S,S–2,8-diazabicyclo[4.3.0]non-8-yl radical (CAS: 4aS-*cis*-octahydro-pyrrolo[3.4-b]pyridine) synthesized by us proved to be more effective than the corresponding active compounds with an R,R–2,8-diazabicyclo[4.3.0] non-8-yl substituent. Some examples of these compounds are shown in Tab. 14.5.

Tab 14.5 MIC-values of various 7-(S,S–2,8-diazabicyclo[4.3.0]non-8-yl)-quinolones ($\mu g\ mL^{-1}$).

Compound	X	R^1	R^5	E. coli Neumann	S. aureus 133	B. fragilis ES 25
38	C-F	C$_2$H$_5$	H	0.03	0.125	32
39	C-H	(—N$_3$)	H	1	> 128	> 32
40	C-F	(F, F)	H	0.06	0.25	8
41	C-F	(F, F)	H	1	4	> 32
42	C-H	(*tert*-butyl)	H	0.125	0.25	32
43	C-H	(allyl)	H	0.06	0.25	4
44 BAY Y 3114	C-H	(cyclopropyl)	H	0.008	0.06	2
36 BAY X 8843	C-F	(cyclopropyl)	H	0.004	0.03	0.5
45 BAY Y 3118	C-Cl	(cyclopropyl)	H	0.004	0.015	≤0.015
46	C-Br	(cyclopropyl)	H	≤0.015	≤0.015	0.5
47 BAY 12–8039	C-OCH$_3$	(cyclopropyl)	H	≤0.015	0.03	1
48	C-OCHF$_2$	(cyclopropyl)	H	0.03	0.06	2
49	C-CH=CH$_2$	(cyclopropyl)	H	0.06	0.125	1
50	C-C≡CH	(cyclopropyl)	H	≤0.015	0.03	0.06

Tab 14.5 Continued.

x HCl

Compound	X	R¹	R⁵	Organism		
				E. coli Neumann	S. aureus 133	B. fragilis ES 25
51	C-CF$_3$		H	0.06	0.25	4
52	N		H	≤0.015	0.125	0.05
53	C-F		F	0.015	0.06	1
54	C-F		NH$_2$	≤0.015	≤0.015	0.5
55	C-OCH$_3$		NH$_2$	0.06	0.06	4
56	C-H		CH$_3$	0.03	0.06	2
57	N		CH$_3$	0.03	0.125	
58	C-H		H	2	4	32
59	C-Cl		H	≤0.015	0.25	4
60	C-H		H	0.03	0.125	> 32
61	C-H	trans, rac	H	0.03	0.125	16
62	N	cis, rac	H	0.03	0.25	16
63	N	trans, rac	H	0.03	1	16

Tab 14.5 Continued.

x HCl

Compound	X	R¹	R⁵	Organism		
				E. coli Neumann	S. aureus 133	B. fragilis ES 25
64	C-Cl	cis, rac (cyclopropyl-F)	H	≤0.015	≤0.015	0.5
65	C-Cl	trans, rac (cyclopropyl-F)	H	≤0.015	0.25	4
66	C-F	trans, rac (cyclopropyl-F)	F	0.03	1	16
67	C-F	trans, rac (cyclopropyl-F)	NH₂	≤0.015	0.25	8
68	C-OCH₃	cis, rac (cyclopropyl-F)	H	≤0.015	0.125	2
69	C-Br	trans, rac (cyclopropyl-F)	H	0.06	0.25	2
70	C-F	(difluorophenyl)	(allyl)	0.5	0.125	1
71	N	(difluorophenyl)	H	≤0.015	0.125	> 32
72	(C-O-isobutyl)		H	0.06	0.25	2

In addition to BAY X 8843 **36**, its 8-chlorine analogue 8-chloro-1-cyclopropyl-7-(*S*,*S*–2,8-diazabicyclo[4.3.0]non-8-yl)-6-fluoro-1,4-dihydro-4-oxo-3-quinolinecarboxylic acid hydrochloride BAY Y 3118 **45** was also an important developmental candidate. Due to its broad activity spectrum, which included anaerobic bacteria [129–136], it had the potential for becoming a "problem-solver". Unfortunately, during the Phase I trials phototoxic side effects occurred [137] which were due to the chlorine substituent in the 8-position, and these led to an abandonment of the development of this compound.

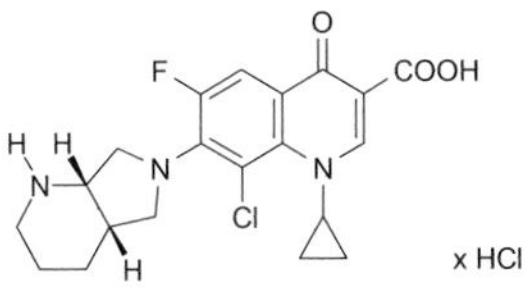

x HCl

45 BAY Y 3118

The corresponding 8-methoxyquinolonecarboxylic acid BAY 12–8039 (moxifloxacin hydrochloride) **47** had already been reported at the 32nd ICAAC, when BAY Y 3118 was first presented, as a comparative product to the latter, in the form of its betaine **9** (BAY Y 6957) [138]. Although it proved to be slightly less active *in vitro* compared to BAY Y 3118, it was – in contrast to the 8-chlorine analogue – photo-resistant. The excellent activity of this 8-methoxyquinolone, in particular against Gram-positive pathogens and anaerobic bacteria, its advantageous pharmacokinetic properties and its high tolerance without any concomitant phototoxicity, were the reasons why the decision was taken to develop this compound clinically. Due to its superior solubility [34], the hydrochloride BAY 12–8039 **47** (24 mg mL^{-1} in water at 25 °C) was given priority over the corresponding tosylate BAY 11–6371 **73** (1.5 mg mL^{-1} in water at 25 °C).

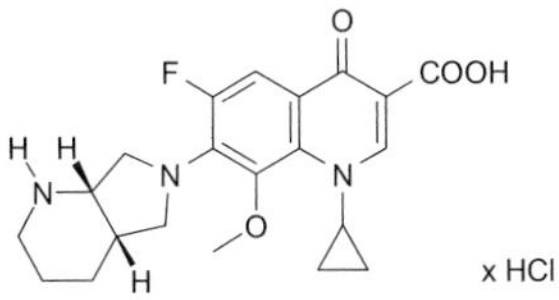

x HCl

x TsOH

47 BAY 12-8039
moxifloxacin hydrochloride

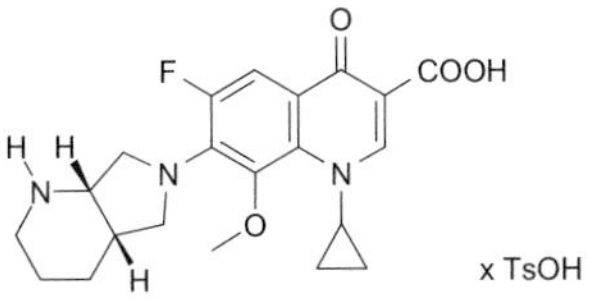

73 BAY 11-6371
moxifloxacin tosylate

14.4
Synthetic Routes

14.4.1
S,S–2,8-Diazabicyclo[4.3.0]nonane

The synthesis of the racemic *cis*–2,8-diazabicyclo[4.3.0]nonane **80** is based on pyridine-2,3-dicarboxylic acid **74**, which is converted with benzylamine via the anhydride **75** into pyridine-2,3-dicarboxylic acid *N*-benzylimide **77**. The pyridine ring is hydrogenated catalytically in the presence of Pd-C, both rings being *cis*-bonded in the resulting bicyclically substituted succinimide **78**. The subsequent reduction of the imide **78** is carried out with LiAlH₄ to give *cis*–8-benzyl-2,8-diazabicyclo[4.3.0]-nonane **79**, which is debenzylated hydrogenolytically to form *cis*–2,8-diazabicyclo[4.3.0]nonane **80** (Scheme 14.1) [103,139].

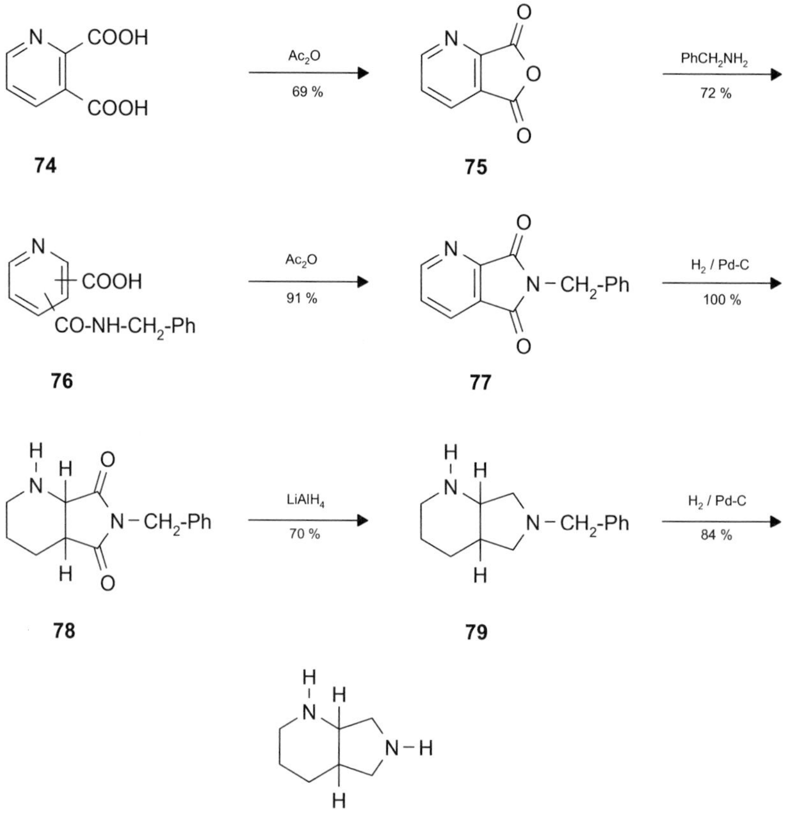

Scheme 14.1 Synthesis of the racemic *cis*–2,8-diazabicyclo[4.3.0]nonane **80**.

The enantiomerically pure amines **82** and **84** are for example prepared by resolution of the racemic 8-benzyl-*cis*–2,8-diazabicyclo[4.3.0]nonane **79** using natural *R,R*(+)-tartaric acid, whereupon the diastereomerically pure *R,R*-tartrate of the *R,R*-enantiomer is crystallized from dimethylformamide (DMF) and can be purified by recrystallization from methoxyethanol. The target *S,S*-enantiomer contained in the mother liquor is first converted into the free base, which is then, for the purpose of further purification, precipitated with *S,S*(–)-tartaric acid to give the diastereomerically pure *S,S*-tartrate. The *S,S*-enantiomer **83** is then liberated with sodium hydroxide solution. The *R,R*-enantiomer **81** is obtained in the same way. Separation of the enantiomers can also be carried out with high optical yields in an aqueous/alcoholic solution [140]. The hydrogenolytic debenzylation of **81** and **83** produces the corresponding pure *R,R*– and *S,S*-2,8-diazabicyclo[4.3.0]nonanes **82** and **84** (Scheme 14.2) [129].

An enzymatic method for resolution of the racemic *cis*–2,8-diazabicyclo[4.3.0]-nonane has also been described, in which reaction of the *S,S*-enantiomer with ethyl acetate in the presence of lipase produces the *S,S*-diacetyl derivative under the reaction conditions and the *R,R*-enantiomer is only reacted to form the mono-acetyl derivative. The components can be readily separated from the resulting reaction mixture, and then deacetylated [141].

14.4.2
Preparation of BAY X 8843 36 and its Analogues

The bicyclic amine **84** can be regioselectively reacted at the pyrrolodine nitrogen atom with 1-cyclopropyl-6,7,8-trifluoro-1,4-dihydro-4-oxo-3-quinolinecarboxylic acid **85** [142] in a mixture of acetonitrile and DMF in the presence of an auxiliary base, such as for example 1,4-diaza[2.2.2]octane (DABCO), without the piperidine nitrogen atom having to be previously masked by a protective group. The active compound is first isolated in the form of the betaine **86**, which is then converted into the hydrochloride **36** using dilute hydrochloric acid (Scheme 14.3) [129]. Other quinolones containing the *S,S*-2,8-diazabicyclo[4.3.0]non-8-yl radical are obtained analogously (see Tab. 14.5).

14.4.3
Preparation of Moxifloxacin Hydrochloride 47

Two methods are known for the production of moxifloxacin hydrochloride (CAS-name: 1-cyclopropyl-6-fluoro-1,4-dihydro-8-methoxy-7-[(4a*S*,7a*S*)-octahydro-6*H*-pyrrolo[3,4-b]pyridin-6-yl]-4-oxo-3-quinolinecarboxylic acid monohydrochloride) (Scheme 14.4).

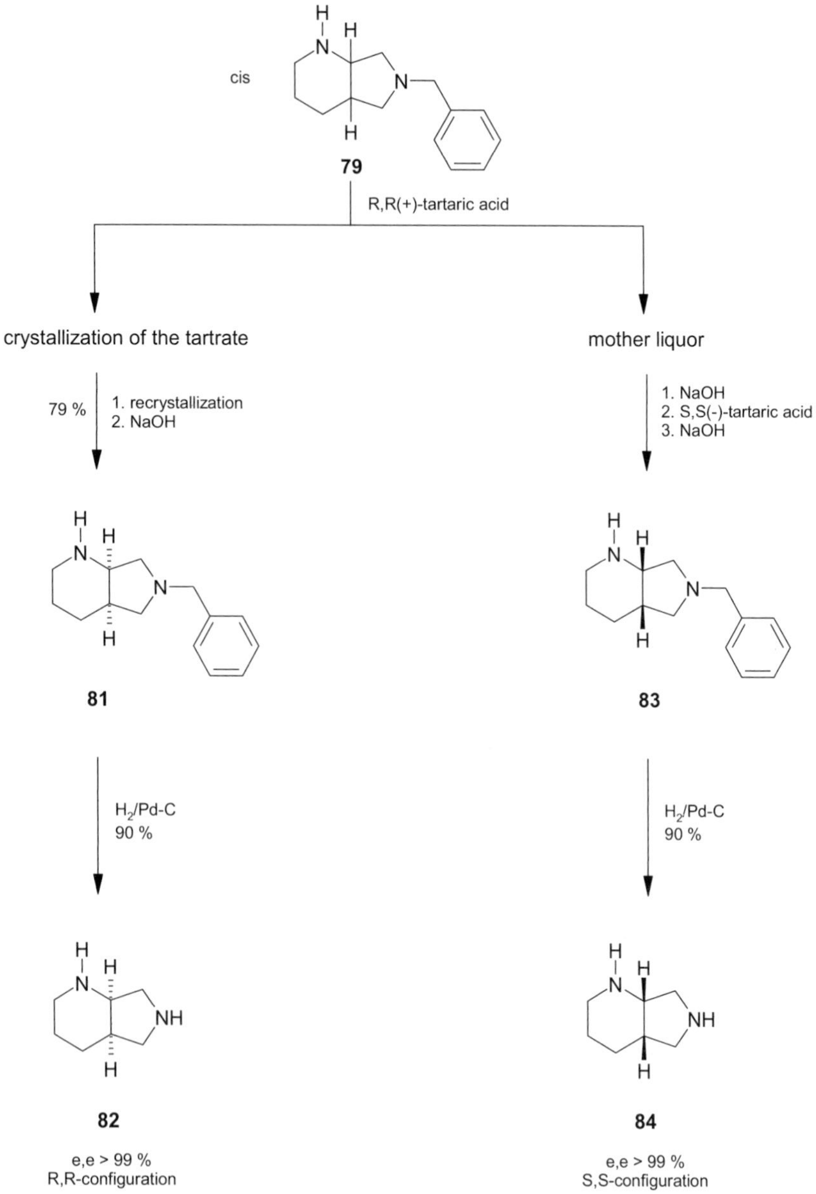

cis

79

R,R(+)-tartaric acid

crystallization of the tartrate

mother liquor

79 % | 1. recrystallization
2. NaOH

1. NaOH
2. S,S(-)-tartaric acid
3. NaOH

81

83

H₂/Pd-C
90 %

H₂/Pd-C
90 %

82

84

e,e > 99 %
R,R-configuration

e,e > 99 %
S,S-configuration

Scheme 14.2 The synthesis of enantiomerically pure 2,8-diazabicyclo[4.3.0]nonanes.

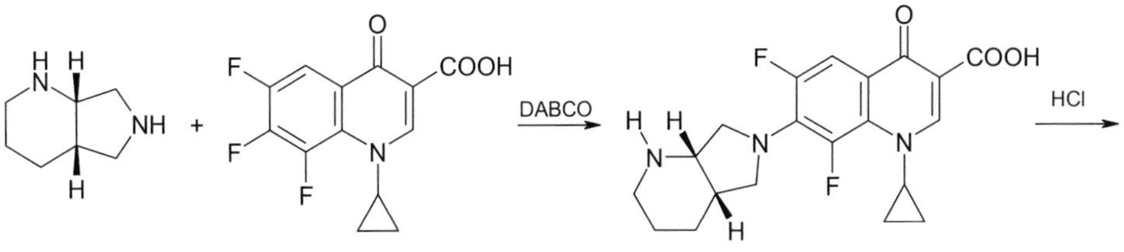

84   **85**   **86**

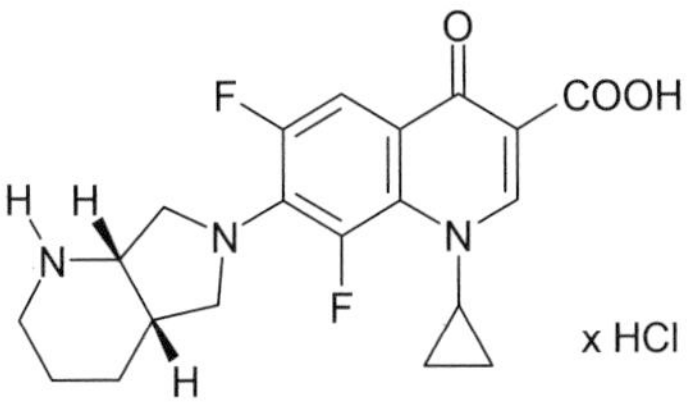

36 BAY X 8843
(S,S-configuration)

The following compounds were prepared analogously:

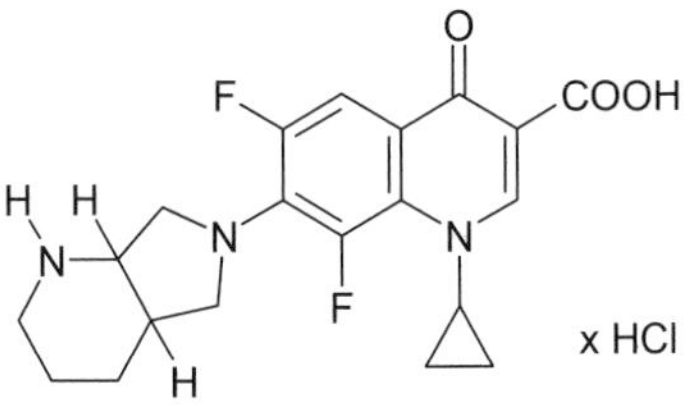

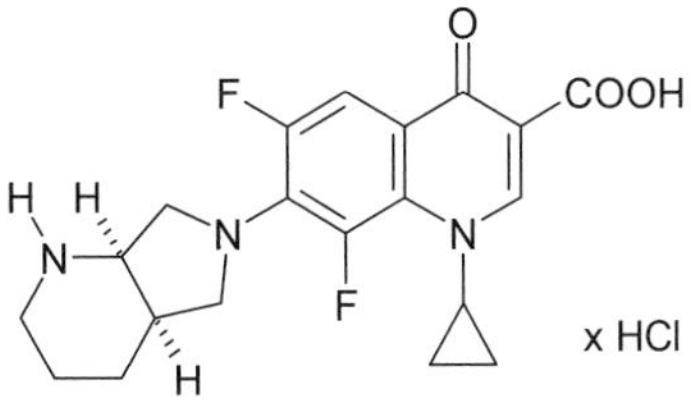

22 BAY W 8801
(racemate)

35 BAY X 8842
(R,R-configuration)

Scheme 14.3 Synthesis of BAY W 8801, BAY X 8842 and BAY X 8843.

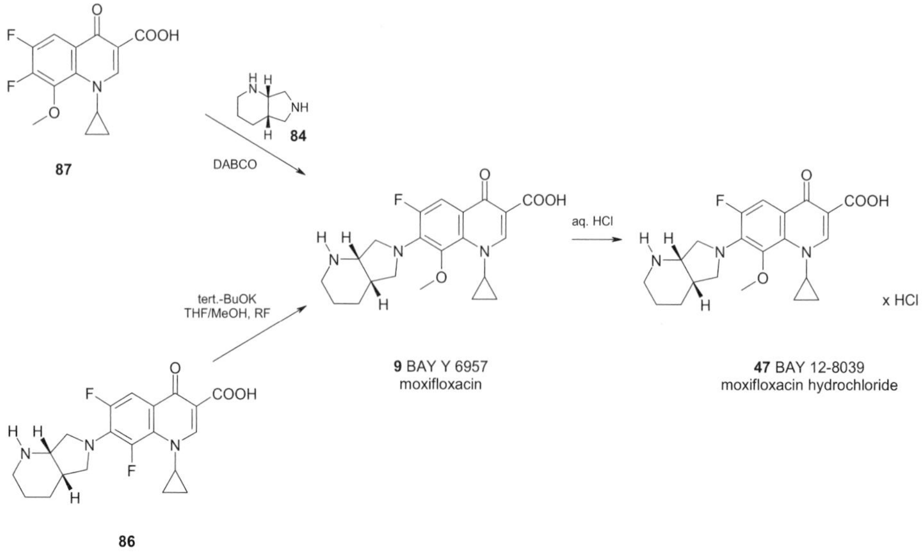

Scheme 14.4 Synthesis of moxifloxacin hydrochloride **47**.

One method of preparation is carried out analogously to the synthesis of BAY X 8843 by the regioselective reaction of 1-cyclopropyl-6,7-difluoro-1,4-dihydro-8-methoxy-4-oxo-3-quinolinecarboxylic acid **87** [143,144] with *S,S*–2,8-diazabicyclo[4.3.0]nonane **84** to give the betaine **9**, which is then converted into the hydrochloride with dilute hydrochloric acid **47** [34,103,139]. In this reaction, the hydrochloride crystallizes out in the form of the monohydrate in an acicular form; the monohydrate can, however, also be isolated in the form of prisms under special conditions [145].

A further process is based on **86**, the betaine of BAY X 8843 **36**. This product is treated with potassium *tert.*-butylate in tetrahydrofuran/methanol, with fluorine being replaced by methoxy in the 8-position to give **9**, which is then converted into the hydrochloride **47** [146].

14.5
The Physicochemical Properties of Moxifloxacin

Moxifloxacin has a distribution coefficient of log $P_{o/w}$ = –1.9 (water)/–0.6 (buffer, pH = 7). The pK_a values are pK_{a1} = 6.4 and pK_{a2} = 9.5, and its isoelectric point is pH 7.95. Moxifloxacin has high thermal and hydrolytic stability, and is also relatively stable in the presence of oxidizing agents. Compared with the former quinolones, moxifloxacin has excellent photostability both in the solid state and in solution. It decomposes only slightly under drastic irradiation conditions (Tab. 14.6) [34].

As was found by determining the molecular weight of moxifloxacin by analytical ultracentrifugation (54 000 rpm., 16 h), when moxifloxacin is dissolved at a concentration of 0.01% in an aqueous solution (phosphate buffer, pH 7, with 0.9% NaCl) at room temperature, it is present in the form of a dimer. In contrast, when it is dissolved at a concentration of 0.16%, it is present in the form of a mixture of associated molecules, presumably consisting of di- and tetramers and having an average degree of association of 3.5 [147]. Analytical ultracentrifugation using the method of "sedimentation diffusion equilibrium" is a reliable method of determining absolute molecular weights, and is particularly suitable for examining association–dissociation equilibria. This method was used for the first time for examining this aspect of quinolones.

Tab 14.6 The stability of moxifloxacin hydrochloride under various conditions.

Thermal stability	2 years at 30 °C; 6 months at 60 °C	Stable
Hydrolytic stability		
0.1% solution in water	1 week, 90 °C	Almost stable, approx. 3% decomposition
0.1% solution in 0.1 M HCl	1 week, 90 °C	Approx. 9% decomposition
0.1% solution in phosphate buffer (p. 7)	1 week, 90 C	Approx. 15% decomposition
0.1% solution in 0.1M NaOH	1 week, 90 C	Almost stable, approx. 6% decomposition
Effects of light		
Solid substance	6 h xenon light (150 kLux; 300–800 nm; approx.	Chemically stable, discoloration of the surface
0.1% solution in water	613 Wh m^{-2})	Approx. 5% decomposition
Stability to oxidants		
Air	4 weeks, 90 °C; open storage	Stable
0.1% solution in water, addition of 3% H$_2$O$_2$	1 week, RT	Approx. 1% decomposition
0.1% solution in a buffer (pH 7), addition of 3% H$_2$O$_2$	1 week, RT	Approx. 16% decomposition

RT: room temperature.

14.6
The Microbiological and Clinical Properties of Moxifloxacin

Moxifloxacin has a broad spectrum of activity against a large number of Gram-positive and Gram-negative pathogens and anaerobes. An overview of the *in vitro* activity of moxifloxacin against some selected pathogens compared with other fluoroquinolones is provided in Tab. 14.7. These data show moxifloxacin to be one of the most effective compounds against *Staphylococcus* spp. and *Enterococcus* spp. Moxifloxacin's activity against Gram-negative bacteria, excluding *Pseudomonas aeruginosa*, is similar to that of ciprofloxacin.

Moxifloxacin is more effective against anaerobes than ciprofloxacin or sparfloxacin, but has approximately the same activity as clindamycin (Tab. 14.8). In 2005 it was approved by the FDA as the first quinolone for the treatment of complicated intra-abdominal infections.

The rapid bacterial action of moxifloxacin was demonstrated by *in vitro* investigations of its killing kinetics against *S. pneumoniae* and *S. aureus* compared with sparfloxacin, ofloxacin, ciprofloxacin, penicillin G, amoxicillin, amoxicillin/clavulanic acid, cefuroxime, and clarithromycin. Moxifloxacin was found to have the most powerful bactericidal activity and dose-dependent killing kinetics over its entire concentration range. By contrast, the comparative substances only produced maximum killing rates under specific concentrations, above which no further increase occurred. There was also no correlation between these killing rates and the minimum inhibitory concentration (MIC) values [33,148].

Due to its powerful specific activity against commonly isolated community-acquired respiratory tract pathogens [33,149–158], including penicillin-sensitive and -resistant *Streptococcus pneumoniae*, methicillin-sensitive *Staphylococcus aureus*, *Haemophilus* spp., *Moraxella catarrhalis* and atypical pathogens such as *Mycoplasma pneumoniae*, *Chlamydia pneumoniae* and *Legionella pneumophila* and *Klebsiella pneumoniae* and anaerobic bacteria [159–162], moxifloxacin was developed as a respiratory tract anti-infective [163–168].

Of the above-mentioned pathogens, *S. pneumoniae* is the most common bacterial cause of CAP [169], acute bacterial sinusitis (ABS) [170], acute otitis media [171], and meningitis. *S. pneumoniae* also plays a major role in the acute exacerbation of chronic bronchitis (AECB) [172]. Pneumonia is the sixth most common cause of death in the USA [173]. The increased resistance of *S. pneumoniae* to penicillins and macrolides, and the occurrence of strains with reduced sensitivity to third-generation cephalosporins, highlighted the urgent need for new antibiotics with powerful activity against *S. pneumoniae* [174,175]. Moxifloxacin is more effective *in vitro* against *S. pneumoniae* than ciprofloxacin, norfloxacin, sparfloxacin, ofloxacin, levofloxacin, lomefloxacin, and grepafloxacin [150,151,176–179], and has comparable activity to trovafloxacin [149,180,181]. Also, its *in vitro* activity against pneumococci, and in particular penicillin-resistant strains, is greater than that of azithromycin, clarithromycin, amoxicillin/clavulanic acid and oral β-lactams such as cefuroxime [149,176,177, 182]. Moxifloxacin's powerful activity against *S. pneumoniae* is evidently due both to its increased action against DNA gyrase and topoisomerase IV and to the reduced efflux from the bacterial cell [183].

Tab 14.7 Antibacterial activity of moxifloxacin hydrochloride compared to other quinolones [34].

Organism		MIC [mg L^{-1}]								
		Moxifloxacin	Ciprofloxacin	Trovafloxacin	Tosufloxacin	Gatifloxacin	Ofloxacin	Lomefloxacin	Clinafloxacin	Sparfloxacin
E. coli	Neumann	0.008	0.004	0.004	0.008	0.008	0.015	0.004	0.015	0.015
E. coli	ATCC 25922	0.008	0.004	0.004	0.008	0.008	0.015	0.004	0.015	0.015
K. pneumoniae	8085	0.015	0.008	0.015	0.008	0.008	0.015	0.015	0.008	0.008
K. pneumoniae	63	0.03	0.015	0.03	0.015	0.015	0.015	0.015	0.008	0.008
E. cloacae	2427	0.03	0.03	0.03	0.03	0.03	0.03	0.06	0.03	0.03
E. aerogenes	ICB 5240	1	1	1	1	1	1	1	1	1
Morganella morganii	932	0.03	0.03	0.015	0.03	0.03	0.06	0.06	0.03	0.06
Providencia sp.	12012	0.06	0.03	0.06	0.03	0.06	0.12	0.12	0.06	0.12
Providencia sp.	12052	0.5	0.25	0.5	0.25	0.5	1	1	0.5	1
Serratia marcescens	16040	1	0.5	1	0.5	1	0.5	0.5	0.25	1
Micrococcus luteus	9341	0.25	0.25	0.25	0.25	0.25	0.25	0.25	0.25	0.25
Staph. aureus	ICB 25701	1	16	1	1	1	16	16	1	1
Staph. aureus	ATCC 29213	0.03	0.5	0.03	0.03	0.12	0.5	0.5	0.03	0.03
Staph. aureus	133	0.03	0.5	0.03	0.03	0.25	0.5	0.5	0.03	0.03
E. faecalis	27101	0.06	0.25	0.12	0.06	0.06	0.5	0.5	0.25	0.06
E. faecalis	9790	0.06	0.25	0.12	0.06	0.12	0.5	0.5	0.25	0.06
Acinetobacter sp.	14068	0.015	0.12	0.03	0.03	0.03	0.25	0.25	0.03	0.015
P. aeruginosa	Walter	0.5	0.06	0.1	0.1	0.25	1	1	0.1	0.1
P. aeruginosa	ATCC 27853	0.5	0.06	0.06	0.06	0.25	1	1	0.1	0.1

Tab 14.8 Antibacterial activity of moxifloxacin hydrochloride against anaerobic bacteria compared to other antibacterial agents [34].

Organism		MIC [mg L^{-1}]				
		Moxifloxacin	Sparfloxacin	Ciprofloxacin	Metronidazole	Clindamycin
Bacteroides fragilis	ES 25	0.25	2	8	4	0.25
Bacteroides fragilis	GA 1–0558	0.125	2	4	1	0.25
Bacteroides fragilis	ATCC 25285	0.5	2	nt	nt	0.5
Bacteroides fragilis	DSM 2151	0.25	1	4	1	0.25
Bacteroides fragilis	4661	0.5	2	8	0.5	>64
Bacteroides thetaiotaomicron	DSM 2079	1	2	8	1	1

nt: not tested

Moxifloxacin hydrochloride was approved for the peroral treatment of AECB, CAP and ABS, and is also available as an infusion solution for the sequential treatment of CAP. In 2004, it was additionally approved by the FDA as the first intravenously and perorally administrable antibiotic for treating CAP, which is caused by multidrug-resistant *S. pneumoniae* (MDRSP). MDRSP is understood to include those strains of *S. pneumoniae* that are resistant to two or more of the following classes of antibiotics: penicillins and second-generation cephalosporins such as cefuroxime, macrolides, tetracyclines and trimethoprim/sulfamethoxazole.

A very comprehensive multicenter, randomized, double-blind study of two parallel treatment arms (the MOSAIC study; MOxifloxacin compared to Standard therapy in Acute Infectious exacerbations of Chronic infections) demonstrated the powerful clinical activity of moxifloxacin for the treatment of AECB. Five-day treatment with moxifloxacin (400 mg, once daily for 5 days) was found to produce clinical cure rates that were superior to those achieved with 7-day treatment with a standard antibiotic (amoxicillin 500 mg three times daily for 7 days; clarithromycin 500 mg twice daily for 7 days; cefuroxime 250 mg twice daily for 7 days) [184].

In ABS caused by penicillin-resistant *S. pneumoniae*, moxifloxacin is recommended as the agent of first choice. A study has shown that a cure rate of 97% is obtained with 7-day treatment, and a cure rate of 100% with 10-day treatment [185].

The effectiveness of moxifloxacin (MIC$_{90}$ 0.06–0.125 mg L^{-1}) against methicillin-sensitive *S. aureus* (MSSA) is four- to eight-fold greater than that of ciprofloxacin, ofloxacin, and levofloxacin, although its *in vitro* activity against methicillin-resistant *S. aureus* (MRSA) is less powerful (MIC$_{90}$ 1–8 mg L^{-1}) [33,149,150, 182,186].

Moxifloxacin is also highly active against Gram-negative respiratory tract pathogens such as *H. influenzae* (MIC$_{90}$ 0.03–0.06 [0.125] mg L^{-1}) [33,149–151,182],

H. parainfluenzae (MIC_{90} 0.03–0.25 mg L^{-1}) [150], M. catarrhalis (MIC_{90} 0.03–0.125 mg L^{-1}) [149,151,182] and K. pneumoniae [33,182,187,188], although its activity against specific bacteria is less powerful than that of ciprofloxacin [188]. When moxifloxacin was used in combination with cefepime or piperacillin/tazobactam against clinical isolates of K. pneumoniae, Enterobacter cloacae and Acinetobacter baumannii, an increase in bactericidal activity was obtained compared to that of the individual compounds; this was due to synergistic effects between the different compounds [189].

The powerful activity of the more recent fluoroquinolones moxifloxacin, gatifloxacin and levofloxacin against clinical isolates of S. pneumoniae, H. influenzae and M. catarrhalis isolated in various European countries in 2000–2001 and the low rates of resistance to these compounds have been demonstrated in a clinical study [190].

14.6.1
Mycobacterium tuberculosis

In the treatment of tuberculosis, resistant strains of M. tuberculosis (multidrug-resistant tuberculosis, MDRTB) present a growing problem, so that new antituberculotic agents are required which act according to a different mechanism to that of standard agents such as isoniazid, rifampicin, pyrazinamide, and ethambutol. The more modern fluoroquinolones are of particular interest, and in particular moxifloxacin, which has powerful in vitro and in vivo activity and, in contrast to sparfloxacin and clinafloxacin, is not phototoxic [191].

Moxifloxacin's MIC_{90} value of 1 mg L^{-1} means that it has the same in vitro activity against M. tuberculosis as levofloxacin, and is more effective than ofloxacin (MIC_{90} = 2 mg L^{-1}) and ciprofloxacin (MIC_{90} = 4 mg L^{-1}) [192–194]. A combination of moxifloxacin and isoniazid proved to be more effective in vivo than the individual compounds [195,196], whereas a combination with ethambutol was less effective [196]. Based on the mutant prevention concentration (MPC), which is a parameter for the selection of resistant pathogens during antibiotic treatment, moxifloxacin was found to be the most effective fluoroquinolone against M. tuberculosis [197].

14.6.2
Skin Infections

Fluoroquinolones are the first orally administrable anti-infectives having activity against the most important Gram-positive and Gram-negative pathogens causing skin and skin structure infections (SSSI) [198]. Moxifloxacin was also approved for the treatment of bacterial skin infections. The growing resistance of Gram-positive pathogens such as S. aureus or S. pyogenes, which, in addition to other bacteria, such as Streptococcus spp., coagulase-negative Staphylococcus, Enterococcus spp., Escherichia coli, Pseudomonas aeruginosa, Enterobacter spp., Proteus mirabilis, Klebsiella pneumoniae and Bacteroides fragilis, are considered to be the main cause of SSSI, presents increasing therapeutic problems. Due to their effective tissue penetration [199] and broad antibacterial spectrum, fluoroquinolones such as moxifloxacin provide welcome additional therapeutic alternatives for the treatment of dermatological infections [198,200–203]. Compared to the other antibiotics, moxifloxacin has powerful *in vitro* activity against aerobic and anaerobic pathogens isolated from infected bite wounds (excluding the relatively resistant *Fusobacterium* spp. and a *Prevotella loeschii* strain) [159] as well as in the peroral treatment of uncomplicated bacterial skin infections (*S. aureus, Streptococcus* spp.) [168,204,205].

In animal experiments, moxifloxacin is highly active against *M. leprae*, in particular in combination with rifapentine/minocycline [206].

14.6.3
Ophthalmology

In 2003, Alcon obtained FDA approval for a 0.5% solution of moxifloxacin (Vigamox™ ophthalmic solution) for the treatment of bacterial conjunctivitis caused by moxifloxacin-sensitive pathogens. Its advantages for this indication are also considered to be its effective tissue penetration and reduced resistance compared with earlier quinolones [207,208]. In *in vitro* tests on keratitis isolates, the more modern quinolones were found to have advantages over their predecessors [209–211]; in endophthalmitis isolates, moxifloxacin was found to have the greatest activity against Gram-positive pathogens compared with gatifloxacin, ciprofloxacin, levofloxacin, and ofloxacin, whereas its activity against Gram-negative pathogens was more or less the same [212].

14.6.4
Dental Medicine

Due to its effective tissue penetration and powerful activity against relevant bacteria, moxifloxacin is also effective in the field of dental medicine [213–216]. Thus, endodontic diseases, parodontopathic conditions and oral mucosa bone wounds were successfully treated by topical treatment [217].

14.7
Pharmacokinetics/Pharmacodynamics of Moxifloxacin

A comparative study of the pharmacokinetic and pharmacodynamic properties of the more modern fluoroquinolones [218,219], as well as overviews on the pharmacokinetics [163,220] and pharmacodynamics [221] of moxifloxacin, can be found in the literature.

When administered orally, moxifloxacin is absorbed rapidly and almost completely, and has a bioavailability of approx. 91% and a plasma half-life of approx. 12 h, which allows once-daily dosing. In the dosage range of 50 to 800 mg, the peak plasma concentration (C_{max}) and the area under the plasma concentration/

time curve (AUC) are linearly related to dose [222]. Upon intravenous administration of 400 mg, C_{max} is increased by 26% and AUC by approx. 10% compared with oral administration.

The plasma protein binding value of 40–42% is low [223]. Moxifloxacin penetrates rapidly into the tissues and body fluids and forms high concentrations in the target tissue; these remain higher than the MIC values for the most important pathogens over the entire dosage period of 24 h [199,224,225].

Slight to moderate renal dysfunction does not significantly affect the pharmacokinetics of moxifloxacin and its metabolites; hence, there is no need for dosage adjustment under such circumstances [226]. Although slight to moderate liver dysfunction tends to reduce the plasma concentrations and renal secretion of moxifloxacin, these do not affect the drug's safety or efficacy [227].

Moxifloxacin's pharmacokinetics are neither age- nor sex-dependent [228], although special measures may be required when treating older patients with restricted organ function with fluoroquinolones [229].

In order to assess the clinical efficiency of anti-infectives, pharmacodynamic parameters such as the AUIC = AUC_{24}/MIC_{90} and C_{max}/MIC_{90} are used for the corresponding pathogens; the values for AUIC should be higher than 100, and for C_{max}/MIC_{90} higher than 10 [221,230]. Comparative values of the activity of 10 fluoroquinolones against *S. pneumoniae*, of which moxifloxacin, clinafloxacin and gemifloxacin were the most efficacious, are shown in Tab. 14.9 [231, 232].

Tab 14.9 Comparative pharmacodynamics of 10 quinolones against *S. pneumoniae* [231].

Compound	AUC/MIC_{90}	C_{max}/MIC_{90}
Lomefloxacin	<5	<1
Ciprofloxacin Ofloxacin	5–25	1–5
Levofloxacin Grepafloxacin Gatifloxacin	25–27	5–10
Trovafloxacin	75–250	10–20
Moxifloxacin Clinafloxacin Gemifloxacin	>250	>20

On analyzing the clinical data so far obtained, moxifloxacin has been found to be highly valuable in practice due to its excellent pharmacokinetics and pharmacodynamics, as well as its high level of safety and tolerance [233].

14.8
Development of Resistance to Moxifloxacin

The resistance mechanisms that cause the inactivation of penicillins, cephalosporins, aminoglycosides, macrolides and tetracyclines do not apply to fluoroquinolones, and there is therefore no cross-resistance between quinolones and other antibiotics.

There are two basic resistance mechanisms involved [10–12,67,234–237]:

1.2 Spontaneous chromosomal mutation of the target enzymes DNA gyrase (GyrA and GyrB) and topoisomerase IV (ParC and ParE), which can gradually produce an increase in resistance to quinolones.

2. An increased expression of efflux pumps, which prevents accumulation of the quinolone in the bacterial cell. Thus, the efflux pump NorA is responsible for the reduced sensitivity of *S. aureus* to hydrophilic fluoroquinolones such as norfloxacin, ciprofloxacin and ofloxacin, whereas quinolones such as sparfloxacin, trovafloxacin and moxifloxacin are not as greatly affected thereby.

Plasmid-mediated resistance has so far evidently played no role in the spread of quinolone resistance and has seldom been detected [238], although plasmid-mediated resistance of *K. pneumoniae* and *E. coli* has been reported in China and the USA [239,240].

Despite the cross-resistance which exists between the quinolones, some quinolone-resistant Gram-positive and anaerobic pathogens are still sensitive to moxifloxacin. The extremely low resistance rate of Gram-positive pathogens to moxifloxacin (10^{-7} to 10^{-10}) compared to ciprofloxacin, which is due to spontaneous mutations, is an important feature of moxifloxacin. On repeated exposure of *S. aureus* 133 to subinhibitory concentrations of moxifloxacin, ofloxacin, levofloxacin, trovafloxacin, and sparfloxacin over a period of 8 days, the increase in MIC value was considerably lower for moxifloxacin than for the comparative quinolones (Fig. 14.14). Neither was any increase in MIC value detected when suboptimum doses of moxifloxacin were used in animal models [33,241,242]. Moxifloxacin retained a high degree of activity against clinical isolates that were resistant to ciprofloxacin and ofloxacin [243]. As was shown by a comparison between gatifloxacin and the analogous quinolone with no substituents in the 8-position, increased activity against quinolone-resistant mutants is due to the effect of the 8-methoxy group [244].

The relatively low development of resistance is also reflected in the low MPC [245], which defines the threshold concentration of anti-infectives sufficient for preventing the development of resistant populations. For clinical use, the MPC must be below the serum concentration or the concentration in the infected tissue. This requirement is met by the 8-methoxy quinolones moxifloxacin and gatifloxacin [246]. Tests carried out on 1-cyclopropyl-7-piperazinyl-quinolones showed

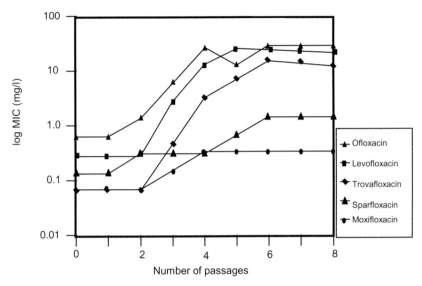

Fig. 14.14 The development of resistance in *S. aureus* 133.

that the 8-methoxy substituent significantly reduced the MPC against *M. tuberculosis*, *M. bovis* BCG, *M. smegmatis* and *S. aureus* [246–248]. The same also applies to moxifloxacin, for which the MPC of 0.6 μg mL^{-1} against *S. aureus* is lower than the MPC of the analogous quinolone without the 8-methoxy group (**44**, BAY Y 3114; 1.7 μg mL^{-1}) [249]. In a comparative test with five fluoroquinolones, in which their potential for suppressing the development of resistance of *S. pneumoniae* was examined, the following descending order was determined: moxifloxacin > trovafloxacin > gatifloxacin > grepafloxacin > levofloxacin [250]. Moxifloxacin is four times more effective than levofloxacin in suppressing the selection of resistant mutants of *S. pneumoniae* [251]. In a more recent study, other sequences were determined for MSSA based on the MIC$_{90}$ value [gemifloxacin (0.063 mg L^{-1}) = moxifloxacin (0.063 mg L^{-1}) > gatifloxacin (0.05 mg L^{-1}) = levofloxacin (0.25 mg L^{-1})] and the MPC values [moxifloxacin (0.25 mg L^{-1}) > gemifloxacin (0.5 mg L^{-1}) > gatifloxacin (1 mg L^{-1}) = levofloxacin (1 mg L^{-1})] [252]. In order to prevent the development of a level of resistance which correlates with the quinolone concentration obtainable *in vivo* [219,232], this concentration should be higher than the MIC value (t > MIC) for as long a period as possible. Moxifloxacin is thus present for the longest period in sufficiently high serum concentrations which exceed the MPC$_{90}$ value: moxifloxacin (>24 h) > levofloxacin (>18 h) > gatifloxacin (12 h) > gemifloxacin (9 h) [252].

14.9
Safety and Tolerability of Moxifloxacin

Whereas some of the earlier quinolones (enoxacin, pefloxacin, fleroxacin, tema-floxacin, lomefloxacin, sparfloxacin, tosufloxacin, grepafloxacin) were associated with major side effects and some of them had to be withdrawn from the market [68,69,253], the more recent fluoroquinolones can be considered to be safe and well-tolerated [12,99,254,255]. The side effects of moxifloxacin [253,256–258], gatifloxacin [253,259,260] and gemifloxacin [261] are comparable to those of conventional quinolones, and most frequently include gastrointestinal events (e.g., abdominal pain, nausea, vomiting, diarrhea, dyspepsia) and CNS side effects (e.g., headache, dizziness, insomnia). Apart from nausea and diarrhea, all of the side effects of moxifloxacin have been observed with a frequency of <3%. Tables comparing the frequencies of the side effects of various quinolones can be found in the literature [12, 262]. In early 2004, however, it was announced that the sale of gatifloxacin in Europe had been discontinued due to the occurrence of hypoglycemic side effects.

In the electrophysiological *ex vivo* model referred to earlier, in which slices of rat hippocampus were used, moxifloxacin displayed a comparable potential to ciprofloxacin for inducing CNS side effects [113].

The phototoxicity displayed particularly by the 8-halogenoquinolones (fleroxacin, sparfloxacin, lomefloxacin, clinafloxacin and BAY Y 3118) [263] plays virtually no role in the 8-methoxy-substituted quinolones moxifloxacin [237,256,264] and gatifloxacin.

As with some other quinolones, moxifloxacin also prolongs the QT_c interval [265], although the prolongation time of 4–6 ms (i.e., 1.4–1.6% of the starting interval) is relatively minimal. For safety reasons, the treatment of patients with QT interval prolongation and certain cardiac diseases is therefore contraindicated. Other medicaments with a potential for prolonging the QT interval may not be administered simultaneously with moxifloxacin. These include anti-arrhythmic drugs of class IA (e.g., quinidine, hydroquinidine, disopyramide) and III (e.g., amiodarone, sotalol, dofetilide, ibutilide), intravenous erythromycin, tricyclic anti-depressives, and cisapride etc.

Quinolones are generally associated with chondrotoxicity and the occurrence of tendopathies, although the risks vary considerably from one quinolone to the other [266,267]. Pathogenetically, this is possibly linked to the magnesium-chelating properties of quinolones. On administering four times the therapeutic dose of moxifloxacin to young beagle dogs (the animals species and strain most susceptible to this side effect), lesions occurred in the articular cartilage subjected to strain [256]. It is for this reason that moxifloxacin is contraindicated in growing children and juveniles, and also during pregnancy and lactation.

Significant drug interactions with theophylline [268], itraconazole [269], warfarin, β-acetyldigoxin, ranitidine, probenecid and oral contraceptives have not been reported with moxifloxacin [12,237].

It is important to note that, due to the chelating properties of quinolones with specific metal ions such as Mg^{2+}, Fe^{2+}, Zn^{2+} and Al^{3+}, moxifloxacin's bioavailability may be reduced when it is administered together with drugs containing these cations. Typical drugs include aluminum-containing preparations such as maalox and sucralfate [12,237]. The absorption of moxifloxacin is however hardly affected by Ca^{2+}; hence, no dosage adjustments are required on the simultaneous administration of Ca^{2+} [270].

Animal experiments in rats, dogs, and monkeys have shown that, in contrast to trovafloxacin, treatment with 400 mg moxifloxacin does not present any risks of hepatoxicity in humans [271].

14.10 Metabolism, Excretion and Biodegradability of Moxifloxacin

When moxifloxacin is administered orally or intravenously, approximately 45% is excreted in an unchanged form (ca. 20% via the urine and 25% via the feces), and approximately 52% via pharmacologically inactive conjugates in the form of its sulfate **88** or glucuronide **89**, (Scheme 14.5) [223,272].

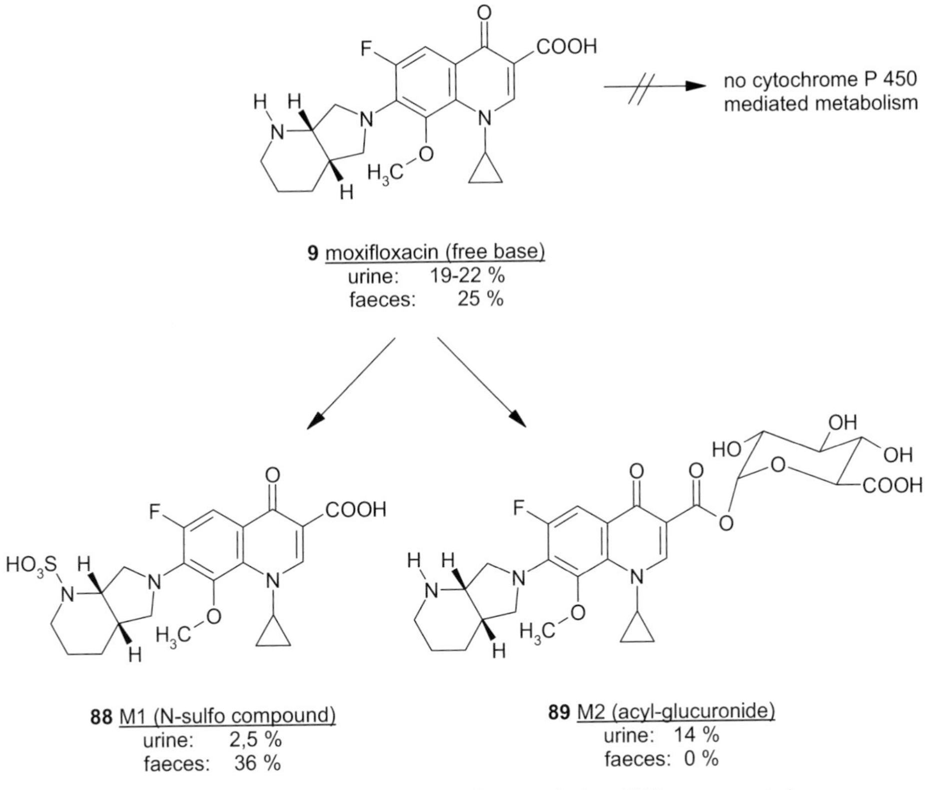
no cytochrome P 450 mediated metabolism
9 moxifloxacin (free base)
urine: 19-22 %
faeces: 25 %
88 M1 (N-sulfo compound)
urine: 2,5 %
faeces: 36 %
89 M2 (acyl-glucuronide)
urine: 14 %
faeces: 0 %

Scheme 14.5 Metabolism of moxifloxacin in humans after a single dose (400 mg, p.o. or i.v.).

The sulfate conjugate (M1, **88**) comprises approx. 36% of the dose, and is found predominantly in the feces, whereas approx. 14% of the moxifloxacin dose is excreted exclusively via the urine in the form of its glucuronide (M2, **89**). The maximum plasma concentrations of the metabolites were much lower than those of moxifloxacin after oral or intravenous application.

With regard to drug interactions, it is important to note that the cytochrome P-450 system is not involved in the metabolism of moxifloxacin. *In vitro* studies using cytochrome P-450 enzymes have shown that moxifloxacin does not inhibit the enzymes CYP 1A2, CYP 3A4, CYP 2C9, CYP 2C19, and CYP 2D6. Moxifloxacin's metabolic profile makes it highly unlikely that it will alter the pharmocokinetics of drugs affected by these enzymes. In rhesus monkeys, oxidative phase I biotransformations do play an additional role, however [272,273].

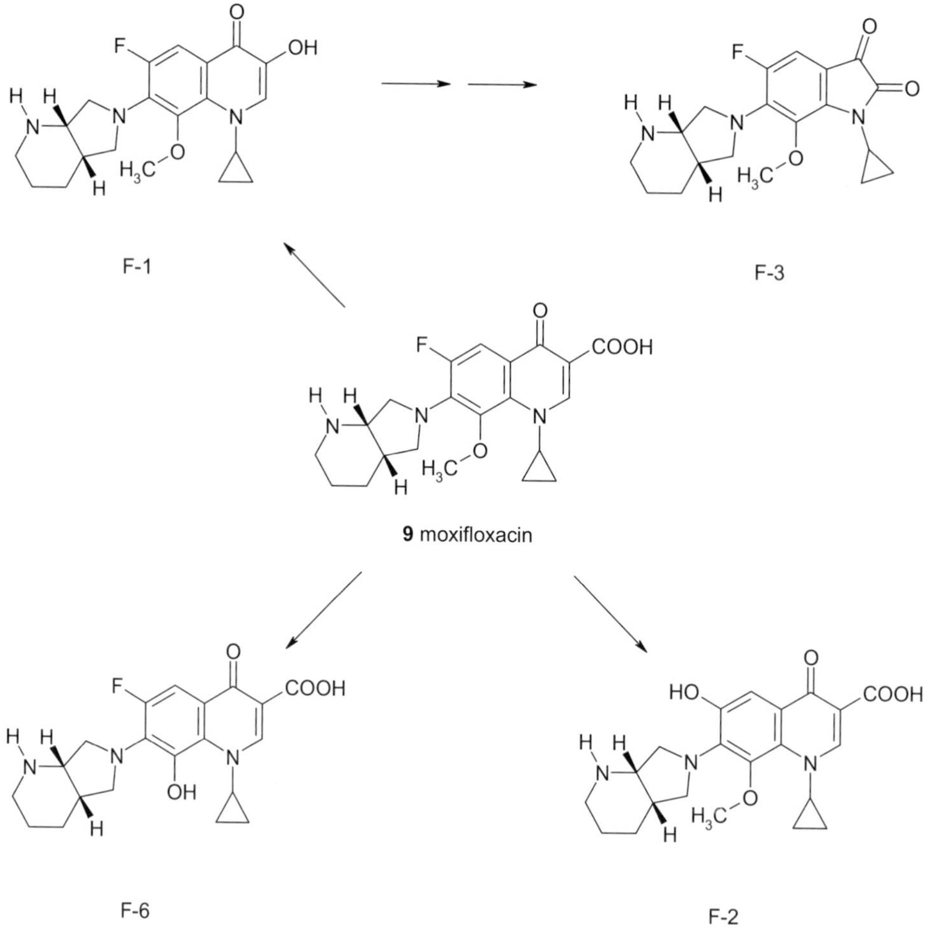

Scheme 14.6 Degradation of moxifloxacin by *G. striatum* [278].

In the above connection, the question of moxifloxacin's degradation on being released into the environment is also important. Although extensive investigations have been carried out on the biodegradability of ciprofloxacin [274] and the veterinary quinolone enrofloxacin [275,276], to date only one laboratory study has been conducted on the degradation of moxifloxacin. In this study, moxifloxacin was efficiently mineralized by the wood-destroying basidiomycete *Gloeophylum striatum* via the same degradation scheme as ciprofloxacin and enrofloxacin [277]. The initially obtained degradation products had no (F-1, F-3), or only very slight (F-2, F-6), residual antibacterial activity (Scheme 14.6) [278].

14.11
Future Prospects for Quinolones

Today, fluoroquinolones represent a valuable advance in the treatment of bacterial infections. The more recently discovered DNA gyrase inhibitors such as the 2-pyridones [3] or 6-desfluoroquinolones (DFQs) and the nonfluorinated quinolones (NFQs) [66,279–281], which are completely free of fluorine, are still undergoing development. They provide additional information on the structure–activity relationships of DNA gyrase inhibitors [282]. Of the 6-desfluoroquinolones, garenoxacin 90 [4], which has an aryl side chain in the 7-position, has been most widely investigated: it is highly active, well-tolerated and, compared to the fluoroquinolones, is less chondrotoxic in juvenile dogs and rats [10,66,267,283].

90 garenoxacin   **91** pradofloxacin   **92** premafloxacin

New quinolones which contain a substituted pyrrolidinyl substituent as the amine radical in the 7-position are also currently being developed for use in veterinary medicine. These include pradofloxacin **91** with an *S,S*–2,8-diazabicyclo[4.3.0]non-8-yl radical in the 7-position and a novel 8-cyano group [284,285], and premafloxacin **92** with a 3-(1-methylaminoethyl)-1-pyrrolidinyl radical in the 7-position and an 8-methoxy group [286,287].

14.12
Summary and Conclusions

The evolution of the quinolones, which began with nalidixic acid and has produced the modern fluoroquinones moxifloxacin, gatifloxacin and gemifloxacin, was based not only on modifications of the basic quinolone structure but also on the development of new cyclic amines for the 7-position. The cyclopropyl radical, which was introduced for the first time in ciprofloxacin, remains the most effective substituent for the 1-position. Moxifloxacin is an 8-methoxy fluoroquinolone with a novel enantiomerically pure S,S–2,8-diazabicyclo[4.3.0]non-8-yl radical in the 7-position.

Moxifloxacin has a broad spectrum of activity which includes Gram-positive cocci, atypical pathogens and anaerobic bacteria responsible, *inter alia*, for infections of the respiratory tract. Moreover, moxifloxacin is one of the most effective fluoroquinolones against pneumococci, including the penicillin- and macrolide-resistant strains. The development of resistance to moxifloxacin is slower than that of the other fluoroquinolones.

Moxifloxacin has excellent pharmacokinetics, is well-tolerated, nonphototoxic, and penetrates rapidly into the body tissues and fluids. Its powerful bactericidal activity and rapid tissue penetration and diffusion to the site of infection produce a rapid onset of action. Due to its long plasma half-life of 12 h, a once-daily dose of 400 mg is sufficient for treatment.

Moxifloxacin has been approved as an anti-infective for the treatment of respiratory tract infections (CAP, AECB, ABS) and skin and soft tissue infections, and was launched in 1999. In the ophthalmological sector, it is used to treat bacterial eye infections.

Among the more recently developed quinolone products, garenoxacin, a 6-desfluoroquinolone, is the first highly active compound without a fluorine atom in the 6-position and with an aryl side-chain replacing the diamine in the 7-position.

Acknowledgments

The quest for new quinolones, and the development of moxifloxacin, has involved many colleagues, whom I thank for their successful collaboration: Chemistry (S. Bartel, K. Grohe, R. Grosser, H. Hagemann, T. Himmler, T. Jaetsch, A. Krebs, F. Kunisch, N. Lui, M. Matzke, B. Mielke, T. Philipps, M. Ruther, T. Schenke, M. Schriewer); microbiology (K.-D. Bremm, A. Dalhoff, R. Endermann, I. Haller, H. Labischinski, K. G. Metzger, H.-O. Werling, H. J. Zeiler); toxicology (H. J. Ahr, E. von Keutz, B. Herbold, M. Pickel, G. Schlüter, M. Schmidt, G. Schmuck, H.-W. Vohr, G. Wasinska-Kempka); analytics (R. Fröde, A. Grunenberg, W. Heilmann, H. Heitzer, W. Karl, H.-F. Mahler, V.-R. Muschalek); pharmacokinetics (C. Kohlsdorfer, J. Petersen-von Gehr, H.-M. Siefert); clinical pharmacology (D. Kubitza, H. Staß); metabolism and isotope chemistry (W. Kanhai, A. Kern, U. Pleiß); galenics (P. Bosché, B. Kühn, T. Laich); drug safety (C. Reiter); project management (P.-K. Bauer, B. Graefe, M. Wegner); Pharmacology (R. Groß, F. Seuter); product development (H.-D. Heilmann, C. Krasemann, M. Springsklee); process development (H. Diehl, P. Fey, R. Gehring, K. Mohrs, P. Naab); production (M. Bock, F. Dürholz , E. Hammerschmidt, H. Lahr, N. Lui, O. Neuner, S. Rosche); marketing (A. Domdey-Bette, Z. Ecarius-Kunze, L. Forstmeier, H. Geritz, M. Golan, R. Kubin, G. Range, J. Talbutt, S. Wegener, A. Westwood).

References

1 Lesher GY, et al. 1,8-Naphthyridine derivatives. A new class of chemotherapeutic agents. *J. Med. Pharm. Chem.*, **1962**, 91, 1063–1065.

2 Bozdogan B, Appelbaum PC. Oxazolidinones: activity, mode of action, and mechanism of resistance. *Int. J. Antimicrob. Agents*, **2004**, 23, 113–119.

3 Li Q, Mitscher LA, Shen LL. The 2-pyridone antibacterial agents: bacterial topoisomerase inhibitors. *Med. Res. Rev.*, **2000**, 20, 231–293.

4 Takahata M, et al. In vitro and in vivo antimicrobial activities of T-3811ME, a novel des-F(6)-quinolone. *Antimicrob. Agents Chemother.*, **1999**, 43, 1077–1084.

5 Roca J. The mechanisms of DNA topoisomerases. *Trends Biochem. Sci.*, **1995**, 20, 156–160.

6 Drlica K, Zhao X. DNA gyrase, topoisomerase IV, and the 4-quinolones. *Microbiol. Mol. Biol. Rev.*, **1997**, 61, 377–392.

7 Maxwell A, Critchlow SE. Mode of action. In: *Quinolone Antibacterials*. Kuhlmann J, Dalhoff A, Zeiler H-J (Eds.) Springer: Berlin, Heidelberg, New York, Barcelona, Budapest, Hong Kong, London, Milan, Paris, Santa Clara, Singapore, Tokyo, **1998**, pp. 119–166.

8 Hooper DC. Mode of action of fluoroquinolones. *Drugs*, **1999**, 58 (Suppl. 2), 6–10.

9 Drlica K. Mechanism of fluoroquinolone action. *Curr. Opin. Microbiol.*, **1999**, 2, 504–508.

10 Dougherty TJ, Beaulieu D, Barrett JF. New quinolones and the impact on resistance. *Drug Discov. Today*, **2001**, 6, 529–536.

11 Heisig P, Wiedemann B. Actio und Reactio – Wirkungs- und Resistenzmechanismen der Chinolone. *Pharm i u Zeit*, **2001**, 30, 382–393.

12 Zhanel GG, et al. A critical review of the fluoroquinolones: focus on respiratory infections. *Drugs*, **2002**, 62, 13–59.

13 ICI: BE 564 863, **1954**.

14 Kaminsky D, Meltzer RJ. Quinolone antibacterial agents. Oxolinic acid and related compounds. *J. Med. Chem.*, **1968**, 11, 160–163.

15 Scavone JM, Gleckman RA, Fraser DG. Cinoxacin: mechanism of action, spectrum of activity, pharmacokinetics, adverse reactions, and therapeutic indications. *Pharmacotherapy*, **1982**, 2, 266–272.

16 Minami S, Shono T, Matsumoto J. Pyrido[2,3-d]pyrimidine antibacterial agents. II. Piromidic acid and related compounds. *Chem. Pharm. Bull.*, **1971**, 19, 1426.

17 Matsumoto J, Minami S. Pyrido(2,3-d)pyrimidine antibacterial agents. 3. 8-Alkyl- and 8-vinyl-5,8-dihydro-5-oxo-2-(1-piperazinyl) pyrido(2,3-d)pyrimidine-6-carboxylic acids and their derivatives. *J. Med. Chem.*, **1975**, 18, 74–79.

18 Gerster JF. **1975**. Minnesota Mining: US Patent 3 917 609 [C. A. 84, 30 930 (1976)].

19 Koga H, et al. Structure–activity relationships of antibacterial 6,7- and 7,8-disubstituted 1-alkyl-1,4-dihydro-4-oxoquinoline-3-carboxylic acids. *J. Med. Chem.*, **1980**, 23, 1358–1363.

20 Goueffon Y. et al. A new synthetic antibacterial: 1-ethyl-6-fluoro-7-(4-methyl-1-piperazinyl)-4-oxo-1, 4-dihydroquinoline-3-carboxylic acid (1589 R.B.). *C. R. Seances Acad. Sci. III*, **1981**, 292, 37–40.

21 Matsumoto J, et al. Pyridonecarboxylic acids as antibacterial agents. 2. Synthesis and structure–activity relationships of 1,6,7-trisubstituted 1,4-dihydro-4-oxo-1,8-naphthyridine-3-carboxylic acids, including enoxacin, a new antibacterial agent. *J. Heterocycl. Chem.*, **1984**, 27, 292–301.

22 Bremner DA, Dickie AS, Singh KP. In-vitro activity of fleroxacin compared with three other quinolones. *J. Antimicrob. Chemother.*, **1988**, 22 (Suppl. D), 19–23.

23 Grohe K, Zeiler HJ, Metzger K. 1-Cyclopropyl-6-fluoro-1,4-dihydro-4-oxo-7-piperazinoquinoline-3-carboxylic acids and an antibacterial agent containing them. **1981**. Bayer AG: DE 3 142 854 [C. A. 99, 53 790 h (**1983**)].

24 Zeiler HJ, Grohe K. The in vitro and in vivo activity of ciprofloxacin. *Eur. J. Clin. Microbiol.*, **1984**, 3, 339–343.

25 Wise R, Andrews JM, Edwards LJ. In vitro activity of Bay O 9867, a new quinoline derivative, compared with those of other antimicrobial agents. *Antimicrob. Agents Chemother.*, **1983**, 23, 559–564.

26 Sato K, et al. In vitro and in vivo activity of DL-8280, a new oxazine derivative. *Antimicrob. Agents Chemother.*, **1982**, 22, 548–553.

27 Hayakawa I, Hiramitsu T, Tanaka Y. Synthesis and antibacterial activities of substituted 7-oxo-2,3-dihydro-7H-pyrido[1,2,3-de][1,4]benzoxazine-6-carboxylic acids. *Chem. Pharm. Bull. (Tokyo)*, **1984**, 32, 4907–4913.

28 Atarashi S, et al. Synthesis and antibacterial activities of optically active ofloxacin and its fluoromethyl derivative. *Chem. Pharm. Bull. (Tokyo)*, **1987**, 35, 1896–1902.

29 Grohe K, Heitzer H. Synthese von 4-Chinolon-3-carbonsäuren. *Liebigs Ann. Chem.*, **1987**, 29–37.

30 Grohe K. The importance of the cycloacylation process for the synthesis of modern fluoroquinolones. *J. prakt. Chem.*, **1993**, 335, 397–409.

31 Grohe K. The chemistry of the quinolones: Methods of synthesizing the quinolone ring system. In: *Quinolone Antibacterials.* Kuhlmann J, Dalhoff A, Zeiler H-J (Eds.) Springer: Berlin, Heidelberg, New York, Barcelona, Budapest, Hong Kong, London, Milan, Paris, Santa Clara, Singapore, Tokyo, **1998**, pp. 13–62.

32 Petersen U, Schenke T. The chemistry of the quinolones: Chemistry in the periphery of the quinolones. In: *Quinolone Antibacterials.* Kuhlmann J, Dalhoff A, Zeiler H-J (Eds.) Springer: Berlin, Heidelberg, New York, Barcelona, Budapest, Hong Kong, London, Milan, Paris, Santa Clara, Singapore, Tokyo, **1998**, pp. 63–118.

33 Dalhoff A, Petersen U, Endermann R. In vitro activity of BAY 12–8039, a new 8-methoxyquinolone. *Chemotherapy (Basel)*, **1996**, 42, 410–425.

34 Petersen U, et al. The synthesis and in vitro and in vivo antibacterial activity of moxifloxacin (BAY 12–8039), a new 8-methoxyquinolone. In: *Moxifloxacin in practice, Vol. 1.* Adam D, Finch R (Eds.) Maxim Medical: Oxford, **1999**, pp. 13–26.

35 Bay 12–8039, Moxifloxacin hydrochloride. *Drugs Fut.*, **1999**, 24, 193–197.

36 Moxifloxacin hydrochloride, Bay 12–8039, Avelox®, Avalox®, Actira®, Octegra®, Proflox®. *Drugs Fut.*, **2000**, 25, 211–219.

37 Moxifloxacin hydrochloride, Avelox®, Avalox®, Actira®, Octegra®, Proflox®. *Drugs Fut.*, **2001**, 26, 189–194.

38 Hosaka M, et al. In vitro and in vivo antibacterial activities of AM-1155, a new 6-fluoro-8-methoxy-quinolone. *Antimicrob. Agents Chemother.*, **1992**, 36, 2108–2117.

39 AM 1155. *Drugs Fut.*, **1993**, 18, 203–205.

40 AM-1155. *Drugs Fut.*, **1994**, 19, 280–281.

41 AM-1155. *Drugs Fut.*, **1995**, 20, 293.

42 Perry CM, et al. Gatifloxacin: a review of its use in the management of bacterial infections. *Drugs*, **2002**, 62, 169–207.

43 Gatifloxacin, Tequin®, Gatiflo®. *Drugs Fut.*, **2000**, 25, 311–317.

44 Gatifloxacin, Tequin®, Gatiflo®. *Drugs Fut.*, **2001**, 26, 297–301.

45 Naber KG, Adam D. Classification of fluoroquinolones. *Int. J. Antimicrob. Agents*, **1998**, 10, 255–257.

46 King DE, Malone R, Lilley SH. New classification and update on the quinolone antibiotics. *Am. Fam. Physician*, **2000**, 61, 2741–2748.

47 Albrecht R. Development of antibacterial agents of the nalidixic acid type. In: *Progress in Drug Research – Fortschritte der Arzneimittelforschung*, Vol. 21. Jucker E (Ed.) Birkhäuser: Basel, Stuttgart, **1977**, pp. 9–104.

48 Bouzard D. Recent advances in the chemistry of quinolones. In: *Recent Progress in the Chemical Synthesis of Antibiotics*. Lukacs G (Ed.) Springer-Verlag, Berlin, Heidelberg, **1990**, pp. 249–283.

49 Chu DTW. Fluoroquinolone carboxylic acids as antibacterial drugs. In: *Organofluorine Compounds in Medicinal Chemistry and Biomedical Applications*. Filler R, Kobayashi Y, and Yagupolskii LM (Eds.). Elsevier: Amsterdam, London, New York, Tokyo, **1993**, pp. 165–207.

50 Chu DTW, Fernandes PB. Recent developments in the field of quinolone antibacterial agents. *Adv. Drug Res.*, **1991**, 21, 39–144.

51 Lesher GY. Nalidixic acid and other quinolone carboxylic acids. In: *Encyclopedia of Chemical Technology*, 3rd edn. Kirk-Othmer (Ed.) John Wiley & Sons: New York, Chichester, Brisbane, Toronto, **1978**, pp. 782–789.

52 Leysen DC, et al. Synthesis of antibacterial 4-quinolone-3-carboxylic acids and their derivatives. Part 1. *Pharmazie*, **1991**, 46, 485–501.

53 Leysen DC, et al. Synthesis of antibacterial 4-quinolone-3-carboxylic acids and their derivatives. Part 2. *Pharmazie*, **1991**, 46, 557–572.

54 Mitscher LA, et al. Recent advances on quinolone antimicrobial agents. In: *Horizons on Antibiotic Research*. Davis BD, et al. (Eds.) Japan Antibiotics Research Association: Tokyo, **1988**, pp. 166–193.

55 Rádl S, Bouzard D. Recent advances in the synthesis of antibacterial quinolones. *Heterocycles*, **1992**, 34, 2143–2177.

56 Mitscher LA. Bacterial topoisomerase inhibitors: quinolone and pyridone antibacterial agents. *Chem. Rev.*, **2005**, 105, 559–592.

57 Andriole VT. *The Quinolones*. Academic Press: London, San Diego, New York, Berkeley, Boston, Sydney, Tokyo, Toronto, **1988**.

58 Asahina Y, Ishizaki T, Suzue S. Recent advances in structure activity relationships in new quinolones. In: *Fluorinated quinolones – new quinolone antimicrobials; Progress in Dug Research – Fortschritte der Arzneimittelforschung*. Jucker E (Ed.), Vol. 38. Mitsuhashi (Ed.) Birkhäuser: Basel Stuttgart, **1992**, pp. 57–106.

59 Domagala JM. Review: Structure–activity and structure–side-effect relationships for the quinolone antibacterials. *J. Antimicrob. Chemother.*, **1994**, 33, 685–706.

60 Mitscher LA, Devasthale PV, Zavod RM. Structure–activity relationships of fluoro-4-quinolones. In: *The 4-Quinolones: Antibacterial Agents in Vitro*. Crumplin GC (Ed.). Springer: London, Berlin, Heidelberg, New York, Paris, Tokyo, Hong Kong, **1990**, pp. 115–146.

61 Mitscher LA, Devasthale P, Zavod R. Structure–activity relationships. In: *Quinolone Antimicrobials Agents*, 2nd edn. Hooper DC, Wolfson JS (Eds.) American Society for Microbiology: Washington, **1993**, pp. 3–51.

62 Rádl S. Structure–activity relationships in DNA gyrase inhibitors. *Pharmacol. Ther.*, **1990**, 48, 1–17.

63 Rosen T. The fluoroquinolone antibacterial agents. In: *Progress in Medicinal Chemistry*, Vol. 27. Ellis GP, West GB (Eds.). Elsevier: Amsterdam, New York, Oxford, **1990**, pp. 235–295.

64 Schentag JJ, Domagala JM. Structure–activity relationships with the quinolone antibiotics. *Res. Clin. Forums*, **1985**, 7, 9–13.

65 Gootz TD, Brighty KE. Fluoroquinolone antibacterials: SAR mechanism of action, resistance, and clinical aspects. *Med. Res. Rev.*, **1996**, 16, 433–486.

66 Domagala J, Hagen SE. Structure–activity relationships of the quinolone antibacterials in the new millennium: Some things change and some do not. In: *Quinolone Antimicrobial Agents.* Hooper DC, Rubinstein E (Eds.) ASM Press: Washington, DC, **2003**, pp. 3–18.

67 Schmitz FJ, et al. Activity of quinolones against gram-positive cocci: mechanisms of drug action and bacterial resistance. *Eur. J. Clin. Microbiol. Infect. Dis.*, **2002**, 21, 647–659.

68 Bertino JJ, Fish D. The safety profile of the fluoroquinolones. *Clin. Ther.*, **2000**, 22, 798–817.

69 Rubinstein E. History of quinolones and their side effects. *Chemotherapy*, **2001**, 47 (Suppl. 3), 3–8.

70 Petersen U, et al. 7-Amino-1-cyclopropyl-6,8-dihalogen-1,4-dihydro-4-oxo-3-chinolincarbonsäuren, Verfahren zu ihrer Herstellung sowie diese enthaltene antibakterielle Mittel. **1984**. Bayer AG: EP 167 763 [C.A. 104: 186 447v].

71 Irikura T, et al. Quinolonecarboxylic acid derivative and process for its preparation. **1985**. Kyorin Pharmaceutical Co: EP 195 316 [C.A.: 106, 50068f].

72 AM-1091, Clinafloxacin, PD-127391, Bay-V 3545, CI-960. *Drugs Fut.*, **1992**, 17, 938–940.

73 Clinafloxacin, AM-1091, PD-127391, Bay-V 3545, CI-960 (as HCl). *Drugs Fut.*, **1993**, 18, 962.

74 Clinafloxacin. *Drugs Fut.*, **1994**, 19, 950–960.

75 Cohen MA, et al. In-vitro activity of clinafloxacin, trovafloxacin, and ciprofloxacin. *J. Antimicrob. Chemother.*, **1997**, 40(2), 205–211.

76 Clinafloxacin. *Drugs Fut.*, **1999**, 24, 1131–1134.

77 Petersen U, et al. Enantiomerenreine 7-(3-Amino-1-pyrrolidinyl)-chinolon-und -naphthyridoncarbonsäuren. **1989**. Bayer AG: EP 26 744.

78 Humphrey GH, et al. Pharmacokinetics of clinafloxacin enantiomers in humans. *J. Clin. Pharmacol.*, **1999**, 39, 1143–1150.

79 Benkert K, Wiegand H. Stellungnahme der Firma Gödecke Parke-Davis. *Chemother. J.*, **2000**, 9, 100–101.

80 Prous J, Graul A, Castañer J. DU-6859. *Drugs Fut.*, **1994**, 19, 827–834.

81 DU-6859, Sitafloxacin. *Drugs Fut.*, **1997**, 22, 1035ff.

82 Sitafloxacin, DU-6859. *Drugs Fut.*, **1998**, 23, 1047.

83 Sitafloxacin hydrate, DU-6859 a. *Drugs Fut.*, **1999**, 24, 1040–1042.

84 Sitafloxacin hydrate. *Drugs Fut.*, **2000**, 25, 993–995.

85 Kawakami K, et al. DK-507k, a New 8-Methoxyquinolone: Synthesis and Biological Evaluation of 7-[(3-amino-4-substituted)pyrrolidin-1-yl] Derivatives. In: 41st Interscience Conference on Antimicrobial Agents and Chemotherapy, Abstract F-546. Chicago, **2001**.

86 Otani T, et al. DK-507k, a New 8-Methoxyquinolone: In Vitro and In Vivo Antibacterial Activities. In: 41st Interscience Conference on Antimicrobial Agents and Chemotherapy, Abstract F-547. Chicago, **2001**.

87 Takahashi Y, et al. In vitro activity of HSR-903, a new quinolone. *Antimicrob. Agents Chemother.*, **1997**, 41, 1326–1330.

88 Yoshizumi S, et al. In vivo activity of HSR-903, a new fluoroquinolone, against respiratory pathogens. *Antimicrob. Agents Chemother.*, **1998**, 42, 785–788.

89 Yoshizumi S, et al. The in vivo activity of olamufloxacin (HSR-903) in systemic and urinary tract infections in mice. *J. Antimicrob. Chemother.*, **2001**, 48, 137–140.

90 Sun J, et al. Protonation equilibrium and lipophilicity of olamufloxacin (HSR-903), a newly synthesized fluoroquinolone antibacterial. *Eur. J. Pharm. Biopharm.*, **2003**, 56, 223–229.

91 Domagala JM, et al. 1-Ethyl-7-[3-[(ethyl-amino)methyl]-1-pyrrolidinyl]-6,8-difluoro-1,4-dihydro-4-oxo-3-quinoline-carboxylic acid. New quinolone antibacterial with potent gram-positive activity. *J. Med. Chem.*, **1986**, 29, 445–448.

92 Brighty KE, Gootz TD. The chemistry and biological profile of trovafloxacin. *J. Antimicrob. Chemother.*, **1997**, 39 (Suppl. B), 1–14.

93 Lode H. Potential interactions of the extended-spectrum fluoroquinolones with the CNS. *Drug Safety*, **1999**, 21, 123–135.

94 Hong CY, et al. Novel fluoroquinolone antibacterial agents containing oxime-substituted (aminomethyl)pyrrolidines: synthesis and antibacterial activity of 7-(4-(aminomethyl)-3-(methoxyimino)-pyrrolidin-1-yl)-1-cyclopropyl-6-fluoro-4-oxo-1,4-dihydro[1,8]naphthyridine-3-carboxylic acid (LB 20304). *J. Med. Chem.*, **1997**, 40, 3584–3593.

95 SB-265805/LB-20304a, Gemifloxacin mesilate. *Drugs Fut.*, **1999**, 24, 1281–1286.

96 Gemifloxacin mesilate, SB-265805/LB-20304a, Factive®. *Drugs Fut.*, **2000**, 25, 1194–1202.

97 Blondeau JM, Missaghi B. Gemifloxacin: a new fluoroquinolone. *Expert Opin. Pharmacother.*, **2004**, 5, 1117–1152.

98 Hong CY. Discovery of gemifloxacin (Factive, LB20304a): a quinolone of a new generation. *Farmaco*, **2001**, 56, 41–44.

99 Saravolatz LD, Leggett J. Gatifloxacin, gemifloxacin, and moxifloxacin: the role of 3 newer fluoroquinolones. *Clin. Infect. Dis.*, **2003**, 37, 1210–1215.

100 Yun HJ, et al. In vitro and in vivo antibacterial activities of DW286, a new fluoronaphthyridone antibiotic. *Antimicrob. Agents Chemother.*, **2002**, 46, 3071–3074.

101 Birkofer L, Feldmann H. Über bicyclische Dilactame. *Liebigs Ann. Chem.*, **1964**, 677, 154–157.

102 Schenke T, Petersen U. Diazabicyclo[3.3.0]octane und ein Verfahren zu ihrer Herstellung. **1989**. Bayer AG: EP 393 424.

103 Petersen U, et al. 7-(1-Pyrrolidinyl)-3-chinolon- und naphthyridoncarbon-säure-Derivate, Verfahren zu ihrer Herstellung sowie substituierte mono- und bicyclische Pyrrolidinderivate als Zwischenprodukte zu ihrer Herstellung und sie enthaltende antibakterielle Mittel und Futterzusatzstoffe. **1988/1989**. Bayer AG: EP 350 733.

104 Petersen U, et al. 7-(2,7-Diazabicyclo[3.3.0]octyl)-3-chinolon- und naphthyridoncarbonsäure-Derivate. **1990**. Bayer AG: EP 481 274.

105 Laborde E, Schroeder M. 7-Substituted quinolones and naphthyridones as antibacterial agents. **1992**. Warner-Lambert Co.: US 5 342 844.

106 Kitzis MD, et al. SYN987, SYN1193, and SYN1253, new quinolones highly active against gram-positive cocci. *J. Antimicrob. Chemother.*, **1995**, 36, 209–213.

107 Klopman G, et al. Anti-*Mycobacterium avium* activity of quinolones: in vitro activities. *Antimicrob. Agents Chemother.*, **1993**, 37, 1799–1806.

108 Petersen U, et al. Quinolone- and naphthyridone carboxylic acid derivatives, process for their production, antibacterial compositions and feed additives containing them. **1988**. Bayer AG: US 5 173 484.

109 Ogata M, et al. Synthesis and antibacterial activity of new 7-(aminoazabicycloalkanyl)quinolonecarboxylic acids. *Eur. J. Med. Chem.*, **1991**, 26, 889–906.

110 Petersen U, et al. 7-(Aminomethyl-oxa-7-aza-bicyclo[3.3.0]oct-7-yl)-chinolon-und naphthyridoncarbonsäure-Derivate. **1992**. Bayer AG: EP 589 318.

111 Petersen U, et al. Quinolone- and naphthyridonecarboxylic acid derivatives. **1994**. Bayer AG: EP 705 828.

112 Dimpfel W, et al. Hippocampal activity in the presence of quinolones and fenbufen in vitro. *Antimicrob. Agents Chemother.*, **1991**, 35, 1142–1146.

113 Schmuck G, Schürmann A, Schlüter G. Determination of the excitatory potencies of fluoroquinolones in the central nervous system by an in vitro model. *Antimicrob. Agents Chemother.*, **1998**, 42, 1831–1836.

114 Petersen U, et al. Verwendung von 7-(1-Aminomethyl-2-oxa-7-aza-bicyclo[3.3.0]oct-7-yl)-chinoloncarbonsäure- und -naphthyridoncarbonsäure-Derivaten zur Therapie von *Helicobacter pylori*-Infektionen und den damit assoziierten gastroduodenalen Erkrankungen. **1996**. Bayer AG: DE 19 652 219.

115 Matzke M, et al. Verwendung von 7-(2-Oxa-5,8-diazabicyclo[4.3.0]non-8-yl)-chinoloncarbonsäure- und -naphthyridoncarbonsäure-Derivaten zur Therapie von *Helicobacter pylori*-Infektionen und den damit assoziierten gastroduodenalen Erkrankungen. **1996**. Bayer AG: DE 19 652 239.

116 Kim JH, et al. In vitro and in vivo antibacterial efficacies of CFC-222, a new fluoroquinolone. *Antimicrob. Agents Chemother.*, **1997**, 41, 2209–2213.

117 Kim JH, et al. In-vitro and in-vivo antibacterial activity of CFC-222, a new fluoroquinolone. *J. Antimicrob. Chemother.*, **1998**, 41, 223–229.

118 Yatsunami T, et al. Isoindoline derivative. **1988**. Wakunaga Seiyaku Kabushiki Kaisha: EP 343 560.

119 Petersen U, et al. 7-Azaisoindolenyl-chinolon- und naphthyridoncarbonsäure-Derivate. **1991**. Bayer AG: EP 520 277.

120 Petersen U, et al. 1-(2-Fluorcyclopropyl)-chinolon- und naphthyridoncarbonsäure-Derivate. **1993**. Bayer AG: EP 653 425.

121 Petersen U, et al. 7-Isoindolinyl-chinolon- und naphthyridoncarbonsäure-Derivate. **1991**. Bayer AG: DE 4 120 646.

122 Okura A, et al. Pyridonecarboxylic acid derivative. **1990**. Banyu Pharmaceutical Co.: WO92/12146.

123 Petersen U, et al. Chinolon- und Naphthyridoncarbonsäure-Derivate. **1995**. Bayer AG: EP 721 948.

124 Petersen U, et al. Chinolon- und Naphthyridoncarbonsäure-Derivate. **1995**. Bayer AG: EP 704 443.

125 Philipps T, et al. 7-Isoindolenyl-quinolone derivatives and 7-isoindolenyl-naphthyridone derivatives. **1996**. Bayer AG: US 5 739 339.

126 Schenke T, Krebs A, Petersen U. Enantiomerenreine 2-Oxa-5,8-diazabicyclo[4.3.0]nonane sowie Verfahren zu ihrer Herstellung. **1992**. Bayer AG: DE 4 200 415.

127 Petersen U, et al. The synthesis and biological properties of 6-fluoro-quinolone-carboxylic acids. *Bull. Chim. Soc. Belg.*, **1996**, 105, 683–699.

128 Kim WJ, et al. Novel quinolone carboxylic acid derivatives and process for preparing the same. **1994**. Korea Res. Inst. Chem. Technology: EP 622 367.

129 Petersen U, et al. Chinolon- und Naphthyridoncarbonsäure-Derivate als antibakterielle Mittel. **1992**. Bayer AG: EP 550 903.

130 Bremm KD, et al. In vitro evaluation of BAY Y3118, a new full-spectrum fluoroquinolone. *Chemotherapy*, **1992**, 38, 376–387.

131 Wise R, Andrews JM, Brenwald N. The in-vitro activity of Bay y 3118, a new chlorofluoroquinolone. *J. Antimicrob. Chemother.*, **1993**, 31, 73–80.

132 Garcia-Rodriguez JA, et al. In vitro activity of BAY y 3118, and nine other antimicrobial agents against anaerobic bacteria. *J. Chemother.*, **1995**, 7, 189–196.

133 Aldridge KE. Increased activity of a new chlorofluoroquinolone, BAY y 3118, compared with activities of ciprofloxacin, sparfloxacin, and other antimicrobial agents against anaerobic bacteria. *Antimicrob. Agents Chemother.*, **1994**, 38, 1671–1674.

134 Fass RJ. In vitro activity of Bay y 3118, a new quinolone. *Antimicrob. Agents Chemother.*, **1993**, 37, 2348–2357.

135 Bauernfeind A. Comparative in-vitro activities of the new quinolone, Bay y 3118, and ciprofloxacin, sparfloxacin, tosufloxacin, CI-960 and CI-990. *J. Antimicrob. Chemother.*, **1993**, 31, 505–522.

136 Sirgel F, Venter A, Heilmann HD. Comparative in-vitro activity of Bay y 3118, a new quinolone, and ciprofloxacin against *Mycobacterium tuberculosis* and *Mycobacterium avium* complex. *J. Antimicrob. Chemother.*, **1995**, 35, 349–351.

137 Schmidt U, Schlüter G. Studies on the mechanism of phototoxicity of BAY y 3118 and other quinolones. *Adv. Exp. Med. Biol.*, **1996**, 387, 117–120.

138 Petersen U, et al. BAY Y 3118, a novel 4-quinolone: Synthesis and in vitro activity. In: 32nd Interscience Conference on Antimicrobial Agents and Chemotherapy, Abstract No. 642. Anaheim, **1992**.

139 Petersen U, et al. Synthesis and in vitro activity of BAY 12–8039, a new methoxyquinolone. In: 36th Interscience Conference on Antimicrobial Agents and Chemotherapy, Abstract No. F1. New Orleans, **1996**.

140 Fey P. *Verfahren zur Herstellung von (S,S)-Benzyl-2,8-diazabicyclo[4.3.0]nonan.* **1999**, Bayer AG: DE 19 821 039.

141 Dreisbach, C., *Verfahren zur Racematspaltung von cis- und trans-Pyrrolopiperidin.* **1997**, Bayer AG: DE 19 735 198.

142 Grohe, K., et al., *7-Amino-1-cyclopropyl-6,8-difluor-1,4-dihydro-4-oxo-3-chinolincarbonsäuren, Verfahren zu ihrer Herstellung sowie diese enthaltende antibakterielle Mittel.* **1983**, Bayer AG: EP 126 355.

143 Masuzawa, K., et al., *8-Alkoxyquinolonecarboxylic acid and salts thereof excellent in the selective toxicity and process of preparing the same.* **1986**, Kyorin Pharmaceutical Co.: EP 230 295.

144 Iwata, M., et al., *Quinoline-3-carboxylic acid derivatives, their preparation and use.* **1986**, Ube Industries: EP 241 206.

145 Grunenberg, A. and P. Bosché, *A new crystal modification of 1-cyclopropyl-7-([S,S]-2,8-diazabicyclo[4.3.0]-non-8-yl)-6-fluoro-1,4-dihydro-8-methoxy-4-oxo-3-quinolinecarboxylic acid hydrochloride (CDCH), a process for its production and pharmaceutical preparations containing same.* **1995**, Bayer AG: DE 19 546 249; US 5 849 752.

146 Gehring, R., et al., *Verfahren zur Herstellung von 8-Methoxy-chinoloncarbonsäuren.* **1999**, Bayer AG: DE 19 751 948.

147 Müller HG. Unpublished results. Bayer AG, **1999**.

148 Lister PD, Sanders CC. Pharmacodynamics of moxifloxacin against *Streptococcus pneumoniae* in an in vitro pharmacokinetic model. In: 38th Interscience Conference on Antimicrobial Agents and Chemotherapy, Abstract A-21. San Diego, CA, **1998**.

149 Woodcock JM, et al. In vitro activity of BAY 12–8039, a new fluoroquinolone. *Antimicrob. Agents Chemother.*, **1997**, 41(1), 101–106.

150 Bauernfeind A. Comparison of the antibacterial activities of the quinolones Bay 12–8039, gatifloxacin (AM 1155), trovafloxacin, clinafloxacin, levofloxacin and ciprofloxacin. *J. Antimicrob. Chemother.*, **1997**, 40, 639–651.

151 Brueggemann AB, Kugler KC, Doern GV. In vitro activity of BAY 12–8039, a novel 8-methoxyquinolone, compared to activities of six fluoroquinolones against *Streptococcus pneumoniae, Haemophilus influenzae*, and *Moraxella catarrhalis*. *Antimicrob. Agents Chemother.*, **1997**, 41, 1594–1597.

152 Boswell FJ, Andrews JM, Wise R. Pharmacodynamic properties of BAY 12–8039 on gram-positive and gram-negative organisms as demonstrated by studies of time-kill kinetics and postantibiotic effect. *Antimicrob. Agents Chemother.*, **1997**, 41, 1377–1379.

153 Bébéar CM, et al. In vitro activity of BAY 12–8039, a new fluoroquinolone against mycoplasmas. *Antimicrob. Agents Chemother.*, **1998**, 42, 703–704.

154 Biedenbach DJ, et al. BAY 12–8039, a novel fluoroquinolone. Activity against important respiratory tract pathogens. *Diagn. Microbiol. Infect. Dis.*, **1998**, 32, 45–50.

155 Donati M, et al. Comparative in-vitro activity of moxifloxacin, minocycline and azithromycin against *Chlamydia* spp. *J. Antimicrob. Chemother.*, **1999**, 43, 825–827.

156 Ruckdeschel G, Dalhoff A. The in-vitro activity of moxifloxacin against *Legionella* species and the effects of medium on susceptibility test results. *J. Antimicrob. Chemother.*, **1999**, 43 (Suppl. B), 25–29.

157 Blondeau JM, Felmingham D. In vitro and in vivo activity of moxifloxacin against community respiratory tract pathogens. *Clin. Drug Invest.*, **1999**, 18, 57–78.

158 Miyashita N, et al. In-vitro activity of moxifloxacin and other fluoroquinolones against *Chlamydia* species. *J. Infect. Chemother.*, **2002**, 8, 115–117.

159 Goldstein EJ, et al. In vitro activity of Bay 12–8039, a new 8-methoxyquinolone, compared to the activities of 11 other oral antimicrobial agents against 390 aerobic and anaerobic bacteria isolated from human and animal bite wound skin and soft tissue infections in humans. *Antimicrob. Agents Chemother.*, **1997**, 41, 1552–1557.

160 MacGowan AP, et al. Bay 12–8039, a new 8-methoxy-quinolone: comparative in-vitro activity with nine other antimicrobials against anaerobic bacteria. *J. Antimicrob. Chemother.*, **1997**, 40, 503–509.

161 Aldridge KE, Ashcraft DS. Comparison of the in vitro activities of Bay 12–8039, a new quinolone, and other antimicrobials against clinically important anaerobes. *Antimicrob. Agents Chemother.*, **1997**, 41, 709–711.

162 Ackermann G, et al. Comparative activity of moxifloxacin in vitro against obligately anaerobic bacteria. *Eur. J. Clin. Microbiol. Infect. Dis.*, **2000**, 19, 228–232.

163 MacGowan A. Moxifloxacin (Bay 12–8039): a new methoxy quinolone antibacterial. *Expert Opin. Invest. Drugs*, **1999**, 8, 181–199.

164 Nightingale CH. Moxifloxacin, a new antibiotic designed to treat community-acquired respiratory tract infections: a review of microbiologic and pharmacokinetic-pharmacodynamic characteristics. *Pharmacotherapy*, **2000**, 20, 245–256.

165 Ball P. Moxifloxacin (Avelox): an 8-methoxyquinolone antibacterial with enhanced potency. *Int. J. Clin. Pract.*, **2000**, 54, 329–332.

166 Blondeau JM, Hansen GT. Moxifloxacin: a review of the microbiological, pharmacological, clinical and safety features. *Expert Opin. Pharmacother.*, **2001**, 2, 317–335.

167 Caeiro JP, Iannini PB. Moxifloxacin (Avelox): a novel fluoroquinolone with a broad spectrum of activity. *Expert Rev. Anti-Infect. Ther.*, **2003**, 1, 363–370.

168 Keating GM, Scott LJ. Moxifloxacin: a review of its use in the management of bacterial infections. *Drugs*, **2004**, 64, 2347–2377.

169 Bartlett JG, Mundy LM. Community-acquired pneumonia. *N. Engl. J. Med.*, **1995**, 333, 1618–1624.

170 Gwaltney JMJ. Acute community-acquired sinusitis. *Clin. Infect. Dis.*, **1996**, 23, 1209–1223; quiz 1224–1225.

171 Yagupsky P, et al. In vitro activity of novel fluoroquinolones against *Streptococcus pneumoniae* isolated from children with acute otitis media. *Chemotherapy*, **2001**, 47, 354–358.

172 Ball P, Make B. Acute exacerbations of chronic bronchitis: an international comparison. *Chest*, **1998**, 113 (3 Suppl.), 199S–204S.

173 Bartlett JG, et al. Practice guidelines for the management of community-acquired pneumonia in adults. Infectious Diseases Society of America. *Clin. Infect. Dis.*, **2000**, 31, 347–382.

174 Cubbon MD, Masterton RG. New quinolones – a fresh answer to the pneumococcus. *J. Antimicrob. Chemother.*, **2000**, 46, 869–872.

175 Esposito S, Noviello S, Ianniello F. Activity of moxifloxacin and twelve other antimicrobial agents against 216 clinical isolates of *Streptococcus pneumoniae*. *Chemotherapy*, **2001**, 47, 90–96.

176 Klugman KP, Capper T. Concentration-dependent killing of antibiotic-resistant pneumococci by the methoxyquinolone moxifloxacin. *J. Antimicrob. Chemother.*, **1997**, 40, 797–802.

177 Visalli MA, Jacobs MR, Appelbaum PC. Antipneumococcal activity of BAY 12–8039, a new quinolone, compared with activities of three other quinolones and four oral beta-lactams. *Antimicrob. Agents Chemother.*, **1997**, 41, 2786–2789.

178 Saravolatz L, et al. Antimicrobial activity of moxifloxacin, gatifloxacin and six fluoroquinolones against *Streptococcus pneumoniae*. *J. Antimicrob. Chemother.*, **2001**, 47, 875–877.

179 Dalhoff A, et al. Penicillin-resistant *Streptococcus pneumoniae*: review of moxifloxacin activity. *Clin. Infect. Dis.*, **2001**, 32 (Suppl. 1), S22–S29.

180 Blondeau JM. A review of the comparative in-vitro activities of 12 antimicrobial agents, with a focus on five new 'respiratory quinolones'. *J. Antimicrob. Chemother.*, **1999**, 43 (Suppl. B), 1–11.

181 Blondeau JM. Expanded activity and utility of the new fluoroquinolones: a review. *Clin. Ther.*, **1999**, 21, 3–40; discussion 1–2.

182 Fass RJ. In vitro activity of Bay 12–8039, a new 8-methoxyquinolone. *Antimicrob. Agents Chemother.*, **1997**, 41, 1818–1824.

183 Pestova E, et al. Intracellular targets of moxifloxacin: a comparison with other fluoroquinolones. *J. Antimicrob. Chemother.*, **2000**, 45, 583–590.

184 Wilson R, et al. Short-term and long-term outcomes of moxifloxacin compared to standard antibiotic treatment in acute exacerbations of chronic bronchitis. *Chest*, **2004**, 125, 953–964.

185 Church D, et al. Efficacy of moxifloxacin in the treatment of acute bacterial sinusitis caused by penicillin-resistant *Streptococcus pneumoniae*. In: 40th Interscience Conference on Antimicrobial Agents and Chemotherapy, Abstract 833. Toronto, **2000**.

186 Souli M, Wennersten CB, Eliopoulos GM. In vitro activity of BAY 12–8039, a new fluoroquinolone, against species representative of respiratory tract pathogens. *Int. J. Antimicrob. Agents*, **1998**, 10, 23–30.

187 Lemmen SW, et al. Comparison of the bactericidal activity of moxifloxacin and levofloxacin against *Staphylococcus aureus*, *Staphylococcus epidermidis*, *Escherichia coli* and *Klebsiella pneumoniae*. *Chemotherapy*, **2003**, 49, 33–35.

188 Aktas Z, et al. Moxifloxacin activity against clinical isolates compared with the activity of ciprofloxacin. *Int. J. Antimicrob. Agents*, **2002**, 20, 196–200.

189 Jung R, et al. Synergistic activities of moxifloxacin combined with piperacillin-tazobactam or cefepime against *Klebsiella pneumoniae*, *Enterobacter cloacae*, and *Acinetobacter baumannii* clinical isolates. *Antimicrob. Agents Chemother.*, **2004**, 48, 1055–1057.

190 Jones ME, et al. In vitro susceptibility of *Streptococcus pneumoniae*, *Haemophilus influenzae* and *Moraxella catarrhalis*: a European multicenter study during 2000–2001. *Clin. Microbiol. Infect.*, **2003**, 9, 590–599.

191 Ji B, et al. In vitro and in vivo activities of moxifloxacin and clinafloxacin against *Mycobacterium tuberculosis*. *Antimicrob. Agents Chemother.*, **1998**, 42, 2066–2069.

192 Rodriguez JC, et al. In vitro activity of four fluoroquinolones against *Mycobacterium tuberculosis*. *Int. J. Antimicrob. Agents*, **2001**, 17, 229–231.

193 Rodriguez JC, et al. In vitro activity of moxifloxacin, levofloxacin, gatifloxacin and linezolid against *Mycobacterium tuberculosis*. *Int. J. Antimicrob. Agents*, **2002**, 20, 464–467.

194 Tortoli E, Dionisio D, Fabbri C. Evaluation of moxifloxacin activity in vitro against *Mycobacterium tuberculosis*, including resistant and multidrug-resistant strains. *J. Chemother.*, **2004**, 14, 334–336.

195 Miyazaki E, Chaisson RE, Bishai WR. In vivo activity of BAY12–8039, a new 8-methoxy-quinolone, in a mouse model of tuberculosis. In: 38th Interscience Conference on Antimicrobial Agents and Chemotherapy, Abstract E-210, **1998**.

196 Lu T, Drlica K. In vitro activity of C-8-methoxy fluoroquinolones against mycobacteria when combined with anti-tuberculosis agents. *J. Antimicrob. Chemother.*, **2003**, 52, 1025–1028.

197 Rodríguez JC, et al. Mutant prevention concentration: comparison of fluoroquinolones and linezolid with *Mycobacterium tuberculosis*. *J. Antimicrob. Chemother.*, **2004**, 53, 441–444.

198 Nichols RL. Optimal treatment of complicated skin and skin structure infections. *J. Antimicrob. Chemother.*, **1999**, 44 (Suppl. A), 19–23.

199 Müller M, et al. Penetration of moxifloxacin into peripheral compartments in humans. *Antimicrob. Agents Chemother.*, **1999**, 43, 2345–2349.

200 Sadick NS. Systemic antibiotic agents. *Dermatol. Clin.*, **2001**, 19, 1–21.

201 Karchmer AW. Fluoroquinolone treatment of skin and skin structure infections. *Drugs*, **1999**, 58 (Suppl. 2), 82–84.

202 Alam MR, Hershberger E, Zervos MJ. The role of fluoroquinolones in the treatment of skin and soft tissue infection. *Curr. Infect. Dis. Rep.*, **2002**, 4, 426–432.

203 Blondeau JM. The role of fluoroquinolones in skin and skin structure infections. *Am. J. Clin. Dermatol.*, **2002**, 3, 37–46.

204 Parish LC, et al. Moxifloxacin versus cephalexin in the treatment of uncomplicated skin infections. *Int. J. Clin. Pract.*, **2000**, 54, 497–503.

205 Muijsers RB, Jarvis B. Moxifloxacin in uncomplicated skin and skin structure infections. *Drugs*, **2002**, 62, 967–973; discussion 974–975.

206 Consigny S, et al. Bactericidal activities of HMR 3647, moxifloxacin, and rifapentine against *Mycobacterium leprae* in mice. *Antimicrob. Agents Chemother.*, **2000**, 44(10), 2919–2921.

207 Mah FS. Fourth-generation fluoroquinolones: new topical agents in the war on ocular bacterial infections. *Curr. Opin. Ophthalmol.*, **2004**, 15, 316–320.

208 Hwang DG. Fluoroquinolone resistance in ophthalmology and the potential role for newer ophthalmic fluoroquinolones. *Surv. Ophthalmol.*, **2004**, 49 (Suppl. 2), S79–S83.

209 Kowalski RP, et al. Gatifloxacin and moxifloxacin: an in vitro susceptibility comparison to levofloxacin, ciprofloxacin, and ofloxacin using bacterial keratitis isolates. *Am. J. Ophthalmol.*, **2003**, 136, 500–505.

210 Callegan MC, et al. Antibacterial activity of the fourth-generation fluoroquinolones gatifloxacin and moxifloxacin against ocular pathogens. *Adv. Ther.*, **2003**, 20, 246–252.

211 Dajcs JJ, et al. Effectiveness of ciprofloxacin, levofloxacin, or moxifloxacin for treatment of experimental *Staphylococcus aureus* keratitis. *Antimicrob. Agents Chemother.*, **2004**, 48, 1948–1952.

212 Mather R, et al. Fourth generation fluoroquinolones: new weapons in the arsenal of ophthalmic antibiotics. *Am. J. Ophthalmol.*, **2002**, 133, 463–466.

213 Müller HP, et al. In vitro antimicrobial susceptibility of oral strains of *Actinobacillus actinomycetemcomitans* to seven antibiotics. *J. Clin. Periodontol.*, **2002**, 29, 736–742.

214 Sobottka I, et al. In vitro activity of moxifloxacin against bacteria isolated from odontogenic abscesses. *Antimicrob. Agents Chemother.*, **2002**, 46, 4019–4021.

215 Eick S, Seltmann T, Pfister W. Efficacy of antibiotics to strains of periodontopathogenic bacteria within a single species biofilm – an in vitro study. *J. Clin. Periodontol.*, **2004**, 31, 376–383.

216 Eick S, et al. In vitro antibacterial activity of fluoroquinolones against *Porphyromonas gingivalis* strains. *J. Antimicrob. Chemother.*, **2004**, 54, 553–556.

217 Schulz H-H, Schlimbach G. Use of chemotherapeutic agents. **1999**. WO/01/45679 A2.

218 Aminimanizani A, Beringer P, Jelliffe R. Comparative pharmacokinetics and pharmacodynamics of the newer fluoroquinolone antibacterials. *Clin. Pharmacokinet.*, **2001**, 40, 169–187.

219 Dalhoff A, Schmitz FJ. In vitro antibacterial activity and pharmacodynamics of new quinolones. *Eur. J. Clin. Microbiol. Infect. Dis.*, **2003**, 22, 203–221.

220 Wise R. A review of the clinical pharmacology of moxifloxacin, a new 8-methoxyquinolone, and its potential relation to therapeutic efficacy. *Clin. Drug Invest.*, **1999**, 17, 365–387.

221 MacGowan AP. Pharmacodynamics of Moxifloxacin. In: *Moxifloxacin in Practice*, Vol. 2. Adam D, Finch R (Eds.). Maxim Medical: Oxford, **1999**, pp. 5–15.

222 Stass H, et al. Pharmacokinetics, safety, and tolerability of ascending single doses of moxifloxacin, a new 8-methoxy quinolone, administered to healthy subjects. *Antimicrob. Agents Chemother.*, **1998**, 42, 2060–2065.

223 Stass H, Kubitza D. Pharmacokinetics and elimination of moxifloxacin after oral and intravenous administration in man. *J. Antimicrob. Chemother.*, **1999**, 43 (Suppl. B), 83–90.

224 Andrews J, et al. Penetration of moxifloxacin into bronchial mucosa, epithelial lining fluid and alveolar macrophages following a single, 400-mg oral dose. In: 38th Interscience Conference on Antimicrobial Agents and Chemotherapy, Abstract No. A 29. San Diego, **1998**.

225 Stass H. Distribution and tissue penetration of moxifloxacin. *Drugs*, **1999**, 58 (Suppl. 2), 229–230.

226 Stass H, et al. Pharmacokinetics of moxifloxacin, a novel 8-methoxy-quinolone, in patients with renal dysfunction. *Br. J. Clin. Pharmacol.*, **2002**, 53, 232–237.

227 Stass H, Kubitza D, Wensing G. Pooled analysis of pharmacokinetics, safety and tolerability of single oral 400 mg moxifloxacin (MFX) doses in patients with mild and moderate liver cirrhosis (LC). In: 40th Interscience Conference on Antimicrobial Agents and Chemotherapy, Abstract No. 2269. Toronto, **2000**.

228 Sullivan JT, et al. The influence of age and gender on the pharmacokinetics of moxifloxacin. *Clin. Pharmacokinet.*, **2001**, 40 (Suppl. 1), 11–18.

229 Stahlmann R, Lode H. Fluoroquinolones in the elderly: safety considerations. *Drugs Aging*, **2003**, 20, 289–302.

230 Scheld W. Maintaining fluoroquinolone class efficacy: review of influencing factors. *Emerg. Infect. Dis.*, **2003**, 9, 1–9.

231 MacGowan AP. In vitro models of infection: pharmacokinetic/pharmacodynamic correlates. In: *First International Moxifloxacin Symposium*. Mandell L (Ed.). Springer-Verlag: Berlin, **1999**, pp. 104–110.

232 Wise R. Maximizing efficacy and reducing the emergence of resistance. *J. Antimicrob. Chemother.*, **2003**, 51 (Suppl. 1), 37–42.

233 Lode H, Garau J. Improving care for patients with respiratory tract infections. *J. Chemother.*, **2002**, 14 (Suppl. 2), 22–28.

234 Everett MJ, Piddock LJV. Mechanisms of resistance to fluoroquinolones. In: *Quinolone Antibacterials*. Kuhlmann J, Dalhoff A, Zeiler H-J (Eds.) Springer: Berlin, Heidelberg, New York, Barcelona, Budapest, Hong Kong, London, Milan, Paris, Santa Clara, Singapore, Tokyo, **1998**, pp. 259–296.

235 Piddock LJ. Mechanisms of fluoroquinolone resistance: an update 1994–1998. *Drugs*, **1999**, 58 (Suppl. 2), 11–18.

236 Wise R. A review of the mechanisms of action and resistance of antimicrobial agents. *Can. Respir. J.*, **1999**, 6 (Suppl. A), 20A–22A.

237 Barman Balfour JA, Lamb HM. Moxifloxacin: a review of its clinical potential in the management of community-acquired respiratory tract infections. *Drugs*, **2000**, 59, 115–139.

238 Martinez-Martinez L, Pascual A, Jacoby GA. Quinolone resistance from a transferable plasmid. *Lancet*, **1998**, 351 (9105), 797–799.

239 Wang M, et al. Plasmid-mediated quinolone resistance in clinical isolates of *Escherichia coli* from Shanghai, China. *Antimicrob. Agents Chemother.*, **2003**, 47, 2242–2248.

240 Wang M, et al. Emerging plasmid-mediated quinolone resistance associated with the qnr gene in *Klebsiella pneumoniae* clinical isolates in the United States. *Antimicrob. Agents Chemother.*, **2004**, 48, 1295–1299.

241 Dalhoff A. Dissociated resistance among fluoroquinolones. In: 2nd European Congress of Chemotherapy, Abstract T 160. Hamburg, **1998**.

242 Dalhoff A. Lack of in vivo emergence of resistance against BAY 12–8039 in *S. aureus* and *S. pneumoniae*. In: 8th International Congress on Infectious Diseases, Poster No. 47.003. Boston, **1998**.

243 Pong A, et al. Activity of moxifloxacin against pathogens with decreased susceptibility to ciprofloxacin. *J. Antimicrob. Chemother.*, **1999**, 44, 621–627.

244 Lu T, Zhao X, Drlica K. Gatifloxacin activity against quinolone-resistant gyrase: allele-specific enhancement of bacteriostatic and bactericidal activities by the C-8-methoxy group. *Antimicrob. Agents Chemother.*, **1999**, 43, 2969–2974.

245 Drlica K, Malik M. Fluoroquinolones: action and resistance. *Curr. Top. Med. Chem.*, **2003**, 3, 249–282.

246 Sindelar G, et al. Mutant prevention concentration as a measure of fluoroquinolone potency against mycobacteria. *Antimicrob. Agents Chemother.*, **2000**, 44, 3337–3343.

247 Dong Y, et al. Fluoroquinolone action against mycobacteria: effects of C-8 substituents on growth, survival, and resistance. *Antimicrob. Agents Chemother.*, **1998**, 42, 2978–2984.

248 Dong Y, et al. Effect of fluoroquinolone concentration on selection of resistant mutants of *Mycobacterium bovis* BCG and *Staphylococcus aureus*. *Antimicrob. Agents Chemother.*, **1999**, 43, 1756–1758.

249 Zhao X, Drlica K. Restricting the selection of antibiotic-resistant mutant bacteria: measurement and potential use of the mutant selection window. *J. Infect. Dis.*, **2002**, 185, 561–565.

250 Blondeau JM, et al. Mutant prevention concentrations of fluoroquinolones for clinical isolates of *Streptococcus pneumoniae*. *Antimicrob. Agents Chemother.*, **2001**, 45, 433–438.

251 Li X, Zhao X, Drlica K. Selection of *Streptococcus pneumoniae* mutants having reduced susceptibility to moxifloxacin and levofloxacin. *Antimicrob. Agents Chemother.*, **2002**, 46, 522–524.

252 Metzler K, et al. Comparison of minimal inhibitory and mutant prevention drug concentrations of 4 fluoroquinolones against clinical isolates of methicillin-susceptible and -resistant *Staphylococcus aureus*. *Int. J. Antimicrob. Agents*, **2004**, 24, 161–167.

253 Sprandel KA, Rodvold KA. Safety and tolerability of fluoroquinolones. *Clin. Cornerstone*, **2003**, Suppl. 3, S29–S36.

254 Ball P. Future of the quinolones. *Semin. Respir. Infect.*, **2001**, 16, 215–224.

255 Talan DA. Clinical perspectives on new antimicrobials: focus on fluoroquinolones. Clin. Infect. Dis., **2001**, 32 (Suppl. 1), S64–S71.

256 von Keutz E, Schlüter G. Preclinical safety evaluation of moxifloxacin, a novel fluoroquinolone. *J. Antimicrob. Chemother.*, **1999**, 43 (Suppl. B), 91–100.

257 Culley CM, et al. Moxifloxacin: clinical efficacy and safety. *Am. J. Health Syst. Pharm.*, **2001**, 58, 379–388.

258 Iannini PB, et al. Reassuring safety profile of moxifloxacin. *Clin. Infect. Dis.*, **2001**, 32, 1112–1114.

259 Lode H, Vogel F, Elies W. Grepafloxacin: a review of its safety profile based on clinical trials and postmarketing surveillance. *Clin. Ther.*, **1999**, 21, 61–74.

260 Lode H, Vogel F, Elies W. Clinical safety profile of grepafloxacin. *Drugs*, **1999**, 58 (Suppl. 2), 395–396.

261 Ball P, et al. A new respiratory fluoroquinolone, oral gemifloxacin: a safety profile in context. *Int. J. Antimicrob. Agents*, **2004**, 23, 421–429.

262 Strehl E, Hofmann C. Struktur-Wirkungs-Beziehungen und Nebenwirkungen von Chinolon-Derivaten. *Krankenhauspharmazie*, **2001**, 22, 359–366.

263 Stahlmann R, Lode H. Toxicity of quinolones. *Drugs*, **1999**, 58 (Suppl. 2), 37–42.

264 Vohr H-W, Wasinska-Kempka G, Ahr HJ. An investigation into the phototoxic potential of moxifloxacin. In: *Moxifloxacin in Practice*, Vol. 2. Adam D, Finch R (Eds.). Maxim Medical: Oxford, **1999**, pp. 83–90.

265 Ball P. Quinolone-induced QT interval prolongation: a not-so-unexpected class effect. *J. Antimicrob. Chemother.*, **2000**, 45, 557–559.

266 Stahlmann R, Shakibaei M. Fluorchinolon-induzierte Tendopathien – klinische und experimentelle Aspekte. *Chemother. J.*, **2000**, 9, 140–147.

267 Stahlmann R. Clinical toxicological aspects of fluoroquinolones. *Toxicol. Lett.*, **2002**, 127, 269–277.

268 Stass H, Kubitza D. Lack of pharmacokinetic interaction between moxifloxacin, a novel 8-methoxyfluoroquinolone, and theophylline. *Clin. Pharmacokinet.*, **2001**, 40 (Suppl. 1), 63–70.

269 Stass H, et al. Lack of influence of itraconazole (ITR) on moxifloxacin (MFX) pharmacokinetics in healthy male subjects. In: 40th Interscience Conference on Antimicrobial Agents and Chemotherapy, Abstract No. 2270. Toronto, **2000**.

270 Stass H, et al. Effect of calcium supplements on the oral bioavailability of moxifloxacin in healthy male volunteers. *Clin. Pharmacokinet.*, **2001**, 40 (Suppl. 1), 27–32.

271 von Keutz E, Schlüter G. Moxifloxacin and the liver: results of preclinical investigations. In: *Moxifloxacin in Practice*, Vol. 2. Adam D, Finch R (Eds.). Maxim Medical: Oxford, **1999**, pp. 1–4.

272 Stass H, Kern A. Moxifloxacin – review of clinical pharmacokinetics: metabolism and excretion. In: 6th International Symposium on New Quinolones; Abstract No. 132. Denver, Colorado, **1998**.

273 Kern A, et al. BAY 12–8039, a new 8-methoxyquinolone: metabolism in rat, monkey and man. In: 36th Interscience Conference on Antimicrobial Agents and Chemotherapy, Abstract. No. F-023. New Orleans, USA, **1996**.

274 Wetzstein H-G, et al. Degradation of ciprofloxacin by basidiomycetes and identification of metabolites generated by the brown rot fungus *Gloeophyllum striatum*. *Appl. Environ. Microbiol.*, **1999**, 65, 1556–1563.

275 Martens R, et al. Degradation of the fluoroquinolone enrofloxacin by wood-rotting fungi. *Appl. Environ. Microbiol.*, **1996**, 62, 4206–4209.

276 Wetzstein H-G, Schmeer N, Karl W. Degradation of the fluoroquinolone enrofloxacin by the brown rot fungus *Gloeophyllum striatum*: identification of metabolites. *Appl. Environ. Microbiol.*, **1997**, 63, 4272–4281.

277 Wetzstein H-G, et al. Residual antibacterial activity of metabolites derived from the veterinary fluoroquinolone enrofloxacin. In: 100th General Meeting of the American Society for Microbiology. Los Angeles, **2000**.

278 Wetzstein H-G, Dalhoff A, Karl W. BAY 12–8039, A new 8-methoxyquinolone, is degraded by the brown rot fungus *Gloephyllum striatum*. In: 37th Interscience Conference on Antimicrobial Agents and Chemotherapy, Abstract F 157. Toronto, **1997**.

279 Gray JL, et al. Synthesis and testing of non-fluorinated quinolones (NFQs). A study on the influence of the C6 position. In: 40th Interscience Conference on Antimicrobial Agents and Chemotherapy, Abstract No. 1506. Toronto, **2000**.

280 Barry AL, Fuchs PC, Brown SD. In vitro activities of three nonfluorinated quinolones against representative bacterial isolates. *Antimicrob. Agents Chemother.*, **2001**, 45, 1923–1927.

281 Ledoussal B, et al. Discovery, structure–activity relationships and unique properties of non-fluorinated quinolones (NFQs). *Current Med Chem -Anti-Infective Agents*, **2003**, 2, 13–25.

282 Kim OK, Ohemeng K, Barrett JF. Advances in DNA gyrase inhibitors. *Expert Opin. Investig. Drugs*, **2001**, 10, 199–212.

283 Kastner M, et al. Concentrations of the des-F(6)-quinolone garenoxacin in plasma and joint cartilage of immature rats. *Arch. Toxicol.*, **2004**, 78, 61–67.

284 Bartel S, et al. Possibly substituted 8-cyano-1-cyclopropyl-7-(2,8-diazabicyclo[4.3.0]nonan-8-yl)-6-fluoro-1,4-dihydro-4-oxo-3-quinolinecarboxylic acids and their derivatives. **1996**. Bayer AG: WO 97/31001.

285 Himmler T, et al. Synthesis and in vitro activity of pradofloxacin, a novel 8-cya-nofluoroquinolone. In: 42nd ICAAC Abstracts, American Society for Micro-biology, p. 188, Abstract: F-566. San Diego, CA, **2002**.

286 Zerva L, Marshall SA, Jones RN. Prema-floxacin: a projected veterinary use fluoro-quinolone with significant activ-ity against multi-resistant gram-positive human pathogens. *J. Antimicrob. Chemother.*, **1996**, 38, 742–744.

287 Watts JL, et al. In vitro activity of prema-floxacin, a new extended-spectrum fluoroquinolone, against pathogens of veterinary importance. *Antimicrob. Agents Chemother.*, **1997**, 41, 1190–1192.

15
The Development of Bisphosphonates as Drugs

Eli Breuer

The story of bisphosphonates (BPs) is typical for the development of drugs in the semi-rational era, when actions were based on seemingly logical assumptions. Although the actions luckily led to eventual success, the initial assumptions turned out to be greatly oversimplified, if not incorrect.

15.1
Historical Background

The first authentic synthesis of a bisphosphonic acid, 1-hydroxyethane-1,1-bisphosphonic acid (etidronic acid), was published by von Baeyer and Hofmann in 1897 [1], although it is possible that the same compound had already been obtained in 1865 by Menschutkin [2]. The synthetic approach to BPs employed today differs only little from that used over a hundred years ago, and it is still based on the reaction of a carboxylic acid with phosphorus trichloride with the addition of some water [3]. (Note: BPs are usually represented as the free bisphosphonic acids; the generic names given refer, however, to their salt forms.)

Ever since the discovery of the steam engine, the precipitation of insoluble calcium salts in boilers and pipes has been causing severe problems. Thus, the discovery that low concentrations of phosphoric, pyrophosphoric or bisphosphonic acids can prevent scale had enormous industrial importance. This information, combined with the characteristic ability of BPs to chelate virtually every heavy metal, laid the basis for their wide industrial importance [4]. As a result, bisphosphonates are in current use in a wide variety of industrial areas, as scale inhibitors, for the treatment of metal and other surfaces, in water treatment, in the textile industry, in the detergent and cosmetic industry, in the polymer industry, and several others [4].

Analogue-based Drug Discovery. IUPAC, János Fischer, and C. Robin Ganellin (Eds.)
Copyright © 2006 WILEY-VCH Verlag GmbH & Co. KGaA, Weinheim
ISBN: 3-527-31257-9

15.2
Discovery of the Biological Activity of Pyrophosphate and of Bisphosphonates

The development of BPs for medical use began with the realization by Fleisch and his colleagues that inorganic pyrophosphate (PPi), a naturally occurring compound in the body, and present also in the serum and urine, can prevent calcification by binding to newly forming crystals of hydroxyapatite [5]. Consequently, it was postulated that PPi might be the agent that normally prevents calcification of soft tissues, and regulates bone mineralization, and that pathological disorders might be linked to disturbances in PPi metabolism. Attempts to exploit these concepts by using pyrophosphate and polyphosphates to inhibit ectopic calcification in blood vessels, skin, and kidneys in laboratory animals were successful only when the compounds were injected. Orally administered pyrophosphate and polyphosphates were inactive, due to the hydrolysis of PPi in the gastrointestinal tract, probably by mucosal brush-border phosphatases [6]. During the search for more stable analogues of PPi that might have similar antimineralization properties, but that would be resistant to hydrolysis, several different chemical classes were studied. The BPs (at that time called diphosphonates, and known since the 19th century), structural analogues of pyrophosphate (see Fig. 15.1), were among those studied. Like PPi, BPs had a high affinity for bone mineral and were found to prevent calcification both *in vitro* and *in vivo*. However, unlike PPi, they were also able to prevent pathological calcification when given orally to rats [7]. Perhaps the most important step towards the future use of BPs occurred when they were found to inhibit the dissolution of hydroxyapatite (HAP, the inorganic constituent of bone) crystals, and to inhibit bone resorption, both in tissue culture and *in vivo* [8].

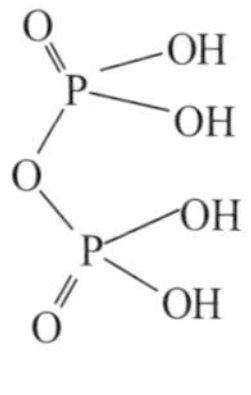

Pyrophosphoric acid 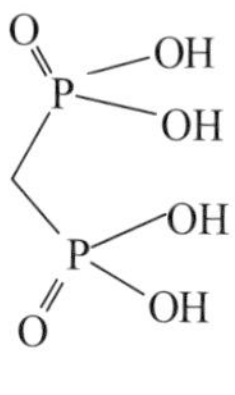   Methanebisphosphonic acid

Fig. 15.1 Bisphosphonates are structural analogues of pyrophosphate.

15.3
Bone-Related Activity of Bisphosphonates

15.3.1
Overview

Bisphosphonates are currently in clinical use for the following indications:
• As bone markers in nuclear medicine (as technetium complexes) for the diagnosis of bone metastases and other bone lesions. This

property follows from the dual ability of BPs, namely to chelate technetium whilst simultaneously retaining bone affinity. The most frequently used are complexes of radioactive technetium, ^{99m}Tc, with methanebisphosphonic acid (medronate, MDP) or its hydroxy derivative (hydroxymethanebisphosphonic acid, oxidronate; see Fig. 15.2). The gamma-emitting Tc-BP complex is injected intravenously, and images are collected after a few hours when the excess complex has been cleared. This method is uniquely suited for the investigation of skeletal pathology, because of its great sensitivity and rapidity [9].

- Osteolytic bone diseases or bone resorption: the most important clinical application of BPs is as inhibitors of osteolytic bone diseases, including osteoporosis, tumor-related bone destruction, and Paget's disease. These are discussed in detail in the next section.
- Inhibition of abnormal calcification: initials explorations of BPs as inhibitors of calcification showed promise, and early applications of etidronate included use in myositis ossificans, as well as in patients who had undergone total hip replacement surgery, in order to prevent subsequent heterotopic ossification and to improve mobility [10].

15.3.2
Osteolytic Bone Diseases

15.3.2.1 Osteoporosis

Osteoporosis is a disease characterized by a reduction in bone mass and a deterioration in bone microarchitecture, which leads to enhanced fragility. This is a very common disorder, and it is becoming more common with the increase of life expectancy; thus, it is a major public health issue. Osteoporosis is caused by an increase in bone loss and at the same time a decrease in bone formation. With increasing age, bone formation is slowed down, especially in women after the menopause, although it is also seen sometimes in men, usually aged over 70 years (senile osteoporosis).

15.3.2.2 Osteolytic Tumors

Osteolytic tumors may induce bone destruction either through local invasion, or by a secondary metastatic bone disease. The most frequent types of primary tumors that develop into metastatic bone disease are, in the order of prevalence: breast, prostate, thyroid, kidney, and bronchial tumors, whereas esophageal, gastrointestinal and rectal tumors are much less metastatic [11]. Very often, the destruction of bone in a metastatic bone disease leads to hypercalcemia of malignancy, which also responds to BPs. In addition, hematological cancers such as

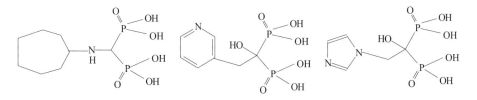

Medronic acid

Oxidronic acid

Clodronic acid

Etidronic acid

Tiludronic acid

Pamidronic acid

Alendronic acid

Ibandronic acid

Neridronic acid

Incadronic acid

Risedronic acid

Zoledronic acid

Fig. 15.2 Bisphosphonates in current clinical use.

multiple myeloma and lymphomas may also induce hypercalcemia, which may result in severe clinical symptoms, including death, depending on the serum calcium level. In contrast, non-tumor-induced hypercalcemia (e.g., hyperparathyroidism) is less responsive to BP treatment. In the treatment of bone problems associated with malignancy, the skeletal complications are reduced with BP therapy. In addition, there is emerging evidence indicating that nitrogen-containing BPs possess anticancer activity in bone, and occasionally in other organs [12].

15.3.2.3 Paget's Disease

Paget's disease is a localized disorder of bone remodeling characterized by the development of deformations and changes in the shapes, sizes and directions of the diseased bones, such as the skull, spine, arms, and legs [13].

15.3.3
Structure–Activity Relationships

15.3.3.1 The Molecular Skeletons of Bisphosphonates

All bone-active BPs possess the P-C-P structure, which is the basis of the structural analogy to pyrophosphates (see Fig. 15.1). This P-C-P moiety has been called the "bone hook", namely it is the primary structural feature that endows the molecule its affinity and targets it to the bone. Molecules in which the two phosphonic functions are farther apart are biologically mostly inactive, although there are reports of some long-chain bisacylphosphonates having calcium-related biological activity [14].

15.3.3.2 Phosphonic Acid Groups

Another requirement for good activity is the presence of four ionizable POH groups, although molecules having only three OH groups have also shown some activity (see Fig. 15.3).

15.3.3.3 The Geminal Hydroxy Group

The presence of an hydroxyl group on the central (geminal) carbon is important for strong affinity of the bisphosphonates to HAP, the inorganic constituent of bone, presumably because this allows the formation of a tridentate mode of Ca binding. However, this hydroxyl group has no influence on the potency of the BP in bone resorption, and there are potent BPs devoid of such hydroxyl groups.

15.3.3.4 Nitrogen-Containing Side Chain

The greatest improvements in the biological activities of BPs have, however, resulted from the introduction of a basic nitrogen atom into the side chain R^2 on the geminal carbon. The first such compound was pamidronate, and further jumps in the activity in bone resorption were obtained either by transposing the amine group farther from the "bone hook" or by alkylating the primary amine to a tertiary one. Finally, the highest activity attained to date has been with the nitrogen-containing heterocyclic BPs, risedronate and zoledronate.

15.3.3.5 Structure–Activity Relationships of BPs: A Summary

The least potent antiresorptive BPs *in vivo* are those that most closely resemble PPi. These include BPs with the simplest R^1 and R^2 substituents, such as medronate ($R^1 = R^2 = H$), oxidronate ($R^1 = H$ and $R^2 = OH$), clodronate ($R^1 = R^2 = Cl$), and etidronate ($R^1 = CH_3$ and $R^2 = OH$) (see Figs. 15.2 and 15.3). The potency of the first two compounds is so low that their use is limited as complexants of technetium. Etidronate was found to be less potent than clodronate both *in vitro* and *in vivo*, even though it had a higher affinity for bone mineral, presumably due to the presence of a hydroxyl group in the R^1 position. This discrepancy between antiresorptive potency and affinity for calcium was one of the indicators that BPs might inhibit bone resorption by cellular effects on bone-resorbing cells, rather than by acting as crystal poisons that simply prevent HAP crystal dissolution by a physico-chemical mechanism. The structures of BPs that have gained approval for clinical use are shown in Fig. 15.2.

Following the initial discovery that BPs which closely resembled PPi could inhibit bone resorption, new BPs were discovered that were up to 1000 to 10 000-fold more potent at inhibiting bone resorption *in vivo*. Such increases in potency could be achieved by various means, for example by increasing the length of the R^2 side chain attached to the geminal carbon, from a single methyl (-CH_3) group (as in etidronate) to a longer alkyl chain, while retaining the hydroxyl (-OH) group in the R^1 position to maximize the affinity for bone mineral.

BPs having $R^1 = OH$ have higher affinity to hydroxyapatite

BPs having basic nitrogen containing R^2 are of enhanced potency and common mechanism

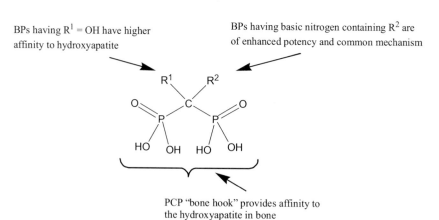

PCP "bone hook" provides affinity to the hydroxyapatite in bone

Fig. 15.3 Structure–activity relationships in geminal bisphosphonates.

A crucial point was reached in BP research when it was found that the antiresorptive potency of a BP increased significantly by the attachment of an amino (-NH$_2$) group at the terminus of the R^2 alkyl chain. BPs with -C$_2$H$_4$NH$_2$, -C$_3$H$_6$NH$_2$, or -C$_5$H$_{10}$NH$_2$ aminoalkyl groups in the R^2 side chain position (as in pamidronate, alendronate, and neridronate, respectively) were found to be up to 1000-fold more potent than etidronate *in vivo*. The optimal length of the aminoalkyl chain appears to be C4 (including the geminal carbon atom), as in alendronate. Potency could be increased further by modifying the primary amine to a tertiary amine, while leaving the R^1 = OH substituent unchanged. Thus, the replacement of the primary amino group of pamidronate by the tertiary methylpentylamino group to form ibandronate increased *in vivo* potency about 300-fold. BPs containing a secondary amine group in the R^2 side chain also have been found to be more potent than those containing a primary amine. For example, incadronate, which contains a cycloheptyl ring in the R^2 side chain attached to the geminal carbon by a secondary amine (NH) group, was found to be about 30-fold more potent than pamidronate *in vivo*. BPs containing a hydroxyl group at the R^1 side chain and a tertiary nitrogen within ring structures in the R^2 side chain appear to be the most potent antiresorptive BPs discovered to date. Examples include risedronate (which contains a 3-pyridyl group) and zoledronate (containing an imidazole ring); these are up to 10 000-fold more potent than etidronate *in vivo* in rodent models of bone resorption.

15.3.4
Inhibition of Bone Resorption: The Mechanisms of Action

Contrary to the original assumption which linked the inhibition of bone resorption solely to physico-chemical adsorption of the BP onto the bone surface, there is new evidence indicating cellular mechanisms of BP action, although a prerequisite of the biological activity is still the adsorption. The bisphosphonates deposited on the bone surface poison the bone-resorbing osteoclasts after being internalized by them.

Bisphosphonates are classified into two main groups, each with a different mechanism of osteoclast poisoning. Those BPs that most closely resemble pyrophosphate (e.g., clodronate, etidronate, tiludronate) are metabolized to form methylene-containing (AppCp-type) cytotoxic analogues of ATP (Fig. 15.4) [15].

In contrast, more potent nitrogen-containing BPs inhibit the intracellular mevalonate pathway [16]. This biosynthetic pathway is responsible for the production of cholesterol via isoprenoid intermediates such as isopentenyl diphosphate (IPP), farnesyl diphosphate (FPP), and geranylgeranyl diphosphate. The structures of carbocation transition states involved in the biosynthesis of cholesterol, and the structures of some nitrogen-containing BPs, are illustrated in Fig. 15.5. The similarity between the structures suggests that the N-containing BPs may be transition-state analogues of isoprenoid diphosphates. Conversely, clodronate and etidronate do not affect cholesterol synthesis. In addition, there seem to exist other mechanisms that may contribute to the bone-protecting effects of BPs.

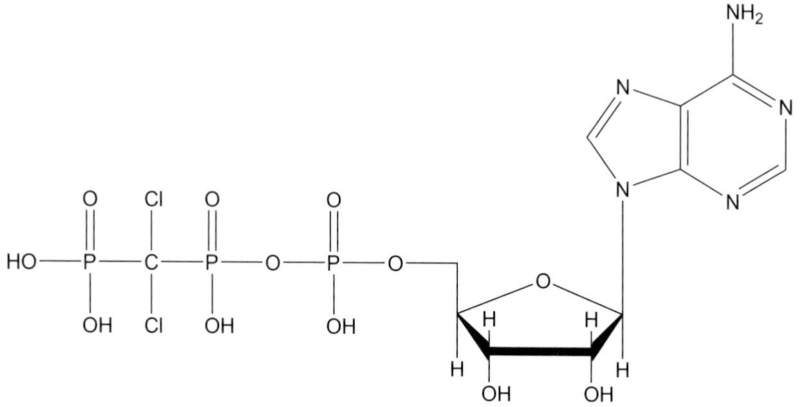

Fig. 15.4 Dichloromethylene-containing (AppCp-type) cytotoxic analogue of ATP.

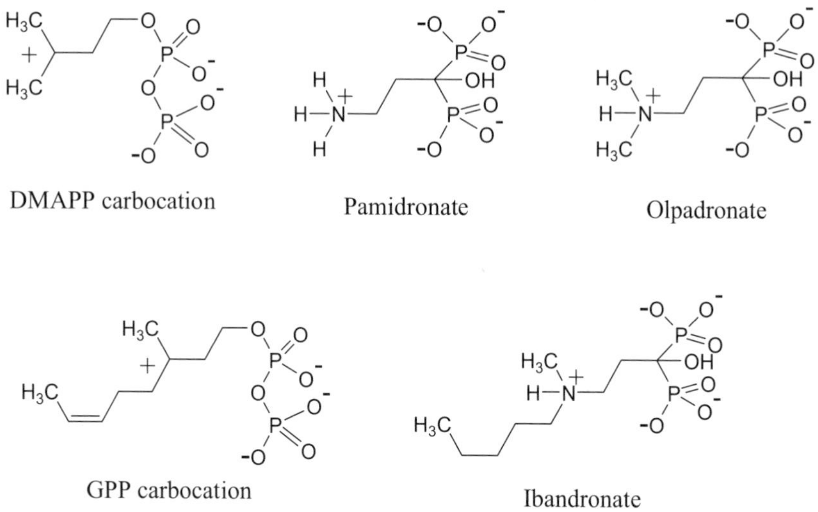

DMAPP carbocation

Pamidronate

Olpadronate

GPP carbocation

Ibandronate

Fig. 15.5 Some isoprenoid diphosphate carbocation intermediates involved in cholesterol biosynthesis, along with some nitrogen containing bisphosphonates, their proposed transition-state-analogues. (DMAPP = dimethylallyl diphosphate, GPP = geranyl diphosphate).

15.3.5
Clinical Pharmacology of Bisphosphonates

Bisphosphonates are very hydrophilic compounds, and are ionized to a high degree at physiological pH (e.g., etidronate has $pK_1 = 1.35$; $pK_2 = 2.87$; $pK_3 = 7.03$; $pK_4 = 11.3$). Therefore, their clinical pharmacology is characterized by very poor intestinal absorption. On the other hand, they have high affinity for bone mineral, which results in highly selective localization in bone, especially in areas of metabolic activity, such as bone formation and bone resorption, and long skeletal retention (half-life measured in years) [17]. Oral absorption is even lower when the drugs are taken with food. Significant side effects are minimal, but sometimes severe esophageal erosions or ulcerations are caused with orally administered BPs. These problems have been overcome successfully by intermittent dosing (e.g., once weekly alendronate, 70 mg [18] or risedronate, 35 mg [19]). Even less frequent regimens have been suggested for highly potent BPs such as ibandronate (either orally 100–150 mg once monthly [20], or 2–3 mg intravenously or subcutaneously administered, once in 2–3 months [21]), or zoledronate (a single dose of 4 mg once a year, or divided into smaller doses, 3 or 6 months apart) [22]. These dosing regimens are adequate to guarantee the presence of sufficient amounts of BPs on the bone to poison the osteoclasts. In this way, better-defined doses can be given with absolute compliance, and therefore it is likely that such regimens will increase in popularity.

15.3.6
Bisphosphonates in Clinical Use

Bisphosphonates that have been approved for clinical use, as reported in the *Annual Report of Medicinal Chemistry*, are listed in Tab. 15.1. It should be noted that although a BP may have gained initial approval for a specific clinical indication, its subsequent use in various countries may not be necessarily limited to that indication.

15.4
Miscellaneous Biological Aspects of Bisphosphonates

15.4.1
Bisphosphonates as Vehicles for Delivering Drugs to Bone

The exceptional affinity of BPs to bone mineral has led to the exploration of their utility for targeting pharmacological agents to bone. Among those explored for bone delivery are the above-mentioned diagnostic radioisotopes, antineoplastic drugs, drugs intended for the augmentation of systemic bone mass, and anti-inflammatory drugs and proteins [23].

15.4.2
Bisphosphonates as Potential Drugs for other Indications

Because of its analogy to PPi and an ability to chelate metals, the BP structure has served as a pharmacophore for drug design in various areas. In addition, as detailed in Chapter I–1, "*...in some cases a new clinical activity observed for an old drug is sufficiently potent and interesting to justify the immediate use of the drug in the new indication...*". Some of the cases that follow fall into this class, and include especially antiparasitic, anti-arthritic, and antirestenosis applications.

Tab 15.1 Bisphosphonates approved for clinical use.

Name (introduced by)	Main clinical use	Approximate relative potency in bone resorption	First introduced	Reference (Ann. Rep. Med. Chem., Vol., p.)
Medronate	^{99m}Tc-complex as diagnostic aid (radioactive imaging agent)	NA	?	–
Oxidronate	^{99m}Tc-complex as diagnostic aid (radioactive imaging agent)	NA	?	–
Etidronate (Procter& Gamble)	Paget's disease (Later hypercalcemia heterotopic ossification)	1	≤1982	
Clodronate (Oy Star)	Hypercalcemia Tumor osteolysis	~10	1986	22, 319
Pamidronate (Ciba Geigy)	Hypercalcemia of malignancy Tumoral osteolysis Paget's disease Osteoporosis	~100	1987	25, 312
Alendronate (MS&D)	Osteoporosis Paget's disease	~100> <1000	1993	29, 332
Tiludronate (Sanofi Winthrop)	Paget's disease Hypercalcemia of malignancy Tumoral osteolysis Osteoporosis	~10	1995	31, 350
Ibandronate (Boehringer Mannheim)	Hypercalcemia of malignancy Tumoral osteolysis OsteoporosisPaget's disease	~1000–~10 000	1996	32, 309
Incadronate (Yamanouchi)	Osteoporosis	~100–~1000	1997	33, 335
Risedronate (Procter& Gamble)	Osteoporosis Paget's disease	~1000–~10 000	1998	34, 330
Zoledronate (Novartis)	Hypercalcemia of malignancy	>10 000	2000	36, 314
Neridronate (Abiogen)	Osteogenesis imperfecta Hypercalcemia	100< – <1000	2002	38, 361

NA: not applicable.

15.4.2.1 Antiviral Drugs

In parallel with the development of BPs for calcium-related disorders, several studies were directed for the development of antiviral drugs involving "pyrophosphate analogues". These analogues included, in addition to clodronate and its dibromo analogue, phosphonoacetic acid (PAA) and phosphonoformic acid (PFA). These two phosphonocarboxylic acids are sometimes also called "pyrophosphate analogues". Only PFA has become a clinically useful antiviral drug (Foscarnet® [24]), while the two dihalomethylenebisphosphonates and phosphonoacetic acids were abandoned, due to their high affinity for bone. The initial rationale for testing in these activities was the inhibitory potency on some viral enzymes, presumably due to metal chelation [25].

15.4.2.2 Bisphosphonate Inositol-Monophosphatase Inhibitor: A Potential Drug for Bipolar Disorders

Lithium has been hypothesized to exert its therapeutic effects in the treatment of bipolar disorder by inhibiting inositol monophosphatase (IMPase), thereby causing a depletion of intracellular inositol. This in turn results in a reduction of the synthesis of the phosphoinositol required to sustain this signaling pathway. Several alicyclic and aromatic bisphosphonates, putative analogues of inositol 1-phosphate, were synthesized and tested, and some showed significant *in-vitro* potency [26].

15.4.2.3 Hypocholesterolemic Bisphosphonates (Squalene Synthase Inhibitors)

Inhibitors of squalene synthase have recently received considerable attention in the search for new antihypercholesterolemic agents. BP-based inhibitors were designed as mimetics of the putative carbocation intermediate formed from FPP in the initial step of the squalene synthase reaction. (See also mechanism of action of nitrogen-containing BPs and Fig. 15.5.) The BP function served as a stable surrogate for the pyrophosphate moiety [27].

15.4.2.4 Antiparasitic Drugs

As mentioned previously, BPs such as pamidronate, alendronate, and risedronate, act by inhibiting enzymes of the mevalonate pathway, and also the formation of farnesyl pyrophosphate, a compound involved in protein prenylation. It was not surprising therefore, that BPs have activity against the *in-vitro* proliferation of *Trypanosoma cruzi*, *Trypanosoma brucei rhodesiense*, *Leishmania donovani*, *Toxoplasma gondii*, and *Plasmodium falciparum*. Moreover, promising results have been found in animal models, with several parasites including parasitological cures with two *Leishmania* species [28]. Because the bisphosphonates mentioned are already approved for use in humans, they represent an attractive class of chemotherapeutic agents for development for clinical use.

15.4.2.5 Anti-Inflammatory and Anti-Arthritic Bisphosphonates

Bisphosphonates (particularly clodronate) have been shown to have anti-inflammatory effects in animal models of rheumatoid arthritis (RA), as well as in arthritis in humans. In adjuvant- and antigen-induced arthritis in rats, clodronate suppresses the inflammatory articular lesions in the inflamed joints [29], whilst in human RA, clodronate decreases the levels of interleukin (IL)-1, tumor necrosis factor-alpha (TNFα and β-microglobulin in the circulation [30]. *In vitro*, clodronate inhibits cytokine and nitric oxide (NO) release and inducible nitric oxide synthase (iNOS) expression in macrophage-like cells.

15.4.2.6 Cardiovascular Applications of Bisphosphonates

Systemically administered liposome-encapsulated bisphosphonates (e.g., clodronate or alendronate) have been shown to be effective, in rat and rabbit models, in the prevention of restenosis. This complication frequently occurs following percutaneous coronary interventions. Thus, these compounds show considerable promise for clinical application in this area [31].

15.5
Conclusions

The group of bisphosphonates complies in, the strictest sense, with the definition of analogues. These compounds have a common pharmacophore, the P-C-P function, and the only variation among the different members of the family is in one substituent linked to that pharmacophore. Moreover, the analogy extends also to the biological activity, which is largely common to all members. Finally, the case study described by this chapter provides excellent proof of the success of the analogue approach, since it demonstrates how the potency of drugs in one family of close structural and pharmalogical analogues could be improved ten thousand-fold, simply by changing one substituent in the molecule.

References

1 Von Baeyer H, Hofmann K. A. Acetodiphosphorige Säure. *Ber. Dtsch. Chem. Ges.*, **1897**, 30, 1973–1978.

2 Menschutkin N. Über die Einwirkung des Chloroazetyls auf phosphorige Säure. *Ann. Chem. Pharm.*, **1865**, 133, 317–320.

3 Lecouvey M, Leroux Y. Synthesis of 1-hydroxy-1,1-bisphosphonates. *Heteroatom Chem.*, **2000**, 11, 556–561.

4 Blomen LJMJ. History of the bisphosphonates. Discovery and history of the non-medical uses of bisphosphonates. In: *Bisphosphonate on bones*. Bijvoet OLM, Fleisch HA, Canfield RE, Russell RGG (Eds.). Elsevier, **1995**, pp. 111–124.

5 (a) Fleisch H, Bisaz S. Isolation from urine of pyrophosphate, a calcification inhibitor. *Am. J. Physiol.*, **1962**, 203, 671–675; (b) Fleisch H, Russell RGG, Straumann F. Effect of pyrophosphate on hydroxyapatite and its implications in calcium homeostasis. *Nature*, **1966**, 212, 901–903.

6 Fleisch H, Russell RGG, Bisaz S, Muhlbauer RC, Williams DA. The inhibitory effect of phosphonates on the formation of calcium phosphate crystals in vitro and on aortic and kidney calcification in vivo. *Eur. J. Clin. Invest.*, **1970**, 1, 12–18.

7 Francis MD, Russell RGG, Fleisch H. Diphosphonates inhibit formation of calcium phosphate crystals in vitro and pathological calcification *in vivo*. *Science*, **1969**, 165, 1264–1266.

8 Fleisch H, Russell RGG, Francis MD. Diphosphonates inhibit hydroxyapatite dissolution in vitro and bone resorption in tissue culture and in vivo. *Science*, **1969**, 165, 1262–1264.

9 (a) Fogelman I, Bessent RG, Turner JF, Citrin DL, Boyce IT, Greig WR. The use of whole-body retention of Tc-99m diphosphonate in the diagnosis of metabolic bone disease. *J. Nucl. Med.*, **1978**, 19, 270–275; (b) Francis MD, Fogelman I. 99mTc-diphosphonate uptake mechanisms on bone. In: *Bone Scanning in Clinical Practice*. Fogelman I (Ed.). Springer-Verlag, **1987**, pp. 1–6.

10 Fleisch H. *Bisphosphonates in Bone Disease. From the Laboratory to the Patient.* 3rd edn. New York: The Parthenon Publishing Group, **1997**, pp. 145–150.

11 Bijvoet OLM, Fleisch HA, Canfield RE, Russell RGG (Eds.). *Bisphosphonate on Bones*. Elsevier Science B.V.: Amsterdam, **1995**, p. 337.

12 Green JR. *Cancer*, **2003**, 97 (3 Suppl.), 840–847.

13 Delmas PD, Meunier PJ. The management of Paget's disease of bone. *N. Engl. J. Med.*, **1997**, 336, 558–66.

14 Golomb G, Schlossman A, Saadeh H, Levi M, Van Gelder JM, Breuer E. Bisacylphosphonates inhibit hydroxyapatite formation and dissolution *in vitro* and dystrophic calcification *in vivo*. *Pharm. Res.*, **1992**, 9, 143–148.

15 Frith JC, Mönkkönen J, Blackburn GM, Russell RGG, Rogers MJ. Clodronate and liposome-encapsulated clodronate are metabolised to a toxic ATP analog, adenosine $5'$-($\beta\beta,\gamma\gamma$-dichloromethylene) triphosphate, by mammalian cells in vitro. J. Bone Miner. Res., **1997**, 12, 1358–1367.

16 Fisher JE, Rogers MJ, Halasy JM, Luckman SP, Hughes DF, Masarachia PJ, Wesolowski G, Russell RGG, Rodan GA, Reszka AA. Mechanism of action of alendronate: geranylgeraniol, an intermediate of the mevalonate pathway, prevents inhibition of osteoclast formation, bone resorption and kinase activation in vitro. *Proc. Natl. Acad. Sci. USA*, **1999**, 96, 133–138.

17 Fleisch H. *Bisphosphonates in Bone Disease. From the Laboratory to the Patient.* 3rd edn. New York: The Parthenon Publishing Group, **1997**, pp. 60–62.

18 Schnitzer T, Bone HG, Crepaldi G, Adami S, McClung M, Kiel D, Felsenberg D, Recker RR, Tonino RP, Roux C, Pinchera A, Foldes AJ, Greenspan SL, Levine MA, Emkey R, Santora AC, Kaur A, Thompson DE, Yates J, Orloff JJ. Therapeutic equivalence of alendronate 70 mg once-weekly and alendronate 10 mg daily in the

treatment of osteoporosis. Alendronate Once-Weekly Study Group. *Aging (Milan)*, **2000**, 12, 1–12.

19 Brown JP, Kendler DL, McClung MR, Emkey RD, Adachi JD, Bolognese MA, Li Z, Balske A, Lindsay R. The efficacy and tolerability of risedronate once a week for the treatment of postmenopausal osteoporosis. *Calcif. Tissue Int.*, **2002**, 71, 103–111.

20 Schimmer R, Bauss F. Effect of daily and intermittent use of ibandronate on bone mass and bone turnover in postmenopausal osteoporosis: a review of three phase II studies. *Clin. Ther.*, **2003**, 25, 19–34.

21 Adami S, Felsenberg D, Christiansen C, Robinson J, Lorenc RS, Mahoney P, Coutant K, Schimmer RC, Delmas PD. Efficacy and safety of ibandronate given by intravenous injection once every 3 months. *Bone*, **2004**, 34, 881–889.

22 Reid IR, Brown JP, Burckhardt P, Horowitz Z, Richardson P, Trechsel U, Widmer A, Devogelaer J-P, Kaufman J-M, Jaeger P, Body J-J, Meunier P-J. Intravenous zoledronic acid in postmenopausal women with low bone mineral density. *N. Engl. J. Med.*, **2002**, 346, 653–661.

23 Uludag H. Bisphosphonates as a foundation of drug delivery to bone. *Current Pharm. Design*, **2002**, 8, 99–110.

24 Wagstaff AJ, Bryson HM. Phosphonoformate. *Drugs*, **1994**, 48, 199–226.

25 Hutchinson DW. Metal chelators as potential antiviral agents. *Antiviral Res.*, **1985**, 5, 193–205.

26 Atack JR. Inositol monophosphatase inhibitors – Lithium mimetics? *Med. Res. Rev.*, **1997**, 17, 215–224.

27 Ciosek CP, Jr., Magnin DR, Harrity TW, Logan JVH, Dickson JK, Jr., Gordon EM, Hamilton KA, Jolibois KG, Kunselman LK, Lawrence RM, Mookhtiar KA, Rich LC, Slusarchyk DA, Sulsky RB, Biller SA. Lipophilic 1,1-bisphosphonates are potent squalene synthase inhibitors and orally active cholesterol-lowering agents *in vivo*. *J. Biol. Chem.*, **1993**, 268, 24832–24837.

28 Martin MB, Grimley JS, Lewis JC, Heath HT, Bailey BN, Kendrick H, Yardley V, Caldera A, Lira R, Urbina JA, Moreno SNJ, Docampo R Croft SL, Oldfield E. Bisphosphonates inhibit the growth of *Trypanosoma brucei, Trypanosoma cruzi, Leishmania donovani, Toxoplasma gondii*, and *Plasmodium falciparum*: a potential route to chemotherapy. *J. Med. Chem.*, **2001**, 44, 909–916.

29 Osterman T, Kippo K, Lauren L, Hannuniemi R, Sellman R. Effect of clodronate on established adjuvant arthritis. *Rheumatol. Int.*, **1994**, 14, 139–147.

30 Cantatore FP, Introsso AM, Carrozzo M. Effects of bisphosphonates on interleukin 1, tumor necrosis factor, and micro-2 globulin in rheumatoid arthritis. *J. Rheumatol.*, **96**, 23, 1117–1118.

31 Danenberg HD, Fishbein I, Gao J, Mönkkönen J, Reich R, Gati I, Moerman E, Golomb G. Macrophage depletion by clodronate-containing liposomes reduce neointimal formation following balloon injury in rats and rabbits. *Circulation*, **2002**, 106, 599–605.

16
Cisplatin and its Analogues for Cancer Chemotherapy

Sándor Kerpel-Fronius

16.1
Introduction

The development of the platinum (Pt) -containing cytotoxic drugs for the treatment of solid tumors is a major success story of analogue-based drug discovery. Very soon after the introduction of the first Pt complex into clinical practice, three main goals for the development of subsequent derivatives were formulated:
- the reduction of dose-limiting, severe, frequently lethal nephro-
  toxicity;
- the broadening of the tumor spectrum; and
- oral administration.

Although all three aims have been achieved to various degrees during the past years, none of the orally administrable Pt analogues has been proven to be sufficiently reliable for broad clinical application. Hence, only the characteristics of the two clinically successful Pt analogues, exemplifying fulfillment of the first and second goals, will be compared to the originator molecule, cisplatin.

16.2
Cisplatin

16.2.1
Discovery

Cisplatin was serendipitously discovered by Rosenberg and his coworkers in 1965 [1] who, while investigating the effect of electric current on bacteria, observed growth inhibition accompanied by a filamentous clustering of the cells around the Pt electrode. The phenomenon was found to be due to the interaction of the growth medium and the Pt electrode, which led to the formation of a Pt complex, *cis*-diamminedichloroplatinum(II). Fortunately, the investigators concluded that this Pt compound might exert also antitumor activity. Their hypothesis was con-

Analogue-based Drug Discovery. IUPAC, János Fischer, and C. Robin Ganellin (Eds.)
Copyright © 2006 WILEY-VCH Verlag GmbH & Co. KGaA, Weinheim
ISBN: 3-527-31257-9

firmed in tumor cell cultures and transplantable murine tumor models. Since the compound has already been described in the 19th century under the name Peyrone's chloride, it could never be fully patented, but eventually it received an application patent for anticancer treatment.

16.2.2
Structure

Cisplatin is a neutral square planar coordination complex with two chloride and ammonium ligands in the cis-configuration. The compounds with trans-configuration are mostly devoid of activity. Pt exists in two main oxidation states, namely Pt^{2+} [Pt(II) and Pt^{4+} Pt(IV)]. In the latter, two of the ligands are located axially directly above and below the square plane, which results in an octahedral configuration. Although many of the Pt(IV) derivatives exhibit anticancer activity, only the Pt(II) complexes with cis-configuration have become widely applied in the clinical practice [2].

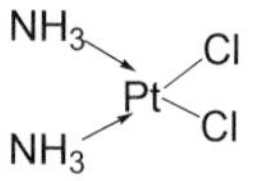

Fig. 16.1 Cisplatin.

16.2.3
Mechanism of Action

The reactivity of the Pt compounds depends on the relative displacement reactions of the various ligands. The nitrogen bonds are essentially stable under physiological conditions. Consequently, any chemical modification at the nitrogen atoms will influence the size and shape of the Pt complex bound to the DNA, but not the chemical reactivity of the compound. The stability of the Pt–halogen bond is much lower. Indeed, approximately 100 mmol chloride concentration is required to suppress the formation of mono- and diaquo species. Therefore, cisplatin must be administered in a saline solution to maintain the chemical neutrality of *cis*-diamminedichloroplatinum(II) which is essential for rapid penetration of the compound into the cells. In the intracellular milieu, the formation of the much more labile, positively charged mono- and diaquo derivatives proceeds rapidly, and these are considered to be the active intermediates. These, in turn, can react with many types of nucleophilic groups located on the various macromolecules. It seems that cisplatin binds less avidly to RNA than to DNA, and has even less affinity to proteins. For anticancer effectiveness, the crucial binding occurs on the N-7 position of the imidazole ring of guanine (G) and adenine (A) of DNA. First, monofunctional binding occurs, though many of the bonds are later converted to bifunctional bindings forming intrastrand and interstrand cross-links, respectively. Around 60–65% of the adducts are formed between adjacent guanines (GpG) or adjacent adenine and guanine (ApG, 20–25%) on the same DNA strand. The binding of adenines or guanines separated by one nucleotide also occurs, though rarely (GpXpG or ApXpG). Although the number of interstrand bonds is small, they seem to play an essential role in cell killing.

16.2.4
Pharmacokinetics

Descriptions of the pharmacokinetic properties of the Pt compounds are usually based on the measurement of total Pt, which includes both Pt bound to plasma proteins and ultrafiltrable Pt. Only the latter is considered to be pharmacologically active [3,4]. The protein binding of cisplatin is extensive (>90%), due to its high chemical reactivity. Consequently, the alpha and beta half-lives are much shorter for ultrafiltrable than for total Pt. The final excretion half-life is several days in length, and is due to the elimination of Pt which has remained bound to the catabolic products of macromolecules (Tab. 16.1). It is interesting to note that due to the tight tissue binding, small amounts of Pt can be detected in the plasma of patients with testicular cancer for up to 10 years after the termination of cisplatin treatment [5].

Tab 16.1 Clinical pharmacokinetics of cisplatin and its analogues.

Parameter	Cisplatin	Carboplatin	Oxaliplatin
$t_{1/2}$ alpha [h]			
Total Pt	0.23–0.82	0.2–1.6	0.43
Ultrafiltrate	0.15–0.5	0.13–1.45	0.35
$t_{1/2}$ beta [h]			
Total Pt	0.7–4.6	1.3–1.7	–
Ultrafiltrate	0.7–0.8	1.7–5.9	–
$t_{1/2}$ gamma [h]			
Total Pt	24–127	8.2–40	38–47
Ultrafiltrate	–	–	–
Protein binding [%]	>90	24–50	85
Urinary excretion [%]	23–50	54–82	>50

16.2.5
Clinical Efficacy

Cisplatin was found to be outstandingly active against testicular and ovarian cancers, and also to exhibit useful clinical efficacy in head and neck, bladder and lung cancers. Over the years, it has become one of the most broadly used cytotoxic agents. Unfortunately, the outstanding antitumor effect is accompanied by dose-limiting, frequently lethal, nephrotoxicity (Tab. 16.2).

Tab 16.2 Clinical toxicity of cisplatin and its analogues.

	Cisplatin	Carboplatin	Oxaliplatin
Myelotoxicity	+	++	++
Nephrotoxicity	++		
Neurotoxicity	++	+	++
Ototoxicity	++		
Nausea/vomiting	++	+	+

16.2.6
Adverse Effects

The pathologic changes affect primarily the distal convoluted tubules and collective ducts, and consist of acute tubular necrosis and the formation of casts. Due to the kidney damage, blood urea nitrogen and creatinine levels are increased and some patients die as a result of renal failure. The full exploitation of the antitumor potency of the drug using 100–120 mg m^{-2} cisplatin every 3–4 weeks was made possible only by the development of forced diuresis using both extensive pre- and posthydration and/or osmotic diuresis elicited by mannitol [6]. Some authors have recommended the use of furosemide for maintaining high urine output. It is believed that these methods reduce the renal damage by preventing the extensive immediate binding of the highly reactive cisplatin to tubular proteins. With the use of hypertonic saline to minimize the aquation of Pt in the kidneys, further protection of the renal tubules could be achieved. Then, the dose-limiting toxicity becomes severe sensory neuropathy. Therefore, hypertonic saline-supported high-dose cisplatin treatment has found only limited acceptance in the medical practice.

In addition to nephrotoxicity, cisplatin causes severe, cumulative neurotoxicity in around 85% of the patients above a total dose of 300 mg m^{-2}. The toxicity includes peripheral sensory neuropathy, though autonomic functions can also be damaged. Hearing loss is evident mainly in the high-frequency region, but in some cases other frequencies are also affected which may interfere with normal social function or even cause complete deafness. The sensory neuropathy is only partly reversible, and recovery may take several weeks or months. Finally, cisplatin stimulates directly the vomiting center, causing extremely severe nausea and vomiting which cannot be effectively influenced by conventional anti-emetic agents. The problem was only solved following the discovery of the key role of the serotoninergic pathway in the pathophysiology of chemotherapy-related nausea and vomiting. This led to the introduction of a new class of antivomiting agents, the 5-HT$_3$ antagonists ondansetron and its analogues, which can substantially suppress this side effect.

16.3
Carboplatin

16.3.1
Development

When considering all the above-described very severe and frequently irreversible toxicities of cisplatin, the search for a similarly effective but less toxic analogue became the primary goal of further drug development. Since the kidney damage could be linked with high probability to the ease of displacement of the halogen atoms, it seemed logical to seek analogues with a more stable bonding. Among the many hundreds of analogues investigated, carboplatin [cis-diammine-1,1-cyclobu-tane dicarboxylate platinum (II)] became the unchallenged winner (Fig. 16.2). The cytotoxic aquated products of cisplatin and carboplatin are the same, but the rate constant of their formation is up to 100 times slower in the latter. In addition, the concentration of carboplatin necessary to achieve the same degree of cellular effect is at least 10 times higher. However, the DNA–platinum adducts formed by the two drugs are similar, and consequently the two compounds show a high degree of cross-resistance.

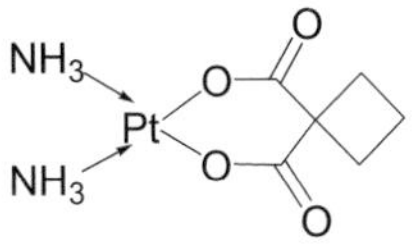

Fig. 16.2 Carboplatin.

16.3.2
Administration and Pharmacokinetics

Due to the lower reactivity of carboplatin it is not necessary to administer it in saline. Furthermore, the binding of Pt to plasma proteins is lower than that of cisplatin, usually less than 50% (see Tab. 16.1). As a consequence, the half-lives of protein-bound and ultrafiltrable Pt are close. The urinary elimination of unchanged carboplatin becomes the dominant feature of carboplatin pharmacokinetics [3,4]. Myelotoxicity (especially thrombocytopenia), the major dose-limiting, cumulative toxicity of carboplatin, is directly related to the plasma level of unchanged carboplatin. Due to this very close relationship, it is much safer to define the optimal anticancer dose in terms of the area under the curve (AUC) of carboplatin, the optimal range being between AUC 5 to 7. Calculating the dose on the basis of either body weight or body surface results in substantial toxicity in patients with impaired renal function. On the contrary, suboptimal anticancer plasma levels are frequently observed in young patients with testicular cancer having a high glomerular filtration rate (GFR). Usually, the dosing formula of Calvert is used:

$$\text{Dose (mg)} = \text{Target AUC (mg mL}^{-1} \times \text{min)} \times [\text{GFR (mL min}^{-1}) + 25]$$

The GFR can be measured most precisely using the ^{51}Cr-EDTA clearance method, but for everyday practice the determination based on nomograms using serum creatinine values provides an adequately accurate estimate. It should be emphasized, however, that the myelotoxicity is also influenced by the general condition of the patient, especially by the extent of bone marrow reserve. Therefore, even when using an AUC-based administration scheme the seriousness of myelotoxicity cannot be accurately predicted.

16.3.4
Adverse Effects

The nephrotoxicity of carboplatin is negligible compared to that of cisplatin, and usually even prolonged treatments do not cause clinical problems. Forced diuresis is not required for carboplatin administration. Nausea and vomiting also occur much less frequently, and 20% of patients do not have any emetogenic side effects. In general, conventional anti-emetic agents are sufficient to control the mild to moderate nausea and vomiting, and 5-HT$_3$ receptor antagonists are needed only occasionally. Sensory neuropathy can be diagnosed in only about 3% of patients.

16.3.5
Clinical Efficacy

Carboplatin shows in general an antitumor activity spectrum similar to that of cisplatin. However, the dose required for similar *in-vitro* efficacy is about four- to 10-fold higher. This difference is also evident in human dosing. In spite of many comparative trials, there is no unequivocal evidence available either to prove or disprove the equal clinical efficacy of the two drugs. In some trials, the remission rate after cisplatin administration was shown to be slightly higher, though without any survival benefit. Thin controversy is due to the fact that the results vary according to the Pt sensitivity of the tumor type, and also according to the stage of the disease studied. The selection between cisplatin and carboplatin is usually made by clinicians considering the palliative or curative intent of the therapy, the general condition of the patients, and the available budget, since carboplatin is more expensive.

16.4
Oxaliplatin

16.4.1
Development

The search for an analogue with a broader tumor spectrum led to the group of Pt derivatives in which different chemical groups were attached to the ammine ligands. From hundreds of such analogues, oxaliplatin, *cis*-[oxalate (trans-*λ*-1,2-DACH) platinum] became internationally registered for clinical use in which 1,2-diaminocyclohexane (DACH) replaces the original ammine ligands and the halogens are substituted by an oxalate group (Fig. 16.3). The DACH group is carried with the Pt even after binding to its target on the macromolecules. This bulky attachment results in a steric configuration of the DNA–Pt adduct which interferes with DNA repair mechanisms and leads to cell death in some tumors hitherto known to be resistant against Pt compounds [7]. Indeed, the NCI tumor panel screen identified oxaliplatin as an analogue distinctly different from cisplatin and carboplatin. In particular, its remarkable activity against colon cancer cell lines was noted, and this was later corroborated also in the clinical setting. Cells made resistant to cisplatin were found to be cross-resistant to carboplatin, but not to oxaliplatin.

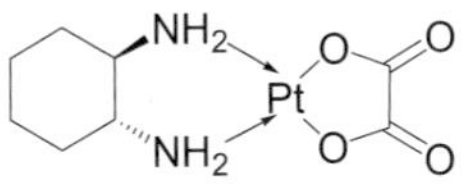

Fig. 16.3 Oxaliplatin.

16.4.2
Cellular Resistance to Various Pt Analogues

Several mechanisms exist which confer cellular resistance to Pt. These are decreased cellular uptake of Pt compounds, increased inactivation by intracellular glutathione and metallothionein, increased nucleotide excision repair and increased replicative bypass and, finally, the defect in mismatch repair. These mechanisms affect all types of Pt compounds, except for the two last resistance mechanisms which appear to be much less (or even not functioning at all) in the case of oxaliplatin. Enhanced replicative bypass permits the DNA replication to proceed beyond the attached Pt derivative after cisplatin or carboplatin exposure, thus avoiding apoptosis. Cell division can proceed, leading to daughter cells carrying multiple DNA lesions. Mismatch repair is a very complex mechanism involving many different enzymes. During mismatch repair, the DNA strand synthesized opposite to the Pt–DNA adduct is removed, and this results in a replication failure beyond the lesion. The continuous presence of the Pt adduct provokes repeated futile cycles of excision, which finally leads to cell death. The mismatch repair enzymes can bind to the Pt–DNA adducts of cisplatin and carboplatin, but not to those with the attached large DACH group. Hence, in cells with the DNA–Pt-(DACH) adduct, the deadly mismatch repair cycles can not be initiated. Increased replicative bypass ability and deficiency in mismatch repair confer only a relatively low additional level of resistance to the cells against the classical Pt derivatives. The increased activation of these mechanisms have been proven also in tumor biopsies taken from patients who developed resistance against cisplatin or carboplatin. Some of these patients could be effectively treated with oxaliplatin, which demonstrated that the inability of these resistance mechanisms to overcome oxaliplatin poisoning confers a pharmacologically appreciable advantage to treat with oxaliplatin patients resistant to the classical Pt drugs.

16.4.3
Metabolism and Pharmacokinetics

Oxaliplatin gives rise to several metabolites, but ultimately the DNA is attacked by the diaquo product, similar to the other Pt analogues. Displacement of the oxalate group by H_2O is a slower process than that of the halogens in cisplatin, and consequently oxaliplatin is less reactive. Protein binding is about 85%, and over 50% of administered oxaliplatin are excreted in the urine (see Tab. 16.1). Oxaliplatin does not accumulate in the plasma during repeated cycles. Elimination is mainly through the kidneys, consequently compromised renal function leads to increased drug levels. Since many other inactive metabolites are also formed, the correlation of the plasma level of the parent compound with the GFR is not as strong as in the case of carboplatin.

16.4.4
Adverse Effects

Oxaliplatin causes myelosuppression similar to most cytotoxic agents. It is devoid of nephrotoxicity, and can be administered without forced diuresis. Nausea and vomiting are moderate to severe, and conventional and/or $5\text{-}HT_3$ antagonists easily control any emesis. Mild to moderate diarrhea is observed in about half of the patients, but this becomes severe only in a few cases. The major, dose-limiting toxicity is cumulative sensory neuropathy with both an acute and a late component. The early mild phase becomes apparent within minutes after initiation of the oxaliplatin infusion and resolves spontaneously within minutes to few hours (occasionally days). It is manifested as distal or perioral dysesthesia, and may be due to the chelating properties of oxaliplatin, which interfere with calcium-dependent sodium channels. The later-developing dysesthesia and paresthesia developing on the extremities are cumulative in nature. Following the administration of $1500\text{--}1700$ mg m^{-2} oxaliplatin, 80–90% of patients develop grade 2–3 neurotoxicity causing functional impairment due to the loss of fine sensory motor coordination. However, as opposed to cisplatin, the recovery is predictable and is usually completed within 20 weeks.

16.4.5
Clinical Efficacy

The clinical tumor spectrum for oxaliplatin is not yet fully characterized, but oxaliplatin is considered to be the most effective agent beside 5-fluorouracil (5-FU) and irinotecan against colorectal cancer. It also shows a promising activity in breast cancer, which is only marginally sensitive to the conventional Pt compounds. Oxaliplatin is also active in the known Pt-sensitive tumors, such as ovarian, head and neck, and lung cancers. It is of great clinical significance that patients with ovarian cancer, who became resistant to cisplatin or carboplatin can be effectively treated with oxaliplatin [7–9].

16.5
Summary

The three Pt complexes discussed above constitute one of the most effective classes of cytotoxic agents. The full therapeutic potency of the Pt complexes could be realized only by the logical, stepwise development of analogues with less nephrotoxicity and/or with an extended tumor. Many thousands of additional analogues were synthesized, many hundreds of which underwent preclinical – and a few of them also clinical testing – but, unfortunately, most of them failed. Nedaplatin [cis-diammine (glycolato)platinum] is a cisplatin-like drug registered in Japan. Satraplatin [bis-acetato ammine dichloro(cyclohexylamine) platinum(IV)] (JM216) is an orally available Pt drug with carboplatin-like activity which unfortunately never reached the market [10]. Recently even the main dogma of structure–activity relationships of platinum complexes, stating that trans-platinum complexes are inactive, was successfully challenged. In spite of promising preclinical efficacy, none of the active trans-platinum complexes has yet entered clinical practice. Nonetheless, the search for new Pt analogues continues, supported by the enthusiasm and optimism of researchers who never give up hope of developing a clinically more useful analogue.

References

1 Rosenberg B, Van Camp L, Krigas T. Inhibition of cell division in *Escherichia coli* by electrolysis products from platinum electrode. *Nature*, **1965**, 205, 698–699.

2 Judson I, Kelland LR. Cisplatin and analogues. In: *Textbook of Oncology*. Souhami RL, Tannock I, Hohenberger P, et al. (Eds.). Oxford University Press, **2002**, 655–662.

3 O'Dwyer PJ, Stevenson JP, Johnson SW. Clinical pharmacokinetics and administration of established platinum drugs. *Drugs*, **2000**, 59 (Suppl. 4), 19–27.

4 Go RS, Adjei AA. Review of the comparative pharmacology and clinical activity of cisplatin and carboplatin. *J. Clin. Oncol.*, **1999**, 17, 409–422.

5 Gieterna JA, Meinardi MT, Messerschmidt J, et al. Circulating plasma platinum more than 10 years after cisplatin treatment for testicular cancer. *Lancet*, **2000**, 355, 1075–1076.

6 Hayes DM, Cvitkovic E, Golbey RB, et al. High dose cis-platinum diammine dichloride. Amelioration of renal toxicity by mannitol diuresis. *Cancer*, **1977**, 39, 1372–1381.

7 Raymond E, Chaney SG, Taamma A, et al. Oxaliplatin: a review of preclinical and clinical studies. *Ann. Oncol.*, **1998**, 9, 1053–1071.

8 Massari C, Brienza S, Rotarski M, et al. Pharmacokinetics of oxaliplatin in patients with normal versus impaired renal function. *Cancer Chemother. Pharmacol.*, **2000**, 45, 157–164.

9 Wiseman LR, Adkins JC, Plosker GL, et al. Oxaliplatin. A review of its use in the management of metastatic colorectal cancer. *Drugs Aging*, **1999**, 14, 459–475.

10 Becouarn Y, Agostini C, Trufflandier N, et al. Oxaliplatin: available data in non-colorectal gastrointestinal malignancies. *Crit. Rev. Oncol. Hematol.*, **2001**, 40, 265–272.

11 Judson I, Kelland LR. New developments and approaches in the platinum arena. *Drugs*, **2000**, 59 (Suppl. 4), 29–36.

17
The History of Drospirenone

Rudolf Wiechert

17.1
General Development

Drospirenone (6β,7β;15β,16β-dimethylen-3-oxo-17α-pregn-4-en-21,17-carbolactone (Fig. 17.1) [1–3] is a progestin with a novel, advantageous biological profile. It originates from a project, the goal of which was the discovery of a new aldosterone antagonist.

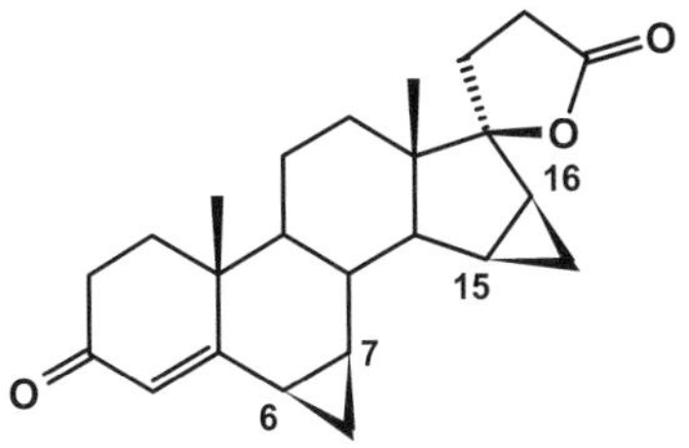

Fig. 17.1 Drospirenone.

Aldosterone (Fig. 17.2) [4] is a highly active mineralocorticoid, which controls potassium and sodium levels, and thereby the water milieu, in organisms.

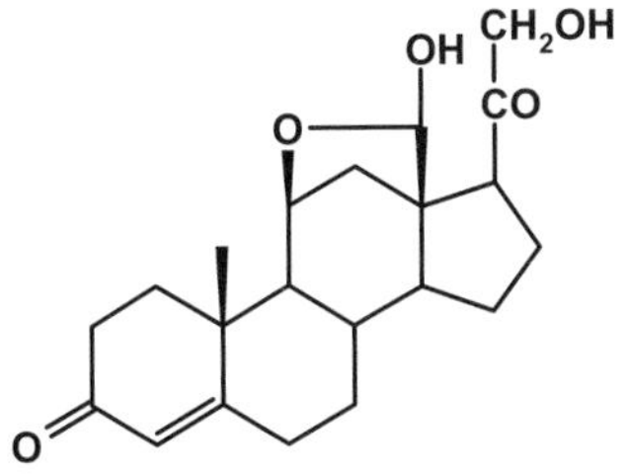

Fig. 17.2 Aldosterone.

Analogue-based Drug Discovery. IUPAC, János Fischer, and C. Robin Ganellin (Eds.)
Copyright © 2006 WILEY-VCH Verlag GmbH & Co. KGaA, Weinheim
ISBN: 3-527-31257-9

During the 1970s and 1980s, the synthetic aldosterone antagonist spironolactone (Fig. 17.3) [5] was used extensively (among others) as a diuretic agent for the treatment of edema, liver cirrhosis and certain cardiac diseases.

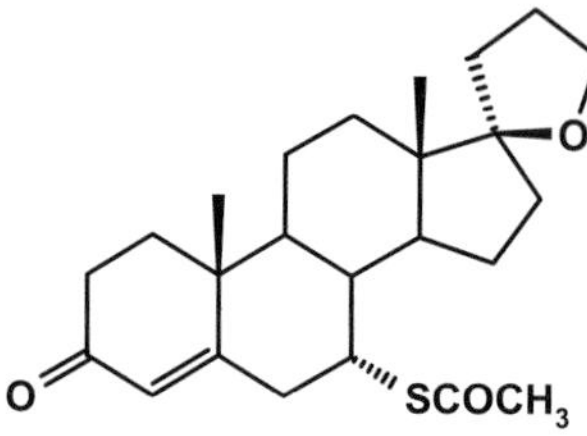

Fig. 17.3 Spironolactone.

In 1986, the total production of steroidal agents was about 393 tons, of which 125 tons was spironolactone [6]. Subsequently, however, the amount produced annually fell as the consequence of a reduction in the pharmaceutical application of spironolactone.

Between the mid-1970s and the mid-1980s, Schering AG carried out a project to develop a compound which had a much higher antagonistic action than that of spironolactone, as well as a lesser sex organ-specific action, since chronic treatment and high doses of spironolactone had caused hormonal disorders in both women and men.

During the course of this project, over 600 new compounds with anti-aldosterone activity were synthetized. Some of these has a fused cyclopropane ring in place of a double bond because of the known electronic similarity under certain circumstances. The test models were developed and the anti-aldosterone activity tests were performed by Losert and coworkers [7]. Sex hormone activity investigations were conducted at Schering's endocrine pharmacology unit, and spirorenone ($6\beta,7\beta;15\beta,16\beta$-dimethylen-3-oxo-$17\alpha$-pregna-1,4-dien-21,17-carbolactone) (Fig. 17.4) [3,8] was finally chosen as the candidate for development.

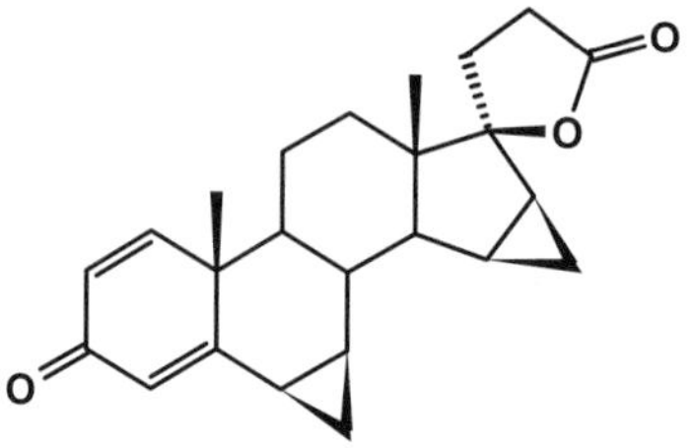

Fig. 17.4 Spirorenone.

In animal tests, the biological profile of spirorenone achieved the proposed goal, with the compound displaying five-fold higher anti-aldosterone activity compared to spironolactone, but with far fewer hormonal side effects. This outcome was deemed highly successful, especially as since 1957 – the year in which spironolactone was produced – no compound with higher anti-aldosterone activity had been developed, despite worldwide research efforts.

Initially, when spirorenone was tested in low doses on a male human, an unambiguous reduction of the subject's testosterone level was detected. The reason for this change was provided by pharmacokinetic monitoring, wherein it was realized that in humans and in monkeys, unlike other species, spirorenone was metabolized exclusively to 1,2-dihydrospirorenone (drospirenone), by the action of a Δ^1-hydrase (Fig. 17.5).

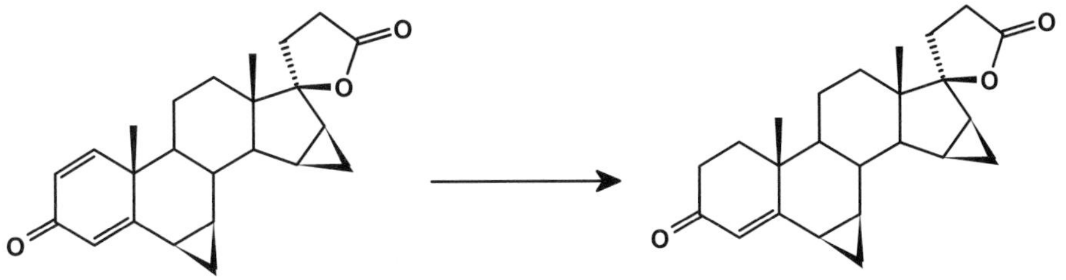

Fig. 17.5 Formation of drospirenone from spirorenone in human and in monkey by Δ^1-hydrase.

Consequently, drospirenone underwent careful pharmacological characterization, and was found to possess a very interesting biological profile. Drospirenone is an orally active progestin with notable anti-androgen and antimineralocorticoid properties which qualitatively match those of natural progesterone. The suggestion was also made that drospirenone, with its antimineralocorticoid component, might also possess certain advantages as a contraceptive. Indeed, on the basis of these pharmacological properties, the recommendation was made to develop the compound as a contraceptive since, if its activity profile could be verified clinically, this would represent a remarkable step forward in the field of oral contraception.

17.2
Syntheses

The synthesis of drospirenone (see Fig. 17.1) was unusually complicated, with the new structural elements being methylene groups at positions $6\beta,7\beta$ and $15\beta,16\beta$. Originally, in 1966, the $15\beta,16\beta$-methylene group [9,10] had been introduced into the steroid skeleton in the process of another Schering project, and good yields were obtained. The major problem in the synthesis was stereoselective $6\beta,7\beta$-methylenation; the key compound for this was the 5β-hydroxy-6-ene structure A, from which the $6\beta,7\beta$-methylene compound B was synthetized stereoselectively by means of a Simmons-Smith reaction (Zn-Cu/CH$_2$Br$_2$) (Fig. 17.6).

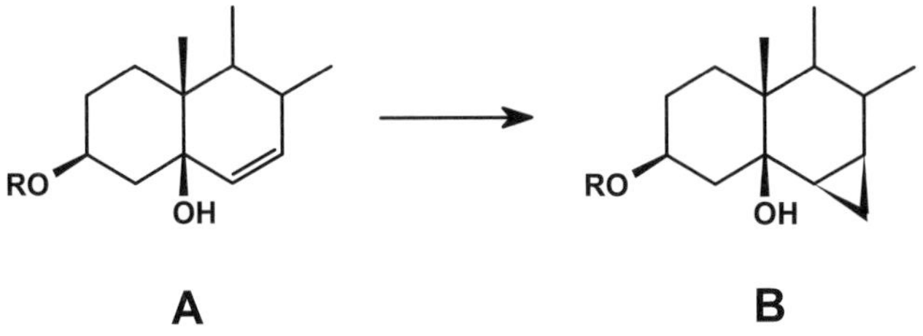

Fig. 17.6 6β,7β-methylenation by the Simmons-Smith reaction.

Further possibilities investigated for this 6β,7β-methylenation proved ineffective for several reasons. Initially, a number of starting materials were used and the methylenation methods were varied [11].

The synthetic studies were therefore concentrated on a rational synthesis of 5β-hydroxy-6-ene structure A. The result was the synthesis briefly outlined in Fig. 17.7, which led to drospirenone (9.6% yield) in a 12-step process from androstenolone (3β-hydroxyandrost-5-en-17-one). These investigations were carried out at Schering's pharmaceutical chemistry unit, under the supervision of H. Laurent.

Androstenolone, **1**, can be transformed microbiologically to the 7a,15a-dihydroxy derivative **2** by the action of *Colletotrichium lini*. During formation of the acetal (**3**), inversion takes place on C-7. Acidic cleavage of **3** results in the triol, **4**, which can also be produced by direct acidic catalysis from **2** [12,13]. After selective protection to the 3β,15a-dipivalate (**5**), the 15β,16β-methylene compound, **6**, can be synthetized, and then stereoselectively transformed to the 5β,6β-epoxide, **7**. This reacts with triphenylphosphine and tetrachloromethane in pyridine to produce the 7a-chloro derivative, **8**. On treatment with zinc and acetic acid, **8** can be converted to the key compound **9**, which has a 5β-hydroxy-6-ene structure. Compound **9** can then be methylenated stereoselectively in the 6β,7β position by the Simmons-Smith method. The last three steps – **10** → **11** → **12** → drospirenone – include the build-up of the spironolactone ring, after which water is lost from the molecule and oxidation affords drospirenone.

About 100 g drospirenone was synthetized in several charges at the pharmaceutical chemistry unit in order to conduct the pharmacological tests.

In November 2000, drospirenone in combination with ethinylestradiol was introduced by Schering AG to the drug market in Germany as a contraceptive pill under the tradename Yasmin®. Today, it is available almost worldwide, and is considered to be the most successful oral contraceptive of the past decade.

Due to the unique biological similarity of drospirenone (e.g., mild anti-aldosterone and anti-androgenic activity) to the natural progesterone, Yasmin® has important advantages over the known oral contraceptives [14]. For example, the use of Yasmin® does not initially cause any increase in body weight, and it is also effective in treating mild to moderate acne. Likewise, menstrual cycle control is excellent.

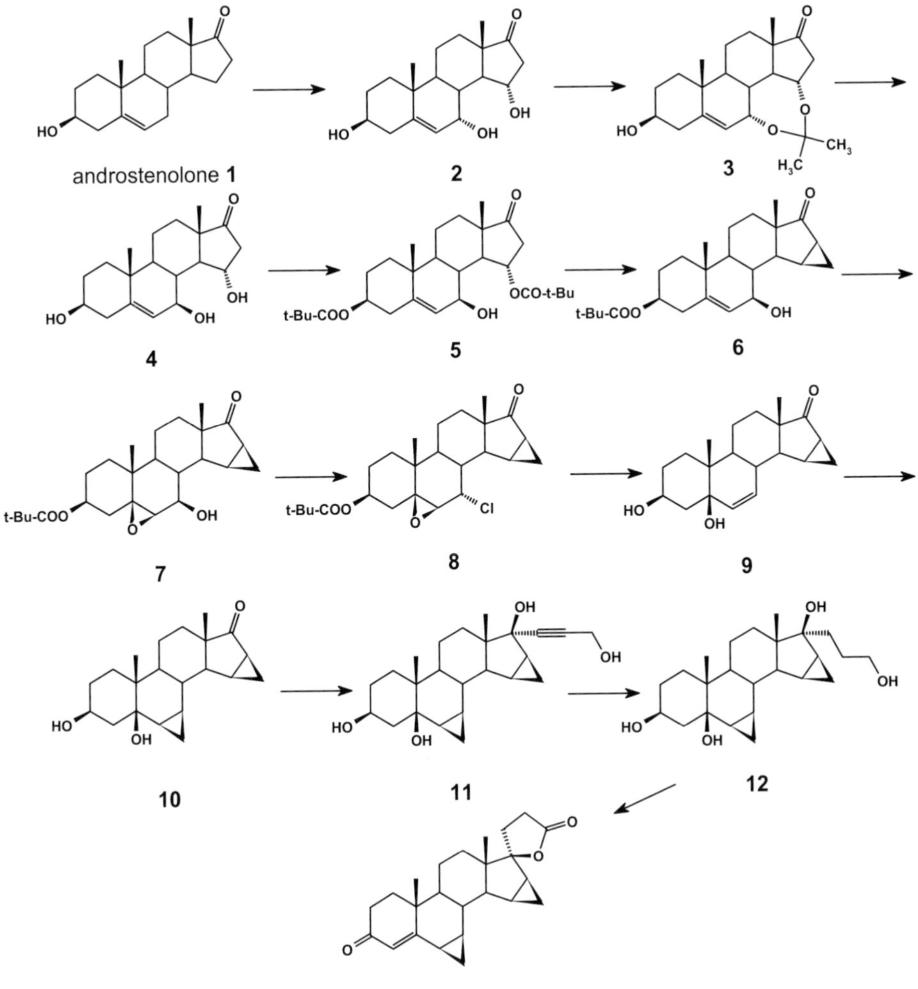

androstenolone 1   2   3

4   5   6

7   8   9

10   11   12

drospirenone

Fig. 17.7 Synthesis of drospirenone from androstenolone.

References

1 Bittler D. Laboratory book number 34 (1-10-1976-30-9-1977).
2 Wiechert R, Bittler D, Kerb U, Casals-Stenzel J, Losert W. DE 2,652,761 (priority date: 16.11.1976).
3 Bittler D, Hofmeister H, Laurent H, Nickisch K, Nickolson R, Petzoldt K, Wiechert R. *Angew. Chem.*, **1982**, 94, 718; *Angew. Chem. Int. Ed.*, **1982**, 21, 696.
4 *Fieser and Fieser: Steroids.* Verlag Chemie, Weinheim/Bergstrasse, **1961**.
5 Cella JA, Kagawa CM. *J. Am. Chem. Soc.*, **1957**, 79, 4808; Kagawa CM. *Endocrinology*, **1960**, 67, 125.
6 Breitfeld R. Pharma Chemikalien Marketing, Schering AG report of 27-4-1989 to Wiechert R.
7 Losert W, Bittler D, Busse M, Casals-Stenzel J, Haberey M, Laurent H, Nickisch K, Schillinger E, Wiechert R. *Arzneim.-Forsch.*, **1986**, 36, 1583.
8 Wiechert R, Bittler D, Kerb U, Prezewowsky K, Casals-Stenzel J, Losert W. DE 2,922,500 (priority date: 31.05.1979).
9 Schmidt O, Prezewowsky K, Wiechert R. DE 1,593,500 (priority date: 02.04.1966).
10 Schmidt O, Prezewowsky K, Schulz G, Wiechert R. *Chem. Ber.*, **1968**, 101, 939.
11 Laurent H, Bittler D, Hofmeister H, Nickisch K, Nickolson R, Petzoldt K, Wiechert R. *J. Steroid Biochem.*, **1983**, 19, 771.
12 Petzoldt K, Laurent H, Wiechert R. EP 0,075,189 (priority date: 21.09.1981).
13 Petzoldt K, Laurent H, Wiechert R. *Angew. Chem.*, **1983**, 96, 413; Angew. Chem. Int. Ed., **1983**, 22, 406.
14 Keam SJ, Wagstaff AJ. *Treat. Endocrinol.*, **2003**, 2, 49–70.

18

Histamine H₁ Blockers: From Relative Failures to Blockbusters Within Series of Analogues

Henk Timmerman

18.1
Introduction

The story of histamine and histamine blocking agents starts in the early years of the 20th century, when histamine [2-(imidazol-4-yl)ethylamine] was first synthetized [1].[1)] The famous pharmacologist Dale showed that histamine possesses potent contractile properties on smooth muscle, as well as blood pressure-lowering effects [2,3]. The compound was subsequently shown to have a wide distribution in nature, both in animals and plants. Initially, it was considered solely as the decarboxylation product from L-histidine from bacteria, but later the compound was shown also to be present in noninfected tissue, though its physiological role was not recognized. In the Foreword to the volume of *Handbook of Experimental Pharmacology*, devoted to histamine and antihistamines, Sir Henry Dale describes how, during the 1910s and 1920s, the pharmacological profile of histamine was observed to closely resemble the pathology seen in shock (vasodilatation, permeability of blood vessels) [4]. At that time, there was much interest in the so-called "H substance", a compound which was still hypothetical and thought to be released from injured cells, as well as for the phenomenon of the anaphylactic shock. It is beyond the scope of this chapter to discuss the interesting story of how, in the course of time, the H substance was found to be a complex of compounds (of which histamine was one) that played a major role in anaphylaxis. Histamine was found to occur in many tissues, and especially in the skin and lungs. Its role in allergic reactions (vasodilatation, constriction of airways, skin reactions due to increased permeability of blood vessel) became increasingly accepted, and it was during the 1930s that the search for H₁ blockers began. Originally, it was thought that such derivatives would be very useful in treating allergic patients, including asthmatics.

---

1) Note: the term H₁ blockers, H₁ antagonists and antihistamines are used to denote the same meaning.

Analogue-based Drug Discovery. IUPAC, János Fischer, and C. Robin Ganellin (Eds.)
Copyright © 2006 WILEY-VCH Verlag GmbH & Co. KGaA, Weinheim
ISBN: 3-527-31257-9

18.2
The First Antihistamines

In the early days, when the role of histamine in allergy had been accepted, much discussion took place on the question of whether or not histamine reacts on an "own" receptor. In fact, this question was important, because it underlined whether or not it made sense to search for antihistamines. At the time, specific pharmacological models were not available, but one pioneer in the field was Staub, who defined an "antihistaminic index" (AHI). This was the number of LD_{50} doses blocked by 5 mg of the compound under investigation. In a model (which for ethical reasons would be completely unacceptable today), the test compound was dosed at 5 mg to mice (i.v.) and subsequently histamine was injected at one, two, three, etc., times the i.v. LD_{50} dose [5]; ultimately, the parameter "measured" was lethality.

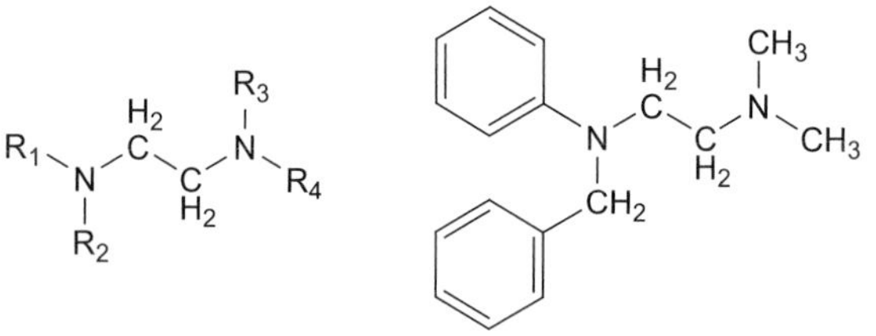

Fig. 18.1 General structure of the diamines. Right: phenbenzamine.

The first useful compounds found by Staub were ethylene diamines (Fig. 18.1). One highly active compound – phenbenzamine – reached the market during the early years of World War II, and may be considered as the archetype of all classic antihistamines. A close analogue of phenbenzamine – in which the phenyl group has been replaced by a 2-pyridyl nucleus, tripelennamine – is still in use today as an anti-itching medicament for skin applications.

The success of phenbenzamine led many research groups to initiate programs to develop new derivatives, with a major success being achieved by Rieveschl, a young lecturer at the Cincinnati University, who synthesized diphenhydramine (Fig. 18.2), an analogue of the first ethers published by Staub.

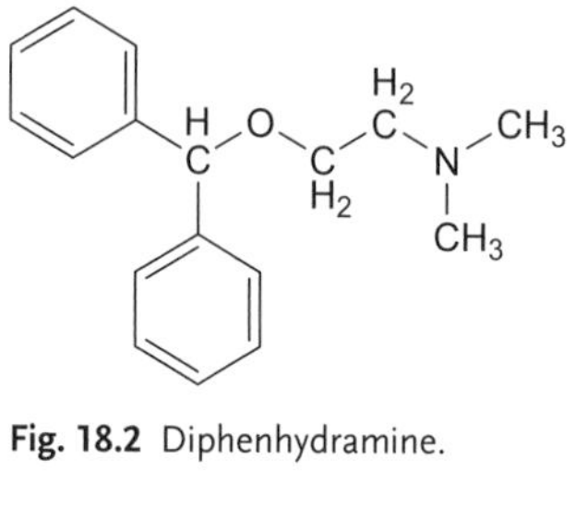

Fig. 18.2 Diphenhydramine.

This new antihistamine reached the market shortly after the World War II, and became a great success. Subsequently, antihistamines became popular drugs and many companies began research program to identify new derivatives. The new antihistamines were, for their structure, all inspired by the very early compounds; the nitrogen of phenbenzamine, already changed in a way in diphenhydramine into an oxygen, was subsequently replaced by a CH_2 (pheniramine) or an unsaturated carbon chain (triprolidine). The benzhydryl group could be bridged, as for example in fenethazine, while the dimethylamino function could be changed into other amino groups (e.g., cyclizine and triprolidine) (Fig. 18.3).

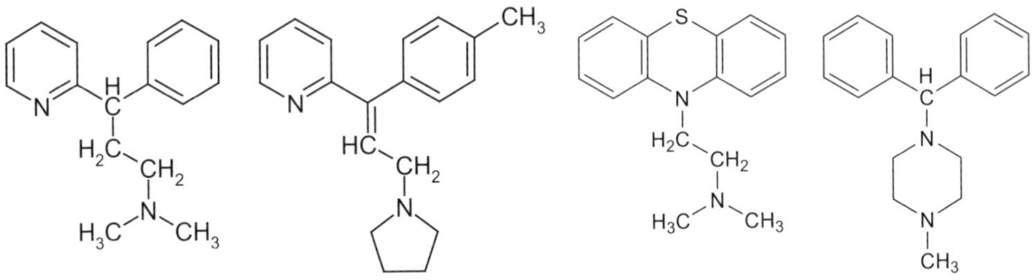

Fig. 18.3 Diphenhydramine analogues.

18.3
Diphenhydramine as a Skeleton for Antihistamines

By far the largest majority of all known H_1 blocking compounds are (closely) related to the classical class of diarylalkylamines, and are excellent examples of analogues of each other (Fig. 18.4).

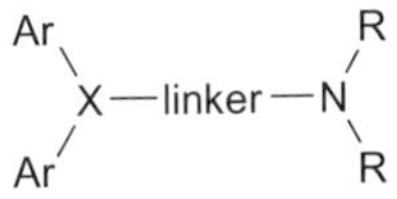

Fig. 18.4 General structure of diphenhydramine analogues.

It is virtually impossible to list all of the analogues, and several reviews have dealt with the structure–activity relationships (SAR) of antihistamines [6–9]. Together, these reviews provide an almost complete listing of the classic (first-generation) and the so-called nonsedating (second-generation) antihistamines.

In applying the analogue principle, the basic structure as shown has been modified in all four elements of the skeleton: the diaryl group; the X moiety; the linker; and the amino function. These four types of alteration will be discussed in the following sections.

18.3.1
The Diaryl Group

A very large number of diphenhydramine analogues has been synthesized (Fig. 18.5), and the compounds obtained subjected to intensive quantitative structure–activity relationship (QSAR) studies by Rekker et al. The case of the diphenhydramines is seen as being representative of the entire class.

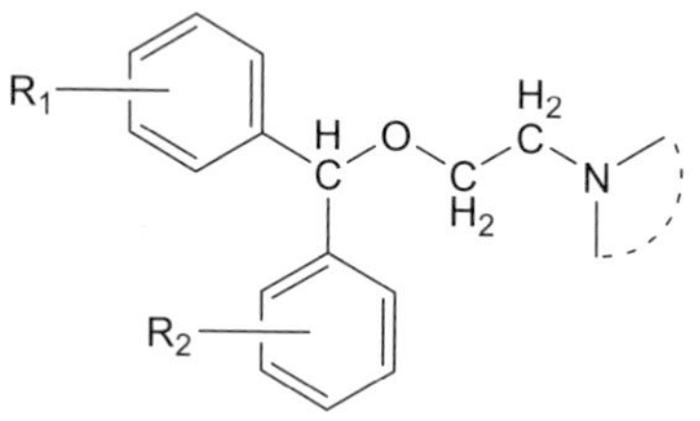

Fig. 18.5 Diphenhydramine derivatives.

The influence of the substitution pattern in the benzhydryl group follows very general patterns. In very broad terms, it may be said that:

- *Ortho*-substitution decreases the antihistamine effects, but increases the antimuscarinic properties; the *ortho*-methyl analogue of diphenhydramine (orphenadrine) was successfully introduced as an anti-Parkinson medicament.
- The introduction of small groups at one *para*-position (methyl, halogen; but not *tert.*-butyl) increases the antihistaminic efficacy, whereas the antimuscarinic effects are reduced.
- The introduction of two *para*-substituents reduce strongly the antihistamine properties, indicating that the two phenyl nuclei have different functions in binding to the receptor. This observation is confirmed by the large difference in activity between the enantiomers of 4, 4^1-CH_3-diphenhydramine.

These general observations explain why the benzhydryl function may be replaced by comparable groups in which two aromatic functions should be present. In the following scheme, a number of striking examples is shown, in which the benzhydryl part is changed, keeping the remainder of the skeleton constant (Fig. 18.7). The activities are relative to that of diphenhydramine and approximate (taken from the cited review articles).

From these examples, the prominent role of the two (!) aromatic groups becomes very clear.

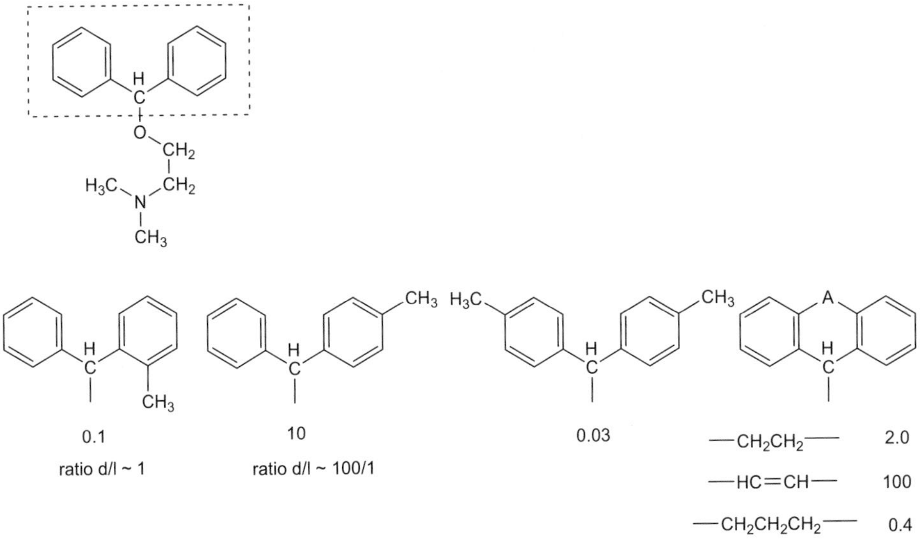

Fig. 18.6 Changing the diphenhydramine skeleton.

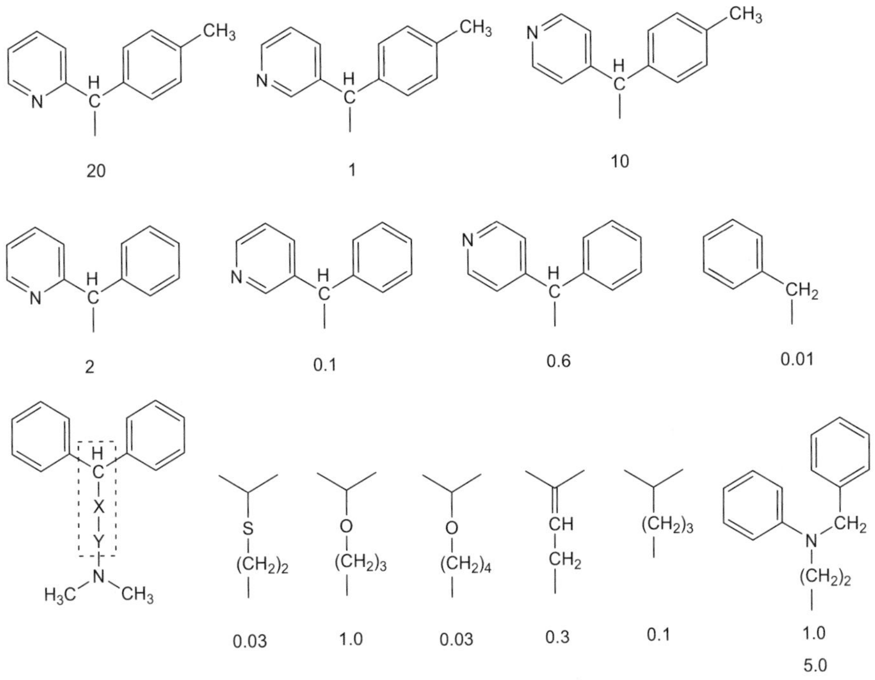

Fig. 18.7 Changing the diphenhydramine nucleus and skeleton, continued.

18.3.2
The Linker

The linker is also of importance, as becomes clear from the next examples. It seems that an oxygen or nitrogen attached to the benzhydryl group leads to better results than either an alkyl or alkylene function. The replacement of an oxygen by a sulfur reduces antihistamine activity drastically (though the influence of this latter change has a positive influence on the antimuscarinic efficacy of these derivatives) (Fig. 18.7).

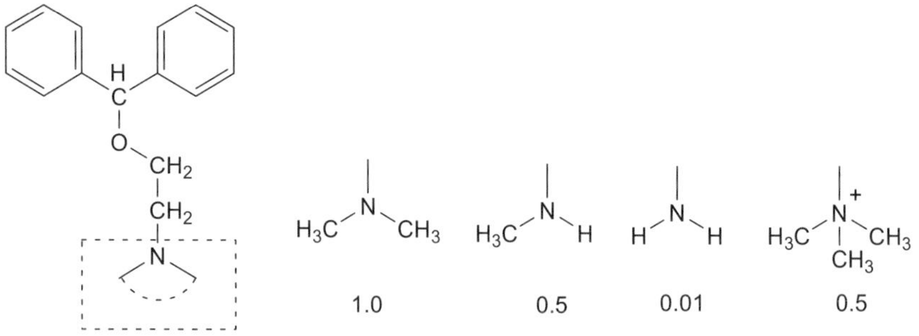

Fig. 18.8 Changing the diphenhydramine skeleton, continued.

18.3.3
The Basic Group

The third part of the skeleton, the amino function, is also of great importance. This is, of course, to be expected as it is known that the amino function binds – in its quaternary state – to an aspartate of the receptor protein (Fig. 18.8). Many different amino functions have been used in the antihistamines. Several examples are shown in the next scheme (Fig. 18.9), and clearly the shape of the substituents is important, as is the pK_a value of the base.

One remarkable derivative is that in which the ammonium function has been replaced by a sulfonium moiety, with an activity of 0.3 of the ammonium analogue of p-CH$_3$-diphenhydramine (Fig. 18.10).

In general terms, it may be said that quaternary compounds (trimethyl) have an activity which does not differ very much from that of the tertiary analogue. The diphenhydramines have been investigated by multiple regression analysis intensively by the group of Rekker [7,8,10]. These studies confirmed the role of the three parts of the diphenhydramine skeleton, with conclusions being reached that the major characteristics for activity in this class included: a high lipophilicity of an aromatic group with two nuclei; spatial orientation of these groups; a given distance between the hydrophobic region and the observed quaternary function; and a relatively large "bulk acceptance" being observed around the quaternary moiety.

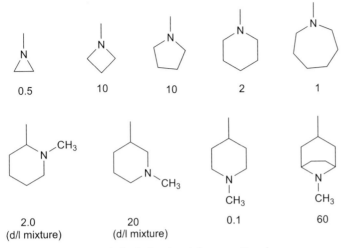

0.5 10 10 2 1

2.0 (d/l mixture) 20 (d/l mixture) 0.1 60

Fig. 18.9 Changing the diphenhydramine skeleton; continued.

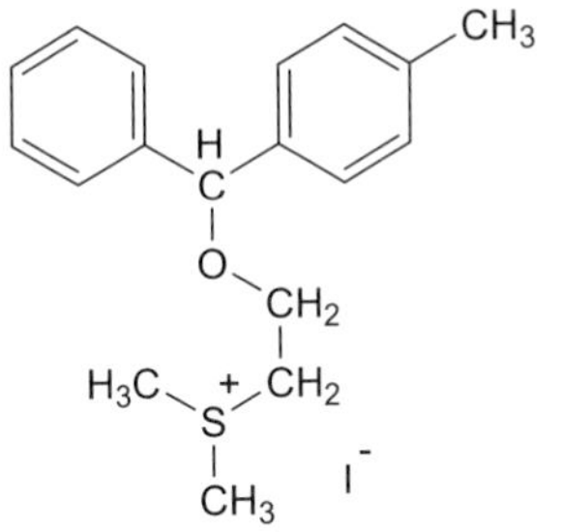

Fig. 18.10 An unusual quaternary antihistamine.

18.3.4
The Analogue Principle

Throughout the years, a very large number of antihistamines have been identified, and many of these have been introduced into therapy. In some ways, the analogue approach has been used by intuition with subsequent QSAR studies following, and this has led to general conclusions. In most – if not all – antihistamines, those elements seen in the skeleton of diphenhydramine are present.

The analogue approach of identifying antihistamines has been successful, and this class of compounds is often used to explain the success of medicinal chemists during the post-World War II years. The reason why antihistamines did not become major medicines was not due to a lack of antihistamine efficacy, as the compounds were known to be effective in treating affectations in which histamine plays a strong role – for example, itching caused by an insect sting or an allergic skin disease. The pathology of allergic asthma however, is too complex to be treated successfully solely with an antihistamine. Moreover, the incidence of side effects (especially drowsiness) were of such a nature that broad application was made impossible.

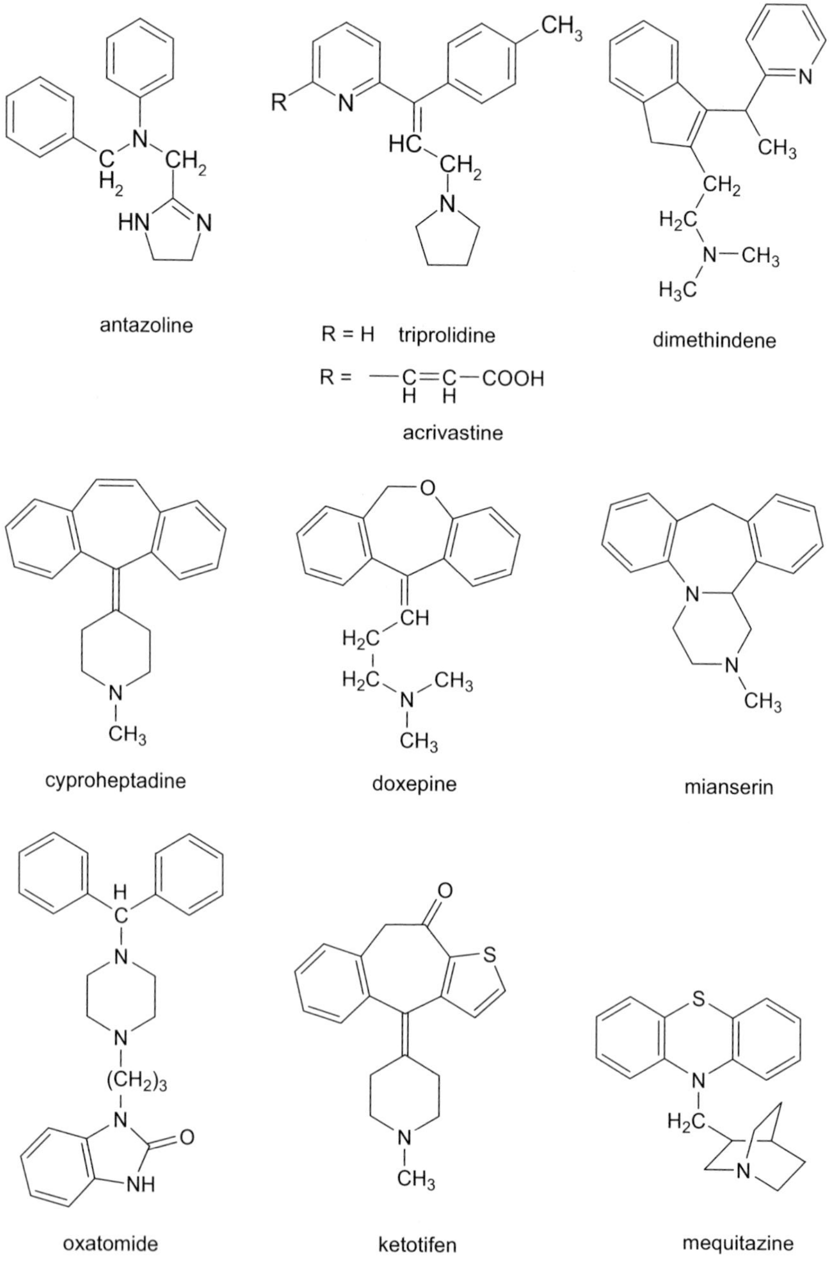

antazoline

R = H   triprolidine

R =   —C=C—COOH
          H   H

acrivastine

dimethindene

cyproheptadine

doxepine

mianserin

oxatomide

ketotifen

mequitazine

Fig. 18.11 Serveral chemically different antihistamines.

By inspecting the structure of several antihistamines, it is clear that in fact they are all analogues, and the conclusions of the QSAR studies of Rekker et al. [10] appear valid in a broad sense.

A number of examples are provided in Fig. 18.11, although it is again impossible to show all examples. Thus, a selection has been made which attempts to include some remarkable structures.

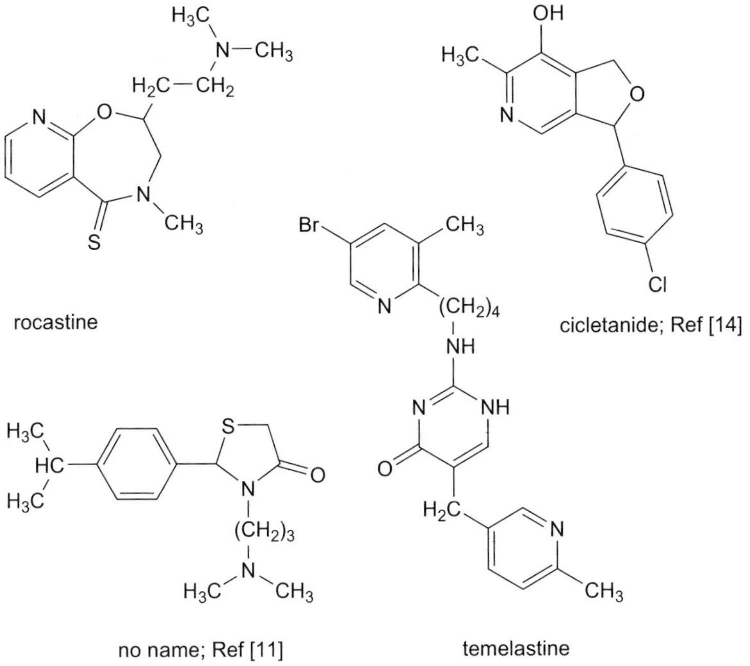

rocastine

cicletanide; Ref [14]

no name; Ref [11]

temelastine

Fig. 18.12 Antihistamines not fitting the analogue principle.

For an extensive review of the structures of antihistamines, the reader should consult Ref. [9]. Although the analogue principle is acceptable, not all antihistamines fall easily into the principle. The examples shown in Fig. 18.12 are matters in case; such compounds may be considered as "pharmacological" rather than "chemical" analogues.

18.3.5
The Analogue Principle in Perspective

The findings of the intensive QSAR analysis conducted by Rekker et al. [10] led to the conclusion that antihistaminic activity may be expected in compounds characterized by the presence of a hydrophobic group (the diaryl moiety) at a certain distance (the linker) from a positive charge (the basic function). The hydrophobic moiety is by far the most critical part for antihistaminic activity. In fact, the often-used term "hydrophobic pocket" is rather misleading, as the part of the receptor to which the diaryl group binds is far from "hydrophobic" as it only concerns a region of the receptor that allows very specific interactions. The following observations provide evidence in this respect:

- The two aromatic groups play very different roles. This is clear from the high enantioselectivity of the chiral antihistamines and the Z:E ratio for activity of compounds of the alkylene series. A good example of this is the case of clemastine and its diastereoisomers (Tab. 18.1) [11]. The results show that that chirality at the benzhydryl carbon plays by far the larger role. For an analysis of the different roles of the two aromatic nuclei, see Ref. [9]. These authors have been able to identify the role of these groups in a large group of antihistamines, showing a high incidence of comparability in all subclasses.
- The interesting difference in the binding mode of triprolidine and acrivastine (see Fig. 18.11). A lysine (Lys200) in the pharmacophore of the H_1 receptor was found – by mutation to alanine – to be involved in binding of acrivastine (a non-sedating antihistamine), but not of triprolidine; the arginine mutant showed the same behavior [12]. Apparently, the carboxyl group of acrivastine is involved in binding to the receptor. However, from QSAR studies [10] it appeared that, in general, a substitution on the position of the carboxyl group in acrivastine is unfavorable. This example shows simultaneously and clearly the limitations of QSAR studies and multiple regression analysis.

Tab 18.1 The antihistamine activity of a series of stereoisomers.

Isomer	pA$_2$*
Clemastine (R,R)	9.45
R,S-isomer	9.40
S,S-isomer	7.99
S,R-isomer	8.57

A negative logarithm of the dissociation constant of the antagonist–receptor complex is indicative of affinity.

18.4
The New Antihistamines

The original expectations of the potential of the antihistamines were not met, as the compounds proved to be effective in those patients with certain allergic affections (skin allergy, rhinitis), but much less so in other patients. In particular, the efficacy in asthma is rather limited, which may be explained by the rather complex nature of the pathology leading to asthma. The major reason for a lack of success of the class was the high incidence of drowsiness induced by all (!) active antihistamines. It is remarkable that it took so many years before the compounds' antihistaminic property was seen to be responsible for the sedative effect. Previously, it was thought that a histamine had no CNS effects and that sedation was caused by other properties (e.g., anticholinergic). If this were true, it should have been relatively simple to derive a nonsedating antihistamine, but this was not the case. Some antihistamines are even used as sleep-promoting agents. Diphenhydramine (and some piperazine analogues) also found use as anti-emetic medicines (against motion sickness) although how this relates to the histamine blocking effect remains unclear.

Elegant experiments conducted by the group of Schwartz in Paris [15] identified a clear relationship between the level of occupation of H_1 receptor in the CNS and sleep induction. When Nicholson et al. [16] showed that, for two pairs of enantiomers of chiral antihistamine, only the active (antihistaminic) enantiomer caused sleep in man, it became clear that the antihistaminic properties are responsible for drowsiness.

Subsequently, a new interest in antihistaminic nonsedating derivatives emerged. Early approaches applying quaternization of the amino function to reduce CNS penetration had failed, and research groups [17,18] began to explore the possibility that there might be differences between central and peripheral receptors. The present author's research group reported clear pharmacological evidence [19] that such differences do not exist – a finding later confirmed when the receptor gene was cloned.

The only remaining possibility of obtaining nonsedating antihistamines was to identify compounds with no (or poor) capacity to pass the blood–brain barrier (BBB), and it is difficult to see why no such compound was identified until the 1980s. The first useful antihistamine without serious sedative effects was terfenadine (Fig. 18.13). Remarkably, this compound was not developed as an antihistamine but rather for other purposes, notably for CNS effects (dopamine antagonism) and calcium channel-blocking properties. Although the CNS effects were absent, the Ca^{++} entry-blocking effects were present however, and subsequently antihistaminic properties, though restricted to peripheral systems, were also observed. This presence of antihistamine properties should not have been too surprising, in fact, as the compound could be considered a diphenhydramine derivative. Clearly, the analogue principle had been successful.

Although terfenadine was the first member of a new class of so-called nonsedating or second-generation antihistamines, several similar compounds were to follow. Terfenadine is pharmacologically an intriguing compound, as the chiral center neither plays a role in H_1 blocking, nor is involved in receptor binding. It

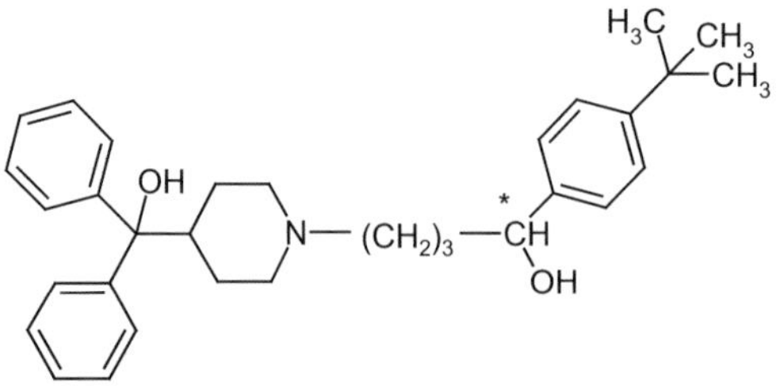

Fig. 18.13 Terfenadine, the first nonsedating antihistamine.

has been shown that the analogue in which OH was replaced by a proton is equi-effective to terfenadine (Fig. 18.13) [20]. Moreover, *in vivo* it is the metabolite fexo-fenadine (Fig. 18.14) which causes the antihistamine effects and is responsible for the poor CNS penetration.

The serious effects of terfenadine on K^+-HERG (human ether-a-go-go related gene) channels leading to cardiac arrhythmia are not seen with fexofenadine, and this is why terfenadine – despite widespread use – was subsequently replaced by fexofena-dine. This serious adverse effect is also seen for astemizole, another second-genera-tion antihistamine. This problem is not confined to second-generation antihista-mines, as some older first-generation antihistamines produce similar effects [21].

The second-generation antihistamines have become a major marketing success, and indeed some representatives are true "blockbusters". On inspection, the struc-ture of several second-generation antihistamines causes no surprises as they may be considered analogues of their first-generation counterparts, with terfenadine being a classical antihistamine with an extended side chain. In Fig. 18.15 there is a selection of second-generation compounds, including ebastine, a diphenhydra-mine analogue.

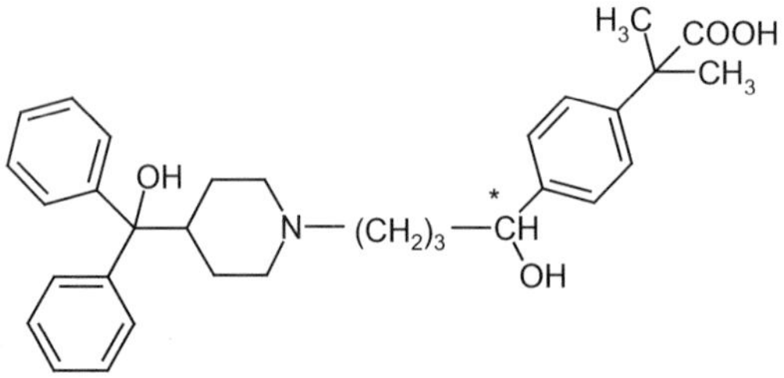

Fig. 18.14 Fexofenadine, the active metabolite of terfenadine.

An inspection of the structures listed (this list is not comprehensive) makes it clear that the analogue principle has also been applied for the nonsedating com-pounds. Most of these compounds have been identified by chance. For example, loratadine (Fig. 18.15) – a derivative of another highly sedating antihistamine, aza-

ebastine

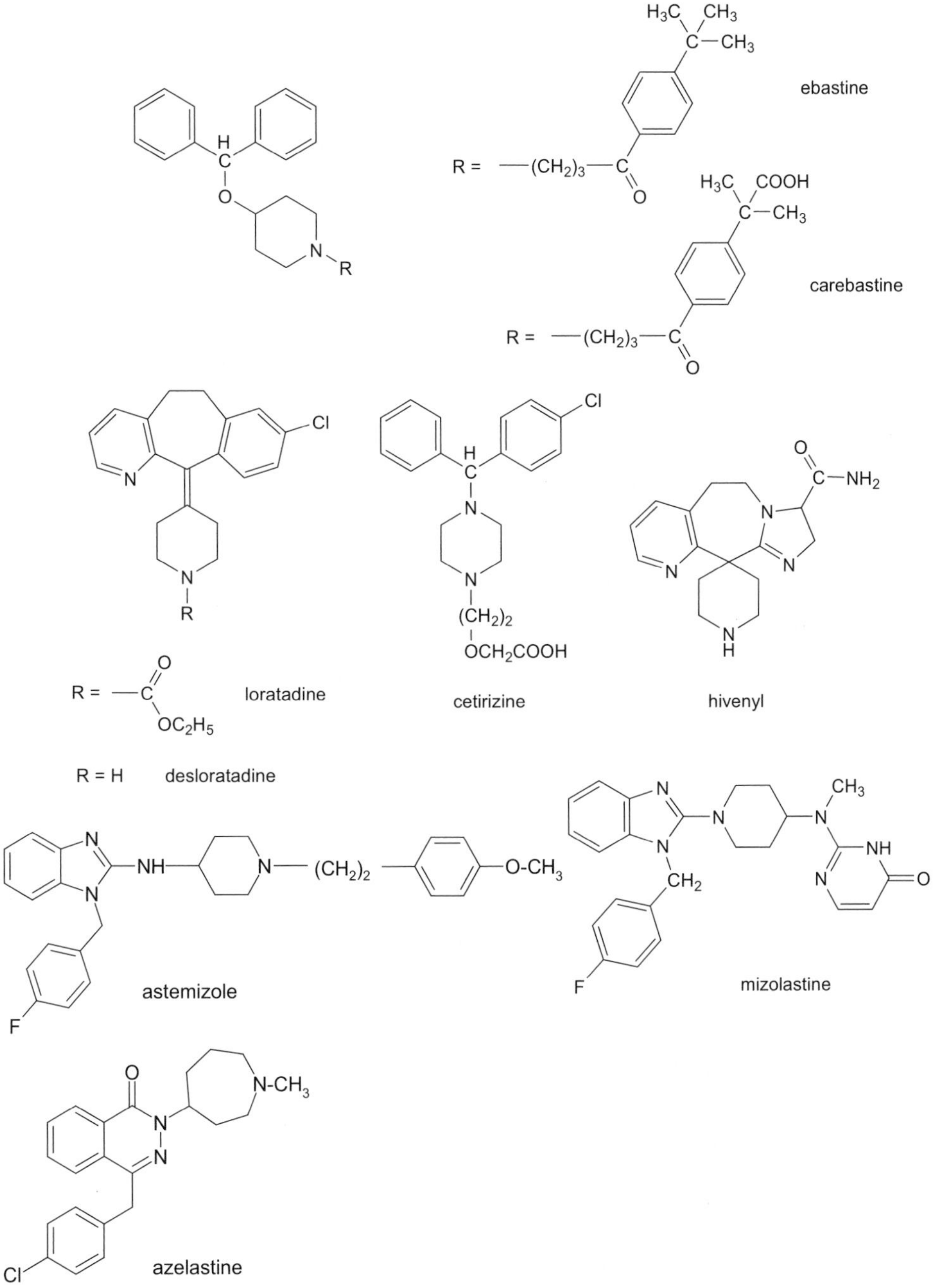

carebastine

loratadine

R = (with structure) OC₂H₅

R = H   desloratadine

cetirizine

hivenyl

astemizole

mizolastine

azelastine

Fig. 18.15 Several nonsedating antihistamines.

tadine – was designed such that the N-substituent was introduced to reduce BBB passage. A second group member, cetirizine, is the metabolite of the potent CNS-active (strong sedatory) compound, hydroxyzine. Strangely, it was later found that a metabolite of loratadine, without the carboxyethyl group originally introduced to avoid CNS-penetration, was responsible for the *in-vivo* antihistaminic activity of this compound (Fig. 18.16).

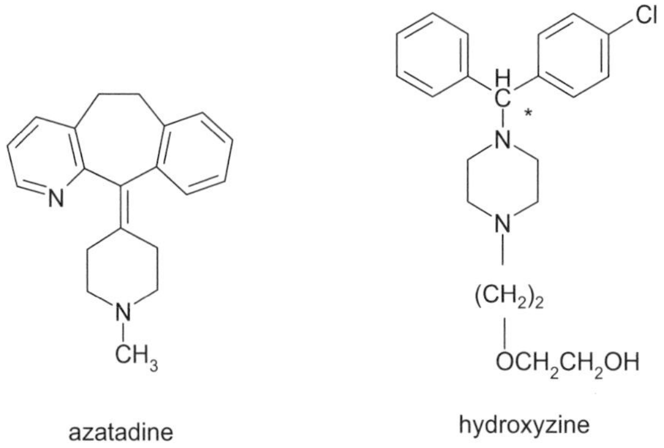

azatadine

hydroxyzine

Fig. 18.16 The sedating analogues of loratadine and cetirizine.

It is perhaps not surprising that because these compounds were mostly found, rather than designed (to avoid BBB passage), it is difficult to understand why they penetrate the CNS so poorly. In textbooks and reviews [22], it is often said that this is because of "high hydrophilicity", but it is clear that the lipophilicity of several compounds would permit brain penetration. Moreover, this parameter does not provide any explanation why azatadine and hydroxyzine penetrate the CNS while loratadine and cetirizine do not.

Here, the so-called Δlog P approach as used for H_2 ligands by Young et al. [23] was applied to determine CNS-penetrating capacity for a series of nonsedating antihistamine [24]. From the results of these studies it became clear that Δlog P (Δlog $P_{octanol/water}$ – log $P_{apolar\ solvent/water}$), which explains a high capacity to form hydrogen bonds and therefore protein binding of compounds with a high Δlog P value, causing reduced BBB passage, does not fully explain the low brain concentrations of these compounds. In a qualitative sense, it may be said that either a low log $P_{octanol-water}$ value (log P <0) or a high value (log P >3) leads to poor entry to the brain; for compounds with log P value between 0 and 3, the Δ log P approach works, namely that there is good penetration at log P values <2, and poor penetration when log P >4.

It seems, however, that other factors may play a more important role in the process of BBB passage. In 1999, Reichel et al. showed that terfenadine had an affinity for the P-glycoprotein (Pgp) efflux pump, but did not indicate whether the compound was a substrate or a blocker of the enzyme [25]. In the case of fexofenadine (the active metabolite of terfenadine), CNS concentrations were found to be much increased in Pgp-knock-out mice, as compared to wild-type mice [26].

Some years later Chen et al. studied the role of Pgp in a number of sedating and nonsedating antihistamines with regard to their brain-penetrating capacity

[27]. The sedating agents hydroxyzine, diphenhydramine and triprolidine were found not to be substrates of Pgp, whereas nonsedating agents such as cetirizine, loratadine and desloratadine were [26]. To date, no convincing SAR have been published for Pgp substrates, but for the antihistamines it is clear that the SAR for antihistaminic activity differs from that for the same compounds as Pgp substrates. Here the analogue principle does not appear to work. It should also be borne in mind that exporting pumps other than Pgp may play comparable roles.

18.5
Antihistamines for Which the Analogue Principle Does not Seem to Work

Few compounds exist with appreciable antihistamine activity that could not easily be considered as analogues of the many compounds developed as antihistamines. Some selected examples are presented in the following section.

It is remarkable that, for histamine, very few compounds of plant origin have been found to have H_1 (or H_2, H_3, H_4) -blocking properties (Fig. 18.17). α-Mangostin [28] is an example, although remarkably this compound (a competitive H_1 antagonist) does not carry a basic function. A basic tertiary amino group is also absent in cicletanide, an intriguing compound that has a chiral center and in which the antihistaminic activity resides in the levo isomer (Fig. 18.17). Rocastine and temelastine are two other antihistaminic compounds with unusual structures (Fig. 18.12).

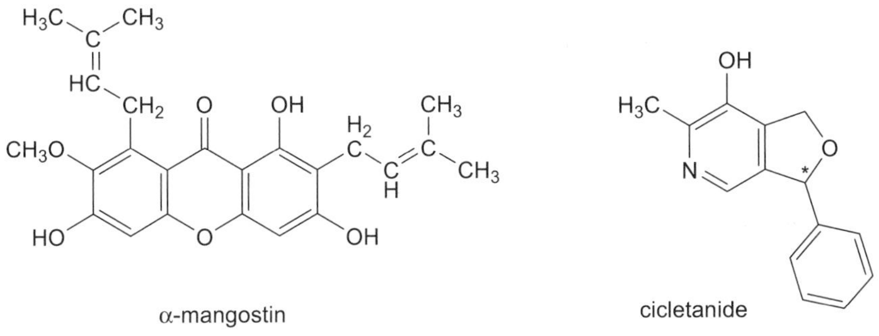

Fig. 18.17 Some unusual structures for H_1 antagonists.

18.6
Inverse Agonism

Virtually all antihistamines of both the first and second generations have been identified as being inverse agonists [29]. This observation led us to a search for compounds possessing classical, neutral antagonizing properties, but to date no compound which could be considered as an analogue of the existing class can be classified as a (neutral) antagonist. However, in another class compounds were found with antihistamine activity, albeit only moderate [30].

The derivative histaprodifen (Table 18.2) is a potent and selective, partial H_1 agonist, the activity being at the same level as seen for histamine. The two higher homologues histabudifen and histapendifen have comparable affinities, but no – or almost no – intrinsic activity (Tab. 18.2). Subsequently, it was shown that hista-budifen antagonizes both the effects of the agonist histamine and the inverse ago-nist mepyramine [30]. Although histabudifen behaves as a classical antagonist at the H_1 receptor, the (limited) data available showed that, when introducing sub-stituents in the benzhydryl group, a SAR was found different from that seen for classical antihistamines carrying such a benzhydryl nucleus. The benzhydryl group in both classes clearly behaves in different ways, showing that in this case the analogue principle is not applicable.

Tab 18.2 Neutral antagonists of the H_1 receptor.

Compound	n	pK_i^a	pEC_{50}^b	α^c
Histaprodifen	2	5.7	6.4	+0.69
Histabudifen	3	5.8	<4	+0.02
Histapendifen	4	5.9	<4	−0.09
Histamine	–	4.2	6.8	+1.0

a. Affinity; COS-7 cells exp. H_1.
b. Activity; COS-7 cells; reportergen, NFκB, luciferase.
c. Intrinsic activity.

"hista-alkyl-difen"

18.7
A Further Generation of Antihistamines?

Due to the success of agents such as cetirizine, (des)loratadine, fexofenadine and, to a lesser extent other second-generation antihistamines (e.g., ebastine, mezolas-tine), the pharmaceutical companies became – for marketing reasons especially – interested in a new, third-generation class of antihistamines. In various marketing campaigns, desloratadine and fexofenadine have been termed "third-generation" antihistamines, although both products are simply metabolites of previously used second-generation medicines. The L-isomer of cetirizine, which was also intro-duced as a new anti-allergic medicine, has also been called a third-generation compound. A consensus group of experts in the field (both pharmacologists and clinicians) concluded that, for any of these compounds, there is no justification whatsoever to use the indication "third generation" [31].

The same consensus group also raised the question of whether any new class (third-generation) antihistamine was feasible [31]. The conclusion was that in the-ory this would be possible, although the new-generation compounds should have a highly specific profile that would allow them to be considered as a class rather than as an isolated compound with an individual profile. For example, a new generation could be formed by a group of neutral antagonists, or of compounds with a beneficial property in addition to antihistamine effects that would provide a clinical advantage. This "extra" property might include anti-LTB_4 effects, anti-inflammatory properties and/or mediator release-reducing properties. The group stressed that the advantage should be demonstrated clinically. Too often, only pharmacological animal data are available and the concentrations used are too high. Anti-PAF data were considered to have only minor relevance. Special attention was given to the combination in a (class of) compounds of H_1- and H_4-blocking activities; the newly detected H_4 receptor was most likely involved in pro-inflammatory effects induced by histamine, and these are not blocked by H_1-, H_2- or H_3-blocking agents.

18.8
Conclusions

During the process which led to the development of both first- and second-generation H_1-blocking antihistamines, the "analogue-principle" was followed intensively. In reality, there is no chemically useful antihistamine available which does not fall within the class of analogues.

Recent developments that have led to the identification of inverse agonists (all classical antihistamines) and neutral antagonists might mean that a totally new group of H_1 receptor-blocking compounds will shortly become available.

The distinction between first- and second-generation antihistamine is both useful and meaningful, as the members of both classes fall within the same group of analogues. The absence – or presence – of cardiac toxicity (as seen in both first- and second-generation compounds) does not mean that they belong to a new class, as the observed cardiotoxicity is not considered to be a class effect.

In theory, a third generation of antihistamines is possible, wherein all such members should have, in addition to H_1-blocking properties, an extra feature (perhaps H_4-antagonizing effects) that would provide an extra, clinically relevant benefit.

References

1 Windaus A, Vogt W. *Chem. Ber.*, **1907**, 40, 3691.
2 Dale HH, Laidlaw PP. *J. Physiol.*, **1910**, 41, 318.
3 Dale HH, Laidlaw PP. (), *J. Physiol.*, **1911**, 43, 182.
4 Dale HH. Foreword. In: *Handbook of Exp. Pharmacol. XVIII-1, Histamine and anti-Histamines.* Rocha e Silva M (Ed.) Springer-Verlag, Berlin, **1966**.
5 Staub AM. *Ann. Inst. Pasteur Paris*, **1939**, 63, 400, 420, 485.
6 Casy AF. Chemistry and structure–activity relationship of synthetic anti-histamines. In: *Histamine and anti-Histaminics. II, Handbook of Experimental Pharmacology.* Rocha e Silva M (Ed.). Springer, Berlin, **1978**, p. 175.
7 Nauta WTh, Rekker RF. Structure–activity relationships of H$_1$ receptor antagonists. In: *Histamine and anti-Histaminics. II, Handbook of Experimental Pharmacology.* Rocha a Silva M (Ed.). Springer, Berlin, **1978**, p. 215.
8 Harms AF, Hespe W, Nauta WTh, Rekker RF, Timmerman H, Vries J. de. Diphenhydramine derivatives: through manipulation toward design. In: *Drug Design. Vol. VI.* Ariëns EJ (Ed.). Academic Press, New York, **1975**, p. 2.
9 Zhang MQ, Leurs R, Timmerman H. Histamine H$_1$ receptor antagonists. In: *Burger's Medicinal Chemistry and Drug Discovery.* Wolff ME (Ed.). John Wiley & Sons, **1997**.
10 Rekker RF. The predictive merits of antihistamine QSAR studies. In: *Strategy in Drug Research.* Keverling Buisman JA (Ed.). Elsevier, Amsterdam, **1982**, p. 315.
11 Diurno MV, Mazzaroni O, Piscopo E, Calignano F, Bolognese A. *J. Med. Chem.*, **1992**, 35, 2910.
12 Wieland K, Ter Laak AM, Smit NJ, Kühne A, Timmerman H, Leurs R. *J. Biol. Chem.*, **1999**, 274, 29994.
13 Ebenother A, Weber HP. *Helv. Chem. Acta*, **1976**, 59, 2462.
14 Schoefter P, Ghysel-Burton J, Cabanie M, Godfraind T. *Eur. J. Pharmacol.*, **1987**, 139, 235.
15 Awack TT, Duchemin AM, Rise C, Schwartz JC. *Eur. J. Pharmacol.*, **1979**, 60, 391.

16 Nicholson AN, Pascoe PA, Turner C, Ganellini CR, Greengrass PN, Casy AF, Mercer AD. *Br. J. Pharmacol.*, **1991**, 104, 270.
17 Barbe J, Andrews PR, Lloyd EJ, Browant P, Soyfer JC, Galy JP, Galy AM. *Eur. J. Med. Chem.*, **1983**, 18, 1531.
18 Ahn HS, Barett A. *Eur. J. Pharmacol.*, **1986**, 127, 153.
19 ter Laak AM, Donné-Op den Kelder GM, Bast A, Timmerman H. *Eur. J. Pharmacol.*, **1993**, 232, 199.
20 Zhang MQ, ter Laak AM, Timmerman H. *Biorg. Med. Chem. Lett.*, **1991**, 1, 387.
21 Taglialatela M, Timmerman H, Annunziata L. *Trends Pharmacol. Sci.*, **2002**, 21, 52.
22 Passalacqua G, Bousquet J, Bachert C, Church MK, Bindslev-Jensen C, Nagy L, Szemere P, Davies RJ, Durham SR, Horak F, Kontou-Fili K, Malling H-J, van Cauwengberge P, Canonica GW. *Allergy*, **1991**, 51, 666.
23 Young RC, Ganellin GR, Griffith R, Mitchell RC, Parsons ME, Saunders D, Sore NE. *Eur. J. Med. Chem.*, **1993**, 28, 201.
24 ter Laak AM, Tsai RS, Donné-op den Kelder GM, Carrupt P-A, Testa B, Timmerman H. *Eur. J. Pharm. Sci.*, **1994**, 2, 373.
25 Reichel A, Siva J, Abbott NJ, Begley DJ. *J. Physiol.*, **1999**, 51, 58.
26 Cvethovic M, Leake B, Fronn MF, Wilkinson CR, Kim RB. *Drug Metab. Dispos.*, **1999**, 27, 866.
27 Chen C, Hanson E, Watson JW, Lee JS. *Drug Metab. Dispos.*, **2003**, 31, 312.
28 Chairungsrilerd N, Furukawa K-I, Ohta T, Nozoe S, Ohizumi Y. (1996), *Eur. J. Pharmacol.*, **1996**, 314, 351–356.
29 Bakker RA, Schoonus SB, Smit MJ, Timmerman H, Leurs R. *Mol. Pharmacol.*, **2001**, 60, 1133.
30 Bruijsters M, Pertz HH, Teunissen A, Bakker RA, Gkillard M, Chatelain P, Schumach W, Timmerman H, Leurs R. *Eur. J. Pharmacol.*, **2004**, 487, 55.
31 Holgate ST, Canonica GW, Simons FER, Tagliatatela M, Tharp M, Timmerman H, Yanai K. *Clin. Exp. Allergy*, **2003**, 33, 1305.

19

Corticosteroids: From Natural Products to Useful Analogues

Zoltán Tuba, Sándor Mahó, and Csaba Sánta

19.1
Introduction

The wide biological spectrum characterizing both natural and synthetic steroids has made this group of compounds one of the most important classes of biologically active materials. As a result of an almost unprecedented number of structural modifications, it has become possible to introduce highly active specific dependable and last, but not least, economically producible derivatives into therapy.

In J.N. Hogg's view, two important discoveries have been responsible for the tremendous acceleration of research into steroids that occurred during the 1950s and beyond the so-called "golden age" of steroids [1]:

- The announcement from the Mayo Clinic in 1949, by Kendall, Hench and their associates, of cortisone's dramatic effectiveness in the treatment of rheumatoid arthritis. This discovery was reported in the *New York Times* on April 20, 1949 and the Proceedings of Staff Meetings of Mayo Clinic [2], and in 1950 resulted in the award of the Nobel Prize in Physiology and Medicine for Kendall and Hench and Reichstein from ETH Zürich. Understandably, the discovery of the exciting potential of cortisone stimulated tremendous interest in steroid chemistry, endocrinology and related areas of medicine, and virtually every leading pharmaceutical company and many research and medical centers worldwide began active programs in these areas.
- The second discovery, by Pincus et al. at the Worcester Foundation, was the demonstration that the female hormone progesterone acts as contraceptive agent in animals [3].

This chapter provides a short survey of the history and progress in steroidal research, focusing both on clinically important compounds and on those at an advanced stage of development.

Analogue-based Drug Discovery. IUPAC, János Fischer, and C. Robin Ganellin (Eds.)
Copyright © 2006 WILEY-VCH Verlag GmbH & Co. KGaA, Weinheim
ISBN: 3-527-31257-9

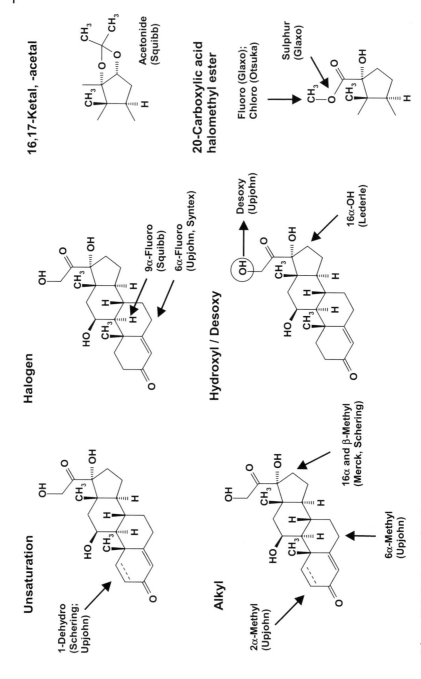

Scheme 19.1 Corticosteroid analogues.

19.2
Corticosteroid Analogues

Corticosteroids are the most effective drugs in the treatment of acute and chronic inflammatory diseases. They were introduced in 1949, since when they have enjoyed a remarkable "career". Numerous synthetic corticosteroids have been developed by modifying the chemical structure of cortisone and hydrocortisone. Indeed, the scientific literature shows that there is not one of the carbon atoms of the sterane structure that has remained unmodified in an endeavor to achieve a favorable biological effect.

Important structural modifications to the cortisone and hydrocortisone have been made, including halogenation, methylation, hydroxylation, and the introduction of a double bond at the C-1 position of the steroid skeleton (Schemes 19.1 and 19.2). Esterification of the 17a and/or 21-hydroxy groups and replacement of the 21-hydroxy group with a halogen atom have also been shown to enhance the anti-inflammatory activity of corticosteroids by improving penetration into the skin and by increasing affinity for glucocorticoid receptor.

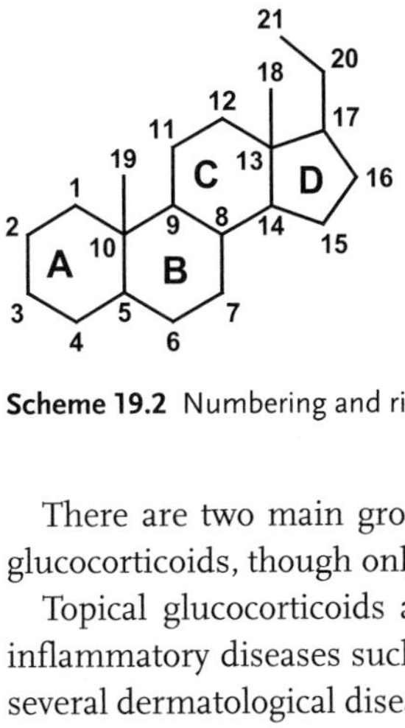

Scheme 19.2 Numbering and ring letters of the pregnane skeleton.

There are two main groups of corticosteroids, namely mineralocorticoids and glucocorticoids, though only the latter group can act locally and systemically.

Topical glucocorticoids are used extensively for the treatment of a variety of inflammatory diseases such as rhinitis, inflammatory bowel disease, asthma, and several dermatological diseases.

Since 1952, about 40 new topically active anti-inflammatory corticosteroids have been introduced into therapy.

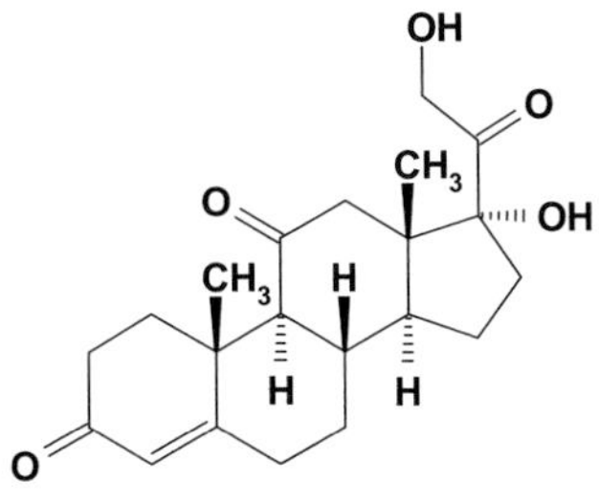

Fig. 19.1 Cortisone.

Oral glucocorticoids such as dexamethasone and prednisolone are still used in patients with severe asthma, though these agents are associated with adverse systemic effects such as growth retardation, osteoporosis, and the suppression of hypothalamic-pituitary adrenal function and the immune system.

19.2.1
Cortisone

As mentioned in the introduction, the announcement of the dramatic efficacy of cortisone (Fig. 19.1) in the treatment of rheumatoid arthritis stimulated tremendous interest in steroidal chemistry, endocrinology and related areas of medicine. This effort was remarkably successful and resulted in the introduction of several important pioneering drugs. The cortisone era had a profound impact on drug discovery as it also led to the logical application of steric and electronic concepts to medicinal chemistry. Last, but not least, the cortisone era taught medicinal chemists many important lessons about drug–receptor interactions.

It is for this reason that we can say that cortisone is the "pioneer drug" in the series of corticosteroids.

Cortisone acetate in readily absorbed from the gastrointestinal tract, and the cortisone is rapidly converted in the liver to its active metabolite, hydrocortisone (cortisol) (Fig. 19.2). The biological half-life of cortisone itself is only about 30 min.

Cortisone acetate is rapidly effective when given orally, and more slowly by intramuscular injection. Cortisone acetate has been used in the treatment of many of the allergic and inflammatory disorders for which corticosteroid therapy is helpful but prednisolone or other synthetic glucocorticosteroids are generally preferred.

Cortisone is a corticosteroid secreted by the adrenal cortex. It has glucocorticoid activity, as well as appreciable mineralocorticoid activity; 25 mg cortisone acetate is equivalent in anti-inflammatory activity to about 5 mg prednisolone.

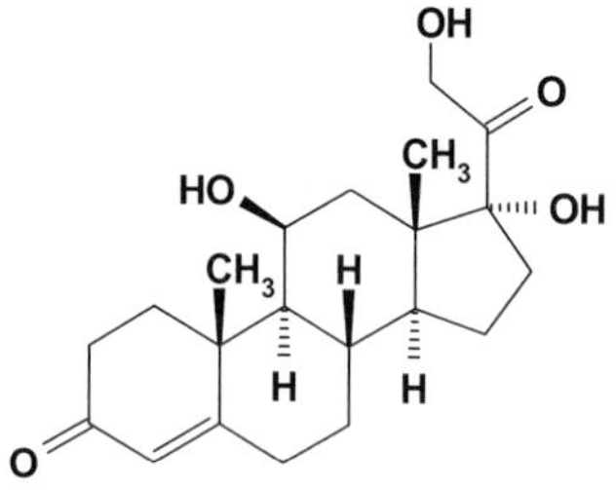

Fig. 19.2 Hydrocortisone.

19.2.2
Hydrocortisone

Biochemical studies conducted by Bush and coworkers [4,5] suggested that cortisol (hydrocortisone), the 11β-hydroxy analogue of cortisone, is the functional hormone and that cortisone owes its biological activity to enzymic conversion to cortisol *in vivo*. Therefore, the first chemical synthesis of cortisol was an important achievement. This was accomplished by Wendler et al. [6] in 1950, and an improved synthesis was described by Wendler, Huang Minlon, and Tishler [7] during the following year.

In 1951–1952, Sulzberger et al. showed that systemically – but not topically – applied cortisone acetate was effective in the treatment of eczema and other dermatitides, whereas hydrocortisone was effective with both routes of administration [8,9]. Because the topical route promised a relative freedom from troublesome systemic side effects (e.g., salt retention). Schering recognized the need for an effective synthesis of the latter.

Hydrocortisone is readily absorbed from the gastrointestinal tract, and peak blood concentrations are attained in about 1 h. The plasma half-life is about 100 min, and hydrocortisone is more than 90% bound to plasma proteins. Following intramuscular injection, absorption of the water-soluble sodium phosphate and sodium succinate ester is rapid, whilst absorption of free alcohol or lipid-soluble esters is slower.

Hydrocortisone is metabolized in the liver and most body tissues to hydrogenated and degraded forms, for example, tetrahydrocortisone and tetrahydrocortisol. These are excreted in the urine, mainly conjugated as glucuronides, together with a very small proportion of unchanged hydrocortisone.

Hydrocortisone is a corticosteroid with both glucocorticoid and, to a lesser extent, mineralocorticoid activity.

Esterification generally alters the potency of hydrocortisone, and compounds with equivalent hydrocortisone content may not have equivalent clinical effects.

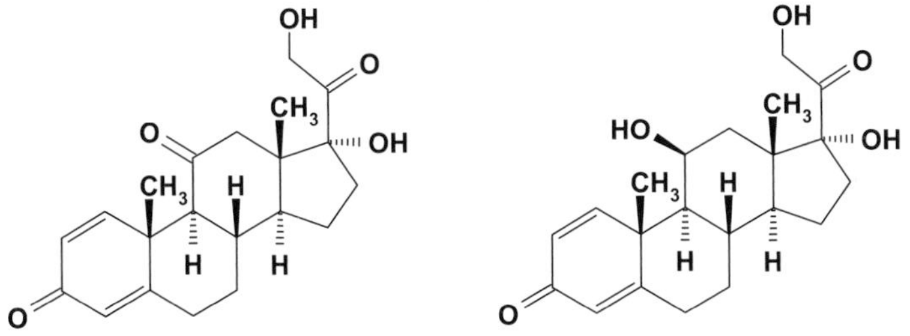

Fig. 19.3 Prednisone and prednisolone.

19.2.3
Prednisone and Prednisolone

Enhanced glucocorticosteroid activity (three to five times that of cortisone) and reduced mineralocorticoid activity at the projected therapeutic dose were demonstrated by the collaborators of Schering Corporation using appropriate assays. Anti-arthritic activity and the absence of any significant associated salt retention were quickly demonstrated for both prednisone and prednisolone.

Prednisone was launched for the first time in March 1955 for use in arthritis and related conditions characterized by inflammation. Prednisolone was introduced soon thereafter.

Although at least five other pharmaceutical companies claimed priority of the invention (Merck, Squibb, Pfizer, Upjohn and Syntex), none was able to convince the U.S. Patent Office. A patent was finally issued to Schering Corp. in 1964 (10 years after the initial filing). To this day, prednisone and prednisolone remain the standard of systemic corticosteroid therapy throughout the world.

Prednisolone and prednisone are both readily absorbed from the gastrointestinal tract, but whereas prednisolone exists in a metabolically active form, prednisone must be converted in the liver to its active metabolite prednisolone.

Peak plasma concentrations of prednisolone are reached 1–2 h after oral dosing, and the plasma half-life is 2–4 h.

Prednisolone is extensively bound to plasma proteins, although less so than hydrocortisone. Prednisolone is a corticosteroid with mainly glucocorticoid activity; in fact, 5 mg prednisolone is equivalent in anti-inflammatory activity to ca. 25 mg cortisone acetate.

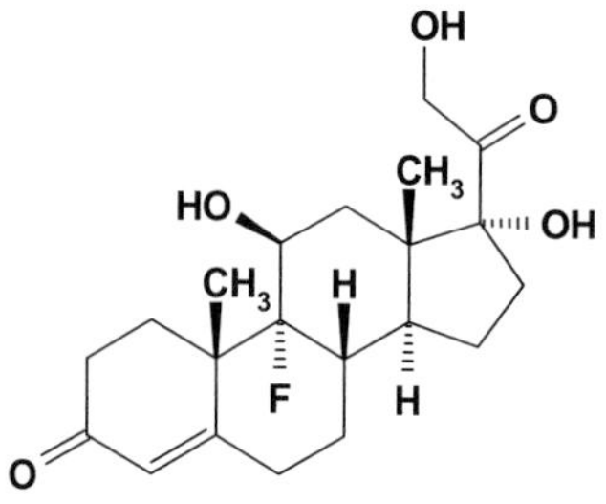

Fig. 19.4 Fludrocortisone.

19.2.4

Fludrocortisone

In 1953, Fried and Sabo announced the synthesis of the highly potent steroid analogue fludrocortisone (9-fluoro-hydrocortisone), formally accenting the beginning of the analogue era [10–13].

Today, it is well known that the introduction of a chlorine or fluorine atom at the C-9 position of the natural adrenal substances cortisone and hydrocortisone (and of their dehydro-analogues) markedly enhances the anti-inflammatory activity of these agents, and is accompanied by a striking increase in both salt and water retention [14]. These undesirable side effects, which are manifested generally by 9-halo-steroids, preclude their use systemically in the management of disorders that are normally responsive to adrenocortical steroid therapy.

This earliest of the important enhancement discoveries was in itself not of therapeutic value, due to salt retention, yet it became a component in nearly all of the ultrapotent corticoids to be marketed. Indeed, it is why the discovery of the potent 9-fluoro-hydrocortisone was an event that had far-reaching significance.

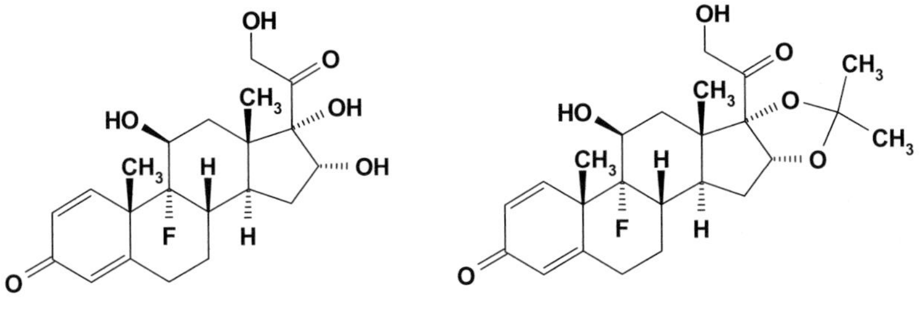

Fig. 19.5 Triamcinolone and triamcinolone acetonide.

19.2.5
Triamcinolone and Triamcinolone Acetonide

Biological studies on 16a-hydroxy-hydrocortisone [15] have demonstrated that this type of corticoid derivative still maintains a considerable activity in the usual types of assays (glycogen deposition, thymus involution and anti-inflammatory tests). It was therefore of interest to prepare the 16a-hydroxy derivatives of the more potent 9a-halo steroids.

Introduction of the 16a-hydroxyl group into 9-fluoro-hydrocortisone and 9-fluoro-prednisolone has been shown by Bernstein et al. [16] to result in complete suppression of the salt-retaining properties of these steroids, without appreciably impairing their glucocorticoid activity. Moreover, studies in man have demonstrated the anti-arthritic activity of 9-fluoro-16a-hydroxy-prednisolone and have confirmed its lack of salt-retaining activity. Subsequently, Lederle reported that the anti-inflammatory potency was lowered, but salt retention was eliminated.

Triamcinolone is reported to have a half-life in plasma of about 2 h to over 5 h. It is bound to plasma albumin to a much smaller extent than hydrocortisone. The acetonide, diacetate, and hexacetonide esters of triamcinolone are only very slowly absorbed from injection sites.

A 4-mg dose of triamcinolone is equivalent in anti-inflammatory activity to about 5 mg prednisolone. It is used, either in the form of the free alcohol or in one of the esterified forms in the treatment of conditions for which corticosteroid therapy is indicated.

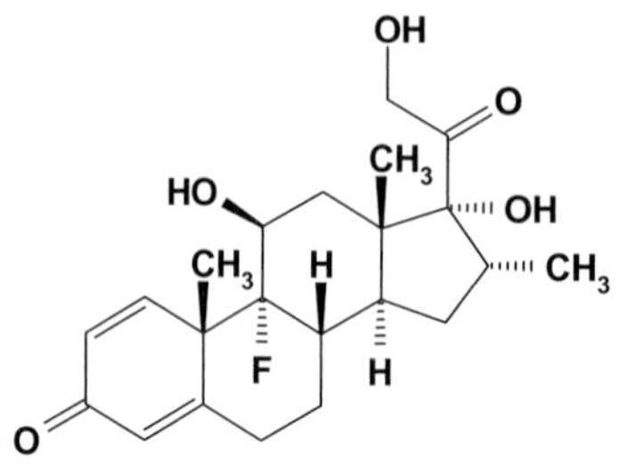

Fig. 19.6 Dexamethasone.

19.2.6
Dexamethasone

Since the discovery that cortisone was effective in the treatment of rheumatoid arthritis [2], intensive efforts have been expended towards modifying the structure of this hormone in the hope of finding a related compound with superior therapeutic properties.

It has been established that in addition to reductions in the A-ring (see Scheme 19.2), two pathways of metabolic inactivation of cortisone or hydrocortisone involve reduction at C-20 [17,18] to an alcohol and scission of the side chain [19] to 17-keto compounds.

On the hypothesis that the side chain could perhaps be stabilized against such metabolic inactivation by an appropriately placed but otherwise chemically inert substituent, some analogues of cortisone containing a 16α-methyl group were synthesized by Arth et al. [20] in the expectation that the 16α-methyl substituent would enhance potency by protecting the 20-ketone from metabolism. Ultimately, these authors prepared 9-fluoro-16α-methylprednisolone which was the first derivative of fluorocortisol with markedly increased anti-inflammatory activity that was devoid of mineralocorticoid activity. This latter fact supports the concept that it is the presence of 16α-substituents that prevents binding to the mineralocorticoid receptor. Dexamethasone has become Merck's principal marketed anti-inflammatory steroid.

Dexamethasone is a corticosteroid with mainly glucocorticoid activity; in fact, 750 µg dexamethasone is equivalent in anti-inflammatory activity to ca. 5 mg prednisolone.

Dexamethasone has been used either in the form of the free alcohol, or in one of the esterified forms in the treatment of conditions for which corticosteroid therapy is indicated. Its return to mineralocorticoid properties makes dexamethasone particularly suitable for treating conditions where water retention would be a disadvantage.

For parenteral administration in intensive therapy or in emergencies, the sodium phosphate ester may be given intravenously by injection or infusion, or intramuscularly by injection. Dexamethasone sodium phosphate is also used in the treatment of cerebral edema caused by malignancy. The sodium phosphate ester is given by intra-articular, intra-lesional or soft tissue injection. Dexamethasone acetate may be given by intramuscular injection in conditions where corticosteroid treatment is indicated, but a prompt response of short duration is not required.

For ophthalmic disorders or for topical application in the treatment of various skin disorders, either dexamethasone or its esters may be used.

For allergic rhinitis and other allergic or inflammatory nasal conditions, a nasal spray containing dexamethasone isonicotinate is available; the acetate, phosphate, sodium phosphate and sodium metasulfobenzoate have also been used.

Dexamethasone is given intravenously and orally for the prevention of nausea and vomiting induced by cancer chemotherapy.

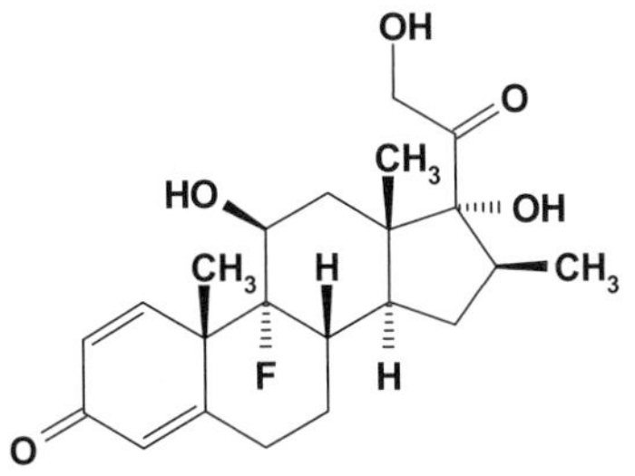

Fig. 19.7 Betamethasone.

19.2.7
Betamethasone

Research teams at Glaxo then undertook the synthesis of derivatives of betamethasone that might afford superior local anti-inflammatory and anti-allergic effectiveness. Using McKenzie and Stoughton's [21] new human-based pharmacologic test that could identify with ease the relative topical potency of steroid inflammatory compounds, a series of 17-esters of betamethasone prepared by Elks [22] was evaluated. This resulted in compounds with new standards of topical potency such as triamcinolone acetonide and fluocinolone acetonide. It was then discovered, that potency peaked with betamethasone-17-valerate and betamethasone-17,21-dipropionate, which were between four- and ten-fold more potent than the standard.

The anti-inflammatory activity of 750 μg betamethasone is equivalent to ca. 5 mg prednisolone. It has been used either in the form of the free alcohol or in one of the esterified forms in the treatment of conditions for which corticosteroid therapy is indicated. Betamethasone has little or no effect on sodium and water retention.

For topical application in the treatment of various skin disorders, the dipropionate and valerate esters are extensively used. Betamethasone valerate has also been used by inhalation for the prophylaxis of asthma.

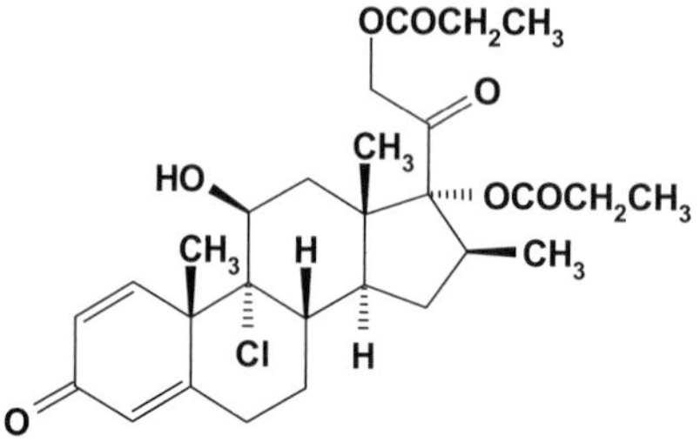

Fig. 19.8 Beclomethasone dipropionate.

19.2.8

Beclomethasone Dipropionate

Beclomethasone dipropionate is the 9-chloro-analogue of betamethasone dipropionate. It was developed by Glaxo and introduced in the United States by Schering Corp. for topical use in the treatment of asthma and allergic rhinitis by the inhalation of a freon-propelled aerosol.

Beclomethasone dipropionate is stated to be readily absorbed from sites of local application and rapidly distributed to all body tissues. It is metabolized principally in the liver, but also in other tissues including the gastrointestinal tract and the lung; enzymatic hydrolysis rapidly produces the monopropionate, and more slowly, the free alcohol, which is virtually devoid of activity.

Beclomethasone dipropionate is a corticosteroid with mainly glucocorticoid activity that is started to exert a topical effect on the lungs, without significant systemic activity. It is used by inhalation, generally from a metered-dose aerosol, for the prophylaxis of asthma. Inhalation of nebulized beclomethasone dipropionate has also been used in the management of asthma in children.

Beclomethasone dipropionate is also used as a nasal spray in the prophylaxis and treatment of allergic and nonallergic rhinitis, and is used topically in the treatment of various skin disorders. It is generally applied as a cream or ointment. Beclomethasone salicylate has also been used topically.

Corticosteroids and β_2 adrenoceptor agonists form the cornerstone of the management of asthma. Patients requiring very occasional relief from symptoms may be managed with β_2 adrenoceptor agonists alone, but an inhaled corticosteroid such as beclomethasone is added if symptomatic relief is needed more than once daily.

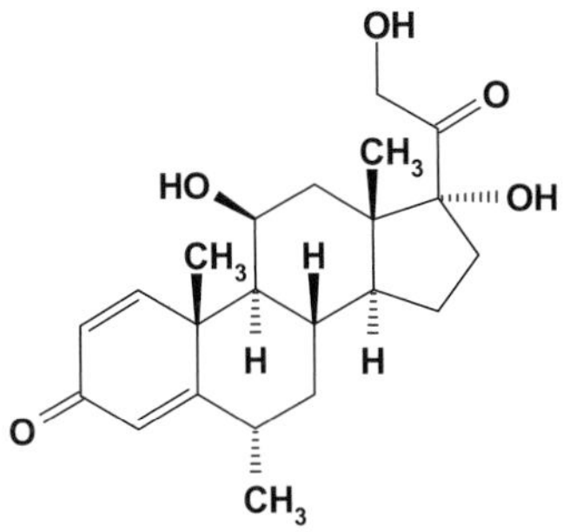

Fig. 19.9 Methylprednisolone.

19.2.9
Methylprednisolone

The synthesis of modified cortical steroids has received particular interest since the report of greatly enhanced activity in the series of 9-halogen derivatives of hydrocortisone [23]. Further modifications have included the introduction of the C-1 double bond [24] and the 2-methyl group [25]. The increased activity due to these latter changes has been attributed to protection of A-ring against inactivation via enzymatic reduction. [26] It seemed reasonable to suppose that a 6a-methyl group would be similarly effective.

A 4-mg dose of methylprednisolone is equivalent in anti-inflammatory activity to about 5 mg prednisolone.

Methylprednisolone may be slightly less likely than prednisolone to cause sodium and water retention. It is fairly rapidly distributed following oral administration, with a plasma half-life of 3.5 h or more. The tissue half-life is reported to range from 18 to 36 h.

The sodium succinate ester is rapidly absorbed following intramuscular administration, with peak plasma concentrations obtained in 2 h. For parenteral administration in intensive or emergency therapy, methylprednisolone sodium succinate may be given by intramuscular or intravenous injection or by intravenous infusion. The intravenous route is preferred for its more rapid effect in emergency therapy.

6a-Methylprednisolone is more active than prednisolone, and has a better therapeutic index with respect to salt retention. Together with its 21-succinate, this became Upjohn's most important corticoid product [27].

19.2.10
Fluocinolone Acetonide, Flunisolide, Fluocortin-21-Butylate and Flumetasone

Further research in this area led to the development of 6a-fluoro derivatives of other known biologically active corticoids. Since the 6a-fluoro substituent appears to be the single most active potentiating group, and since the 6a,9a-difluoro corti-

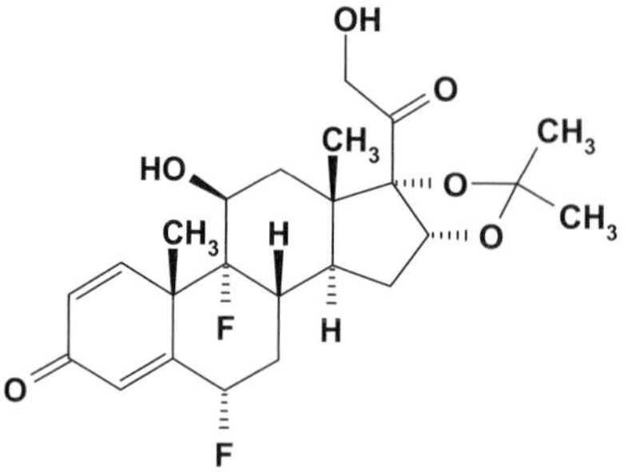

Fluocinolone acetonide

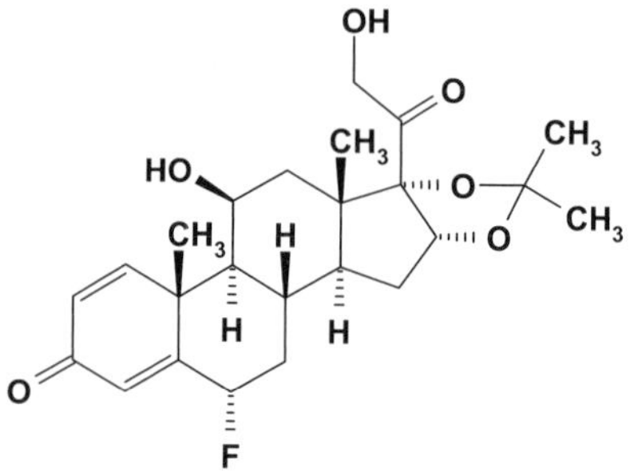

Flunisolide

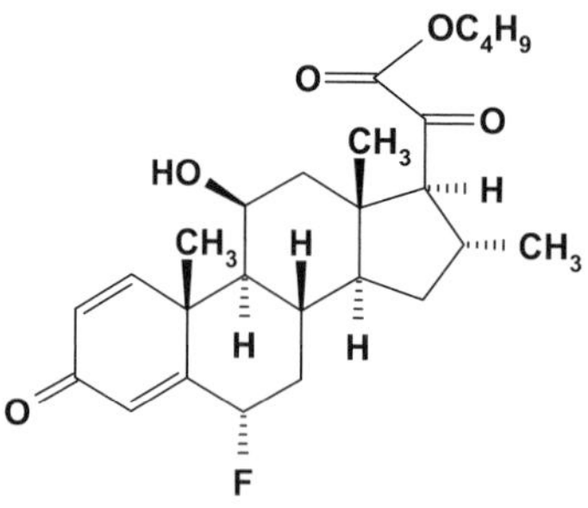

Fluocortin butylate

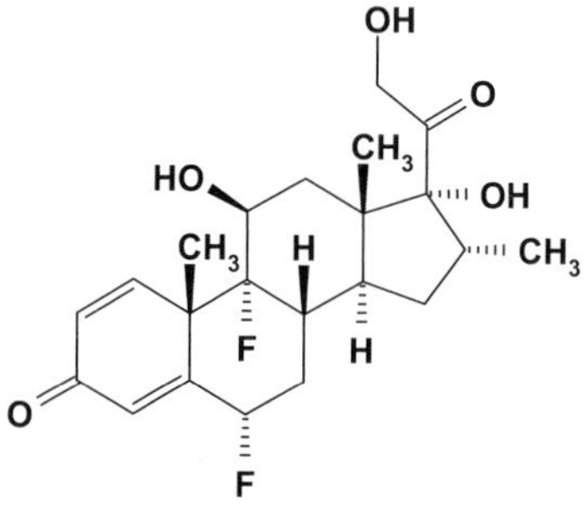

Flumetasone

Fig. 19.10 Fluocinolone acetonide, flunisolide, fluocortin-21-butylate and flumethasone.

coids [28–30] are still not devoid of sodium retention, an important extension was the preparation of 6a,9a-difluoro-16a-hydroxy and 6a,9a-difluoro-16a-methyl corticoids [31].

The introduction of a 16a-hydroxy group into 6a,9-difluoro-prednisolone led to a compound (fluocinolone) with the anticipated favorable biological spectrum, namely high anti-inflammatory activity (35-fold that of hydrocortisone, seven-fold that of triamcinolone) and – in contrast to the C-16 unsubstituted compound – no retention of sodium. The corresponding 16,17-acetonide (fluocinolone acetonide) exhibited 100-fold the anti-inflammatory activity of hydrocortisone, with no sodium retention. In clinical trials, 6a,9-difluoro-16a-hydroxyprednisolone was found to be a potent suppressor of inflammatory conditions such as rheumatoid arthritis, as well as allergic conditions such as asthma, whilst its acetonide proved to be highly effective as topical corticoid.

As a nasal spray, flunisolide is effective in allergic rhinitis in clinical use [32–34], and their protective effects in patients with bronchial asthma have also been studied. The efficacy and safety of a new metered-dose inhaler of flunisolide has recently been reported [35].

Clinical data showing good a protective effect of fluorocortin butyl ester in patients with bronchial asthma are available [36].

Flumetasone pivalate is a corticosteroid used topically for its glucocorticoid activity in the treatment of various skin disorders. It is usually employed as a cream, ointment, or lotion. Flumetasone pivalate is also used in ear drops.

Flumetasone [37] can be used as an intermediate for the preparation of fluticasone.

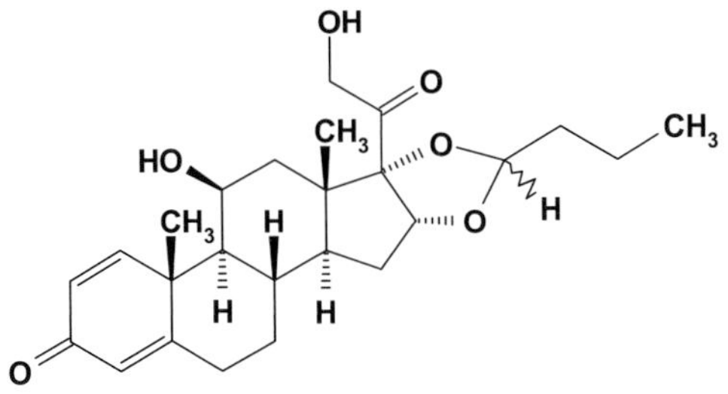

Fig. 19.11 Budesonide.

19.2.11
Budesonide

One of the probable prerequisites for obtaining high topical activity is a balance between the lipophilic and hydrophilic properties of the steroid molecule [38, 39]. The lipophilicity of the molecule can be increased by masking one or two of the hydroxy substituents by, for example, esterification or acetalization. Thus, triamcinolone acetonide, but not triamcinolone, has a high topical activity. Thalèn et al. [40] found that the topical anti-inflammatory potency could be increased without a corresponding rise in systemic glucocorticoid activity by using various aldehydes instead of acetone to mask the 16a,17a-hydroxy groups in 16a-hydroxyprednisolone. In this way, it has been possible to design a nonhalogenated corticoid with very high topical anti-inflammatory activity.

The epimeric mixture of 16a(R,S),17-[butylidenebis(oxy)]-11β,21-dihydroxy-pregna-1,4-diene-3,20-dione was shown to have a local anti-inflammatory potency comparable to that of fluorocinolone acetonide. In contrast, its systemic glucocorticoid activity was found to be four to seven times lower than that of fluocinolone acetonide.

Budesonide was originally developed by Astra Zeneca, but is now a generic compound. It exhibits prolonged residence in the lung tissues as a consequence of its increased lipophilic character resulting from the acetal group.

Budesonide is rapidly and almost completely absorbed following administration by mouth, but has poor systemic availability (ca. 10%) due to extensive first-pass metabolism in the liver, mainly by the cytochrome P450 isoenzyme CYP3A4. The major metabolites 6β-hydroxybudesonide and 16a-hydroxyprednisolone have less than 10% of the glucocorticoid activity of unchanged budesonide. Budesonide is reported to have a terminal half-life of about 2–4 h.

Budesonide is indicated for the treatment of chronic obstructive pulmonary disease, rhinitis, and general allergies.

Local formulations of budesonide are used in the management of inflammatory bowel disease. In mild to moderate Crohn's disease it is given by mouth as modified release capsules intended for a topical effect on the gastrointestinal tract. This approach offers new hope for Crohn's disease sufferers.

Budesonide is also given by inhalation as a nebulized solution in the management of childhood croup. It is also used topically in various skin diseases as a cream or ointment.

Budesonide nasal spray is available for the treatment of seasonal and perennial allergic rhinitis in adults and children.

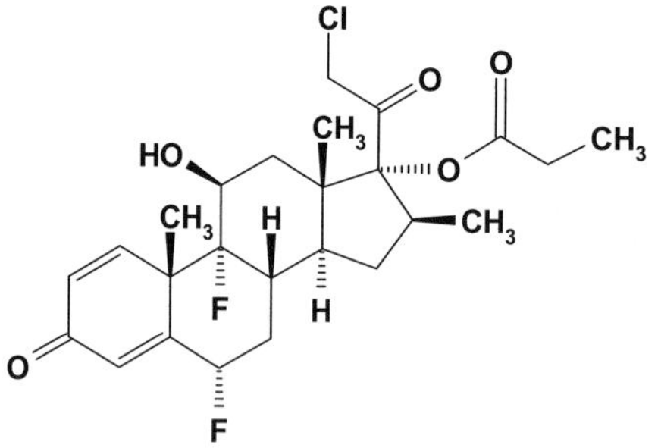

Fig. 19.12 Halobetasol propionate.

19.2.12
Halobetasol Propionate

Scientists at Ciba-Geigy synthesized a series of polyhalogenated corticosteroids, and one compound in the series containing halogen atoms at three positions – halobetasol propionate – was selected for further evaluation [41,42].

The anti-inflammatory and antihypoproliferative activities of this compound were compared to those of other compounds in several animal models. The compound was 1300-fold more potent than hydrocortisone, and 11-fold more potent than clobetasol propionate, respectively, in suppressing the erythemal responses in a guinea pig model of UV-induced inflammation, and it was 10-fold more potent than clobetasol in the cotton pellet granuloma assay in rats.

Halobetasol propionate is a class I topical corticosteroid, or an ultrapotent corticosteroid according the Stoughton classification system which is widely used in United States. Halobetasol propionate has been approved in the United States for the treatment of corticosteroid-responsive dermatoses. In a number of clinical studies, halobetasol propionate cream and ointment have both been tested against vehicle and compared with other ultrapotent corticosteroids [43].

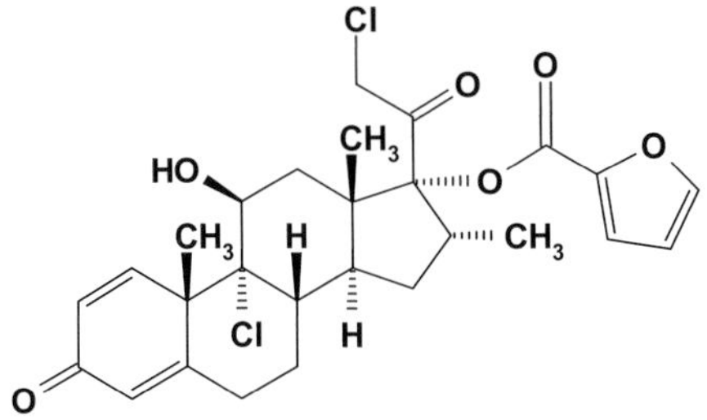

Fig. 19.13 Mometasone furoate.

19.2.13
Mometasone Furoate

The new class of corticosteroids consists of heterocyclic ester derivatives involving the 17a-hydroxy function. These include furoyl, thiophenoyl, and pyrrolylcarbonyl esters. Shapiro devised momethasone furoate as the first 16a-methyl-17a-hydro-xysteroid esterified at C-17 to possess great topical potency relative to the plethora of topical steroid drugs already available [44–48].

Anti-inflammatory potencies were measured in mice by a 5-day modification of the Tonelli croton oil ear assay. The most potent topical anti-inflammatory compounds were 17a-heterocyclic esters in the 16a-methyl series where the 21-substituent was either chloro or fluoro. Thus, [21-chloro-17-(2'-furoate)] was eight-fold more potent than betamethasone valerate, while [21-fluoro-17-(2'-furoate)], [21-chloro-17-(2'-thenoate)] and 6a-fluoro-21-chloro-17-(2'-furoate)] were three-fold as potent as betamethasone valerate.

Exceptional safety through the effective separation of topical from systemic activity (via absorption through skin) was demonstrated in studies, and the product has, since its introduction in 1987, enjoyed steadily increasing use in human medicine.

Schering-Plough has developed a once-daily nasal spray formulation of momethasone furoate for the treatment and prevention of seasonal and perennial allergic rhinitis. Inhalation formulations have also been developed. Schering-Plough has developed an inhaled powder formulation for asthma maintenance treatment.

Momethasone furoate has been launched in the US (as an ointment and a cream) for once-daily use in the treatment of corticosteroid responsive dermatoses and as a lotion for the scalp. It has also been launched in many countries for the control and management of persistent asthma in patients aged >12 years.

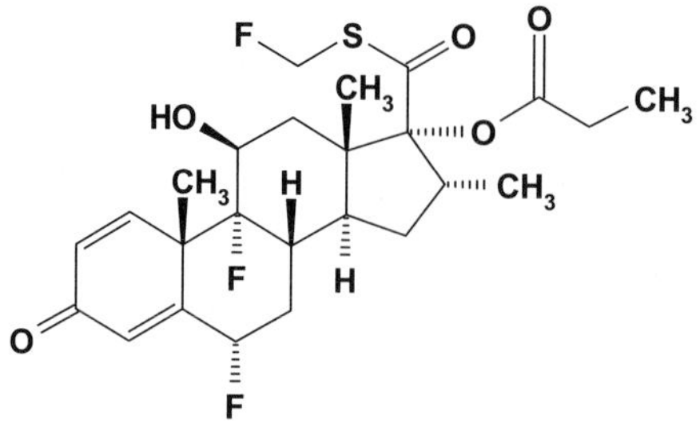

Fig. 19.14 Fluticasone propionate.

19.2.14
Fluticasone Propionate

In order to avoid even the minimal systemic side effects seen with currently available corticosteroids, scientists at Glaxo synthesized a series of androstane 17β-carboxylates and carbothiolates. It was found that halomethyl carbothiolates showed the highest topical activity as assessed by the inhibition of croton oil-induced ear inflammation in mice. The highest activity was found in the fluoromethyl carbothiolate-17-propionate (fluticasone propionate) synthesized from flumethasone [49–51].

Fluticasone propionate, when administered topically, about ten-fold more active, whereas beclomethasone dipropionate is five-fold more active than fluocinolone acetonide using the vasoconstriction assay in man.

Fluticasone propionate (GlaxoSmithKline) is currently the only marketed inhaled corticosteroid with patent protection. It shows very low oral bioavailability, due to a combination of low solubility and oxidative liver metabolism of the C(20) thioester to the corresponding acid which shows much reduced *in-vitro* activity.

Over 100 million people worldwide suffer from asthma, and the most effective anti-inflammatory treatment of the disease is provided by glucocorticoids.

Oral glucocorticoids such as dexamethasone and prednisolone are still used in patients with severe asthma, though these agents are associated with adverse systemic effects. Inhaled glucocorticoid therapy was introduced in 1972 with beclomethasone dipropionate, which dramatically reduced systemic effects. Fluticasone propionate (launched in 1993) is very efficiently inactivated in the liver, and exhibits low oral bioavailability, which in turn leads to a further reduction in systemic exposure.

A recent development has been the advent of combination therapy delivering a corticosteroid and a long-acting β_2 adrenergic agonist in the same inhalation device; this has been shown to provide synergistic benefits in clinical end-points in asthma. The only currently approved combination of an inhaled corticosteroid with β_2-agonist activity is fluticasone propionate with salmeterol (GlaxoSmithKline).

Fluticasone propionate is poorly absorbed from the gastrointestinal tract, and undergoes extensive hepatic first-pass metabolism; as a result, the oral bioavailability is reported to be only about 1%.

Fluticasone propionate is administered by nasal spray in the prophylaxis and treatment of allergic rhinitis, but is applied topically in the treatment of various skin disorders (as creams and ointments).

Fluticasone propionate is also available in some countries as a powder or aerosol inhalation for the treatment of chronic obstructive pulmonary disease.

Fluticasone propionate, when given by mouth, has produced variable results in the treatment of Crohn's disease and ulcerative colitis.

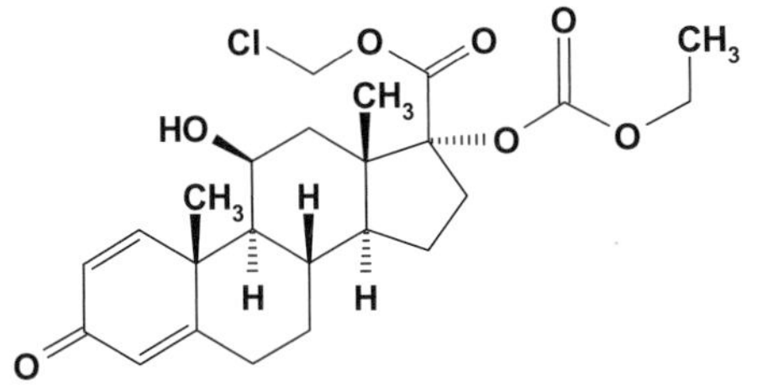

Fig. 19.15 Loteprednol etabonate.

19.2.15
Loteprednol Etabonate

Loteprednol etabonate is a so-called "soft drug" that was designed using a concept developed by Bodor [52,53]. According to this concept, the synthesis of a soft drug is achieved by starting with a known inactive metabolite of known active drug. The inactive metabolite is then modified to an active form that, after having achieved its therapeutic role, will undergo a predictable and controllable one or two-step transformation *in vivo* back to the inactive metabolite via known processes of enzymic deactivation [54,55].

In this case, unlike prednisolone a metabolically labile ester function occupies the 17β-position. The ester is hydrolyzed to an inactive carboxylic acid, Δ^1-cortienic acid etabonate and then into the lead compound Δ^1cortienic acid in the biological system.

Loteprednol etabonate has good local activity and high therapeutic index, being devoid of systemic side effects. As mentioned above, loteprednol etabonate is a soft corticosteroid that is rapidly deactivated after reaching the general circulation. Loteprednol etabonate was co-developed by Bausch and Lomb and Pharmos, and launched in 1998 as an ophthalmic suspension for the treatment of allergic conjunctivitis and postoperative ocular inflammation following ocular surgery.

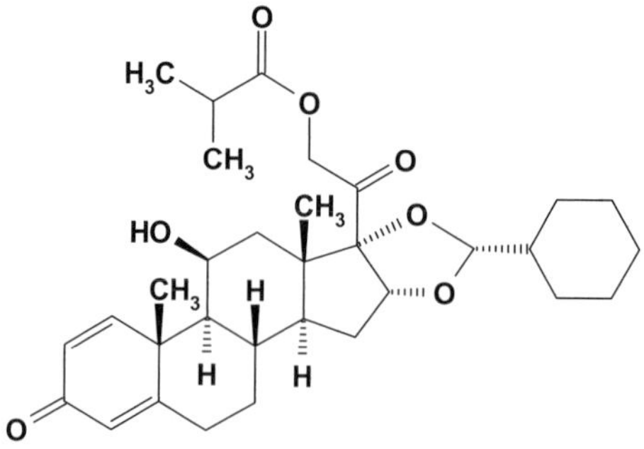

Fig. 19.16 Ciclesonide.

19.2.16
Ciclesonide

The ester prodrug ciclesonide is a new-generation inhaled nonhalogenated glucocorticoid with high local anti-inflammatory properties developed by Altana
Pharma. The advantages of on-site activation include targeted activation in the
lung, minimal systemic adverse effects, and minimal oropharyngeal side effects
[56].

Ciclesonide is an agent that is inactive until it reaches its target site – the lung
– where it is converted to its active metabolite des-isobutyryl-ciclesonide, catalyzed
by esterases in the lung, whereas further metabolism of des-isobutyryl-ciclesonide
to inactive metabolites occurs in the liver and is catalyzed by cytochrome P450
enzymes, especially by CYP 3A4.

Ciclesonide, which is licensed for the prophylactic treatment of persistent
asthma in adults, has a number of pharmacological properties which allow it to be
taken only once a day. According to Philip Ind (Hammersmith Hospital), "... Ciclesonide's high lipid solubility, high protein binding, high lung deposition and
long-long retention make once-daily dosing a reality".

Ind also explained that, since ciclesonide is activated in the lung, it has an
improved systemic safety profile compared with other corticosteroids, and
patients should be less likely to suffer from local side effects.

As mentioned above, ciclesonide is a nonhalogenated glucocorticoid-ester
which exists as (R)- and (S)-epimers, each with different receptor affinities [57].
The (R)-epimer, which has a 5.2-fold greater affinity for the glucocorticoid receptor
than the (S)-epimer, was the sole isomer selected for clinical development.

The epimeric stability by ciclesonide has been demonstrated in healthy volunteers, in which there is little conversion of the (R)-epimer to the less potent
(S)-epimer. This suggests that drug deposited in the lungs is the active metabolite,
and that little or none of it is converted to the inactive (S)-epimer or its inactive
metabolites [58,59].

19.3
Summary

We can say that, in the series of corticosteroids, the pioneer molecule is cortisone and that it owes its biological activity to enzymic conversion to cortisol *in vivo*.

The initial breakthrough came with the introduction of a 1,2-double bond in the A-ring of the steroid nucleus, giving prednisone and prednisolone which have enhanced activity (three- to five-times that of cortisone) and reduced mineralocorticosteroid activity at the projected therapeutic dose.

Following observations by Fried and Sabo that the introduction of C-9 halo-substituents (particularly fluoro) into the steroid nucleus markedly increased glycogenic anti-inflammatory and mineralocorticoid potency, a considerable effort has been devoted to the synthesis of cortical hormone analogues bearing substituents at other sites of the molecule, in the hope of obtaining high anti-inflammatory activity without undesirable sodium retention. This search has led to the preparation, among others, of 6α-methyl, 6α-fluoro, 16α-hydroxy, 16α-methyl, 16β-methyl hormone analogues with the 6- and 16-substituted compounds appearing to be of particular interest.

Introduction of the 6α-methyl group into prednisolone produced a compound which is more active than prednisolone, and with a better therapeutic index with respect to salt retention.

Thus, whilst the 6α-methyl, -chloro and -fluoro groups potentiate anti-inflammatory activity and promote sodium excretion, a single modification is incapable of completely overcoming the profound sodium retention induced by the 9α-fluoro atom.

The introduction of a 16α-hydroxy-group into 9α-fluoro-prednisolone leads to a compound which is devoid of sodium retention but has anti-inflammatory activity considerably lower than that of the parent compound. The 16α– and 16β-methyl-9α-fluoro-prednisolones, on the other hand, are more potent than the parent compound yet free of sodium retention. Peculiarly, the conversion of 16α-hydroxy-9α-fluoro-prednisolone to the 16α,17-acetonide markedly increases topical activity, without significantly affecting oral anti-inflammatory activity.

The 9α-fluoro substituent appears to be the single most active potentiating group, yet since the 6α,9α-difluoro corticoids are still not devoid of sodium retention, an important extension was the preparation of 6α,9α-difluoro-16-methyl corticoids and the 6α,9α-difluoro-16α-hydroxy derivatives.

The introduction of a 16α-hydroxyl group into 6α,9-difluoro-prednisolone led to 6α,9-difluoro-16α-hydroxy-prednisolone with an anticipated favorable biological spectrum.

The corresponding 16,17-acetonide exhibited 100-fold the anti-inflammatory activity of hydrocortisone, with no sodium retention.

Introduction of the 16α-methyl group into 6α,9-difluoro-prednisolone led to flumethasone, which is used topically in the treatment of various skin disorders.

The high topical activity of the nonhalogenated budesonide was based on the balance between the lipophilic and hydrophilic properties of the steroid molecule.

The lipophilicity of the molecule was increased by replacement of the 16a,17-hydroxy groups with an acetal group.

A new generation of nonhalogenated glucocorticoids with high local anti-inflammatory properties is represented by the "ester prodrug", ciclesonide. This is inactive until it reaches its target site in the lung, where it is converted to the active metabolite by esterases. The advantages of this approach include targeted activation in the lung and a prolonged duration of action.

A new class of corticosteroids consists of heterocyclic ester derivatives of the 17a-hydroxy function. The most potent topical anti-inflammatory compounds were 17a-heterocyclic esters in the 16a-methyl series, where the 21-substituent was either chloro or fluoro. Exceptional safety through the effective separation of topical from systemic activity has been demonstrated in studies.

In order to further minimalize the systemic side effects, scientists at Glaxo synthesized a series of androstane 17β-carboxylates and 17β-carbothioates. The carbothioates showed the highest topical activity, with the highest activity being found in the fluoromethyl-carbothioate-17-propionate synthetized from flumethasone.

Synthesis of the "soft drug", loteprednol etabonate, was achieved by starting with a known inactive metabolite of a known active drug. The inactive metabolite was then modified to an active form which, after having achieved its therapeutic role, was subsequently metabolized to an inactive species.

It has taken a vast synthetic effort over a number of years to exploit the advantages of steroids as potent and effective anti-inflammatory agents. Much of the success that has been achieved has resulted from significant and innovative chemical transformations that allowed both regio- and stereoselective modifications within the steroidal skeleton to be accomplished. As a result of this research, several useful analogues of cortisone have been discovered and developed.

References

1 Hogg JA. Steroids, 1992, 57, 593–616.
2 Hench PS, Kendall EC, Slocumb CH, Polley HF. Proc. Staff Meet. Mayo Clin., 1949, 24, 181–197.
3 Maisel AQ. The Hormone Quest. Random House, 1965, Chapter 7.
4 Bush JE, Sandberg AA. J. Biol. Chem., 1953, 205, 783–793.
5 Bush JE. Experientia, 1956, 12, 325–331.
6 Wendler NL, Graber RP, Jones RE, Tishler M. J. Am. Chem. Soc., 1950, 72, 5793–5794.
7 Wendler NL, Huang Minlon, Tishler M. J. Am. Chem. Soc., 1951, 73, 3818–3820.

8 Sulzberger MB, Sauer GC, Herman F, Baer RL, Milberg IL. J. Invest. Dermatol., 1951, 16, 323–327.
9 Sulzberger MB, Vitten VH. J. Invest. Dermatol., 1952, 19, 101–102.
10 Fried J, Sabo EF. J. Am. Chem. Soc., 1953, 75, 2273–2274.
11 Fried J, Sabo EF. J. Am. Chem. Soc., 1953, 76, 1455–1456.
12 Fried J, Thoma RW, Gerke JR, Herz JE, Donin MN, Perlman D. J. Am. Chem. Soc., 1952, 74, 3962–3963.
13 Fried J., et al. Recent Prog. Hormone Res., 1955, 11, 149.
14 Liddle GW. Ann. N.Y. Acad. Sci., 1959, 82, 854.

15 Allen WS, Bernstein S. *J. Am. Chem. Soc.*, **1956**, 78, 1909.

16 Bernstein S, Lenhard RH, Allen WS, Heller M, Litter L, Sholar M, Feldman LI, Blank RH. *J. Am. Chem. Soc.*, **1956**, 78, 5693.

17 Caspi E, Levy H, Hechter OM. *Arch. Biochem. Biophys.*, **1953**, 45, 169.

18 Glenn EM, Stafford RO, Lyster SC, Bowman BJ. *Endocrinology*, **1957**, 61, 128.

19 Burstein S, Savard K, Dorfman RJ. *Endocrinology*, **1953**, 52, 448.

20 Arth GE, Fried J, Johnston DBR, Hoff DR, Sarett LH, Stoerk HC, Winters CA. *J. Am. Chem. Soc.*, **1958**, 80, 3161–3163.

21 McKenzie AW, Stoughton RB. *Arch. Dermatol.*, **1962**, 86, 608–610.

22 Elks J, Bailey EJ. US Patent 3,376,193, **1968**.

23 Fried J, Sabo EF. *J. Am. Chem. Soc.*, **1953**, 75, 2273.

24 Herzog HL, Nobile A, Tolksdorf S, Charney W, Hershberg EB, Perlman PL, Pechet MM. *Science*, **1955**, 121, 176.

25 Hogg JA, Lincoln RW, Jackson RW, Schneider WP. *J. Am. Chem. Soc.*, **1955**, 77, 6401.

26 Ely RS, Done AK, Kellery VC. *Proc. Soc. Exp. Biol. Med.*, **1956**, 91, 503.

27 Spero GB, Thompson JL, Magerlein BJ, Hanze AR, Murray HC, Sebek OK, Hogg JA. *J. Am. Chem. Soc.*, **1956**, 78, 6213–6214.

28 Hogg JA, Spero GB, Thomson JL, Magerlein BJ, Schneider WP, Peterson DH, Campbell JA. *Chem. Ind.*, **1958**, 1002.

29 Edwards JA, Ringold HJ, Djerassi C. *J. Am. Chem. Soc.*, **1959**, 81, 3156.

30 Bowers A, Denot E, Sánchez MB, Ringold HJ. *Tetrahedron*, **1959**, 7, 153 (1959)

31 Schneider WP, Lincoln FH, Spero GB, Murray HC, Thomson JL. *J. Am. Chem. Soc.*, **1959**, 81, 3167.

32 Tasaka K. *Drugs Today*, **1986**, 22(3), 101–133.

33 Jain P, Golish JA. *Drugs*, **1996**, 52(6), S1–S11.

34 Hollmann H, Derendorf H, Barth H, Meibohm B, Wagner M, Krieg M, Weisser H, Knoller J, Mollman A,

Hockhaus G. *J. Clin. Pharmacol.*, **1997**, 37, 893–903.

35 Corren J, Tashkin DP. *Clin. Ther.*, **2003**, 25, 776–798.

36 Chen-Waldon H. *Z. Erkr. Atmungsorgane*, **1981**, 157(3), 287–290.

37 Edwards JA, Ringold HJ, Djerassi C. *J. Am. Chem. Soc.*, **1960**, 82, 2318–2322.

38 Popper TL, Watnick AS. In: Anti-inflammatory agents, Vol. 1. Scherrer RA, Whitehouse MW (Eds.). Academic Press, New York, **1974**, pp. 245–294.

39 Schlagel CA. *J. Pharm. Sci.*, **1965**, 54, 335.

40 Brattsand RL, Thuressonat A, Ekenstam B, Claeson KG, Thalèn A. U.S. Patent 3,929,768, **1975**.

41 Kaldova J, Anner G. Ciba-Geigy AG; CH 631,185.

42 Kaldova J, Anner G. Ciba-Geigy AG; CH 632,000; DE 2,743,069.

43 Robinson CP. *Drugs Today*, **1991**, 27(5), 304–311.

44 Shapiro EL, Genthes MJ, Tiberi R, Popper T, Bechenkopf J, Lutsky B, Watnick AS. *J. Med. Chem.*, **1987**, 30, 1581–1588.

45 Shapiro EL. US Patent 4,472,393.

46 Shapiro EL, Gentles MJ, Tiberi RL, Popper TL, Berkenkopf J, Lutsky B, Watnick AS. *J. Med. Chem.*, **1982**, 30, 1068.

47 Shapiro EL. US Patent 4,472,393.

48 Shapiro EL, Gentles MJ, Titeri RL, Popper TL, Bergenkopf J, Lutsky B, Watnick AS. *J. Med. Chem.*, **1987**, 30, 1068–1073.

49 Philips GH, Bain BH, Williamson C, Steeples JP, Laing SB. (Glaxo Group Ltd.); GB 2,088,877, NL 8,100,707.

50 Phillip GH. *Resp. Med.*, **1990**, 84 (Suppl. A), 19–23.

51 Edwards JA. Syntex USA Inc., US Patent 4,187,301.

52 Bodor N. A novel approach to safer drugs. Cristophers E, et al. (Eds.). Raven Press, New York, **1988**, pp. 13–25.

53 Bodor NS. Otsuka Pharm. Co. Ltd.; BE 0,889,563.

54 Leibowitz HM, Kupferman A, Ryan WJ, Reares TA, Howes J. *Invest. Ophthalmol. Visual Sci.*, **1991**, 32(4), Abs. 346.

55 Bodor N, Murakami T, Wu WM. *Pharm. Res.*, **1995**, 12, 869–874.

56 Dietzel K, Engelstatter R, Keller A. *Prog. Respir. Res.*, **2001**, 31, 91–93.

57 Nave R, Drollmann A, Mayer W, Zech K. *Am. J. Respir. Crit. Care Med.*, **2003**, 167, A771.

58 Calatayud J, Conde JR, Luna M. (Byk Elmu SA); BE 1,005,876, CH 683,343, DE 4,129,535, GB 2,247,680, JP 199,2257,599, US 5,482,934.

59 Amscler H, Flockerzi D, Gutterer B. (Byk Gulden Lomberg) Chem. F. GmbH; DE 19,635,498, PCT WO 98/09982.

Part III
Table of Selected Analogue Classes

Analogue-based Drug Discovery. IUPAC, Janos Fischer, and C. Robin Ganellin (Eds.)
Copyright © 2006 WILEY-VCH Verlag GmbH & Co. KGaA, Weinheim
ISBN: 3-527-31257-9

Table of Selected Analogue Classes*

Erika M. Alapi and János Fischer

Analogues have been collected from the list of the most frequently used drugs (IMS) which have been classified and completed with the help of ATC system, but in the case of structural analogues, i.e. analogues with biological activities that are different from the lead, a remark (R) indicates that the ATC code is different. For a given class, the analogues are listed in the order of their discovery dates.

Abbreviations

MW   molecular weight (without considering a salt or a solvate)
P   first patent(s) or journal article (year and originator)
L   first launch year (originator)
S   salt(s)
R   referring to a different subclass of ATC in case of a structural analogue and/or other remarks (eg. withdrawal)

---

* This book was carefully produced. Nevertheless, editor and publishers do not warrant the information contained therein to be free of errors. Readers are advised to keep in mind that statements, data, illustrations, procedural details, or other items may inadvertently be inaccurate.

Analogue-based Drug Discovery. IUPAC, Janos Fischer, and C. Robin Ganellin (Eds.)
Copyright © 2006 WILEY-VCH Verlag GmbH & Co. KGaA, Weinheim
ISBN: 3-527-31257-9

A
Alimentary Tract and Metabolism

A02BA
H$_2$-receptor antagonists
(cimetidine, ranitidine, roxatidine, famotidine, nizatidine)

Cimetidine

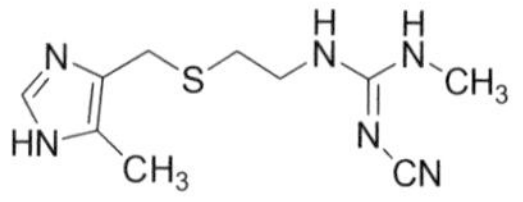

MW: 252.35
P: US 3,950,333
(1971, Smith Kline & French Lab.)
L: 1977 (Smith Kline & French Lab.)
S: –

Ranitidine

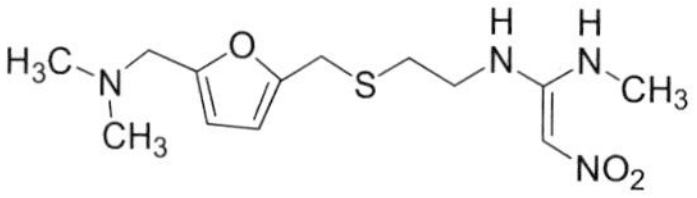

MW: 314.42
P: DE 2,734,070 (1976, Allen & Hanburys Ltd.)
L: 1981 (Glaxo, GSK)
S: HCl, bismuth citrate

Roxatidine

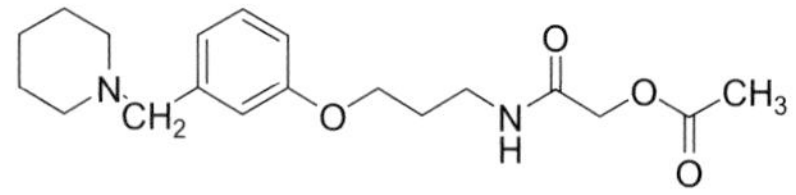

MW: 348.44
P: EP 24,510 (1979, Teikoku Hormone)
L: 1986 (Teikoku Hormone)
S: HCl, acetate

Famotidine

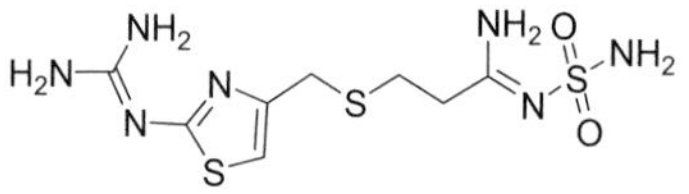

MW: 337.45
P: DE 2,951,675 (1979, Yamanouchi)
L: 1985 (Yamanouchi)
S: –

Nizatidine

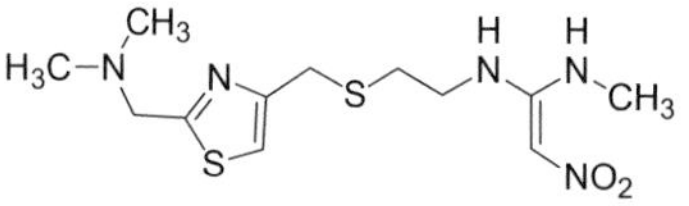

MW: 331.46
P: EP 49,618 (1980, Eli Lilly)
L: 1987 (Eli Lilly)
S: –

A02BC
Proton Pump Inhibitors
(omeprazole, pantoprazole, lansoprazole, rabeprazole, esomeprazole)

Omeprazole

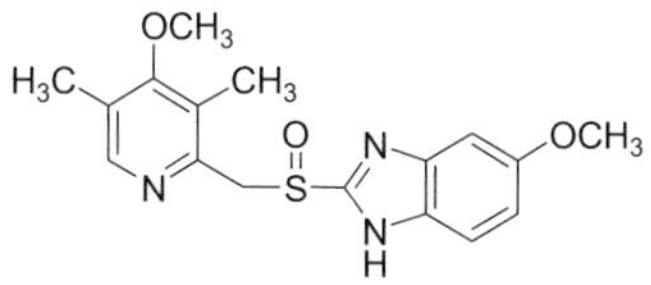

MW: 345.42
  P: EP 5,129 (1978, Hässle)
  L: 1988 (Hässle, AstraZeneca)
  S: ½ Mg, Na

Pantoprazole

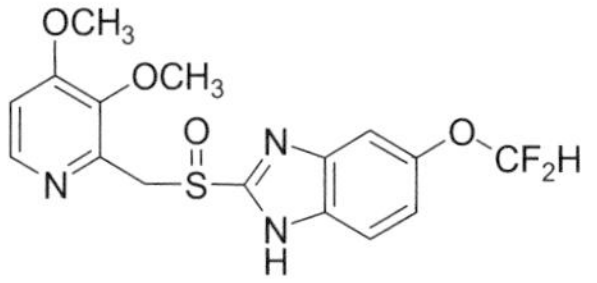

MW: 383.37
  P: EP 166,287 (1984, Byk Gulden Lomberg)
  L: 1994 (Byk Gulden Lomberg, ALTANA
     Pharma)
  S: Na

Lansoprazole

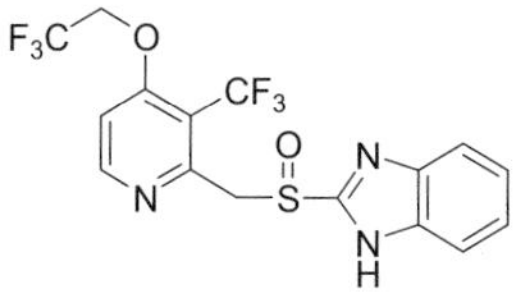

MW: 369.37
  P: EP 174,726 (1984, Takeda)
  L: 1992 (Takeda)
  S: −

Rabeprazole

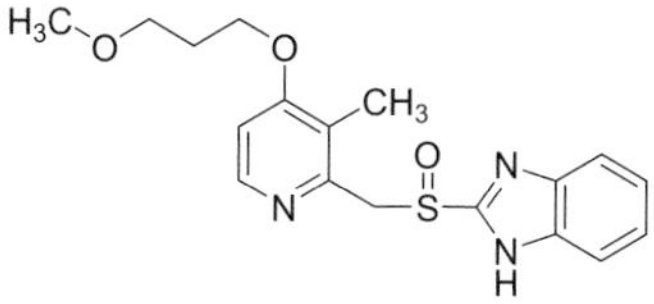

MW: 359.45
  P: EP 268,956 (1986, Eisai)
  L: 1997 (Eisai)
  S: Na

Esomeprazole

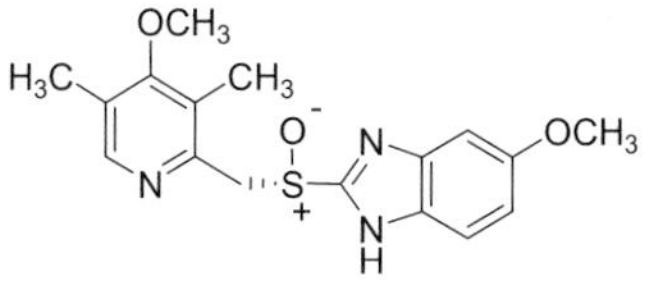

MW: 345.42
  P: WO 94/27988 (1993, AstraZeneca)
  L: 2000 (AstraZeneca)
  S: ½ Mg

A03BB

**Drugs for functional gastrointestinal disorders
(methylscopolamine bromide, butylscopolamine bromide)
and structural analogues
(trospium chloride, ipratropium bromide, oxitropium bromide, tiotropium bromide)**

Methylscopolamine bromide

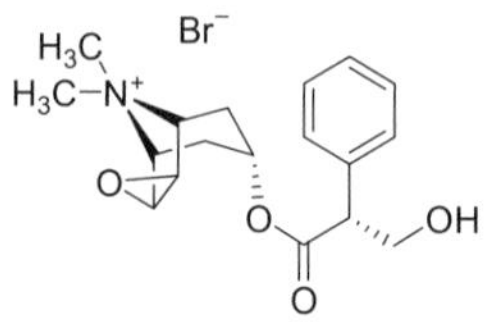

MW: 398.30
P: DE 145,996 (1902, E. Merck AG.)
L: 1947 (Boehringer Ingelheim)
S: bromide

Butylscopolamine bromide

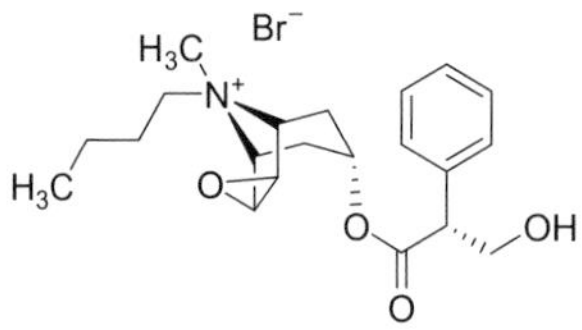

MW: 440.38
P: DE 856,890 (1950, Boehringer Ingelheim)
L: 1951
S: bromide

Trospium chloride

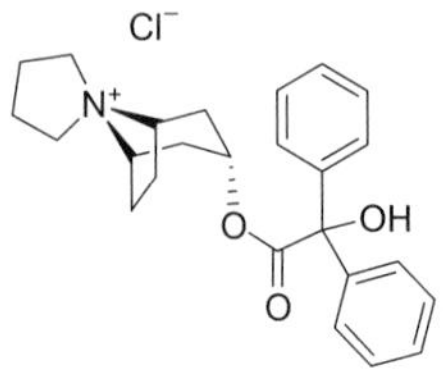

MW: 427.97
P: DE 1,194,422 (1963, Pfleger)
L: 1990 (Madaus)
S: –
R: urinary antispasmodic
(ATC: G04BD)

Ipratropium bromide

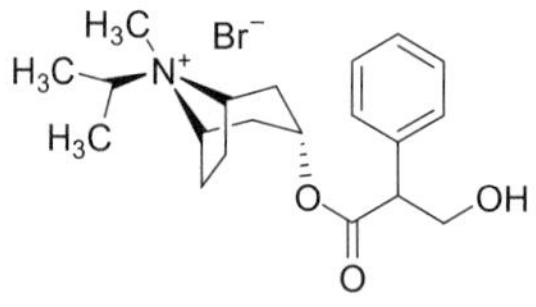

MW: 412.37
P: DE 1,670,177
(1966, Boehringer Ingelheim)
L: 1974 (Boehringer Ingelheim)
S: bromide
R: obstructive airway disease
(ATC: R03BB)

Oxitropium bromide

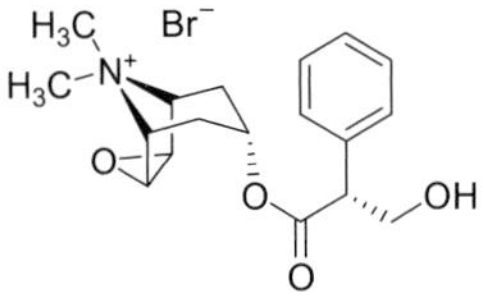

MW: 412.32
- **P:** DE 1,795,818
  (1966, Boehringer Ingelheim)
- **L:** 1983
- **S:** bromide
- **R:** obstructive airway disease
  (ATC: R03BB)

Tiotropium bromide

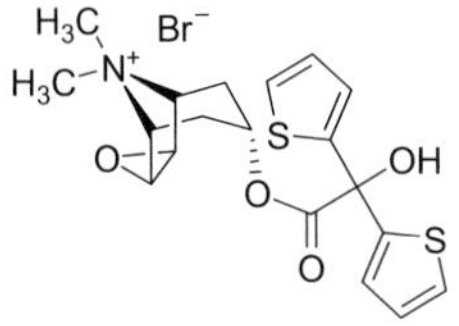

MW: 472.42
- **P:** DE 3,931,041
  (1989, Boehringer Ingelheim)
- **L:** 2002 (Boehringer Ingelheim)
- **S:** bromide
- **R:** obstructive airway disease
  (ATC: R03BB)

A04AA
Serotonin (5HT₃) antagonists
(tropisetron, ondansetron, granisetron, dolasetron, alosetron)

Tropisetron

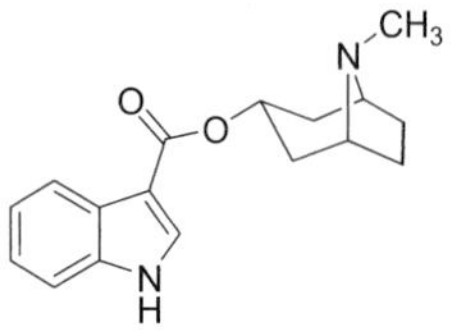

MW: 284.36
P: DE 3,322,574 (1982, Sandoz)
L: 1992 (Novartis)
S: –

Ondansetron

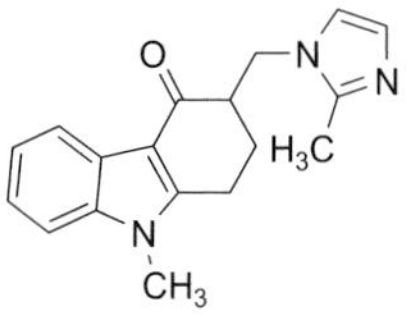

MW: 293.37
P: DE 3,502,508 (1984, Glaxo Group)
L: 1990 (Glaxo, GSK)
S: HCl
R: first marketed drug in the class

Granisetron

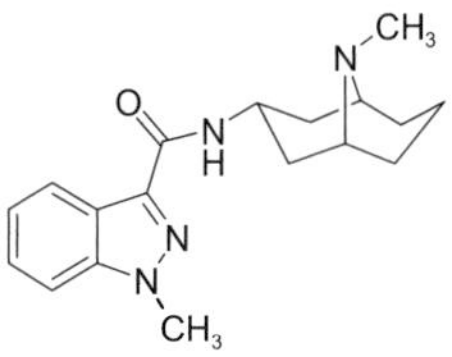

MW: 312.42
P: EP 200,444 (1985, Beecham)
L: 1991 (Hoffmann-La Roche)
S: HCl

Dolasetron

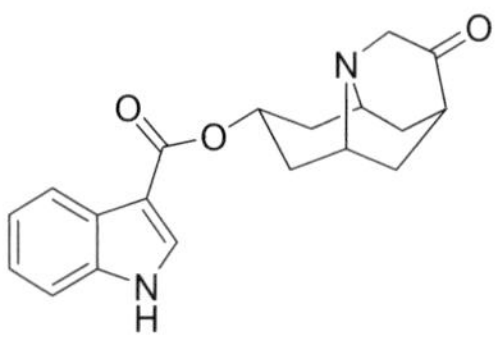

MW: 324.38
P: EP 266,730 (1986, Merrell Dow Phama)
L: 1997 (Aventis Pharma, Sanofi-Aventis)
S: mesylate

Alosetron

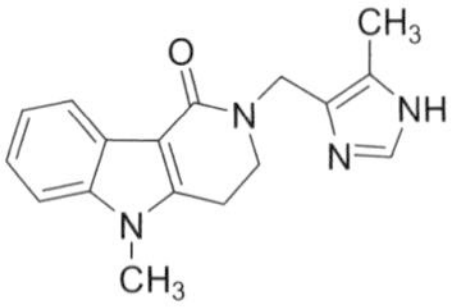

MW: 294.36
P: EP 306,323 (1987, Glaxo)
L: 2002 (GlaxoSmithKline)
S: HCl
R: shortly after the introduction (2000)
withdrawn, then reintroduced in 2002

A10BB
Sulfonamides, urea derivatives
(carbutamide, gliclazide, glipizide, glimepiride)

Carbutamide

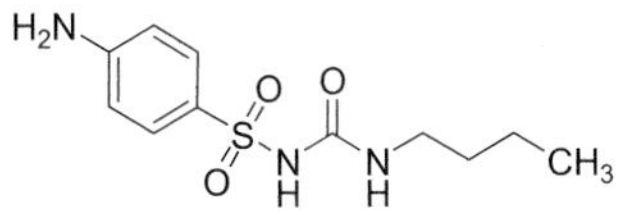

MW: 271.33
P: DE 1,117,103
(1953, Boehringer Mannheim)
L: 1956 (Boehringer Mannheim, Hoffmann-La Roche)
S: –
R: withdrawn

Gliclazide

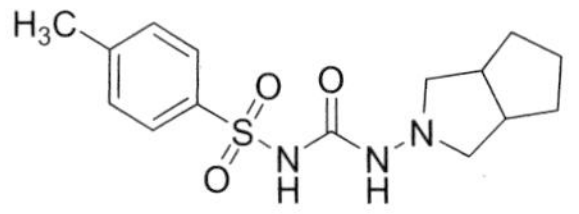

MW: 323.42
P: US 3,501,495 (1966, Science Union et Cie)
L: 1972 (Servier)
S: –

Glipizide

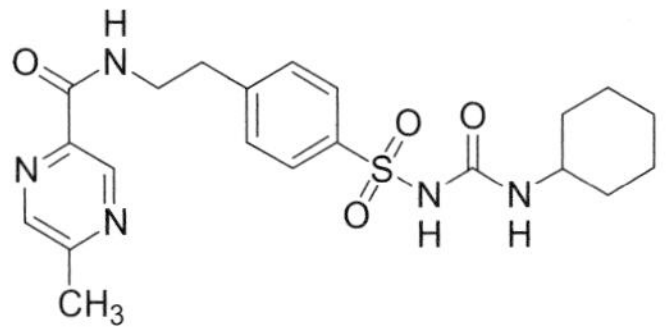

MW: 445.54
P: DE 2,012,138 (1969, Carlo Erba)
L: 1971 (Pfizer)
S: –

Glimepiride

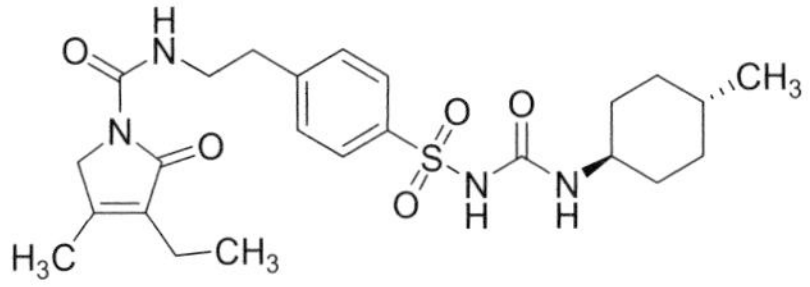

MW: 490.63
P: DE 3,067,390 (1979, Hoechst AG)
L: 1995 (Aventis Pharma, Sanofi-Aventis)
S: –

A10BG
Thiazolidinediones
(troglitazone, pioglitazone, rosiglitazone)

Troglitazone

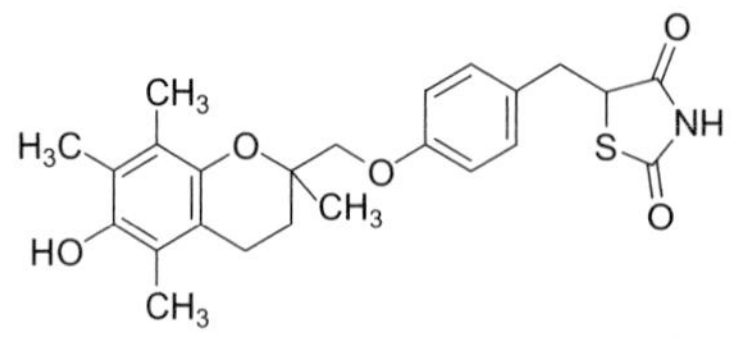

MW: 441.55
  P: EP 139,421 (1983, Sankyo)
  L: 1997 (Sankyo)
  S: –
  R: withdrawn (1997)

Pioglitazone

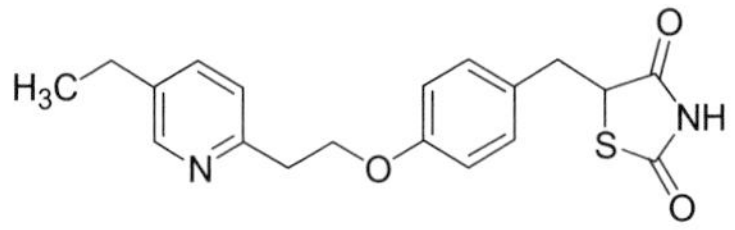

MW: 356.45
  P: EP 193,256 (1985, Takeda)
  L: 1999 (Takeda)
  S: HCl

Rosiglitazone

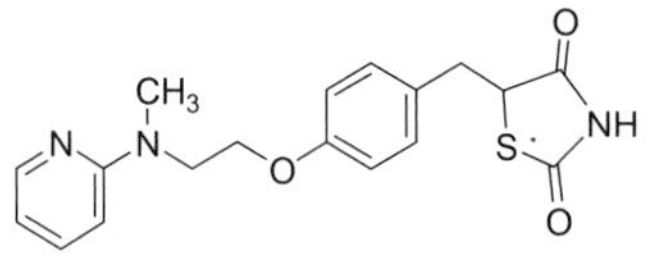

MW: 357.43
  P: EP 306,228 (1987, Beecham)
  L: 1999 (GlaxoSmithKline)
  S: maleate

A11CC
Vitamin D and analogues
(ergocalciferol, cholecalciferol, alfacalcidol, calcitriol, calcipotriol, paricalcitol)

Ergocalciferol (Vitamin D$_2$)

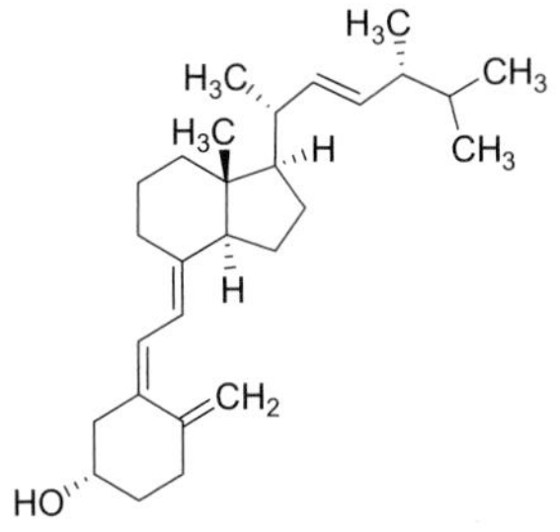

MW: 396.66

   P: A.Windaus et al.: Annalen 1936,
      521, 160–175

   L: –

   R: in multivitamin preparations

Cholecalciferol (Vitamin D$_3$)

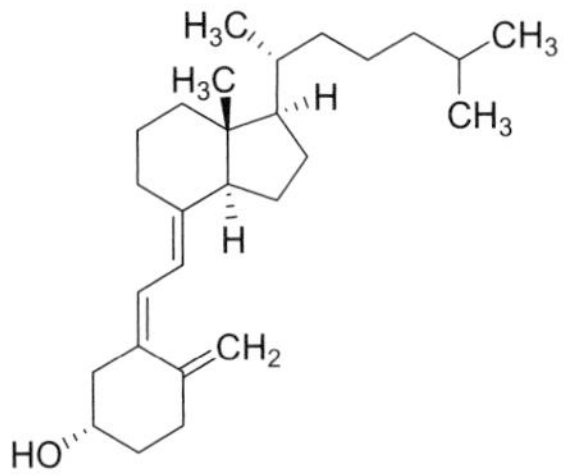

MW: 384.65

   P: Windaus A. et al.: Z. Physiol.
      Chem, 1936, 241, 100

   L: –

   R: in multivitamin compositions

Calcitriol

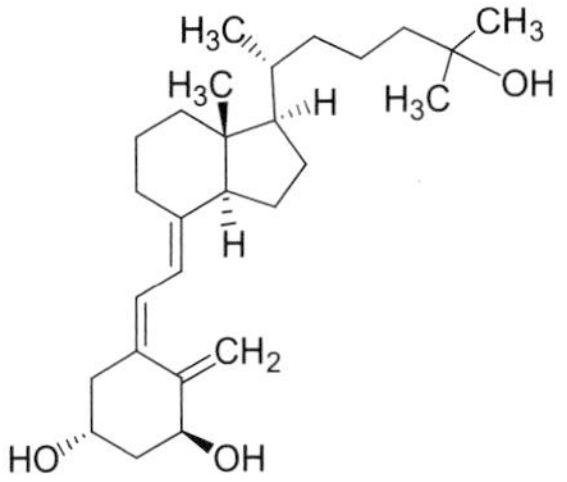

MW: 416.65

   P: Blunt, J. W. et al.: Biochemistry 1968, 7,
      3317,
      US 3,993,675
      (1974, Hoffmann-La Roche)

   L: 1978 (Hoffmann-La Roche)

Alfacalcidol

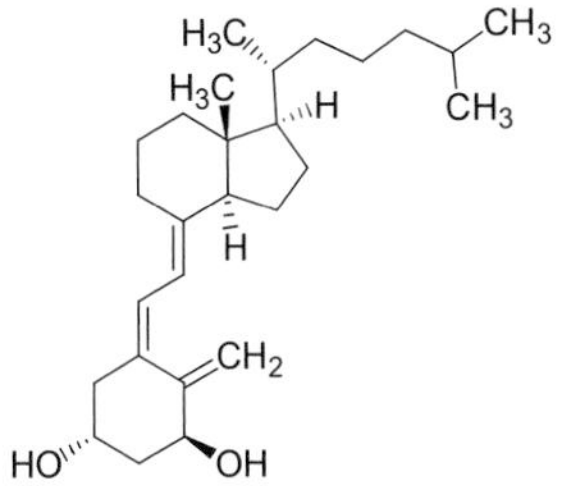

MW: 400.65

   P: US 3,741,996 (1971, Wisconsin Alumni Res.
      Found.)

   L: 1978 (Leo Pharm. Prod.)

Calcipotriol

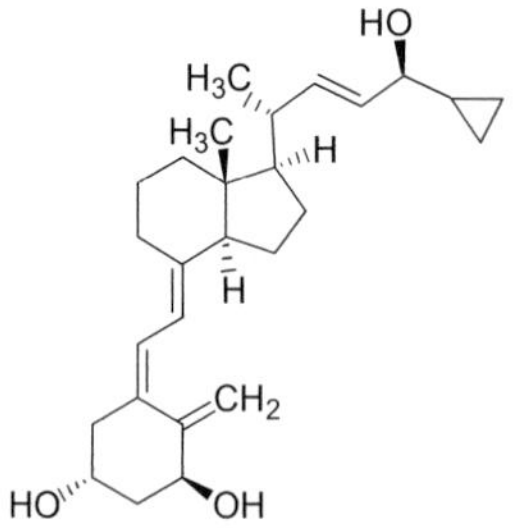

MW: 412.61
P: EP 227,826
    (1985, Leo Pharm. Prod.)
L: 1991 (Leo Pharm. Prod.)

Paricalcitol

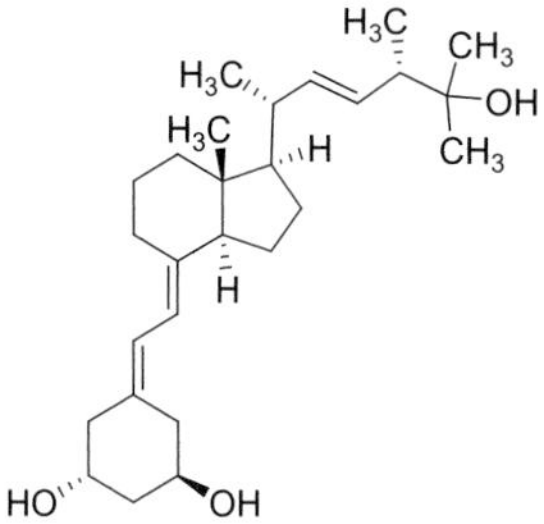

MW: 416.65
P: EP 387,077 (1989, Wisconsin Alumni Res.
    Found.)
L: 1998 (Abbott)

B
Blood

B01AC
Platalet aggregation inhibitors
(ticlopidine, clopidogrel)

Ticlopidine

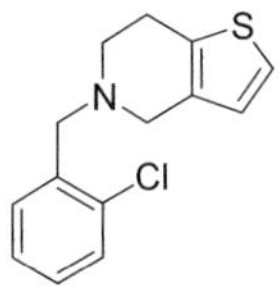

MW: 263.79
P: US 4,051,141 (1973, Centre Etd. Ind. Pharm.)
L: 1978 (Sanofi-Aventis)
S: HCl

Clopidogrel

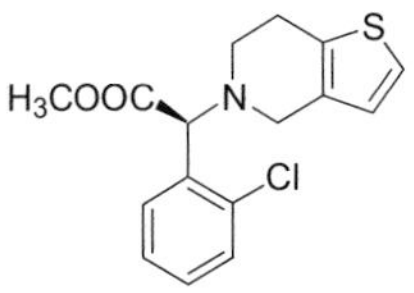

MW: 321.82
P: EP 99,802 (1982, Sanofi)
L: 1998 (Sanofi-Aventis)
S: ½ H_2SO_4

C
Cardiovascular System

C01DA
Organic Nitrates
(isoamyl nitrite, glyceryl trinitrate, isosorbide dinitrate, pentaerythrityl tetranitrate, isosorbide mononitrate)
and structural analogue
(nicorandil)

Isoamyl nitrite

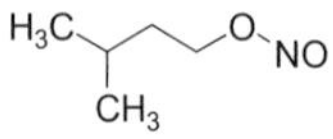

MW: 117.15
P: Balard, M. C.: R. Acad. Sci., 1844, 19, 634,
   Brunton, T. L.: Lancet, 1857, 97
L: 1867

Glyceryl trinitrate

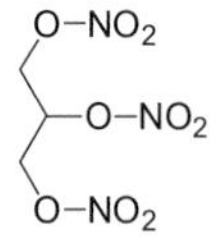

MW: 227.09
P: Sobrero, A.: Mem. R. Acad.,
   Torino 1846, 10, 195,
   Murrell, W.: Lancet, 1879, 80
L: –

Isosorbide dinitrate

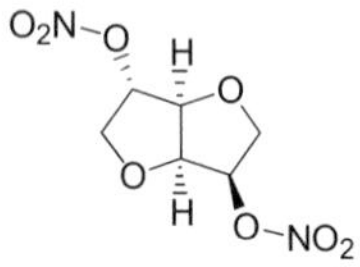

MW: 236.14
P: Krantz, J. C. et al.: J. Pharm. Exp. Ther.
   1939, 67, 187
L: –

Pentaerythrityl tetranitrate

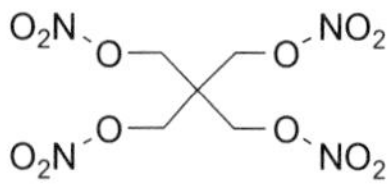

MW: 316.14
P: US 2,370,437 (1943, Du Pont de Nemours)
L: 1946 (Pfizer)

Isosorbide mononitrate

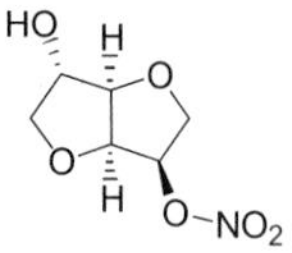

MW: 191.14
P: DE 2,221,080 (1971, American Home
   Products)
L: 1981

Nicorandil

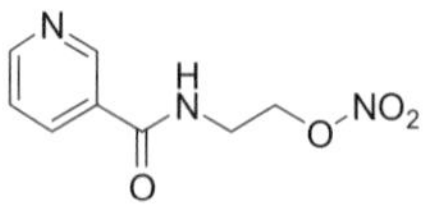

MW: 166.18
P: US 4,200,640 (1976, Chugai)
L: 1983 (Chugai, Hoffmann-La Roche)
S: –
R: other vasodilator ATC: C01DX

C02CA
Alpha-adrenoreceptor antagonists
(prazosin, doxazosin)
other applications (beningn prostatic hypertrophy)
(terazosin, alfuzosin)

Prazosin

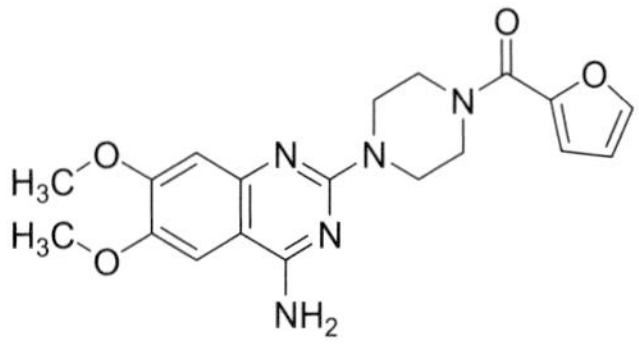

MW: 383.41
P: US 3,511,836 (1965, Pfizer)
L: 1974 (Pfizer)
S: HCl

Terazosin

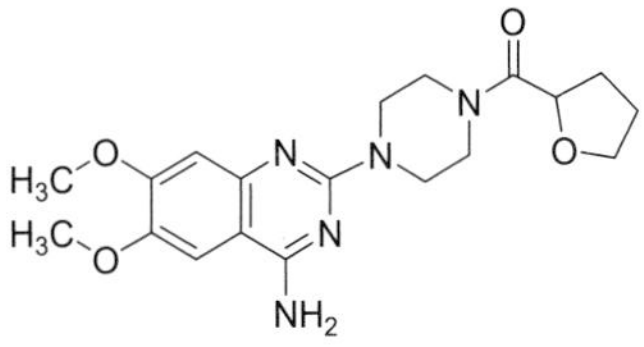

MW: 387.44
P: US 4,026,894 (1975, Abbott)
L: 1985 (Abbott)
S: HCl
R: **benign prostatic hypertrophy** ATC: G04CA

Doxazosin

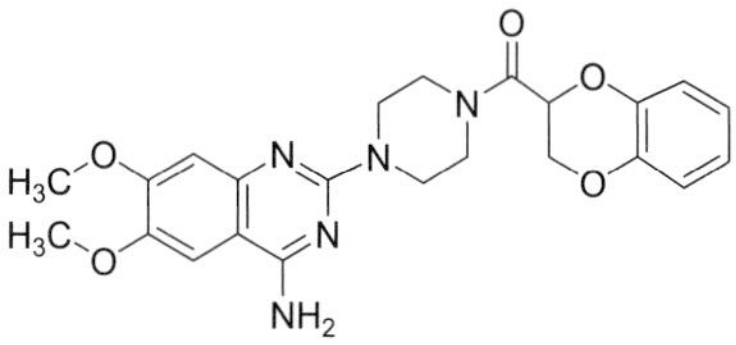

MW: 451.48
P: US 4,188,390 (1977, Pfizer)
L: 1988 (Pfizer)
S: mesylate

Alfuzosin

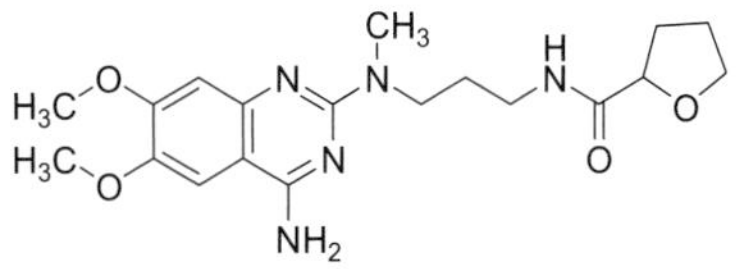

MW: 389.46
P: US 4,315,007 (1978, Synthelabo)
L: 1988 (Sanofi-Synthelabo, Sanofi-Aventis)
S: HCl
R: **benign prostatic hypertrophy** ATC: G04CA, G04CB

C03AA
Thiazides
(chlorothiazide, bendroflumethazide, hydrochlorothiazide)

Chlorothiazide

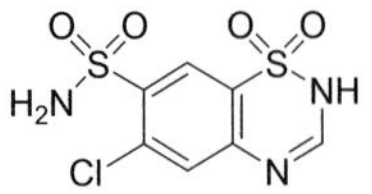

MW: 295.73
  P: US 2,809,194 (1956, Merck & Co.)
  L: 1958 (Merck & Co.)
  S: Na

Bendroflumethiazide

MW: 421.42
  P: US 3,392,168 (1958, Lovens Kemiske Fabrik)
  L: 1960 (Bristol-Myers Squibb)
  S: –

Hydrochlorothiazide

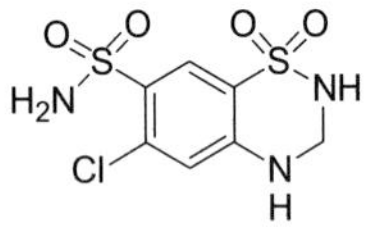

MW: 297.74
  P: US 3,163,645 (1958, Ciba)
  L: 1959 (Ciba, Novartis)
  S: –

C03BA
**Low-ceiling diuretics
(chlortalidone, metolazone, indapamide)**

Chlortalidone

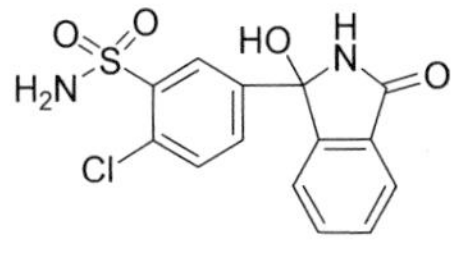

MW: 338.77
P: US 3,055,904 (1957, Geigy)
L: 1960 (Novartis)

Metolazone

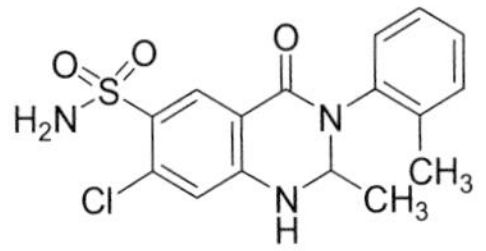

MW: 365.84
P: DE 1,620,740 (1966, Pennwalt)
L: 1974 (Celltech)
S: –

Indapamide

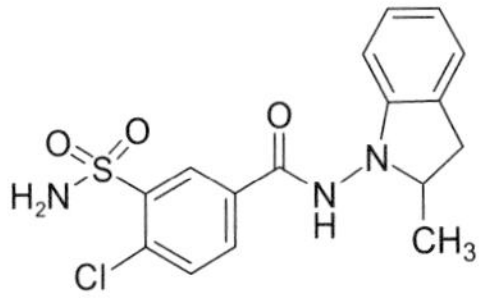

MW: 365.84
P: US 3,565,911 (1968, Science Union et Cie)
L: 1977 (Servier)

C03CA
High-ceiling diuretics, sulfonamide
(furosemide, bumetanide, torasemide, piretanide)

Furosemide

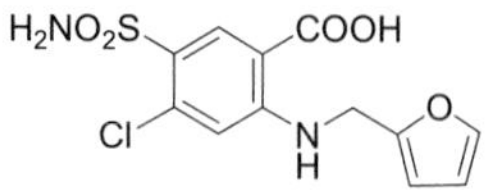

MW: 330.75
  P: US 3,058,882 (1959, Hoechst)
  L: 1964 (Aventis Pharma,
    Sanofi-Aventis)
  S: –

Bumetanide

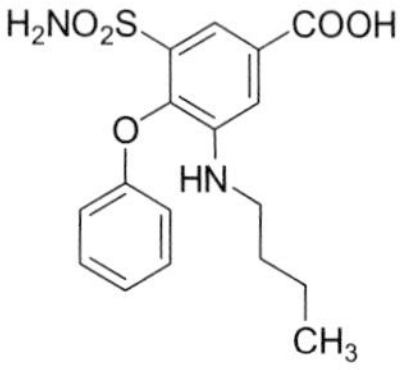

MW: 364.42
  P: US 3,806,534 (1968, Lovens Kem. Fabrik)
  L: 1972 (Hoffmann-La Roche)
  S: –

Torasemide

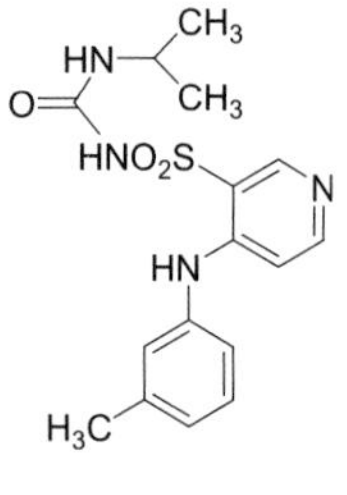

MW: 348.43
  P: US 4,018,929 (1974, A. Christiaens
    Soc. Anonyme)
  L: 1993 (Hoffmann-La Roche)
  S: –

Piretanide

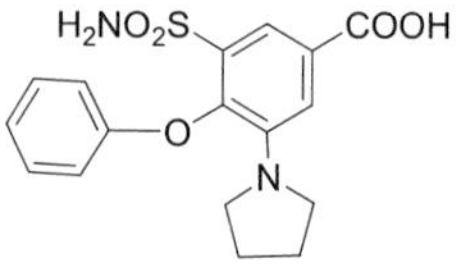

MW: 362.41
  P: DE 2,419,970 (1974, Hoechst)
  L: 1981 (Aventis Pharma, Sanofi-Aventis)
  S: –

C03DA
**Diuretics, aldosterone antagonist
(spironolactone, eplerenone)
and
structural analogue:
(drospirenone)**

Spironolactone

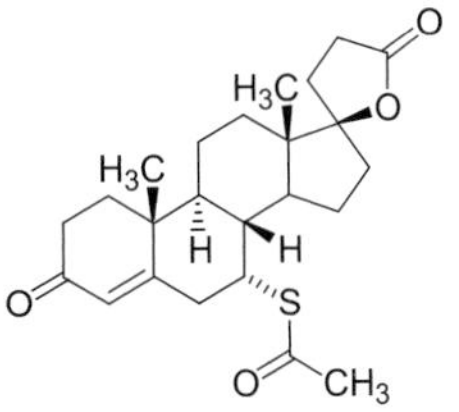

MW: 416.58
  P: DE 1,121,610 (1958, Searle)
  L: 1959 (Pharmacia)

Drospirenone

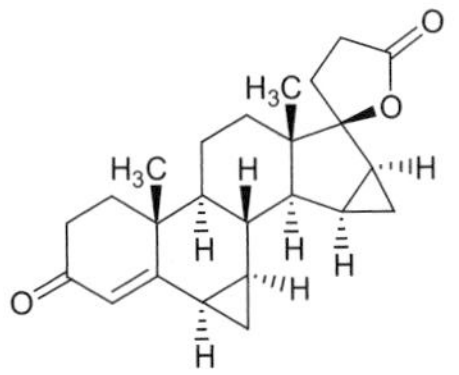

MW: 366.49
  P: DE 2,652,761 (1976, Schering AG)
  L: 2000 (Schering AG)
  R: contraceptive ATC: G03AA

Eplerenone

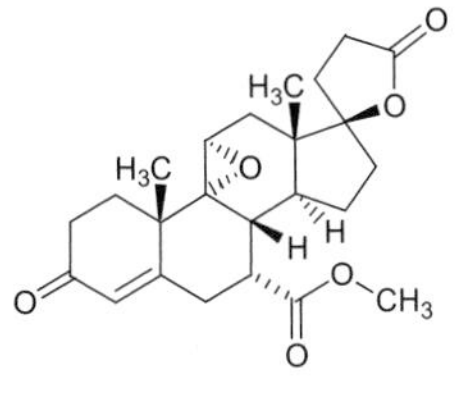

MW: 414.50
  P: EP 122,232 (1983, Ciba-Geigy)
  L: 2004 (Pfizer)

C07AA

Beta Blocking Agents, non-selective
(propranolol, sotalol, timolol, nadolol, carteolol)

Propranolol

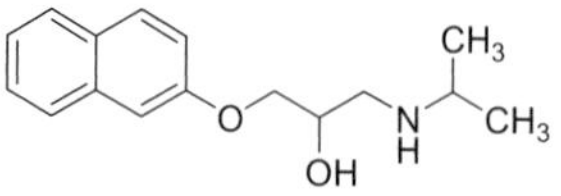

MW: 259.35
 P: US 3,337,628 (1962, Imperial
  Chemical Industries, ICI)
 L: 1964
 S: HCl

Sotalol

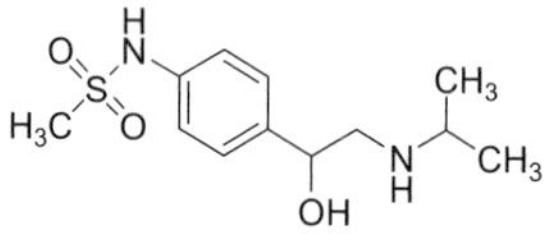

MW: 272.37
 P: Larsen, A. A.: Nature 1964, 203,
  1283
 L: 1974
 S: HCl

Timolol

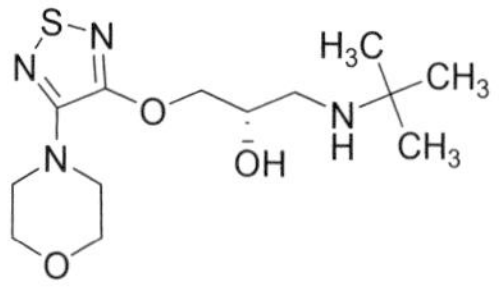

MW: 316.43
 P: US 3,655,663 (1968, C. E. Frosst & Co.)
 L: 1974
 S: maleate
 R: antiglaucoma agent (ATC: S01ED01))

Nadolol

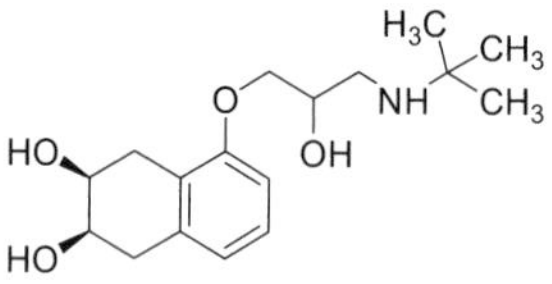

MW: 309.40
 P: US 3,935,267 (1970, Squibb)
 L: 1978 (Bristol-Myers Squibb)
 S: –

Carteolol

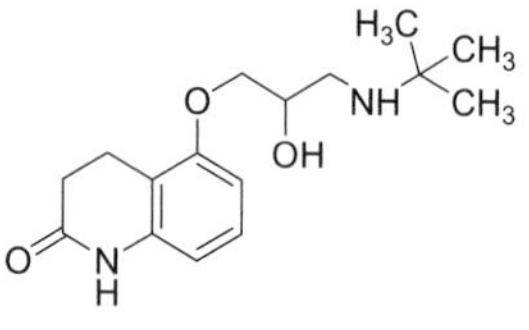

MW: 292.38
 P: US 3,910,924 (1972, Otsuka)
 L: 1980 (Otsuka)
 S: HCl

C07AB
Beta Blocking Agents, selective
(acebutolol, atenolol, metoprolol, celiprolol, betaxolol, bisoprolol, esmolol, nebivolol)

Acebutolol

MW: 336.43
P: GB 1,247,384 (1967, May & Baker)
L: 1973 (Aventis Pharma, Sanofi-Aventis)
S: HCl

Atenolol

MW: 266.34
P: US 3,663,607 (1969, Imperial
   Chemical Industries, ICI)
L: 1975 (ICI, AstraZeneca)
S: –

Metoprolol

MW: 267.37
P: DE 2,106,209 (1970, Hässle)
L: 1975 (Hässle, AstraZeneca)
S: tartrate

Celiprolol

MW: 379.50
P: US 4,034,009 (1973, Chemie Linz)
L: 1982 (Aventis Pharma, Sanofi-Aventis)
S: HCl

Betaxolol

MW: 307.43
P: DE 2,649,605 (1975, Synthelabo)
L: 1983 (Sanofi-Syntélabo, Sanofi-Aventis)
S: HCl

Bisoprolol

MW: 325.45
P: DE 2,645,710 (1976, Merck KGaA)
L: 1986 (Merck KGaA)
S: fumarate

Esmolol

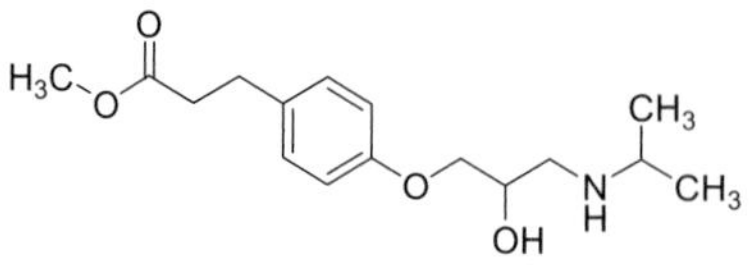

MW: 295.38
P: EP 41,491 (1980, Hässle)
US 4,387,103
(1983, American Critical Care)
L: 1987 (Baxter)
S: HCl

Nebivolol

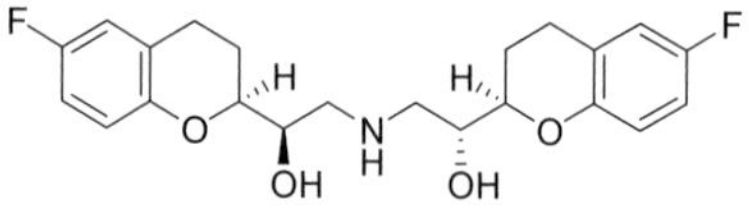

MW: 405.44
P: EP 145,067 (1983, Janssen Pharma)
L: 1997 (Menarini)
S: HCl

C07AG
**Alpha and Beta Blocking Agents
(labetalol, carvedilol)**

Labetalol

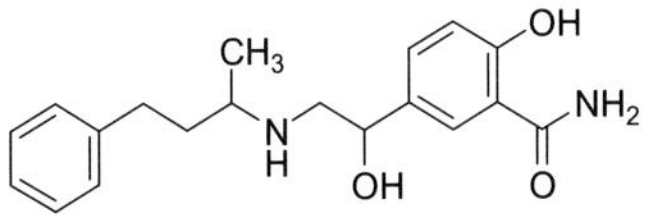

MW: 328.41
P: US 3,705,233 (1966, Allen & Hanburys)
L: 1977 (Genta)
S: HCl

Carvedilol

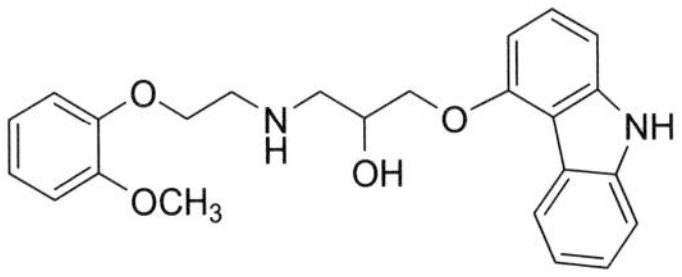

MW: 406.48
P: DE 2,815,926 (1978, Boehringer Mannheim)
L: 1991 (Hoffmann-La Roche)
S: –

C08CA

Dihydropyridine derivatives

(nifedipine, nimodipine, nitrendipine, nicardipine, nisoldipine, felodipine, isradipine, benidipine, manidipine, amlodipine, cilnidipine, lercanidipine, lacidipine)

Nifedipine

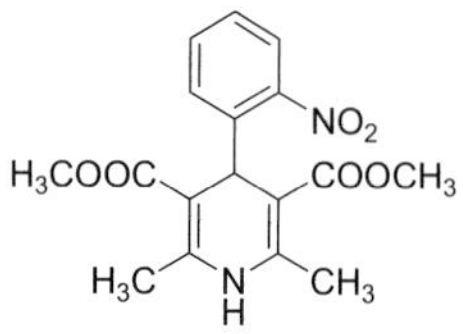

MW: 346.34
P: GB 1,173,862 (1967, Bayer)
L: 1975 (Bayer)

Nimodipine

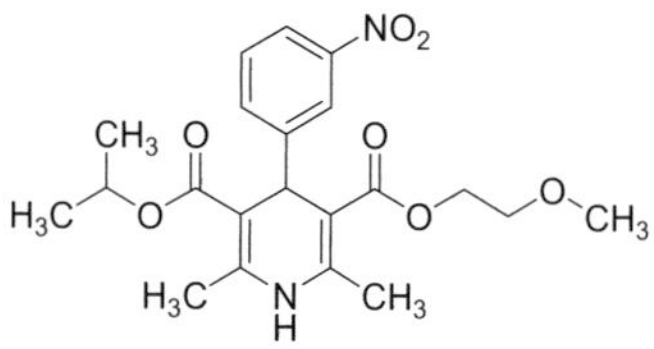

MW: 418.45
P: DE 2,117,573 (1971, Bayer)
L: 1985 (Bayer)

Nitrendipine

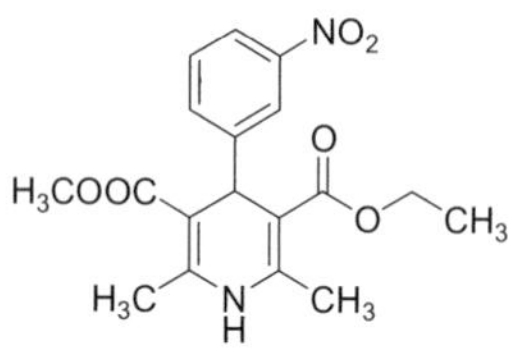

MW: 360.37
P: US 3,799,934 (1971, Bayer)
L: 1985 (Bayer)

Nicardipine

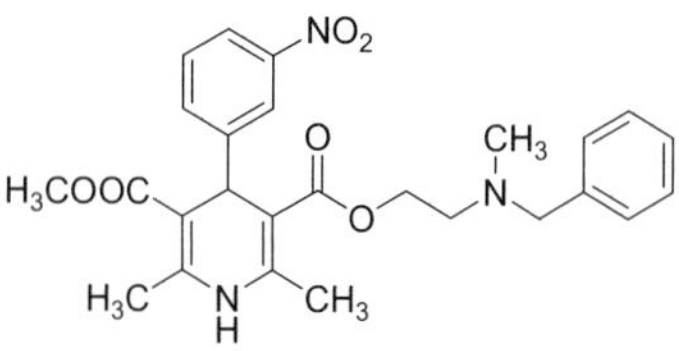

MW: 479.53
P: DE 2,407,115 (1973, Yamanouchi)
L: 1981 (Yamanouchi)
S: HCl

Nisoldipine

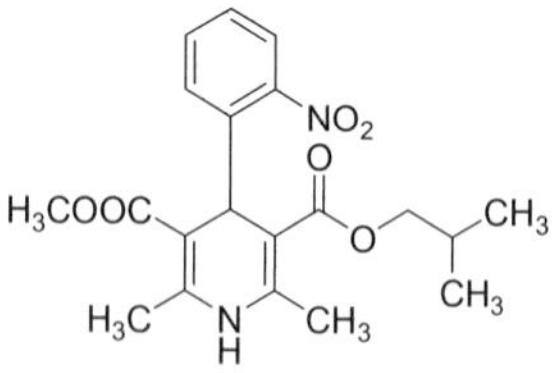

MW: 388.42
P: US 4,154,839 (1975, Bayer)
L: 1990 (Bayer)

Felodipine

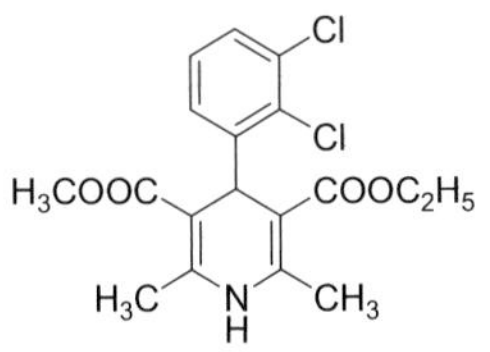

MW: 384.26
  P: EP 7,293 (1978, Hässle)
  L: 1988 (Hässle, AstraZeneca)

Isradipine

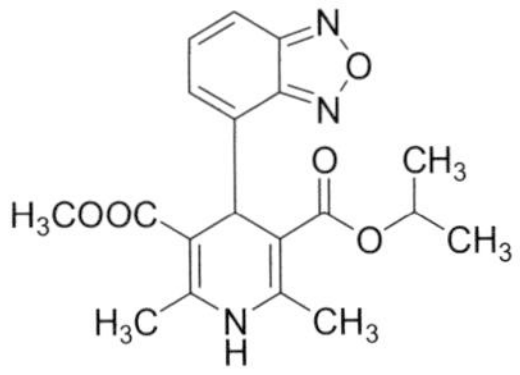

MW: 371.39
  P: DE 2,949,491 (1978, Sandoz)
  L: 1989 (Sandoz, Novartis)

Benidipine

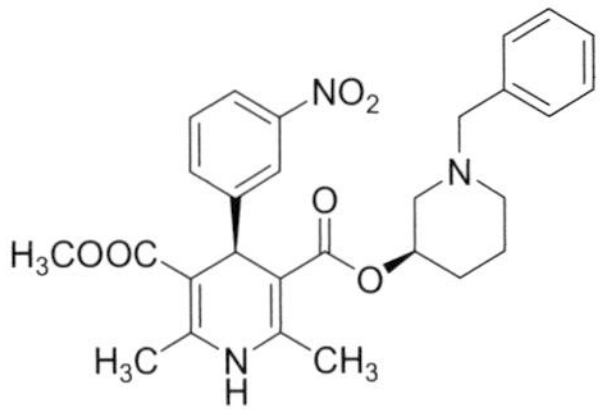

MW: 505.57
  P: EP 63,365 (1981, Kyowa Hakko)
  L: 1991 (Kyowa Hakko)
  S: HCl

Manidipine

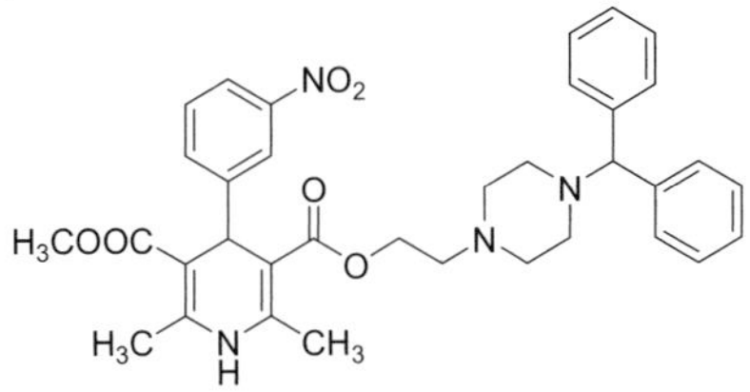

MW: 610.71
  P: EP 94,159 (1982, Takeda)
  L: 1990 (Takeda)
  S: 2 HCl

Amlodipine

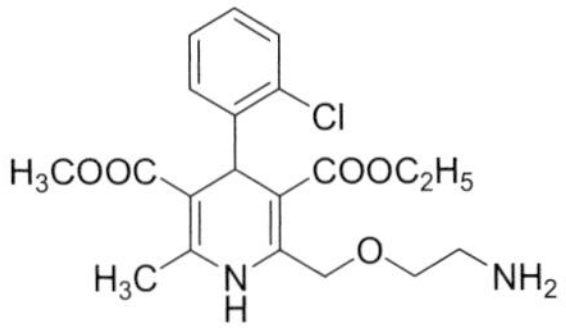

MW: 408.88
  P: EP 89,167 (1982, Pfizer)
  L: 1990 (Pfizer)
  S: besylate

Cilnidipine

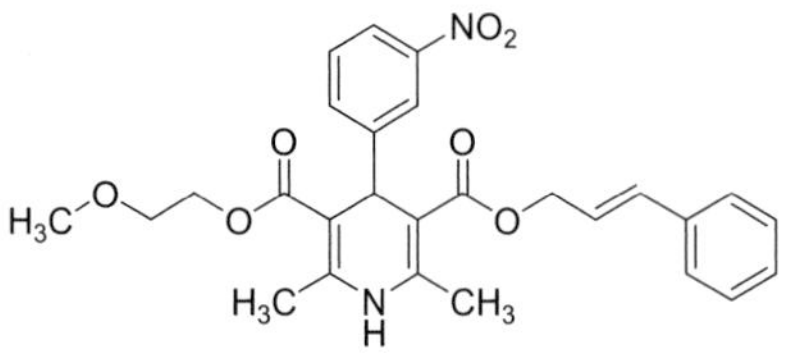

MW: 492.53
  P: EP 161,877 (1984, Fujirebio)
  L: 1995 (Union Chimique Belge, UCB Japan)

Lercanidipine

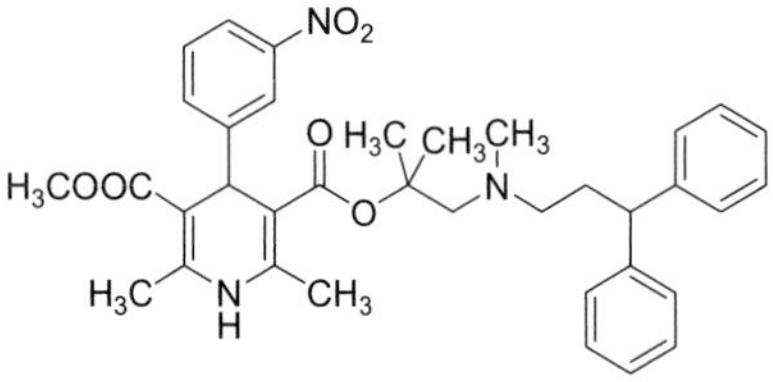

MW: 648.20
  P: EP 153,016
    (1984, Recordati Chem. Pharm.)
  L: 1997 (Recordati Chem. Pharm.)
  S: HCl

Lacidipine

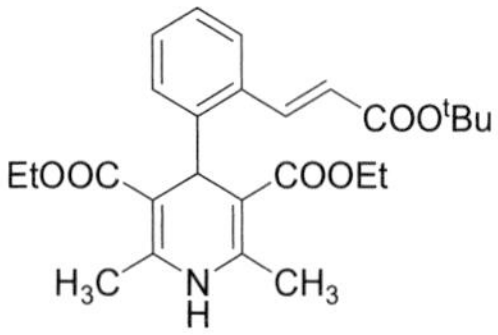

MW: 455.55
  P: DE 3,529,997 (1984, Glaxo)
  L: 1991 (Zambon)

C09AA

ACE inhibitors

(captopril, enalapril, lisinopril, zofenopril, perindopril, quinapril, moexipril, fosinopril, spirapril, benazepril, ramipril, trandolapril, cilazapril, imidapril, temocapril)

Captopril

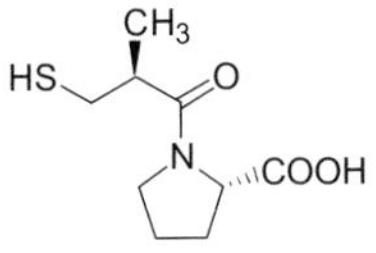

MW: 217.29
  P: DE 2,703,828 (1976, Squibb)
  L: 1980 (Bristol-Myers Squibb)
  S: –

Enalapril

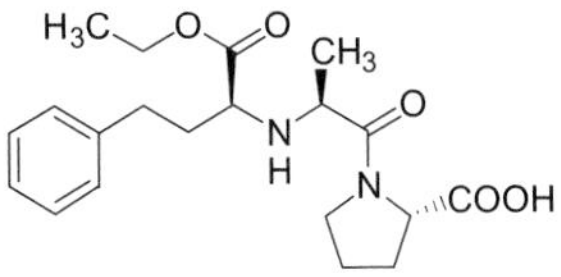

MW: 376.45
  P: US 4,374,829 (1978, Merck & Co.)
  L: 1984 (Merck & Co.)
  S: hydrogen maleate

Lisinopril

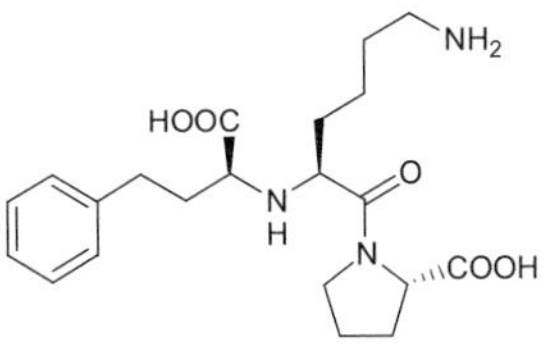

MW: 405.50
  P: US 4,374,829 (1978, Merck & Co.)
  L: 1987 (Merck & Co.)
  S: –

Zofenopril

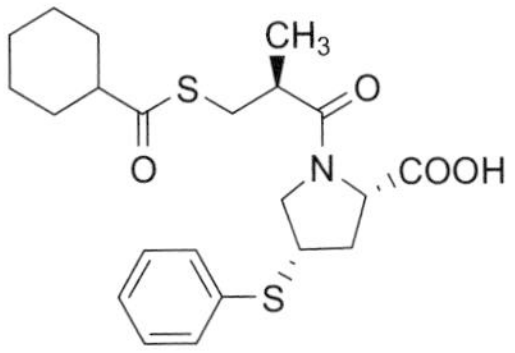

MW: 435.60
  P: US 4,316,906 (1978, Bristol-Myers Squibb)
  L: 2000 (FIRMA, Guidotti, Menarini)
  S: Ca

Perindopril

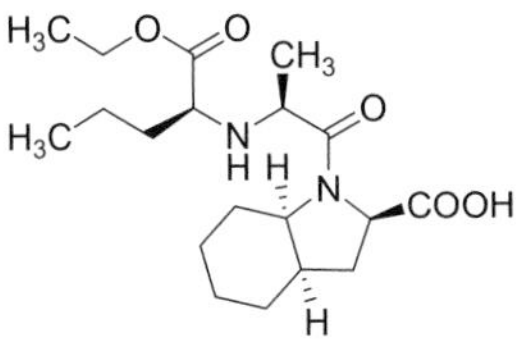

MW: 368.47
  P: EP 49,658 (1980, ADIR)
  L: 1988 (Servier)
  S: erbumine

Quinapril

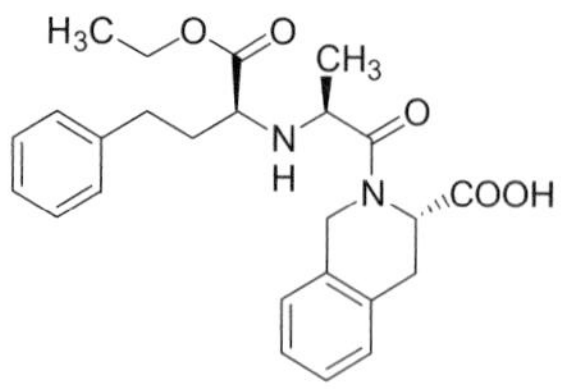

H₃C — O — CH₃ — N H — COOH

MW: 438.52
P: EP 49,605 (1980, Warner-Lambert)
L: 1989 (Pfizer)
S: HCl

Moexipril

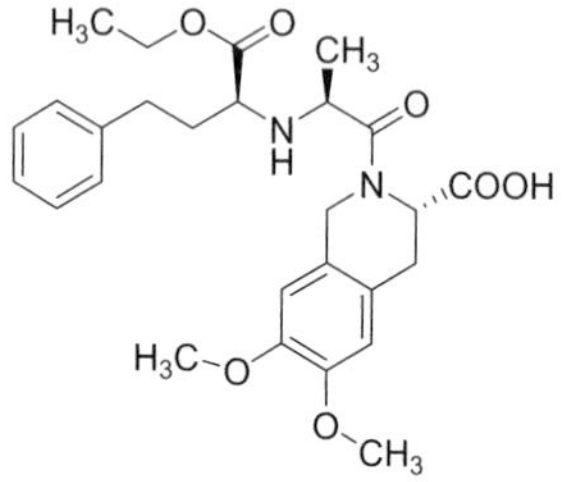

H₃C — O — CH₃ — N H — COOH — H₃C — O — O — CH₃

MW: 498.58
P: EP 49,605 (1980, Warner-Lambert)
L: 1995 (Schwarz)
S: HCl

Fosinopril

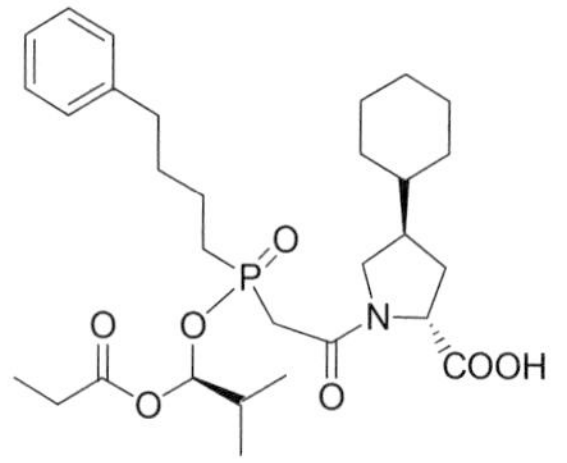

P — O — COOH — O — O

MW: 563.67
P: US 4,337,201 (1980, Squibb)
L: 1991 (Bristol-Myers Squibb)
S: Na

Spirapril

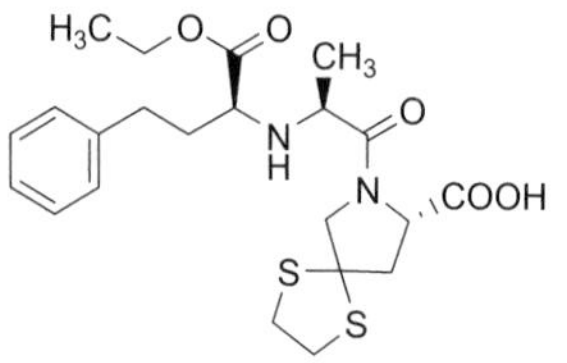

H₃C — O — CH₃ — N H — COOH — S — S

MW: 466.61
P: US 4,470,972 (1980, Schering Corp.)
L: 1995 (Pfizer)
S: HCl

Benazepril

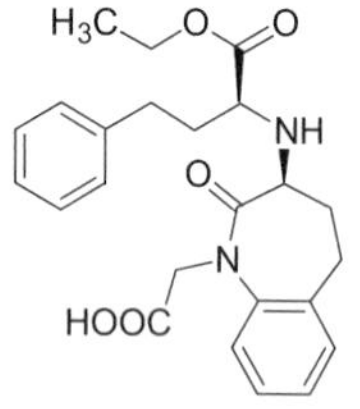

H₃C — O — O — NH — O — N — HOOC

MW: 424.49
P: US 4,410,520 (1981, Ciba-Geigy)
L: 1990 (Novartis)
S: HCl

Ramipril

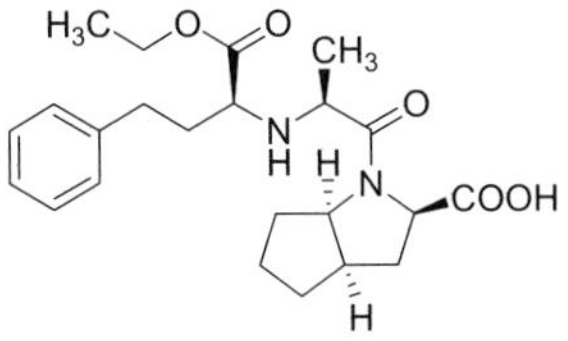

MW: 416.51
P: EP 79,022 (1981, Hoechst)
L: 1989 (Aventis Pharma)
S: –

Trandolapril

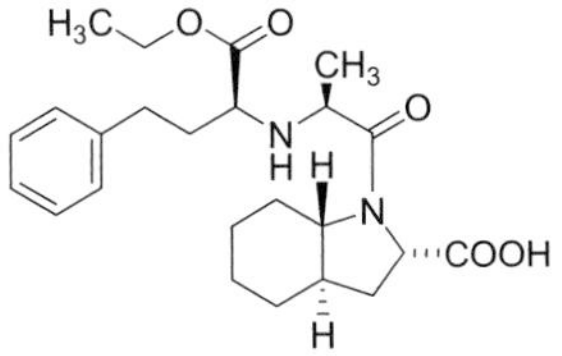

MW: 430.55
P: DE 3,151,690 (1981, Hoechst)
L: 1993 (Abbott)
S: –

Cilazapril

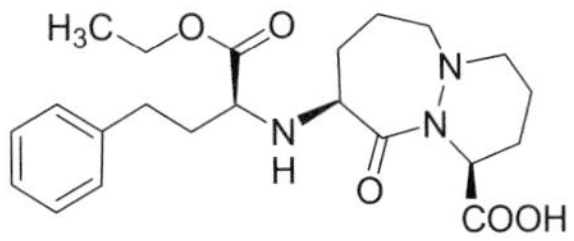

MW: 417.51
P: US 4,512,924 (1982, Hoffmann-La Roche)
L: 1990 (Hoffmann-La Roche)
S: –

Imidapril

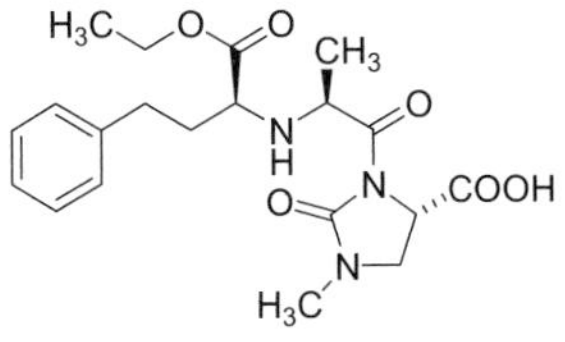

MW: 405.45
P: EP 95,163 (1982, Tanabe Seiyaku)
L: 1993 (Tanabe Seiyaku)
S: HCl

Temocapril

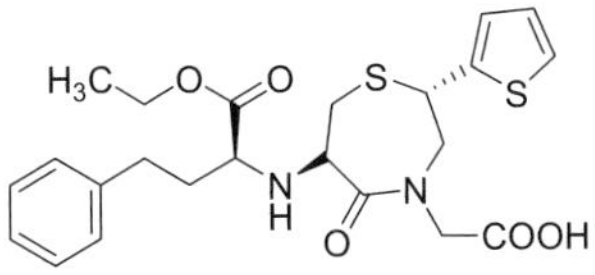

MW: 476.62
P: EP 161,801 (1984, Sankyo)
L: 1994 (Sankyo)
S: HCl

C09CA

Angiotensin II antagonists
(losartan, eprosartan, valsartan, irbesartan, candesartan cilexetil, telmisartan, olmesartan medoxomil)

Losartan

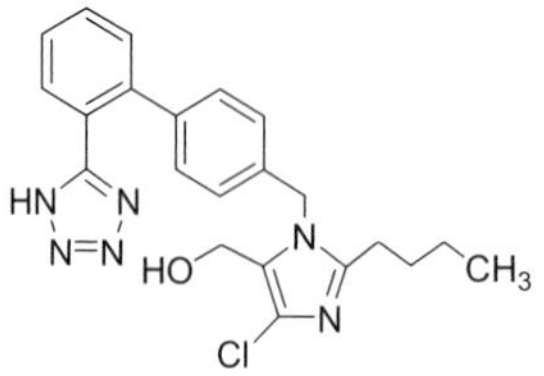

MW: 422.16
  P: EP 253,310 (1986, Du Pont)
  L: 1994 (Merck)
  S: K

Eprosartan

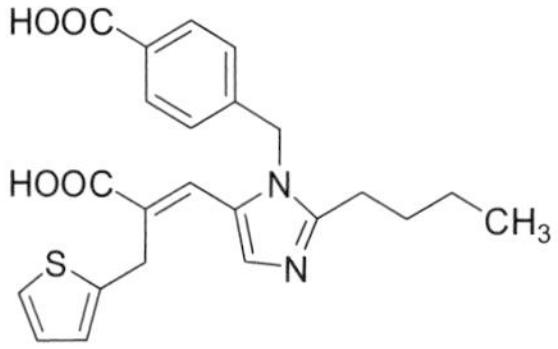

MW: 424.52
  P: EP 403,159 (1989, SmithKline Beecham)
  L: 1997 (GlaxoSmithKline)
  S: mesylate

Valsartan

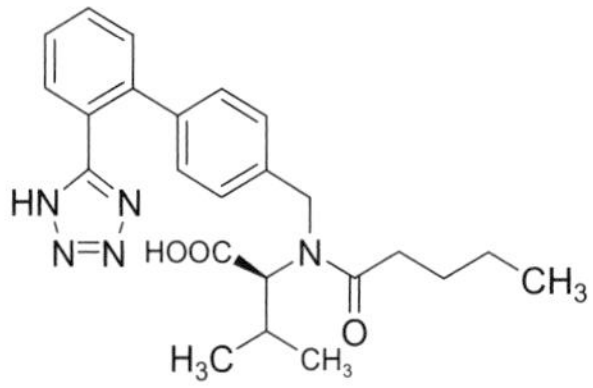

MW: 435.52
  P: EP 443,983 (1990, Ciba-Geigy)
  L: 1996 (Novartis)
  S: –

Irbesartan

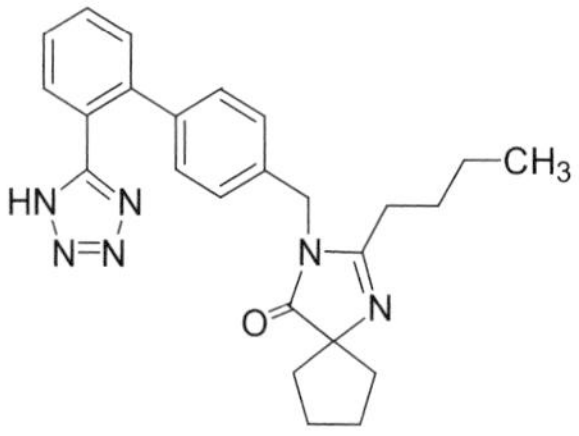

MW: 428.53
  P: EP 454,511 (1990, Sanofi)
  L: 1997 (Sanofi-Synthélabo)
  S: –

Candesartan cilexetil

MW: 610.66
  P: EP 459,136 (1990, Takeda)
  L: 1997 (Takeda)
  S: –

Telmisartan

MW: 514.63
  P: EP 502,314 (1991, Thomae)
  L: 1999 (Boehringer Ingelheim)
  S: –

Olmesartan medoxomil

MW: 558.60
  P: EP 503,785 (1991, Sankyo)
  L: 2002 (Sankyo)
  S: –

C10AA
HMG CoA reductase inhibitors
(lovastatin, simvastatin, pravastatin, fluvastatin, atorvastatin, pitavastatin, rosuvastatin)

Lovastatin

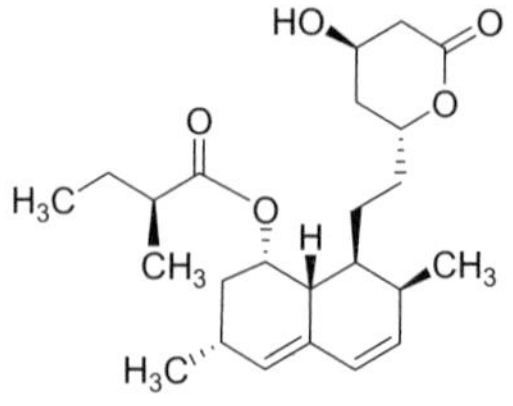

MW: 404.55
P: GB 2,046,737 (1979, Sankyo),
US 4,231,938 (1979, Merck & Co.)
L: 1987 (Merck Sharp & Dohme)
S: –

Simvastatin

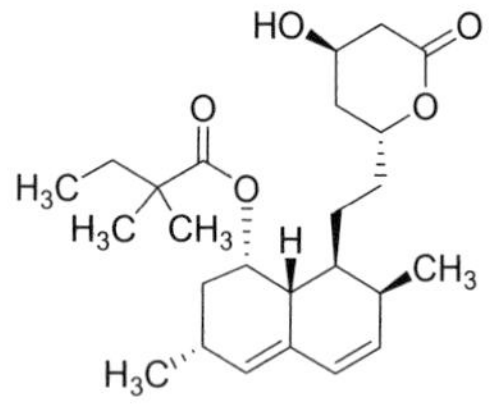

MW: 418.57
P: US 4,444,784 (1980, Merck & Co.)
L: 1988 (Merck)
S: –

Pravastatin

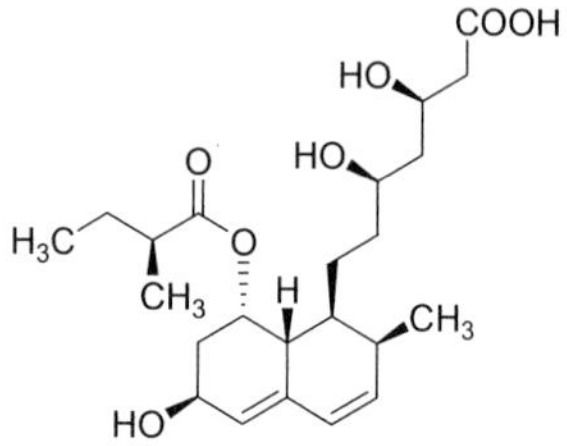

MW: 424.53
P: DE 3,122,499 (1980, Sankyo)
L: 1989 (Sankyo)
S: Na

Fluvastatin

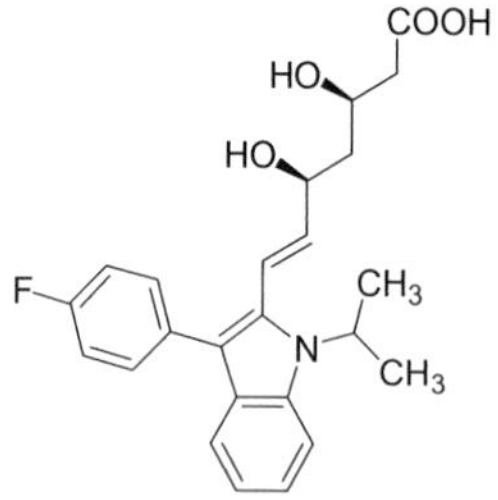

MW: 411.47
P: EP 114,027 (1982, Sandoz)
L: 1994 (Novartis)
S: Na

Atorvastatin

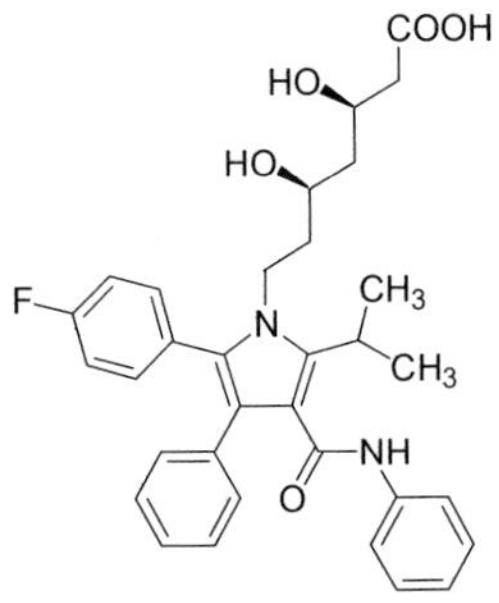

MW: 558.65
   P: US 4,681,893 (1986, Warner-Lambert)
   L: 1997 (Pfizer)
   S: ½ Ca

Pitavastatin

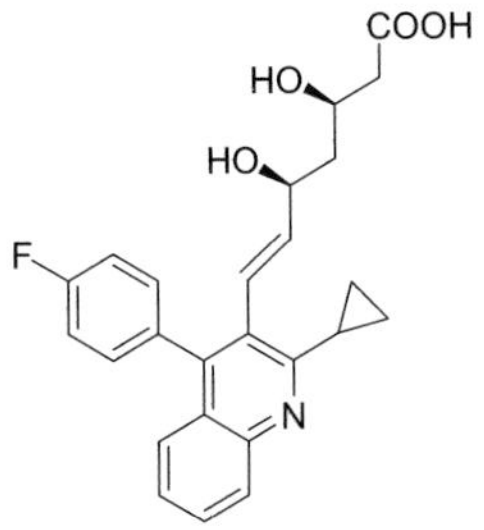

MW: 421.47
   P: US 5,102,888 (1987, Nissan Chemical)
   L: 2003 (Nissan Chemical)
   S: ½ Ca

Rosuvastatin

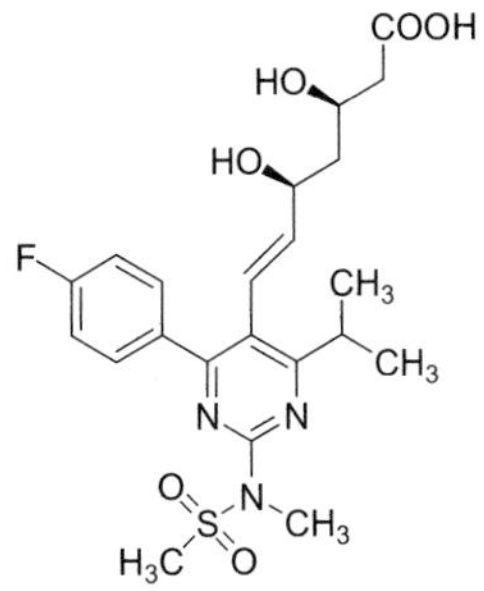

MW: 481.54
   P: EP 521,471 (1991, Shionogi)
   L: 2003 (AstraZeneca)
   S: ½ Ca

C10AB
Fibrates
(clofibrate, gemfibrozil, fenofibrate, bezafibrate, ciprofibrate)

Clofibrate

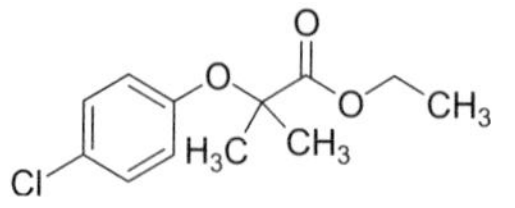

MW: 242.70
P: US 3,262,850 (1958, Imperial
    Chemical Industries, ICI)
L: 1963 (Wyeth Pharmaceuticals)

Gemfibrozil

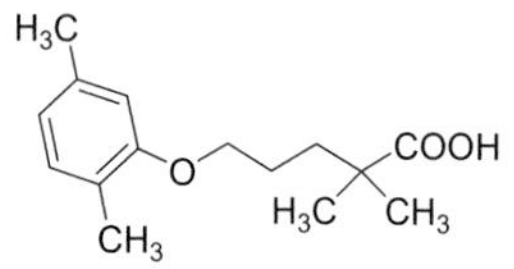

MW: 250.34
P: DE 1,925,423 (1968, Parke Davis)
L: 1982 (Pfizer)
S: –

Fenofibrate

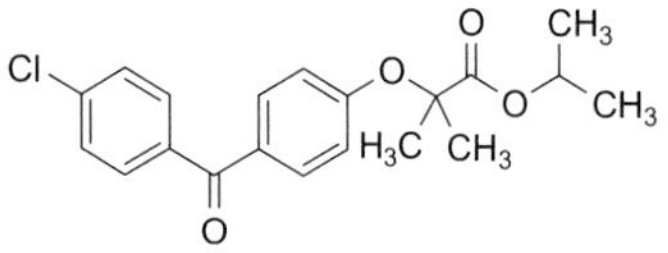

MW: 360.84
P: US 4,058,552 (1969, Orchimed)
L: 1975 (Fournier)

Bezafibrate

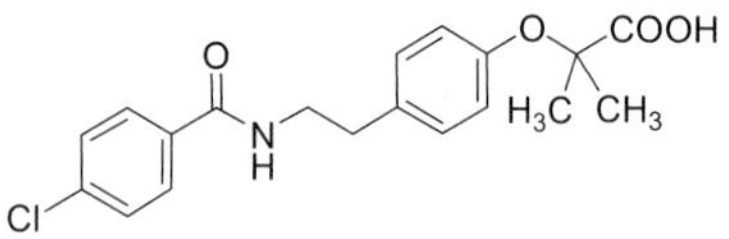

MW: 361.83
P: DE 2,149,070 (1971, Boehringer Mannheim)
L: 1978 (Boehringer Mannheim)
S: –

Ciprofibrate

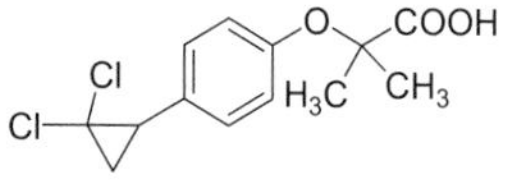

MW: 289.16
P: US 3,948,973 (1972, Sterling Drug)
L: 1985 (Sanofi-Synthelabo)
S: –

D
Dermatologicals

D04AB
Anesthetics for Topical Use
(benzocaine, tetracaine)

Benzocaine

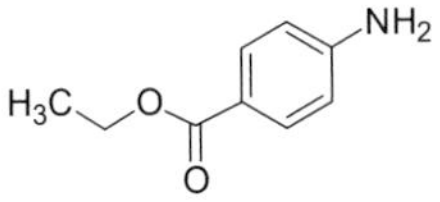

MW: 165.19
P: Salkowski, Ber. 1895, 28, 1921,
  DE 181,324 (1904, AG. f. Anilin Fabr. Berlin)
L: 1902 (Hoechst)
S: –

Tetracaine

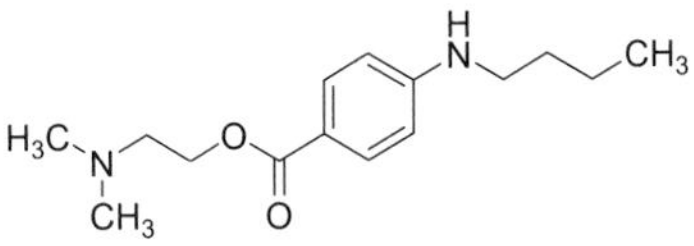

MW: 264.37
P: US 1,889,645 (1930, Winthrop)
L: 1941 (Sanofi-Synthelabo)
S: HCl

D10AD

**Retinoids for Topical Use in Acne
(retinol, tretinoin, isotretinoin)**

Retinol (Vitamin A)

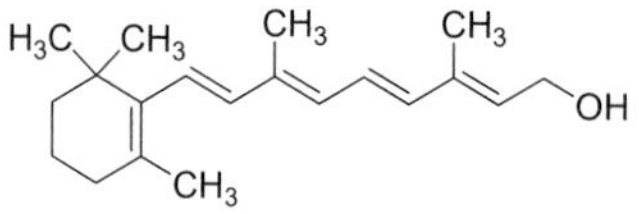

MW: 286.46
P: Karrer, P. et al. Helv. Chim. Acta:
1931, 14, 1036–1040
L: 1947 (Hoffmann-La Roche)
S: –

Tretinoin

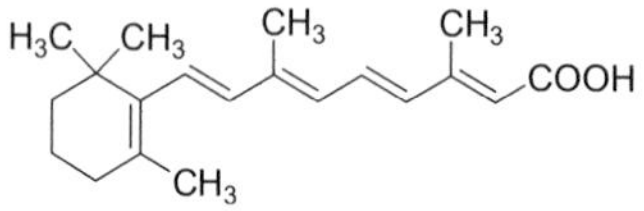

MW: 300.44
P: DE 1,035,647 (1957, BASF)
L: 1962
S: –

Isotretinoin

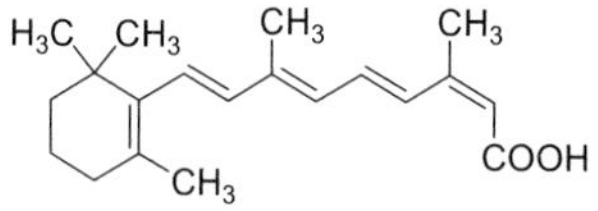

MW: 300.44
P: US 3,746,730 (1969, Hoffmann-La Roche)
L: 1982 (Hoffmann-La Roche)
S: Na, K

G
Genito Urinary System and Sex Hormones

G01AA
Antibiotics
(nystatin, amphotericin B)

Nystatin

MW: 926.12
P: US 2,797,183 (1952, Research Corp.)
L: 1954 (Bistrol-Myers Squibb)
S: H_2SO_4
R: Nystatin A_1

Amphotericin B

MW: 924.09
P: US 2,908,611 (1954, Olin Mathieson)
L: 1958 (Bistrol-Myers Squibb)
S: –

G03A

Progestogens
(progesterone, norethisterone, etynodiol, medroxyprogesterone, chlormadinone, megestrol,
levonorgestrel, norgestrel, norgestimate, desogestrel, etonogestrel, gestodene)

Progesterone

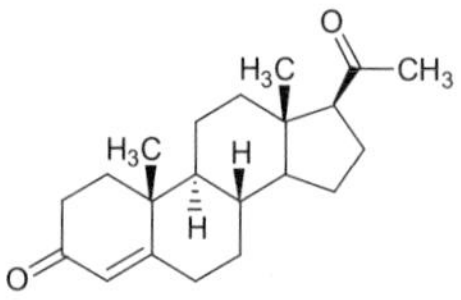

MW: 314.47
   P: Butenandt A. et al.: Ber. **67**, 1611, 1934
   L: –

Norethisterone

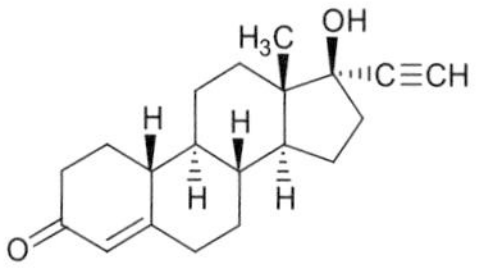

MW: 474.59
   P: US 2,744,122 (1951, Syntex)
   L: 1957
   R: hormonal contraceptive
      (ATC: G03AC)

Etynodiol acetate

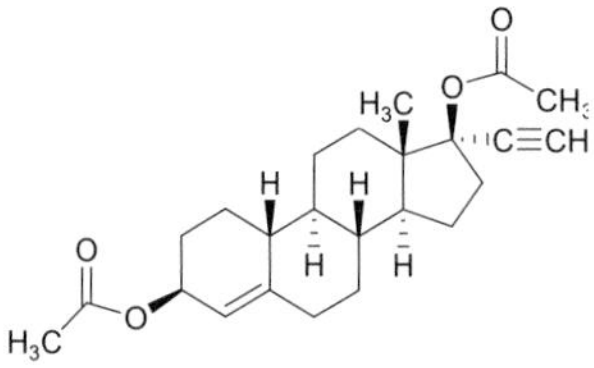

MW: 300.45
   P: US 2,843,609 (1955, Searle)
   L: –
   S: –
   R: hormonal contraceptive
      (ATC: G03AC)

Medroxyprogesterone acetate

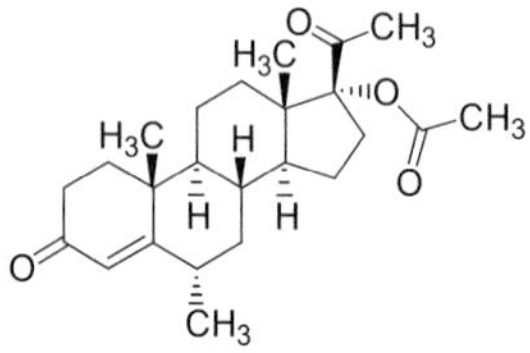

MW: 344.48
   P: US 3,147,290 (1958, Upjohn)
   L: 1958
   S: –
   R: hormonal contraceptive
      (ATC: G03AC)

Chlormadinone

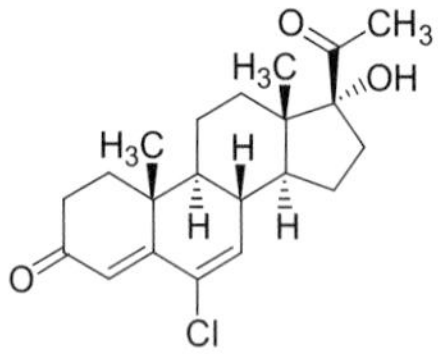

MW: 362.89
   P: DE 1,075,114 (1958, E. Merck)
   L: 1963
   S: –
   R: hormonal contraceptive in
      combination with estrogen
      (ATC: G03AB)

Megestrol acetate

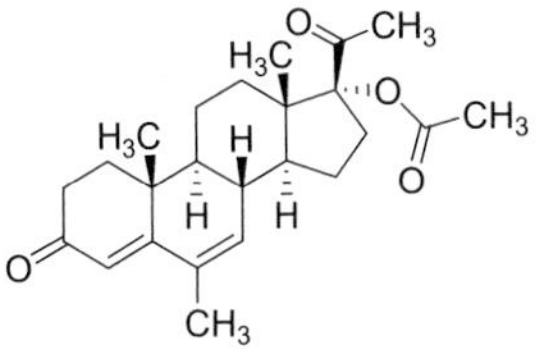

MW: 384.52
  P: US 2,891,079 (1959, Searle)
  L: 1963
  S: –
  R: hormonal contraceptive
    (ATC: G 03AC)

Levonorgestrel

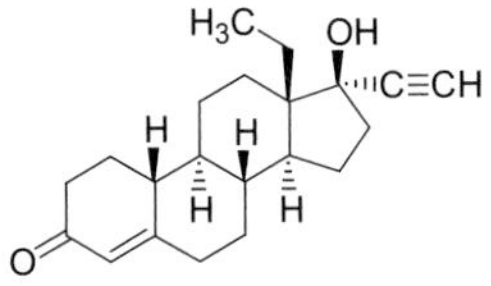

MW: 312.45
  P: US 3,959,322 (1960, Herchel Smith)
  L: 1978 (Wyeth Labs.)
  S: –
  R: hormonal contraceptive
    (ATC: G 03AC)

Norgestrel

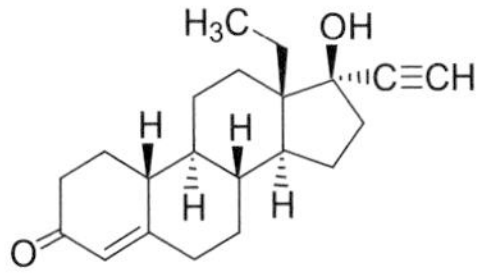

MW: 312.45
  P: GB 1,041,279 (1961, Herchel Smith)
  L: 1968
  S: –
  R: racemate
    hormonal contraceptive in
    combination with estrogen
    (ATC: G03AB)

Norgestimate

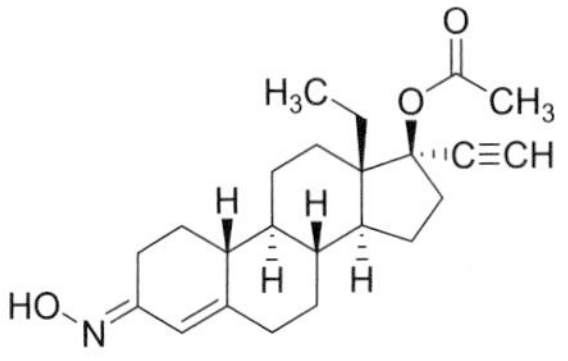

MW: 369.52
  P: GB 1,123,104 (1965, Ortho Pharma)
  L: 1986
  S: –
  R: hormonal contraceptive in
    combination with estrogen
    (ATC: G03AB)

Desogestrel

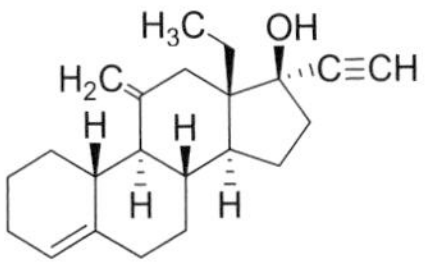

MW: 310.48
  P: DE 2,361,120 (1972, Akzo)
  L: 1982
  S: –
  R: hormonal contraceptive
    (ATC: G 03AC)

Etonogestrel

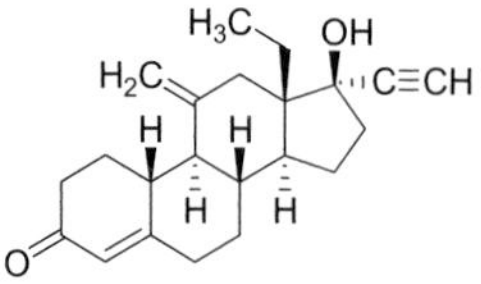

MW: 324.46

P: US 3,927,046 (1972, Akzona Incorp.)

L: –

S: –

R: hormonal contraceptive
(ATC: G 03AC)

Gestodene

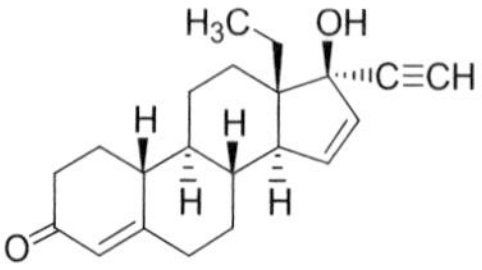

MW: 310.44

P: DE 2,546,062 (1975, Schering AG)

L: 1987

S: –

R: contraseptive in combination
with estrogen
(ATC: G03AB)

G03BA

Androgens: 3-oxoandrosten (4) derivatives
(testosterone, methyltestosterone)

Testosterone

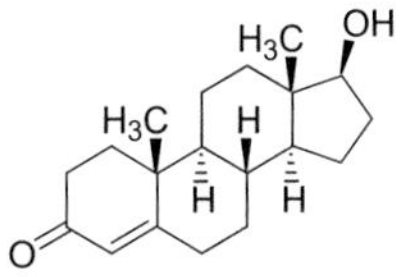

MW: 288.43
 P: US 2,308,833 (1935, Ciba)
 L: 1939

Methyltestosterone

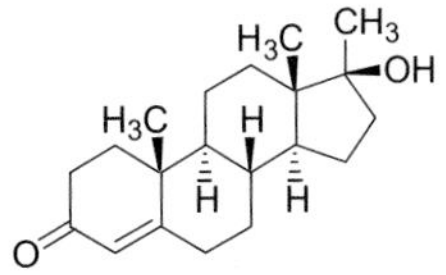

MW: 302.46
 P: US 2,143,453 (1935, Society of Chem. Ind.
    Basle)
 L: 1939

G03CA

**Natural and Semisynthetic Estrogens
(estradiol, ethinylestradiol)**

Estradiol

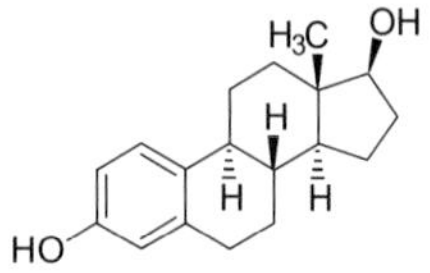

MW: 272.39
  P: US 2,096,744 (1932, Schering AG)
  L: 1938

Ethinylestradiol

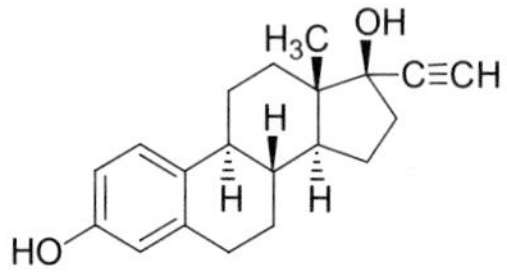

MW: 296.41
  P: US 2,243,887 (1935, Schering
     AG),Inhoffen, H. H. et al.
     Naturwiss. 1938, 26, 96
  L: 1938
  R: combination with progestogens

G04CB
**Testosterone-5-alpha reductase inhibitors
(finasteride, dutasteride)**

Finasteride

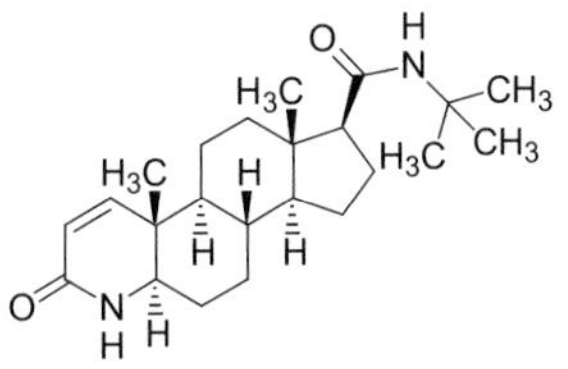

MW: 372.55
  P: US 4,760,071 (1984, Merck & Co.)
  L: 1992 (Merck Sharp & Dohme)

Dutasteride

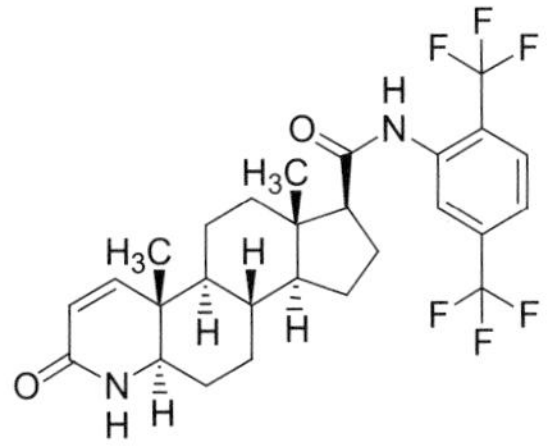

MW: 528.54
  P: EP 719,277 (1993, Glaxo Wellcome)
  L: 2003 (GlaxoSmithKline)

H
Systemic Hormonal Preparations

H02AB
Systemic hormonal preparations
(cortisone, hydrocortisone, fludrocortisone, prednisolone, prednisone dexamethasone,
betamethasone, triamcinolone, methylprednisolone, deflazacort)
Other applications (dermatology, respiratory diseases)
(flumetasone, flucinolone acetonide, beclomethasone, flunisolide, clobetasol, fluocortin butylate,
budesonide, halobetasol propionate, fluticasone, loteprednol etabonate, mometasone furoate,
ciclesonide)

Cortisone

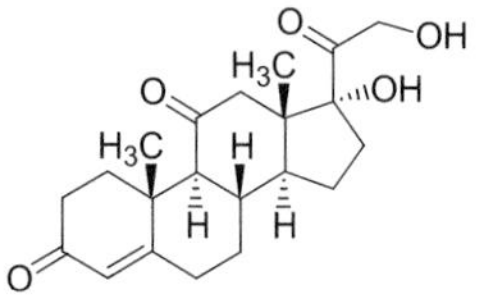

MW: 360.45
  P: Hensch, P.S. et al.: Proc. Staff Meet. Mayo
  Clin. **1949**, 24, 181
  L: –

Hydrocortisone

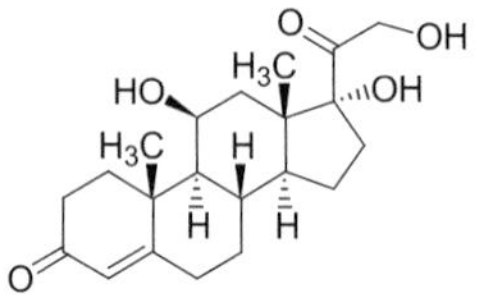

MW: 362.47
  P: US 2,183,589 (1936, Roche-Organon)
  L: 1941

Flumetasone

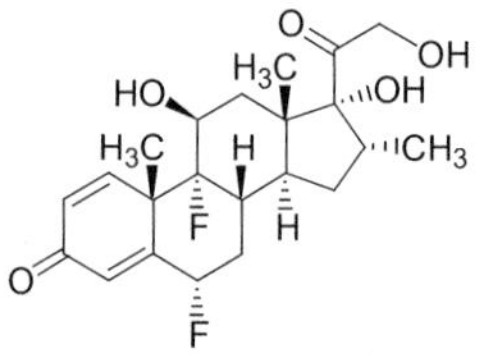

MW: 410.46
  P: US 2,671,752 (1951, Syntex)
  L: 1964
  R: dermatology (ATC: D07AB)

Fludrocortisone

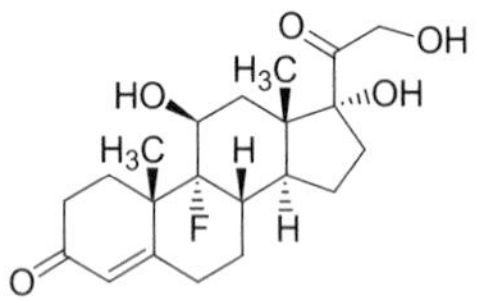

MW: 380.46
  P: US 2,771,475 (1953, Upjohn)
  L: –

Prednisolone

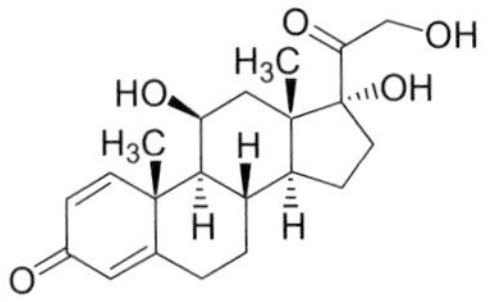

MW: 360.45
  P: US 2,897,216 (1954, Schering Corp.)
  L: 1959

Prednisone

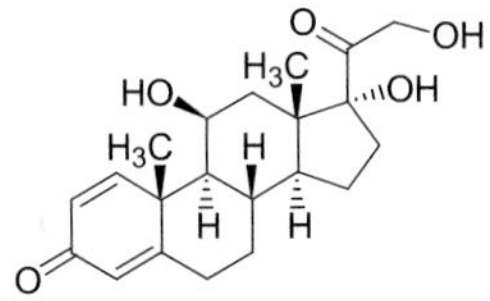

MW: 358.43
  P: US 2,897,216 (1954, Schering Corp.)
  L: 1954 (Schering-Plough)

Dexamethasone

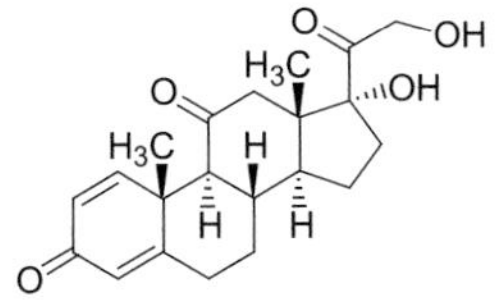

MW: 392.47
  P: DE 1,113,690 (1957, Merck & Co.)
  L: 1958

Fluocinolone acetonide

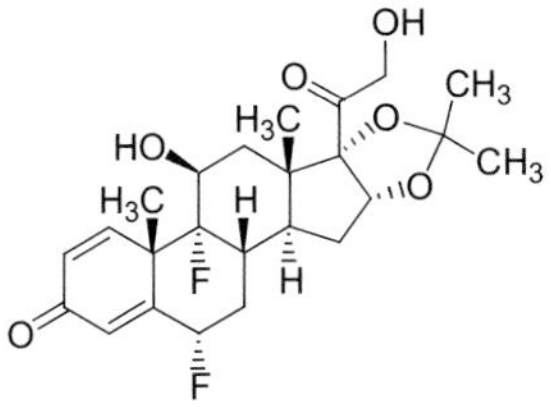

MW: 452.49
  P: US 3,197,469 (1958, Pharmaceutical
    Research Prod.)
  L: 1961 (Medicis)
  R: **dermatology** (ATC: D07AB)

Betamethasone

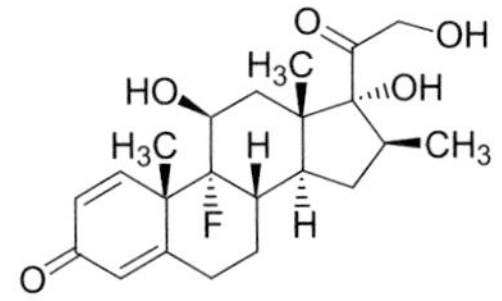

MW: 392.47
  P: US 3,053,865 (1958, Merck)
  L: 1961
  R: **OAD*** (ATC: R03BA)

*OAD = Obstructive Airway Disease

Triamcinolone

MW: 394.44
P: US 2,789,118 (1956, American Cyanamid)
L: 1958
R: also in OAD* (ATC: R03BA)

Methylprednisolone

MW: 374.48
P: US 2,897,218 (1956, Upjohn)
L: 1955

Beclomethasone

MW: 408.92
P: GB 901,093 (1957, Schevico)
L: 1972
R: dermatology (ATC: D07AB)
OAD (ATC: R03BA)

Flunisolide

MW: 434.50
P: US 3,126,375 (1958, Syntex)
L: 1978
R: nasal preparation (ATC: R01AD)
OAD (ATC: R03BA)

Deflazacort

MW: 441,52
P: GB 1,077,393 (1965, Lepetit)
L: 1985 (Aventis Pharma)

Clobetasol propionate

MW: 466.98
P: DE 1,902,340 (1968, Glaxo Lab.)
L: 1978 (GlaxoSmithKline)
R: **dermatology** (ATC: D07AB)

Fluocortin butylate

MW: 446,56
P: DE 2,150,268 (1971, Schering AG)
L: 1977
R: **dermatology** (ATC: D07AB)

Budesonide

MW: 430.54
P: US 3,929,768 (1972, Bofors)
L: 1981 (AstraZeneca)
R: **nasal preparation** (ATC: R01AD)
OAD (ATC: R03BA)

Halobetasol propionate

MW: 484.96
P: DE 2,743,069 (1975, Ciba-Geigy)
L: 1991 (Bristol-Myers Squibb)
R: **dermatology** (ATC: D07AB)

Fluticasone propionate

MW: 500.58
P: BE 887,518 (1980, Glaxo)
L: 1990 (GlaxoSmithKline)
R: **nasal preparation** (ATC: R01AD)
OAD (ATC: R03BA)

Loteprednol etabonate

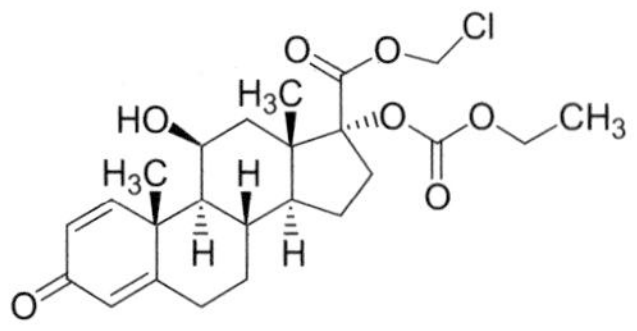

MW: 466.95
P: US 4,996,335 (1980, Soft Drugs)
L: 1998 (Bausch & Lomb)
R: **topical anti-inflammatory**

Mometasone furoate

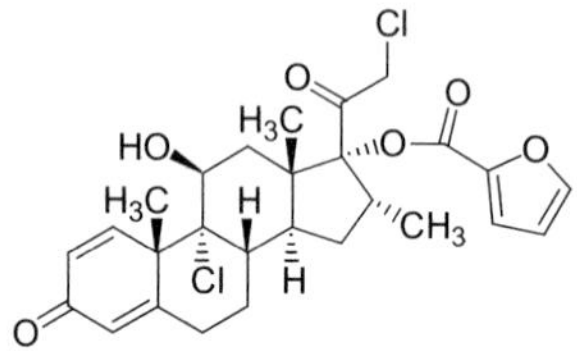

MW: 521.44
P: EP 57,401 (1981, Schering Corp.)
L: 1987 (Schering-Plough)
R: **nasal preparation** (ATC: R01AD)
   OAD (ATC: R03BA)

Ciclesonide

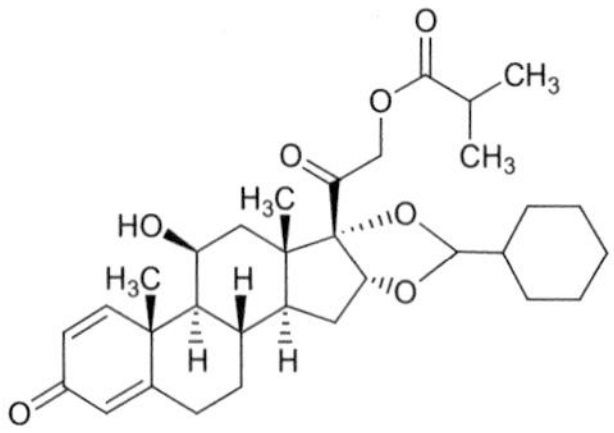

MW: 540.69
P: DE 4,129,535 (1990, Byk Elmu SA)
L: 2005 (Altana Pharma)
R: **OAD** (ATC: R03BA)

J
Antiinfectives for Systemic Use

J01AA
Tetracyclines
(oxytetracycline, tetracycline, doxycycline, minocycline, doxorubicin)

Oxytetracycline

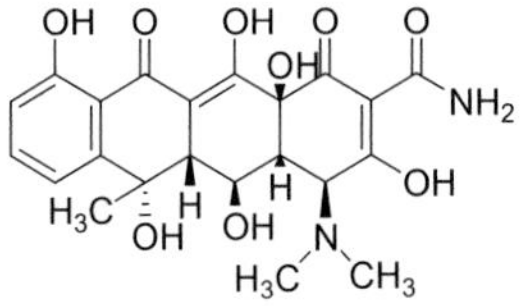

MW: 460.44
   P: US 2,516,080 (1949, Pfizer)
   L: 1950
   S: –

Tetracycline

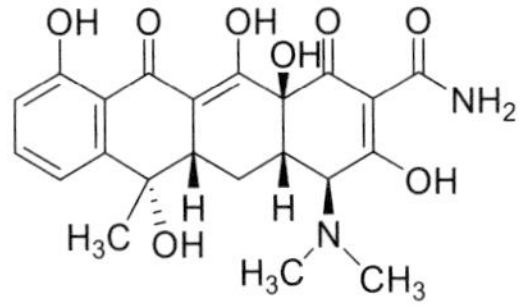

MW: 444.44
   P: US 2,699,054 (1953, L. H. Conover)
   L: 1978 (Procter & Gamble)
   S: HCl

Doxycycline

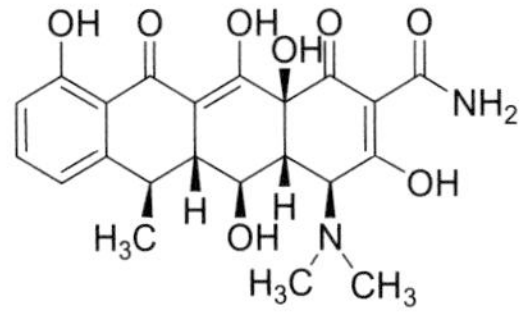

MW: 444.44
   P: DE 1,082,905 (1957, American Cyanamid)
   L: 1967
   S: HCl, hyclate

Minocycline

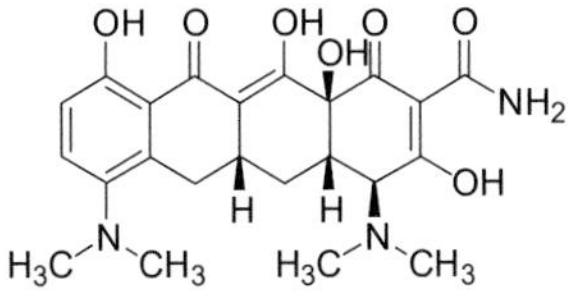

MW: 457.48
   P: US 3,148,212 (1961, American Cyanamid)
   L: 1971 (Wyeth-Lederle)
   S: HCl

J01C
Beta-Lactam Antibacterials, Penicillins
(benzylpenicillin, ampicillin, amoxicillin, cloxacillin, oxacillin, dicloxacillin, flucloxacillin, ticarcillin, piperacillin, sultamicillin)

Benzylpenicillin

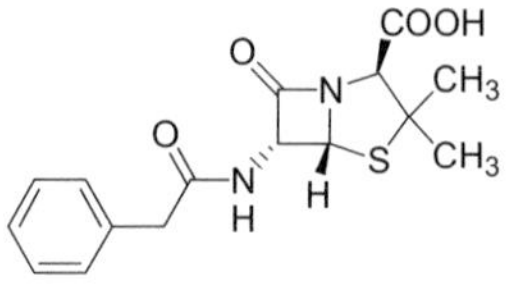

MW: 334.40
P: Fleming, A.: Brit. J. Exp. Pathol. **1929**, 10, 226
L: 1942
S: Na, K, procaine

Ampicillin

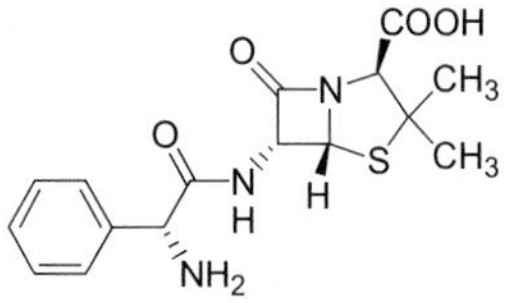

MW: 349.41
P: US 2,985,648 (1958, Beecham)
L: 1961
S: Na

Amoxicillin

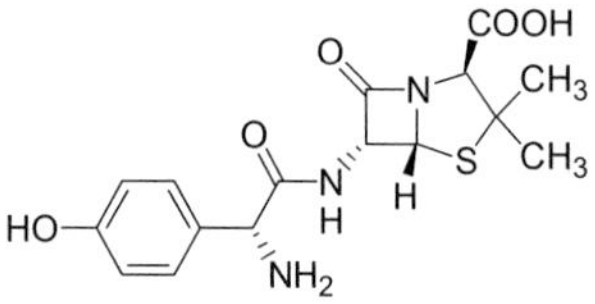

MW: 365.41
P: GB 978,178 (1958, Beecham)
L: 1964 (Beecham)
S: –

Cloxacillin

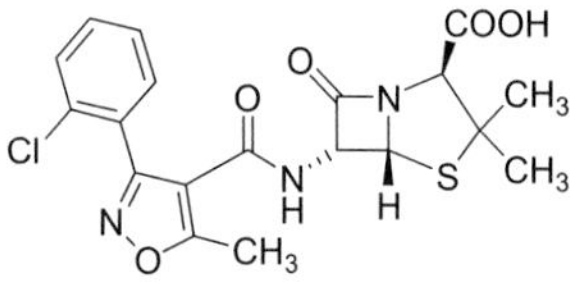

MW: 435.89
P: US 2,996,501 (1960, Beecham)
L: 1965
S: Na

Oxacillin

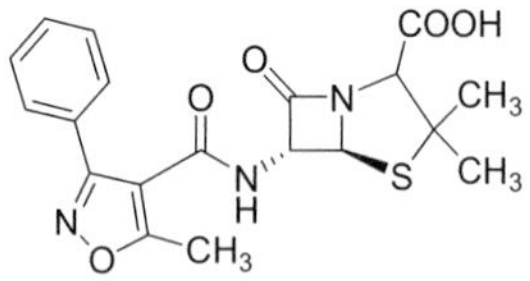

MW: 401.44
P: US 2,996,501 (1960, Beecham)
L: 1962 (Beecham)
S: Na

Dicloxacillin

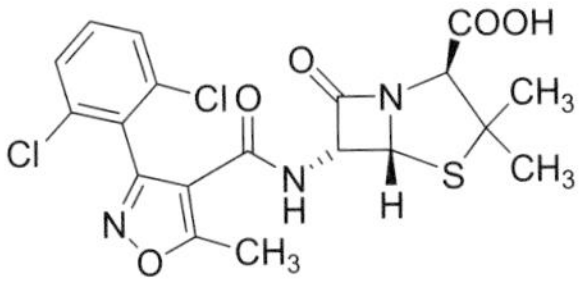

MW: 470.33
P: GB 978,299 (1961, Beecham)
L: 1968
S: Na

Flucloxacillin

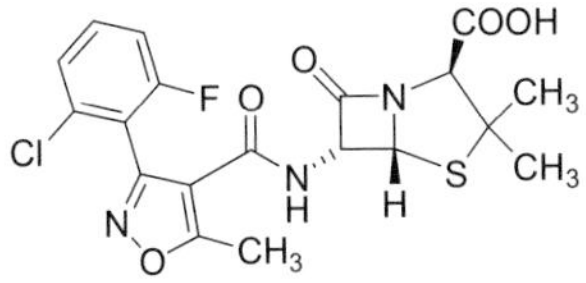

MW: 453.88
P: GB 978,299 (1961, Beecham)
L: GlaxoSmithKline
S: Na

Ticarcillin

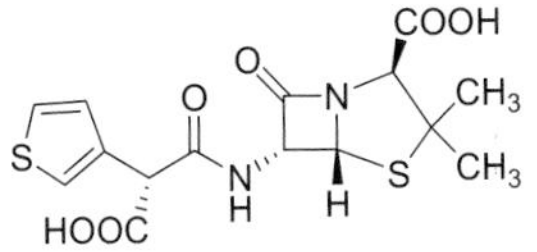

MW: 384.43
P: DE 1,295,558 (1963, Beecham)
L: GlaxoSmithKline
S: 2 Na

Piperacillin

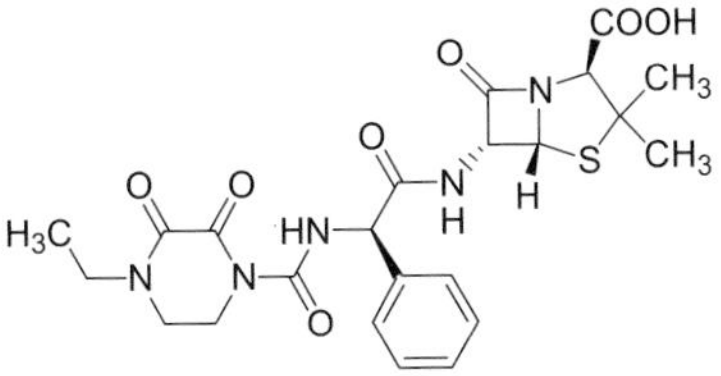

MW: 517.56
P: GB 1,508,062 (1974, Toyama)
L: 1981
S: Na

Sultamicillin

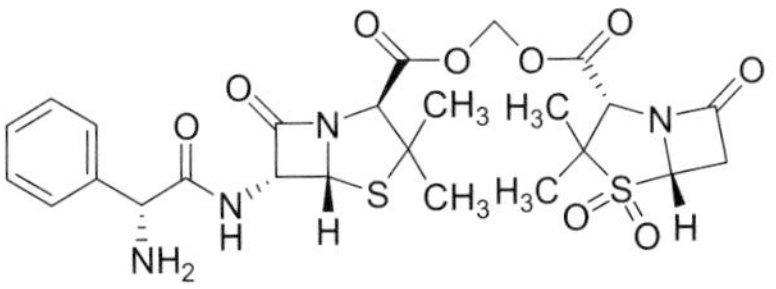

MW: 594.67
P: US 4,342,772 (1979, Leo Pharm. Prod.)
L: 1987 (Pfizer)
S: tosylate

J01CG
Beta-Lactamase Inhibitors
(clavulanic acid, sulbactam, tazobactam)

Clavulanic acid

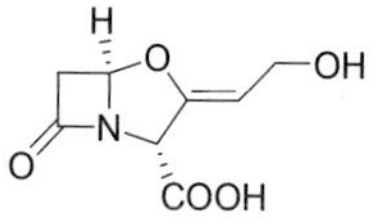

MW: 199.16
P: US 4,529,720 (1974, Beecham)
L: –
S: K
R: combination with beta-lactam
antibacterials

Sulbactam

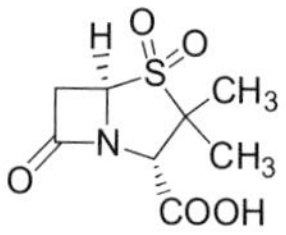

MW: 233.24
P: US 4,234,579 (1977, Pfizer)
L: 1986 (Pfizer)
S: Na
R: combination with beta-lactam
antibacterials

Tazobactam

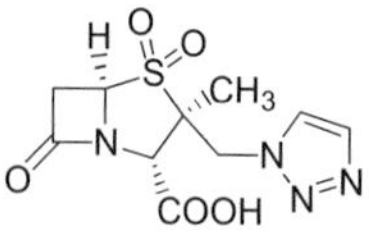

MW: 300.29
P: EP 97,446 (1982, Taiho Pharmaceutical)
L: 1992 (Otsuka)
S: Na
R: combination with beta-lactam
antibacterials

J01DA
Cephalosporins and Related Substances
(cefalexin, cefazolin, cefadroxil, cefuroxime, cefaclor, cefotiam, cefoperazone, cefotaxime, cefuroxime, ceftazidime, ceftriaxone, cefixime, cefdinir, cefprodoxime, cefepime, flomoxef, cefditoren, cefcapene, cefprozil)

Cefalexin

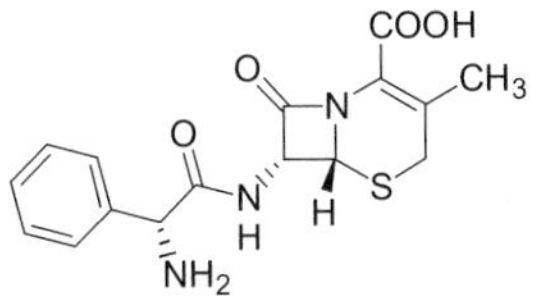

MW: 347.40
P: US 3,507,861 (1966, Eli Lilly)
L: 1970 (Eli Lilly)
S: HCl

Cefazolin

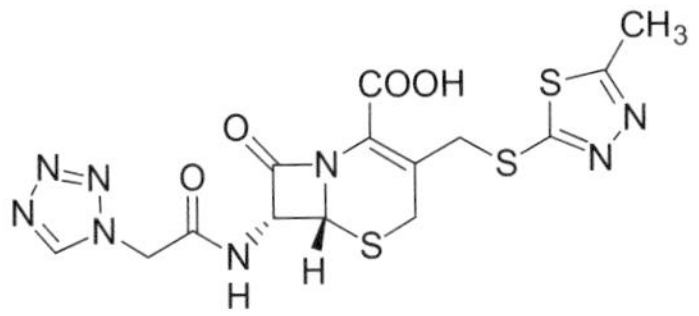

MW: 454.52
P: DE 1,770,168 (1967, Fujisawa)
L: 1971
S: Na

Cefadroxil

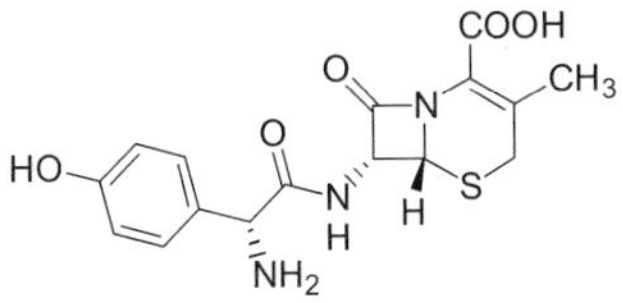

MW: 363.39
P: GB 1,240,687 (1967, Bristol-Myers)
L: 1978
S: –

Cefuroxime

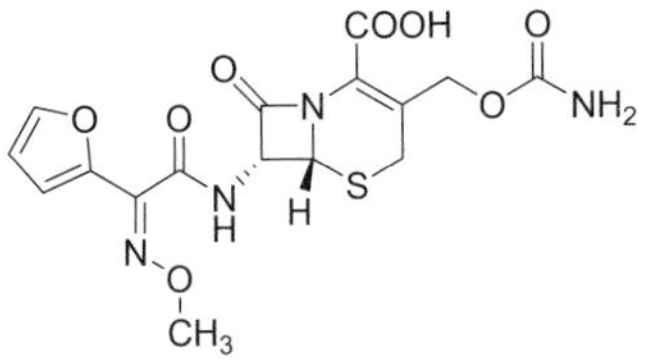

MW: 424.39
P: DE 2,223,375 (1971, Glaxo)
L: 1977 (GlaxoSmithKline)
S: Na

Cefaclor

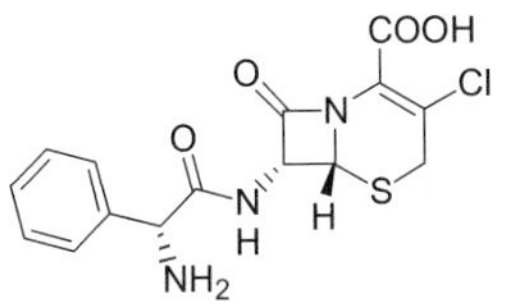

MW: 367.81
P: US 3,925,372 (1973, Eli Lilly)
L: 1979 (Eli Lilly)
S: –

Cefotiam

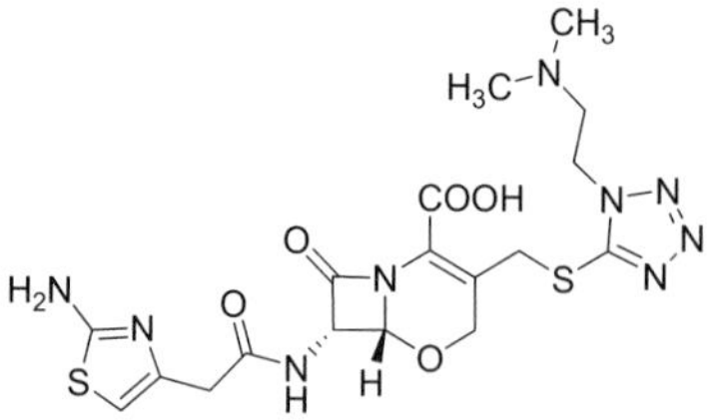

MW: 525.64
   P: DE 2,462,736 (1973, Takeda)
   L: 1981 (Takeda)
   S: 2 HCl

Cefoperazone

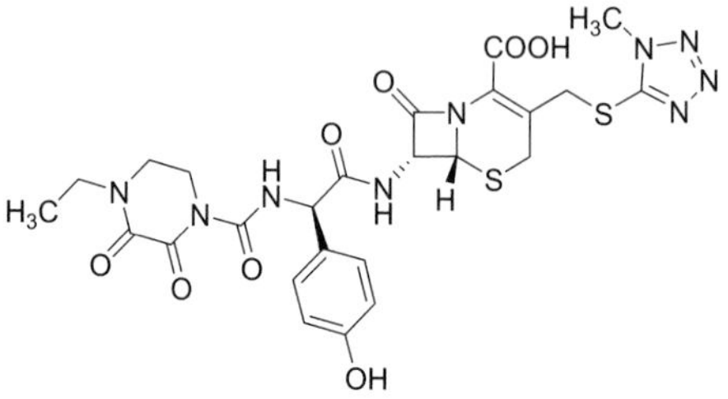

MW: 645.68
   P: GB 1,508,062 (1974, Toyama Chemical)
   L: 1981 (Toyama, Pfizer)
   S: Na

Cefotaxime

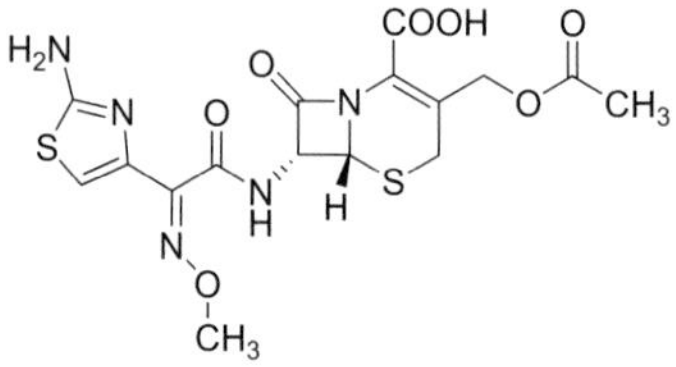

MW: 455.47
   P: DE 2,702,501 (1976, Roussel-Uclaf)
   L: 1980 (Aventis)
   S: Na

Cefuroxime axetil

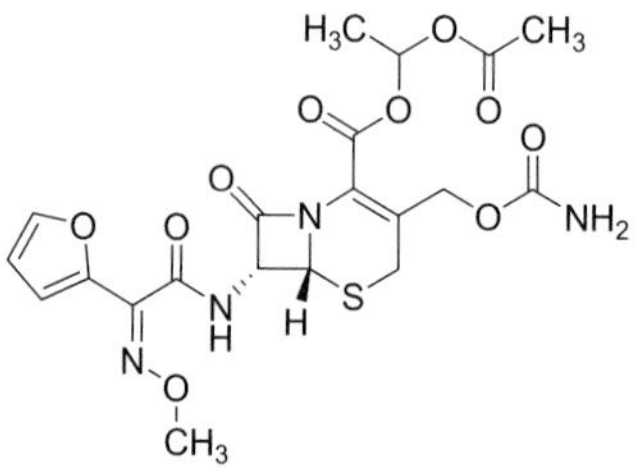

MW: 510.48
   P: US 4,267,320 (1976, Glaxo)
   L: 1987 (GlaxoSmithKline)
   S: –

Ceftazidime

MW: 546.59
   P: GB 2,025,398 (1978, Glaxo)
   L: 1984 (GlaxoSmithKline)
   S: 2HCl

Ceftriaxone

MW: 554.59
   P: DE 2,922,036 (1978, Hoffmann-La Roche)
   L: 1982 (Hoffmann-La Roche)
   S: 2 Na

Cefixime

MW: 453.46
   P: EP 30,630 (1979, Fujisawa)
   L: 1987 (Fujisawa)
   S: –

Cefdinir

MW: 395.42
   P: US 4,559,334 (1979, Fujisawa)
   L: 1991 (Fujisawa)
   S: –

Cefprodoxime proxetil

MW: 557.60
   P: EP 49,118 (1980, Sankyo)
   L: 1989 (Sankyo)
   S: –

Cefepime

MW: 480.57
   P: DE 3,307,550 (1982, Bristol-Myers)
   L: 1994 (Bristol-Myers Suibb)

Flomoxef

MW: 496.47
   P: DE 3,345,989 (1982, Shionogi)
   L: 1988
   S: Na

Cefditoren pivoxil

MW: 620.73
   P: EP 175,610 (1984, Meiji Seika Kaisha)
   L: 1994 (Meiji Seika)
   S: –

Cefcapene pivoxil

MW: 567.64
   P: BE 904,517 (1985, Shionogi)
   L: 1997
   S: HCl

Cefprozil

MW: 389.43
   P: US 4,520,022 (1983, Bristol-Myers)
   L: 1992 (Bristol-Myers Squibb)
   S: –
   R: Z/E isomeric mixture

J01DH
Carbapenems
(imipenem, meropenem)

Imipenem

MW: 299.35
P: US 4,194,047 (1975, Merck & Co.)
L: 1985
S: –
R: with enzyme inhibitor

Meropenem

MW: 383.47
P: EP 126,587 (1983, Sumitomo Chemical Co.)
L: 1994 (AstraZeneca)
S: –

J01FA

Macrolides

(erythromycin, clarithromycin, roxithromycin, azithromycin, telithromycin)

Erythromycin

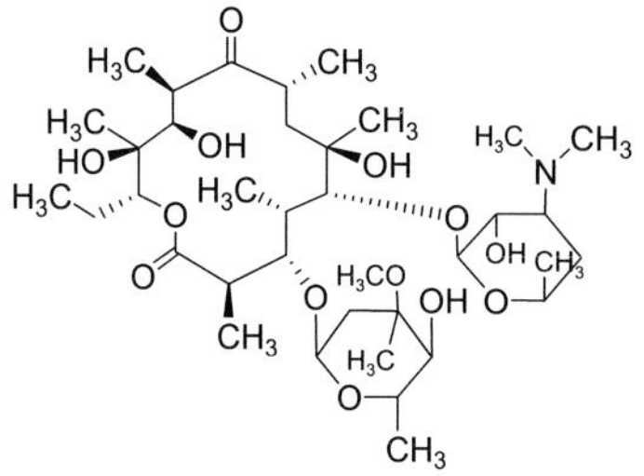

MW: 733.93
  P: US 2,653,899 (1952, Eli Lilly)
  L: 1952
  S: –

Clarithromycin

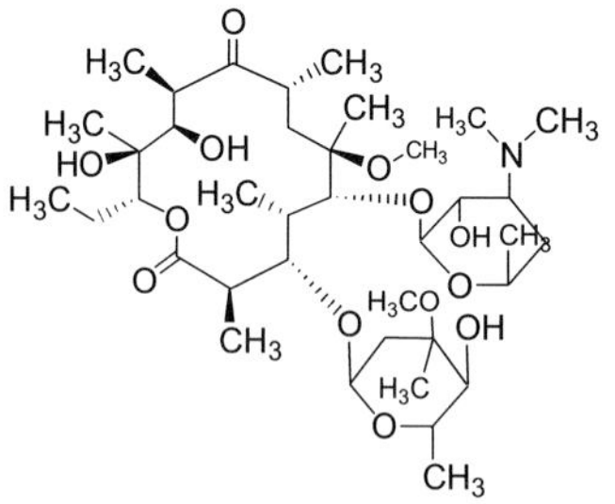

MW: 747.95
  P: EP 41,355 (1980, Taisho Pharma)
  L: 1990
  S: –

Roxithromycin

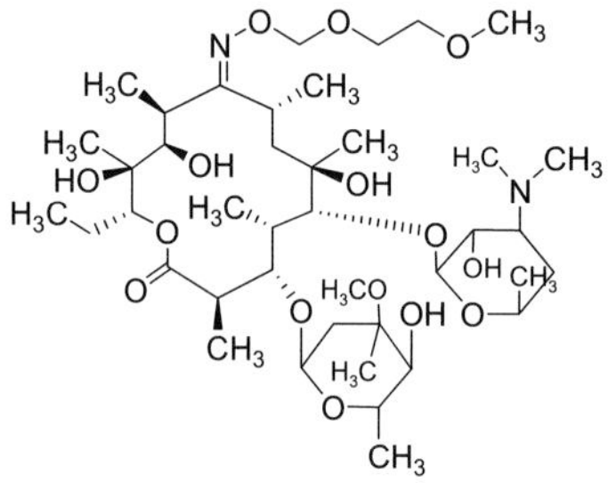

MW: 837.06
  P: EP 33,255 (1980, Roussel Uclaf)
  L: 1987 (Aventis Pharma)
  S: –

Azithromycin

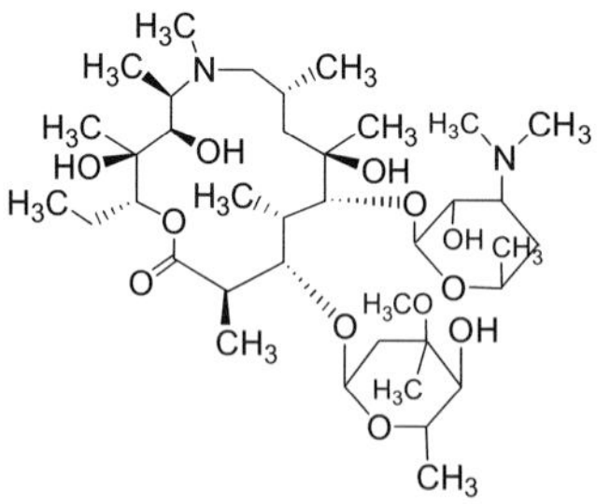

MW: 748.99
  P: DE 3,140,449 (1981, Pliva Pharm. & Chem. Works)
  L: 1988 (Pliva Pharm. & Chem. Works)
  S: HCl

Telithromycin

MW: 812.01
P: EP 680,967 (1994, Roussel Uclaf)
L: 2001 (Aventis Pharma)
S: –

J01MA

Quinolone Antibacterials
(norfloxacin, ofloxacin, ciprofloxacin, lomefloxacin, levofloxacin, sparfloxacin, moxifloxacin, gatifloxacin)

Norfloxacin

MW: 319.34
P: DE 2,804,097 (1977, Kyorin Seiyaku)
L: 1983 (Merck Sharp & Dohme)
S: –

Ofloxacin

MW: 361.37
P: EP 47,005 (1980, Daiichi Seiyaku)
L: 1985 (Daiichi Seiyaku)
S: HCl

Ciprofloxacin

MW: 331.34
P: EP 49,355 (1980, Bayer AG)
L: 1986 (Bayer AG)
S: HCl

Lomefloxacin

MW: 351.35
P: EP 140,116
       (1983, Hokuriku Pharmaceutical)
L: 1989 (Abbott)
S: HCl

Levofloxacin

MW: 361.37
P: EP 206,283 (1985, Daiichi Seiyaku)
L: 1993 (Daiichi Seiyaku)
S: –

Sparfloxacin

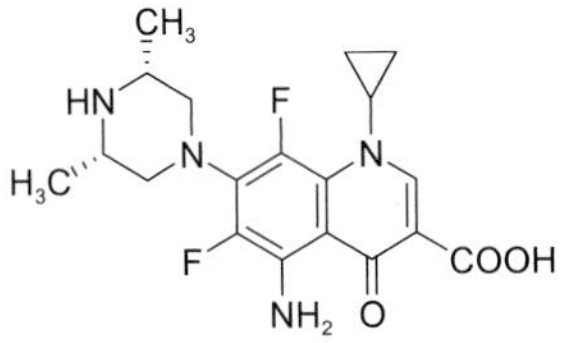

MW: 392.41
  P: EP 221,463 (1985, Dainippon)
  L: 1993 (Dainippon)
  S: –

Moxifloxacin

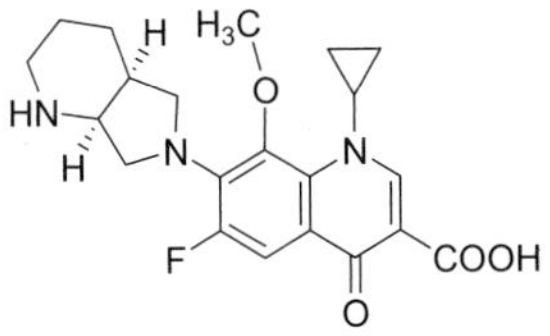

MW: 401.44
  P: EP 350,733 (1988, Bayer)
  L: 1999 (Bayer)
  S: HCl

Gatifloxacin

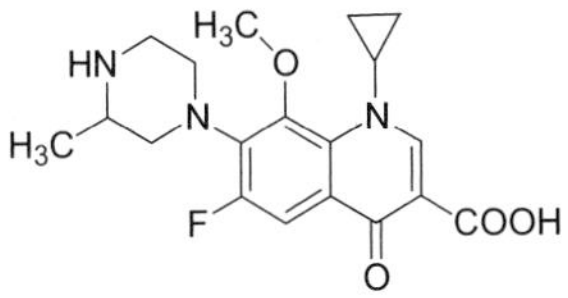

MW: 375.39
  P: EP 230,295 (1986, Kyorin Seiyaku)
  L: 1999 (Bristol-Myers Squibb)
  S: –

J02AB-AC
Antimycotics for Systemic Use
(miconazole, ketoconazole, itraconazole, fluconazole, voriconazole)
Antifungals for Topical Use
(econazole, isoconazole, bifonazole, oxiconazole, tioconazole)

Miconazole

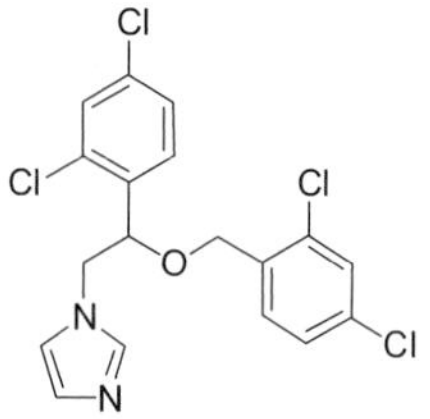

MW: 416.14
P: US 3,717,655 (1968, Janssen)
L: 1971
S: nitrate

Econazole

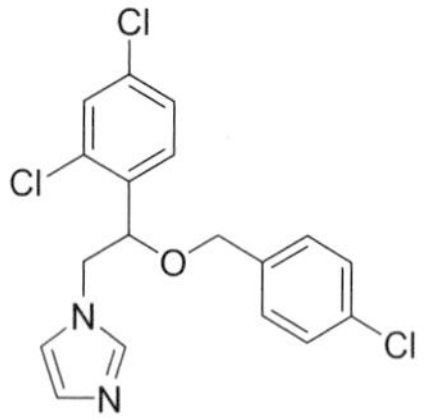

MW: 381.69
P: DE 1,940,388 (1968, Janssen)
L: 1974 (Janssen)
S: nitrate
R: dermatology (ATC:D01AC)

Isoconazole

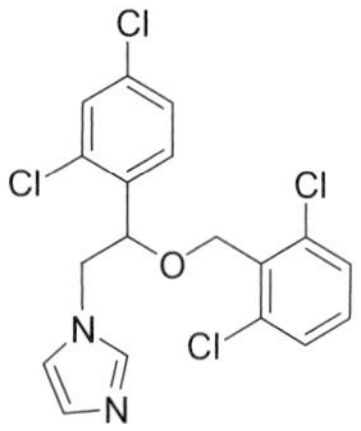

MW: 416.14
P: DE 1,940,388 (1968, Janssen)
L: 1979 (Janssen)
S: nitrate
R: dermatology (ATC:D01AC)

Bifonazole

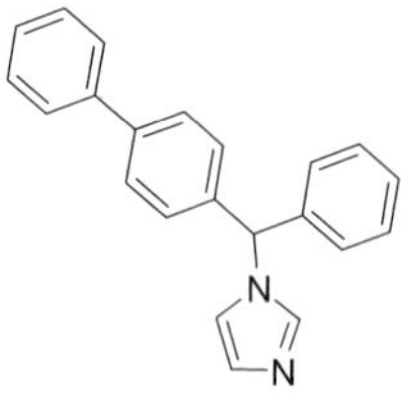

MW: 310.40
P: DE 2,461,406 (1974, Bayer)
L: 1983 (Bayer)
S: –
R: dermatology (ATC:D01AC)

Oxiconazole

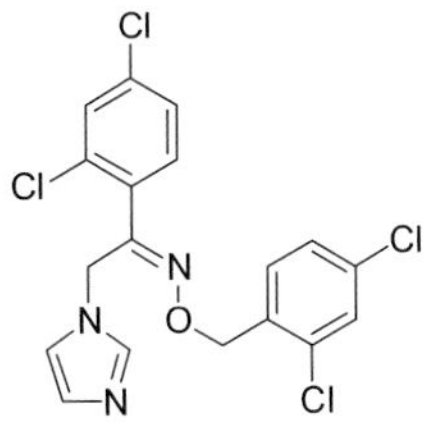

MW: 429.13
P: DE 2,657,578 (1975, Siegfried AG)
L: 1983 (Roche)
S: nitrate
R: dermatology (ATC:D01AC)

Tioconazole

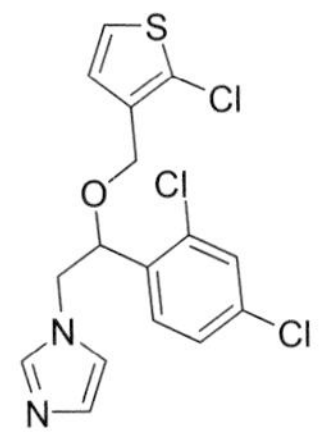

MW: 387.72
P: DE 2,619,381 (1975, Pfizer)
L: 1982 (Eli Lilly)
S: –
R: dermatology (ATC:D01AC)

Ketoconazole

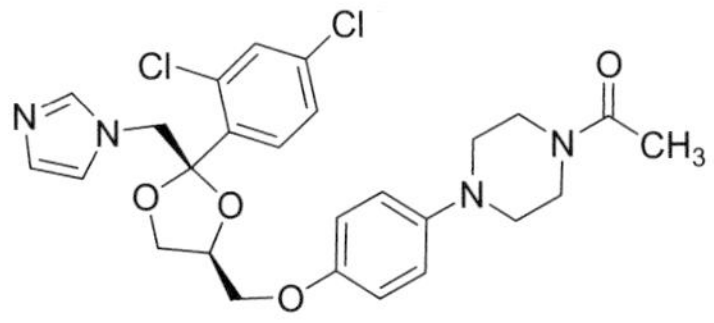

MW: 531.44
P: DE 2,804,096 (1977, Janssen)
L: 1981 (Janssen)
S: –

Itraconazole

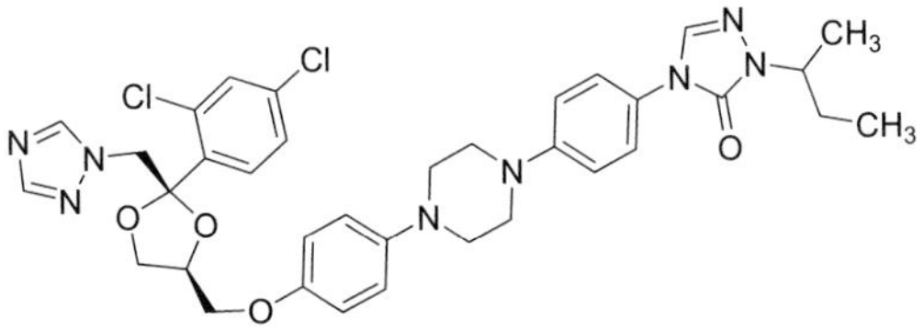

MW: 705.65
P: US 4,267,179 (1978, Janssen)
L: 1988 (Janssen)
S: –

Fluconazole

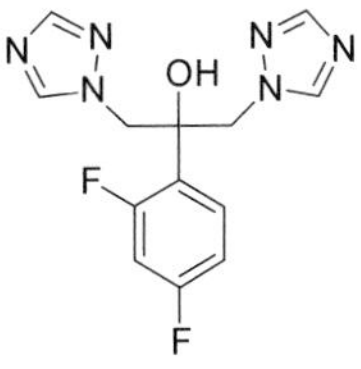

MW: 306.28
P: GB 2,099,818 (1981, Pfizer)
L: 1988 (Pfizer)
S: –

Voriconazole

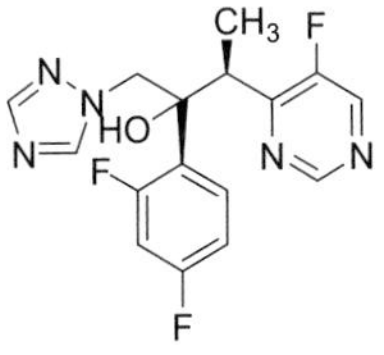

MW: 349.31
P: EP 440,372 (1990, Pfizer)
L: 2002 (Pfizer)
S: –

J05AB
Nucleosides and Nucleotides excl. Reverse Transcriptase Inibitors
(ribavirin, aciclovir, ganciclovir, famciclovir, valaciclovir)

Ribavirin

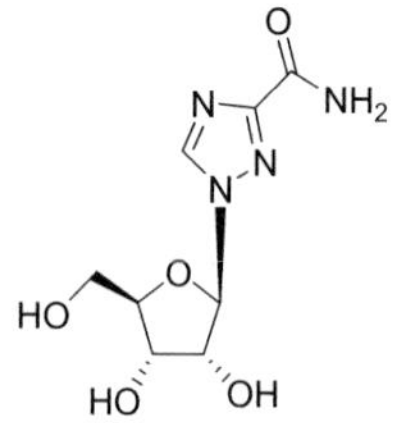

MW: 244.21
  P: US 3,976,545 (1971, ICN Pharmaceutical)
  L: 1986
  S: –

Aciclovir

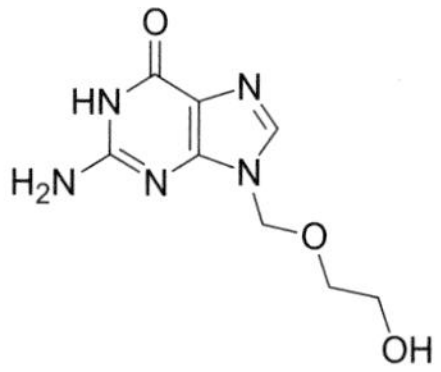

MW: 225.21
  P: DE 2,539,963 (1974, Wellcome)
  L: 1981 (GlaxoSmithKline)
  S: –, Na

Ganciclovir

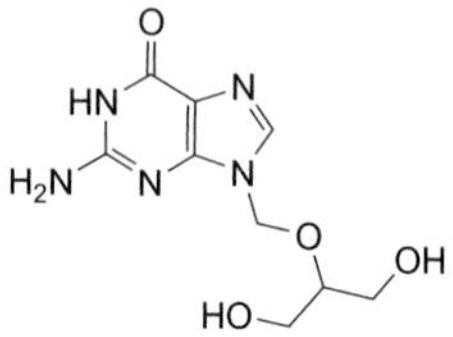

MW: 255.23
  P: EP 49,072 (1980, Ens Biologicals)
  L: 1988 (Roche)
  S: –, Na

Famciclovir

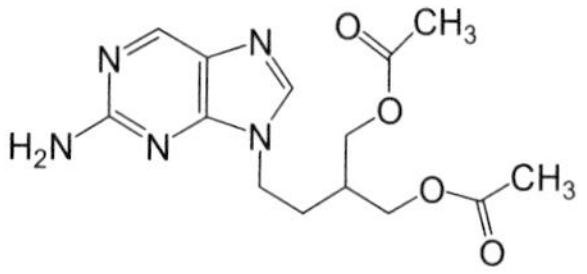

MW: 321.34
  P: EP 182,024 (1983, Beecham Group)
  L: 1994 (Novartis)
  S: –

Valaciclovir

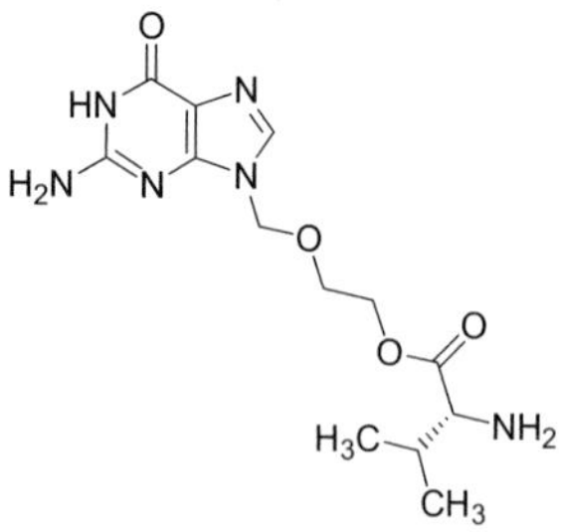

MW: 324.34
  P: EP 308,065 (1987, Wellcome Found. Ltd)
  L: 1995 (GlaxoSmithKline)
  S: HCl

J05AF
Nucleoside Reverse Transcriptase Inhibitors
(zidovudine, didanosine, stavudine, abacavir, lamivudine, tenofovir disoproxil)

Zidovudine

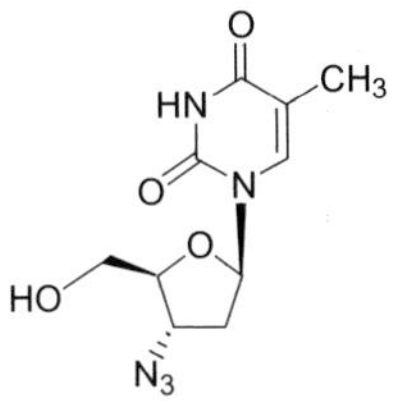

MW: 267.24
  P: Horwitz, J. P. et al.: J. Org. Chem.
     1964, 29, 2076,
     DE 3,608,606 (1985, Wellcome
     Found.)
  L: 1987 (GlaxoSmithKline)

Stavudine

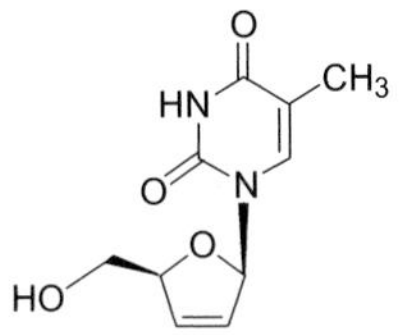

MW: 224.21
  P: Horwitz J. P. et al.: J. Org. Chem.
     1966, 31, 205
  L: 1994 (Bristol-Myers Squibb)

Didanosine

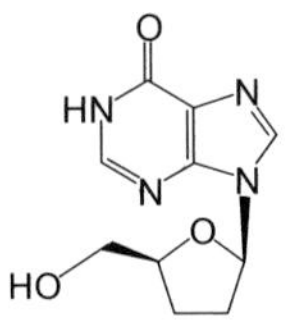

MW: 236.23
  P: Plunkett, W. et al.: Cancer Res. 1975, 35,
     1547, EP 216,510 (1985, US Com-
     merce)
  L: 1991 (Bristol-Myers Squibb)

Tenofovir disoproxil

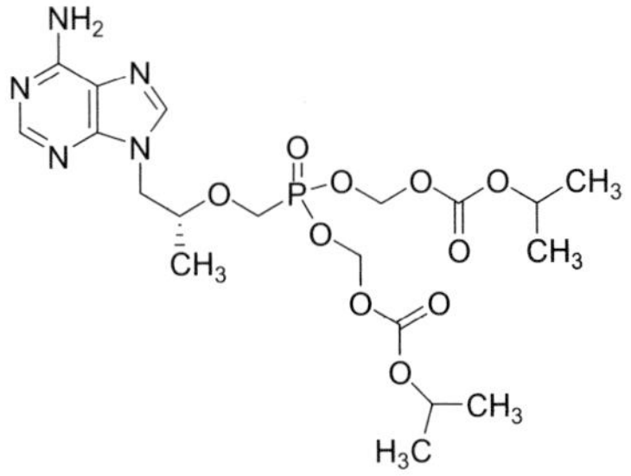

MW: 519.44
  P: EP 915,894 (1996, Gilead Sciences)
  L: 2001 (Gilead)
  S: fumarate

Abacavir

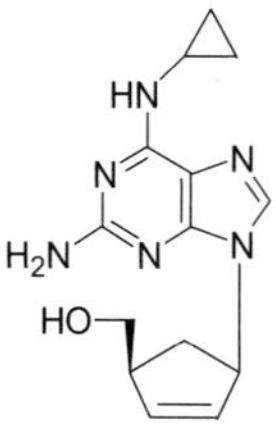

MW: 286.34
  P: US 5,034,394 (1988, Borroughs Welcome)
  L: 1999 (GlaxoSmithKline)
  S: succinate, H_2SO_4

Lamivudine

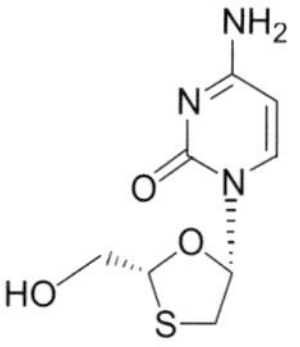

MW: 229.26
  P: EP 382,526 (1989, IAF Biochem. Int.)
  L: 1995 (Shire BioChem)
  S: −

J01GB
Aminoglycosides
(neomycin B, gentamicin, tobramycin, amikacin, netilmicin, isepamicin)

Neomycin B

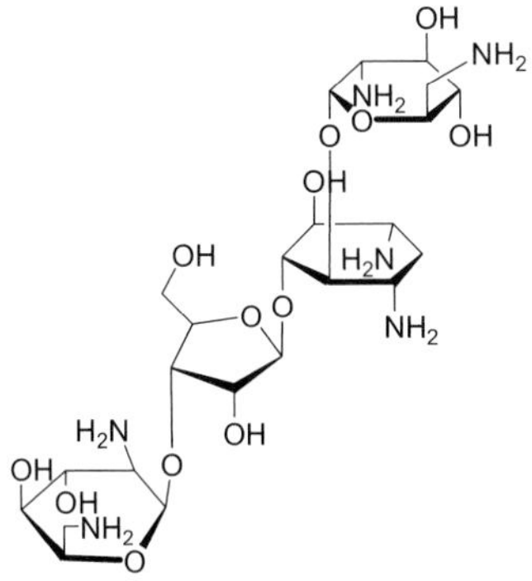

MW: 614.65
P: US 2,848,365 (1950, Upjohn)
L: 1952
S: 1/3 H_2SO_4

Gentamicin

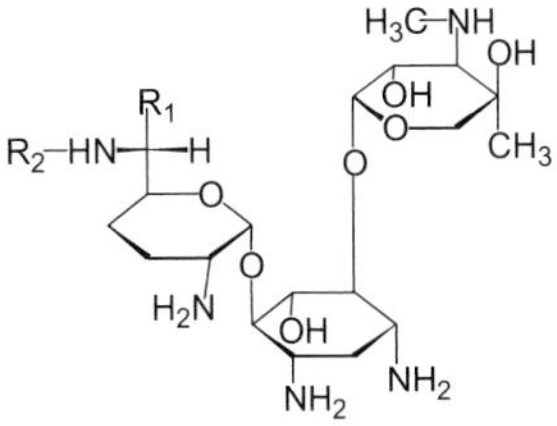

C_1 R_1: CH_3, R_2: CH_3
C_2 R_1: CH_3, R_2: H
C_{1a} R_1: H, R_2: H

MW: unspecified
P: US 3,091,572 (1962, Schering-Plough)
L: 1964 (Schering-Plough)
S: H_2SO_4

Tobramycin

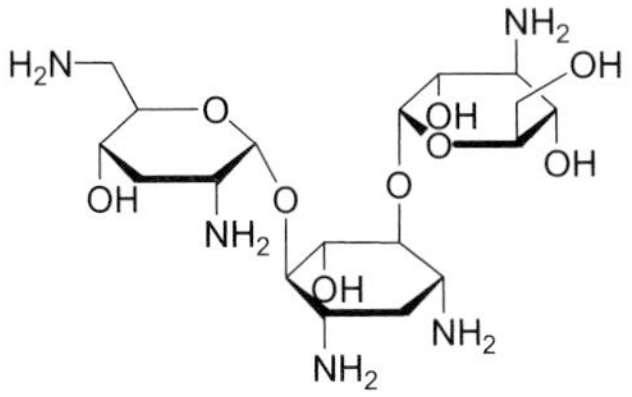

MW: 467.52
P: US 3,691,279 (1965, Eli Lilly)
L: 1974 (Eli Lilly)
S: H_2SO_4

Amikacin

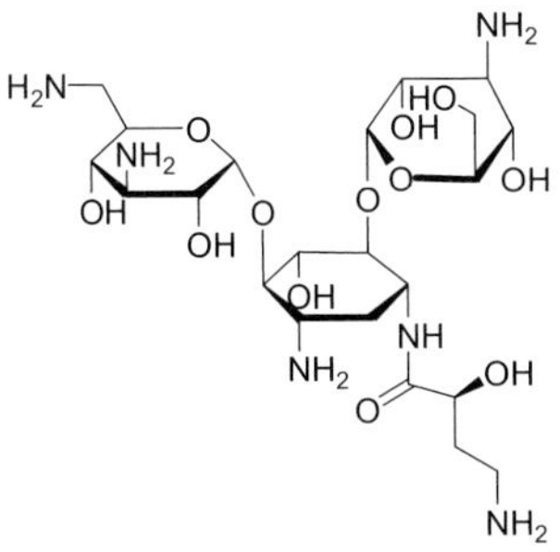

MW: 585.61
P: GB 1,401,221 (1971, Bristol Myers)
L: 1976 (Bristol-Myers Squibb)
S: –

Nelfinavir

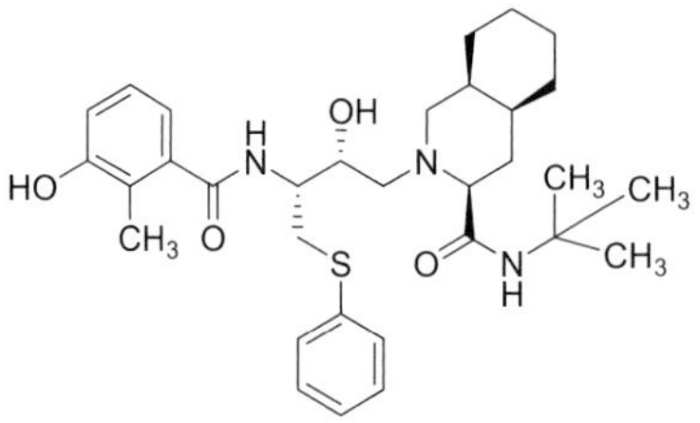

MW: 567.31
P: US 5,484,926
 (1992, Agouron Pharmaceutical)
L: 1997 (Agouron)
S: mesylate

Lopinavir

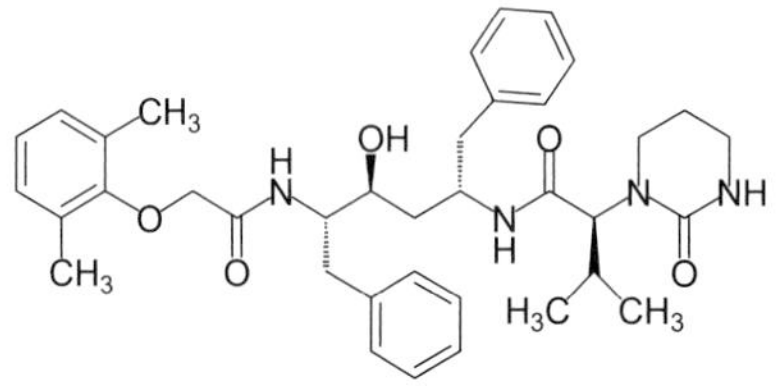

MW: 628.82
P: WO 97/21683 (1995, Abbott Labs.)
L: 2000 (Abbott)
S: –

L
Antineoplastic and Immunomodulating Agents

L01BC
Pyrimidine analogues
(fluorouracil, cytarabine, tegafur, gemcitabine, capecitabine)

Fluorouracil

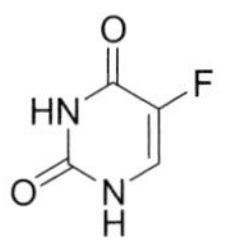

MW: 130.08
P: US 2,802,005 (1956, C. Heidelberger)
L: 1962 (Roche)
S: –

Cytarabine

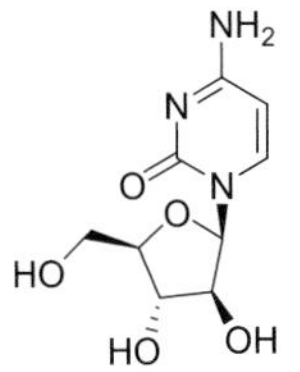

MW: 243.22
P: US 3,116,282 (1960, Upjohn)
L: 1969 (Pfizer)
S: –

Tegafur

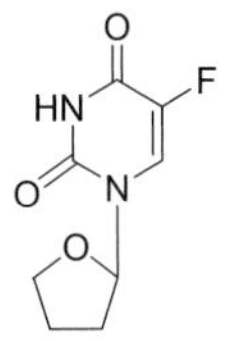

MW: 200.17
P: US 3,635,946 (1967, S.A. Giller)
L: 1972
S: Na

Gemcitabine

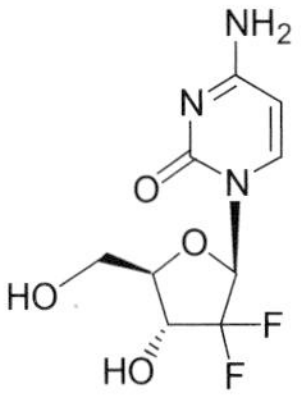

MW: 263.19
P: GB 2,136,425 (1983, Eli Lilly)
L: 1995 (Lilly)
S: HCl

Capecitabine

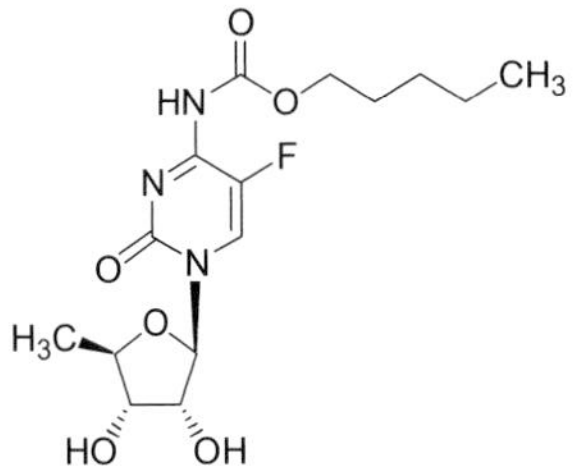

MW: 359.35
P: EP 602,454 (1992, Hoffmann-La Roche)
L: 1998 (Hoffmann-La Roche)
S: –

L01CD
**Taxanes
(paclitaxel, docetaxel)**

Paclitaxel

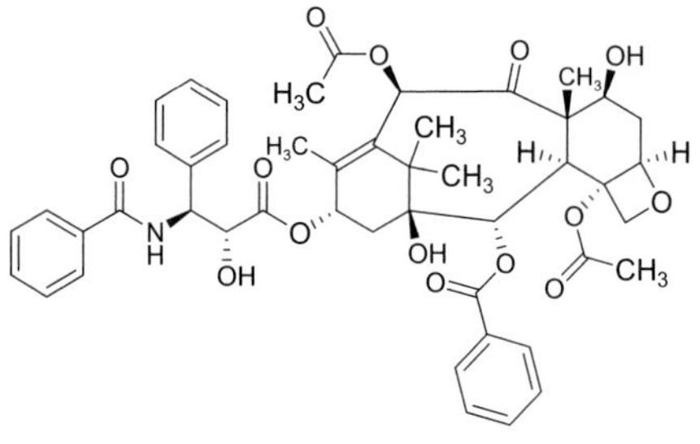

MW: 853.92
P: Mansukhlal C. Wani et al.: J. Am.
Chem. Soc. 1971, 93, 2325
L: 1993 (Bristol-Myers Squibb)

Docetaxel

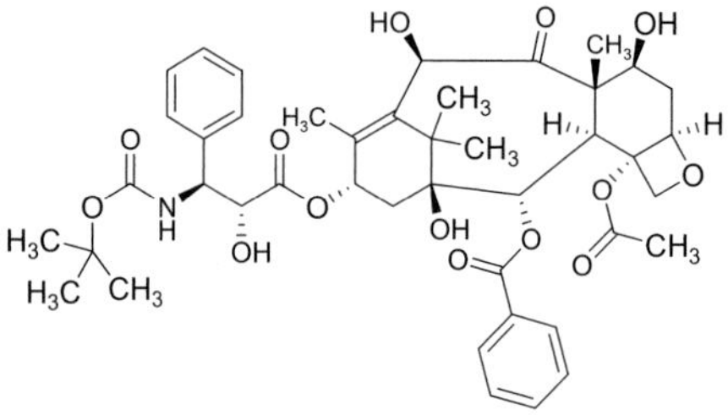

MW: 807.89
P: EP 253,738 (1986, Rhône-Poulenc Sante)
L: 1995 (Aventis Pharma)

L01XA
**Platinum Compounds
(cisplatin, carboplatin, oxaliplatin)**

Cisplatin

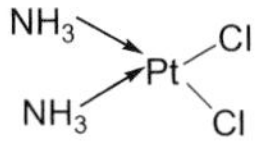

MW: 300.05
   P: Peyrone, M.: Ann. 1845, 51, 1
      Rosenberg, B.: Nature 1965, 205,
      698
   L: 1979 (Bristol-Myers Squibb)

Carboplatin

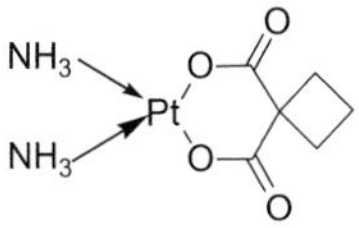

MW: 371.25
   P: DE 2,329,485 (1972, Research Corp.)
   L: 1986 (Bristol-Myers Squibb)

Oxaliplatin

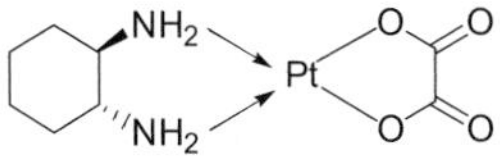

MW: 397.29
   P: JP 53,031,648 (1976, Y. Kidani & K. Inagaku)
   L: 1996 (Sanofi-Synthélabo)

L02AE

Gonadotropin Releasing Hormone Analogues
(leuprorelin, buserelin, triptorelin, goserelin)

Leuprorelin

MW: 1209.42
  P: US 4,005,063 (1973, Abbott Labs.)
  L: 1984 (Takeda)
  S: acetate

Buserelin

MW: 1239.45
  P: DE 2,438,350 (1974, Hoechst)
  L: 1985 (Aventis Pharma)
  S: acetate

Triptorelin

MW: 1311.47
  P: US 4,010,125 (1975, Tulane Univ.)
  L: 1986 (Ferring, Ipsen)
  S: pamoate

Goserelin

MW: 1269.43
  P: US 4,100,274 (1976, Imperial Chemical
    Industries, ICI)
  L: 1987 (AstraZeneca)
  S: acetate

L02BB
**Anti-androgenes
(flutamide, bicalutamide)**

Flutamide

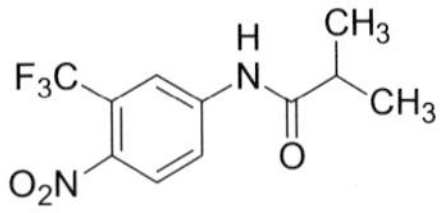

MW: 276.21
   P: Baker, J. W. et al.: J. Med. Chem.
   1967, 10, 93, DE 2,130,450
   (1971, Scherico)
   L: 1983 (Schering-Plough)

Bicalutamide

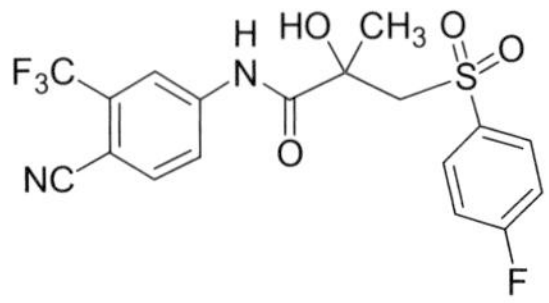

MW: 430.38
   P: EP 100,172 (1982, Imperial Chemical
   Industries, ICI)
   L: 1995 (AstraZeneca)

L02BG

Antineoplastic and immunomodulating agents
(letrozole, anastrozole)

Letrozole

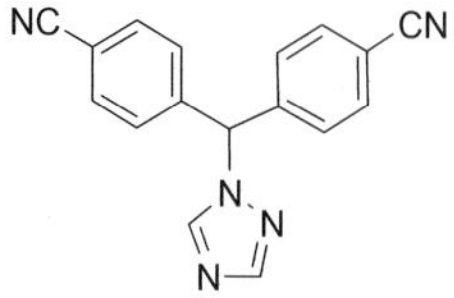

MW: 285.31
P: EP 236,940 (1986, Ciba-Geigy)
L: 1996 (Novartis)

Anastrozole

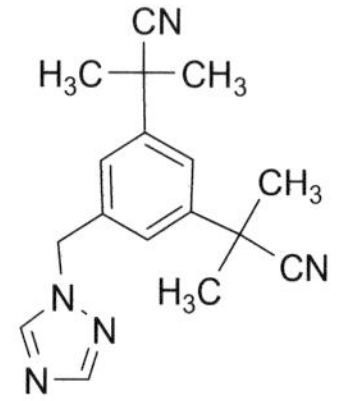

MW: 293.37
P: EP 296,749 (1987, Imperial Chemical
Industries, ICI)
L: 1995 (AstraZeneca)

M
Musculo-Skeletal System

M01AB
Acetic Acid Derivatives and Related Substances
(indometacin, diclofenac, sulindac, etodolac, aceclofenac)
and structural analogue
(lumiracoxib)

Indometacin

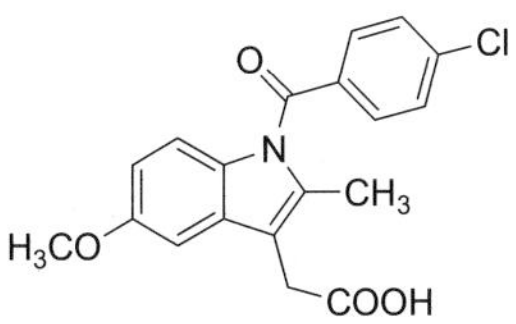

MW: 357.79
P: DE 1,620,031 (1961, Merck & Co.)
L: 1963
S: Na

Diclofenac

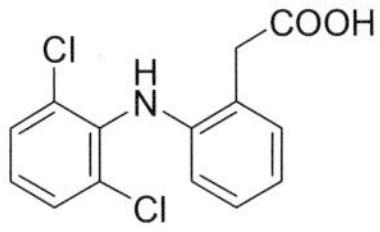

MW: 296.15
P: DE 1,793,592 (1965, Ciba-Geigy)
L: 1974 (Ciba-Geigy, Novartis)
S: Na

Sulindac

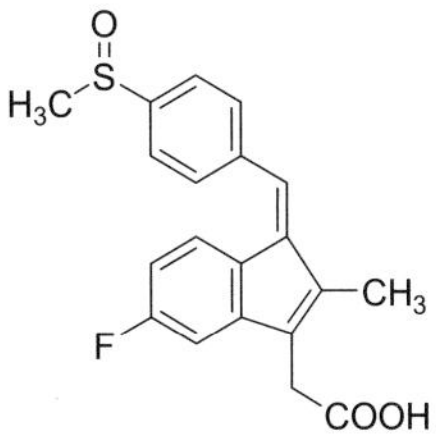

MW: 356.42
P: US 3,654,349 (1969, Merck & Co)
L: 1976 (Neopharmed)
S: −

Etodolac

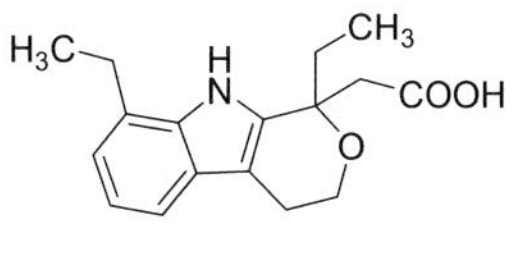

MW: 287.36
P: US 3,939,178 (1971, American Home
      Products)
L: 1985 (Roberts)
S: −

Aceclofenac

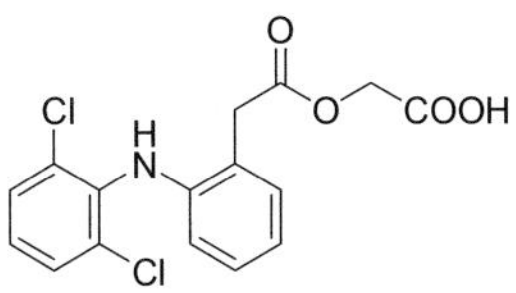

MW: 354.19
P: EP 119,932 (1983, Prodes)
L: 1992 (Almirall Prodesfarma)
S: −

Lumiracoxib

MW: 293.72
P: US 6,291,523 (1997, Novartis AG)
L: 2003 (Novartis)
S: –
R: selective COX-2 inhibitor
(ATC:M01AH)

M01AC
Oxicams
(piroxicam, tenoxicam, meloxicam, lornoxicam)

Piroxicam

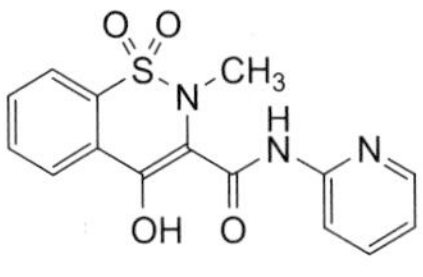

MW: 331.35
  P: US 3,591,584 (1968, Pfizer)
  L: 1979 (Pfizer)
  S: –

Tenoxicam

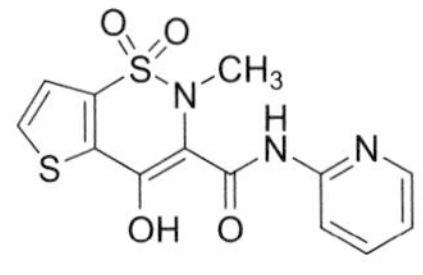

MW: 337.38
  P: US 4,076,709 (1974, Hoffmann-La Roche)
  L: 1987 (Hoffmann-La Roche)
  S: –

Meloxicam

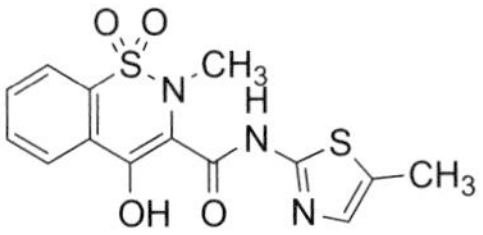

MW: 351.41
  P: DE 2,756,113 (1977, Thomae GmbH)
  L: 1996 (Boehringer Ingelheim)
  S: –

Lornoxicam

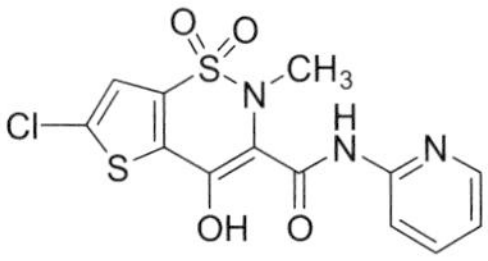

MW: 371.83
  P: DE 2,838,851 (1977, Hoffmann-La Roche)
  L: 1997 (Nycomed Pharma)
  S: –

M01AE
Propionic Acid derivatives
(ibuprofen, flurbiprofen, naproxen, ketoprofen, oxaprozin, tiaprofenic acid, loxoprofen, ketorolac)

Ibuprofen

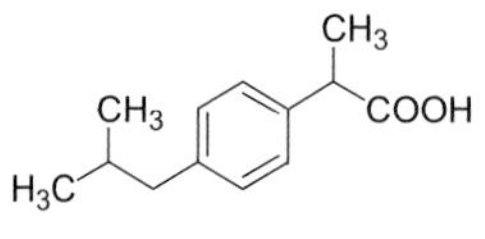

MW: 206.29
  P: DE 1,443,429 (1961, Boots Pure Drug)
  L: 1969 (BASF)
  S: lysinate

Flurbiprofen

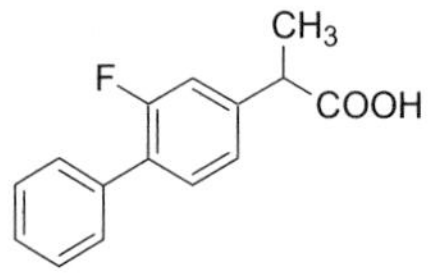

MW: 244.27
  P: DE 1,518,528 (1964, Boots Pure Drug)
  L: 1987 (BASF)
  S: Na

Naproxen

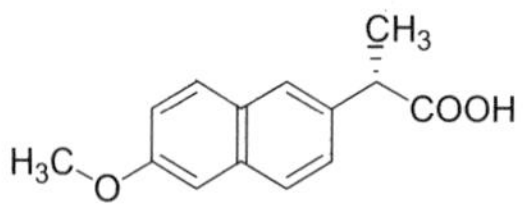

MW: 230.26
  P: US 3,896,157 (1967, Syntex)
  L: 1973 (Roche)
  S: Na

Ketoprofen

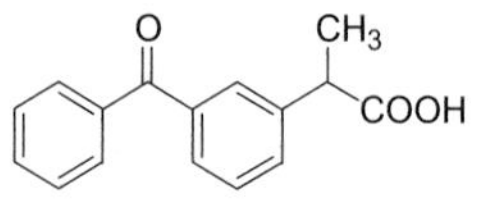

MW: 254.29
  P: US 3,641,127 (1967, Rhône-Poulenc)
  L: 1980 (Sanofi-Aventis)
  S: lysinate

Oxaprozin

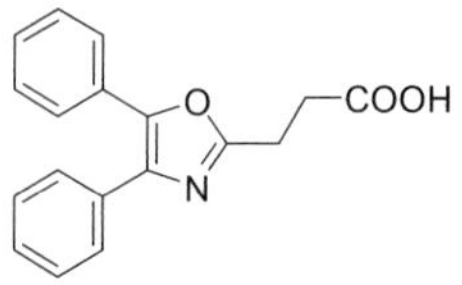

MW: 293.32
  P: US 3,578,671 (1967, Wyeth)
  L: 1983 (Wyeth)
  S: –

Tiaprofenic acid

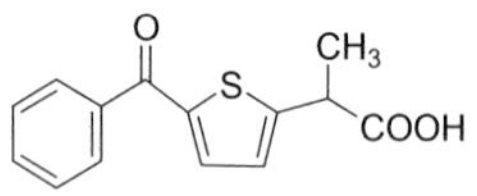

MW: 260.31
  P: FR 2,068,425 (1969, Roussel-Uclaf)
  L: 1981 (Sanofi-Aventis)
  S: –

Ketorolac

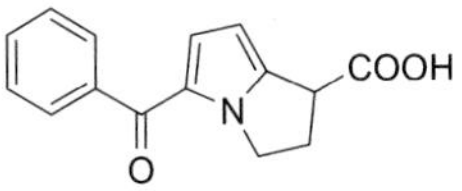

MW: 255.27
  P: US 4,089,969 (1976, Syntex)
  L: 1991 (Hoffmann-La Roche)
  S: tromethamine, Na

Loxoprofen

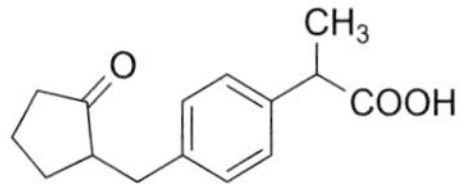

MW: 246.31
  P: DE 2,814,556 (1977, Sankyo)
  L: 1986 (Sankyo)
  S: Na

M01AH
Coxibs
(rofecoxib, celecoxib, valdecoxib, parecoxib, etoricoxib)

Rofecoxib

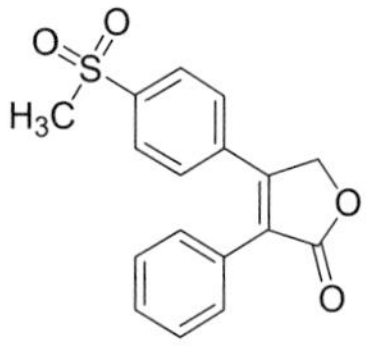

MW: 314.36
  P: EP 705,254 (1993, Merck Frosst)
  L: 1999 (Merck & Co.)
  S: –
  R: withdrawn (2004)

Celecoxib

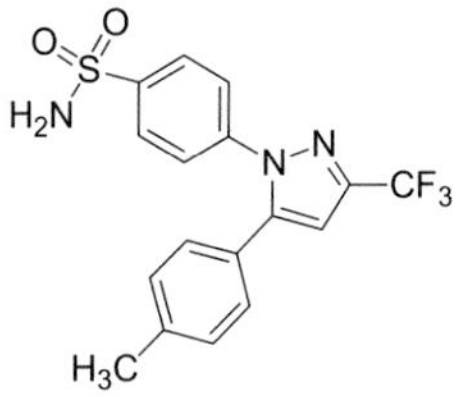

MW: 381.38
  P: WO 95/15316 (1993, Searle)
  L: 1999 (Searle)
  S: –

Valdecoxib

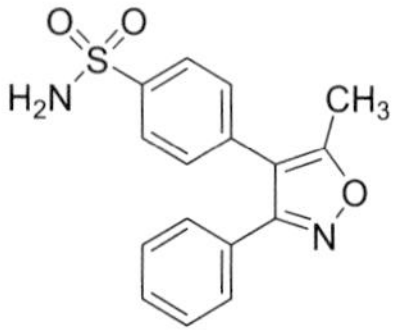

MW: 314.37
  P: US 5,633,272 (1995, Searle)
  L: 2002 (Pfizer)
  S: –
  R: withdrawn (2005)

Parecoxib

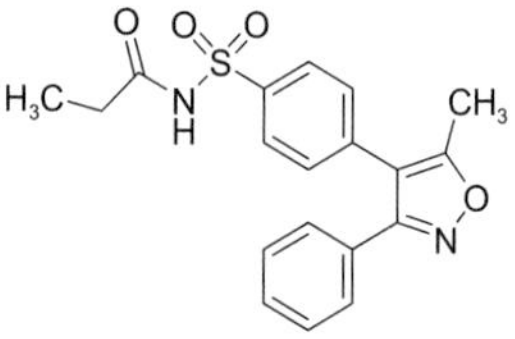

MW: 370.43
  P: US 5,932,598 (1996, Searle)
  L: 2002 (Pfizer)
  S: Na
  R: prodrug of valdecoxib

Etoricoxib

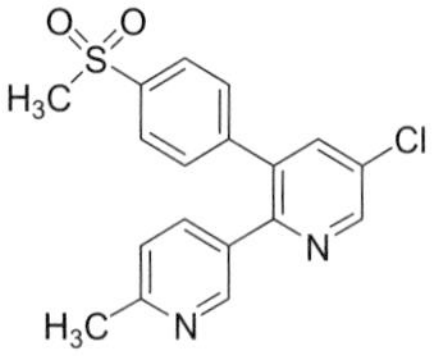

MW: 358.85
  P: US 5,861,419 (1996, Merck Frosst)
  L: 2002 (Merck & Co)
  S: –

M05BA

Bisphosphonates
(etidronic acid, pamidronic acid, alendronic acid, risedronate, ibandronic acid, zoledronic acid)

Etidronic acid

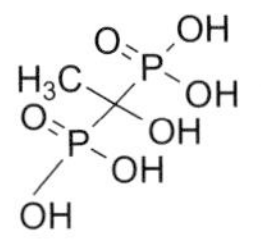

MW: 206.03
- **P:** Hofmann, K. A. et al.: Ber. 1897, 30, 1973.
  FR 1,531,913 (1966, Procter & Gamble)
- **L:** 1977 (Procter & Gamble)
- **S:** 2 Na

Pamidronic acid

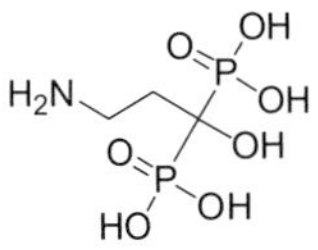

MW: 235.07
- **P:** DE 2,130,794 (1971, Benckiser)
- **L:** 1987 (Gador)
- **S:** 2 Na

Alendronic acid

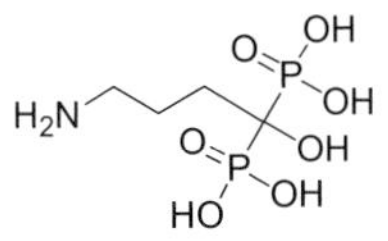

MW: 249.10
- **P:** Izv. Akad. Nauk. SSR, Ser. Khim 1978, 2, 433.
  DE 3,016,289 (1980, Henkel K.G.);
  GB 2,118,042 (1982, Gentili)
- **L:** 1993 (Merck & Co.)
- **S:** Na

Risedronate

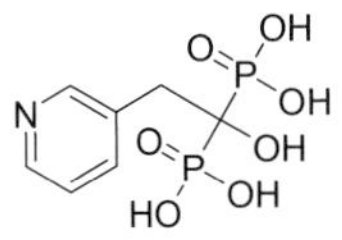

MW: 305.10
- **P:** EP 186,405 (1984, Procter and Gamble Co.)
- **L:** 1998 (Procter and Gamble Co.)
- **S:** Na

Ibandronic acid

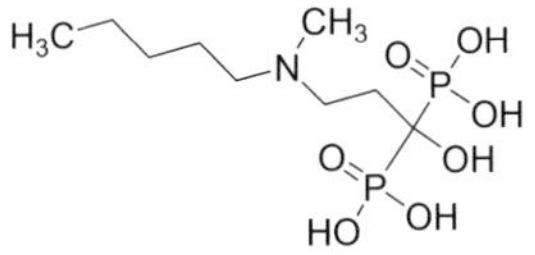

MW: 319.23
- **P:** EP 252,504 (1986, Boehringer Mannheim)
- **L:** 1996 (Hoffmann-La Roche)
- **S:** Na

Zoledronic acid

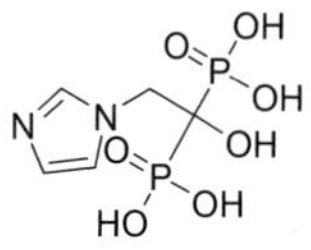

MW: 272.09

P: US 4,939,130 (1986, Ciba-Geigy)

L: 2000 (Novartis)

S: –

N
Nervous System

N02A
Opioids
(morphine, codeine, heroin, hydromorphone, hydrocodone, oxycodone, dihydrocodeine, levorphanol, pethidine, methadone, dextromethorphan, oxymorphone, pentazocine, naloxone, fentanyl, tramadol, nalbuphine, naltrexone, buprenorphine, butorphanol)

Morphine

MW: 285.34
P: Sertürner, F.W.: Trommsdorff's J. Pharm., 1805, 13, 234
L: –
S: HCl, sulfate

Codeine

MW: 299.37
P: Robiquet, P.J.: Ann. Chim., 1832, 51, 259
L: –
S: phosphate
R: **cough suppressant** (ATC: R 05DA)

Heroin

MW: 369,16
P: Dreser, H. Monatsh. 1898, 12, 509
L: 1898 (Bayer)
S: –, HCl

Dihydrocodeine

MW: 301.39
P: Skita, F.: Ber. 1911, 44, 2862
L: 1948
S: hydrogen tartrate
R: **cough suppressant** (ATC: R 05DA)

Hydromorphone

MW: 285.34
  P: DE 365,683 (1921, Knoll)
  L: Abbott
  S: HCl

Hydrocodone

MW: 299.37
  P: DE 415,097 (1923, E.Merck)
  L:
  S: R,R- tartrate, HCl
  R: **cough suppressant** (ATC: R 05DA)

Oxycodone

MW: 315.37
  P: DE 411,530 (1923, E.Merck AG)
  L: 1917
  S: HCl

Pethidine

MW: 247,34
  P: US 2,167,351 (1937, Winthrop Chemical)
  L: 1943 (Hoechst, Sanofi-Aventis)
  S: HCl

Methadone

MW: 309.45
  P: DE 865,314 (1941, Farbw. Hoechst)
  L: 1947 (Roxane)
  S: HCl
  R: **opioid dependence** (ATC: N 07BC)

Levorphanol

MW: 257.38
  P: Grewe, Naturwiss. 1946, 33, 333
      CH 280,674 (1949, Hoffmann-La Roche)
  L: 1951
  S: tartrate

Dextromethorphan

MW: 271.40
  P: US 2,676,177 (1949, Hoffmann – La Roche)
  L: 1953
  S: HBr
  R: cough suppressant (ATC: R 05DA)

Oxymorphone

MW: 301.34
  P: US 2,806,033 (1955, M.J. Lewenstein and
      U. Weiss)
  L: 1959 (Endo)
  S: HCl

Pentazocine

MW: 285.43
  P: GB 1,000,506 (1960,Sterling Drug)
  L: 1964
  S: HCl

Naloxone

MW: 327,38
  P: US 3,254,088 (1961, M.J. Lewenstein)
  L: 1969
  S: HCl
  R: antidote (ATC: V 03AB)

Fentanyl

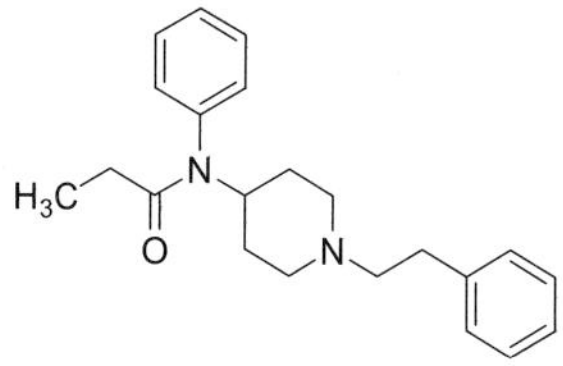

MW: 336.48
 P: US 3,164,600 (1961, Janssen)
 US 3,141,823 (1962, Janssen)
 L: 1963
 S: citrate

Tramadol

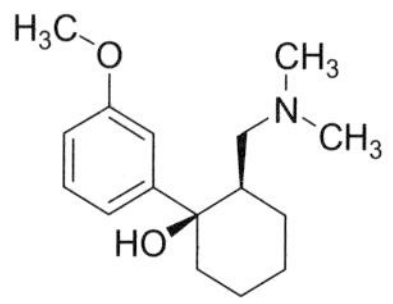

MW: 263.38
 P: GB 997,399 (1963, Gruenenthal Chemie)
 L: 1977 (Gruenenthal)
 S: HCl

Nalbuphine

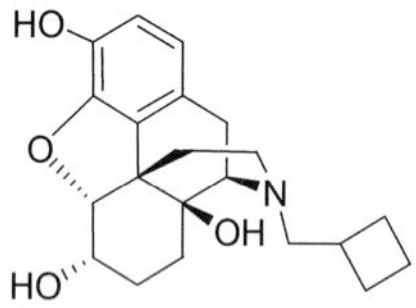

MW: 357.45
 P: US 3,332,950 (1963, Endo Laboratories)
 L: 1980 (Endo Laboratories)
 S: HCl

Naltrexone

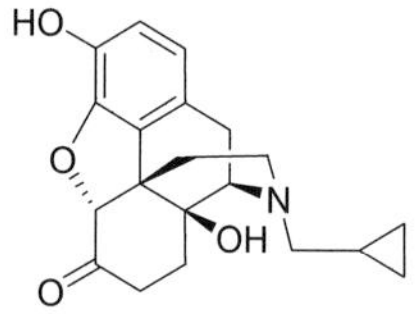

MW: 341,41
 P: US 3,332,950 (1963, Endo Laboratories)
 L: 1984
 S: HCl
 R: alcohol dependence
 (ATC: N 07BB)

Buprenorphine

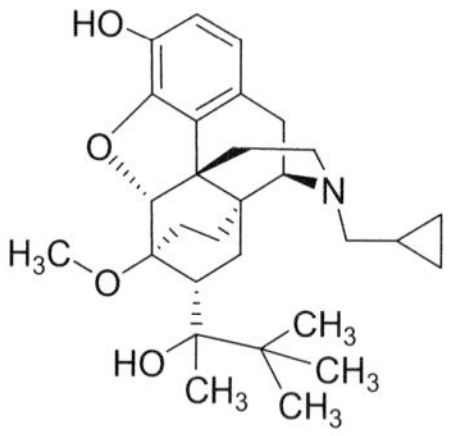

MW: 467,65
 P: DE 1,620,206 (1965, Reckitt & Sons)
 L: 1978 (Reckitt & Sons)
 S: HCl

Butorphanol

MW: 327.47
P: US 3,819,635 (1971, Brystol-Myers)
L: 1979
S: tartrate

N02BB
**Analgesics, Pyrazolones
(metamizole, propyphenazone)**

Metamizole

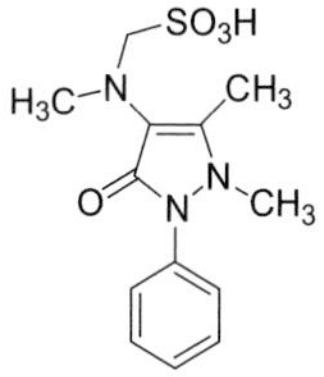

MW: 311.36
  P: DE 476,663 (1922, I.G. Farben)
  L: –
  S: Na

Propyphenazone

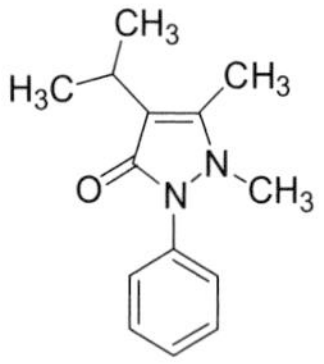

MW: 230.31
  P: DE 565,799 (1931, Hoffmann-La Roche)
  L: –

N02CC
Analgesics, Selective Serotonin (5HT₁) Agonists
(sumatriptan, naratriptan, zolmitriptan, rizatriptan, almotriptan)

Sumatriptan

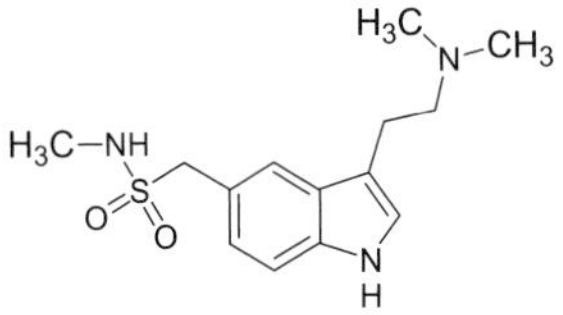

MW: 295.41
  P: GB 2,124,210 (1982, Glaxo)
  L: 1991 (GlaxoSmithKline)
  S: succinate

Naratriptan

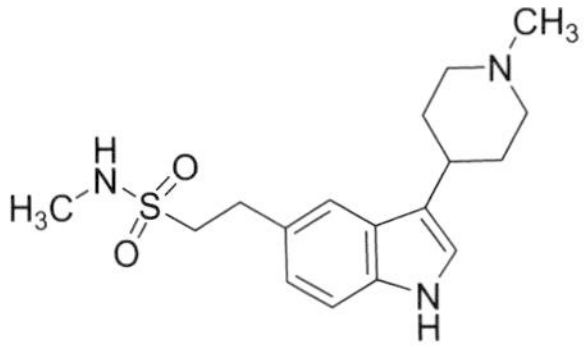

MW: 335.47
  P: EP 303,507 (1987, Glaxo)
  L: 1997 (GlaxoSmithKline)
  S: HCl

Zolmitriptan

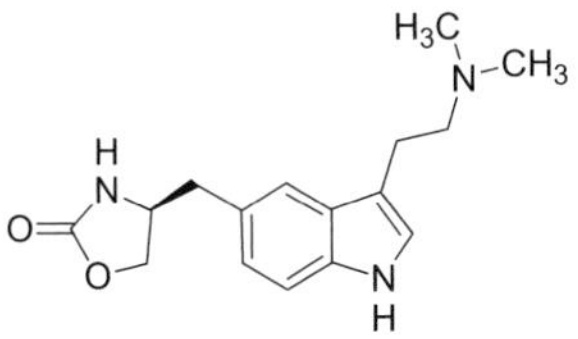

MW: 287.36
  P: EP 486,666 (1990, Wellcome Foundation)
  L: 1997 (AstraZeneca)
  S: HCl

Rizatriptan

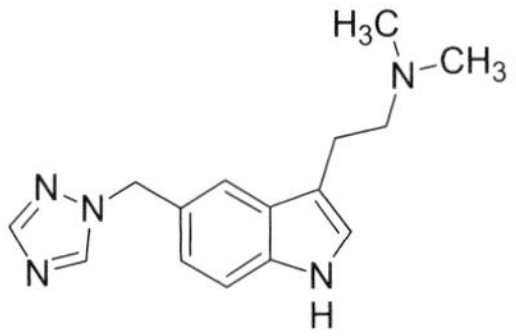

MW: 269.35
  P: EP 497,512 (1991, Merck Sharp & Dohme)
  L: 1998 (Merck & Co.)
  S: benzoate, H₂SO₄

Almotriptan

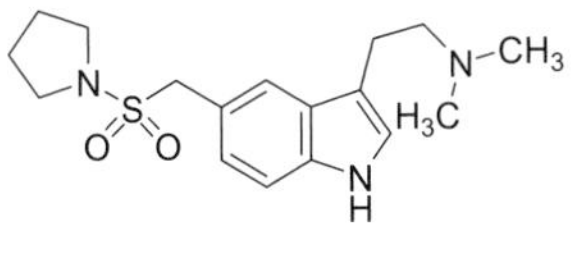

MW: 335.47
  P: EP 605,697 (1992, Almirall S. A.)
  L: 2000 (Almirall Prodespharma)
  S: maleate

N03AF
Antiepileptics, Carboxamide Derivatives
(carbamazepine, oxcarbamazepine)

Carbamazepine

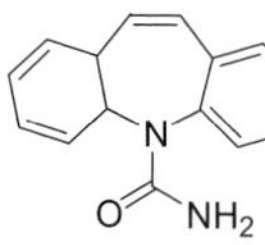

MW: 236.27
   P: US 2,948,718 (1957, Geigy Chemical)
   L: 1963

Oxcarbazepine

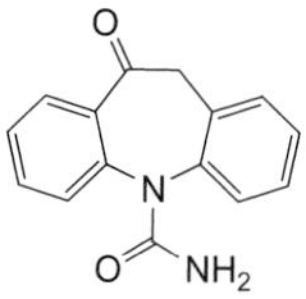

MW: 252.27
   P: DE 2,011,087 (1969, Geigy)
   L: 1990 (Novartis)

N04BC
Anti-Parkinson Drugs, Dopamine Agonists
(bromocriptine, pergolide, cabergoline)

Bromocriptine

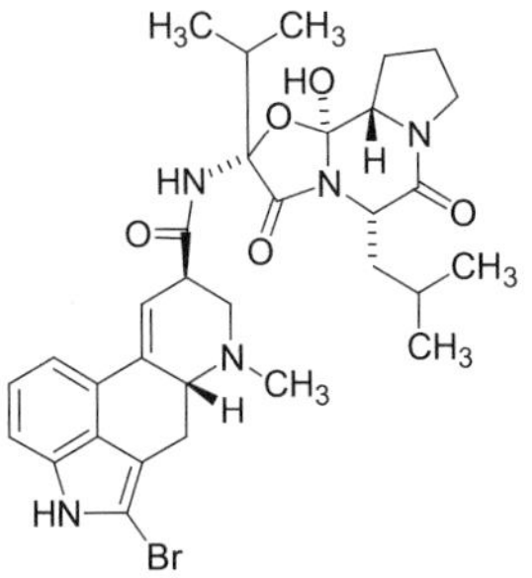

MW: 654.61
  P: US 3,752,814 (1968, Sandoz)
  L: 1975
  S: mesylate

Pergolide

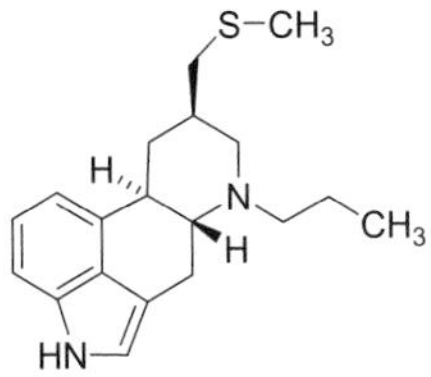

MW: 314.50
  P: US 4,166,182 (1978, Lilly)
  L: 1989
  S: mesylate

Cabergoline

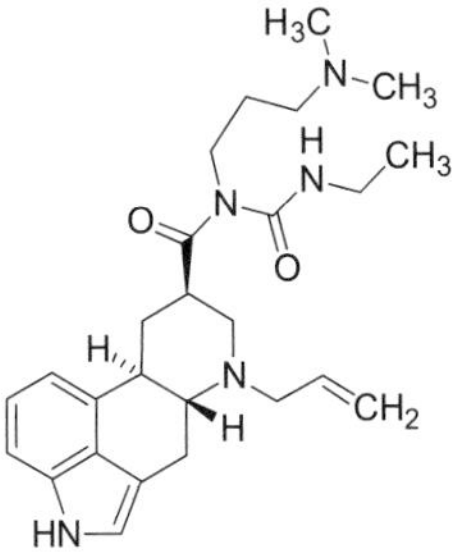

MW: 451.62
  P: GB 2,074,566 (1980, Farmitalia Carlo Erba)
  L: 1993 (Pharmacia)
  S: –

N05AH

**Psycholeptics, Diazepines, Oxazepine and Thiazepines
(clozapine, quetiapine, olanzapine)**

Clozapine

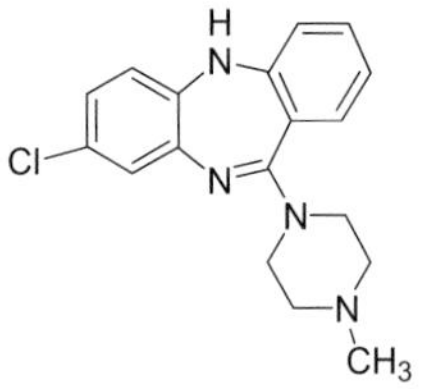

MW: 326.83
  P: GB 980,853 (1960, Wander)
  L: 1972 (Wander)
  S: –

Quetiapine

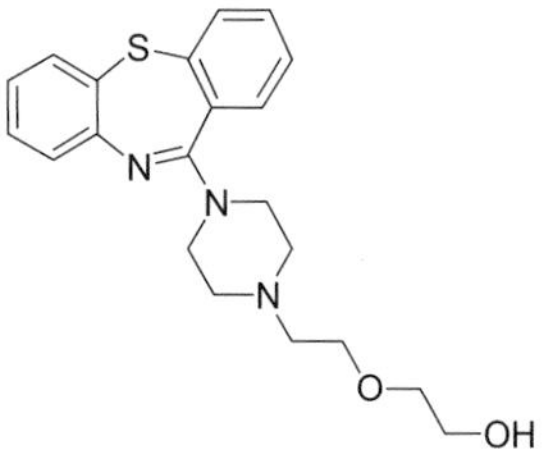

MW: 383.52
  P: EP 240,228 (1986, Imperial Chemical
    Industries, America, ICI)
  L: 1996 (AstraZeneca)
  S: fumarate

Olanzapine

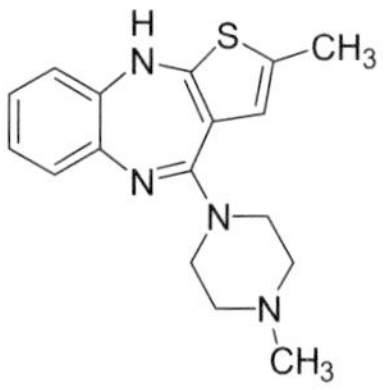

MW: 312.44
  P: EP 454,436 (1990, Lilly)
  L: 1996 (Lilly)
  S: –

N05BA

Anxiolytics, Benzodiazepine Derivatives
(chlordiazepoxide, diazepam, clonazepam, bromazepam, lorazepam, oxazepam, clorazepate, etizolam, alprazolam, /midazolam/)

Chlordiazepoxide

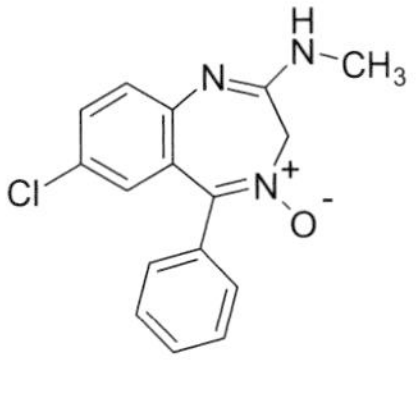

MW: 299.76
> **P:** US 2,893,992 (1958, Hoffmann-La Roche)
> **L:** 1960 (Hoffmann-La Roche)

Diazepam

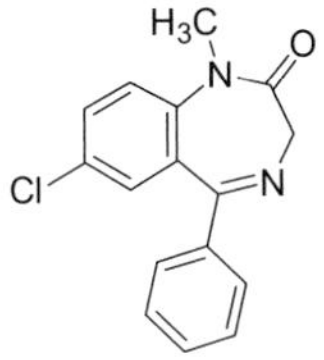

MW: 284.75
> **P:** DE 1,136,709 (1959, Hoffmann-La Roche)
> **L:** 1961 (Hoffmann-La Roche)

Clonazepam

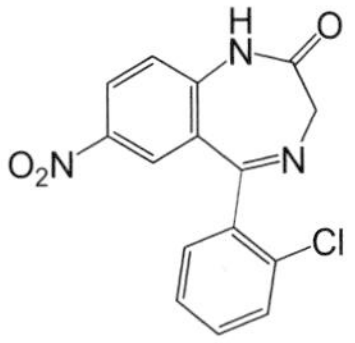

MW: 315.72
> **P:** US 3,121,114 (1960, Hoffmann-La Roche)
> **L:** 1973
> **S:** –
> **R: antiepileptic** (ATC: N03AE)

Bromazepam

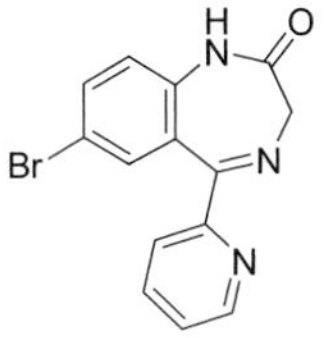

MW: 316.16
> **P:** US 3,100,770 (1961, Hoffmann-La Roche)
> **L:** 1974

Lorazepam

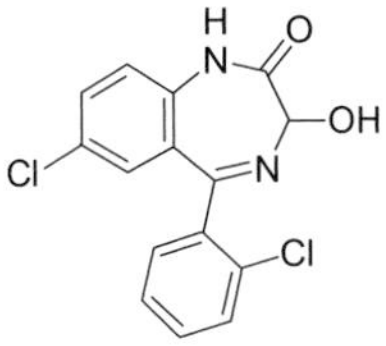

MW: 321.16
> **P:** US 3,296,249 (1962, American Home Products)
> **L:** 1971

Oxazepam

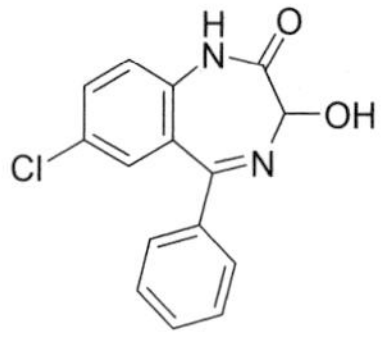

MW: 286.72
P: DE 1,645,904 (1962, American Home
   Products)
L: 1964 (Wyeth)
S: –

Clorazepate

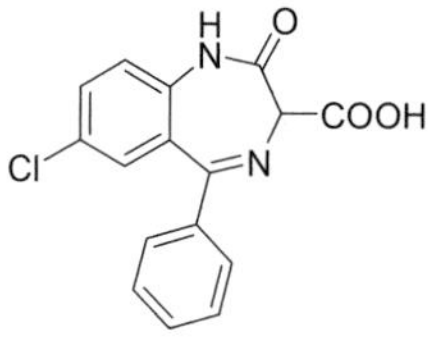

MW: 314.73
P: DE 1,518,764 (1965, Clin.Byla Ind.)
L: 1967 (Sanofi-Synthelabo)
S: 2 K

Etizolam

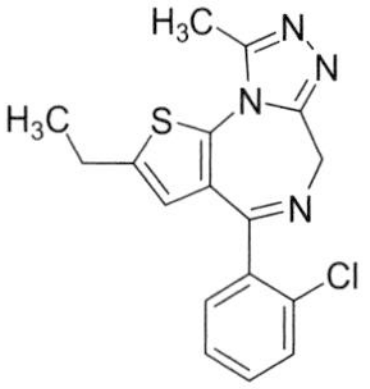

MW: 342.85
P: US 3,904,641 (1971, Yoshitomi
   Pharmaceut.)
L: 1984 (Mitsubishi Pharma)
S: –

Alprazolam

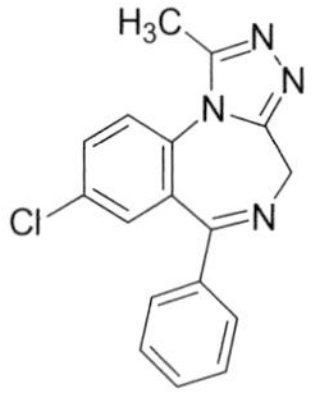

MW: 308.77
P: DE 2,203,782 (1971, Upjohn)
L: 1983 (Pharmacia, Pfizer)
S: –

N05CD

Hypnotics and Sedatives
Benzodiazepine Derivatives
(nitrazepam, lormetazepam, temazepam, flunitrazepam, flurazepam, estazolam, triazolam, quazepam, doxefazepam, brotizolam, midazolam, loprazolam, cinolazepam)

Nitrazepam

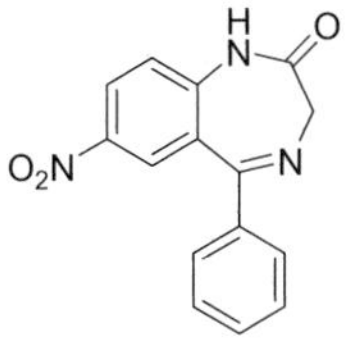

MW: 281.27
P: US 3,109,843 (1961, Hoffmann-La Roche)
L: 1965 (Hoffmann-La Roche)
S: –

Lormetazepam

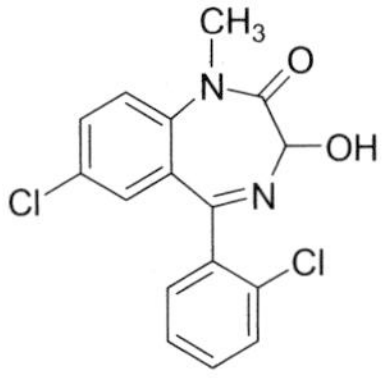

MW: 335.19
P: US 3,295,249 (1961, American Home Products)
L: 1980 (Wyeth)
S: –

Temazepam

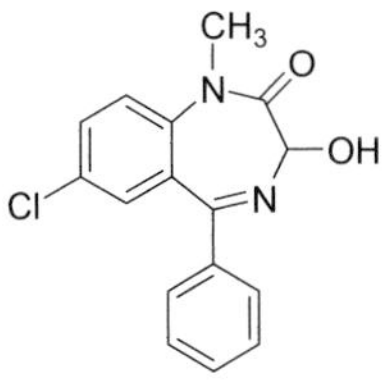

MW: 300.75
P: GB 1,022,642 (1962, American Home Products)
L: 1969
S: –

Flunitrazepam

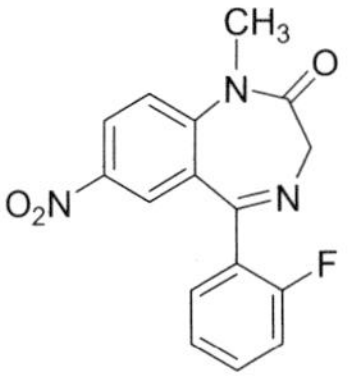

MW: 313.29
P: US 3,116,203 (1962, Hoffmann-La Roche)
L: 1974 (Hoffmann-La Roche, Chugai)
S: –

Flurazepam

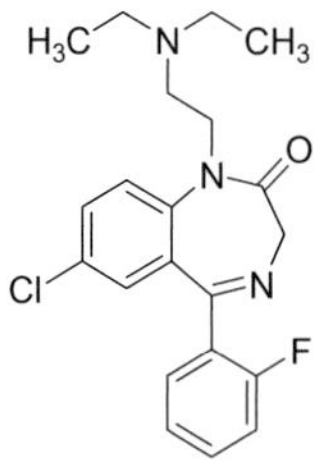

MW: 387.89
P: US 3,567,710 (1968, Hoffmann-La Roche)
L: 1968 (Hoffmann-La Roche)
S: 2 HCl

Estazolam

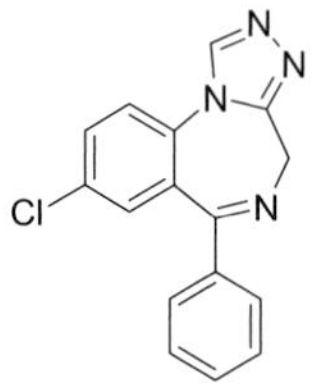

MW: 294.75
P: US 4,116,956 (1968, Takeda)
L: 1975 (Takeda)
S: –

Triazolam

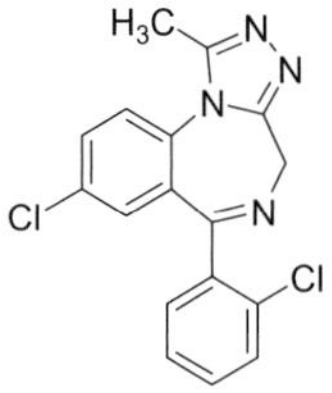

MW: 343.22
P: DE 2,012,190 (1969, Upjohn)
L: 1978 (Upjohn, Pfizer)
S: –

Quazepam

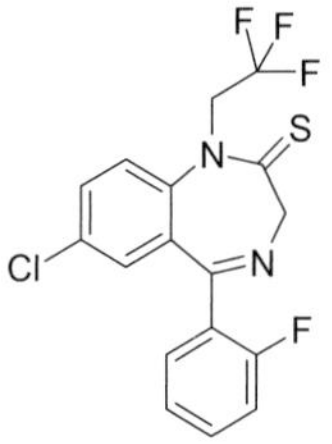

MW: 386.80
P: US 3,845,039 (1970, Schering Corp.)
L: 1985 (Schering-Plough)
S: –

R
Respiratory System

R01BA
Nasal Decongestants for Systemic Use Sympathomimetics (pseudoephedrine, phenylpropanolamine, phenylephrine) and structural analogues (epinephrine, ephedrine)

Pseudoephedrine

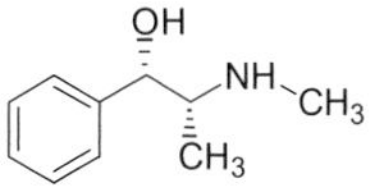

MW: 165.26
P: Schmidt, E. et al. Arch.Pharmaz. **1909**, 247, 141
L: –
S: HCl, ½ H_2SO_4
R: combinations with H_1 antagonists

Phenylephrine

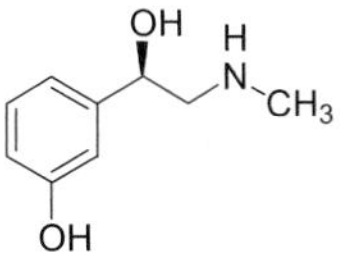

MW: 167.21
P: US 1,932,347
(1927, Frederick Stearns & Co.)
L: 1938
S: HCl

Phenylpropanolamine

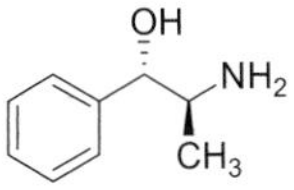

MW: 151.21
P: US 2,151,517 (1938, Jonas Kamlet)
L: –
S: HCl

Epinephrine (Adrenaline)

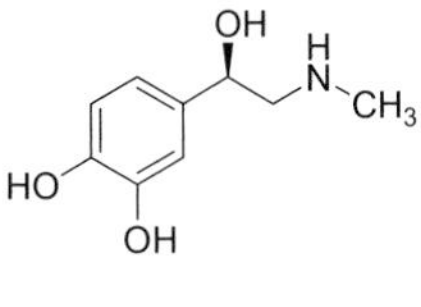

MW: 183.21
P: DE 152,814 (1903, Hoechst)
L: 1905
S: tartrate, HCl
R: hormone, **cardiac stimulant (C01C),**
combinations with local anesthetics (N01BB)

Ephedrine

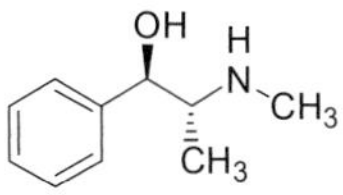

MW: 165.26
P: US 1,956,950 (1930, E. Bilhuber)
L: 1926
S: H_2SO_4, HCl
R: **adrenergic for systemic use (R03CA)**

R03AC
Drugs for Obstructive Airway Diseases
Selective Beta-2-adrenoreceptor Agonists
(fenoterol, reproterol, terbutaline, salbutamol, clenbuterol, pirbuterol, formoterol, procaterol, salmeterol)

Fenoterol

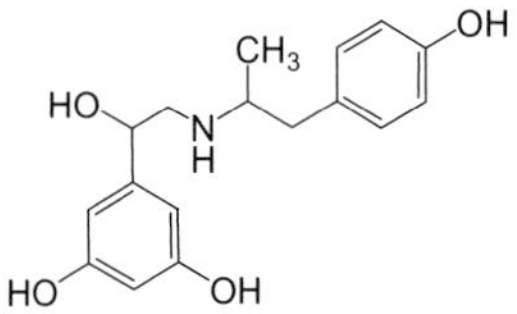

MW: 303.36
P: DE 1,286,047 (1962, Boehringer Ingelheim)
L: 1971
S: HBr

Reproterol

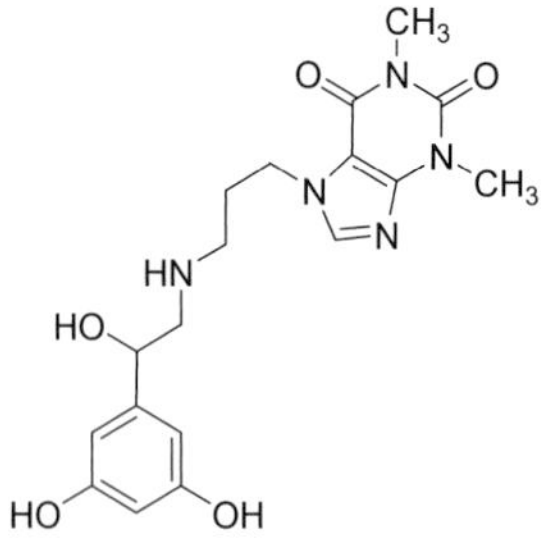

MW: 389.41
P: DE 1,545,725 (1965, Degussa)
L: 1977 (Asta Medica)
S: HCl

Terbutaline

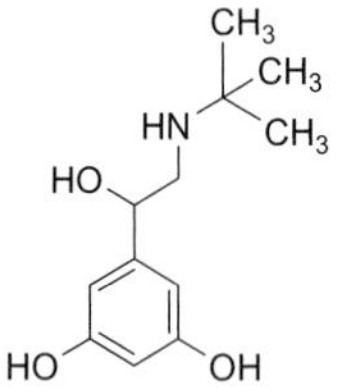

MW: 225.29
P: GB 1,199,630 (1966, Draco, Lund)
L: 1970
S: H_2SO_4

Salbutamol

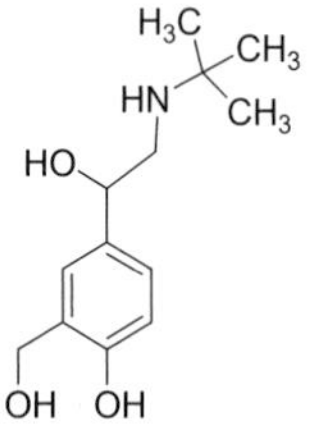

MW: 239.32
P: GB 1,200,886 (1966, Allen & Hanburys)
L: 1968 (Glaxo)
S: H_2SO_4

Clenbuterol

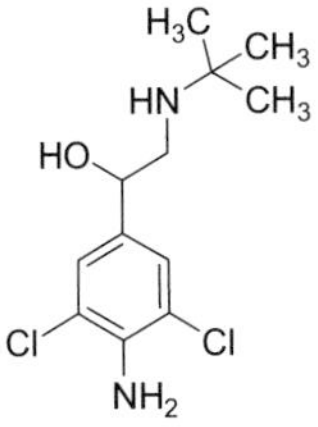

MW: 277.20
  P: DE 1,793,416 (1967, Thomae)
  L: 1977
  S: HCl

Pirbuterol

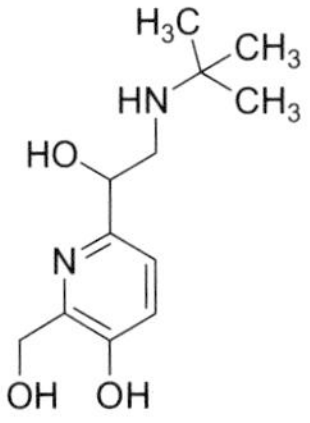

MW: 240.30
  P: US 3,700,681 (1971, Pfizer)
  L: 1983 (Pfizer)
  S: acetate

Formoterol

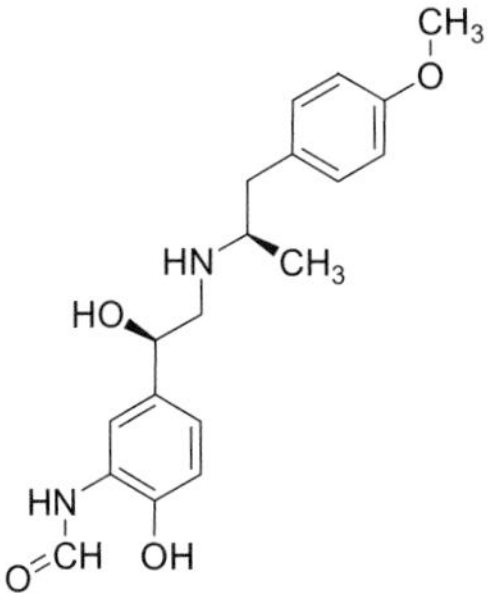

MW: 344.41
  P: DE 2,305,092 (1972, Yamanouchi)
  L: 1986 (Yamanouchi)
  S: fumarate (2:1)

Procaterol

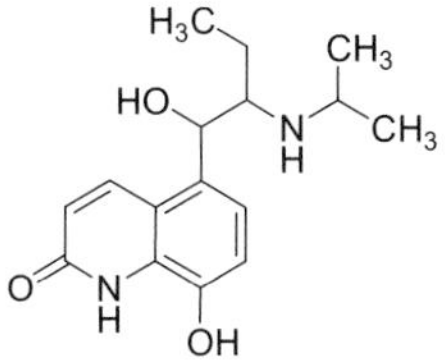

MW: 290.36
  P: DE 2,461,596 (1974, Otsuka)
  L: 1980 (Otsuka)
  S: HCl

Salmeterol

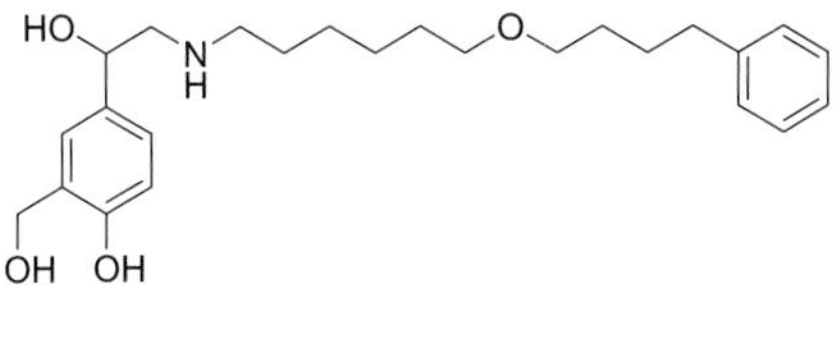

MW: 415.57
  P: DE 3,414,752 (1983, Glaxo)
  L: 1990 (GlaxoSmithKline)
  S: xinafoate

R05CB
Cough and Cold Preparations
Mucolytics
(acetylcysteine, bromhexine, carbocisteine, ambroxol)

a. Cysteine derivatives

Carbocisteine

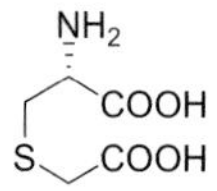

MW: 179.20
P: Armstrong, L. et al.: J. Org. Chem.
   1951, 16, 749
L: 1960
S: –

Acetylcysteine

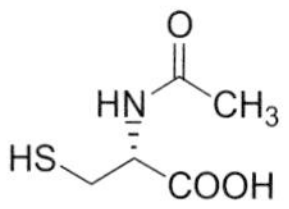

MW: 163.20
P: US 3,091,569 (1960, Mead Johnson)
L: 1968 (Zambon)
S: Na

---

b. Bromhexine and its active metabolite

Bromhexine

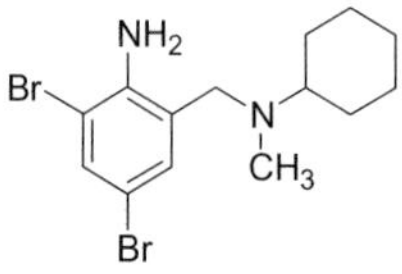

MW: 376.14
P: DE 1,169,939 (1961, Thomae)
L: 1966
S: HCl

Ambroxol

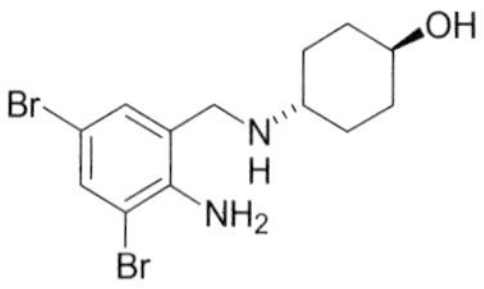

MW: 378.11
P: US 3,536,713 (1966, Thomae)
L: 1979 (Boehringer Ingelheim)
S: HCl

R06AA–AX
Antihistamines for Systemic Use
(diphenhydramine, mepyramine, promethazine, carbinoxamine, chlorpheniramine, triprolidine, bromphenyramine, pheniramine, doxylamine, meclozine, dimetindene, cyproheptadine, clemastine, dexchlorphenyramine, mequitazine, ketotifen, azelastine, oxatomide, fexofenadine, epinastine, loratadine, cetirizine, ebastine, desloratadine, olopatadine)

Diphenhydramine

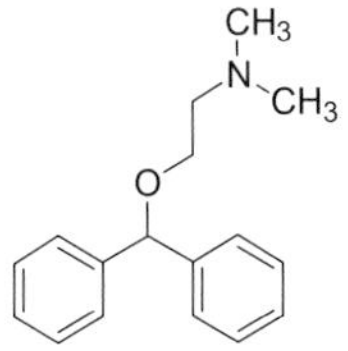

MW: 255.36
P: US 2,567,351 (1946, Parke, Davis)
L: 1946 (Parke, Davis)
S: HCl
R: **topically use** (ATC: D04A)

Mepyramine

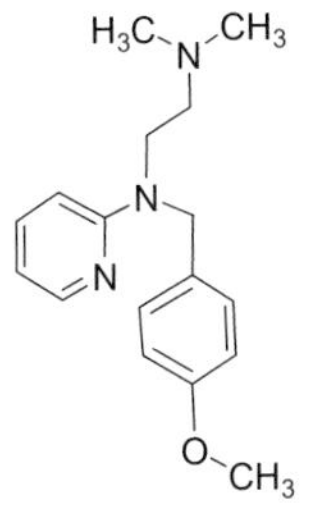

MW: 285.35
P: US 2,502,151 (1943, Rhône-Poulenc)
L: 1949
S: –
R: **topically use** (ATC: D04A)

Promethazine

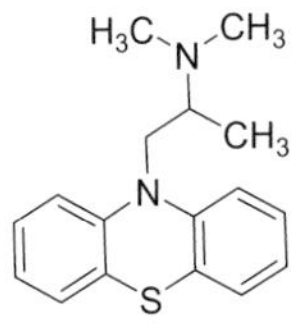

MW: 284.43
P: US 2,530,451 (1946, Rhône-Poulenc)
L: 1949
S: HCl
R: **topically use** (ATC: D04A)

Carbinoxamine

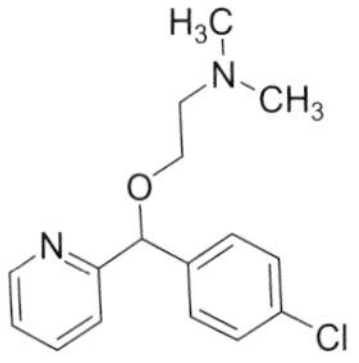

MW: 290.79
P: US 2,606,195 (1947, Merrell)
L: 1953
S: –

Chlorphenamine

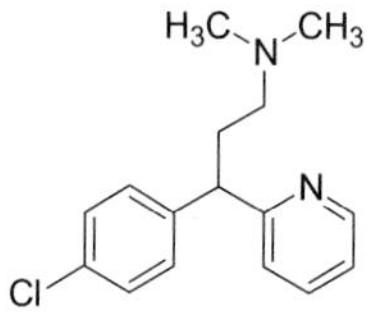

H₃C–N–CH₃ structure (4-chlorophenyl pyridine propanamine)

MW: 274.80
  P: US 2,567,245 (1948, Schering Corp.)
  L: 1949 (Schering-Plough)
  S: maleate

Triprolidine

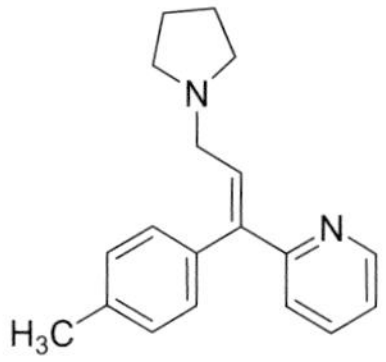

Pyrrolidine structure with 4-methylphenyl and pyridine

MW: 278.40
  P: US 2,712,020 (1948, Burroughs Wellcome)
  L: 1953 (GlaxoSmithKline)
  S: HCl

Brompheniramine

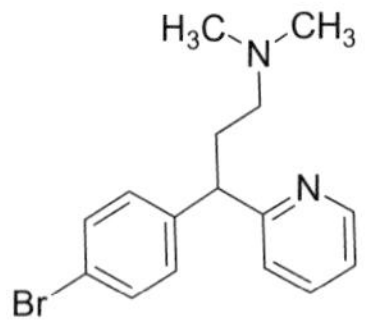

H₃C–N–CH₃ structure (4-bromophenyl pyridine propanamine)

MW: 319.25
  P: US 2,567,245 (1948, Schering Corp.)
  L: 1955 (Schering-Plough)
  S: –

Pheniramine

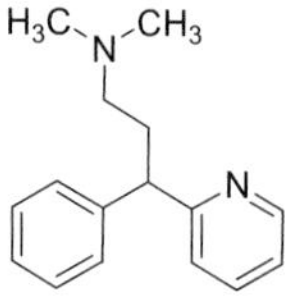

H₃C–N–CH₃ structure (phenyl pyridine propanamine)

MW: 240.35
  P: US 2,567,245 (1948, Schering Corp.)
  L: –
  S: maleate

Doxylamine

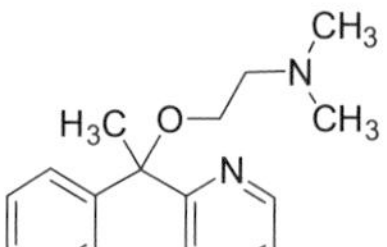

structure with O, N(CH₃)₂, phenyl and pyridine

MW: 270.38
  P: Tilford, C. H. et al.: J. Am. Chem. Soc.
    1948, 70, 4001
  L: –
  S: succinate

Meclozine

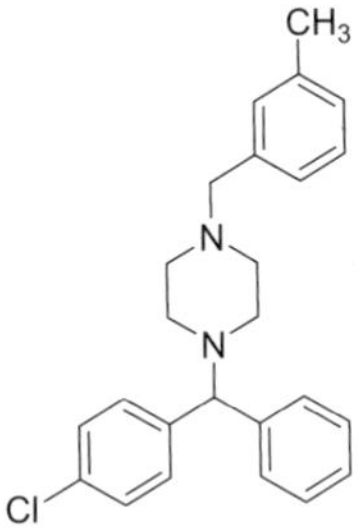

MW: 390.96
P: US 2,709,169 (1951, Union Chimique Belge, UCB)
L: 1953 (Pfizer)
S: HCl

Dimetindene

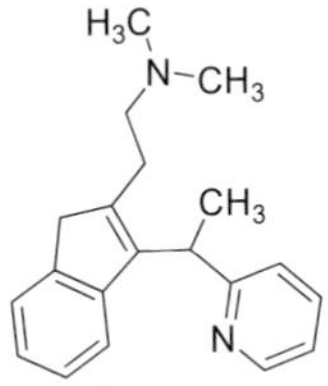

MW: 292.43
P: US 2,970,149 (1958, Ciba)
L: 1960
S: maleate
R: **topically use** (ATC: D04A)

Cyproheptadine

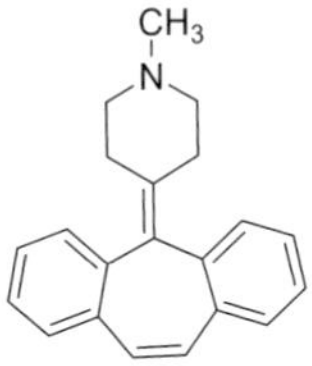

MW: 287.41
P: US 3,014,911 (1959, Merck & Co)
L: 1961 (Merck & Co)
S: HCl

Clemastine

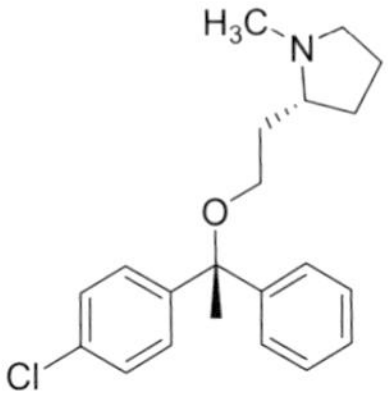

MW: 343.90
P: GB 942,152 (1960, Sandoz)
L: 1967 (Novartis)
S: hydrogen fumarate
R: **topically use** (ATC: D04A)

Dexchlorpheniramine

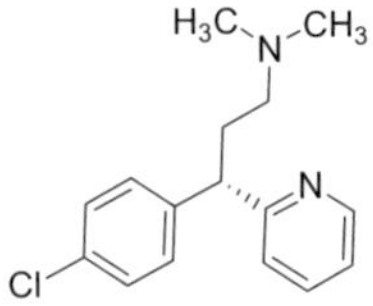

MW: 274.80
P: US 3,061,517 (1962, Schering Corp.)
L: 1959
S: maleate

Mequitazine

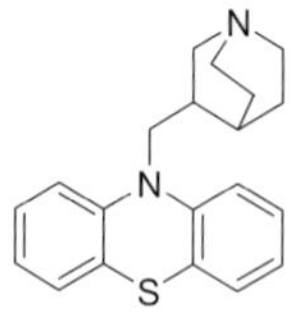

MW: 322.48
P: DE 2,009,555 (1969, Sogeras)
L: 1976 (Aventis)
S: –

Ketotifen

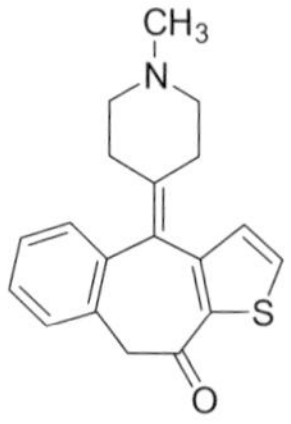

MW: 309.43
P: DE 2,111,071 (1970, Sandoz)
L: 1978
S: fumarate

Azelastine

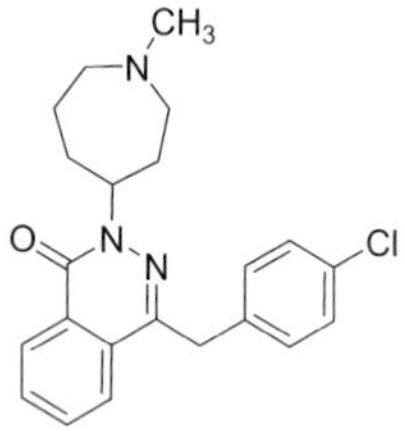

MW: 381.91
P: DE 2,164,058 (1971, ASTA-Werke)
L: 1986 (Eisai)
S: HCl

Oxatomide

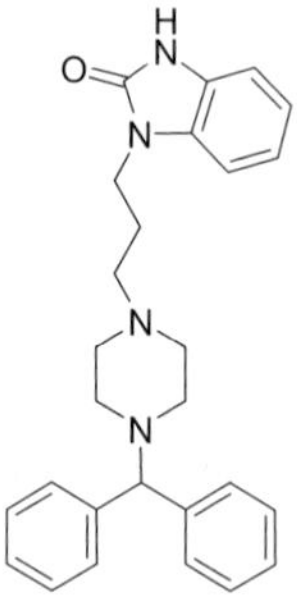

MW: 426.56
P: BE 852,405 (1976, Janssen)
L: 1981 (Janssen)
S: –

Fexofenadine

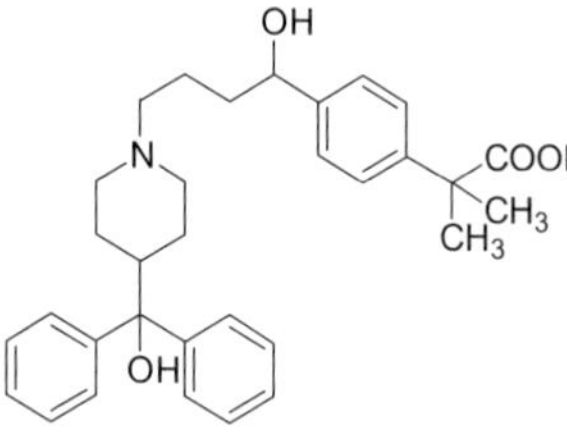

MW: 501.67
P: US 4,254,129 (1979, Richardson-Merrell)
L: 1996 (Aventis Pharma)
S: HCl

Epinastine

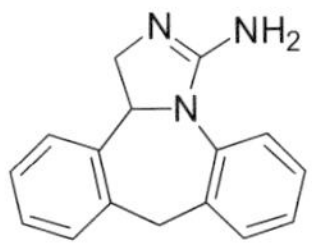

MW: 249.32
P: DE 3,008,944 (1980, Boehringer Ingelheim)
L: 1994 (Nippon Boehringer Ingelheim)
S: HCl

Loratadine

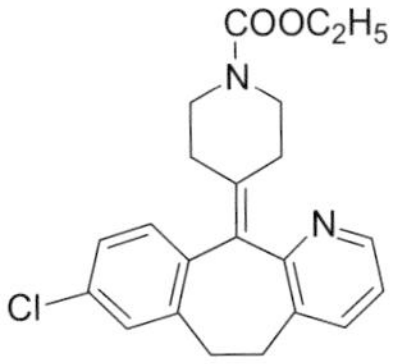

MW: 382.89
P: US 4,282,233 (1980, Schering Corp.)
L: 1988 (Schering-Plough)
S: –

Cetirizine

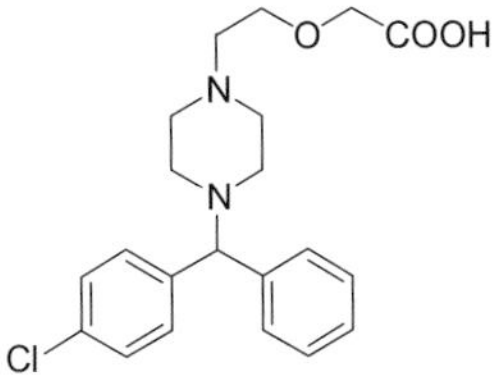

MW: 388.90
P: EP 58,146 (1981, Union Chimique Belge, UCB)
L: 1987 (Union Chimique Belge, UCB)
S: 2 HCl

Ebastine

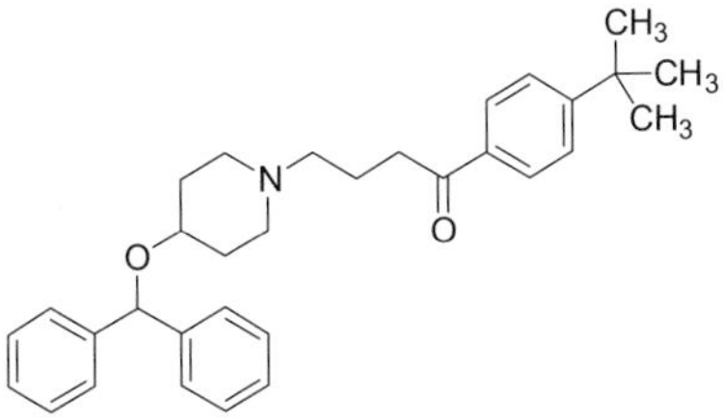

MW: 469.67
P: EP 134,124 (1983, Fordonal)
L: 1990 (Almirall Prodesfarma)
S: –

Desloratadine

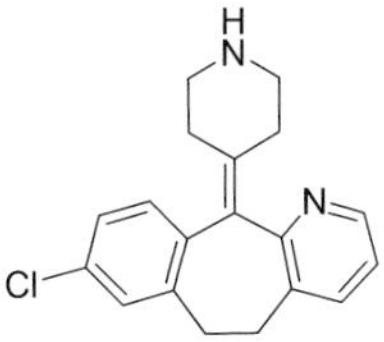

MW: 310.83
P: EP 152,897 (1984, Schering Corp.)
L: 2001 (Schering-Plough)
S: –

Olopatadine

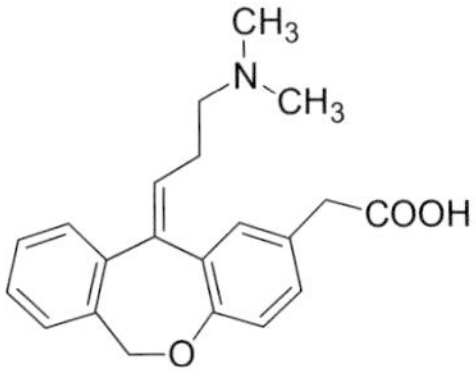

MW: 337.41
P: EP 235,796 (1986, Kyowa Hakko Kogyo Co.)
L: 1997 (Alcon)
S: HCl

S
Sensory Organs

S01EA
Antiglaucoma Preparations and Miotics
Sympathomimetics in glaucoma therapy
(clonidine, brimonidine)

Clonidine

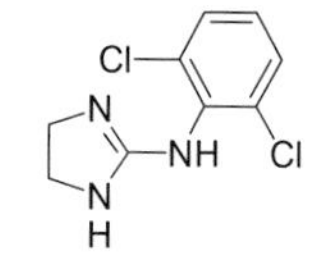

MW: 230.10
- **P:** DE 1,303,141 (1961, Boehringer Ingelheim)
- **L:** 1966 (Boehringer Ingelheim)
- **S:** HCl
- **R:** centrally acting antiadrenergic agent
  (ATC:C02AC)
  migraine (ATC:N02CX)

Brimonidine

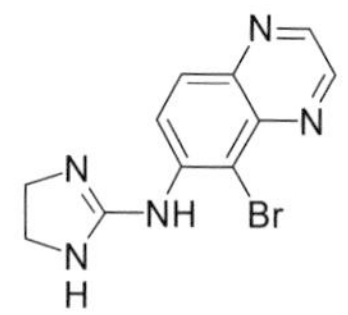

MW: 292.14
- **P:** DE 2,309,160 (1972, Pfizer)
- **L:** 1996 (Allergan)
- **S:** tartrate (1:1)

S01FA
Mydriatics and Cycloplegics
Anticholinergics
(scopolamine, atropine)

Scopolamine

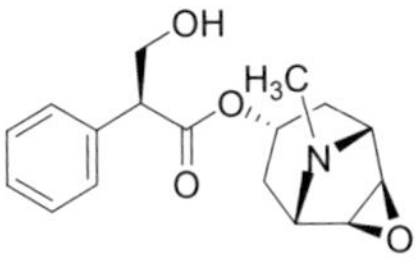

MW: 303.36
  P: Landenburg, A.:Ann. **1881**, 206, 274
  L: 1947
  S: borate, HBr

Atropine

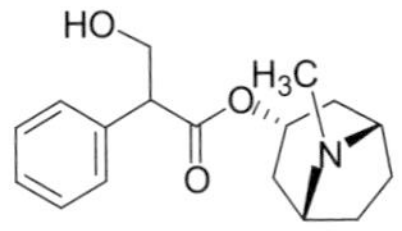

MW: 289.38
  P: Landenburg, A.: Ann. **1883**, 217, 75
  L: –
  S: ½ H_2SO_4

S01GA

Decongestants and Antiallergics
Sympathomimetics Used as Decongestants
(naphazoline, tetryzoline, xylomethazoline, tramazoline)

Naphazoline

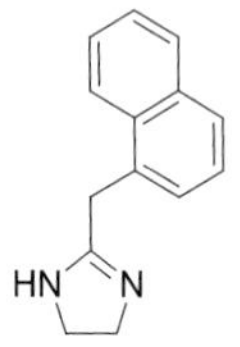

MW: 210.28
  P: US 2,161,938 (1934, Soc.Chem. Ind. Basle)
  L: 1942
  S: HCl, nitrate

Tetryzoline

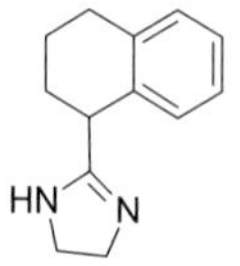

MW: 200.29
  P: US 2,731,471 (1954, Sahyun Labs.)
  L: 1956
  S: –

Xylometazoline

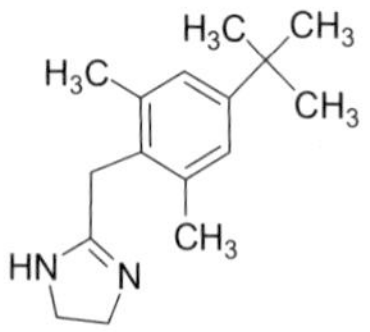

MW: 244.38
  P: US 2,868,802 (1956, Ciba)
  L: 1959
  S: HCl

Tramazoline

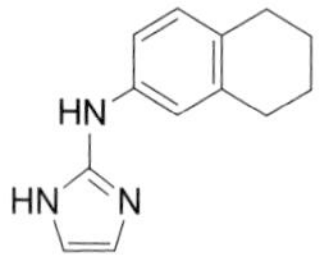

MW: 215.30
  P: DE 1,173,904(1961, Thomae)
  L: 1962 (Thomae, Boehringer Ingelheim)
  S: HCl

Index

a

Abacavir 506
ABDD (Analogue-based Drug Discovery)
  XXIX, 25, 193, 195 f., 201, 206 f., 211 ff.,
  217 f., 226 f., 233 ff., 242 f., 256, 385
ABS (acute bacterial sinusitis) 344, 346
absolute molecular weight 343
absorption 205, 215, 220, 291, 319, 348, 353,
  379, 424, 431
abuse 263, 268 ff.
ACE-independent pathways 160
ACE inhibitors 4, 13, 59, 161
acebutolol 196, 207, 208, 211 ff., 461
aceclofenac 517
acetaminophen (paracetamol) 264, 266
acetylcholine (ACh) 3, 26, 55, 91, 115, 278,
  280 f., 283
acetylcholine agonists and antagonists 3, 57
acetylcholinesterase (AChE) 278, 280 f.,
  285 f., 290 ff.
acetylcholinesterase gene 289
acetylcholinesterase inhibitors 17, 279, 284,
  286, 288 ff., 292
acetylcysteine 544
β-acetyldigoxin 352
6-acetylmorphine 263
aciclovir 504
acid rebound 115
Acinetobacter baumannii 347
Acinetobacter sp. 347
acne 398
acrivastine 408, 410
active metabolite 39, 126 f., 129, 133, 163 f.,
  177, 263, 411, 414, 422, 424, 436, 544
acute bacterial sinusitis (ABS) 344, 346
acute coronary syndrome (ACS) 149
acute exacerbation of chronic bronchitis
  (AECB) 323, 344, 346
acute mania 308

acute otitis media 344
acute renal failure 175
addictive 262 f., 267 f., 270 f., 274
adefovir 27, 36
adefovir dipivoxil 27, 36 f.
ADMET 193, 195, 201, 203, 205, 210, 214 ff.,
  220 f.
adrenal cortex 422
adrenaline (epinephrine) 60, 193, 541
α-adrenergic agonist 60
β-adrenergic agonist 60
β_2 adrenergic agonist 60, 428
α,β-adrenergic agonist 60, 193
α_1-adrenergic antagonism 306
β-adrenergic antagonist 60, 182
α_1-adrenergic receptor 59, 61, 301 f., 306 f.
α_2-adrenergic receptor 61
α_{2A}-adrenergic receptor 301
α_{2B}-adrenergic receptor 301
α_{2C}-adrenergic receptor 301
α- and β-adrenergic receptors 59
adrenergic burst-prompted trauma 242
adrenocortical steroid therapy 425
adsorption of BPs 377
AECB 323, 344, 346
aerobic pathogens 348
AIDS, treatment of 37
agranulocytosis 297, 300, 307, 310
alanine aminotrasferase 152
aldicarb 286 f.
aldosterone 62, 160, 174, 221, 395
aldosterone antagonist 62. 395 f., 459
alendronic acid (alendronate) 374, 377,
  379 f., 382, 523
alfacalcidol 451
alfuzosin 455
aliskerin XXI
alkaloids 263, 277 ff.
allergy 323, 401, 402, 407, 430, 432

Analogue-based Drug Discovery. IUPAC, János Fischer, and C. Robin Ganellin (Eds.)
Copyright © 2006 WILEY-VCH Verlag GmbH & Co. KGaA, Weinheim
ISBN: 3-527-31257-9

allergic conjunctivitis   435
allergic rhinitis   427 f., 430, 432 f., 435
allo-pregnanolone   62
N-allylnormorphine   264 f.
almotriptan   26, 43 ff., 531
alosetron   26, 448
alprazolam   536
alprenolol   196
Alzheimer's disease (AD)   277, 288 ff., 293
ambroxol   544
amikacin   507
amiloride   204
aminoglycosides   316, 350
amiodarone   12, 152, 352
aminolevulinic acid   27
amisulpride   298, 305
amitriptylline   32 f.
amlodipine   162, 165, 185, 187, 204, 465
amoxapine   298 f.
amoxicillin   344, 346, 490
amoxicillin/clavulanic acid   344
amphetamine   60
amphetamine-induced hyperlocomotion
   (AHL)   302 f.
amphetamine-induced disruption of
   prepulsive inhibition   302
amphotericin B   477
ampicillin   490
amprenavir   509
amrubicin   27
β-amyloid plaques   288
anabolic   62
anaerobic bacteria   319 ff., 337, 344, 348
analgesics   259, 261 ff., 268 f., 273 f.
analogue - IUPAC definition   XXIII
analogue types   XXIII, XXIV
analogue optimization   87
anaphylactic shock   401
anastrozole   516
androgen   62
androstenolone   398 f.
anesthetic   62, 273
angina pectoris   185 f., 194 f., 200, 203, 210 f.,
   216, 221, 247, 248, 252, 256
angioedema   175
angiogenesis   160
angiogenesis inhibition   14
angioneurotic edema   176
angiotensin-converting enzyme (ACE)   169,
   173
angiotensin-converting enzyme (ACE)
   inhibitors   XXI, 13, 157 f., 160, 169 ff.,
   173 f., 177, 179, 204

angiotensinogen   169
angiotensin I   39, 169
angiotensin II   39, 157 ff., 169
angiotensin II receptor blockers (ARBs)   13,
   39, 40, 56, 157, 160 ff., 179, 204, 470
angiotensin II receptor (AT$_1$)   160 f., 163, 166
angiotensin II receptor (AT$_2$)   160
4-anilidopiperidine class   273
ankle edema   186 f.
antacids   71, 82, 115, 176
antazoline   408
anti-aldosterone activity   396 ff.
anti-allergic drugs   417, 422, 427
anti-amebic   62
anti-androgen   62, 72 f., 397 f.
anti-anginal drug   64, 195, 200 f., 247, 250,
   255
anti-arrhythmic   195, 203, 210
anti-arrhythmic drugs of class IA, QT
   prolongation of   352
anti-arrhythmic drugs of class III, QT
   prolongation of   352
anti-arthritic   379, 424, 425
antibiotics   315, 350
anticancer activity   375, 386
anticholelithogenic   62
anticholesterolemic BPs   381
anticholinergics   73, 277, 279
anticholinergic toxicity   310, 411
anticonvulsants   57
antidepressants   6, 8, 9, 16, 55, 57 f., 293, 299
antidote   264, 279, 285
anti-emetic drugs   388, 411
anti-estrogen   62
antifungals   152
antihistamine effects   404, 406, 410 ff., 415
antihistamines   401 ff., 406 ff.
antihistaminic index (AHI)   402
anti HIV-1 effects   37
anti-infectives   315
anti-inflammatory effects   382, 417, 421 f.,
   425 ff., 429 f., 434, 437 f.
anti-itching medicament   402
anti-LTD$_4$ effects   417
antimalarial drug   13
antimineralocorticoid   397
antimuscarinic drugs   119, 121, 133
antimuscarinic effects   404, 406
antineoplastic drugs   379
antinoniceptive effect   56
antioxidant   10, 191, 221
antiparasitic drugs   379, 381
anti-Parkinson medicament   404

antiprogestogen 62
antipsychotics 12, 43, 45 f., 59, 299, 302, 307 ff., 310
antiparkinson drugs 57
antipsychotics 297, 300, 302
antirachitic 62
antiresorptive BPs 377
antirestenosis 379
antisecretory effect 86 f.
antitumor activity 385, 388, 390
antitussive 263 f., 266, 269, 274
antiulcer agents 71
antiviral drugs 381
anxiolytic activity 43, 200
aplastic anemia 289
apomorphine antagonism 299
apomorphine-induced climbing (APC) 302 f.
apoptosis 160
appetite suppressant 60
aprepitant 28
aripiprazole 27
aromatase inhibitor 62
arrhythmias 200
artemisinin 26
artemotil 26
arterial smooth muscle 184, 186
arthritis 424
arylcarbamates 280
aryloxypropanolamines 195 f., 200, 206
aspartate aminotransferase 152
aspartyl proteases 58 f.
Aspergillus terreus 139
aspertame 83
aspirin 266
aspirin, low-dose 176, 182
astemizole 412 f.
asthma 203, 401, 411, 421 f., 428, 430, 433 f., 436
asymmetric oxidation 109
atazanavir 28
ATC XVII, 247
atenolol 162, 191, 196, 201 ff., 210 f., 213, 216 f., 221, 241, 461
atherosclerosis 160, 182, 183, 185, 191, 204
atomoxetine 28, 33 ff.
atorvastatin 42 f., 139, 141 ff., 149 ff., 473
atrial fibrillation 205, 211
atrial natriuretic peptide (ANP) 64
atropine XX, 271, 278 f., 287, 291, 551
attention deficit hyperactivity disorder (ADHD) 35
attrition rate 309

atypical (second-generation) antipsychotics 298 ff., 304, 307, 310
atypical pathogens 319, 344
AUC (area under the curve) 348 f., 389
AUC-based administration 390
AUIC (AUC_{24}/MIC_{90}) 349
azaquinolones 317
azatadine 32 f., 412, 414
azelastine 413, 548
azelnidipine 28
azithromycin 344, 498

b
bacterial conjunctivitis 348
bacterial infections 315, 355
bacterial skin infections 347
Bacteroides fragilis 347
Bacteroides theataiotaomicron 348
balofloxacin 27, 45, 48 f.
baygon 286
BAY 11-4119 329
BAY 12-8039 (moxifloxacin) 334 f.
BAY 15-7828 326 f.
BAY 35-3397 328
BAY W 8801 329, 332 f., 341
BAY X 8841 329, 333
BAY X 8842 332 f., 341
BAY X 8843 332 ff., 339, 341, 342
BAY X 8507 329
BAY Y 3114 334
BAY Y 3118 334 f., 352
Beagle Pain Syndrome 86
beclomethasone dipropionate 428, 434, 486
benazepril 171 f., 178, 468
benazeprilat 174
bendiocarb 286
bendroflumethazide 456
benidipine 465
benign gastric ulceration 79
benzamide-type antipsychotics 297
benzbromarone 12
benziodarone 12
benzocaine 475
benzodiazepines 57 ff.
benzodiazepine agonist 57
benzodiazepine antagonist 57
benzologue 5
benzomorphans 270
benzylpenicillin 490
bepotastine 26
BETACAR study 211
betaine (zwiterion) forms 337, 339, 342
betamethasone 427 f., 485

betamethasone-17,21-dipropionate   427
betametasone-17-valerate   427, 433
betaxolol   201, 205 ff., 214, 216 f., 461
bevantolol   196, 225
bexarotene   26
bezafibrate   152, 474
biapenem   27
bicalutamide   515
bifonazole   502
bile acids   118
bimatoprost   27
bioavailability   39, 43, 103 ff., 108, 125, 132,
      145, 164, 170 f., 183, 185, 187, 207, 210,
      215 f., 250, 262, 266, 268, 271, 348, 352, 431,
      434 f.
biochemical models   292
bioisosteric   72, 194, 265 f., 317, 322
bipolar disorder   308 f., 381
bisacylphosphonates   375
bisoprolol   461
bisphosphonates (BPs)   371 ff.
bisphosphonates, pK values of   378
bisphosphonic acid   371
bisoprolol   206 f., 216, 221
bladder   279
bladder cancer   387
blockbuster   403, 414
blood-brain barrier   7, 13, 31, 164, 203, 263,
      272, 283, 288 f., 411, 414
blood monitoring   307
blood pressure lowering effects   165, 174,
      181, 401
blood sugar   322
blood urea nitrogen level   175, 388
bone delivery   379
bone destruction   373
bone formation   378
bone formation, decrease in   373
bone formation stimulation   14
bone fragility   373
bone hook   375 f.
bone loss   373
bone markers   372
bone marrow dysfunction   289
bone marrow reserve   390
bone metastases   372
bone microarchitecture   373
bone mineralization   372, 376, 378
bone remodeling   375
bone resorbing osteoclasts   377
bone resorption   373, 375 ff.
bortezomib   28
bosentan   27

bowel   279
BPs, potency of   376
bradycardia   210, 279
bradykinin   161, 169, 174 f.
brain   272, 283, 287, 291, 297
brain cholinergic activity   279
brain nicotinic system   279
breast cancer   373, 393
brimonidine   550
bromazepam   6, 535
bromhexine   544
bromocriptine   533
brompheniramine   546
bronchial asthma   10, 430 f.
bronchial tumor   373
bronchodilator   193
bronchospasm   195
brotizolam   539
bucumolol   196
budesonide   431 f., 437, 487
budipine   57
bufetolol   196
bufuralol   196
bulaquine   26
bumetanide   458
bunitrolol   196
bunolol   196
bupivacaine   26
bupranolol   196
buprenorphine   262, 267, 528
burimamide   XXIII, 71 ff., 76, 115
buserelin   514
buspirone   43, 46, 47
butidrine   196
butocrolol   196
butorphanol   269, 270, 529
butoxamine   196
butylscopolamine bromide   446
butyrophenones   297, 299
butyrylcholinesterase (BChE)   278, 280, 289,
      291
butyrylcholinesterase inhibitors   289, 292

c
cabergoline   533
calabar bean (*Physostigma venenosum*)   277 f.
calcipotriol   38 f., 452
calcitriol   38 f., 451
calcium antagonists   181 ff., 188 f., 204, 411
cAMP (cyclic adenosine monophosphate)   87
cAMP-stimulated acid secretion   87
cancer   264
cancer chemotherapy   427

candesartan cilexetil   40 f., 158 f., 162 ff., 471
CAP   323, 344, 346
CAPE II trial   204
capecitabine   511
captopril   XXI, 27, 39, 162, 170 f., 174, 176 ff., 204, 467
carazolol   196
carbamate intoxication   287
carbamates   277 f., 280, 282, 284 ff., 291 f.
carbamazepine   532
carbaryl   286 f.
carbinoxamine   545
carbocisteine   544
carbofuran   282
carbonic anhydrase inhibitor   XX
carboplatin   387 ff., 513
carbutamide   449
cardiac arrhythmia   412
cardiac depressant   62, 200
cardiac glycoside   62
cardiac hypertrophy   173
cardioselectivity   206 f., 212 f.
cardiotoxicity   417
cardiovascular mortality   204
carebastine   413
carteolol   196, 460
carvedilol   211, 463
catalepcy induction   302
catalepsy inducing dose (CATL)   303
cataleptogenic activity   299
cataract   310
catatonic state   272
catecholamines   193
catechol O-methyl-transferase inhibitor   4
cathepsin   160
CCK$_A$-agonists   56
cefaclor   493
cefadroxil   493
cefalexin   493
cefazolin   493
cefcapene pivoxil   496
cefdinir   495
cefditoren pivoxil   496
cefepime   347, 496
cefixime   495
cefoperazone   494
cefotaxime   494
cefotiam   494
cefprodoxime proxetil   495
cefprozil   496
ceftazidime   495
ceftriaxone   495
cefuroxime   344, 346, 493

cefuroxime axetil   494
celecoxib   28, 30, 522
CELIMENE study   216
celiprolol   201, 205, 207, 211 f., 214 ff., 226, 461
cell death   391
cellular mechanism of BPs   377
central nervous system (CNS)   277, 279, 283, 288, 291, 302, 322, 327, 411, 540
cephaloporins   315, 344, 346, 350
cerebral edema   427
cerebral cortex   288, 291
cerebral vasculature   187
cerivastatin   42 f., 148, 152
cetirizine   27, 413 ff., 549
cetophenicol   6, 7
cevimeline   26
CHARM study   162
chelating heavy metal   371
chemical library   XXI
chemical warfare agents (CWA)   284, 286
chemogenomics   53 f.
chenodiol   62
childhood croup   432
chimase   160
Chlamydia pneumoniae   319, 344
chloramphenicol   6 f., 316
chlordiazepoxide   535
chlormadinone   478
chlorothiazide   456
chlorphenamine   546
chlorpromazine   XX, 6, 12 f., 20, 32 f., 57, 297 ff., 305 ff.
chlorprothixene   6
chlortalidone   457
cholecalciferol (Vitamin D$_3$)   38 f., 62, 451
cholecystokinin-B (CCK-B) antagonist   57 f.
choleretic   62
cholesterol   62, 138, 148, 377
cholic acid   62
cholinergic activity   279
cholinergic agonist   16, 288
cholinergic neurons   288
cholinergic receptor   279
cholinergic synapses   278
cholinergic system   277 f.
chondrotoxicity   352
chronic myelogenous leukemia   63
chronic obstructive pulmonary disease (COPD)   221, 432, 435
ciclesonide   436, 438, 488
cicletanide   409, 415
cilazapril   171 f., 178, 469

cilnidipine 466

cimetidine XXI, 72, 78 f., 83, 115, 118, 124, 444

cincalcet 29

cinolazepam 539

ciprofibrate 474

ciprofloxacin 45, 48 f., 318 ff., 344 ff., 352, 356, 500

circadian rhythm 177

cis- and trans-configuration of Pt compounds 393

cisapride 352

cis-diamminedichloroplatinum (II) 385

cis-diammine-1,1-cyclobutane dicarboxylate platinum(II) 389

cis-[oxalate (trans-*λ*-1,2-diaminocyclohexane) platinum (II) 391

cisplatin 385 ff., 513

cisplatin, DNA cross-links 386 f.

citalopram 27, 65 f.

CLAIM study 165

clanobutin 118

clarithromycin 344, 346, 498

class effect 310, 417

classes of antibiotics 316

clavulanic acid 492

clemastine 410, 547

clenbuterol 543

clinafloxacin 322, 345, 347, 349, 352

clindamycin 344

clobetasol propionate 432, 487

clodronic acid (clodronate) 374, 376 f., 380 ff.

clofibrate 474

clometherone 62

clonazepam 535

clonidine 550

clopidogrel 453

clorazepate 536

clotiapine 298 f.

cloxacillin 490

clozapine 297 ff., 534

clozapine metabolism 307

C_{max}/MIC_{90} 349

CMN-131 83 f., 119 f.

CNS effects 411

CNS penetration 411 f., 414

CNS toxicity 327, 411

codeine 261 ff., 266, 269, 525

codeine-6-glucuronide 263

cognition disorders 14, 288, 305

cognitive decline, prevention of 162 f., 166 f.

cognitive tests 302

Colletotrichium lini 400

colon cancer cell line 391

colorectal cancer 393

combinatorial chemistry 47

competitive antagonists 164, 186, 190

competitive H_1 antagonist 415

compound library 243

computational chemistry 41, 223

community-acquired pneumonia (CAP) 323, 344, 346

conditioned avoidance response (CAR) 302

conessin 62

conformation, favorable 187 f.

congestive heart failure (CHF) 162, 165, 173 f., 177, 179, 210, 216, 221, 247, 248, 251 f., 255 f.,

constipation 261, 271

constriction of airways 401

contraceptive 397 f., 419

convulsant 262

coordination complex 386

coronary artery disease (CAD) 162, 173, 216

coronary heart disease (CHD) 150, 247

corticosteroid analogues 420, 421

corticosteroids 419, 421 f., 426 f., 432 ff., 436 ff.

corticosteroids, combination wirh $β_2$-agonists 428, 434

corticotropin-releasing factor 1 (CRF1) antagonists 7 f., 18

corticosteroid responsive dermatoses 432

cortisol (hydrocortisone) 62, 422, 437

cortisone 419, 421 f., 437, 484

cough 160, 264

[51]Cr-EDTA clearance method 390

creatine kinase (CK) 151

creatinine level 175 f., 388 f.

Crohn's disease 432, 435

cromakalim 9, 251

cromoglycate 64

cromolyn 10

cross-resistance 350, 389, 391

cyclizine 403

cyclooxygenase 1 (COX-1) inhibitor 25, 29, 31

cyclooxygenase 2 (COX-2) inhibitor 25, 29 ff., 161

cyclosporine 152, 176

cymserine 284, 289

cyproheptadine 408, 547

cyproterone acetate 62

CYP1A2 183

CYP2C8 148

CYP2C9 147, 152

CYP2D6   208, 263, 266, 269
CYP3A4   147, 152, 182 f., 431, 436
cytarabine   511
cytochrome P-450 (CYP-450)   75, 109, 130,
    137, 145, 165, 225, 254, 354
cytokine   382
cytosol   241
cytotoxicity   329

d
danazol   62
danofloxacin   320, 321
daunomycin   27
deafness   388
decalin   139
decamethonium   3, 9
deflazacort   486
delta (δ) opioid receptor (DOP)   259 ff., 267
dementia   160, 288
deoxycytidine   28
dependence   261
dermatitis   423
dermatological infections   348
desloratadine   27, 31 ff., 413 ff., 549
desogestrel   479
Dess-Martin oxidation   219
devazepide   57 f.
developmental candidate   89, 326, 333, 337,
    396
dexamethasone   42, 62, 426, 485
R(+)-dexamisole   8 f.
dexchlorpheniramine
dexmedetomidine   26
dextromethorphan   54, 269, 527
dextromethorphan abuse   269
dextrorphan   269
diabetes   162, 174, 179, 204, 310
diabetes, type 2   162
diabetic nephropathy   162, 166, 175, 179
diagnostic radioisotopes   379
dialyzability   177
diarrhea   392
diastolic blood pressure (DBP)   165 f.
diazepam   6, 19, 57 f., 76, 535
dichloroisoprenaline/dichloroisoproterenol
    55, 60, 207
dichloromethylene-containing cytotoxic
    analogue of ATP   377, 378
diclofenac   28, 517
dicloxacillin   491
didanosine   36 f., 505
dideoxyadenosine   36 f.
dideoxynucleosides   37

Diels-Alder reaction   267
6*a*,9*a*-difluoro-16*a*-hydroxyprednisolone
    (flucinolone)   430
digoxin   62, 165, 176, 211
dihydrocodeine   264, 525
dihydromorphine   264
1,4-dihydropyridines   182, 187 f.
dihydroxyvitamin D3   38
diltiazem   XXIV, 181 f., 211
dimethindene   408, 547
dimethylallyl diphosphate (DMAPP)
    carbocation   378
diphenhydramine   55, 57, 402 f., 411, 414,
    545
diphenhydramine analogues   403 ff., 411 f.
direct analogue   XXIII, 10, 20, 141, 170
distribution   205, 216, 241
disuprazole   100
disopyramide   352
dissociation from the D$_2$ receptors   305
distribution   220
diuresis   261
diuretics   204, 396
dizziness   216
DK-507 k   322
DNA gyrase (topoisomerase II)   317, 344,
    350, 355
DNA-platinum adducts   389, 391
dobutamine   60
docetaxel   512
dofetilide   26, 352
dolasetron   448
L-DOPA (levodopa)   4, 26
dopamine   60, 293
dopamine antagonist   57, 411
dopaminergic agonist   60
dopamine D$_1$ receptor   59, 61, 297, 301
dopamine D$_2$ receptor   59, 61, 297, 301 f.,
    304 ff.
dopamine D$_3$ receptor   301, 304 f.
dopamine D$_4$ receptor   59, 61, 300 f.
dopamine metabolite   299
dopamine receptors   302
dopaminergic neuron   272
down-regulation of the breathing center   279
doxazosin   455
doxefazepam   539
doxepine   408
doxycycline   489
doxylamine   546
drofenine   55
drospirenone   XXIV, 26, 395, 397 f., 459

drospirenone, combination with ethinylestradiol 398
drospirenone, synthesis of 398 f.
drowsiness 411
drug analogues 25
drug dependence 200
drug-drug interactions 132, 137, 165, 176, 221, 321, 352
drug-food interaction 177
drug-likeness 14, 25, 47, 53, 193
dry cough 174
duloxetine 29, 34 f.
duodenal ulcer 79, 90, 115
DuP 105 26
DuP 697 28, 31
DuP 721 26
duration of action 161, 164 f., 171, 177, 179, 183, 185, 187 ff., 191, 207, 233 f., 268, 273 f., 288 ff., 292, 438
dutasteride 28, 483
DW-286 323
dyslipidemia 152
dyspepsia 79
dysphoria 261, 265, 268, 271
D_2 antagonism 304
D_2 occupancy 305
D_2/5-HT$_{2A}$ occupancy 305
D_2 presynaptic receptor antagonist 43
D_2/D_3 nonselective antagonist 18
D_2/D_3 partial agonist 18

e
econazole 502
edaravone 27
E. aerogenes (Enterobacter aerogenes) 347
Early Manifest Glaucoma Trial 211
early morning surge 165
early-phase analogues 25, 171
ebastine 412, 413, 416, 549
E1cB mechanism 280
α-ecdysone 62
E. cloacae (Enterobacter cloacae) 347
E. coli (Escherichia coli) 347, 350
ectopic calcification 372
eczema 423
edema 216, 396
edifolone 62
E. faecalis (Enterococcus faecalis) 347
efflux pumps 350
egualen sodium 26
elastase inhibitors 59
eletriptan 27, 43 ff.
elimination 272, 389, 392

ELITE II study 162
ELSA study 191, 204
emergency medication 234, 242, 427, 429
emesis 261, 392
emetogenic side effect 390
emtricitabine 28
enalapril 4, 171, 177 f., 216, 221, 467
enalaprilat 171, 174, 177
enantiomers 8, 184, 218, 269, 271, 274, 279, 282, 291, 319, 322 f., 332 f., 404, 411, 415, 417, 436
endocrine tumors 89
endocrinology 419
endodontic diseases 348
endogenous antioxidant 94
endogenous opioid receptor ligand 259
endogenous ORL-1 receptor ligand 260
endophthalmitis isolates 348
endothelial cell 221
endothelial dysfunction 160, 173
endothelin ET$_A$ receptor ligands 15
endothelium-derived relaxing factor (EDRF) 253
endothelium-derived nitric oxide (NO) 173
enkephalin 259
enoxacin 319 f., 352
enrofloxacin 320 f., 355
entacapone 4 f.
Enterobacter aerogenes 347
Enterobacter cloacae 347
Enterobacter spp. 347
enterochromaffin-like (ECL) cells 90
Enterococcus faecalis 347
Enterococcus spp. 344, 347
enzymatic hydrolysis 280
enzyme 280
ephedrine 60, 541
epilepsy-type convulsions 279
epinastine 549
epinephrine (adrenaline) 193, 541
eplerenone 27, 459
eprosartan 40 f., 158, 160 f., 163, 165, 470
eptasigmine 7, 284, 289, 291
erectile dysfunction 13
ergocalciferol (Vitamin D$_2$) 451
ertapenem 27
erythromycin 498
Escherichia coli 347, 350
escitalopram 27
eserine (physostigmine) 277
eseroline 278 f., 282, 289
esmolol 34 f., 201, 233 f., 236, 240 ff., 462
esomeprazole 26, 81, 105, 107 ff., 133, 445

esophageal erosion   379
esophageal tumor   373
esophageal ulceration   379
esophagitis   103
essential hypertension   181 f.
estazolam   538
estradiol   62, 482
estrogen   62
eszopiclone   29
ethambutol   347
ethinylestradiol   482
etidronic acid (etidronate)   371, 373 f., 376 f.,
   380, 523
etizolam   536
etodolac   517
etonogestrel   480
etoricoxib   28, 30 f., 522
etynodiol acetate   478
euphoria   261, 271
exaprolol   196
excretion   205, 210
exemestane   26
exercise tolerance   221, 251
exercise-induced angina   182
exertional angina   182
existing classes of anti-infectives   315
existing drugs   XIX, 66
EXP-3174   163 f.
EXP-3179   161
EXP-6155   158
EXP-6803   158
EXP-7711   158
extensive metabolizer   107
external ester analogue   236, 243
extrapyramidal symptoms (EPS)   297, 304,
   307, 309
ezetimibe   28, 153

f

factor Xa inhibitors   59
famciclovir   504
famotidine   XXIII, 76 ff., 115, 444
farnesyl diphosphate (FPP)   377
farnesyl transferase inhibitors   58
fast metabolizer   220 f.
fatigue   210, 216
felodipine   183, 187, 465
fendiline   29
fenethazine   403
fenofibrate   148, 152, 474
fenoterol   542
fentanyl   54, 273, 528
fentanyl analogues   273

fexofenadine   412, 414, 416, 548
fibrates   148, 151
fibrinogen receptor   61
finasteride   28, 62, 483
first in class   XX, 31
fleroxacin   319 ff., 352
flomoxef   496
flucinolone   430
flucinolone acetonide   429 ff., 434, 485
flucloxacillin   491
fluconazole   28, 29, 152, 165, 503
fludrocortisone   424, 484
fluid accumulation in the lungs   279
flumazenil   57 f.
flumequine   318
flumetasone   429 ff., 434, 437 f., 484
flumetasone pivalate   431
flunisolide   429 f., 486
flunitrazepam   537
fluocortin-21-butylate   429 ff., 487
9-fluoro-hydrocortisone   424 f.
9-fluoro-16α-hydroxy-prednisolone   425
9-fluoro-16α-methylprednisolone   426
fluoroquinolone antibiotics   45, 317, 327,
   347 ff., 355 f.
5-fluorouracil (5-FU)   392, 511
fluoxetine   28 f., 34 f., 65 f.
flurazepam   538
flurbiprofen   520
flutamide   515
fluticasone   431
fluticasone propionate   434 f.
fluvastatin   29, 42, 43, 139, 141 ff., 149 f., 152,
   472
folk law remedies   XX
food   220
forebrain   288
formestane   62
formethanate   286 f.
formoterol   543
formulation studies   89
fosamprenavir   28
foscarnet   381
fosfluconazole   29
fosinopril   171, 173, 177 f., 468
fosinoprilat   174
Friedel-Crafts acylation   214
Fries rearrangement   214
frovatriptan   26
fudosteine   27
fulvestrant   28
furosemide   388, 458
Fusobacterium spp.   348

g

GABA   29
GABA receptor   58
gabexate   27
ganciclovir   27, 504
garenoxacin   317, 355, 356
gastrin   90 f., 115
gastric acid   71, 75, 81, 91, 116
gastrin antagonists   76, 83 f.
gastric carcinoma   327
gastric fistula dog   83, 87
gastric musoca   91 f., 96, 132, 327
gastric ulcer   115, 327
gastritis   327
gastroesophageal reflux disease (GERD)   79,
   81 f., 90, 115, 132 f.
gastroprotective drugs   118
gastrointestinal tumor   373
gatifloxacin   45, 318 ff., 323, 345, 347 ff., 356,
   501
gefitinib   28
gemcitabine   511
gemfibrozil   137, 148, 152, 474
gemifloxacin   28, 49, 323, 349, 351 f., 356
gene   278, 411
genetic deficiency   263
gentamicin   507
geranyl diphosphate (GPP) carbocation   378
geranylgeranyl diphosphate   377
gestodene   480
glands   279
glaucoma   161, 200, 210 f., 283
glibenclamide   251
gliclazide   449
glimepiride   449
glipizide   449
globular monomeric form (G1) of AChE   291
globular tetrameric form (G4) of AChE   291
glomerular filtration rate (GFR)   389 f., 392
glomerulosclerosis   160
glucocorticoids   62, 421 ff., 426, 428, 434
glucocorticoid receptor   421
glucose transport   179
glutathione   94, 127
glutathione-S-transferase   254 f.
glyceryl trinitrate (nitroglycerin)   204, 221,
   247 ff., 251 ff., 454
glycogen deposition   425
glycopeptides   316
GMC-series   298, 300
goserelin   514
Gould Jacobs process   319
gout   166

G protein-coupled receptors (GPCRs)   53, 222
GPCR ligands   59
GP IIb/IIIa receptor   61, 63
gram-negative bacteria   317, 319, 321, 326,
   344, 347 f.
gram-positive bacteria   317, 319 ff., 323, 326,
   337, 344, 347 f.
granisetron   29, 448
granulocytopenia   XXIII, 72
grapefruit juice   183, 215
grepafloxacin   320 f., 344, 349, 351 f.
growth retardation   422
guaiazulene sulfonate   26
guanilate cyclase   253 ff.
gut   183
gynaecomastia   72

h

Haemophilus influenzae (*H. influenzae*)   346 f.
Haemophilus parainfluenzae
   (*H. parainfluenzae*)   349
Haemophilus spp.   344
halobetasol propionate   432
hallucinogen   60
haloperidol   20, 54, 297 ff., 307
9-halo-steroids   425
Hansch analysis   187
HAP (hydroxyapatite)   372, 375 f.
headache   191, 210
head cancer   387, 393
hearing loss   388
heart failure   211
heart rate   173, 211, 234, 248, 252
Heck reaction   190
Helicobacter felis murine model   327
Helicobacter pylori   80, 132, 327
Helicobacter pylori ferret model   327
hematological cancers   373
heparin   174
hepatic activation   177
hepatic cholesterol synthesis   137
hepatic first-pass metabolism   103, 148, 183,
   185, 187, 203, 209, 212, 214, 221, 249 f., 263,
   267 f., 270 f., 431, 435
hepatic intolerance   322
hepatitis B, treatment of   37
hepatocellular injury   152
hepatotoxicity   353
heroin   262 f., 267, 273, 525
heterotopic ossification   373, 380
hexafluorocalcitriol   27, 37 ff.
hexamethonium   3, 9

high-density lipoprotein-cholesterol (HDL-C) 145 f., 150 f., 153
high hydrophilicity 416
high-throughput screening (HTS) XXI, 47, 244
H. influenzae 346 f.
hip replacement surgery 373
hippocampus 288, 291
hippocampus slice model 327
histamine 31, 55, 76, 115, 401 f., 411, 416
histamine-stimulated gastric acid secretion 75, 119
histamine H_1- and H_4-blockers 417
histamine H_1-blockers 401, 415
histamine H_1- receptor 59, 61, 301 f., 306 f., 411
histamine H_2-receptor 71 ff.
histamine H_4- receptor 417
histabudifen 416
histapendifen 416
histaprodifen analogues 416
HIV antivirals 152
HIV protease 58
HIV protease inhibitor 59
hivenyl 413
H^+/K^+ ATPase (proton pump) 87, 92 f., 96, 111, 133
H^+/K^+ ATPase inhibitors 79, 91 ff., 98, 132, 134
HMG-CoA reductase 138, 148
HMG-CoA reductase inhibitors 13, 41, 137, 472
homatropine 264
homologue 3, 71, 139, 244, 329
hormone 423, 426
hormonal disorders 396
H. parainfluenzae 349
5-HT reuptake inhibitor 34
5-HT_{1A} receptor 301
5-HT_{1A} receptor partial agonist 43
5-HT_{2A} receptor 59, 61, 297, 302, 305 f.
5-HT_{2A} antagonism 304
5-HT_{2A}/D_2 affinity 299 f.
5-HT_{2B} receptor 59, 61, 301, 306
5-HT_{2C} receptor 59, 61, 301
5-HT_{1D} receptor 301
5-HT_{1D} receptor agonists 43
5-HT_3 receptor/ion channel 59 ff., 301
5-HT_3 antagonists 388, 390, 392
5-HT_3 ligand, selective 60
5-HT_4 receptor 60
5-HT_4 ligand, selective 60
5-HT_6 receptor 301, 306
5-HT_7 receptor 301, 306
human brain AChE inhibition 292
human liver microsomes 110
Δ^1-hydrase 399
hydrochlorothiazide 162, 204, 456
hydrocodone 264, 526
hydrocortisone (cortisol) 421 ff., 429 f., 432, 437, 484
hydrolytic stability 343
hydromorphone 264, 526
hydroquinidine 352
hydroxyapatite (HAP) 372, 375 f.
hydroxydione 62
16α-hydroxy-hydrocortisone 425
hydroxymethanebisphosphonic acid (oxidronate) 373
5-hydroxytryptamine-induced head twitch (HTW) 302 f.
hydroxyzine 414
hypercalcemia 373, 375, 380
hypercholesterolemia, treatment of 41
hypergastrinemia 90
hyperglycemia 310
hyperkalemia 174, 176
hyperparathyroidism 375
hyperprolactinemia 305, 310
hypertension, treatment of 39, 183 ff., 195, 210, 216, 221, 252
hypertension and diabetes 162, 173
hypertensive emergency 177, 179
hyperthyroidism 310
hypnotics 58, 299
hypocalcemia 380
hypothalamic-pituitary adrenal funtion 422
hypotension 174, 203
H_1 receptor antagonist XXI, 31 f., 57, 80
H_2 receptor antagonist XXI, 13, 71, 81, 83, 90, 131 ff.

i
ibandronic acid (ibandronate) 374, 377 ff., 523
ibuprofen 520
ibutilide 352
imatinib 27, 63 f.
imidapril 171 f. 469
imipenem 497
imipramine 6, 12, 20, 32, 33, 57
IMS Health XVII
immune cells 261
immune system 261, 422
immunmodulator 13
immunosuppression 261

immunstimulant 8
impaired kidney function 272, 389
incadronic acid 374, 377, 380
indapamide 204, 457
indenolol 196
indinavir 509
indisetron 29
indomethacin 165, 517
inducible nitric oxide synthase (iNOS) 382
infected bite wounds 348
infections 315
inflamed joints 382
inflammatory bowel disease 421, 432
inhalation treatment 248
inhaled glucocorticoid therapy 434, 436
inhaled powder formulation 433
inorganic pyrophosphate (PPi) 372, 375 f.
inositol monophosphatase (IMPase)
    inhibitor 381
insect juvenile hormone 6
insecticides 277, 284, 286, 287
insect-repellents 286
insulin resistance 179
insurmountable antagonists 164
intravascular metabolism 249
integrin antagonists 58, 61
inter-individual variability 110 f.
interleukin (IL)-1 382
internal ester analogue 236
International Union of Pure and Applied
    Chemistry XVII
intestinal worms 86
intestinal absorption 378
intracellular cell adhesion molecules 161
intracellular mevanolate pathway 377
intraocular pressure 211, 279
intravenous administration 321
intrinsic sympathomimetic activity (ISA)
    195, 197, 201 f., 213, 233, 237, 239, 243
intubation 242
inverse agonism of antihistamines 415 ff.
inverse agonists at dopamine receptors 306
inverse agonists at 5-HT$_{2A}$ and 5-HT$_{2C}$
    receptors 306
iodine uptake, inhibition of 86, 88
ipratropium bromide 446
iprocrolol 197
irbesartan 40 f., 158 f., 162 ff., 470
irinotecan 392
iris 279
irreversible brain damage 272
irritable bowel disease 421
ischemic heart disease 150

isepamicin 508
ISIS-1 study 205
isoamyl nitrite 247 f., 252, 454
isoconazole 502
isoelectric point 342
isolated rabbit ear artery 188
isolated systolic hypertension 186
isoniazid 349
isopentenyl diphosphate (IPP) 377
isoprenaline, isoproterenol 55, 60, 193
isopreoid diphosphates 377
isoprenylated proteins 137
isoproterenol (isoprenaline) 193, 240, 242
isosorbide dinitrate 247, 249 f., 252, 254 f.,
    454
isosorbide mononitrate 247, 250 ff., 254, 454
isostere 3, 72
isotretinoin 476
isradipine 183, 187, 465
itraconazole 215, 352, 503
IUPAC XVII

j
jaundice 152
JL-13 298, 300

k
kappa (κ) opioid receptor (KOP) 259 ff., 265,
    267
kappa opioid receptor agonist 57, 268, 270,
    274 f.
keratitis isolates 350
ketoconazole 503
ketoprofen 520
ketorolac 521
ketotifen 408, 548
K$^+$-HERG (human ether-a-go-go related
    gene) 412
kidney damage 388 f.
kidney tumor 373
kidney tubular cells 152
kidney vasculature 187
kinases 63
kinase inhibitors 63
Klebsiella pneumoniae (K. pneumoniae) 344,
    347, 350
K. pneumoniae (Klebsiella pneumoniae) 344,
    347, 350
K$^+$-sparing diuretics 174, 176

l
LAARS study 204
labetalol 197, 463

lacidipine   181, 185, 187 ff., 204, 466
β-lactams   315 f., 319, 344
lafutidine   26
lamivudine   506
lamtidine   77 f.
landiolol   28, 34 f.
Land's classification   195
lansoprazole   101 f., 111, 117, 133, 445
laryngeal edema   175
latanoprost   27
L-class voltage-gated calcium blocker   XXIV
L-DOPA decarboxylase inhibitors 4 f.
LD$_{50}$ 279, 286 f., 290 f.
lead compound   47, 86, 111, 134, 138, 157 f., 170 f., 182
lead optimization   158 f., 189
left venticular function   221
left ventricular hypertrophy (LVH)   160, 162, 166, 204
Legionella pneumophilla   319, 344
Leishmania donovani   381
leminoprazole   101
leprosy, chemotherapy of   13
lercarnidipine   185 ff., 466
letrozole   516
leukemia   289
leukemic cells   289
leuprorelin   514
S-(–)-levamisole   8 f.
levetiracetam   26
levobupivacaine   26
levocetirizine   27
levodopa (L-DOPA)   4,   26
levofloxacin   318 ff., 323, 344, 346 ff., 500
levonorgestrel   479
levorphanol   269, 527
levosimendan   26
lidocain   83
lifelong toxicological studies   89
LIFE study   162, 166, 204
ligand-based drug design   193, 207
light sensitive drugs   182
linear dose-response curve   177
linezolid   26, 317
lipid peroxidation inhibitor   62
lipophilicity   7, 143, 177, 185, 187 f., 190, 195, 209, 289 f., 414, 431, 438
lipoproteins   210
liposome-encapsulated BPs   382
liranaftate   26
lisinopril   171 f., 174, 177 f., 467
lithium   176, 381
liver cirrhosis   396

liver dysfunction   349
local anesthetics   83, 208
Log P   30 ff., 239, 342, 414
lomefloxacin   320 f., 344 f., 349, 352, 500
loperamide   54, 57
lopinavir   26, 510
loprazolam   539
loratadine   27, 32 f., 412 ff., 549
lorazepam   535
lormetazepam   537
lornoxicam   519
losartan   28, 39 ff., 158 f., 161 ff., 204, 470
loss of consciousness   279
loteprednol etabonate   435, 438, 488
lotrafiban   61
lovastatin   20, 41 ff., 139, 143 ff., 150, 152, 472
low blood pressures   176
low-density lipoprotein (LDL)   191
low-density lipoprotein-cholesterol (LDL-C) 137, 146, 149 ff., 153
loxapine   298 ff.
loxitidine   77
loxoprofen   521
lumiracoxib   28, 517 f.
lung cancer   387, 393
lymphomas   375
lysolecithin   118
L-745,870   300

m

maalox   353
maconde sculpture   316
macrolide antibiotics   152, 315 f., 319, 346, 350
maculopapular rash   175
magaldrate   117
MALT lymphoma   327
a-mangostin   415
mania   200, 309
manidipine   186 f., 465
MAO-B inhibitors   292
maprotiline   6
marbofloxacin   320 f.
maxacalcitol   26, 37 ff.
M. bovis BCG (*Mycobacterium bovis* BCG) 351
M. catarrhalis   344, 347
MDMA (Ecstasy)   60
MDP (medronate)   373 f., 376
meclozine   547
medetomidine   26
medronic acid (medronate)   373 f., 376, 380
medroxyprogesterone acetate   478

megestrol acetate   479
melevodopa   26
meloxicam   519
membrane-modulating agent   62
membrane-stabilizing activity (MSA)   195, 197, 200 ff., 213, 233
Mendelson's syndrome   79
meningitis   344
menopause   373
menstrual cycle control   398
meperidine (pethidine)   272 f.
mepindolol   197
mepyramine   416, 545
mequitazine   408, 548
meropenem   27, 497
mesocortial selectivity   306
metabolic instability   189
metamizole   530
metastatic bone disease   373
metered-dose inhaler   430
methadone   273 f., 526
methanebisphosphonic acid (medronate, MDP)   372 f.
methicillin-resistant *S. aureus* (MRSA)   346
methicillin-sensitive *S. aureus* (MSSA)   346
methomyl   286
methylene-containing cytotoxic analogue of ATP   377
methylphenidate   27
D-threo-methylphenidate   27
methylprednisolone   429, 486
methylprednisolone sodium succinate   429
methylscopolamine bromide   446
methyltestosterone   481
metiamide   XXIII, 72 f., 131
metiapine   298 f.
metipranolol   197
metolazone   457
metoprolol   60, 197, 204, 206 ff., 211, 214, 216, 236, 461
mevalonic acid   137 f.
mevastatin   138 f.
mezolastine   416
mianserin   408
MIC values of quinolones   329 ff., 344 ff.
mice tail-flick assay   56
miconazole   502
Micrococcus luteus   347
β-microglobulin   382
MIDAS study   183
midazolam   539
mifepristone   62
miglustat   29

migraine, treatment of   43, 200, 203, 205
minaprine   16, 17, 55
mineralocorticoid   62, 395, 421 ff., 426, 437
miosis   261, 291
miotic   279, 282
miotin   282 f., 289 ff.
mismatch repair enzymes   391
mitiglinide   29
Mitsunobu reaction   219
mixed agonist-antagonist opioid ligands   265, 268 ff.
mizolastine   413
M. leprae (Mycobacterium leprae)   348
modafinil   57
moexipril   171 f., 468
molecular modeling   41, 158
molecular weight   30 ff., 343
mometasone furoate   433, 488
monoamine oxidase (MAO)   214, 225
monoamine oxidase B (MAO-B)   272
monomethyl-carbamates   279
moprolol   197
Moraxella catarrhalis (M. catarrhalis)   344, 347
Morganella morganii   347
morphinan skeleton   266 f., 269 f.
morphine   XX, 54 f., 259 f., 262, 263, 265 ff., 525
morphine analogues   263, 267
morphine-6-glucuronide   263
morphine poisoning   264
MOSAIC study   346
moxifloxacin (BAY 12-8039)   XX, 315, 318 f., 321, 323, 334 f., 342, 344 ff., 356, 501
moxifloxacin betaine form (BAY Y 6957)   337, 342
moxifloxacin combinations   347 f.
moxifloxacin, CNS side effects of   352
moxifloxacin, degradation of   354 f.
moxifloxacin, gastrointestinal side effects of   352
moxifloxacin, metabolism of   353
moxifloxacin salts   337, 339, 342
moxifloxacin synthesis   338 ff.
MPC   XXX, 350 f.
MPPP   272
MPP$^+$ 272
MPTP   272
MRC study   204
M. smegmatis (Mycobacterium smegmatis)   351
MSSA   XXX, 351
M. tuberculosis (Mycobacterium tuberculosis)   319, 347, 351

mucosal brush-border phosphatases 372

multidrug-resistant *S. pneumoniae* (MDRSP) 346

multidrug-resistant tuberculosis (MDRTB) 347

multiple myeloma 375

multiple regression analysis 410

multiple sclerosis 14, 283

multipotent drugs 59

multireceptor profile 300, 302

MOP, μ-opioid receptor 259 ff., 266 f.

muscarine 73

muscarinic activities 279, 306

muscarinic blocker 291

muscarinic group 279

muscarinic acetylcholine receptor (ACh-M) 301

muscarinic M_1 receptor affinity 16 f., 59, 61, 301, 306 f.

muscarinic M_2 receptor 59, 61, 301, 306

muscarinic M_3 receptor 59, 61, 301, 306

muscarinic M_4 receptor 59, 61, 301, 306

muscarinic M_5 receptor 59, 61, 306

muscle relaxants 58, 62

muscle twitching 279

muscle weakness 279, 285

muscular fasciculation 279

mutant prevention concentration (MPC) 347, 350

myalgia 151

myasthenia gravis 277, 283, 288

Mycobacterium bovis BCG *(M. bovis BCG)* 351

Mycobacterium leprae (M. leprae) 348

Mycobacterium smegmatis (M. smegmatis) 351

Mycobacterium tuberculosis (M. tuberculosis) 319, 347, 351

mycophenolate 29

Mycoplasma pneumoniae 319, 344

myelotoxicity 388 ff., 392

myocardial contractility 184

myocardial infarction (MI) 162, 177 ff., 203 ff., 233 f., 252, 253

myocardial ischemia 177, 247, 251

myocardial oxygen supply 247, 252

myocardial oxygen demand 247

myocarditis 310

myopathy 151

myositis ossificans 373

n

nadolol 197, 460

nalbuphine 268 f. 528

nalidixic acid 48 f., 315, 317 f., 356

nalorphine XXIV, 55, 264 f.

naloxone 265, 267 f., 271, 527

naltrexone 268, 528

nandrolone decanoate 62

naphazoline 552

naproxen 520

naratriptan 43 ff., 531

natural products XX

nausea 175, 388, 390, 392, 427

nebivolol 201, 205, 207, 217, 219 ff., 462

NEBIS trial 221

neck cancer 387, 393

necrotizing vasculitis 86, 88, 125

nedaplatin 393

nefopam 57

nelfinavir 510

neointima formation 160

neomycin B 507

neostigmine 283 f., 540

nephropathy, in type 2 diabetes mellitus 166

nephrotoxicity 385, 387 f., 390, 392 f.

neridronic acid (neridronate) 28, 374, 377, 380

nerve gases 281

netilmicin 508

neuramidase inhibitors 3, 4

neurodegenerative disorders 287

neuroleptics 6, 54, 58, 59, 299, 307

neuromodulator 62

neuromuscular blocking agents 10

neuromuscular junction (NMJ) 279, 283

neuron loss 287

neuropathy 388

neuropeptide Y receptor ligand 20

neuroprotective effect 160

neurotoxicity 272, 333, 388

neurotoxin 272

neurotransmitter 278, 279

neurotransmitter pathways 288

neurotransmitter reuptake inhibitor 271

neutral antagonists of H_1 receptor 416, 417

neutropenia 175, 176

niacin 151

nicardipine 28, 464

nicorandil 247, 251 f., 255, 454

(–)-nicotine 56

nicotinic acetlycholine receptors 56

nicotinic activity 279

nicotinic group 279

nicotinic receptors 279, 283

nifedipine XXIV, 181 ff., 186 f., 189, 204, 464

nifenalol 196

nimodipine 184, 187, 464

nipecotic acid 7
nisoldipine 184, 464
nisoxetine 34, 35, 65, 66
nisvastatin 29
nitazoxanide 29
nitisinone 28
nitrates 211, 247 ff., 251 ff., 256
nitrate tolerance 247, 251, 255 f.
nitrazepam 6, 537
nitrendipine 187, 464
nitric oxide 217, 221, 253, 256, 382
nitric oxide synthases (NOSs) 253
nitroglycerine (glyceryl trinitrate) 204, 221, 247 ff., 251 ff., 454
nizatidine XXIII, 76, 78 f., 444
NK-1 receptor antagonists 58
NMDA (*N*-methyl-D-aspartate) receptor 269
nociceptin 260
nolinium bromide 119 f.
NONOates 256
nonparasitized dog 86
nonsedating antihistamine 411 ff.
nonsteroidal anti-inflammatory drugs (NSAIDs) 25, 29, 79, 118, 132, 174, 176
noradrenaline (norepinephrine) 60, 193
noradrenaline reuptake inhibitor 9, 34
norcodeine 265
norelgestromin 29
norepinephrine (noradrenaline) 193
norethisterone 478
norfloxacin 48, 49, 318 ff., 344, 350, 500
norgestimate 479
norgestrel 29, 479
normeperidine 272
normorphine 265
norpseudoephedrine 60
noscapine 261
NO-synthase (NOS) 221
nuclear receptor ligands 61
nucleoside and nucleotide analogues 36 f.
nystatin 477

o

obstructive airway disease (OAD) 200 f.
oculomucocutaneous syndrome 200
ocular pressure 161, 166
ofloxacin 28, 45, 48 f., 318 ff., 344 ff., 500
olamufloxacin 322
olanzapine 59, 61, 298 ff., 534
olanzapine metabolism 307
olmesartan 28, 39 ff., 158 f., 164, 166, 471
olopatadine 549
olpadronate 378

omeprazole 26, 79, 81, 85, 88 ff., 111, 117, 121, 124, 130, 133, 445
omeprazole cycle 95, 96
OMPIMAAL study 162
once-daily administration 321, 348
ondansetron 26, 388, 448
onset time 177, 185, 187, 205, 215, 263, 272 f., 423
opioid abuse potential 269
opioid agonists 266
opioid analgesic activity 272, 274
opioid antagonist 265, 267 f.
μ opioid receptors 259 ff., 274
opioid receptor ligands XX, 259, 262 f., 265, 273
opioids 259
opioid withdrawal agent 267 f., 274
opium poppy 259, 261 f.
optical isomer 3, 8
OP_1 receptor 261
OP_2 receptor 261
OP_3 receptor 261
oral administration 321, 385
oral contraceptives 352, 397 f.
oral mucosa bone wounds 348
orange juice 215
organ trasplantation 152
organic acid transporter polypeptides (OATP) 143 f., 148
organophosphates 284 f., 289
oripavine class 267
ORL-1 receptor 260
orphan diseases 20
orphaninan FQ 260
orphenadrine 404
orthostatic hypotension 297, 310
osteoclasts 377
osteoclast poisoning 377, 379
osteogenesis imperfecta 380
osteolytic bone diseases 373
osteolytic tumors 373
osteoporosis 373, 380, 422
ototoxicity 388
ovarian cancer 387, 393
oxacillin 490
oxaliplatin 387, 388, 391 ff., 513
oxaprozin 520
oxatomide 408, 548
oxazepam 536
oxazolidinones 315, 317
oxcarbazepine 532
oxiconazole 503
oxidronic acid (oxidronate) 373, 374, 376, 380

oxitropium bromide   28, 447
oxmetidine   72 f.
oxprenolol   197
oxycodone   266 f., 526
oxymorphone   266 f., 527
oxytetracycline   489

p
Paclitaxel   512
P. aeruginosa   319, 326, 344, 347
Paget's disease   373, 375, 380
pain   261 ff., 270, 274
palonosetron   29
pamatolol   197
pamidronic acid (pamidronate)   27 f., 374 f., 377 f., 380, 523
pancuronium bromide   62
pantoprazole   99, 100, 102, 111, 115, 117 ff., 123 f., 126, 128 ff., 445
pantoprazole sodium sesquihydrate   130
papaverine   261
Papaver somniferum   261 ff.
paracetamol   264, 266
paradontopathic conditions   348
paralysis   279
parecoxib   28, 30 f., 522
pargolol   197
paricalcitol   38 f., 452
parietal cell   87, 91, 93, 96, 116, 126, 129, 133
Parkinsonian symptoms   306, 307
Parkinson's disease   272, 293
Parkinson's disease model   272
partial β_2-agonist   60
partial H_1-agonist   416
PASS   14, 21
patient compliance   177, 182
Payne rearrangement   223 f., 241
pazufloxacin   28, 45, 49
P-C-P structure   375, 382
peak action time   205
peak plasma concentration   348
penbutolol   197
penicillin G   344
penicillin-resistant strains   344
penicillins   XX, 315 f., 346, 350
Penicillium citrinum   138
pentaerythrityl tetranitrate   454
pentazocine   270, 271, 527
peptic ulcer disease   71, 78 f., 82, 132
perfloxacin   319 ff., 352
pergolide   533
perindopril   171 f., 178, 204, 467
perindoprilat   174

peripheral nervous system (PNS)   277, 279
perlapine   298 f.
permeability of blood vessels   401
pethidine (meperidine)   272, 526
perospirone   26, 43, 46, 47
Peyrone's chloride   388
P-glycoproteine (Pgp) efflux pump   414
Pgp-knock-out mice   414
Pgp substrates   415
pharmacodynamics   110, 163, 274, 321, 348, 349
pharmacokinetics (PK)   110, 132, 161, 163 f., 178, 193, 206, 210, 211, 213 f., 227, 243, 251, 319, 321, 348 f., 387, 389
pharmacological analogues   XXIV, 21, 409
pharmacophore   201, 222, 225, 236 f., 262, 265 f., 274, 382
phenbenzamine   XXI, 402 f.
phencyclidine (PCP)   269
phencyclidine-induced hyperlocomotion   302
pheniramine   403, 546
phenobarbital   165
phenothiazine derivative   297, 299
phenserine   284, 289
phenylephrine   541
phenylethylamine, biological activities of   60
phenylpiperidine class   272 f.
phenylpropanolamine   541
phenylpropylamines   273
phenytoin   57
Philadelphia chromosome   63
phosphodiesterases   64
phosphodiesterase 5 (PDE5) inhibitor   XXI, 5, 13 f. 58, 64 f.
phosphodiesterase 6 (PDE6)   64
phosphonoacetic acid (PAA)   381
phosphonoformic acid (PFA)   381
photostability   342 f.
phototoxicity   322 f., 337, 347, 352
Physostigma venenosum (Calabar bean)   277
physostigmine (eserine)   XX, 7, 277 ff., 281 ff., 285 f., 288 ff.
physostigmine analogues   281 ff.
pibutidine   26
picoprazole   86, 88, 111, 121
pimobendan   26
pinacidil   251
pindolol   197, 204
pioglitazone   450
pioneer drug   XX, 72, 178 f., 193, 195 f., 213, 234 f., 247, 422, 437
pipemidic acid   317, 318, 320
piperacillin   491

piperacillin/tazobactam 347
piracetam 26
pirbuterol 543
pirenzepine 119
piretanide 458
piroxicam 519
pitavastatin 139, 141 f., 144 f., 473
plasma atrial natriuretic peptide 221
plasma half-life 161, 164, 170 f., 177, 182,
    185, 187, 195, 205, 209 f., 212, 216, 239 ff.,
    245, 250, 289, 292, 348, 379, 387, 389,
    422 ff., 429, 432
plasma protein binding 205, 210, 216, 220,
    233, 242, 349, 387, 389, 392, 423 ff., 436
plasmid-mediated resistance 350
Plasmodium falciparum 381
platelet-activating factor (PAF) antagonists
    33
platinum (Pt)-containing cytotoxic drugs 385
pleiotropic effect 163
pneumonia 344
POH groups 375
polymethylenic bis-ammonium compounds
    3
polyphosphates 372
poor metabolizer 108, 110
poppy 259, 262 f.
positional isomer 3
positron emission tomography (PET) 305
postoperative patients 265
postural hypotension 306
potassium channel 251
potassium channel blocker 9
potassium channel modulator 58
potassium channel opener 251
potassium level 176
PPi (inorganic pyrophosphate) 372, 375 f.,
    379
practolol 197, 200, 201, 206 f., 211 ff., 236
pradofloxacin 355
pralidoxime 285, 287
pravastatin 41 ff., 139, 142 ff., 149 ff., 472
prazosin 455
prednisolone 422 ff., 427, 429, 435, 437, 485
prednisone 423 f., 437, 485
pregabalin 29
pregnane skeleton 421
premafloxacin 355
Prevotella loeschii 348
primaquine 26
privileged structures 53, 59
probenecid 352
probucol 10

procaine 234, 235, 243
procaterol 543
procindolol 197
prodrug 26 ff., 31, 37, 41, 89, 96, 106, 120,
    126, 133, 139, 143 ff., 164, 171, 177, 254,
    263, 436, 438
progestin 395, 397
progesterone 62, 397 f., 419, 478
progestogen 62
prolactin 297, 302, 304
promethazine XX, 57, 545
pronethalol XXI, 194, 195, 197, 200, 207
propiverine 57
propranolol 34, 35, 76, 195, 197, 200 f., 203,
    205 ff., 210 f., 213 f., 216 f., 220, 223, 233,
    235, 240 ff., 460
propyphenazone 530
prostaglandin G_2 29
prostate tumor 373
protein kinase C (PKC) inhibitor 63 f.
proteins, β-turns of 57
proteinuria 152, 175
Proteus mirabilis 347
proton pump 87, 96, 111, 116, 127, 129, 132
proton-pump inhibitors (PPIs) 79, 80, 81,
    98 ff., 103, 111, 115, 117 f., 121, 123 f., 125,
    127, 129, 131 ff.
prototype drug 71, 259, 262, 273, 299
Providencia sp. 347
prulifloxacin 28, 45, 48 f.
pseudoephedrine 541
Pseudomonas aeruginosa (P. aeruginosa) 319,
    326, 344, 347
Pseudomonas spp. 317
psychiatry 307
psychostimulants 57
psychotic symptoms 297, 307
Pt(II) complexes 386
pulmonary effects 216
pyrazinamide 347
pyridostigmine 277, 283 ff., 289, 540
pyrophosphates (PPi) 372, 375 ff.
pyrophosphate analogues 381

q
QSAR of antihistamines 404, 407, 409 f.
QT interval prolongation 297, 352
quazepam 538
quetiapine 298 ff., 308 ff., 534
quinacrine 13
quinapril 171 f., 178, 468
quinalaprilat 174
quinidine 352

quinolone antibiotics   315 ff., 334, 350, 352
quinolones, magnesium-chelating properies
   of   352
quinolones, evolution of   356

r

raclopride   305
radioactive imaging agent   380
ramipril   171 f., 177, 178, 469
ramiprilat   174
ranitidine   XXIII, 73 ff., 78 f., 90, 115, 118,
   352, 444
rabeprazole   102, 111, 117, 133
REASON study   204
reboxetine   34 f.
β_1-receptors   213, 239
β_2-receptors   208, 239
μ-receptor agonist   265, 267, 271
μ-receptor antagonist   268, 270
receptor selectivity   265
rectal tumor   373
5a-reductase inhibitor   62
reduction in bone mass   373
reflex tachycardia   173
reflux esophagitis   90
remission rate   390
renal artery stenosis   176
renal elimination   177
renal failure   388
renal impairment   174, 349
renally hypertensive dog   190
renin   169, 175, 221
renin-angiotensin-aldosterone system
   (RAAS)   160
renin-angiotensin system (RAS)   169, 173,
   179
renin inhibitor   XXI, 58
renovascular hypertension   186
reproterol   542
reserpine   XX
resistance to antibiotics   319, 321, 344, 346 f.
resistance against Pt drugs   392 f.
resistance to moxifloxacin   350
respiratory depression   261, 271
respiratory tract infections   319, 344, 356
restenosis   382
retinol (Vitamin A)   476
reversed ester derivative of meperidine   272
reversible inhibitors   134
RGD   61
rhabdomyolysis   137, 147 f., 150 ff.
rheumatoid arthritis (RA)   382, 419, 422, 426,
   430

rhinitis   411, 421, 428, 432
ribavirin   504
rifampicin   165, 347
rifapentine/minocycline, in combination with
   moxifloxacin   348
ring transformations   3, 9
risedronic acid (risedronate)   374, 376 f., 380,
   523
risperidone   298 ff., 305, 307
ritonavir   26, 509
rivastigmine   284, 288 ff., 540
rivastigmine analogues   290, 292
rizatriptan   43 ff., 531
rocastine   409, 415
rodent model of bone resorption   377
rofecoxib   30, 522
rosiglitazone   450
rosuvastatin   29, 41 ff., 139, 141 ff., 150 ff.,
   473
rotatable bonds   41, 49
roxatidine   XXIII, 78, 79, 444
roxithromycin   498
rufloxacin   320
rupatadine   29, 31 ff.

s

safety profile   271, 349, 352
salbutamol   60, 542
salivation   279
salmeterol   434, 543
salt and water retention   425, 428, 429
salt retention   423 ff., 429, 437
salvarsan   315
salvinorin A   274, 275
saquinavir   509
saralasin   157
satraplatin   393
S. aureus   319, 344, 347, 348, 350, 351
scale inhibitors   371
schizophrenia   59, 200, 297 ff., 307 ff.
SCOPE study   162, 166
scopolamine   277, 279, 287, 551
screening   XIX, 202
sebacyl dicholine   10
sedation   261, 268, 306, 411
sedative/hypnotic drug   13
sedimentation diffusion equilibrium   343
seizures   272, 310
β_1-selective agonist   60
β_1-selective blockers   193, 195, 200 f., 208,
   213 f., 217, 226, 233
selective norepinephrine reuptake inhibitor
   33, 35

selegiline   292 f.
semi-rigid analogues   213
senile osteoporosis   373
SENIORS trial   222
sensory neuropathy   390, 392
sequential therapy   321, 346
serine protease inhibitors   59
serotonin (5-HT)   27, 302
serotonin (5-HT) receptor subtypes   304
Serratia marcescens   347
sertindole   298 ff.
severe headache   247 ff.
sex hormone activity   396
Sharpless asymmetric epoxidation   219
Shay rat model   118
shelf stability   89
shock   401
side-effect profile   310
sildenafil   XXI, 13, 29, 64 f.
SILVHIA study   162
Simmons-Smith reaction   397 f.
simvastatin   20, 41 ff., 139, 142 ff., 149 ff., 472
sitafloxacin   322
sivelestat   28
skeletal muscles contraction   279
skin allergy   411
skin and skin structure infections (SSSI)   347
skin infections   347, 348
skin penetration   421
skin rash   323
skin reactions   401
sleep induction   411
slow metabolizer   220
slow-release formulation   182 ff., 187
small arteries   184, 187
smooth muscle contraction   279, 401
S-nitrosothiols   256
sodium nitroprusside (SNP)   254, 256
sodium oxybate   28
sodium retention   430, 437
soft corticosteroid   435
soft drug   242, 435, 438
solid tumors   385
solifenacin   29
solubility   125, 130 f., 182, 239, 243, 264, 337, 434, 436
soraprazan   133 f.
SOSA   14 f., 18, 21, 55
sotalol   197, 352, 460
sparfloxacin   320 f., 344 f., 347, 350, 352, 501
spasmolytics   57
spiralpril   171 f., 177 f., 468
spironolactone   XXIV, 26 f., 62, 176, 396, 459

spirorenone   396 f.
S. pneumoniae (Streptococcus pneumoniae)
   319 f., 344, 347, 349, 351
spontaneously hypertensive rat (SHR)
   188 ff., 208
S. pyogenes (Streptococcus pyogenes)   347
squalene synthase inhibitors   381
SSSI   348
stability to oxidants   343
stable angina pectoris   205, 217, 221, 255
Staph. aureus   319, 344, 347 f., 350 f.
Staphylococcus aureus (Staph. aureus), (S. aureus)   319, 344, 347 f., 350 f.
Staphylococcus spp.   344
statins   137, 140 f., 149, 151, 185
stavudine   505
sterane structure   421
steroids   59, 419, 438
stigmines   277
stimulant   60
STOP-Hypertension trial   204
STOP-Hypertension-2 trial   204
Stoughton classification system   432
Straub tail reaction   271
steam engine   371
Streptococcus pneumoniae (S. pneumoniae)
   319 f., 344, 347, 349, 351
Streptococcus pyogenes (S. pyogenes) 347
Streptococcus spp.   347 f.
streptogramins   316
stress ulceration   79
stroke   160, 162, 166, 177, 179, 184, 204
structural analogues   XXIV, 184, 297, 298, 310, 372, 459, 517 f., 541
sructural and pharmacological analogues
   XVII, XXIII, 382
structure-activity relationship (SAR)   XXIII, 45, 122, 187, 194 f., 197, 200 ff., 206 ff., 213 f., 217 f., 222, 236 f., 243 f., 262, 265 f., 273, 277, 282, 291, 299, 319, 355, 375, 376, 393, 403, 415 f.
structure-based drug design   XIX, XXI , 222
strychnine   262
ST 1460   298, 300
sublingual administration   179
substance P   174
substantia nigra   272
succinyl dicholine   10
succinyl L-proline   171
sucralfate   353
suicidal behavior   307 f.
sulbactam   492
sulfanilamide   XX

sulfathiazole   15 f.
sulfenamide   94 f., 97, 103, 126 ff., 133
sulfisoxalole   15 f.
sulfonamides   54, 316
sulindac   517
sulpiride   18
sultamicillin   494
sumatriptan   26, 43 ff., 531
surmountable antagonist   164
sustained duration of action   191
sweating   279
sydnonimines   256
sympathomimetic, indirect   60
syncope   247
synergistic effect   436
systemic corticosteroid therapy   424
synthetic glucocorticoids   422
syphilis   315
systemic infections   319
systemic safety profile   436
systolic blood pressure (SBP)   165
systolic/diastolic blood pressures   165, 183
S-8307   40, 157 f.
S-8308   40, 158 f.

t

tacalcitol   38 f.
tachycardia   182, 240
tadalafil   29, 64 f.
talopram   65 f.
taltirelin   26
tandospirone   43, 46 f.
tardive dyskinesia   297, 305, 307
targeted activation   436 f.
taste disturbances   175
tazobactam   492
tazolol   197
Tc-BP complex   375
technetium chelate   373
tegafur   511
tegaserod   27
telenzepine   119
telithromycin   499
telmisartan   40 f., 158 f., 163 ff., 471
temafloxacin   320 f., 352
temazepam   537
temelastine   409, 415
temocapril   171 f., 469
tenatoprazole   100 f., 121 f.
tendopathies   352
tenofovir   27, 36 f.
tenofovir disoproxil   27, 36 f., 505
tenonitrozole   29

tenoxicam   519
teprotide   157, 169 f.
teratogenicity   13, 175
terazosin   455
terbutaline   542
terfenadine   411 f., 414
testicular cancer   387, 389
testosterone   62, 397, 481
tetracaine   475
tetracyclines   316, 346, 350, 489
tetryzoline   552
thalidomide   13
thebaine   261 f., 267
thebaine analogues   266
theophylline   76, 352
therapeutic space   66
therapeutic window   153, 289
thermal stability   343
thiamphenicol   6, 7
thioridazine   298, 307
thrombin inhibitors   59
thromboxan-A$_2$ (TXA-2)   161
thymus gland   86
thymus involution   425
thyroid gland   86 f.
thyroid tumor   373
thyrotoxicosis   200
thyrotropin-releasing hormone   26
tiagabine   7
tiaprofenic acid   520
TIBET study   205
ticarcillin   491
ticlopidine   453
tifluadom   57 f.
tiludronic acid (tiludronate)   374, 377, 380
timolol   197, 460
timoprazole   84, 86 ff., 98, 103, 111, 117,
   119 ff., 122, 125, 127 f.
tioconazole   503
tiprenolol   197
tiquinamide   119 f.
tiospirone   26 f., 43, 46 f.
tiotidine   76 f.
tiotropium bromide   28, 447
tirilazad   62
tissue regeneration   160
tobramycin   507
tolamolol   197, 225
tolcapone   4 f.
tolerance   261, 267, 274
toliprolol   197
tolnaftate   26
tomoxetine   33

Tonelli croton oil ear assay   435
tonin   160
topical anti-inflammatory activity   431
topical glucocorticoids   421, 428, 431, 433
topoisomerase II (DNA gyrase)   317, 344
topoisomerase IV   317, 344, 350
torasemide   458
Torpedo californica AChE   280, 292
tosufloxacin   28, 322, 345, 352
toxicity   289
toxogonin   285, 287
Toxoplasma gondii   381
tracheal tissue   213
tramadol   271, 528
trandolapril   171 f., 177 f., 469
tranquilizer   57
transdermal use   273, 288
transition-state analogue   377
tranplantable murine tumor models   385
travoprost   27
treadmill exercise   211, 255
treatment-resistant schizophrenia   308
tremor   200
treprostinil   28
tretinoin   476
triamcinolone   425, 430 f., 486
triamcinolone acetonide   425, 431
triazolam   538
tricyclic antidepressants   352
triglyceride   137, 146, 148, 150 f., 153
trimethoprim/sulfamethoxazole   346
tripelennamine   271, 402
triprolidine   403, 408, 410, 414, 546
triptorelin   514
trithiozine   119 f.
troglitazone   450
tropisetron   29, 448
trospium chloride   446
trough-to-peak ratio   177
trovafloxacin   322 f., 327, 344 f., 349 ff., 353
"truth drug"   277
Trypanosoma brucei rhodesiense   381
Trypanosoma cruzi   381
trypsin inhibitor   59
tuberculosis   347
tumoral osteolysis   380
tumor biopsies   391
tumor cell cultures   386
tumor necrosis factor-alpha (TNFα)   382
tumor-related bone destruction   373
tumor spectrum   385, 392 f.
twin drugs   3

u

typical antipsychotics (neuroleptics)   297 ff., 306, 307, 309

ubiquinone   137
UDP-glucuronosyltransferases   145, 148
UGT1A1 and UGT1A3, inhibition of   148
UGT1A9 and UGT2B7, inhibition of   148
ulcerative colitis   435
ultra-short-acting compound   234 f.
unstable angina   182
uric acid excretion   161, 166
uricosuric activity   12, 167
urinary elimination   389
urinary tract infection   317

v

valaciclovir   504
valdecoxib   28, 30 f., 522
valganciclovir   27
ValHeFT study   163
VALIANT study   162
valsartan   40 f., 158 f., 162 ff., 470
ValSyst study   166
VALUE study   162
vardenafil   29, 64, 65
variant angina   182
vascular media hypertrophy   160
vasoconstriction   160, 169
vasodilation   160 f., 169, 173, 182 f., 186, 190, 401
vasopressin   160
vasopressin receptor antagonists   58
verapamil   XXIV, 181 f.
Verloop parameter   187
veterinary quinolones   355
viloxazine   9, 34, 35
voltage-gated calcium channel, L class of   181, 186
vomiting   388
von Braun reaction   265
voriconazole   28
vinylogue   3
vitamin A (retinol)   476
vitamin D analogue   37 f.
vitamin D nuclear receptor   39
vitamin D_2 (ergocalciferol)   451
vitamin D_3 (cholecalciferol)   451
vitamin D_3, 1,2-dihydroxy   26 f.
vitamin E   191
vitronectin receptor   61, 63
vomiting   388, 390, 392, 427
voriconazole   503

w

warfarin   76, 352
water retention   426
weight gain   300, 306, 310, 398
WHO   XVII
Wittig reaction   190, 219
Wolff-Parkinson-White syndrome   12
wound-healing   160

x

xamoterol   60
ximelagatran   29
xylometazoline   552

y

Y-931   298, 300

z

zaprinast   XXI, 5, 64
zidovudine   505
ziprasidone   27, 43, 46 f., 298 ff.
zofenopril   27, 170, 178, 467
zoledronic acid (zoledronate)   27, 374, 376 f.,
   380, 524
zolenzepine   119
Zollinger-Ellison syndrome   79, 132
zolmitriptan   43 ff., 531
zopiclone   29
zotepine   298, 300